World archaeoastronomy

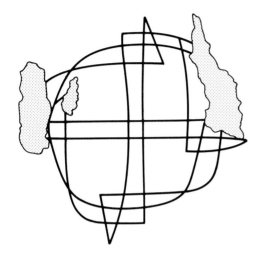

Symbol of the 2nd Oxford International Conference on Archaeoastronomy

Graffito 1 (painted in black, diameter 26 cm), incised into the plaster floor of the central chamber of Str. 1-sub during the 8th C. – a quadripartite diagram that was made from a single line tracing a four-part, complexly articulated circular pattern around a central point. Located in SE corner of Str. 1-sub, Dzibilchaltun. (Description after Coggins, C. (1983) *M.A.R.I. Publ. 49*, pp. 37–9.)

World archaeoastronomy

Selected papers from the
2nd Oxford International Conference on Archaeoastronomy
Held at Merida, Yucatan, Mexico
13–17 January 1986

Edited by A. F. Aveni

Colgate University, Hamilton, USA

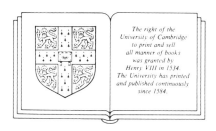

Cambridge University Press

Cambridge
New York   New Rochelle   Melbourne   Sydney

Published by the Press Syndicate of the University of Cambridge
The Pitt Building, Trumpington Street, Cambridge CB2 1RP
32 East 57th Street, New York, NY 10022, USA
10 Stamford Road, Oakleigh, Melbourne 3166, Australia

First published 1989

Printed in Great Britain at The Bath Press, Avon

British Library cataloguing in publication data

Oxford International Conference of Archaeoastronomy
(*2nd: 1986: Merida*) World archaeoastronomy.
1. Astronomy, Prehistoric
I. Title II. Aveni, Anthony F.
930.1   GN799.A8

Library of Congress cataloguing in publication data

Oxford International Conference on Archaeoastronomy (2nd: 1986:
Mérida, Mexico)
World archaeoastronomy : selected papers from the 2nd Oxford
International Conference of Archaeoastronomy, held at Merida,
Yucatan, Mexico, 13–17 January 1986/edited by A. F. Aveni.
    p.   cm.
ISBN 0 521 34180 9
1. Astronomy, Prehistoric–Congresses. 2. Astronomy, Ancient–
Congresses. 3. Indians–Astronomy–Congresses. I. Aveni, Anthony
F. II. Title. --
GN799.A8094 1989
520′.97–dc19   87–35400 CIP

ISBN 0 521 34180 9

BS

Contents

Contents

PART 3　ARCHAEOASTRONOMY: AN INTERDISCIPLINE IN PRACTICE

PART 4　ADDITIONAL ABSTRACTS SUBMITTED　491

Contributors

Carmen Aguilera
Biblioteca del Instituto Nacional de
  Antropología e Historia
Museo Nacional de Antropologia
Reforma y Gandhi
Mexico 5, D.F., Mexico

Anthony F. Aveni
Department of Physics and
  Astronomy
Colgate University
Hamilton, NY 13346, USA

Katalin Barlai
Konkoly Observatory
H-1525
Budapest XII
P.O. 67, Hungary

Harvey Bricker
Department of Anthropology
Tulane University
New Orleans, LA 70118, USA

Victoria Bricker
Department of Anthropology
Tulane University
New Orleans, LA 70118, USA

William Brito Sansores
Department of Literature
University of Yucatan
Merida, Yucatan, Mexico

Johanna Broda
Instituto de Investigaciónes
  Historicas, UNAM
Torre l de Humanidades, 8°piso
Ciudad Universitaria
Delegacion Coyoacan
04510 Mexico, D.F., Mexico

Gordon Brotherston
Department of Literature
University of Essex
Wivenhoe Park
Colchester CO4 3SQ, UK

Edward E. Calnek
Department of Anthropology
University of Rochester
Rochester, NY 14627, USA

David Carrasco
Religious Studies Department, CB
  292
University of Colorado
Boulder, CO 80303, USA

Robert Chadwick
John Abbott College
C.P. 2000
St. Anne de Bellevue
Quebec, Canada

Von Del Chamberlain
Hansen Planetarium
15 S. State Street
Salt Lake City, UT 84111, USA

Michael P. Closs
Department of Mathematics
University of Ottawa
Ottawa K1N 9B4, Canada

Clemency Chase Coggins
Peabody Museum
Harvard University
Cambridge, MA 02138, USA

David Dearborn
Lawrence Livermore National
  Laboratory
P.O. Box 808, L-23
Livermore, CA 94550, USA

R. David Drucker
c/o Krueger Enterprises Inc.
Geochron Lab Division
24 Blackstone Street
Cambridge, MA 02139, USA

R. Dvorak
Universitäts – Sternwarte
Turkenschanzstrasse 17
Vienna, Austria

I. Ecsedy
Orientalist Research Centre
Orszaghaz u. 30 Budapest
Hungary

Munro S. Edmonson
Department of Anthropology
Tulane University
New Orleans, LA 70118, USA

Contributors

Claire Farrer
Anthropology Department
California State University at Chico
Chico, CA 95929, USA

Maria Firneis
Department of Astronomy
University of Vienna
Turkenschanzstrasse 17
A-1180 Vienna, Austria

James A. Fox
Department of Anthropology
Stanford University
Stanford, CA 94305, USA

Jaime Ganot R.
Universidad Juarez de Durango
Av. Universidad y Fanny
Anitua, Durango
CP 34000, Mexico

Owen Gingerich
Harvard-Smithsonian Center for
   Astrophysics
Cambridge, MA 02138, USA

Luis González-Reimann
El Colegio de Mexico
C.E.A.A. Camino al Ajusco 20
01000 D.F., Mexico

N'omi Greber
Cleveland Museum of Natural
   History
Wade Oval
Cleveland, OH 44106, USA

Phillip C. Green
Department of Anthropology
Oregon State University
Corvallis, OR 97331, USA

Roberta L. Hall
Department of Anthropology
Oregon State University
Corvallis, OR 97331, USA

Horst Hartung
University of Guadalajara
Calle Caballo Arete #3065
Club Hipico Tapatio
C.P. 44570 Guadalajara, Mexico

Ronald Hicks
Department of Anthropology
Ball State University
Muncie, IN 47306, USA

Michael Hoskin
Churchill College
Cambridge CB3 0DS, UK

Tom Hoskinson
813 Old Farm Road
1000 Oaks, CA 91360, USA

Stanisław Iwaniszewski
State Archeological Museum
52 Dtuge St
00950 Warsaw, Poland

John S. Justeson
Computer Studies
Stanford University
Stanford, CA 94305, USA

Christian Köberl
Institute for Geochemistry
University of Vienna
Währingerstrasse 38
A-1090 Vienna, Austria

Ulrich Köhler
Institut für Völkerkunde
   der Universität
Werderring 10
7800 Freiburg/Br., FRG

Ed Krupp
Griffith Observatory
2800 East Observatory Road
Los Angeles, CA 90027, USA

Jesus F. Lazalde M.
Universidad Juarez de Durango
Av. Universidad y Fanny
Anitua, Durango
CP 3400, Mexico

Miguel León-Portilla
Instituto de Investigaciónes
   Antropologicas de la UNAM
Ciudad Universitaria/Delegacion
   Coyoacan
04510 Mexico D.F., Mexico

Floyd G. Lounsbury
Department of Anthropology
Yale University
New Haven, CT 06520, USA

Stephen C. McCluskey
Department of History
West Virginia University
Morgantown, WV 26506, USA

J. McKim Malville
University of Colorado
Campus Box 391
Boulder, CO 80309, USA

Alexander Marshack
Peabody Museum–Harvard
   University
4 Washington Square Village
New York, NY 10012, USA

Michael P. Marshall
Cibola Research Consultants
Corrales
New Mexico, USA

William Breen Murray
Faculty of Humanities and Social
   Sciences
University of Monterrey
Monterrey, Mexico

David S. Nivison
Department of Philosophy, Building
   90
Stanford University
Stanford, CA 94305 USA

Alejandro Peschard F.
Universidad Juarez de Durango
Av. Universidad y Fanny
Anitua, Durango
CP 34000, Mexico

Arturo Ponce de Leon
Football 118–2
Col. Country Club
CP 04220 Mexico D.F., Mexico

Allen F. Roberts
Albion College and University of
   Michigan
Center for Afroamerican and African
   Studies
Ann Arbor, MI 48103, USA

Contributors

Jack H. Robinson
Astronomy Program
University of South Florida
3507 Nakora Drive
Tampa, FL 33618, USA

Curt Roslund
Section of Astronomy
Chalmers University of Technology
Gothenburg, Sweden

Clive L. N. Ruggles
Computing Studies Department
University of Leicester
Leicester LE1 7RH, UK

Brad Schaefer
NASA/GSFC
Code 661
Greenbelt, MD 20771, USA

R. Schult
Zentralinstitut für Astrophysik
60 Sonnenberg, GDR

Rolf M. Sinclair
Physics Division
National Science Foundation
Washington, DC 20550, USA

Nathan Sivin
University of Pennsylvania
8125 Roanoke Street
Philadelphia, PA 19118, USA

Anna Sofaer
Solstice Project
P.O. Box 9619
Washington, DC 20016, USA

John Sosa
Department of Anthropology
State University of New York at
    Cortland
Cortland, NY 13045, USA

Carolyn Tate
University of Texas
5007 Placid Place
Austin, TX 78731, USA

Barbara Tedlock
Department of Anthropology
SUNY Buffalo
Buffalo, NY 14260, USA

Dennis Tedlock
Program in Folklore, Mythology
    and Film
Department of English
SUNY Buffalo
Buffalo 14260, USA

Raymond White
Steward Observatory
University of Arizona
Tucson, AZ 85721, USA

Ray A. Williamson
Office of Technology Assessment
US Congress
Washington, DC 20510, USA

M. Jane Young
Department of American Studies
307 Ortega Hall
University of New Mexico
Albuquerque, NM 87106, USA

Michael Zeilik
Department of Physics and
    Astronomy
University of New Mexico
Albuquerque, NM 87131, USA

R. Tom Zuidema
Department of Anthropology
University of Illinois
109 Davenport Hall
607 S. Mathews
Urbana, IL 61801, USA

Preface

What place could have been more suitable than Yucatan to hold the 2nd Oxford International Conference on Archaeoastronomy? There, in the Maya heartland, more than a millennium ago, people developed one of the most orderly, complex and precise means of time reckoning – a scheme tied directly to the observation of celestial bodies and the incorporation of that information into agriculture, religion, political ideology, 'cosmovisión' – indeed, every aspect of life. Little wonder then that we see the workings of the firmament mirrored in the inscriptions, the ethnohistoric documents, the architecture, the sculpture – every segment of the material world.

This is the business of archaeoastronomy: to comprehend the meaning of astronomy to ancient peoples through a study of their texts, both written and unwritten, and the purpose of our meeting was to ask participants from many fields (astronomy, history, anthropology, archaeology, art history) to share openly and constructively their findings, suppositions, hunches, and reactions to one anothers' ideas on this broad subject.

This conference is a continuation of the dialogue, or more appropriately, 'polylogue' that was begun in 1981 at Oxford when Professor Michael Hoskin (with the sponsorship of the International Astronomical Union and the International Union for the History and Philosophy of Science) organized the first conference on an international basis that brought together scholars from a wide variety of disciplines who focused on the common question of the role of astronomy in ancient cultures, both western and non-western. Quite different from the conference on 'The Place of Astronomy in the Ancient World', held in London (under the auspices of the British Academy and the Royal Society of London) some seven years before, which was scientifically more specialized, the 'Oxford' conference aimed at a genuine interdisciplinary focus on questions and problems in ancient astronomy, one that would transcend both culture and method. The two volumes representing the written output of that conference (Aveni, 1982; Heggie, 1982)* remain as evidence by which to judge the success of that meeting.

Using C. P. Snow's old 'island metaphor' as a vehicle for commenting on the compartmentalized nature of the London symposium, D. G. King-Hele stated, '... divisive insularity is still quite strong, not because anyone is being wicked but because it is so tidy and self-perpetuating, both administratively and intellectually to have islands and avoid the traumatic experience of emigration' (King-Hele, 1974, p. 273). As anyone will see who browses through the 'Oxford I' volumes, a number of bridges have now been built between the island disciplines that each have an interest and investment in matters astronomical.

The Oxford conclave, replete with all of its good fellowship and the revelation of common interest among many in places where one might least expect to find it, produced exactly what one might have anticipated – the promise of another conference to be held about five years later to assess the progress or regress in our common, broadly based studies. Thus, 'Oxford II'.

* References to this Preface are listed at the end of Chapter 1.

While many of the papers in the present volume often appear to talk past one another, some communication is beginning to become evident. We base this judgment most directly on the quality and liveliness of the discussions that followed practically every paper that was presented and indirectly on the increased number of papers that deal with data of more than one kind.

No volume can relate fully the amount of activity and the degree of exchange of information and ideas that took place at the 2nd Oxford Conference in the Yucatan, but we trust the text of selected works adequately spans all concerns.

It has been traditional in archaeoastronomy volumes to group the papers by cultural area. Indeed, the collected works from the 1st Oxford International Symposium were divided into two volumes: the Old World and the New World, with contrasting colors applied to their respective covers. In the introductory chapter of this volume I comment on the hidden symbolism of these colors and I criticize this approach to organizing the material. I suggest that archaeoastronomy has progressed in the past five years to the degree that we now might dare to consider arranging the presentations according to the methods and processes employed in the various studies and the common themes they appear to share. Such an unorthodox approach is a far more difficult matter than the rote task of assigning papers by continent, country, or culture – and yet it must be done because such a realignment of scholarship forces us to face several important issues and questions more directly:

(1) What is the source of archaeoastronomical hypotheses, and how can evidence drawn from the allied disciplines be brought to bear on testing these hypotheses?

(2) What kinds of evidence are admissible in a given argument?

(3) What are the deficiencies in studies conducted from the perspective of a single discipline, and how can such studies be improved?

By mixing papers from the culture areas within the text, one hopes to attract a cross-disciplinary readership. One may find the author of a paper on Maya ceremonial building orientations asking the same questions as those of an investigator on the alignment of medieval European churches. Moreover, under the arrangement present in this volume, a reader who is interested in purely text-based astronomical arguments will encounter a wide variety of material that might not ordinarily be confined to the category of 'written text', such as quipus and picture writing. Hopefully, these types of evidence will come to be viewed as part of a literary continuum. These examples and the present arrangement of the Oxford II presentations are intended to stress the fundamental principle that archaeoastronomy is an 'interdiscipline-in-practice' and not simply a multidiscipline, or collection of unrelated investigations, each emanating from different disciplines.

In Part I ('The boundaries of archaeoastronomy: overviews and syntheses'), we present a collection of papers, most of which had been invited as review presentations at the Merida meeting. These range from areal overviews such as Justeson's paper on the Maya, Zeilik's on the US southwest, Sivin's on the Chinese, and Ruggles' on British megalithic astronomy to disciplinary overviews and syntheses that attempt to show the status and contributions of archaeoastronomy as seen from the perspective of the history of religions (Carrasco), the history of science (Gingerich), social theory (Iwaniszewski), and current ethnology (Sosa). While some of these latter papers (e.g. that of Sosa) are confined to a single cultural area, they exemplify the value of the data and forms of inquiry emanating from the established disciplines to the interdisciplinary form of inquiry that makes up archaeoastronomy. Thus, they inform both with regard to content and process in areas where the true interdisciplinary investigator, who works cooperatively with other scholars, needs to be informed.

We entitle Part II 'Archaeoastronomy: the textual basis'. The first of these papers deal with astronomical practice among diverse cultures as it is revealed in the traditional (historic and ethnohistoric) written record. We then move to the papers that display astronomy in the Maya hieroglyphic record, a field that has undergone accelerated development in the past decade. One hopes that the astronomical revelations of such papers, though they emanate from the fringe of what many western scholars would call 'literature', would invite contributions to future volumes from other culture areas that usually appear in publications and journals lodged within a more narrow disciplinary framework. We also include in this section papers that deal with data from even more remote 'written' forms, such as picture writing, quipus and other symbolic representational devices. We include petroglyphs and pictographs, which usually are placed in an entirely separate category that is totally unrelated to the history of astronomy. Thus, we attempt to break the traditional and somewhat confining historic/prehistoric dichotomy by keeping, under the same umbrella, all of the papers that deal with symbolic data on ancient astronomy.

In Part III, 'Archaeoastronomy: an interdiscipline in practice', we bring together those papers that we believe best demonstrate the processes by which the traditional text-based studies can be related to the 'unwritten record' that had made up the sole basis for archaeoastronomical studies during the early stages of its development. While many of these papers dwell on the topic of alignments in ceremonial buildings, architecture and the landscape, nearly all of them employ some other form of evidence. The presentations in this part dramatize the nature of the argumentation in those cultures wherein one has little evidence beyond the standing stones with which to work. Nevertheless, they also show how and under what conditions it is possible to bring the archaeological sciences to bear on an astronomical hypothesis.

Finally, by offering this deliberately unorthodox alignment of papers, one hopes to bring out into the open, rather than hiding under the label of 'exact science', some of the questions the wider readership has been asking about our interdiscipline of archaeoastronomy, beginning with that most frequently asked: What is it?

There is not enough space in this volume to include all of the papers that were read at the conference; therefore this is not a 'proceedings volume' – nor, in fact, were the Oxford I volumes. Rather, the collection is intended to be a representative sampling of papers that were read at the Oxford II meeting. In order to give the reader at least a taste of the wide variety of cultures whose astronomies were discussed, we have decided to include, at the end of the text, abstracts of all the papers that were presented both orally and in the poster sessions and that do not appear in full draft in this text.

The editor wishes to express his gratitude most especially to Luis Felipe Rodriguez of UNAM and Owen Gingerich of Harvard-Smithsonian, who co-organized the conference and co-authored the ultimately successful proposal for its financial support to NSF (USA) and CONACYT (Mexico). We gratefully acknowledge the support of these organizations (Grant Number INT-8514830). We are also indebted to other individuals, including those at the local level who assisted admirably in various phases of the organization process: Lorraine Aveni, Johanna Broda, John Carlson, Lisa Jean Cohen, Victor Garlock, Stanisław Iwaniszewski, Jack Lain, Lucrecia Maupomé, Jeff Santa Ana, Jennifer Schaeffer, Julie Somers and Scott Stowell. Our hosts in Yucatan and in the city of Merida, Joann Andrews, Alfredo Barrera Rubio, Teresa Borge Manzur and the staff of the Hotel Maria del Carmen in Merida, Peter Schmidt, Victor Segovia, Alicia Villar del Blanco of the Department of Tourism of the State of Yucatan and the staff of INAH, Yucatan, were especially cordial. Many conference participants remarked that rarely in their professional careers had they ever been made to feel more welcome. We also thank Irene Pizzie and Lorraine Aveni for editorial assistance.

Our volume is affectionately dedicated to the memories of Alexander Thom and Travis Hudson, two significant contributors to very different aspects of our field, who have recently passed away.

A. Aveni
Hamilton, New York

I

The boundaries of archaeoastronomy: overviews and syntheses

1

Introduction: whither archaeoastronomy?

A. F. Aveni *Colgate University*

The fault, dear Brutus, is not in our stars, But in ourselves, that we are underlings.
Shakespeare, *Julius Caesar*, Act I, Scene II.

1.1 What are the questions being asked in archaeoastronomy? Who is asking them, and why?

A curious dichotomy exists between the subject matter presented in the two volumes of proceedings of the Oxford Archaeoastronomy Conference of 1981. In the Old World (green colored) volume, all 16 contributed articles deal with the possible astronomical orientations of alignments at prehistoric sites in the British Isles and Central Europe consisting of standing stones. But in the New World (brown colored) volume, only three of the nine papers deal substantially with alignments, two of them referring directly to historical and/or ethnohistorical material as a means of establishing explanations. Two-thirds of these papers discuss time reckoning in the context of some sort of written or pictographic record. In fact, one-third of the New World papers make either no reference at all or only an indirect reference to alignments. And yet, both these books carry the word 'archaeoastronomy' in the title; they appear together on the shelves of libraries and in the offices and studies of scholars and laypersons interested in archaeoastronomy, and they are believed by readers to define the domain of this interdisciplinary inquiry.

Judging from the content of the European (megalithic) work (let us call it 'green archaeoastronomy'),

the principal inquiry would appear to consist of determining whether alignments among standing stones,[1] collected carefully, precisely, and (usually) with due regard to the archaeological record, might bear any relationship to astronomical events at the horizon. Indeed, this is the thrust of the discussion commencing at the middle of p. 2 and following all the way to the end of Pedersen's review paper on 'The present position of archaeoastronomy' that appears at the end of the Old World volume.

Statistical discussions of the data abound in this green-colored text and one of the fundamental methodological issues, recently readdressed and updated by Ruggles (1984), is whether so-called 'green archaeoastronomy' has consisted too much of seeking out and over-emphasizing specific examples of astronomical orientations without paying much attention to whether other (non-astronomical) factors could have played a role in determining site planning and orientation.

Much effort has been expended in green archaeoastronomy exploring the evolutionary question of defining the level of absolute intelligence attained by ancient people. For preliterate, low-technology civilizations, the definition of the level of astronomical intelligence is likened to a step on a pyramid. The setting of that level is based upon two fallacious assumptions: (*a*) that there exists a *universal* scale of degrees of difficulty of perception of astronomical phenomena. Thus, recognition of the lunar standstills ranks as a higher achievement than that of the solar extrema, and precession of the equinoxes still higher (see the discussion in Heggie, 1982, pp. 3–4); and (*b*) that levels of intelligence can be related to the degree of precision with respect to which a given phenomenon is observed. This becomes at least a part of the issue in the extensive discussion of and reference to the Kintraw alignment in the first Old World volume (Heggie, 1982, pp. 18, 22, 37, 57, 124, 128, 135–6, 159, 183–90).

The problem of the ways in which astronomy might have influenced architectural orientations also seems to be implicit in many of the Native American investigations, though these deal more often with ceremonial centers, pyramids, doorways of temples, and an assortment of oddly shaped or oriented structures rather than standing stones. (One wonders what Mesoamerican archaeoastronomers might have concluded if only the stelae remained standing at the Mayan sites.) But the nature of the evidence and the method of investigation employed in 'brown archaeoastronomy' papers such as those of Brotherston, Lounsbury and Zuidema, to name only three in the New World proceedings volume, resemble rather closely what Old World scholars would term the history of astronomy. In fact, can it be, asks Pedersen (1982, p. 266), that scholars working on the calendric and astronomical decipherment of New World documents and inscriptions are engaged in a task similar to that tackled by Strassmaier, Kugler and Epping with respect to the Mesopotamian written materials over a century ago?

While there is some overlap in the methodologies employed by Old World historians of astronomy and brown archaeoastronomers, there are nevertheless some significant differences. Let me clarify the distinctions by discussing a few examples from the first Oxford volume on New World Archaeoastronomy. Lounsbury, a linguist and an anthropologist, asks: How was astronomical knowledge employed by the ruling class at the Classic Maya site of Bonampak? Answer: To make war. In seeking to comprehend the relationship between war-ritual and astronomy (Maya Star Wars!) he supplies us with valuable inscriptional material which he interprets as a series of precisely determined time markers associated with aspects of the planet Venus, and he offers us a rationale for why the Maya rulers, during the stage of social and political development one finds in the Yucatan peninsula in the eighth century, fostered astronomy as an instrument of power.

This same anthropological line of inquiry on the connection between astronomical pursuits and social behavior is followed by Schele and others at Palenque and by Tate at Yaxchilan (see Chapter 32). They find that astronomy has meaning only when it is carried along part-and-parcel with certain ritual and religious manifestations. Consider the example of Palenque's Group of the Cross, where we find a triad of buildings, each housing in its inner chamber a bifurcated hieroglyphic text that reflects historical events (on the right half) that are supposed to be re-enactments of the supernatural happenings found on the left side of each text. 'In each text a carefully designed bridge carries the reader from mythological to historical time and space.' (Schele, n.d., p. 5.) The subject matter deals with the transfer of power of rulership from a deceased father to his young son with various gods of the Palenque triad, each of whom may be represented by a planet, being named as agents or instigators of the events relating to this process. There is some evidence that the events depicted in the mythical part of the record may have been enacted metaphorically on the celestial sphere in real time, e.g. by the planets Jupiter and Saturn, whose stationary points, Lounsbury argues, bear a significant level of correlation to event dates appearing on the tablet inscriptions. Indeed, there is no room for ancient astronauts on this informative tablet!

At Palenque, history and astronomy appear welded together, and the alignments found in the skewed, disoriented Group of the Cross become the logical end point in a causal process resulting from supernatural actions recorded in the texts. As if to enframe time itself, the tablets in question are so designed and oriented to be illuminated by the sun on symbolically important dates of the year. Why? 'To make the site itself or perhaps the universe reproduce, for all to see and understand, the cosmological foundation of rulership.' (Schele, 1977, p. 50.) The event portrayed symbolically on the lid of Pacal's sarcophagus – his passage into the underworld – is re-enacted literally on each winter solstice as the sun itself descends into the underworld directly over his tomb as viewed from the Palace. Next day, the rising sun illuminates the Temple of the Sun, in the inner chamber of which one finds a plaque commemorating the transfer of rulership from Pacal to his son Chan Bahlum. Thus, the dying sun enters the underworld through the tomb of the ruler Pacal beyond the west end of the complex only to reappear anew as his son in the east.

Archaeoastronomical studies at Palenque seem to focus more on the issue of *why* people oriented buildings rather than whether they could orient them accurately. While the Group of the Cross does not contain the precise sort of alignments suggested at the Caracol or at Stonehenge, which seem to fascinate both astronomers and green archaeoastronomers, it may well possess greater significance for it yields information not only about the meaning of the act of orienting buildings but also about particular behavioral characteristics of the ancient Maya people.

To stress further differing motives of archaeoastronomers, let us review the content and process exhibited in another of the 1981 New World papers. Zuidema's approach to the problems of the nature of the Inca sidereal lunar calendar is given away in his opening paragraph, and it differs little from that of Lounsbury or Schele. He states quite clearly his desire 'to study characteristics of precision [of the calendar] needed in a state bureaucracy highly concerned with recording and correlating cycles of irrigation, agriculture, husbandry, trade, and warfare' – all operated 'within an intricate system of kinship, age, classes, and sociopolitical organization' (Zuidema, 1982, p. 59).

In the ancient Inca capital the integrative mechanism that bound astronomy to ritual kinship and kingship in a system of hierarchical order was the radial cosmo-political map centered on the 'Temple of the Sun' (Coricancha) that overlies the valley of Cuzco – the ceque system (Zuidema, 1964, 1977, 1982). At the same time, the ceque system also served as a calendar in which the ceque lines and their huacas, or sacred places of worship, functioned like the cords and knots of a quipu by which one could tally the days of the agricultural year.

The study of astronomical alignments in the ceque system offers an object lesson in the danger of attempting to draw firm conclusions on the basis of employing statistical methods alone, a habit one often encounters in green archaeoastronomical studies. If we disregarded the written record provided by the Spanish chroniclers, and if we set out to measure the alignments of the 41 directions to the horizon indicated by the ceque lines, fed the data into a computer and then attempted to correlate the

alignments with astronomical phenomena of conceivable significance at the horizon,[2] we would not have the slightest chance of arriving at the conclusions that, in fact, turn out to be consistent with the historical record (see Aveni, 1981a, for a summary). The valuable Incaic historic material, though offered to us in the garbled tones of a thoroughly confused group of chroniclers, under-educated with respect to indigenous ways of thinking, clearly indicates that only a few ceque lines also functioned as astronomical sight lines. Using alignment data alone, no fieldworker could possibly have deduced that other astronomical sight lines, including the most fundamental one in Cuzco (that concerning the zenith sunrise/anti-zenith sunset positions) were generated by connecting a part of one ceque line to a part of another, these parts being the huacas or sacred places in the local environment which served to delineate the ceques. In Andean archaeoastronomy, ethnohistory does not confirm orientations – it generates the hypotheses and rationale for conceiving of them in the first place.

Astronomy aside, the more we look into Inca planning and their way of thinking, the more we are impressed with the degree to which linearity and in particular radiality insist upon being present. The concept of splitting and bifurcating, the 'coming together' or 'joining at a point', is evident in Andean culture linguistically and grammatically, literally and 'architecturally. This concept is so prevalent in pan-Andean cultures that we have been forced lately to employ it as a way of generating hypotheses about the meaning and function of one morphological class of lines or geoglyphs found etched upon the pampa (elevated desert bordering the valley of the Nazca River in south coastal Peru) (Aveni, 1987). Without digressing too long, our analogical approach to the Nazca line problem has proceeded as follows:

(1) We became interested in the Nazca lines when we recognized from earlier work that many of them exhibited ray patterns that reminded us of ceque lines.

(2) Having measured and analyzed the properties of nearly 800 straight lines on the Nazca pampa, we discovered a pattern of organization: with rare exceptions, the lines appeared to radiate from more than 60 well-defined positions surrounding the pampa.

(3) These focal points turned out to be situated

upon natural hills that flank tributaries or that lie at the intersection of waterways with the principal drainage system (the Nazca River). Many of these centers lie at the lowermost of a series of hills that jut out like a protruding finger onto the pampa as they descend from the high Andes.

Thus, we discovered that all the lines appeared to connect points in the landscape related to the descent of water from the mountains to the sea via the pampa. Our study also disclosed that the major figures (the huge trapezoids and triangles) are preferentially oriented relative to the direction of the flow of water across the pampa.

(4) But we also found, from a careful examination of remains on the surface of the pampa, that the lines were intended to have been walked upon. Perhaps the act of walking was related to a concept of order. The builders may have intended to associate the irrigation process with other concepts of the Nazca world view. Urton's (n.d.) study of post-Conquest documents from Nazca suggests that the assignment of water rights among the coastal peoples, at least at the time of the Conquest, was rather complex, having been based upon principles of kinship. Thus people would have had good reason to walk across the pampa from one valley to another in accordance with prescribed social rules.

(5) Was astronomy involved? One astronomical hypothesis that can be formulated from the association of the lines with water is that there might be a significant correlation between line orientation and the position of sunrise on the day of the late October solar passage across the zenith, a date which can be correlated both astronomically and ethnohistorically in other segments of Andean culture with the time of appearance of the spring rains that precede planting. The statistical data do not conclusively bear out this assertion of a relationship between 'water and zenith sun appearing', but they do offer more than a subtle hint that such an association is possible. In the narrow ecological niche between coastal and montane environments occupied by the Nazca culture, the recognition of a relationship between sun and water expressible through

empirically based, ordered events in the sky–earth environment makes special sense because the water that comes to Nazca rarely falls from the sky. It appears in the rivers only after dark clouds form over the distant mountains at a certain time of the year, and once it arrives it is gone in a flash as it concludes its dramatic, precipitous descent from the high Andes to the Pacific.

In summary, our explanation of the cultural phenomenon of but one category of Nazca lines is a complicated one which has yet to be worked out in detail: the straight lines and the line centers did indeed constitute a system; almost surely that system was associated with water; some of the lines may have been associated with calendar and astronomy; and finally, people walked on them. Ultimately, we believe all of these characteristics must be brought together through the pan-Andean concept of radial hierarchical structure derived from the Cuzco analog.

Methodologically, the study of Nazca line and ceque line orientations stands in marked contrast to the published results on the megalithic sites, where astronomical questions provide the driving force and other factors are given little regard. Having thoroughly reviewed the contents of the 1981 Old World volume on archaeostronomy, I failed to turn up a single discussion of water, mountains, or, for that matter, any environmental parameters or phenomena other than celestial that might have influenced the placement of the standing stones. However, Ruggles (1984, pp. 271–2) has pointed out that such broadly based investigations do exist in megalithic studies. These are confined to the archaeological literature, to which many archaeoastronomers pay little attention. The isolation of culturally based evidence from astronomical argumentation, coupled with the background or corroborative role given such evidence, is viewed as an obstacle in the way of progress toward understanding why ancient people built these sites in the first place.

What, then, do these examples of brown archaeoastronomy from Meso- and South America have in common? Less concerned with the question of astronomical cultural universals, native precision, and purely quantitative matters, they seem to be inquiring: What is the nature of the relationship between astronomical phenomena and cultural behavior? What did astronomical phenomenon X mean to the people who practiced it? Why were they interested in phenomenon X instead of Y? How did they conceive of that phenomenon in their ritual, myth, calendar, religion, architecture and historical chronology (to name a few manifestations of astronomy in Central Mexico (Broda, 1982, pp. 101–2))? What role did it play in shaping their ideology? By what means did people employ phenomenon X as a way of creating order in their cultural reference frame? To turn a phrase of Levi-Strauss (1969, p. 9):

> The real question is not whether [Halley's Comet] caused [the Norman Conquest]. It is rather whether there is a point of view from which [the apparition of Halley's Comet] and [the fall of a kingdom] can be seen as going together and whether some order can be introduced into the universe by means of that pairing.

1.2 Is it science?

Thus far, we have reflected upon the domains of inquiry of Old and New World archaeoastronomy by concentrating mainly on the contents of the New World 1981 conference proceedings volume. But where does traditional history of astronomy fit in?

As we have seen, the foregoing examples of brown archaeoastronomy differ significantly from the work of both the 19th and 20th century historians of astronomy who confine themselves largely to text decipherment and philology. In the New World, we seem to be collectively developing an anthropology of astronomy rather than a history of astronomy. The content and method of the papers I have cited reveal the complex relationship between astronomy and politics, economics, and culture history. Therefore, we seem to be engaged in an endeavor that encompasses traditional history of astronomy.

By contrast, in the Classical Old World astronomy has remained at the level of history, ever the queen of disciplines in the European scholarly community. At a very early stage in the development of these highly specialized studies, historians of astronomy were careful to cull out scientific astronomy in the Greco–Babylonian tradition from other types of astronomical endeavor (e.g. Nilsson, 1920; Dicks, 1970; Aaboe, 1974; Price, 1975). Thus, Dicks speaks of Greek astronomy as the product

of a 'gifted race bequeathed to posterity' (p. 6) and Price of a 'spectacular accident of history' (pp. 14–15):

> Is it not a mystery that, having both essential components of Hellenistic astronomy, they [the Chinese] came nowhere near developing a mathematical synthesis, like the *Almagest*, that would have produced in the fullness of time, a Chinese Kepler, Chinese Newton, and Chinese Einstein?

The generative source of this question, it seems to me, is quite clear. It consists in placing one's own cultural traits at the center of the universe of inquiry and then systematically employing them as a value-base for comparison. Of course, it is inevitable that if we adopt the assumptions of our culture and reject those of the cultures we are studying, we cannot compare those other cultures with our own without finding them wanting. Applied to Maya astronomy, this ethnocentric posture leads to the rejection of astrology and ritual as materials relevant for such studies and this, in turn, results in the characterization, by one historian of Western astronomy, of the Maya calendar as: 'a maze of arithmetical juggling which permeated an entire culture without making it scientific' (Price, 1975, p. 6).

Curiously, this quality of being scientific is rarely defined in discussions regarding cultural contrast, though Aaboe (1974, p. 23) does take the trouble to lay out a number of achievement levels on the road of progress from prescientific to scientific astronomy. The top of the pyramid is crowned by pure scientific astronomy, which: 'gives us control over the irregularities within each period and frees us from constant consultation of observational records'. These statements, while admittedly taken out of context, reflect the ethnocentric, evolutionist underpinnings of traditional western history of astronomy.

In a more liberal discussion of the question: Are they like us? (i.e. Is it science?), McCluskey (1987) has recently singled out some of the criteria that investigators have usually employed to separate scientific from 'primitive' astronomy. Among the traditional reasons given are the static or unchanging nature of primitive astronomy, its lack of openness to theoretical alternatives, the absence of any notion of progress, the absence of a separation of primitive astronomy from its social context and the failure of a given culture to pursue knowledge for its own sake (McCluskey, 1987, p. 206). Such deficiency-oriented, value-based criteria suggest to him that too often we tend to hold up our own way of thinking as the judgmental yardstick by which to measure up all foreign modes of inquiry, which, of course, only in the most perfect of systems could ever be expected to score 100%. Little wonder then that most western-trained scientists would hold that we, and only we, see nature objectively, i.e. 'as it really is', and that we are unaffected by any systems of belief to which we might adhere. For example, how many of us believe the solar system is really, *in fact*, heliocentric and not geocentric? On the other hand, could it be that heliocentricity only seems a preferable alternative to us simply because our scientific culture, at least since the Greeks, has been deeply imbued with concepts of spatiality, orbits, and centricity in general. If queried about his opinion as to the center of things, a hypothetical Mayan Laplace might well have responded, 'we have no need of that idea'.

Following the example of Geertz in anthropology, McCluskey argues that the ultimate goal of archaeostronomy should be the 'deepening of our understanding of culture', a goal which can be achieved only when we understand astronomy as 'an endeavor by which societies seek to make their observations of the heavens intelligible' – in terms 'that are meaningful in that society'. He admirably attempts to describe how to achieve that goal in the study of Hopi astronomy by comparing criticisms of tradition based on direct observation in the Hopi realm with criticisms of theory in western astronomy. Still, it seems to me that like those whom he criticizes, McCluskey ultimately is concerned with 'questions of fact', observations of a 'respectable' level of naked eye precision and other concepts that serve as the defining parameters in the conduct of western science. Indeed, it may be that ultimately we can never hope to make objective cross-cultural comparisons and that perhaps we ought not to consider that task to be one of our scholarly goals.

I believe the separation of Ptolemaic astronomy from peasant astronomy (e.g. as reflected in Hesiod) in ancient Greece has served to narrow and sharpen the probing of the evolution of scientific astronomy in western civilization; this separation also has resulted in an 'isolationist' way of thinking and classifying that has had at least three detrimental effects:

(i) The evolution of Old World astronomy often is treated by some historians of science as if it had been totally removed from its social context. Reading Dicks' (1970) masterful treatise on *Early Greek Astronomy to Aristotle*, one has the impression that all of the developments in the pre-Socratic, Socratic, Platonic and Aristotelian worlds took place in the total absence of historical events. Indeed, despite the recent attempts to reintegrate history of science into the history of western culture, some of us may still imagine that pure scientific thought in the 16th, 18th, or for that matter, 20th century has gone unaffected by events in the world outside scientific institutions.

(ii) The approach that stresses the separation of the study of western scientific thinking from its cultural context is static for it lacks any notion of process or change of culture. It enhances the ethnocentrism already inherent in any trained scientist or historian of science who, whether in an active or passive way, inquires into the astronomy of other cultures. As Ruggles (1984, pp. 284–5) has stated, such individuals usually are interested only in questions of precision and in the age-old 'science vs. ceremonial' debate (which seems to be very important in green and can claim a just importance in certain brown archaeoastronomical studies); normally, high precision is equated with scientific and low precision with ritualistic behavior. Thus, when the western scientist studies the astronomy of other cultures, whether literate or not, he often may seem to be engaged in the narcissistic act of searching for his own history. This predetermined goal of seeking ourselves out in the tattered pages and crumbling walls of other peoples' history is all too evident in many popular works on archaeoastronomy.

(iii) Finally, this separation of the ways of knowing astronomy has had a third effect. The relatively *text*-rich civilizations of the Greco–Romans, Babylonians and Egyptians have offered 19th and 20th century scholars a blinding stream of astronomical information, and yet the number of studies on the possible presence of astronomical phenomena and concepts in the art, iconography, and architecture is exceedingly scant. Are *literary* texts conceived as the only mode in which astronomical information was stored and conveyed? We already know this was not the case in the New World when, given relatively few manuscripts to work with, investigators chose to expand their investigations to interpret the term 'text' to include the archaeological record as well as other forms of evidence.

The redefinition of what constitutes a 'text' has been one of the cardinal contributions of 20th century American anthropology. The Old World humanistic tradition to which classical history of astronomy belongs apparently has not taken this redefinition to heart, a development that is not surprising when we consider how difficult it is for a culture based largely upon written words to accept such a transition. Shortly after the turn of the century, Franz Boas, one of the founders of American anthropology, attempted to create an analogous body of written texts for New World peoples in order that their cultures might be studied on the model of the Greeks and Romans (Stocking, 1977, p. 5):

> The result of anthropological fieldwork carried out in this mode would be a body of material similar to that through which traditional European humanistic scholars studied earlier phases in the cultural history of literate peoples: physical remains of their art and industry; literary materials in which their history and cultural life were described in their own words; and grammatical material derived from the latter – all of them more or less direct expressions of the 'genius' of the people, as free as possible from the 'alternating sounds' imposed by the cultural categories of an outside observer.

Like written texts, architecture and iconography also are capable of yielding information about cultural behavior. And in archaeoastronomy one needs to comprehend the full cultural contexts of alignments and carvings as well as words, for all are subject to interpretation in the light of evidence. Perhaps we have become so acculturated to written words that we tend to assign them a more exalted level of meaning than other categories of evidence. I suggest that archaeology, like ethnohistory, is not simply a body of knowledge to be turned to, as a lesser alternative, when written texts do not happen to be available to back up an astronomical hypothesis. Rather, this human record, like literature, offers us the clay out of which to mold ideas and suppositions that can be tested empirically.[3]

While I have implied that in megalithic studies too much emphasis has been accorded astronomy, I suggest that in other cultures astronomical questions pursued under the banner of the revised definition of 'text' are worthy of far more attention. I believe Greek and Egyptian architecture ought to be submitted to the same organized and systematic interdisciplinary study that has been undertaken in Mesoamerica. That more energy has been expended in the search for astronomical meaning among carved rocks in the south-western USA than in all of the buildings of Classical Greece seems to me a serious imbalance of effort in the field of archaeoastronomy. I believe this imbalance results from two attitudinal preconceptions. First, we continue to perceive astronomical studies of the Old World literate cultures as belonging solely to the domain of the historians of astronomy; and, second, we harbor the collective psychological attitude that archaeoastronomical studies are unworthy of being conducted beyond the confines of one's own neighborhood. Finally, I believe the local working environment is thought to be more convenient and far less threatening to the researcher who elects to do work in archaeoastronomy, which is still considered to be one of the offbeat interdisciplinary branches of academia.

1.3 Changing course

This critical essay on the course of archaeoastronomy has been based on a personal examination of the way in which the interdiscipline has been practiced in recent times, and it has precipitated a number of conclusions. First, we adopt the only sensible definition of what it is we are all doing (in green and brown archaeoastronomy together), i.e. archaeoastronomy is the study of the practice and use of astronomy among the ancient cultures of the world based upon all forms of evidence, written and unwritten.[4]

Employing this definition, green archaeoastronomy, epitomized by the work on the megaliths, was found to utilize what was considered to be the only evidence available: the standing stones, and what was believed to be the only methodological recourse for dealing with that body of data: the statistical approach. Ruggles (1984) has capably described the evolution, levels, insights and pitfalls of green archaeoastronomy (see also Chapter 2). My discussion of the ceque system suggested that the hypothetical application of 'green' techniques to New World material without looking at allied evidence only magnifies how misleading the results can be even though they are conducted with extreme rigor. We have seen that archaeoastronomy of a slightly more 'brown' hue, but nevertheless with greenish overtones, also exists. The study of the role of astronomy in the design of the Nazca lines is a suitable example. But, still, our knowledge of the enduring concepts of the Andean world at the time of the Conquest as well as after the Conquest (e.g. the coastal Peruvian ethnoastronomical work by Urton, 1982), enables us to argue by analogy with greater conviction than is possible in the case of the megalithic studies.[5]

We also reach the conclusion that New World archaeoastronomy as reflected in the brown proceedings volume of the first Oxford conference is not equal to traditional history of astronomy. Rather, it has become an anthropology of astronomy which encompasses the highly focused, well-established discipline of history of astronomy while also embracing a number of other disciplines. Lest the historians of astronomy feel degraded by being forced taxonomically to occupy the same bed (a king-sized one!) as those whom they might tend to view as shabby brethren, let it be clear that the questions we all ask and the methods by which we seek to answer them do exhibit some overlap. Students of the Mesoamerican inscriptions who would aspire to the quality and rigor of the work of a Kugler or a Neugebauer would indeed set formidable goals for themselves. And there have been such scholars in the New World: Proskouriakoff, Teeple and Thompson to name but three.

We must cease carving up the scholarly world and the disciplines into specialties that we persist in believing are unrelated. A review of the content of the articles in the British *Journal for the History of Astronomy* and its *Archaeoastronomy: Supplement to the Journal for the History of Astronomy* only enhances the differences between the two archaeoastronomical courses of a different color that I have traced. Thus, the green version of archaeoastronomy has become isolated from the history of astronomy, while all the brown-type archaeoastronomical studies, whether based upon the study of historical matters or not, are classified as archaeoastronomy. The same sort of division, though more subtle, is nevertheless still apparent in the programs of the Historical Astronomy Division of the American Astronomical Society.

All studies of ancient astronomy should have at least one thing in common: ultimately they should be concerned with people and how they behave. Of course, we need to continue with work in our specialties, but at the same time we also must pay more attention to the questions our other colleagues, those on the other side of the disciplinary gap, are asking. If we intend to make lasting contributions to knowledge, then we must be even more daring in our ventures into these related disciplines – we must *learn* them. Serious archaeoastronomy is not a part-time occupation intended for weekend reflection or relaxation from one's normal travail. Given the breadth and depth of inquiry it spans, archaeoastronomy requires even more attention and focus than the traditional disciplines whose domains of inquiry are well explicated.

We find ourselves in the archaeoastronomy of the mid-1980s in a curious quandary. The questions asked by those trained in the quantitative sciences do not seem to elicit all that much interest in anthropological circles, and those astronomical matters that impress people engaged in the study of culture are given too little attention by the astronomers. It is as if we all pursue that which we find interesting rather than that which was important to the people whose astronomies we are studying. After all, ancient astronomy was what it was and it may have meant something entirely different to its practitioners than either scientist or social scientist can conceive. But the problem with archaeoastronomy lies not only in the different ways it is practiced, but also in how it is perceived. I believe the term 'archaeoastronomy' has become a curse for some of us, for literature carrying that label too often is taken as either simply entertaining or slightly crackpot by large segments of the established disciplines that go into making up its name. Too often it is simply not taken seriously, perhaps rightly so, given the unusual dose of sensationalism that goes with it. Frequently, we employ this term to validate what we are doing as science, when, in fact, archaeoastronomy – though it may employ some of the methods of science – is really a part of culture history rather than an endeavor to be isolated from the domain of the social sciences and moulded artificially into a branch of science.

Astronomy seems to have been a multi-faceted form of inquiry in the ancient mind. And so must we, too, approach it. Unless we change course and institute more interactive, interdisciplinary ways of proceeding, we run the risk of failing to solve some of the really important problems that confront us. And that brings us to the present world congress – a rare forum in which we can openly and critically discuss ideas and procedures, drawn from a wide spectrum of individuals with varied background and training – from all over the world – a union in which the ethnocentrism and disciplinary biases inherent in all of us hopefully will dissolve and

collectively refocus on the central question that concerns us all: What were they up to and why?

And so, whither archaeoastronomy? We have not simply lapsed into the Kuhnian phase of normal science wherein everyone simply carries on with the articulation of the archaeoastronomical paradigm laid down in the revolutionary sixties. On the contrary, I believe our most serious problems still lie ahead of us and, if they are resolved, we could make some of our most significant discoveries.

Acknowledgments

I am indebted to Gary Urton and Curtis Hinsley for their critical commentary on an earlier draft of this paper.

Notes

1 The rationale for the types of alignments sought is no less bizarre than that for the astronomical phenomena searched out to fit alignments. The history of the curious quest for the anticipated lunar horizon extrema in non-literate cultures of the Old and New Worlds is an excellent example, which I have recently traced (Aveni, 1987).

2 I argue that the determination of 'conceivable significance' must be not only a cultural derivative, but also an environmentally dependent one (Aveni, 1981b).

3 Yet ethnohistory remains weak evidence in the minds of many scientists (Ruggles, this volume, p. 24). For some investigators, archaeology even precedes ethnohistory in the pecking order of degree of validity of evidence. Thus, Rowe (1979, pp. 231–2) totally dismisses the internally consistent ethnohistoric arguments about the astronomical use of horizon pillars on the hills surrounding Cuzco by pointing out that these markers were destroyed by the Spaniards and we have not a hope of finding them because the hills have eroded severely since the Conquest. Unfortunately, he also confuses our data-gathering methods with those of the Inca, as is evidenced by his complaint that one does not need a transit to determine the solstices by observing the horizon (p. 231).

4 Unfortunately, it is still the case that many writers outside the interdiscipline view archaeoastronomy as simply the study of alignments; cf. Chippindale (1986).

5 While the significance of ethnoastronomical studies in the New World has recently been recognized, studies of contemporary folk astronomy in Europe are practi-

cally non-existent. Perhaps one assumes Europe has evolved so far beyond the preliterate world of Homer and Hesiod that peasant astronomy is non-existent. Yet, the planting seasons, general ecology, and certainly the skies of the remoter regions have not changed all that much. On a recent visit to Askra, Greece, where Hesiod lived and wrote his *Works and days* (see West, 1978), we were surprised to find the general environmental conditions scarcely different from the way Hesiod and later Pausanias (Jones, 1978) described them.

References

Aaboe, A. (1974). Scientific astronomy in antiquity. *Phil. Trans. Roy. Soc., (London) A* **276**, 21–42.

Aveni, A. (1981a). Horizon astronomy in Incaic Cuzco. In *Archaeoastronomy in the Americas*, ed. R. Williamson, pp. 305–18. Los Altos: Ballena.

Aveni, A. (1981b). Tropical archaeoastronomy. *Science* **213**, 161–71.

Aveni, A. ed. (1982). *Archaeoastronomy in the New World*. Cambridge University Press.

Aveni, A. ed. (n.d.). *The Lines of Nazca*, ed. A. F. Aveni. Philadelphia: American Philosophical Society (in press).

Aveni, A. (1987). Archaeoastronomy in the southwestern U.S.: a neighbor's eye view. In *Astronomy and Ceremony in the Prehistoric Southwest*, eds. J. B. Carlson and J. Judge. Albuquerque: Papers of the Maxwell Museum of Anthropology, no. 2, pp. 9–23.

Broda, J. (1982). Astronomy, cosmovision, and ideology in pre-Hispanic Mesoamerica. In *Ethnoastronomy and Archaeoastronomy in the American Tropics*, eds. A. Aveni and G. Urton, vol. 385, pp. 81–110. New York Academy of Science.

Chippindale, C. (1986). Stonehenge astronomy: anatomy of a modern myth. *Archaeology* **39**, (1), 48–52.

Dicks, D. R. (1970). *Early Greek Astronomy to Aristotle.* London: Thames and Hudson.

Heggie, D., ed. (1982). *Archaeoastronomy in the Old World.* Cambridge University Press.

Jones, W. (1978). *Pausanias' Description of Greece*, (4 vols.). Harvard University Press.

King-Hele, D. G. (1974). Concluding remarks. In *The Place of Astronomy in the Ancient World*, ed. F. R. Hodson. *Phil. Trans. Roy. Soc. (London) A* **276**, 273–5.

Levi-Strauss, C. (1969). *The Raw and the Cooked.* New York: Harper and Row.

McCluskey, S. (1987). Science, society, objectivity, and the astronomies of the southwest. In *Astronomy and Ceremony in the Prehistoric Southwest*, eds. J. B. Carlson and J. Judge. Albuquerque: Papers of the Maxwell Museum of Anthropology, no 2. pp. 205–17.

Nilsson, M. (1920). *Primitive Time Reckoning.* Lund: Gleerup.

Pedersen, O. (1982). The present position of archaeoastronomy. In *Archaeoastronomy in the Old World*, ed. D. Heggie, pp. 265–74. Cambridge University Press.

Price, D. (1975). *Science Since Babylon.* Yale University Press.

Rowe, J. (1979). Archaeoastronomy in Mesoamerica and Peru. *Latin Am. Res. Rev.* **14**, 227–33.

Ruggles, C. L. N. (1984). Megalithic astronomy: the last five years. *Vistas in Astron.* **27**, 231–89.

Schele, L. (1977). Palenque, the house of the dying sun. In *Native American Astronomy*, ed. A. Aveni, pp. 42–56. University of Texas Press.

Schele, L. (n.d.). *The Maya Message: Time, Text, and Image.* Paper presented at Symposium on Art and Communication, Israel Museum, Jerusalem.

Stocking, G. (1977). The aims of Boasian ethnography: creating the materials for traditional humanistic scholarship. *Hist. of Anth. Newsletter* **4**, (2), 4–5.

Urton, G. (1982). Astronomy and calendrics on the coast of Peru. In *Ethnoastronomy and Archaeoastronomy in the American Tropics*, eds. A. Aveni and G. Urton, vol. 385, pp. 231–48. New York Academy of Sciences.

Urton, G. (n.d.). The social historical context of the maintenance of the Nazca lines. In *The Lines of Nazca*, ed. A. F. Aveni. Philadelphia: American Philosophical Society (in press).

West, M. (1978). *Hesiod Works and Days.* Oxford University Press.

Zuidema, R. T. (1964). *The Ceque System of Cuzco.* Leiden: Brill.

Zuidema, R. T. (1977). The Inca calendar. In *Native American Astronomy*, ed. A. Aveni. University of Texas Press.

Zuidema, R. T. (1982). The sidereal lunar calendar of the Incas. In *Archaeoastronomy in the New World*, ed. A. Aveni, pp. 59–108. Cambridge University Press.

2

Recent developments in megalithic astronomy

C. L. N. Ruggles   *University of Leicester*

2.1 Introduction

It was undoubtedly British 'megalithic astronomy', and especially the intense astronomical interest in Stonehenge during the 1960s, that sparked off archaeoastronomy as a field of interest in its own right. Much of the first Oxford conference in 1981 was devoted to discussions of the astronomical significance of megalithic sites in the British Isles, written versions of papers on this topic almost filling an entire one of the two volumes of conference proceedings (Heggie, 1982).

To summarise all that has happened since, especially for an audience who generally have a much wider and more general interest in archaeoastronomy, and many of whom may have very little knowledge of the British material, is a difficult undertaking. The most significant recent event, however, is not hard to recognise: this was the death, in November 1985, of Alexander Thom. Although most eyes were elsewhere in the 1960s, when megalithic astronomy first caught the public imagination, it is without a doubt Thom's very extensive work, starting in the 1930s and continuing until only a year or so before his death, which has inspired and motivated much of the current work in megalithic astronomy – and, indirectly, in archaeoastronomy as a whole.

In this chapter I shall attempt firstly to describe some recent twists in 'classical'-style debates about individual megalithic sites and groups of sites. I shall then proceed to discuss some more recent, and rather different, approaches to particular problems. Having summarised the directions in which

I feel British archaeoastronomy is proceeding, I shall finish by discussing briefly the developing relationship between archaeoastronomical studies in prehistoric Britain and elsewhere.

A wide range of views is still held on many of the topics discussed in this paper. Some fundamental disagreements still remain amongst those actively involved in British archaeoastronomy. While I shall make clear my own overall views and interpretations, I shall try to represent alternative views and controversies as fairly as possible.

2.2 Lines and alignments: the continuing saga

Traditionally, 'megalithic astronomy' has consisted of measuring alignments of standing stones in search of cases of astronomical significance. Ever since the earliest antiquarians roamed the British countryside recording ancient remains, people have occasionally become fascinated by the idea that there might be an astronomical significance in many prehistoric monuments, and they have dedicated large amounts of time and energy in pursuit of this idea. Astronomer Norman Lockyer was perhaps the first to back up astronomical speculations with accurate measurements (Lockyer, 1909); and by far the most influential name of recent decades was that of engineer Alexander Thom, who published three books (Thom, 1967, 1971; Thom and Thom, 1978a) and a great many papers in a diverse collection of journals.

That free-standing megalithic sites – stone circles (more correctly termed stone rings, since many are nowhere near circular), rows of three or more standing stones, stone pairs and single standing menhirs – should have been the focus of almost all of this work is perhaps due to two factors. Firstly, megalithic sites have survived the passage of time peculiarly well: thus investigators particularly interested in astronomy – hardly ever with any formal training in archaeology – have tended to concentrate their attention upon these relatively conspicuous, and consequently well-known, monuments. Secondly, alignments along and between standing stones are easily picked out, and can easily be imagined to be 'indicating' something: the idea that such alignments indicated horizon astronomical events has dominated investigations of prehistoric astronomy, almost to the exclusion of any others.

Thus, over many years, the concept of megalithic astronomy was born, whereupon it rose to become a vigorous field of enquiry as well as one attracting considerable popular interest. There has been much talk of a 'megalithic culture' harbouring astronomical (and geometrical) expertise far in advance of its time and hitherto totally unexpected: 'Megalithic Man', according to Thom (1971, ch. 1), was a competent engineer, had an extensive knowledge of practical geometry, and studied the movements of the sun and moon in great detail.

In fact, the archaeological perspective on the megalithic tradition is rather different. Our current vision of prehistoric societies in Britain depends not only on the evidence of spectacular burial and ritual monuments, into which category the megalithic monuments fall, but also on that from settlement sites, artefacts and environmental data. It is clear that there was no 'megalithic culture' as such, and no 'Megalithic Man': instead megalithic monuments of various types span a period of about four millennia, being built at various stages in a long and complex period of social change, and are related to many independent traditions throughout the northern and western parts of the British Isles. The very term 'megalithic astronomy' is a misleading one, for we should really refer to studies of astronomical practice in prehistoric Britain, 'prehistoric British archaeoastronomy' perhaps; nonetheless the traditional term 'megalithic astronomy' is retained in this chapter.

Despite the diversity of prehistoric culture in Britain, archaeoastronomical enquiries in what one might term the 'classic' mould (that is with studies of alignments at megalithic sites as their principal aim) have continued on two levels. On one level,

there is still considerable debate about a few individual sites, but the emphasis tends to be upon archaeological verification of alignments (some would prefer to use the word 'orientations') of possible astronomical significance. On the other level, there have been fresh investigations of groups of sites with a strong emphasis upon statistical verification. The idea is that, since any alignment of apparent astronomical significance may have arisen through a combination of factors quite unrelated to astronomy, one must seek a sample of alignments from several sites taken together, and look to see whether any orientation trends emerge which *necessitate* an astronomical explanation.

Astronomical alignments at individual sites

Newgrange and Maes Howe   The passage grave at Newgrange in Ireland is perhaps the least contentious example of a monument widely quoted as incorporating an astronomical alignment. Built somewhat before 3000 BC, it is the largest of the vast Boyne Valley passage graves. A 19 m long passage, with ample headroom for the visitor, leads from the entrance to a large central chamber with a corbelled ceiling up to 6 m high. Three side chambers open out from the central one, each housing a stone basin which probably held the cremated remains of the people for whom the tomb was built. The entire passage grave is surrounded by a circle of standing stones.

Newgrange was excavated and restored by O'Kelly (1971). Over the entrance to the passage he discovered a mysterious 'roof-box' which would have continued to admit light into the interior after the entrance was sealed. His own observations, followed up by a survey by Patrick (1974), established that the sun's rays will enter the roof-box, penetrate the entire length of the passage and illuminate the central chamber just after sunrise if the sun's declination lies between about $-23°.0$ and $-25°.9$. Thus, direct sunlight reached the centre of the tomb every morning for about a week on either side of the winter solstice.

Given that south-easterly orientations are quite common amongst passage graves, and may be favoured for other reasons, the probability of a chance solstitial alignment is appreciably high; however, the presence of the roof-box (not readily explicable otherwise) in just the right position for the shaft of light to penetrate the entire passage has made its intentionality almost universally accepted amongst archaeologists and astronomers alike.

Maes Howe was built around 2500 BC and is the greatest of the tombs in the Scottish Orkney islands. Its entrance is oriented south west, and here the setting sun's rays around the midwinter solstice illuminate the rear wall of the central chamber (Moir, 1981). The stone that blocked the passage was not quite tall enough to fit to the top of the entrance, and it has been suggested (Burl, 1981a) that the shortfall was to allow, as at Newgrange, the sun's light to enter the tomb after it had been sealed.

There are some problems remaining at Maes Howe, and certain pitfalls in placing too much emphasis on apparent similarities between the two sites (see Moir, 1981); however, Newgrange and Maes Howe do seem to provide evidence of (low-precision) astronomical alignments being incorporated into monumental architecture as early as about 3000 BC. At the same time the sites were in no sense 'observatories': rather, the light of the midwinter sun was meant, for whatever reason, to fall upon the bones of the dead.

The classic Thom solstitial observatories: Ballochroy and Kintraw Ballochroy, a 5 m long three-stone row situated on the coast of Kintyre, western Scotland, has been a source of debate ever since Thom first mentioned it in an early paper (Thom, 1954; see also Thom, 1967, 1971; MacKie, 1974; Bailey *et al.*, 1975; Burl, 1979). Two indications at the site, along the alignment and across it, appear to provide solstitial indications using distant natural foresights accurate to one or two minutes of arc. A recent twist in the debate has been the discovery by Burl of a sketch of the site made in 1700. This indicates that one of the two high-precision foresights was originally obscured, and implies that this alignment is fortuitous. It does not however necessarily mean that the site itself was not aligned upon the solstitial setting sun, but just that this was not set up to a high precision (Burl, 1983).

The story of Kintraw, some 50 km north of Ballochroy, is one of the best known in megalithic astronomy (Thom, 1969; MacKie, 1974). An archaeological excavation was carried out in order to verify a postulated observing platform perfectly positioned for high-precision solstitial observations

in a distant horizon notch. The excavation sadly proved inconclusive (MacKie, Gladwin and Roy, 1985). Of all the points that have been raised in debate about the site (McGreery, 1980a; MacKie, 1981; Patrick, 1981; McGreery, Hastie and Moulds, 1982) perhaps the most telling, although it is one only mentioned in passing by McGreery *et al.*, is that an observer along the ridge to the south east of the platform certainly would not have been able to view the notch. Yet such observations would have been crucial to setting up the sightline in the first place, i.e. in determining the position of the platform, by making observations of sunsets before and after the solstice (see Thom, 1971, pp. 13–14). This creates severe difficulties for the theory that Kintraw represented a high-precision solstitial sightline.

Minard (Brainport Bay), Argyll The site at Brainport Bay in Argyll, again in western Scotland, which has been uncovered in recent years, is an important one in the context of the megalithic astronomy debate. It is a complex involving an alignment of features – a natural outcrop, which has been paved and terraced, two large boulders and a 'back platform', which has also been artificially modified. The alignment is oriented to the north east towards the only distant horizon visible from the site – at the far end of Loch Fyne. A notch on this distant horizon would mark the upper limb of the rising sun at about a fortnight either side of midsummer solstice.

Excavations by Gladwin, described by MacKie (1981), uncovered two sockets in the main outcrop and, nearby, two small stone slabs into which they fitted well. As viewed from the boulders, dubbed 'observation boulders' by MacKie, the shape of the outcrop frames the distant notch, and the slabs, if erect, would have resembled 'the sights of a huge rifle' pointing at the notch.

Subsequent excavation (MacKie *et al.*, 1985) established carbon 14 dates around 1400 BC, and uncovered more specific evidence of human occupation such as scatters of flint flakes. At the same time the site was not, it seems, primarily intended for domestic, defensive, industrial or funerary use. The 'observation boulders' have, however, been in place since well before Neolithic times. Their presence might be explained by the chance discovery in Neolithic times of a site suitable for rough orientation towards midsummer solstice. MacKie's subsequent work uncovered evidence of a further, equinoctial, alignment at the site, again involving man-made features and a horizon notch.

The site is important because there exist archaeological data which appear to corroborate the hypothesis of constructions oriented upon solar phenomena. MacKie himself sees his work as a successful archaeological test of the hypothesis of high-precision observations using natural foresights. He insists that the idea of high-precision solar observation (and hence Thom's solar calendar) is supported. However, this interpretation rests upon the efficacy of the horizon notches, which are not located exactly at the solstice or at the 'calendrical' equinox (Thom, 1967, ch. 9) and might easily have arisen by chance. The debate continues.

Stonehenge Stonehenge has to be mentioned. Being such a well-known monument, it has attracted a grossly disproportionate amount of astronomical attention, especially during the 1960s. A detailed and balanced appraisal of many of these theories has been given by Heggie (1981) and there is no need to repeat the various arguments here. Thankfully it seems that the tide of unwarranted speculation has abated quite markedly in recent years.

The only reasonably uncontentious statement that can be made about Stonehenge is that the general orientation of the axis of the monument at various stages in its development is towards midsummer sunrise and midwinter sunset, and that this may well have been deliberate. A precision in azimuth of at best a degree or two of arc is involved; the popular notion that the Heel Stone defined the direction of solstitial sunrise more precisely is quite unsupportable, because the supposed observing position (the centre of the monument) cannot be defined precisely enough on the archaeological evidence, and, even it if could be, the Heel Stone is too near to provide an accurate foresight and the horizon behind it is featureless.

The practice most common at Stonehenge has been to identify certain features at the site and to fit a theory to 'explain' them. Even when this is done impartially there are grave dangers in attributing astronomical (and geometrical) frameworks to what is a very limited sample of the original features at the site, namely those which in AD 2000 are superficially obvious, those which happen to have been excavated when large areas of the site are

still unexplored, and so on. These dangers have been amply expounded by Pitts (1981a, b) in the context of the discovery of a possible companion to the Heel Stone.

Some of the most famous astronomical theories at Stonehenge (Hawkins, 1963; Hawkins and White, 1966) depend upon statistical arguments about the number of astronomical alignments between pairs of points chosen as possibly significant. These arguments fall down on many different grounds: lack of *a priori* justification for the points chosen in the first place, and archaeological doubts about some of those that were (Atkinson, 1966); numerical flaws in the probability calculation (see, e.g., Ruggles, 1981a); and, perhaps most importantly, the non-independence of data (e.g. when the precision of alignment being considered is only about one degree and the horizon is flat, a line between a pair of points indicating midsummer sunrise in one direction will inevitably indicate midwinter sunset in the other). It is truly unfortunate that the topic of megalithic astronomy is so coloured by the excessive attention that this one site, and some highly speculative interpretations of it, have received over the years.

Statistically based projects and reappraisals

Reassessments of Thom's analyses In contrast to these discussions of individual sites, arguably the most important evidence presented over the years by A. Thom (and later by A. Thom and A. S. Thom) arises from analyses of groups of sites taken together. The evidence is cumulative in nature. At each stage in the work it consists essentially of one or more analyses of many putative indications from a number of megalithic sites, and at each stage the level of precision of indication being tested is greater than at previous stages. The later work concentrates on lunar sightlines. Four distinct stages have been identified by the current author (Ruggles, 1981b) as follows.

Level 1 In his earliest work Thom measured the declinations indicated by 72 structures at 39 megalithic sites, plotting the results in the form of a probability histogram, or 'curvigram' (Thom, 1955, Fig. 8). This evidence was later extended to 262 structure indications at 145 sites (Thom, 1967, Fig. 8.1). On the basis of these curvigrams and associated statistical tests, Thom suggested the existence of deliberate solar, lunar and stellar alignments set up to a precision of about half a degree. The solar targets concerned were the solar solstices, equinoxes and intermediate declinations representing equal division of the year into eight and maybe 16 parts. The lunar targets were the four declinations which represent the extremes of the moon's limiting monthly motions in its 18.6 year cycle – declinations referred to by Thom as the lunar 'standstills'.

Level 2   In 1967 Thom analysed further those Level 1 indications which fell near the solar solstitial declinations, about 30 of them (Thom, 1967, ch. 9), and found evidence that the upper and lower limbs were preferentially observed; the further analysis of the lunar lines (ch. 10), of which there are about 40, suggests the same thing. This increased the inferred precision to about ten minutes of arc.

Level 3   In his later work Thom considered an effectively distinct hypothesis, namely that distant horizon features such as notches provided natural foresights, and that structures at the sites themselves merely served to identify the observing position and which foresight was to be used. Because of the accuracy of any individual foresight, declinations could now be quoted to a minute or two of arc, and the idea of sightlines of much higher precision could be tested. The analysis of 40 distant horizon features at 23 sites (Thom, 1971, ch. 7) suggests the use of distant foresights for observations accurate to three minutes of arc or better, and to record the 173 day, nine minutes of arc perturbation in the moon's motion (Thom, 1971, ch. 2; see also Morrison, 1980).

Level 4   The analysis at Level 3 took no account of small variable corrections to the declinations, such as variable parallax. At Level 4, represented in three papers by Thom and his son A. S. Thom after 1978 (Thom and Thom, 1978b, 1980; A. S. Thom, 1981), each sightline was considered on its own merits, taking into account the time of year and time of day of presumed use, various small corrections being made accordingly. In addition, the data set was restricted to those few lines the Thoms considered to be most convincing on the ground. The resulting curvigram peaks were more clear-cut than at previous levels, and when the analysis

was reformulated in a more rigorous way as suggested by Morrison (1980), the sightlines seemed so accurate (to better than a minute of arc) that it seemed they could only have been set up at the end of an averaging process lasting some 180 years (A. S. Thom, 1981, p. 38).

There have been a number of critical assessments of the data at different levels (Heggie, 1972; McGreery, 1979, 1980b; Moir, 1980, 1981; Ellegård, 1981). Many authors have pointed out that selection decisions as to which data to analyse in the first place – which sites, which potential indicating devices at each site, and which horizon features taken to be indicated – are subjective and that the particular selection decisions made by Thom may not be justifiable on *a priori* grounds. Others have pointed out that lunar observations of the precision claimed at Levels 3 and 4 simply may not be feasible in practice.

Thorough reassessments of the evidence at each of the four levels have been attempted in recent years by the present author, based upon first-hand examination of each of the sites in question and upon resurveys where appropriate. The reassessments at Levels 2 and 3 (Ruggles, 1981b) were described in the proceedings of the first Oxford conference (Ruggles, 1982a). Those at Level 4 were the subject of two further papers (Ruggles, 1982b, 1983). It was concluded that the argument for very high-precision lunar indications can be convincingly challenged on any one of four different grounds, namely

  (i) the archaeological status of the sites involved;
 (ii) theoretical (astronomical) considerations, i.e. what is actually possible;
(iii) by examining the selection of data (i.e. whether data particularly favourable to the astronomical hypothesis have been preferentially selected and other data preferentially ignored; and
(iv) by reassessing the methods of statistical analysis.

When all four factors are taken together the evidence is overwhelmingly against deliberate lunar indications of very high precision.

A new study of 300 western Scottish sites   At Level 1, the principal questions to be answered in order to reassess Thom's data are to do with the selection of data for consideration, and these questions cannot be adequately answered by simply revisiting those sites considered by Thom. For this reason a surveying programme in western Scotland was begun in 1975; it was finally completed in 1981, after more than 300 sites had been investigated, and the results were published in 1984 (Ruggles, 1984a). A code of practice was laid down both for selecting sites for consideration in the first place, and for selecting potential indications at each site encountered. This code of practice was then strictly adhered to during fieldwork, in order to avoid any accusation of selective bias in the data. As the project progressed, a reassessment of Thom's work was no longer seen as the principal aim of the project; it was more an attempt to lay a new methodological framework for assessing alignments of possible astronomical significance at ancient sites.

The results manifested trends at three levels of precision. At the lowest level, declinations between about $-15°$ and $+15°$ were strongly avoided, suggesting a general preference for structures to be oriented N–S, NW–SE or NE–SW rather than E–W. At the second level, there was a marked preference for southern declinations between $-31°$ and $-19°$, a range which very closely matches (i.e. matches to within about a degree at either end) the range of declinations between the two lunar standstills (i.e. the range of declinations which the moon can reach at the southern limit of its monthly motions). At the most precise level, there was marginal evidence of a preference for six particular declination values to within a precision of one or two degrees. Three of the declinations ($-30°$, $+18°$ and $+27°$) may indicate a specific interest in the lunar standstills, and would imply that organised observations were undertaken over periods of at least 20 years. There is no coherent astronomical explanation for the other three values ($-25°$, $-22°.5$ and $+33°$), but the first of these might indicate an interest in the winter solstice.

Although there was clear evidence of lunar orientation, and marginal evidence of orientation upon the winter solstice, no evidence whatsoever was found for an interest in the summer solstice or equinoxes (indeed, declinations in the vicinity of the equinoxes were strongly avoided). This was in distinct contrast to the earlier conclusions of Thom (1967).

Perhaps the most interesting result to emerge from the project was that a certain group of sites featured predominantly amongst the indications falling in the 'preferred' declination interval between $-31°$ and $-19°$; the three-, four- and five-stone rows of mainland Argyll and the inner Hebridean island of Mull. These small sites, linear in form, are found here in marked preference to the stone rings prevalent in so many other parts of the country. The project motivated the more detailed study of this particular group of sites.

Conclusions reached

The overall picture that emerges from recent studies of lines and alignments is one of low-precision astronomical alignments incorporated in the design of a range of burial and ritual sites. The evidence for high-precision alignments is generally much more shaky.

Early communal tombs like Newgrange and Maes Howe seem to have been orientated so that the rays of the rising or setting sun around the winter solstice could fall upon the bones of the dead. The general orientation of Stonehenge, set up early in the evolution of the site, may also represent an alignment upon the setting midwinter sun in the south-west rather than, as is commonly supposed, one upon the rising midsummer sun in the north-east.

The small megalithic sites of western Scotland – mainly individual standing stones and linear settings – are the product of a very different social structure in a particular geographical area at a somewhat later period of British prehistory. The statistical study described above seems to indicate that these sites are preferentially oriented upon the southern moon; but there is no evidence for high-precision lunar observations or for a precise solar calendar. At individual sites such as Ballochroy the idea that the site, linear in form, is oriented upon the sun or moon setting in the south may be far nearer the truth than the idea that it marks solstitial sunsets to high precision using distant notches. A low-precision interest in the rising sun may also account for the orientation of the linear site at Minard.

2.3 Some new approaches to studying astronomy in prehistoric Britain

The overriding flavour of the megalithic astronomy debate, even in recent years, has been a great deal of talk about the precision of alignments and statistical rigour, and very little about the nature of the sites themselves and the people who built them. The ultimate quest remains, for many investigators, to examine the nature of astronomical observations made in prehistoric times: other archaeological evidence merely serves as an arbiter in the astronomical debate, or else to provide background evidence.

From the point of view of the great majority of archaeologists this whole approach has always been questionable. Why concentrate upon astronomy when so many other factors might also have motivated the construction, design and use of sites about which we know so little? The answer is that this is the approach of the astronomer or the historian of science, and that it is a clear reflection of the backgrounds of the majority of workers in this topic. The archaeologists's approach is inevitably more broad-based, and in recent years a number of archaeological investigations have been undertaken in which astronomy is considered as merely one of a number of possible factors influencing site placement and orientation. These studies have proceeded largely in isolation from those described so far. They have started to uncover evidence that astronomical considerations were a factor influencing the siting and design of a variety of different types of monument at different times in British prehistory, and have started to give us some clues about the role played by astronomy in the lives of the people concerned.

An example is the work, described below, of Fraser (1983) on the chambered cairns of Orkney. He has attempted to draw together many lines of evidence relating to the people who inhabited Orkney during the third and second millennia BC and the land on which they lived. His main source of data is the Orkney chambered cairns: their architectural form, their relationship to other material evidence (other monuments, artefacts and burial evidence), and their spatial relationships.

At the same time, the archaeoastronomers have become more aware of the need to combine their statistical analyses of azimuths and declinations with archaeological evidence which is less tangible: in short to combine numerical skills with more interpretative discussions of more subjective evidence. Some joint enterprises, such as that between the present author and Burl on the eastern Scottish recumbent stone circles (RSCs) (Ruggles, 1984b; Ruggles and Burl, 1985), again discussed below, have attempted to move in this direction. A further attempt has been made by the present author (Ruggles, 1985) in looking more closely at the Argyll and Mull linear sites singled out by the statistical study described above.

Astronomy as a factor in site placement and orientation: the chambered cairns of Orkney

There are at least 76 chambered cairns in the Orkney islands, of which the spectacular Maes Howe, mentioned earlier in this paper, is perhaps the best known. Other monuments of the same age in Orkney include four settlement sites, two henges and a number of individual standing stones. On the basis of the artefactual evidence Fraser (1983) concludes, in contrast to others (e.g. MacKie, 1977), that chambered cairns were places where the ordinary activities of life were carried out, rather than places where activities of special significance, requiring the manufacture and use of special objects (such as ritual objects), took place.

Fraser undertook a locational analysis in an attempt to explore the relationships between the chambered cairns and the landscape of Orkney. Various aspects of the landscape were identified, namely geology, soils, land use capability, topography, vegetation, altitude, visibility from a location, drainage, ease of approach to a location and the nature of the nearest coast. The distribution pattern of the chambered cairns was examined in relation to each of these aspects in turn, and then to all the factors taken together.

Some of the conclusions were as follows:
(i) The accessibility of building stone may have constrained the cairn-builders to prefer land with no superficial deposits (such as a mantle of peat) over land with such deposits.
(ii) The chambered cairns seem preferentially to be located near to soils which were easy to work and freely drained, i.e. on or near the farming land of the builders.
(iii) They were located in a variety of topographic positions (i.e. valley bottoms, coastal plains, hill slopes, hill tops, etc.)
(iv) The distribution of present-day vegetation does not appear to be correlated with the distribution of chambered cairns.
(v) The cairns were not located in places with extensive sectors of restricted visibility.
(vi) They were generally located in positions which are easy to approach.

(vii) They show a pronounced tendency to be located close to the coast: high cliffs were preferred over major beach systems as the nearest coastal type.

Fraser proceeded to examine various relationships and artefacts 'most readily open to interpretation as symbolic'. One of these was symbolism connected with the sky, such as the orientation of cairns with respect to the surrounding land and the visible sky; others were the intervisibility of cairns, the demarcation of space within and around cairns, animal and bird bones found within cairns, and decorations (such as spirals) found on sculpted stones.

Conclusions from studies of the cairn orientations were as follows.

(i) Those cairns which are elongated seem preferentially to be oriented in the NW–SE quadrants.

(ii) Amongst those cairns with elongated chambers, there is no pattern of orientation amongst the chambers.

(iii) As a group, the known chambered cairns of Orkney have passages which are preferentially oriented outwards from the chamber to the south and east.

The azimuth distribution of distant visibility was found to be bimodal, with a major peak around 140° and a minor one around 270°. One major trough is apparent around 0°. The pattern was more pronounced than those for either intermediate or restricted visibility, suggesting that the distant visibility in certain directions, if anything, was the important factor.

The question now is what is special about these particular points of the compass that requires distant visibility. The possibility of local topography influencing orientation of visibility can be discounted since the cairns are widely dispersed over 20 islands. Although astronomical design cannot be tied down very precisely with measurements of this accuracy, it is interesting to note that, at the latitude of Orkney when the cairns were erected, midwinter sunrise occurred at an azimuth of around 140°. Azimuth 270° is, of course, that of the equinoctial setting sun.

Fraser concludes that the neolithic inhabitants of Orkney situated their chambered cairns in places which were chosen so as to make certain solar observations possible. These observations could have been of importance to a farming people in signalling significant times in the agricultural year. He also suggests that the chambered cairn, situated in the physical centre and social heart of the community, would have been the obvious meeting place to rejoice and celebrate the changing of the seasons.

Fraser's work gives a feel for the difference in approach between an archaeologist concerned only in passing with astronomical symbolism and an astronomer or engineer particularly seeking evidence of astronomical achievement.

A more interpretative approach: the recumbent stone circles of eastern Scotland

Situated in and around Aberdeenshire, in eastern Scotland, are about 100 RSCs, about half of which are in a reasonable state of preservation. They are stone rings distinguished by the presence of a heavy recumbent slab flanked by two upright pillars, which seem to be of central importance to the function of the sites. The heavy recumbent was carefully chosen, often transported from some distance and placed in position with great care. The circle stones, in contrast, seem to have been added rather more casually later, and in some cases may never have been added at all. The RSCs seem to be the elegant but simple ritual centres for small, egalitarian groups of subsistence farmers living in Aberdeenshire in the late third and early second millennium BC. Various general features in their placement and design seem to reflect a ritual tradition that was adhered to over a wide area – some 80 × 50 km.

At each site the line joining the centre of the ring and the centre of the recumbent stone provides an obvious principal axis. The orientations of the principal axes are highly clustered in azimuth (between SSE and WSW), a highly significant general trend which cannot be explained either by orientations upon a particular terrestrial feature or by the local topography.

A new study of the Aberdeenshire RSCs was undertaken in 1981 by the author in association with H. A. W. Burl (Ruggles, 1984b; Ruggles and Burl, 1985). Sixty-four sites were visited and examined. Many were surveyed or resurveyed. While great attention was paid to the fair selection of data, no rigorous statistical tests were presented. We tried instead to present the new data and to give some initial, general interpretations which integrated the astronomical and archaeological evidence rather than concentrating upon the astronomy alone. The idea was that further fieldwork and more rigorous statistical analyses could then begin to investigate our initial ideas further.

We concluded that astronomy was undoubtedly a factor, perhaps an important one, in the function of the sites. At the same time, their relatively small diameters and wide recumbents imply that they were not sites of great astronomical precision. From the archaeological evidence it seems likely that the sites had a simple ritual significance.

Although the top of the recumbent always lay below the natural horizon, as viewed from within or across the ring, the flankers usually projected above it – an attempt, perhaps, to reproduce an astronomical 'letterbox' rather like those at Newgrange and Maes Howe. In no case is the horizon above the recumbent nearer than 1 km distant, implying that there was an interest in celestial phenomena near to the horizon occurring behind the recumbent.

There seems, however, to be no simple, direct interpretation of the pattern of indicated declinations in terms of observations of the sun or moon. Burl's earlier suggestion (Burl, 1980) that the sites might have been oriented so that the major standstill moon would pass above the axial line (i.e. the centre of the recumbent stone was viewed from the centre of the site) was ruled out by the large number of declinations found to be somewhat *above* $-30°.0$. There are, however, definite indications of an interest in the rising or setting major standstill moon (and occasionally the minor standstill moon) *generally* over the recumbent stone. This idea seems to be corroborated by the azimuths of cupmarked stones at the sites.

However, there is no simple explanation of all the data, astronomical or otherwise. Instead we find evidence of conflicting concerns: orientation upon the major or minor standstill moon; the presence of conspicuous hilltops within the horizon above the recumbent; a recumbent stone facing due south. These separate goals would rarely have been achievable simultaneously, and we surmise that compromises were often reached. It is very unlikely that the positioning of the recumbent stone was exclusively astronomical.

Several directions emerged for future work. Firstly, at the RSCs themselves it would be of great interest to investigate more thoroughly and objec-

tively conspicuous hilltops occurring within the indicated horizon. Such data might be augmented from sites in a poorer state of repair than those considered in 1981. Secondly excavation, should adequate resources become available, may be able to clarify many of the apparent overall trends and regional variations deduced from the surface evidence. Finally, studies of related groups of sites such as the Clava cairns of neighbouring Invernessshire, which show a similar overall orientation trend to the RSCs, or the other group of RSCs, found in south-west Ireland, would be of the greatest interest.

Combining a statistical with an interpretative approach: the linear stone settings of Argyll and Mull

Following the results of the statistical survey of 300 western Scottish sites, described above, fieldwork was undertaken in 1985 in order to examine more closely the southern indications at the linear stone settings of Argyll and Mull – that is, the three-, four- and five-stone rows, aligned pairs of stones and single, flat slabs (Ruggles, 1985).

Stone rows and aligned pairs with local horizons to the south, excluded from the earlier analysis but surveyed during 1985, were found to fit the general pattern of indicated declinations between about $-31°$ and $-19°$. However, the distribution of indicated declinations within this range was far from that to be expected if sites were merely oriented upon the limiting monthly moon at arbitrary points in the 18.6 year cycle. Instead, a grouping of indications was found within a degree or two of $-30°$ and a second, rather wider, grouping centred upon $-23°$. In the two main concentrations of stone rows and aligned pairs of stones – northern Mull and the Kilmartin area of Argyll – a structure yielding an indication in the second group was invariably situated close to another yielding a declination of around $-30°$. Outside the main geographical concentrations, the rows and aligned pairs of Argyll and Mull seem to fit a more general pattern of orientation between $-30°$ and $-19°$.

The southern declinations obtained from single slabs and menhirs with an obvious longer axis were much more scattered – a fact perhaps hardly surprising in view of the uncertainties inherent in determining an indication from the current disposition of such stones, together with the range of possible purposes for which such stones might have been erected. However, there was some evidence that the orientation pattern observed amongst the rows and aligned pairs extends to certain single standing stones.

It was found that, while the orientations of several sites do reflect the local topography, several others clearly do not. Thus it is eminently possible that the lie of the land was taken into account in choosing a convenient situation for a ceremonial site within the territory available to a particular set of builders, and also when deciding between a rising or setting indication; however, geographical factors such as the lie of the land cannot on their own explain the declination trends.

Thus it was concluded that the orientation of many of the stone rows and aligned pairs of stones in Mull and mainland Argyll was of importance to the builders, and that astronomical, and particularly lunar, alignment was an important factor in determining this orientation. In two areas where particular concentrations of such sites are found – northern Mull and the Kilmartin area of Argyll – it seems that a primary target for orientation was the major standstill moon in the south. There is also some evidence pointing to a habit of secondary orientation, achieved perhaps by the contrary orientation of individual slabs in a longer alignment (as at Kilmartin), or else by setting up a second structure close by and roughly parallel to the first (as at Duncracaig) or else again by setting up a second structure at a short distance from the first (as at Dervaig and Dunamuck). These secondary structures may have been oriented upon the southern limiting moon at another point in the 18.6 year cycle, but then it is perhaps somewhat surprising that no declinations higher than $-21°$ are observed. It is also possible that they were oriented upon the midwinter sun. The northern declinations indicated by these secondary structures seem to manifest less of a consistent pattern. Surveys of the southern profiles indicated by the two wide slabs at Ballochroy, which are oriented across the alignment, may throw further light on the matter. The design of the Kilmartin (Temple Wood) site may even represent the culmination of a local practice, managing to incorporate two different orientations of significance into a single, more complex arrangement of stones. It is interesting to contrast this picture, built up by examining all the sites of similar form in a particular area, with the ideas of Thom about Ballochroy and Temple Wood.

Conclusions reached

The recent work described above points to both solar and lunar significance amongst prehistoric British burial and ritual monuments of different periods.

Some of the early, large megalithic tombs seem to incorporate solar orientations: there are some well-known individual cases of winter solstitial alignments in megalithic tombs, such as those at Newgrange and Maes Howe, and some indication from a group study of the Orkney tombs of solstitial and equinoctial orientation. It is extremely interesting to note that it appears always to be the *winter* solstice that is involved – a fact which may come as no surprise from the ethnographic record, where winter solstice ceremonies are often of great importance, the sun needing to be turned back from its southward movement (Thorpe, 1981; Chamberlain, 1988).

The evidence both from the RSCs in the east of Scotland and the linear stone settings in the west lends considerable support to the idea of low-precision lunar observation, at least of the full moon nearest to the summer solstice. More tentative evidence from the linear settings seems to indicate a 'primary' orientation upon the most southerly (major standstill) moon, a practice which would have required organised lunar observations over a period of at least 20 years.

Burl has drawn attention to independent archaeological evidence supporting the idea of an interest in the moon (Burl, 1980, 1981b, 1983). Nonetheless it is pertinent to ask to what extent even those who do not accept Thom's idea of high-precision lunar alignments have unwittingly been guided by this earlier idea in finding evidence of lunar orientation, albeit at a lower precision.

That observations of the moon might have been important to people in prehistoric Britain comes as little surprise in view of the ethnographic record. The cycle of lunar phases is the most obvious cycle in the sky after the daily one; it demarcates time periods of a convenient length; and it closely matches natural cycles such as the human menstrual cycle. Knowledge of the lunar month is almost universal, and many seasonal calendars are lunar-based (Baity, 1973; Thorpe, 1981; Chamberlain,

1988). On the other hand – and this is perhaps more worrying – ethnographic instances of horizon lunar observations are unknown, at least to the present author. One possible explanation for their importance in prehistoric Scotland would be the particular latitude, whereby the major standstill moon is seen to scrape along the northern or southern horizon – a rare and spectacular event which could well have assumed great importance. Clearly it would greatly support the case for the reality of the horizon lunar observations in prehistoric Britain if evidence were uncovered of similar practices from communities in similar latitudes.

Regarding the interpretation of astronomical practice in prehistoric Britain in its social context, the 'science vs. ceremonial' debate smoulders on, fuelled by those who insist that evidence in favour of high-precision observations becomes evidence in favour of 'scientific' astronomy, while refutation of high-precision indications and evidence in favour of low-precision ones becomes evidence in favour of 'ceremonial' astronomy. Strongly on the side of 'scientific' astronomy is, for example, MacKie (1981; MacKie *et al.*, 1985); in the other camp are authors such as Burl (1980, 1981b, 1983), Barnatt and Pierpoint (1983) and Fraser (1984). In fairness, all these authors derive support for their view from independent archaeological evidence. Thus MacKie (1977) believes that a 'two-tier' social system was in operation in some megalith-building societies, the top tier being an élite of astronomer priests undertaking intricate astronomical observations. Burl, on the other hand, cites the funerary association of much ancient astronomy in support of the idea that it was a ritual, not a scientific, practice.

In fact, the rigid dichotomy between science and ceremonial is undoubtedly misleading; we need to consider first the existence and precision of intentional orientations, astronomical and otherwise, and then to interpret them using far more than two crude (and actually very vaguely defined) categories. As Renfrew (1981) has pointed out, to insist on such a rigid and ultimately subjective dichotomy between science and non-science serves only to obscure the nature of what surely ranks as an interesting development in the evolution of scientific thought.

2.4 Discussion

Lessons learned

British archaeoastronomers have learned the hard way about the dangers of protracted discussion about individual sites. Burl (1981b) has pointed out some poignant examples where convincing astronomical alignments are present at individual sites, but when these sites are seen as part of an archaeological group the trend is not repeated and the alignment then appears to have come about by chance. The present author was disillusioned about the solstitial interpretation of Ballochroy when he first visited a similar three-stone row at Duachy, which was not solstitially oriented.

Hence, studies of groups of sites have recently come into prominence. It is surely from the study of large, integral groups of sites, such as the Scottish RSCs, that the most persuasive evidence is likely to come in the early stages of archaeoastronomical research. There is, however, a considerable problem in marrying together the 'statistical' approach of the numerate scientist and the 'interpretative' approach of the archaeologist or ethnographer.

The advantage of the statistical approach is to pick out trends amongst large quantities of diverse data and focus one's attention upon interesting subsets of the data. It also serves, importantly, to counter unwarranted speculation on the purpose of structure orientations, and to prevent the accumulation of evidence arrived at merely by preferentially selecting oriented structures which appear to support a favoured idea and excluding others which do not.

The more interpretative approach, however, is also necessary in order to take account of cultural diversity and human perversity. It would be wholly unreasonable to expect the design and function of a complete group of superficially similar sites to have conformed exactly to any overriding 'master plan' or purpose.

How, then, do we merge the two approaches? In the present author's view statistical rigour must precede interpretative reasoning. The former serves to isolate any general trends which show up above the background 'random noise' in the data and to focus one's attention upon interesting subsets of the data. The latter then enables one to examine this part of the data and these trends in more detail and in their full archaeological and cultural context.

In the work on the western Scottish linear stone settings (Ruggles, 1985) the present author has attempted to make explicit the transition from the original analysis with its rigid selection criteria, through the further investigation of apparent trends using predetermined (though rather more subjective) criteria, to a more interpretative presentation of the data which is based upon a reappraisal of the archaeological status and current state of repair of the sites, but which also takes into account retrospectively the indicated declinations that were obtained.

A futher way to view the study of lines and alignments at prehistoric sites is as one in which the surface record can give us valuable information in advance of excavation. In archaeology as a whole, it is possible to design investigations (for example an excavation strategy) in order to test particular hypotheses rather than merely to collect data. This becomes increasingly important for the overall progress of archaeology as resources become increasingly scarce. If we can use the surface data to set up viable hypotheses in areas of importance, then we have both a working hypothesis in advance of any excavation, and also a strategy which should save resources if ever we are in a position to undertake one. If corroborative evidence on prehistoric astronomy is sought from excavation, then survey work can isolate groups of sites where excavation might preferentially be undertaken, together with hypotheses (solar and lunar association) which might be tested. The stone rows of Mull and mainland Argyll, and their possible lunar association, are in the author's view a particularly exciting example which has cropped up in recent years, and it is hoped that the excavation of one of these sites will proceed soon.

Prehistoric Britain in the context of world archaeoastronomy

The basic problem in studying prehistoric British archaeoastronomy is that our material record is very thin compared, say, with that encountered when dealing with later pre-Columbian cultures in Mesoamerica. There is precious little independent archaeological evidence (such as radiocarbon dating, relative chronology and excavated artefacts) relating to these sites, and of course no evidence to compare with the ethnohistorical and written evidence available to many workers in the Americas.

The paucity of our prehistoric cultural record does, however, force us to face up to certain problems. In particular it necessitates a rigorous approach to the unbiased selection and statistical analysis of alignment data, in order to avoid a situation in which almost anything can be 'proved' by simply selecting the right data in a certain way and ignoring the rest (the plethora of spectacular theories about Stonehenge makes this point only too clearly). This in turn forces us to try to marry a rigorous statistical approach together with the interpretative approach needed in order to consider the data in their full cultural context.

While such problems may seem very distant to archaeoastronomers blessed with a far richer cultural record, they are none the less crucial. There appears to be a distinct danger that in such situations workers may indulge in collecting alignment data, and arriving at conclusions with little or no rigorous justification, but which are supported by rather weak data from other sources such as the ethnohistoric (for example after having scanned the codices somewhat uncritically for scraps of evidence which seem to support the conclusions). If one is not careful the result may be little better than many of the totally unsupportable theories about Stonehenge.

British archaeoastronomers have often been accused by their American colleagues of lavishing too much attention on statistical rigour at the expense of cultural context; it may soon be the turn of the British to persuade their colleagues that statistical rigour has a crucial rôle for a great many archaeoastronomers, and to teach them how it can be integrated into an approach which quite properly considers the great diversity of cultural evidence which may be available.

Acknowledgments

I am grateful to Pergamon Press Ltd. for permission to reproduce passages from a review paper entitled 'Megalithic astronomy: the last five years', which appeared in the journal *Vistas in Astronomy* in 1984 (Ruggles, 1984c).

References

Atkinson, R. J. C. (1966). Moonshine on Stonehenge. *Antiquity* **40**, 212–16.

Bailey, M. E., Cooke, J. A., Few, R. W., Morgan, J. G. and Ruggles, C. L. N. (1975). Survey of three megalithic sites in Argyllshire. *Nature* **253**, 431–3.

Baity, E. C. (1973). Archaeoastronomy and ethnoastronomy so far. *Current Anthrop.* **14**, 389–449.

Barnatt, J. and Pierpoint, S. (1983). Stone circles: observatories or ceremonial centres? *Scottish Archaeological Review* **2**, 101–15.

Burl, H. A. W. (1979). *Rings of Stone*. London: Frances Lincoln.

Burl, H. A. W. (1980). Science or symbolism: problems of archaeoastronomy. *Antiquity* **54**, 191–200.

Burl, H. A. W. (1981a). *Rites of the Gods*. London: Dent.

Burl, H. A. W. (1981b). 'By the light of the cinerary moon': chambered tombs and the astronomy of death. In *Astronomy and Society in Britain During the Period 4000–1500 BC*, eds. C. L. N. Ruggles and A. W. R. Whittle, pp. 243–74. Oxford: British Archaeological Reports 88.

Burl, H. A. W. (1983). *Prehistoric Astronomy and Ritual*. Aylesbury: Shire.

Chamberlain, von Del, ed. (1988). *Proceedings of the First International Ethnoastronomy Conference*, Washington, DC (in press).

Ellegård, A. (1981). Stone age science in Britain? *Curr. Anthrop.* **22**, 99–125.

Fraser, D. (1983). *Land and Society in Neolithic Orkney*, 2 vols. Oxford: British Archaeological Reports 117.

Fraser, D. (1984). In support of festive astronomy. *Scottish Archaeological Review* **3**, 16–18.

Hawkins, G. S. (1963). Stonehenge decoded. *Nature* **200**, 306–8.

Hawkins, G. S. and White, J. B. (1966). *Stonehenge Decoded*. London: Souvenir Press.

Heggie, D. C. (1972). Megalithic lunar observatories: an astronomer's view. *Antiquity* **46**, 43–8.

Heggie, D. C. (1981). *Megalithic Science*. London: Thames and Hudson.

Heggie, D. C., ed. (1982). *Archaeoastronomy in the Old World*. Cambridge University Press.

Lockyer, N. (1909). *Stonehenge and Other British Stone Monuments Astronomically Considered*, 2nd edn. London: Macmillan.

McGreery, T. (1979). Megalithic lunar observatories – a critique, Part 1. *Kronos* **5**, (1), 47–63.

McGreery, T. (1980a). The Kintraw stone platform. *Kronos* **5**, (3). 71–9.

McGreery, T. (1980b). Megalithic lunar observatories – a critique, Part 2. *Kronos* **5**, (2), 6–26.

McGreery, T., Hastie, A. J. and Moulds, T. (1982). Observations at Kintraw. In *Archaeoastronomy in the Old World*, ed. D. C. Heggie, pp. 183–90. Cambridge University Press.

MacKie, E. W. (1974). Archaeological tests on supposed astronomical sites in Scotland. *Philosophical Transactions of the Royal Society of London A* **276**, 169–94.

MacKie, E. W. (1977). *Science and Society in Prehistoric Britain*. London: Paul Elek.

MacKie, E. W. (1981). Wise men in antiquity? In *Astronomy and Society in Britain During the Period 4000–1500 BC*, eds. C. L. N. Ruggles and A. W. R. Whittle, pp. 111–52. Oxford: British Archaeological Reports 88.

MacKie, E. W., Gladwin, P. F. and Roy, A. E. (1985). A prehistoric calendrical site in Argyll. *Nature* **314**, 158–61.

Moir, G. (1980). Megalithic science and some Scottish site plans: Part 1. *Antiquity* **54**, 37–40.

Moir, G. (1981). Some archaeological and astronomical objections to scientific astronomy in British prehistory. In *Astronomy and Society in Britain During the Period 4000–1500 BC*, eds. C. L. N. Ruggles and A. W. R. Whittle, pp. 221–41. Oxford: British Archaeological Reports 88.

Morrison, L. V. (1980). On the analysis of megalithic lunar sightlines in Scotland. *Archaeoastronomy* **2**, S65–S77.

O'Kelly, M. J. (1971). *An Illustrated Guide to Newgrange*, 2nd edn. Wexford: John English.

Patrick, J. D. (1974). Midwinter sunrise at Newgrange. *Nature* **249**, 517–19.

Patrick, J. D. (1981). A reassessment of the solstitial observatories at Kintraw and Ballochroy. In *Astronomy and Society in Britain During the Period 4000–1500 BC*, eds. C. L. N. Ruggles and A. W. R. Whittle, pp. 211–19. Oxford: British Archaeological Reports 88.

Pitts, M. W. (1981a). Stones, pits and Stonehenge. *Nature* **290**. 46–7.

Pitts, M. W. (1981b). The discovery of a new stone at Stonehenge. *Archaeoastronomy Bulletin* **4**, (2), 16–21.

Renfrew, A. C. (1981). Comment on Ellegård (1981). *Curr. Anthrop.* **22**, 120–1.

Ruggles, C. L. N. (1981a). Archaeoastronomical anomalies. *Nature* **294**, 485–6.

Ruggles, C. L. N. (1981b). A critical examination of the megalithic lunar observatories. In *Astronomy and Society in Britain During the period 4000–1500 BC*, eds. C. L. N. Ruggles and A. W. R. Whittle, pp. 153–209. Oxford: British Archaeological Reports 88.

Ruggles, C. L. N. (1982a). Megalithic astronomical sightlines: current reassessment and future directions. In *Archaeoastronomy in the Old World*, ed. D. C. Heggie, pp. 83–105. Cambridge University Press.

Ruggles, C. L. N. (1982b). A reassessment of the high precision megalithic lunar sightlines, 1: Backsights, indicators and the archaeological status of the sightlines. *Archaeoastronomy* **4**, S21–S40.

Ruggles, C. L. N. (1983). A reassessment of the high precision megalithic lunar sightlines, 2: Foresights and the problem of selection. *Archaeoastronomy* **5**, S1–S36.

Ruggles, C. L. N. (1984a) (principal author: contributions by P. N. Appleton, S. F. Burch, J. A. Cooke, R. W. Few, J. G. Morgan and R. P. Norris). *Megalithic Astronomy: A New Archaeological and Statistical Study of 300 Western Scottish Sites*. Oxford: British Archaeological Reports 123.

Ruggles, C. L. N. (1984b). A new study of the Aberdeenshire recumbent stone circles, 1: Site data. *Archaeoastronomy* **6**, S55–S79.

Ruggles, C. L. N. (1984c). Megalithic astronomy: the last five years. *Vistas in Astronomy* **27**, 231–89.

Ruggles, C. L. N. (1985). The linear settings of Argyll and Mull. *Archaeoastronomy* **9**, S105–S132.

Ruggles, C. L. N. and Burl, H. A. W. (1985). A new study of the Aberdeenshire recumbent stone circles, 2: Interpretation. *Archaeoastronomy* **8**, S25–S60.

Thom, A. (1954). The solar observatories of Megalithic Man. *Journal of the British Astronomical Association* **64**, 396–404.

Thom, A. (1955). A statistical examination of the megalithic sites in Britain. *Journal of the Royal Statistical Society* A **118**, 275–95.

Thom, A. (1967). *Megalithic Sites in Britain*. Oxford University Press.

Thom, A. (1969). The lunar observatories of Megalithic Man. *Vistas in Astronomy*, **11**, 1–29.

Thom, A. (1971). *Megalithic Lunar Observatories*. Oxford University Press.

Thom, A. and Thom, A. S. (1978a). *Megalithic Remains in Britain and Brittany*. Oxford University Press.

Thom, A. and Thom, A. S. (1978b). A reconsideration of the lunar sites in Britain. *Journal for the History of Astronomy* **9**, 170–9.

Thom, A. and Thom, A. S. (1980). A new study of all megalithic lunar lines. *Archaeoastronomy* **2**, S78–89.

Thom, A. S. (1981). Megalithic lunar observatories: an assessment of 42 lunar alignments. In *Astronomy and Society in Britain During the Period 4000–1500 BC*, eds. C. L. N. Ruggles and A. W. R. Whittle, pp. 13–61. Oxford: British Archaeological Reports 88.

Thorpe, I. J. (1981). Ethnoastronomy: its patterns and archaeological implications. In *Astronomy and Society in Britain During the Period 4000–1500 BC*, eds. C. L. N. Ruggles and A. W. R. Whittle, pp. 275–88. Oxford: British Archaeological Reports 88.

3

Exploring some anthropological theoretical foundations for archaeoastronomy

Stanisław Iwaniszewski *State Archaeological Museum, Warsaw, and Instituto de Investigaciónes Antropológicas de la UNAM*

3.1 Introduction

About two decades ago a new paradigm (Kuhn, 1970) emerged that proposed that both literate and illiterate peoples had developed relatively elaborate calendrical and astronomical systems. This paradigm became the base of the archaeoastronomical approach that has attracted many scholars. Starting in the mid-1960s, archaeoastronomical studies have tended to show an exponential growth (S. Iwaniszewski (n.d.)). However, it was not until the 1970s that Invisible Colleges (Price, 1961; Crane, 1972), international symposia, and specialized periodicals appeared. This constituted a basic step toward normal science (Price, 1961, 1963).

Nevertheless, it is difficult to say that archaeoastronomy has achieved the status of a discipline (Zeilik, 1983). As Aveni (1981, p. 32) points out:

> While it is easy to outline a general methodology for fieldwork, tangible results occur only with the understanding of a true interdisciplinary study of an ancient culture's astronomy and with the perception of astronomical activity within the context of that civilization.

This chapter presents a reconsideration of the anthropological theoretical foundations of archaeoastronomy. It is divided into three parts: in the first (Sections 3.2 and 3.3), I consider theoretical concepts useful to archaeoastronomy; the second part (Sections 3.4 and 3.5) presents preliminary results of some archaeological applications of the reconsidered concepts; and the third part (Section 3.6) presents a list of possible functions of astronomical and calendrical activities that need to be tested.

3.2 Theoretical orientation

The intellectual core of anthropology is the study of culture. Recently, Keesing (1974) observed two distinct concepts of culture: behavioral–ecological and ideational–symbolic. The former approach views a culture as a system that relates human populations to their environments. The latter describes a culture as a cognitive (e.g. Goodenough), structural (e.g. Levi-Strauss), or symbolic (e.g. Dumont, Geertz, Schneider) system.

Culture as an open system (Clarke, 1968, p. 88) maintains feedback relations with somatic–genetic and environmental systems and is basic to man's adaptation. Cultural adaptation is the process that a cultural system undertakes in order to establish and maintain a relationship with both other systems (Kirch, 1978, p. 105). The term 'adaptation' implies, in anthropology, the existence of the interrelationship of the following factors: variability (cultural memory + innovations − loss); selective mechanisms; and transmission (retention or propagation; Kirch, 1980, pp. 108–9). Some authors (Brykczyński, 1979; Wierciński, 1979, 1983) add rhythmicity to these.

However, culture is rather an abstract entity. Anthropologists study culture through sociocultural systems. According to Keesing (1974, p. 82): 'sociocultural systems represent the social realizations or enactments of ideational designs-for-living in particular environments'.

A sociocultural system may be viewed as a cybernetic system which consists of the three components: matter, energy, information.

Matter and energy are not destructible according to the law of conservation. Humans can only transform one into another, changing the relations between them, or in other words, their structure. I define structure as the temporal–spatial placement of matter and energy.

Contrasting with matter and energy, information may be created, stored, or destroyed. Cultural transmission of new adaptive information is supposed to be significant for human evolution (Dunnell, 1978, p. 198); in view of the three entities mentioned, it is the only one that can be responsible for the patterning of human behavior (Fritz, 1978, pp. 38–9; van der Leeuw, 1981, p. 234). All living systems process information (Flannery, 1972, p. 400) but cultural systems are information systems (Clarke, 1968, p. 660). In anthropology, this position has focused attention on the organizational aspects of sociocultural systems. As Hole and Heizer (1973, p. 315) state: 'It is becoming more and more apparent that the principles of organization are the basic keys in our understanding of any class of phenomena, including people'.

The following strategy for the study of organization has been recently adopted. All cultural products such as artifacts, rules of social interactions, native cognitive systems, and so on are viewed in the light of the behavior that created them. This behavior is carried out and determined by the context of a given sociocultural system. Cultural products can be considered as the results of culturally patterned behavior. Patterning is present in cultural products because the behavior that produces them is patterned. Through the careful examination and observation of the assemblages of cultural products, anthropologists can make inferences about culturally patterned behavior.

Both behavioral and cognitive aspects are present in cultural adaptation (Keesing, 1974; Alland, 1975; Dunnell, 1978; Kirch, 1980); however, concrete patterned behavior derives ultimately from the cognitive processes. New adaptive information is derived from the interrelation between the human brain and the perception and organization of environmental (natural, sociocultural and psychological)

facts. The selective mechanism operates on three levels (Alland, 1975, p. 69):

(a) new adaptive information is limited by cognitive and decision-making centers from the human brain (somatic–genetic system) and by brain-based 'culturally realized cognitive structures';

(b) the second level tests a new trait against the system's structure;

(c) on the third level the selective mechanism tests new information against environmental pressure. Resulting behavior interrelated with somatic–genetic, sociocultural and environmental systems is perceived again in a continuing dialectic (Kirch, 1980, p. 112).

Within the view presented, archaeoastronomy is a perfect field in which the ideational–symbolic and behavioral–ecological approaches can be used in conjunction with one another.

3.3 Foundations of archaeoastronomy

Astronomy, if isolated from the society, cannot be understood totally. Forming a subsystem within the culture system of ideology, astronomy should be studied as another cultural subsystem. When dealing with astronomical traits, we describe and reconstruct patterns of astronomical behavior which are elements of cultural cognitive systems. Even a scientific cognitive system may be treated as a cultural subsystem (Latour and Woolgar, 1979, pp. 19–37).

Archaeoastronomy should deal then with both behavioral and cognitive aspects of astronomy. Briefly it may be defined as: *the study of astronomy in its sociocultural context.*[1]

In this vein, I propose the following definition of astronomical artifact, following Wierciński's (1983, p. 28) definition of cultural product: *an astronomical product or artifact is any element of the environment related to astronomical phenomena which has been transformed through the conscious work by man in order to utilize its adaptive value.*

I also define astronomical phenomenon in the following way: *every kind of natural phenomenon that occurs outside of the Earth or occurring in the earthly atmosphere is of extraterrestrial origin.*

Although the adaptive value of astronomy has long been recognized (e.g. Pannekoek, 1951), few studies have actually explored that side of astronomy; in fact, the study of astronomy was dominated by the history of the discipline which emphasized the intellectual achievements over the sociocultural ones.

This adaptive aspect was superficially investigated in the social sciences. Adopting a passage from Cuvier which related the Egyptians' need of astronomy to the annual rises of the Nile and seasonal agriculture, Marx (1973, Vol. 1) states that astronomy served as a base for economic dominance in the sense that astronomical knowledge offers a way to control the means of production.

Even today this statement has a great impact in the social sciences. For example, Wittfogel (1957, pp. 29–30, 382) follows Marx stating that calendar making and astronomy used in water scheduling were instrumental in the development of differentiated leadership. However, this simplified point of view cannot be further sustained.

Also, astronomy was taken into account in the field of ethnology and the history of religions. However, this produced some scientific deviations known as 'astralistics', 'solar mythology', 'pan-Egyptianism' or 'pan-Babylonianism'. On the other hand, Berthelot's theory of so-called 'astrobiology' remains almost totally forgotten.

Within the birth of archaeoastronomy, several studies (Renfrew, 1973a; Reichel-Dolmatoff, 1976; Reyman, 1976, 1979; MacKie, 1977a,b; Wood, 1978) offered new approaches to the study of astronomical activities and/or systems of knowledge.

The fact that a lot of astronomical information is embodied in magico-religious systems of the world and that the art of making and/or using calendars is present commonly in human cultures, no matter how advanced, suggests naturally that astronomy possesses a certain adaptive value. If it is found so frequently in different sociocultural systems, it must possess a special adaptive value for man. Nevertheless, saying that astronomy is present in a given sociocultural system because it has an adaptive value (Reyman, 1979) leads to the same kind of tautology that Alland (1975, pp. 75–6) observes. *Instead of asking why astronomy is present in the cultures of the whole world* (the answer will be, because of its common adaptive value for man), *we should ask what different functions it may play in particular sociocultural systems.* Based on case histories we will be able to deduce the place of astronomy in kinds of sociocultural systems.

Since not all observable astronomical phenomena are equally important, their importance will be determined by cultural preference. It may be deduced from the above that the somatic–genetic system, the sociocultural patterns and the nature of the environment will play a decisive role in the cultural selection and transmission of certain astronomical phenomena. *Then, in order to understand the cultural significance of astronomy, it must be matched against the set of other cultural variables.* These may be language affinities, the type of environment, the basic subsistence economy, the kind of social structure, the changes in stylistic variability of cultural products, the patterns of attitudes of man's efforts at self interpretation, and so on. We also must keep in mind the *history* of a given sociocultural system because of the importance of cultural memory.

3.4 Astronomy at Teotihuacan

In my previous work on astronomy at Teotihuacan (Iwaniszewski, 1984) I investigated the importance of astronomical and calendrical activities through ecological–behavioral and ideational–symbolic approaches. It was possible to relate some solar dates pointed out by the city's general orientation, with meteorological and biological cycles being of vital importance for the success of a basic subsistence economy (see Figure 3.1). The possibility that 'pecked cross' designs also served to encode some astronomical rhythms has also been considered. But the most striking feature of astronomical activities was that the production of astronomical artifacts showed similar trends to other cultural variables, such as population growth, changes in the city's area and rhythms in obsidian and pottery production (see Figure 3.2). All these variables showed a relatively rapid growth until the Tlamimilolpa Phase (AD 200–450). After reaching peaks, stagnation during the rest of Teotihuacan existence was observed (Cowgill, 1977). Also, a low precision in the elaboration of pecked cross figures (e.g. TEO 19 and TEO 22) is observed in the Tlamimilolpa Tardío (Late) Phase (AD 300–450). With these data, several preliminary suggestions may be offered.

The phase of rapid growth probably refers to the emergence of the Early State at Teotihuacan. Pacific, nonviolent displacement of population from small settlements in the Valley of Mexico to the growing city in the Valley of Teotihuacan suggests

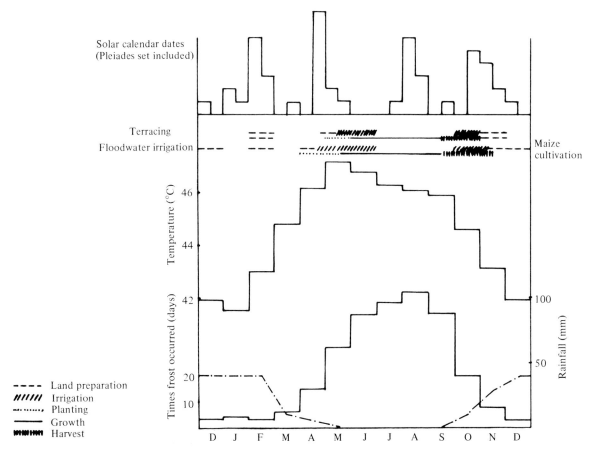

Figure 3.1 Annual relations among ecological cycles, agricultural activities and solar calendar dates indicated by the Teotihuacan astronomical products.

that the political unification of the Basin of Mexico involved religious motivation (Parsons, 1982, pp. 87–9). In fact, the emergence of ceremonial architecture may denote a change in the ideological system: the shift is from the rites performed in the sacred cave towards ceremonies practiced at the pyramids (Heyden, 1975, 1981). The rapid growth of astronomical artifacts related to that architecture suggests that the city builders possessed rather advanced astronomical knowledge. At present, we can detect the presence of solar, lunar, Venusian and Pleiades cycles.

City planning shows a certain preference of alignments. An orientation of 15 to 17 degrees east of north at Teotihuacan replaced the nearly east–west orientation at Cuicuilco. In part, this orientation can be associated with important economic cycles for the Teotihuacanos (e.g. new intensive

agricultural techniques). On the other hand, the shift in basic orientation may indicate a change in ideological pattern. If astronomy formed a significant part of the religious system, then the shift in orientation would signify a religious change.

This orientation is embodied in ceremonial architecture. The architecture which is built for long-term use demands previous programming (Wierciński, 1977, p. 88). Thus, the architecture at Teotihuacan should have served as an essential component of the culture memory, with references in stone and space (Fritz, 1978, pp. 55–6; Cowgill, 1983, pp. 329–30; see also Wheatley, 1971, pp. 411–76; Eliade, 1973, pp. 56–9). Other astronomical information might be revealed to Teotihuacanos through the special light-and-shadow effect (e.g. the equinox effect at the Moon Pyramid, El Adosado), thus reinforcing the existing social and

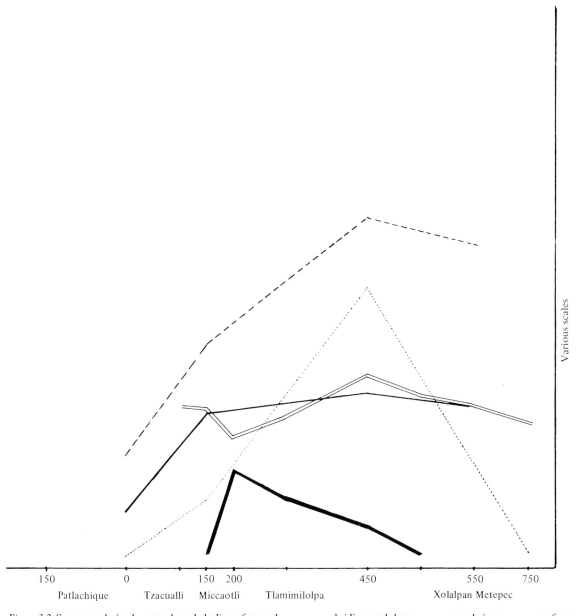

Figure 3.2 Some trends in the growth and decline of several cultural traits at Teotihuacan (after Cowgill, 1977; Millon, 1981; Spence, 1981; Iwaniszewski, 1984; Rattray, n.d.).

...... obsidian workshops; – – – – population; ———— surface; ===== pottery forms; ━━━━ astronomical artefacts.

religious order. Then, the ceremonial architecture not only defined a given socioeconomic system (which may be decoded by the archaeoastronomical approach) but also gave religious sanction to those who had controlled city planning. Astronomy, as a significant element of the urbanization at Teotihuacan, could have played an important role in the

legitimization of the power of the decision-making center.

Some sort of standard measuring units must have been used since large-scale social labor requires both accuracy and cooperation (Childe, 1951, pp. 193–4; Hudson, 1972). This rudimentary mathematical knowledge could also be used for astronomical and calendrical purposes. The possible presence of lunar, solar and Venusian cycles in the pecked cross designs suggests that a kind of numerological speculation related to astronomical cycles was developed; some full-time functionaries must have been involved. Thus, during the initial development of social stratification at Teotihuacan, some specialists in calendrics and astronomy were present.

Elsewhere I have suggested (Iwaniszewski, 1984) that pecked cross designs were symbolic-numerical representations of the world in which rhythmically repeated astronomical, biological and socioeconomic events were mutually synchronized. Pecked cross designs and urban planning at Teotihuacan, sharing the same orientational preference, do indeed reflect the ancient ideational system (Coggins, 1980 and Mansfield, 1981). It may be deduced that astronomy played an active role in the formation of the world-view at Teotihuacan.

The symbolic significance of ceremonial architecture at Teotihuacan refers to political or religious acts (Millon, 1981, p. 235) and should be viewed in the light of the political unification of the Basin of Mexico. Religion might contribute as a coercive or integrative factor in the process of state formation (Webster, 1976), although it does not necessarily imply the presence of a theocracy. Thus, ceremonial structures served as stable symbols for Teotihuacanos (Cowgill, 1983, p. 330). I suggest that, during this transitional period, the integration and stability needed for the maintenance of social equilibrium were reinforced by the incorporation of stable astronomical rhythms into the dominating worldview.

3.5 Astronomy at the Puuc sites

It is well known (*Dresden Codex*) that the Maya used astronomical and calendrical knowledge for ceremonial, agricultural and divinatory purposes. Thus, solar calendar dates pointed out by alignments of Mayan ceremonial architecture may coincide with ritual and agricultural cycles (Aveni and Hartung, 1984). In fact, a kind of relationship can be observed (Figure 3.3) between meteorological phenomena registered at Mérida and solar calendar dates resulting from the orientation of Puuc Terminal Classic architecture.

The growth of astronomically significant architecture during the Late Classic (AD 600–800) and Terminal Classic (AD 800–1000) times demonstrates the same trends as the growth of the whole of Puuc architecture (see Figure 3.4). Also, in plotting all pan-Maya structures, it is possible to observe a similar, exponential growth of astronomically oriented architecture in the same period (see Figure 3.5; however, I cannot present here a histogram with the behavior of pan-Maya architecture as a whole).

Some buildings at Uxmal encode astronomical information (e.g. the Nunnery; see Lamb, 1980), while others employ astronomically significant visual lines that link one structure with another (Aveni, 1980, 1982; Aveni and Hartung, 1984). Visual-symbolic elements embodied in urban planning at Uxmal could have reinforced an existing ideological system that supported the sacrosanct status of the ruling group (Barrera Rubio, 1984).

I shall consider this aspect more generally. Undoubtedly, a kind of rudimentary astronomy was practiced at La Venta, Tres Zapotes and other Olmec sites (Coe, 1977, pp. 189–90, 194). But it was the Maya who developed both calendrics and astronomy and made them very complex. Very early in the Classic Period astronomical events were associated with historical ones. Apart from its utility for ecological adaptation and for divinatory and ceremonial purposes, astronomical phenomena were used to insure the success[2] of various kin linkages (Schele, 1977; Tate, 1984) and regulated ritual wars and captives' sacrifices (Dütting, 1981; Lounsbury, 1982; Carlson, 1984).

The maintenance of the ritual wars could also reinforce interior social cohesion and offer justification for the existing social order (Webster, 1977, pp. 363–6). On the other hand, astronomy strengthened religious sanctions, contributing to the conservation of the theocratic character of Maya culture.

I would also suggest that the increase in the number of astronomically significant structures (which perhaps is followed by the increasing number of stelae with historical and astronomical events; see Figure 3.5) reflects growing social differentiation and the strong position of ruling kin families during Classic Maya times.

Solar calendar
dates of Puuc
Terminal
Classic

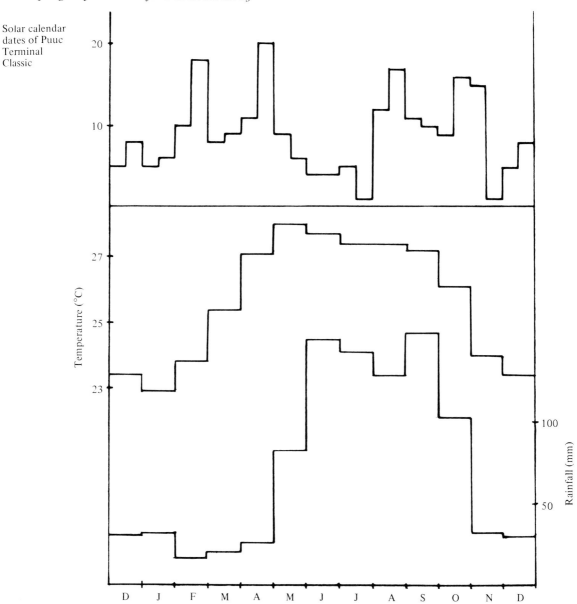

Figure 3.3 Annual relations between ecological cycles and solar
calendar dates indicated by the Puuc Terminal Classic architecture (after Aveni and Hartung, 1984).

3.6 Some suggestions

Mesoamerica offers some examples of Level Two
astronomy in Aaboe's (after Renfrew, 1973b) classification. This roughly corresponds to the astrobio-
logical stage in Berthelot's (1949) taxonomy. Both
taxonomic systems do not explain the cultural pro-

cesses leading from one stage to another. This
chapter also does not intend to do that: I merely
suggest what strategy should be developed in order
to establish processual models. Below, I offer a tentative list of possible functions of astronomical and
calendrical activities. I believe that examining them
in particular sociocultural systems will enable us

to make inferences about the nature of astronomical behavior.

I *Level One: Astronomical practices and natural environment*
  (1) The importance of astronomical activities for ecological adaptation.
  (2) 'Cosmovisión' as a device for ecological equilibrium.

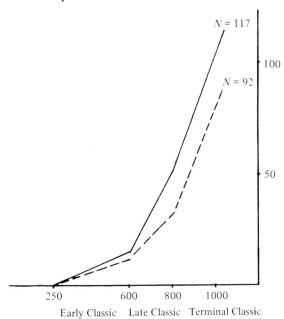

Figure 3.4 The growth of astronomically significant architecture and the whole Puuc architecture during the Late Classic (AD 600–800) and Terminal Classic (AD 800–1000) times. (Data after Pollock, 1980; Gendrop, 1983; and Aveni and Hartung, 1984). ——— Puuc architecture; – – – – Puuc astronomically significant architecture.

II *Level Two: Astronomical practices and society*
  (1) Astronomy, calendars, and socioeconomic rhythms.
  (2) 'Cosmovisión' as an element of basic social norms.
  (3) 'Cosmovisión' and social integration.
  (4) The place of the concept of astrobiology in ideology: the legitimization and justification of the existing social order.
  (5) Astronomy and cultural evolution:
    (*a*) astronomical and calendrical knowledge as a social endogenous synchronizer which offers to the social system a higher level of autonomy in relation to natural environment;

    (*b*) astronomy and the development of social stratification;
    (*c*) astronomy and the general evolution of sociocultural systems.

III *Level Three: Astronomical practices and ideational culture*
  (1) Astronomy and the formation of culturally recognized conceptual models of the world.
  (2) The evolution of ideational–symbolic elements of astronomical subsystems.
  (3) Mathematical and astronomical tools and concepts.

IV *Level Four: Astronomy and the imposition of a collective ideational system on the psychodynamics of the individual*
  (1) Myths representing psychic processes that secondarily employ astronomical pictures.

3.7 Conclusions

In sum, I would say that we still do not understand well the importance of ancient astronomical and calendrical behavior for cultural evolution in Mesoamerica. In particular, we cannot make excessive inferences about ancient world-views. However, it is important to stress that astronomical subsystems reveal the same developmental trends as do other sociocultural subsystems.

I have also tried to point out that archaeoastronomy should deal with both cognitive and behavioral aspects of culture. The archaeoastronomical approach provides archaeology with new tools. This is of crucial importance for archaeology since no adequate strategy to study the cognitive part of the ancient cultures has been developed.

However, archaeoastronomy is not well prepared to study cultural conceptual models of the world. So far, most archaeoastronomical effort has been involved in the study of astronomical alignments, calendrical practices and patterns of local world-views. Nevertheless, it should be stressed that all of these are the products of cultural evolution in which somatic–genetic, environmental and cultural systems are mutually engaged. In this chapter I have attempted to establish, as a working framework, a set of possible cultural functions of astronomical lore in order to investigate first how the processes of building a conceptual model of the world can affect different spheres of human behavior, and, second, how they are dependent on the cultural trajectory of a system. I hope that in this way more

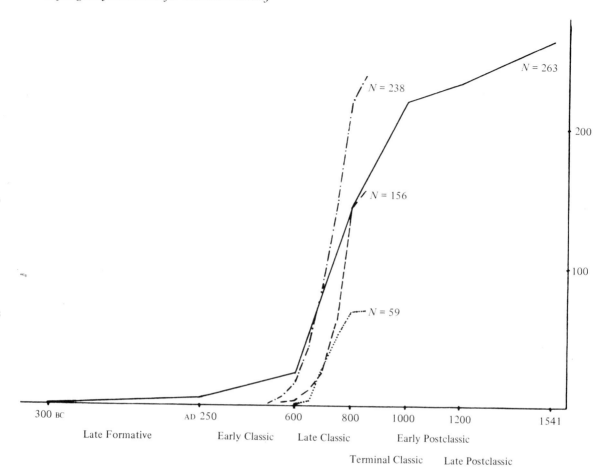

Figure 3.5 The growth of astronomically oriented architecture and the growth of the stelae beginning with Katun 9 (after Morley, 1975; Rands, 1977; and Aveni and Hartung, 1984). ——— astronomically significant architecture, pan-Maya; —·—·— dates recorded in Katun 9 (general, after Morley); ...... dates recorded in Katun 9 from northwestern lowlands; —–—– dates recorded in Katun 9 at upper Usumacinta sites.

efficient strategies with insights into the ancient ideational processes may be offered to archaeology.[3]

Acknowledgments

I am indebted to Matthew Wallrath for discussing some of the problems presented. Valuable comments on earlier versions of this chapter were also made by Arturo Ponce de León and Luis González Reimann. Special thanks are owed to Anthony F. Aveni who kindly permitted me to utilize his data from Maya sites. I also wish to express my personal thanks to Ann Cyphers G. for the revision and the discussion of the English version of this text. However, responsibility for the ideas, as well as any errors which may appear, are my own.

Notes

1 Of course, the name 'archaeoastronomy' can lead to some misunderstanding. According to it, the new discipline should deal with astronomy in the *past*. However, I see no reason to limit our investigation to ancient or contemporary ethnic sociocultural systems. Why do we exclude the study of peasantry? What of the astronomical lore in urban societies? Or the astronomical lore among scientists? On the other hand, archaeology itself does not deal with the past only, for example the excavations at the Nazi concentration camps in

Poland, or the Garbage Project by Rathje conducted at Tucson.

2 I owe that term to Stephen McCluskey.

3 After the Conference I had an opportunity to talk to Alexander Marshack and read his earlier article (Marshack, 1985). It is interesting that our approaches to archaeoastronomy coincide in many ways, although our points of departure differ greatly (see Chapter 24 of this volume).

References

Alland, A., Jr. (1975). Adaptation. *Annual Review of Anthropology* **4**, 59–73.

Aveni, A. F. (1980). *Skywatchers of Ancient Mexico*. Austin: University of Texas Press.

Aveni, A. F. (1981). Archaeoastronomy. In *Advances in Archaeological Method and Theory*, ed. M. B. Schiffer, vol. 4, pp. 1–77. New York: Academic Press.

Aveni, A. F. (1982). Astronomy in the Maya region: 1970–1980. In *Archaeoastronomy in the New World*, ed. A. F. Aveni, pp. 1–30. Cambridge University Press.

Aveni, A. F. and Hartung, H. (1984). Archaeoastronomy and the Puuc sites. Paper submitted to *Proceedings of the Symposium: 'Archaeoastronomy and Ethnoastronomy in Mesoamerica'*. Mexico City: National University of Mexico.

Barrera R. A. (1984). Settlement patterns in the Uxmal Area, Yucatan, Mexico. *Indiana (Berlin)* **10**, (2), 227–35.

Berthelot, R. (1949). *La penseé de l'Asie et l'astrobiologie*. Paris: Payot.

Brykczyński, P. (1979). Rytmy naturalnego otoczenia a socjorytmy. (Rhythms of the natural environment and social rhythms). In *Bioelektronika*, ed. W. Sedlak, pp. 103–18. Lublin: Towarzystwo Naukowe Katolickiego Uniwersytetu Lubelskiego.

Carlson, J. (1984). Venus and ritual warfare in Ancient Mesoamerica: new light from the Maya Grolier Codex. Paper presented at the *Symposium: Archaeoastronomy and Ethnoastronomy in Mesoamerica*. Mexico City: National University of Mexico.

Childe, G. V. (1951). *Man Makes Himself*. London: Watts and Co.

Clarke, D. A. (1968). *Analytical Archaeology*. London: Methuen.

Coe, M. D. (1977). Olmec and Maya: a study of relationships. In *The Origins of Maya Civilization*, ed. R. E. W. Adams, pp. 183–95. Albuquerque: University of New Mexico Press.

Coggins, C. (1980). The shape of time: some political implications of a four-part figure. *American Antiquity* **45**, (4), 727–39.

Cowgill, G. L. (1977). Processes of growth and decline at Teotihuacan: the city and the state. In *Los procesos de cambio. XV Mesa Redonda de la Soc. Mex. de Antrop.*, vol. 1, ed. N. Castillo, pp. 183–91. Guanajuato.

Cowgill, G. L. (1983). Rulership and the Ciudadela: political inferences from Teotihuacan architecture. In *Civilization in the Ancient America: Essays in Honor of Gordon Willey*, eds. R. M. Leventhal and A. L. Kolata, pp. 313–43. Cambridge, MA: Harvard University.

Crane, D. (1972). *Invisible Colleges*. University of Chicago Press.

Dunnell, R. C. (1978). Style and function: a fundamental dichotomy. *American Antiquity* **43**, (2), 192–201.

Dütting, D. (1981). Life and death in Mayan hieroglyphic inscriptions. *Zeitschrift für Ethnologie* **106**, (1–2), 185–228.

Eliade, M. (1973). *Lo sagrado y lo profano*. Madrid: ed. Guadarramas.

Flannery, K. V. (1972). The cultural evolution of civilizations. *Ann. Rev. Ecology & Systematics* **3**, 399–426.

Fritz, J. M. (1978). Paleopsychology today: ideational systems and human adaptation in prehistory. In *Social Archaeology: Beyond Subsistence and Dating*, eds. C. L. Reedman, M. J. Berman, E. V. Curtin, W. T. Langhorne, Jr., N. M. Versaggi and J. C. Wauser, pp. 37–59. New York: Academic Press.

Gendrop, P. (1983). *Los estilos Río Bec, Chenes y Puuc en la arquitectura maya*. Mexico City: Universidad Nacional Autónoma de México.

Heyden, D. (1975). An interpretation of the cave underneath the Pyramid of the Sun in Teotihuacan, Mexico. *American Antiquity* **40**, (2), 131–47.

Heyden, D. (1981). Caves, gods, and myths: world-view and planning in Teotihuacan. In *Mesoamerican Sites and World-Views*, ed. E. P. Benson, pp. 1–39. Washington, DC: Dumbarton Oaks Trustees for Harvard University.

Hole, F. and Heizer, R. F. (1973). *An Introduction to Prehistoric Archaeology*, 3rd edn. New York: Holt, Rinehart and Winston.

Hudson, D. T. (1972). Anasazi Measurement Systems at Chaco Canyon, New Mexico. *The Kiva* **38**, (1), 27–42.

Iwaniszewski, S. (1984). La arqueología y la astronomía en Teotihuacan. Paper submitted to *Proceedings of the Symposium: Archaeoastronomy and Ethnoastronomy in Mesoamerica*. Mexico City: National University of Mexico.

Iwaniszewski, S. (n.d.). 'Algunos aspectos del desarollo de la arqueoastronomia desde 1951 hasta 1980' (1985, unpublished ms.).

Keesing, R. M. (1974). Theories of Culture. *Annual Review of Anthropology* **3**, 73–97.

Kirch, P. V. (1978). Agricultural adaptation in the humid tropics. In *Explorations in Ethnoarchaeology*, ed. Richard A. Gould, pp. 103–25. Albuquerque: University of New Mexico Press.

Kirch, P. V. (1980). The archaeological study of adaptation: theoretical and methodological issues. In *Advances in Archaeological Method and Theory*, ed. M. B. Schiffer, vol. 3, pp. 101–56. New York: Academic Press.

Kuhn, T. S. (1970). *The Structure of Scientific Revolutions*. University of Chicago Press.

Lamb, W. (1980). The Sun, Moon and Venus at Uxmal. *American Antiquity* 45, (1), 79–86.

Latour, B. and Woolgar, S. (1979). *Laboratory Life. The Social Construction of Scientific Facts*. Beverly Hills and London: Sage Publications.

Leeuw, S. E. van der (1981). Information flows, flow structures and the explanation of change in human institutions. In *Archaeological Approaches to the Study of Complexity*, ed. S. E. van der Leeuw, pp. 230–329. Amsterdam: Universiteit van Amsterdam.

Lounsbury, F. G. (1982). Astronomical knowledge and its uses at Bonampak, Mexico. In *Archaeoastronomy in the New World*, ed. A. F. Aveni, pp. 143–68. Cambridge University Press.

MacKie, E. W. (1977a). *The Megalith Builders*. Oxford: Phaidon.

MacKie, E. W. (1977b). *Science and Society in Prehistoric Britain*. London: Paul Elek.

Marshack, A. (1985). A lunar–solar year calendar stick from North America. *American Antiquity* 50, (1), 27–51.

Mansfield, V. N. (1981). Mandalas and mesoamerican pecked circles. *Curr. Anthrop.* 22, (3), 269–84.

Marx, K. (1973). *El Capital*. Mexico: Editorial Cartago.

Millon, R. (1981). Teotihuacan: city, state, and civilization. In *Supplement to the Handbook of Middle American Indians*, eds. J. A. Sabloff and P. A. Andrews, pp. 193–243. Austin: University of Texas Press.

Morley, S. G. (1975). *An Introduction to the Study of the Maya Hieroglyphs*. New York: Dover Publications.

Pannekoek, A. (1951). The origin of astronomy. *Monthly Notices of the Royal Astronomical Soc.* 111, 347–56.

Parsons, J. R. (1982). The Chalco–Xochimilco region in Central Mexico *and* Overall conclusions. In *Prehistoric Settlement Patterns in the Southern Valley of Mexico. The Chalco-Xochimilco Region*, eds. J. R. Parsons, E. Brumfiel, M. H. Parsons and D. J. Wilson, pp. 365–84. Ann Arbor: University of Michigan Press.

Pollock, H. E. D. (1980). *The Puuc. An Architectural Survey of the Hill Country of Yucatan and Northern Campeche, Mexico*. Memoirs of the Peabody Museum, vol. 19. Cambridge, MA: Harvard University.

Price, D. de Solla (1961). *Science since Babylon*. New Haven: Yale University Press.

Price, D. de Solla (1963). *Little Science, Big Science*. New York: Columbia University Press.

Rands, R. L. (1977). The Classic Maya collapse: Usumacinta Zone and the Northern Periphery. In *The Classic Maya Collapse*, ed. T. P. Culbert, pp. 165–205. Albuquerque: University of New Mexico Press.

Rattray, E. (n.d.). The Teotihuacan ceramic chronology: early Tzacualli to Metepec phases. In *Urbanization at Teotihuacan*, ed. R. Millon. Austin: University of Texas Press (in press).

Reichel-Dolmatoff, G. (1976). Cosmology as ecological analysis: a view from the rain forest. *Man* 11, (3), 307–18.

Renfrew, C. (1973a). *Before Civilization*. London: Jonathan Cape.

Renfrew, C. (1973b). The place of astronomy in the ancient world: a Royal Society – British Academy conference. *Archaeology* 26, (3), 222–3.

Reyman, J. E. (1976). Astronomy, architecture, and adaptation at Pueblo Bonito. *Science* 193, (4257), 957–62.

Reyman, J. E. (1979). Some observations on archaeology and archaeoastronomy. *Archaeoastronomy. The Bulletin of the Center for Archaeoastronomy* 2, (2) 11–13.

Schele, L. (1977). Palenque: the House of Dying Sun. In *Native American Astronomy*, ed. A. F. Aveni, pp. 42–56. Austin: University of Texas Press.

Tate, C. (1984). Ritual observation of summer solstice and zenith passage dates at Yaxchilan, Chiapas. Paper presented at the *Symposium: Archaeoastronomy and Ethnoastronomy in Mesoamerica*. Mexico City: National University of Mexico.

Webster, D. L. (1976). On theocracies. *American Anthropologist* 78, (4), 812–28.

Webster, D. L. (1977). Warfare and the evolution of Maya civilization. In *The Origins of Maya Civilization*, ed. R. E. W. Adams, pp. 335–72. Albuquerque: University of New Mexico Press.

Wheatley, P. (1971). *The Pivot of the Four Quarters*. Chicago: Aldine Publishing Company.

Wierciński, A. (1977). Time and space in the Sun Pyramid from Teotihuacan. *Polish Contributions in New World Archaeology*, vol. 1, pp. 87–103. Kraków: Polska Akademia Nauk.

Wierciński, A. (1979). Biorytmy a bioplazma. (Biorhythms and bioplasma). In *Bioelektronika*, ed. W. Sedlak, pp. 87–102. Lublin: Towarzystwo Naukowe Katolickiego Uniwersytetu Lubelskiego.

Wierciński, A. (1983). An anthropological vision of culture and cultural evolution. *Ethnologia Polona* 9, 23–31.

Wittfogel, K. A. (1957). *Oriental Despotism. A Comparative Study of Total Power*. New Haven: Yale University Press.

Wood, J. E. (1978). *Sun, Moon and Standing Stones*. Oxford University Press.

Zeilik, M. (1983). One approach to archaeoastronomy: an astronomer's view. *Archaeoastronomy. The Journal of the Center for Archaeoastronomy* 6, (1–4), 4–7.

4

Reflections on the role of archaeoastronomy in the history of astronomy

Owen Gingerich   *Harvard-Smithsonian Center for Astrophysics*

Modern science is a uniquely Western phenomenon that has arisen only once on our planet. As a historian of Renaissance astronomy, I find that one of the most fascinating questions is: to what extent has this flowering of modern, predictive science been inevitable? We do not have any well developed parallel cases for comparison, although the civilization of China, to a lesser extent that of India, and to a still lesser extent that of the Mayas, offer us areas for exploration. In the present context I would like to ask whether archaeoastronomy can offer insights into the origin of science.

4.1 Some characteristics of modern science

The particulars of modern science are molded by cultural constraints to a far greater extent than most scientists would readily admit. Take, for example, Newton's first two laws of motion. The first states that in the absence of any force a mass will continue in its state of rest or motion. The second states that the presence of a force will cause an acceleration proportional to the force and inversely proportional to the inertial mass of the object. Clearly the first law, the case of zero force, is contained in the second and is therefore redundant, as has been pointed out many times. But if we ask *why* there are two laws rather than one at this point in Newton's *Principia*, we will discover that Descartes had already come close to stating the first without the second and also that Huygens had started out with *two* laws.[1] In other words, the historical context had an important formative effect on the details of the presentation.

If we further examine the configuration of the laws of mechanics, we notice that the same predictions could arise from a wholly different scheme, one that never mentioned force and inertia, but which used conservation laws, potentials and fields instead (Gingerich, 1978). Thus we begin to recognize the *Principia* as a unique creative achievement; not only would words and ordering be otherwise if another scientist had attacked the problems of mechanics, but even the physical concepts could have differed.

Despite the fact that Newton's world picture was in many ways a personal creation, it was so thorough and so probable that most Europeans had no difficulty in accepting it as truth. It was so impressive and apparently so complete that it has helped to characterize the scientific way of looking at the world, at least through the 19th century: deductive, mathematical, atomistic, mechanical.

'At the end of the last century,' and here I quote from a most perceptive book by Jacob Bronowski (1978), 'there were physicists who were perfectly willing to say that there was no need to produce another Newton because there was nothing as fundamental as gravitation for another Newton to discover .... Since then,' he continues, 'the world has fallen about our ears.' Why? Because Einstein's general relativity showed us another way of looking at gravitation, and brought the scientific community to the realization that there was no way to know when any scientific theory was final. Einstein's work has forcefully awakened us to the provisional nature of our scientific picture. This is why I have sometimes referred to the scientific world view as a grand tapestry.[2] It is an interlocked and coherent picture, a most workable explanation, but it is not ultimate truth. Robinson Jeffers (1963) puts it well in one of his poems when he says,

> The mathematicians and physics men
> Have their mythology; they work alongside the truth,
> Never touching it; their equations are false
> But the things *work*.

Even though we claim that our scientific world picture is in part uniquely configured by particular historical circumstances, the fact that things work, that events can be predicted (whether by Newton or Einstein), gives science a powerful efficacy in our modern world. Nowhere is this more impressively shown than in Western technology. We have opened a Pandora's box from which there is no turning back.

Philosophers from Hume onwards have been puzzled as to why science has this remarkable predicting ability, but, understood or not, this power has shaped our contemporary culture. Nevertheless, this predictive power, or 'foresight' as Stephen Toulmin (1961) expresses it, is only part of what makes science acceptable. There is also an explanatory part, 'understanding' in Toulmin's succinct coding. Of course, from earliest times part of the challenge of being human was to gain understanding, to express in mythic terms the interrelation between man and nature. To the extent that men have accepted a rational, purposeful creation they have demanded a strong coherence in the explanatory structure.[3] Yet what seems different about the modern scientific world view is the sheer complexity of the connections between its explanative and predictive aspects, as well as the continual reforming of the theoretical structure in the light of observation and experiment. It is these aspects that characterize modern science and set it qualitatively apart from the rudimentary systems of archaeo- and ethnoastronomy.

Hence, when I inquire whether archaeoastronomy can offer any insight into the origin of science, I mean a complex, evolving explanatory/predictive system of the sort that most educated people within Western culture take for granted.

4.2 Archaeoastronomy as a prelude to modern science

Now up until Oxford I, most practitioners of classical archaeoastronomy – Aveni's 'greens' or those who work with sticks and stones – were well rooted within Western science and engineering, and they firmly believed that their interdisciplinary field had something to say about early astronomy, the oldest of the sciences, and therefore about the roots of science itself. If modern science evolves through measurement and numeracy, why should there not also be stone age Einsteins who could theorize and abstract basic truths from persistent observations of the sky? There was a tacit assumption that the monument builders were not merely looking to the heavens, but they were also establishing the rudiments of science – as Euan MacKie's (1977) title, *Science and Society in Prehistoric Britain* suggests by its explicit reference to 'science'.

But to what extent is the 'science' of these ancient cultures an anticipation of modern science as I have described it? Can archaeoastronomy tell us anything about the origin of science itself? I shall argue that, paradoxically, archaeoastronomy may have more to say about the *non*-origin of science than about its origins, and that to look for the origins of science here may lead to misleading or anachronistic conclusions.

When at Merida I posed the question 'Is science inevitable?' and suggested that a modern predictive physical science with something like Newtonian mechanics might *not* inevitably arise, I expected shouts of outrage. I was nonplussed by the lack of reaction. Perhaps the audience was exhausted at the end of a strenuous conference, or confused as to what I had in mind, or possibly even baffled by a question they had never thought about. But another explanation might be that the audience included a majority of neoclassical archaeoastronomers, Aveni's 'browns' with their links to anthropology, who were accustomed to strangely different cultures and open to the idea that elsewhere there could be an entirely different approach to the natural world or a science that would not at all resemble Newtonian physics.

Now this is not the reaction that I would have got from a typical group of physicists or astronomers. Their response can be surmised from the current popularity of the idea of intelligent life on other worlds. Evolution would inevitably lead to intelligence, and intelligence to technology and hence to communication, so the reasoning runs, and I have heard this notion argued persuasively by some of the virtuosi of our day. Behind this logic lies the idea that science will inevitably converge on something similar in every alien civilization, so that we can exchange scientific notes when the radio links are established.

To a certain extent archaeoastronomy and ethnoastronomy introduce us to aliens – perhaps not as foreign as the putative creatures inhabiting distant solar systems – but to societies at least isolated from our own cultural history. The possibility that the prehistory of science as probed by archaeoastronomers might offer some insight as to whether science arises inevitably and in a form similar to our own is one of the reasons why historians of science necessarily have a stake in this discipline.

4.3 Motivations for the development of astronomy

Let us consider the possible motivations for doing astronomy, which may have stirred different societies along parallel lines.

There are, it seems to me, at least two compelling concerns that would have brought about an examination of the heavens:

(i) cosmology, and

(ii) astrology.

By cosmology, I mean something broader than the modern ideas of the expanding universe, although this concept represents the most sophisticated form of humankind's search for our place in time and space. A quite different form of the search is eloquently summarized in Chapter 16 of this volume by Leon-Portilla. Both cases represent what may be a universal curiosity and longing to know where we stand within the cosmos, the interrelation between man and nature. It is a pervasive desire, and in our culture the language and literature is permeated with vestiges of this ongoing search.

Closely bound with the task of finding one's place in the universe is astrology, well expressed in the motto of Tycho Brahe, *suspiciens despicio*, 'By looking up I understand what is below' – an adage whose history goes back to the Middle Ages and whose roots are in antiquity. In other words, here is the idea that an understanding of the heavens will in turn lead to an understanding of mundane affairs – a prescientific notion if it involves some mysterious synchronicity, or a pioneering scientific concept if it supposes cause and effect. In the West the astrological vision was developed in a scheme wherein the microcosm reflected the macrocosm; although this-

was an interpretation elaborated by a particular culture, we find the belief in connectivity between the heavens and earth in China as well as in Mesoamerica, and it seems plausible to expect some form in other cultures as well.

The borderline between cosmology and astrology is vague, but together they provide an almost subconscious impetus for recording the rhythms and even the discontinuities of the starry vault. Let me be more specific about rhythms and discontinuities. The daily and yearly solar cycles, and their connections with the seasons of hunting, planting and harvesting, are clearly rhythmic and predictable. Earthquakes, great floods, whirlwinds and droughts as well as comets, novae, sunspots and solar eclipses are seemingly random, discontinuous, and evil. No wonder that comets are almost universally seen as bad omens. It is hard to have a *theory* of discontinuity – that is, it is comparatively difficult to have an explanatory scheme for unpredictable events – but a society might provide a partial protection by diligent observation and associated rituals. The wonderful sequence of Chinese observations that are now admired so much in the West (because in retrospect they are important for modern astrophysics) arose from such a search for omens in the discontinuities of the sky.

The cyclic aspects, on the other hand, cry out if not for a theory, at least for a calendar (which is a theory of sorts). This then brings me to two rather more specific applications that have motivated astronomical practice:

(i) calendar, and

(ii) navigation.

These were perceived to be useful, particularly in more sophisticated civilizations. The need for a calendar, especially, helped build numeracy in widely different cultures, but curiously enough, by our own cultural standards, calendars need not imply numeracy.

The study of archaeoastronomy and ethnoastronomy shows clearly that, in one society after another, some sort of calendar develops, apparently quite independently. Almost universally, early calendars seem to be lunar as well as seasonal. This does not mean, however, that they are necessarily numerical or that the seasonal connections are directly linked with solar astronomy. The problem of intercalation – arising from the fact that some seasonal years have 12 lunations and others have

13 – has been solved by a variety of terrestrial observations dependent on the season without invoking the solar movement. The Trobriand Islanders, for example, keep their lunar calendar in step with the seasons by the requirement that the palolo worm, a tasty marine annelid, should surface in the sea near the full moon in the lunar month of Milamala (around our November) (Leach, 1950). If the palolo worm fails to appear, the Trobrianders simply add a second Milamala, thereby effectively placing a 13th lunation into the seasonal year. 'That the system should work at all is almost a paradox,' writes E. R. Leach (1954). 'It does so only because the logic with which the natives approach calendarial time is vague in the extreme and even inconsistent.' Elsewhere, the Mursi in Ethiopia keep their count of lunations in approximate agreement with the seasonal year by starting the sequence with the maximal flooding of the River Omo (Turton and Ruggles, 1978). Though they recognize that the appearance or disappearance of certain star patterns correlate with the vegetation, this is not particularly formative in keeping the calendar in step with the seasons. Similarly, in the old Hebrew calendar, the lunar months were synchronized with the seasonal year by the Sanhedrin's judgment on the signs of spring vegetation, and not until the fourth century of our era was the calendar placed onto the astronomically based Metonic cycle.

Even when the astronomical correlations are roughly understood, only a minimum number of observations or measurements are needed. Most of the leverage for establishing accurate cycles comes from a long span of time rather than precision of positions. Asger Aaboe (1980) has shown with considerable ingenuity how little accuracy of observation was required by the Babylonians to determine their relatively accurate parameters. This emphasizes the value of tradition or of record keeping, rather than devising highly accurate instruments. (This is partly why I have always been suspicious of the Thoms' claims for precision megalithic observatories, but more of that later.)

The more complex calendars, such as the intricately and rigidly dovetailed Mayan system, certainly entail a high level of numeracy (Kelley, 1978). Two remarks seem in order here. First, even if a society has the ritual use of a fairly complex calendar, this does not mean that the typical member of the group has much appreciation of its numerical

or observational basis. Our own Western calendar is bound into an ancient seven-day cycle, an approximate lunar monthly cycle, and a fine-tuned lunar/solar calendar that sets the date of Easter and Passover. Yet few who celebrate these special days have any clue to their astronomical origin. We can expect that in some societies the calendar will be their highest mathematical achievement, far beyond the general numeracy within the society.

Second, the willingness to use and manipulate numbers seems a necessary prerequisite to the development of a Newtonian-type science. A necessary condition, yes, but is it a sufficient condition? Just because a few individuals in a society make the occasional critical observation required for a calendar, this does not guarantee a further impetus toward scientific modeling and the development of causal physical relationships. In China, besides the well publicized ominal observations, there exists a long series of solstitial as well as eclipse observations to be used for the correction of the calendar, but there was no autonomous movement toward what we would recognize as a 'Newtonian revolution.'[4] The apparent dead end that the Mayan mathematical culture reached seems suggestive evidence for the non-inevitability of science, a case for the non-origin of science, if you will.

4.4 On the non-inevitability of modern science

I recognize that my argument is open to the criticism that I am mixing apples and oranges: those who argue that, given enough time, a predictive, Newtonian-type science will inevitably arise *somewhere* are not claiming that it will arise *everywhere*. The evolutionary process is full of blind alleys and dead ends. But this is just my point: we cannot assume that the megalith makers or the Mayans or the Anasazi were on the escalator to nuclear astrophysics or grand unified field theories.

This leads me to say that I suspect that archaeoastronomy has been considered too much a subdivision of history of science. If it is seen exclusively as a branch of history of *science*, we may tacitly assume that these investigations must illuminate the origins of modern science. Accordingly, we have tended to judge these past cultures by the yardstick of Western science, looking for prehistoric Einsteins and precision measurements of the kind that energize modern science. As long as we assume that our science is inevitable we can scarcely prevent a warped perspective and false set of expectations. Even Ptolemy has not escaped the judgmental scorn that imputes modern motivations and standards to ancient ages.[5]

Probably in no other area can we slip into contemporary ethnocentrism more easily than in the interpretation of the role of observations and measurement. Observations are so essential to today's scientific ethos that we are almost disconcerted to learn that *no* new observations or measurements were required for Huygens' discovery of the rings of Saturn or Copernicus' formulation of the heliocentric system (Van Helden, 1974). Similarly, when we think in terms of the modern role of observations, it is hard to know what to make of the giant stone instruments in Delhi or Jaipur. Raised by Jai Singh early in the 18th century, these observatories postdated the invention of the telescope by a century. But whether or not they had telescopic sights makes little difference. If one examines closely, for example, the slotted bowl-shaped instruments that Jai Singh had himself so proudly invented, we see that there is no way, with or without telescopic sights, that useful observations in any modern celestial mechanical sense could have been made with these marble instruments. They were splendid demonstration devices, perhaps comparable to a Zeiss planetarium projector in our contemporary culture, a way to capture the cosmos. Cosmology, yes, but stepping stones to the origins of modern science, no.

By the same sort of argument, I would likewise say that the Chinese, for all their long series of ominal observations, and despite their occasional specifically quantitative measurements, were not demonstratively on the road to modern science. Accurate calendars, yes, but there is no evidence that they were prepared on their own to abandon their holistic view of nature for the reductionist, mathematical approach of post-Renaissance Western science. But Nathan Sivin, who has stressed this viewpoint, has also urged me to consider how little evidence exists that the West in pre-Galilean Europe was about to take on the awesome mental reorientation leading to Western science and the strong cleavage between the scientific and religious ways of looking at nature.

4.5 Precautions and scepticisms

These arguments do not deny the possibility of precision observations in prehistoric or ancient cul-

tures. Rather, I question the *expectation* of finding them, an expectation based on our modern notions of what it takes to do science. But *if* precise measurements are postulated in other cultures, then I would require some plausible motivation *within that culture* and not from our own. In other words, since precision observations can sometimes be made today from the Scottish sites of Kintraw, or Ballochroy, or Brogar, it would be stupid to deny the possibility of such precision in neolithic times, but one can certainly maintain a healthy scepticism in the absence of any understanding as to what that culture possibly would have used such observations for. I can well believe that a people who recognized (and expressed architecturally) the extreme northern swing of the sun might also have been curious about the northernmost swing of the moon (or Venus), and I find it not implausible that the post holes north of the Heel Stone at Stonehenge might be related to an attempt to establish just such a limit. It is not unreasonable to argue that they may have developed eclipse warnings from an understanding of the moon's apparent path. Nevertheless, it boggles my mind to suppose that the Newtons among those people laid out stones to interpolate lunar positions to an accuracy where nutation becomes a significant consideration.

By the same token, I have never been convinced by any of the claims for alignments related to the minor lunar standstill, whether they are in Britain, Brittany, or on Fajada Butte. Perhaps I am making an impossible request in demanding some ethnographic or anthropological context. And the other side of the coin is that, since I cannot rule out the possibility that the alignments were deliberately set on the lunar minor standstill, I might be wholly mistaken in my prejudice. A case in point could involve the Great North Road north of Pueblo Bonito and reported in Chapter 29 of this volume. Within this Chaco Canyon area stands the magnificent Fajada Butte, and on it are found a remarkable series of petroglyphs. No small number of these markings are oriented so that the noon-day sunlight creates a distinctive play of light and shadow across them. Were these deliberate time-keeping markers? From what little we know about the ancient peoples in this area, or what we can deduce from the customs of the more recent inhabitants, it would seem *a priori* unlikely that the makers of the petroglyphs would wish to establish noon so precisely. On the other hand, if the Great North Road really did run accurately straight and due north out of that site, it would offer striking evidence that the north–south cardinal direction had some as-yet-unexplained significance in that culture. Thus, even in the absence of ethnographic data, mute archaeological evidence could make it reasonable to suppose that noon markers (*de facto*, south markers) had relevance to the Anasazi peoples, even though it might have seemed previously implausible. At first I found the new evidence presented by Sofaer *et al.* tantalizing and highly suggestive but, after I had the opportunity to examine aerial views of the meandering road, I lapsed into my earlier scepticism.

4.6 Where is the place of archaeoastronomy?

It was with some trepidation that I accepted the task of making some remarks on the role of archaeoastronomy in the history of science in pre-Galilean Europe. My own arena has been modern astronomy and its historical precedents in that uniquely Western phenomenon we call the scientific revolution. As a historian of astronomy, I am more interested in helping to explain the reasons for scientific change than the change itself. The more cultural units we can examine, the more grist for the mill, because the differences and even the lack of change may help clarify why changes happened in far different contexts.

In any event, the role of archaeoastronomy in the history of science is clear: archaeoastronomy gives us still another tool for probing the past, with the unusual primary use of monuments and artifacts rather than the texts familiar to historians. But in another sense archaeoastronomy transcends the history of science, because it studies not only the roots of science but also how people have coped with the natural world in ways that do not necessarily lead to science. To the extent that archaeoastronomy also studies the *non*-origin of science, it is not so much history of science as part of something far broader, the history of culture itself. It studies what are literally the roots of civilization (to borrow Alexander Marshack's evocative title). Nathan Sivin, in responding to my conclusion, has remarked that archaeoastronomers might consider their field a subset of an as-yet-uninvented discipline that could be called uranology or 'sky studies', a field as interested in literature and ritual as in eclipse prediction.

Archaeoastronomy is by its nature interdisciplinary, and archaeoastronomers armed with a vision of the broad canvas of intellectual history should not hesitate to appropriate tools where they can find them, whether in linquistics or mathematics, laser theodolites or radioactive dating, astronomy or archaeology. And, in the process, we might just gain insights into that difficult question: Is modern Western science inevitable?

Notes

1 For fuller details see Cohen (1967).
2 I have explored this metaphor in greater depth in Gingerich (1983).
3 Thus it has been argued that the combination of the Judeo–Christian tradition and Greek philosophy provided a particularly felicitous if not indispensable setting for modern science; perhaps the strongest case is made by Hooykaas (1972) in his *Religion and the Rise of Modern Science.*
4 See the thought-provoking discussion by Nathan Sivin on whether the Chinese had a scientific revolution (Sivin, 1984).
5 An anachronistic attack on Ptolemy was launched by Robert R. Newton and embellished in his *The Crime of Claudius Ptolemy* (Newton, 1977); for what we consider to be better balanced views, see Gingerich (1981) and Wilson (1984).

References

Aaboe, A. (1980). Observation and theory in Babylonian astronomy. *Centaurus* **24**, 14–35.
Bronowski, J. (1978). *The Origins of Knowledge and Imagination*, p. 56. New Haven: Yale University Press.
Cohen, I. B. (1967). Newton's second law and the concept of force in the *Principia*. *The Texas Quarterly* **10** (3), 127–57.
Gingerich, O. (1978). Circumventing Newton: a study in scientific creativity. *American Journal of Physics* **46**, 202–6.
Gingerich, O. (1981). Ptolemy revisited: a reply to R. R. Newton. *Quarterly Journal of the Royal Astronomical Society*, **22**, 40–4.
Gingerich, O. (1983). Let there be light: modern cosmogony and biblical creation. In *Is God a Creationist?*, ed. R. M. Frye, pp. 119–37. New York: Scribners.
Hooykaas, R. (1972). *Religion and the Rise of Modern*

Science. Edinburgh: Scottish Academic Press.

Jeffers, R. (1963). The great wound. In *The Beginning and the End and Other Poems*, p. 11. New York: Random House.

Kelley, D. H. (1978). Astronomical identities of Mesoamerican gods. *Archaeoastronomy*, Supplement to *Journal for the History of Astronomy*. Suppl. to vol. **11**, no. 2, pp. S1–S54.

Leach, E. R. (1950). Primitive calendars. *Oceania* **20**, 245–62.

Leach, E. R. (1954). Primitive time-reckoning. In *A History of Technology*, eds. C. Singer, E. J. Holmyard and A. R. Hall, vol. 1, p. 120. Oxford University Press.

MacKie, E. (1977). *Science and Society in Prehistoric Britain.* New York: St. Martin's Press.

Newton, R. R. (1977). *The Crime of Claudius Ptolemy.* Baltimore: Johns Hopkins.

Sivin, N. (1984). Why the Scientific Revolution did not take place in China – or didn't it? In *Transformation and Tradition in the Sciences*, ed. E. Mendelsohn, pp. 531–54. Cambridge University Press, Cambridge.

Toulmin, S. (1961). *Foresight and Understanding: An Enquiry into the Aims of Science.* New York: Harper and Row.

Turton, D. and Ruggles, C. (1978). Agreeing to disagree: the measurement of duration in a southwestern Ethiopian community. *Current Anthropology* **19**, 585–93.

Van Helden, A. (1974). 'Annulo Cingitur': the solution of the problem of Saturn. *Journal for the History of Astronomy* **5**, 155–74.

Wilson, C. (1984). The sources of Ptolemy's parameters (essay review). *Journal for the History of Astronomy* **15**, 37–47.

5

The king, the capital and the stars: the symbolism of authority in Aztec religion

David Carrasco   *University of Colorado*

5.1 Background

In James Thurber's autobiographical work *My Life and Hard Times*, the author describes in his chapter 'University days' a struggle he underwent while attempting to pass his undergraduate course in botany. Thurber tells that he never succeeded in passing botany because, in spite of his instructor's guidance, insistence, and even emotional outbursts he (Thurber) 'never once saw a cell through a microscope'. Even though all the other students did see the mechanics of flower cells through the microscope, Thurber didn't and proclaimed, 'It takes away from the beauty anyway' – a comment which drove his teacher into a fury. You were supposed to see a vivid restless clockwork of sharply defined plant cells, but Thurber didn't, exclaiming 'I see what looks like a lot of milk'. The instructor claimed that this was the result of not having adjusted the microscope properly, so the instructor adjusted the microscope. Thurber states, 'And I would look again and see milk ... a nebulous milky substance – a phenomenon of maladjustment.' This academic crisis of perspective reached a high pitch during the second year in the laboratory. The teacher claimed grimly, 'We'll try it, with every adjustment of the microscope known to man. As god is my witness, I'll arrange this glass so that you see cells or I'll give up teaching.'

Then one day, Thurber thought he saw what the teacher so much wanted him to see. He writes,

> With only one microscopic adjustment known to man did I see anything but blackness or the familiar lacteal opacity, and that time I saw, to

my pleasure and amazement, a variegated constellation of flecks, specks and dots. These I hastily drew. The instructor noting my activity came back from an adjoining desk, a smile on his lips and his eyebrows high in hope. He looked at my cell drawing. 'What's that?' he demanded, with a hint of a squeal in his voice. 'That's what I saw,' I said. 'You didn't, you didn't, you didn't,' he screamed, losing control of his temper instantly, and he bent over and squinted into the microscope. His head snapped up. 'That's your eye!' he shouted. 'You've fixed the lens so that it reflects! You've drawn your eye.' (Thurber, 1956.)

I begin my chapter on the symbolism and the symmetry of the Templo Mayor of Tenochtitlan with a story from an American humorist to set out a problem in the interdisciplinary attempts some of us are making to gain a clearer view of how the practice of astronomy and the sacralized character of this practice influenced the construction and experience of sacred space in Mesoamerica. It is the problem of perspective, or rather perspective in the plural, and how perspectives can be in contact with one another (Eliade, 1969).[1] In a way, seeing is what archaeoastronomy seems to be largely about; seeing stars, seeing how the ancients saw stars, seeing alignments between human order and celestial patterns, and seeing to it – in the case of kings – that alignments between celestial events and human society were created and maintained. In our different attempts to see with a native eye, or see how the natives saw things, or, in the case of ideological arguments, seeing how some natives wanted other natives to see things their way, we are often in the position of either Thurber or his teacher. We work hard trying to get others to see things according to the received ideas of how things ought to be done or we 'fix the lens of our discipline' so that we end up seeing our own eye, our eye and mistaking it for their visions.

In the teacher's case, in some teacher's cases, the gesture of persuasion becomes the gesture of coercion to have younger, or North American, or Mexican, or European scholars see what they are supposed to see – the mechanics of flowers or, in the case of astronomy, the mechanics of stars, of mathematics, or symbolism. At the other extreme, we act like Thurber (who, by the way, lost an eye when he was ten after his brother shot him with an arrow while they were playing William Tell) and only see, in fact insist on seeing, reflections of our own eye, perspective, or discipline. In this case 'we reduce the difficult richness of the necessities before us' (Geertz, 1983a,b).

A simple example in the history of archaeoastronomy will illustrate part of the problem. In the fall of 1984 I had the valuable opportunity to study with the archaeoastronomer Anthony Aveni while he was working in the Mesoamerican Archive in Boulder, Colorado. During our eight-week jointly taught seminar on the history of religions and archaeoastronomy, Aveni (who is not like Thurber's teacher) made several presentations on the history of scholarship about Stonehenge. He pointed out the parallel between cultural fashions and the Stonehenge fashion – how popular intellectual fads determined the way people understood Stonehenge for certain decades. He noted that when the fashion was observatories, or computers, or eclipses, then Stonehenge became an observatory, a computer, an eclipse marker. The point was, in Aveni's words (quoting Jacquetta Hawkes), 'Each generation gets the Stonehenge it deserves'. The best illustration was the day Aveni brought to class that Thurberesque advertisement about the Stonehenge watch you could purchase through the mail made up of little stones on the dial which cast shadows according to the time of day.

It led me to ask, 'Does each generation get the Templo Mayor it deserves, the Jerusalem it deserves, the Benares it deserves, or the Palenque it deserves?' Further, my work with Aveni led me to ponder, to what extent does archaeoastronomy

contain a theoretical framework which enriches our understanding about the relationship of astronomy to religious creativity? This question was recently discussed in Lawrence Sullivan's article 'Astral myths rise again: interpreting religious astronomy'. Summarizing two previous periods in Western scholarship when the tie between astronomy and religion was in vogue, Sullivan noted a single tendency which hindered the construction of a sound theoretical framework for understanding the relationship between religious symbolism and astronomy. This was the tendency to 'impose upon the data an assumption of mutual incompatibility between religious and scientific purposes which appear to be built into the data' (Sullivan, 1983).

The problems caused by this assumption were evident in the 1798 publication of Charles Francois Dupuis' 12-volume *The Origin of All Religious Worship: Universal Religion.* Dupuis argued that all religious ideas and practices originated in fantastic distortions of one process of nature, the regular movement of the stars. Sullivan writes,

> Examining evidence of religious belief and
> practice in Chinese, Siamese, Greek,
> Molluccan, Persian, Philippino, Norse,
> Madagascan, Formosan, and Japanese traditions
> he attributed the origin and order of all myth
> ... to nothing more than the unnecessary veil
> of allegory drawn across the visible events of ...
> the equinox, solstices, seven planets, and 12
> signs of the Zodiac.

Dupuis' point was that religion obstructed the acquisition of true knowledge about the stars. True knowledge of the heavens would be achieved through reasonable and precise astronomic data. It is interesting to note, as an aside about cultural fashions and astronomy, that Dupuis' book was dedicated to the French Revolutionary Assembly of 1795, and that in the intervening 25 years according to Sullivan (1983)

> this last significant mythography of the
> Enlightenment had sold out of every edition of
> its three, eight, twelve and single volume formats
> and had been the center of a raging controversy.
> Joseph Priestley, the one time Unitarian minister
> and scientist, refuted it at length and, in exile
> from England, discussed it vehemently with his
> new found American friends, John Adams and
> Thomas Jefferson.

A second interest in astromythology appeared in

the early 20th century among the Pan Babylonianists whose researches in Babylonian, Assyrian, Summerian and Semitic civilizations argued that nearly all myths were literal statements about the movement of heavenly bodies, especially the 'multiform synodical phases of the moon' (Sullivan, 1983). Again the position argued that religion had a narrow origin in human history and that it was, in part, a distortion of observable nature.

It appears that one of the characteristics of contemporary archaeoastronomy is the interest in discovering intersections, crossroads and conjunctions between the multiple aspects of ecology, the patterns of astronomy, and the complex aspects of the religious imagination. This development in archaeoastronomy finds a conversation partner in the position expressed by Ninian Smart in his work on the future of the history of religions. Professor Smart persuasively argues that scholars in religious studies need intellectual interchanges with scholars from other disciplines. These interchanges would result in not only the usual process of 'imports' of ideas into religious studies but the much needed process of 'exporting' ideas from the history of religions into other disciplines. Also, Professor Smart urges historians of religions to carry out more theoretical work which relates what we call religion to other areas of human experience (Smart, 1985). This attempt at an understanding of how things interrelate and how disciplines need to relate seems crucial in Mesoamerican studies, especially in the light of Kent Flannery's claim that 'Mesoamerican archaeology has absolutely no coherent and consistent theoretical framework by means of which ritual or religious data can be analyzed and interpreted' (Flannery, 1976). But Flannery, along with a handful of specialists, has constructed the outlines of such a framework for the interpretation of formative and classical religious patterns. This framework is contextual in character and 'ties religion to social organization, politics and subsistence rather than leaving it on the ephemeral plane of mental activity' (Flannery, 1976). Still, according to Flannery, when it comes to religion most of us are 'guessing', and, as Flannery states, anyone's guesses are as good as anyone else's.

But the reason for having many guesses is not just because each is as good as the next, but because there are so many levels, angles, alignments, and points of view to guess about and from which to guess. As the poet Ovid noted, there are many heavens, the heaven of the astronomer, the heaven of the historian, the heaven of the priest, the heaven of the lover, and the heaven of the myth-maker. If interdisciplinary work is to take place, a recognition of the abundance of the phenomena and the interrelatedness of the components is necessary.

One of the cases which has enjoyed sustained scholarly focus in recent years has been the Templo Mayor of Tenochtitlan. Recently there has been serious speculation concerning the influence of astronomy on its spatial orientation and symbolism (Townsend, 1982; Broda, Carrasco and Matos Moctezuma, 1987; Aveni, Calnek and Hartung, 1988).[2] This chapter is an attempt to enlarge our understanding of religious symbolism in Tenochtitlan by utilizing the discipline of the history of religions to re-think the category of alignment in relation to myths and ritual. Specifically I will focus on a comparison between a sacred symmetry within the New Fire Ceremony reported in Book VII of Bernardino de Sahagun's *Florentine Codex* (Sahagun, 1975) and in Motolinia's *Historia de los Indios de Nueva España* and a similar arrangement in the description of the equinox sunrise at Templo Mayor reported in Motolinia (1569). With respect to these texts I wish to suggest that the ceremonial order of Aztec life, what has recently been called 'The Aztec Arrangement', was a vigorous interplay between a locative view of the world, in which all things are sacred if they are in their place, and an apocalyptic view of the world, in which the sacred dissolves when things have no place or are out of place. Further, I want to suggest that the Aztecs believed that their power and authority depended upon their capacity to integrate certain major elements of their world including astronomy into a meaningful alignment or symmetry.

5.2 Aztec authority

On an evening in the middle of November in 1507, a procession of fire priests with a captive warrior 'arranged in order and wearing the garb of the gods' processed out of the city of Tenochtitlan towards the ceremonial center on the Hill of the Star. During the days prior to this auspicious night, the populace of the Aztec world participated together in the ritual extinction of fires, the casting of statues and hearth stones into the water, and the clean sweeping of houses, patios, and walkways. Book VII of *The*

Florentine Codex entitled 'The sun, the moon, the stars, and the binding of the years' tells us that in anticipation of this fearful night women were closed up in granaries to avoid their transformation into fierce beasts who would eat men, pregnant women would put on masks of maguey leaves, and children were punched and nudged awake to avoid being turned into mice while asleep (Sahagun, 1975). For on this one night in the calendar round of 18 980 nights the Aztec fire priests celebrated 'when the night was divided in half', the New Fire Ceremony which ensured the rebirth of the sun and the movement of the cosmos for another 52 years. This rebirth was achieved symbolically through the heart sacrifice of a brave warrior specifically chosen by the king. We are told that when the procession arrived 'in the deep night' at the Hill of the Star the populace climbed onto their roofs and 'with unwavering attention and necks craned toward the hill became filled with dread that the sun would be destroyed forever'. It was thought that if fire could not be drawn, the demons of darkness would descend to eat men. As the ceremony proceeded, the priests watched the sky carefully for the movement of a star group known as Tianquiztli or Marketplace, the cluster we call the Pleiades. As it made a meridian transit signalling that the movement of the heavens had not ceased, a small fire was started on the outstretched chest of a warrior. The text reads, 'When a little fire fell, then speedily the priests slashed open the breast with a flint knife, seized the heart, and thrust it into the fire. In the open chest a new fire was drawn and people could see it from everywhere.' The populace cut their ears, even the ears of children in cradles, the text tells us, 'and spattered their blood in the ritual flicking of fingers in the direction of the fire on the mountain'. Then the new fire was taken down the mountain, carried to the pyramid temple of Huitzilopochtli in the center of the city of Tenochtitlan, where it was placed in the fire holder of the statue of the god. Then messengers, runners and fire priests who had come from everywhere took the fire back to the cities where the commonfolk, after blistering themselves with the fire, placed it in their homes, and 'all were quieted in their hearts' (Sahagun, 1975).

In reflecting on this famous passage about the New Fire Ceremony, I want to follow the lead of Giovanni Morelli and use an 'a-centric' perspective. Morelli, a 19th century art historian, developed a successful method for distinguishing original masterpieces from copies by focusing his eyes, not on the most obvious characteristics of a painting in order to identify its master, but 'on minor details, especially those considered least significant in the style typical of the painter's own school' (Ginzberg, 1983). Instead of looking at the smiles of Leonardo's women or the eyes of Perugino's characters, which were usually raised to heaven, Morelli studied the ear lobes, the fingernails, the shapes of fingers and toes. This method, the Morelli method, used minor details to gain a picture of the whole.

This passage, which has only a few variants in 16th century accounts, is extraordinarily thick and complex. It has the obvious meanings related to astronomy, calendars, ritual theatres, human sacrifice, and even child rearing. But running through it all is a thread, actually two threads, partly hidden, which not only tie the description together but provide a clue to the underlying social and symbolic purpose of the ritual. These threads are the flow of Moctezuma's authority through all aspects of the ritual, and the presence of the Templo Mayor as the axis mundi of the New Fire Ceremony.

The presence of these threads is more evident when we retrace *just* the physical actions of the description. The drama begins with Moctezuma in Tenochtitlan, even though in this account he is not mentioned at the start. But, elsewhere in this volume, we are told that, months before the New Fire Ceremony, Moctezuma ordered a captive be found whose name contained the word 'xiuitl' meaning turquoise, grass, or comet – a symbolic name connoting precious time. The procession of deity impersonators move along a prescribed passageway, presumably seen and heard by masses of people before arriving at the Hill of the Star. In Motolinia (1569) we are told that Moctezuma 'had special devotion and reverence for the shrine' and the deity on the sacred hill. Assembled in the ceremonial center the group of priests and lords, sharing a heightened sense of expectation and fear, seek another procession – the procession of the stars through the meridian. Once recognized, the heart sacrifice is carried out, the new fire is lit amid universal rejoicing and bleeding, and the fire is taken to the Templo Mayor, presumably with Moctezuma on hand to see its blaze. Then, in what I see as the most meaningful social and symbolic gesture, messengers, priests and runners who have 'come from all directions' to Templo Mayor take the fire back to the towns and cities of the periphery. In Motolinia (1569) we are told that the fire was taken back to the temples only 'after asking permission from the great chief of Mexico' (Broda, 1970).

By focusing our eyes on the minor details of Moctezuma's role and the Templo Mayor as the shrine to which the New Fire is taken in order to be dispersed to the populace, I see a skillful symmetry reflecting the Aztec commitment to the interconnection of their world. By symmetry I mean the orderly arrangement of symbolic components around an axis. This symmetry consists of five elements: (1) the cosmic mountain (in this text there are two, the Hill of the Star and the Templo Mayor); (2) astronomical events; (3) human sacrifice; (4) agricultural renewal; and (5) sacred kingship. I see the center of this symmetry to be interplay between the king's flow of authority and the axis of Aztec society, the Templo Mayor. This interplay constitutes what the University of Chicago scholar of social thought Ed Shils calls a 'center' by which he means 'the point or points in a society where its leading ideas come together with its leading institutions to create an arena in which events that most vitally affect its members' lives take place' (Geertz, 1983a).[3] What is taking place in the New Fire Ceremony is the integration of the leading idea – Moctezuma's authority, with the leading institutions, symbolized by the Templo Mayor, and with the cosmic renewal integrated by an astronomical event.

This symmetry is an example of what Jonathan Z. Smith, in his book *Map is not Territory: Studies in the History of Religions*, calls a locative view of the world consisting of 'a map of the world that guarantees meaning and value through structures of conjunction and conformity' (Smith, 1978). This locative view, which has been discerned in the traditional societies of Mesopotamia and Egypt, in which everything has value and even sacrality when it is in its place, is an imperial view of the world designed to ensure social and symbolic control on the part of the king and the capital. It is informed by a cosmological conviction consisting of five facets which dominated human society for over 2000 years in the Near Eastern world including (1) there is a cosmic order that permeates every level of reality; (2) this cosmic order is the divine society of the gods;

(3) the structure and dynamics of this society can be discerned in the movement and patterned juxtaposition of the heavenly bodies; (4) human society should be a microcosm of the divine society; and (5) the chief responsibility of priests and kings is to attune human order to the divine order. In the New Fire Ceremony – at least in 1507 in Tenochtitlan – Moctezuma the Second is carrying out his chief responsibility by attuning human order to the divine order through the discernment of an astronomical event.

Perhaps the single best illustration of the locative emphasis of Aztec religion appears in the detailed description of the human body chosen to impersonate the king's patron deity Tezcatlipoca, Lord of the Smoking Mirror. Tezcatlipoca, known alternately as 'He Who Created and Brought Down All Things' and 'The Enemy from Both Sides' was one of the supreme manifestations of the Aztec High God. As the patron of Aztec kings, he created, inspired, guided and, according to the text, could turn and kill one of them if they governed badly. Most Aztec gods had human impersonators, usually captive warriors or slaves who were ritually costumed and processed through the streets, plazas and temples during the festivals of the Aztec calendar. These living, moving cult images symbolized numinous impersonal forces, and according to the figures in the pictorial screenfolds some were so lavishly dressed that their human forms were often obscured to the point of being invisible. Consider this description of the physical microcosm hidden beneath Tezcatlipoca's costume during the festival of Toxcatl (Sahagun, 1975):

Indeed he who was thus chosen was of fair countenance, of good understanding, quick, of clean body, slender, reed-like, long and thin, like a stout cane, like a stone column all over, not of overfed body, not corpulent, nor very small, nor exceedingly tall.

Indeed it became his defect if someone were exceedingly tall. The women said to him 'Tall fellow; tree-shaker; star-gatherer.'

He who was chosen as impersonator was without defects. He was like something smoothed, like a tomato, like a pebble, as if sculptured in wood, he was not curly haired, curly headed; his hair was indeed straight, his hair was long. He was not rough of forehead; he had not pimples on his forehead; he did not have a forehead like a tomato; he did not have a baglike forehead. He was not long-headed; the back of his head was not pointed; his head was not like a carrying net; his head was not bumpy, he was not broad-headed; he was not rectangular-headed; he was not bald; he was not of swollen eyelids; ... he was not swollen cheeked; he was not of injured eyes; ... he was not cloven-chinned; he was not of downcast face; he was not flat-nosed, not crooked nosed, but his nose was averagely placed, he was straight nosed. He was not thick lipped, he was not big lipped, he was not bowl lipped; he was not a stutterer, not ring-tongued, he did not speak a barbarous language, he did not lisp, ... He was not buck-toothed, he was not fang-toothed, he was not yellow toothed, he was not ugly toothed, he was not rotten toothed, his teeth were like sea shells; they lay well, they lay in order, he was not bowl toothed, ... he was not little eyed, he was not tiny eyed, he was not yellow eyed, he was not hollow eyed, not sunken eyed, he was not cup-eyed, he was not round eyed, he was not tomato eyed, he was not of pierced eye, he was not of perforated eye ... nor was he large eared, nor long eared, ... he was not long-handed, he was not fat fingered ... he was not fat, he was not big bellied, he was not of protruding navel, he was not of hatchet shaped navel, he was not of wrinkled stomach, he was not of hatchet shaped buttocks, he was not of flabby buttocks, he was not of flabby thighs.

For him who was thus, who had no flaw, who had no bodily defects, who had no blemish, who had no mark, who had on him no wart, no such small tumor, there was taken the greatest care that he be taught to blow the flute, that he be able to play his whistle, and that at the same time he hold all his flowers and his smoking tube.
At the same time he would go playing the flute, he would go sucking the smoking tube, he would go smelling the flowers.

This paragon of physical human order lived in honor, luxury and ritual esteem for one year prior to his sacrifice during the festival of Toxcatl. At the appropriate time during his tenure of office the king Moctezuma came forward and 'repeatedly adorned him, he gave him gifts, he arrayed him, he arrayed him with great pomp. He had all costly things placed on him, for verily he took him to be his beloved god' (Sahagun, 1975). In this festival, the king adorns the image of perfect order in order to renew his own legitimate authority.

5.3 A context for archaeoastronomy

In my own embryonic study of archaeoastronomy and in particular with Johanna Broda, Edward Calnek and Anthony Aveni, I discovered that an extremely valuable framework for the understanding of how astronomy, kingship, ecology, religion and city interrelate was provided by the urban ecologist Paul Wheatley (1971) in his swollen seed of comparative urban studies *The Pivot of the Four Quarters: A Preliminary Enquiry Into the Nature and Character of the Ancient Chinese City*. Here, and in other works, Wheatley has shown how religious symbolism played a major role in the spatial organization and the integration of ecological complexes of cities in the seven areas of primary urban generation including northern China, Mesopotamia, Egypt, southwestern Nigeria, the Indus Valley, Mesoamerica, and the highlands of Peru. Two sentences from Wheatley's work indicate how a mode of thinking called 'cosmo-magical thought' was utilized by kings and priests to organize the ecological complexes of these cultures. Wheatley (1971) notes,

Underpinning urban form not only in traditional China but also throughout most of the rest of Asia, and with somewhat modified aspect in the New World, was a complex of ideas to which Rene Berthelot has given the name astro-biology ... This mode of thought presupposes an intimate parallelism between the mathematically expressible regimes of the heavens, and the biologically determined rhythms of life on earth, (as manifested conjointly in the succession of the seasons and the annual cycles of plant regeneration) ....

This passage is one of the inspirations for the research carried out in the Mesoamerican Archive and Research Project because it asks us to consider two problems. First, it leads us to ask 'what kinds of parallelisms existed between myth, ecology, astronomy, social order, and architecture in specific locations in Mesoamerica?' Second, it notes that in the New World (that is Mesoamerica and Peru), the pattern of interaction between earth, society and the heavens was distinct and slightly different than in the Old World. We are led to ask, in the present context, 'what is the distinctiveness of Mesoamerican archaeoastronomy, of New World parallelisms embedded in myth, ethnographic data, and in architecture?'

By combining the framework of Wheatley with the research of archaeoastronomy, I discovered a number of new ideas which enlarged my perspective on Mesoamerican religions. The first had to do with the impact which astronomical events on the horizon had on the ceremonial order of Tenochtitlan. Celestial archetypes and the Aztec sky manifest their order largely through horizontality – or horizon astronomy. While the Aztec conception of the sky includes the apprehension of the sky as 'up there', 'on high' and 'everywhere', in fact it is the celestial events on or near the horizon which play the fundamental roles in the spatial organization of capitals and the complex cycles of ritual activity. I find this point to alter in an important way one of the fundamental impressions in the discipline of the history of religions concerning the significance of the sky. From reviewing, for instance, the marvelous chapter on the sacredness of the sky in Mircea Eliade's *Patterns in Comparative Religions* (Eliade, 1963), and from an initial study of 24 articles published in the *History of Religions Journal* since 1961 on the organization of sacred space in comparative religions, one gets the impression that (1) astronomy had almost no significant role in the construction of celestial archetypes and ceremonial cities, and (2) when it does, it is through the influence of a static vertical sky. Almost everywhere we have the dominant view that mountains are sacred because they are up high and therefore are the meeting of earth and sky. But in Mesoamerica hills, mountains and temple pyramids also are sacred because they have been aligned with the appearances of stars along the horizon. They are observation points of horizon celestial events. I call these events horizon archetypes because they become paradigmatic for the ritual orientation of buildings, highways, calendars, even entire cities, as well as the ritual activity of thousands of people.

The second idea which emerges from placing archaeoastronomy in contact with urban ecology runs counter to the impression that a locative view of the world dominated Aztec religion. My reading of the New Fire Ceremony suggests that a sixth facet informed the cosmological conviction of Mesoamerica. The sixth facet was the apocalyptic

fear that the cosmic order was periodically filled with so much tension, threat and instability that it could not be rejuvenated. Other evidence in the ethnographic record suggests this as well. Besides the repeated reference to the fear of the populace and their anxiety concerning cosmic renewal in the New Fire Ceremony, we read in Book II of Sahagun (1975) that each day when the sun rose there 'was uncertainty about his movement', and the people felt danger. Comets were omens announcing the death of rulers, the onset of war or famine. The recent research of Ulrich Kohler (see Chapter 22, this volume) on comets and falling stars in the perception of Mesoamerican Indians reveals that the prognostic significance of many astronomical events was decisively negative. Whether falling stars and meteors were considered as cigar butts of certain gods, or star excrement, or as projectiles sent by stars, they are usually associated with a coming disaster. Also the omens associated with the Conquest indicate the apocalyptic element in Aztec astronomy. Likewise, comets were associated with the great famine of 1545, and even the beginning of the Mexican Revolution of 1910. In my view, the most profound example of the instability factor relating myth to astronomy, and horizon astronomy at that, is the account of the Creation of the Fifth Sun.

We remember that, following the collapse of the fourth cosmic era, the gods gathered in the primordial darkness in Teotihuacan to discover 'who would take it upon himself to be the sun, to bring the dawn'. Following a 52-year period of darkness, Nanahuatzin, the Pimply Faced One, musters his courage and throws himself into the sacred fire. And then the text reads,

> Then the gods sat waiting to see where
> Nanahuatzin would come to rise, he who fell
> first into the fire, in order that he might shine
> as the sun in order that dawn might break. When
> the gods had sat and had been waiting for a long
> time, thereupon began the reddening of the
> dawn in all directions, all around, the dawn and
> light extended. Thereupon the gods fell upon
> their knees in order to await where he who had
> become the sun would come to rise. In all
> directions they looked. Everywhere they peered
> and kept turning about. Uncertain were those
> whom they asked. Some thought that it would
> rise in the north. Some thought that it would
> rise in the south. Some thought that it would

rise in the west. They expected that he might rise in all directions because the light was everywhere.

Quetzalcoatl oriented himself to the eastern horizon and the sun appeared. Then, following this ponderous wait, the striking passage (Sahagun, 1975): 'When the sun came to rise, he looked very red. He appeared to wobble from side to side.' This is part of the cosmic condition facing mankind in the Aztec world. The primordial sun is swaying from side to side, unable to achieve stability, or find its place, or initiate a creative movement. Even at the mythic level, the level at which cosmological order was achieved, the sun has a profound difficulty finding its place and orienting the world.

Perhaps the quest for a locative, even imperial, cosmology is succinctly stated in the important passage from Motolinia (1569) about the equinox sunrise at Templo Mayor. My reading of the passage suggests another example of the fivefold symmetry discerned in the New Fire Ceremony.

We are told that the fiesta of Tlacaxipeualiztli, the Feast of the Flaying of Men, at which men were flayed (first the boys, and then the great men) and their flesh was eaten, took place 'when the sun was in the middle of Uchilobos, which was at the equinox, and because it was a little wrong, Moctezuma wished to pull it down and set it straight' (Aveni, Calnek and Hartung, 1988). The inference here is that Moctezuma was looking east toward the Templo Mayor as the equinox sun rose over Huitzilopochtli's temple. From this perspective, a king's eye view, a symmetry is sought between the horizon equinox sun, the orientation of Templo Mayor, the sacrificial festival, and the king's own authority. There are three aspects of this rare passage which I want to highlight: (1) the actual astronomical alignment of the Templo Mayor, (2) the role of the king in the Festival of the Flaying of Men, and (3) the fact that Moctezuma wanted to tear the temple down and realign it.

When the excavation of the Templo Mayor uncovered seven rebuildings of the basic structure of the temple over a hundred-year period, Anthony Aveni and Sharon Gibbs measured with a surveyor's transit and astronomical fix the alignments of the rebuildings. Their first measurement found the alignment of the building deviating seven degrees and six minutes to the south of a true east–west line. Subsequent measurements of the rebuild-

ing of the temple revealed (Aveni, Calnek and Hartung, 1988)

a remarkable similarity of alignment in the seven rebuildings pervading all the structural stages that make up the Templo Mayor. Scarcely a full angular degree of extreme separation in azimuth could be found ... At first the results appeared to contradict Motolinia's historical statement about the use of the building to register the sun and equinox. But upon further reflection, we realized that if the sun had been observed from ground-level over an elevated horizon one should anticipate an alignment to the south of east because as the sun rises it also drifts slightly southwards, that is to the right as the observer faces east upon ascending into the sky ... In fact, using the supposed location of the Temple of Quetzalcoatl as a hypothetical observing point along with estimates of the dimension of the temple from a number of sources, we discovered that the equinox sun observation could be fulfilled by Motolinia's statement *only* if the Templo Mayor were skewed about seven degrees south of east.'

This indicates a self-conscious attempt by Moctezuma and other Tlatoanis to align the Templo Mayor with the equinox sunrise.

Another component in this search for the symmetry of Aztec religion has been illuminated by the detailed work of Johanna Broda on the ethnohistorical record concerning the Festival of the Flaying of Men (Broda, 1970). Her work shows that the greatest number of victims and the most important prisoners of war were slain at the Festival of Xipe, also that the festival was both an agricultural renewal ceremony and an initiation of warriors demonstrating the connection between agriculture, warfare and astronomy. And, as before, the core of the event consists of the relationship of Moctezuma to his temple. The hierarchical element is suggested when we read elsewhere in Motolinia that on the third day of the festival the ceremony of 'the Bringing Out of the Skins' took place at the great palace where the nobles of the major cities danced. For this festival 'a prestigious personage, a lord' from an enemy city was flayed 'so that his skin might be worn by Moctezuma, the great Lord of Mexico, who danced with great gravity' (Broda, 1970). In this case, it is Moctezuma who has become a deity impersonator. The hierarchical emphasis of the ritual is repeated when we see that 'many people went to see' the king dancing – 'a most marvellous thing' because, in other towns, it was not the lords but other leading men who put on skins of the flayed victims. And all this, the numbers, the prized warriors, the king dancing with skins of sacrificed victims impersonating Xipe Totec, was enacted before the eyes of foreign rulers and nobles who were forced to watch from clandestine and public locations. The text reads, 'and also from cities which were his enemies, from beyond the mountains, those with which they were at war, Moctezuma secretly summoned, secretly admitted as his guests' (Broda, 1970). But the most vivid example of imperial authority and kingly worry appears in the statement that there was a misalignment, like Thurber's microscope, between the king's vision and the temple's orientation. As in the myth of creation, alignment is difficult to achieve; and, if it can't be achieved, the king's authority comes into question and the cosmos is in danger. As in the case of European royalty where 'a woman is not a duchess 100

yards from a carriage' (Geertz, 1983a), Moctezuma is not a king if the temple is out of line. So, in an act of immense responsibility, Moctezuma orders the temple re-oriented between the equinox sun and his own perspective.

In this tentative interpretation of a sacred symmetry in the New Fire Ceremony and the orientation of Templo Mayor, I have explored a strategy for the construction of an approach to religion and astronomy that will enable us to discover conjunctions, correspondences and entity involvements. My guess at this stage is that the Aztec dynasty believed that their power and authority depended on their capacity to integrate major elements of their world into a cohesive symmetry, and that astronomy was the most efficient cosmo-magical system enabling them to establish this alignment.

Perhaps, like Thurber, I am seeing more of my own eye than their symmetry, but I'm working, along with an interdisciplinary group of scholars, at adjusting the lens.

Notes

1 My own concern with interdisciplinary work stems from the creative hermeneutics of Mircea Eliade whose quest for a 'new humanism' involves an intellectual orientation designed to transcend cultural and disciplinary provincialism. Consider this statement by Eliade that it is

useful to repeat that *homo religiosus* represents the 'total man'; hence, the science of religions must become a total discipline in the sense that it must use, integrate, and articulate the results obtained by the various methods of approaching a religious phenomenon. It is not enough to grasp the meaning of a religious phenomenon in a certain culture and, consequently, to decipher its 'message' (for every religious phenomenon constitutes a 'cipher'); it is also necessary to study and understand its 'history', that is, to unravel its changes and modifications and, ultimately, to elucidate its contribution to the entire culture. In past few years a number of scholars have felt the need to transcend the alternative *religious phenomenology* or *history of religions* and to reach a broader perspective in which these two intellectual operations can be applied together. It is toward the integral conception of the science of religions that

the efforts of scholars seem to be orienting themselves today. To be sure, these two approaches correspond in some degree to different philosophical temperaments. And it would be naive to suppose that the tension between those who try to understand the *essence* and the *structures* and those whose only concern is the *history* of religious phenomena will one day be completely done away with. But such a tension is creative. It is by virtue of it that the science of religions will escape dogmatism and stagnation.

Among many important statements about new directions in the history of religions, see especially the essays of Eliade (1985) and Smart (1985). Also consider the final essay in Smith (1978) for a different view of prospects in the history of religions; also see Penner and Yonan (1972). Finally, for a recent example of interdisciplinary contact on the three genres of Mircea Eliade's work, see Carrasco and Swanberg (1985).

2 My own work on Templo Mayor has benefited by the scholarly activity of Eduardo Matos Moctezuma, Edward Calnek, Johanna Broda and Anthony Aveni, whose writing and research have transformed our understanding of the axis mundi of Tenochtitlan. See especially Broda's 'The Templo Mayor as ritual space' in Broda, Carrasco and Matos Moctezuma (1987). Also see 'Myth, environment, and orientation of the Templo

Mayor of Tenochtitlan' by A. Aveni, E. Calnek and H. Hartung (1988) (and see page 493 of this volume for abstract), etc.

3 This quote is actually Clifford Geertz's gloss on Shils' work in Geertz (1983b, p. 122).

References

Aveni, A. F., Calnek, E. and Hartung, H. (1988). On the orientation of the Templo Mayor of Tenochtitlan. *American Antiquity* **53**, 287–309; abstract in the present volume, p. 493.

Broda, J. (1970). Tlacaxipehualiztli: a reconstruction of an Aztec calendar festival from 16th century sources. In *Revista Española de Anthropologia Americana*, vol. 5, pp. 197–274. Madrid.

Broda, J., Carrasco, D. and Matos Moctezuma, E. (1987). *The Great Temple of Tenochtitlan, Center and Periphery in the Aztec World*. University of California Press.

Carrasco, D. and Swanberg, J. M. (1985). *Waiting for the Dawn: Mircea Eliade in Perspective*. Boulder: Westview Press.

Eliade, M. (1963). *Patterns in Comparative Religions*. Cleveland: World Publishing.

Eliade, M. (1969). *The Quest*, pp. 1–13. University of Chicago Press.

Eliade, M. (1985). Homo Faber and Homo Religiosus. In *The History of Religions: Retrospect and Prospect*, ed. J. M. Kitagawa. New York: Macmillan Publishing Co.

Flannery, K. (1976). *The Early Mesoamerican Village*. New York: Academic Press.

Geertz, C. (1983a). Blurred genres: the refiguration of social thought. In *Local Knowledge: Further Essays in Interpretive Anthropology*. New York: Basic Books, Inc.

Geertz, C. (1983b). Centers, kings, and charisma: symbolics of power. In *Local Knowledge: Further Essays in Interpretive Anthropology*. New York: Basic Books, Inc.

Ginzberg, C. (1983). Morelli, Freud and Sherlock Holmes: clues and the scientific method. In *The Sign of Three, Dupin, Holmes, Pierce*. Bloomington: Indiana University Press.

Motolinia, Fray Tòribio de Benavente o (1569). *Historia de los* Indios de la Nueva España. Mexico: Editorial Porrua (published in 1973).

Penner, H. H. and Yonan, E. A. (1972). Is a science of religion possible? *Journal of Religion* **52**, (2), 107–33.

Sahagun, B. de (1975). *The Florentine Codex*. Santa Fe, New Mexico: The School of American Research and the University of Utah, Monographs of The School of American Research.

Smart, N. (1985). The history of religions and its conversation partners. In *The History of Religions: Retrospect and Prospect*, ed. J. M. Kitagawa, pp. 73–87. New York: Macmillan Publishing Co.

Smith, J. Z. (1978). *Map is not Territory: Studies in the History of Religions*. Leiden: Brill.

Sullivan, L. E. (1983). Astral myths rise again: interpreting religious astronomy. *Criterion* **22**, (1), 12–17.

Thurber, J. (1956). *My Life and Hard Times*, quoted in *The Art of Interpretive Speech*, eds. C. H. Woolbert and S. E. Nelson. New York: Appleton-Century-Crofts Inc.

Townsend, R. F. (1982). Pyramid and sacred mountain. In *Ethnoastronomy and Archaeoastronomy in the American Tropics*, eds. A. Aveni and G. Urton, *Annals of the New York Academy of Science*, vol. 385. NY Academy of Science.

Wheatley, P. (1971). *Pivot of the Four Quarters*. Chicago: Aldine Press.

6

Chinese archaeoastronomy: between two worlds

N. Sivin *University of Pennsylvania*

Well over a decade ago I attended a conference on archaeoastronomy at MIT. The participants spoke of ancient astronomy in the 'Old World', by which they evidently meant Europe and the Middle East, and the 'New World', which included the southwestern USA, Mexico and parts of Central America. The larger part of the Eurasian land mass, including India and China, was never mentioned, leaving it presumably on another planet. In 1977, when I was part of an American delegation assessing the state of astronomy in the People's Republic of China, I spent several days in the company of Xia Nai. It was Xia who had kept archaeology going during the Cultural Revolution, had for the first time engaged historians of astronomy systematically in the study of sites and artifacts, and had published some of the most important papers on early astronomy of the 1970s. He had been approached for permission to survey Chinese tombs. Xia was curious about these novel proposals, but perplexed. 'What,' he asked me, 'is archaeoastronomy?'

That mutual inattention is a thing of the past. In 1981 Xi Zezong, the dean of Chinese historians of science, concluded a survey of research in that country by remarking that archaeoastronomy is an empty slate 'that can be filled only by organized effort'. I mention the formerly unworldly status of China in archaeoastronomy because it points to a central problem in how we go about our work.

6.1 China between two worlds

The outworn distinction between Old World and New World astronomy did reflect a real difference in the character of our knowledge and the sources on which we build. What we know about Mesopotamian, Egyptian and European star-knowledge has accreted over centuries of scholarship. Astronomical surveying of remains and interpretation of artifacts began in the Old World.

Sir Norman Lockyer, writing at the beginning of this century on the alignments of Stonehenge, acknowledged that it had been studied as an astronomical observatory as early as 1771 (Lockyer, 1906, pp. 52–3). Willy Hartner's remarkable investigations of constellation images on early Middle Eastern clay seals and Old Norse drinking horns showed how erudition and analytical power can be combined in the study of artifacts (Hartner, 1965, 1969, 1973). Work of this kind was stoutly ignored until a decade or so ago by most scholars of early astronomy, for whom records, above all records containing numbers, defined the significant problems. The large picture of archaic astronomy in the Old World remains based almost entirely on books.

In the New World, on the other hand, what spans the great diversity of dead or dying cultures is the astronomy of site alignments and symbolic images. We have some written records of the most evolved peoples, but their significance is largely contentious, and the scope of the practices they record is uncertain. Some specialists who work with uncompromisingly rigorous quantitative methodologies have not yet admitted that there is anything on the Western side of the Atlantic worth including in the history of astronomy.[1]

The result of these differences in sources was a great difference in investigatorial style. The breakthroughs with respect to the Western hemisphere depended largely on prospecting, on serendipity, on (to use an anthropologist's cliché) bricolage. In the Old World that was also true of occasional bold new departures, for instance Lockyer's generalization from Stonehenge to a flock of megalithic circles that revealed how common astronomical orientations were within their narrow band of latitudes (Lockyer, 1906). But there was a great deal less leeway in studying the old civilizations for hunches that might or might not pan out. If speculations were not backed by considerable expertise, both cultural and technical, chances are that their weaknesses would be revealed very quickly. Innocent astronomers hoping to make a neat little reputation on the side found themselves mauled by historical facts; archaeologists and philologists with *idées fixes* about mythology sooner or later had to face the numbers.

With these differences in mind, some students of early astronomy remember with nostalgia a time when China was not complicating the picture. Its astronomical culture is undeniably inconvenient. Westerners have been told incessantly for nearly 400 years that the Chinese were extremely sophisticated observational and mathematical astronomers. We have been assured, at least until recently, that the Chinese were systematically recording the precise times of solar eclipses since the beginning of the second millennium BC. But that does not leave us badly off. Historians today, more skeptical about datings, would make this claim only for the past couple of thousand years, with occasional useful records as early as *c.* 1450 BC. No matter whether we are concerned with supernovas, long-term variations in the period of Halley's comet, time series of visible sunspots, auroras, or earthquakes, we have learned to turn first to East Asian records, usually detailed and always meticulously dated. We hear that work on the history of astronomy is proceeding on a practically industrial scale in China and Japan.

How does one come to grips with such a massive slice of the pie of the universal history of mankind, bursting with indecipherable plums of information under an impenetrable crust of exoticism? Most archaeo–astrophiles have made an honest stab at volume 3 of *Science and Civilisation in China*, applauded the questions it raised and the depths it hinted at but did not reveal, and threw up their hands.[2] Rare is the student of Old or New World astronomy who has followed publications on China in the 35 years since Needham's book appeared.[3] Rarer still, to the sorrow of us all, is the historian of astronomy willing to master classical Chinese, or the sinologist ready to learn the esoteric rudiments of computational astronomy and error analysis, in order to go hunting in the sources without being forced to shoot in the dark.

There we have ancient Chinese astronomy: a most inconveniently large part of the world picture, almost impossible to place in it.

6.2 China in world astronomy

But placing China in the world picture is hardly optional. Seen from the broadest perspective of world ethnoastronomy, what makes Chinese astronomy inconvenient is precisely what makes it valuable. Its continuity is unbroken, from the crude knowledge reflected in the oracle records of 1450 BC to the 20th century. Another salient characteristic of Chinese astronomy is its elaborate technical resources. That point is so well known that I need not spin it out. Many of the old bromides about Chinese reluctance to approach astronomy theoretically and inability to apply spatial reasoning have been retired by the last generation's scholarship.

Despite a continual passage of scientific ideas and techniques back and forth since the Neolithic, Chinese astronomy evolved with little influence from the Old World. The crude observational records of the mid-second millennium BC, and the rudimentary ephemerides of the late first millennium, do not leave much of a formative role for Mesopotamian influence. We see an evolution practically without gaps from these simple beginnings to the zenith of sophistication in the 13th century AD. Indian astronomers held important technical posts in China from the eighth century on; Muslims from the 13th century on; Europeans from shortly after 1600. All left writings in Chinese that we can read today. What strikes us is, once their computational tasks were done, how little they displaced the main astronomical tradition.

In addition to being elaborate, continuous and independent, Chinese astronomy is verbose. If we are tempted to assume that the lunar conjunction and the first day of the lunar month coincided since time immemorial, a glance at the registers of astrological portents shows us that Hipparchus' Chinese contemporaries (mid-second century BC) considered a conjunction on the last day of the month to be as normal as one on the first. That was no longer true a century later. We do not have to wonder why astronomers in 100 BC or AD 1280 moved off in new directions. We have detailed contemporary prospectuses and committee reports that explain the relation between observation and computation and test the reasoning of the innovators.

Verbosity has its advantages in astronomical traditions. What we know of living cultures indicates that their knowledge of the sky is never merely empirical. Behind patterns of data lie metaphysical abstractions that point to a coherent physical reality, or concrete metaphors that tie sky-order to social order.[4] But for most dead cultures we have nothing but columns of numbers, circles of stones, axes of ruined buildings, or stark, cryptic images. It is easy enough to imagine what they may have meant, but very hard to be sure whether we are right or wrong.

China offers us a trove of cases in which the abstractions are spelled out and the metaphors explained, explained again, and explained yet again. We can explore them for clues that may be pertinent elsewhere. For instance, I showed some years ago that we can dissect out of early techniques for computing eclipses, and contemporary reflections on them, changing assumptions about the character of the underlying reality. By tracing changes in these assumptions it was possible to reconstruct a scientific revolution in a culture that is not supposed to have had scientific revolutions.[5]

We can use all the verbose cultures we can find. Rationales and astral speculations are sparse for Mesopotamia and Egypt, and we know remarkably little about those that underlay Greek astronomy before the time of Plato. The antiquity and continuity of the Chinese tradition give it a special role to play in this quest for general insights.

6.3 Characteristic significances in Chinese astronomy

This is not to say that everything we find in China will be true worldwide. The early invention there of true bureaucracy affected astronomy as well as philosophy. The first complete system for calculating the ephemerides, in 104 BC, reduced this task, including prediction of eclipses and planetary motions, to a step-by-step program for a human computer who was expected only to know basic arithmetic. The Grand Inception system used a

simpler program than that of the *Almagest*. It was also committed to much less subtle assumptions, numerical rather than geometric, about what sorts of celestial motions could be eternal.

To look at another aspect, astronomy and astrology were locked in an interesting tension. Astrology, like that of Babylonia, was judicial, concerned with unpredictable phenomena that warned of danger to the ruling dynasty. Phenomena that were regular and could be computed were not omens. The astronomical official looked for unforeseen events in the sky and interpreted their significance. At the same time he tried to incorporate as many phenomena as possible in a correct ephemeris. The most creative astronomers constantly strove in that way to reduce the range of ominous events, to replace the uncertainty of the sky with a set of computational cycles that would make future stargazing unnecessary as a bureaucratized cosmos no longer challenged the political status quo.

Divination on behalf of individuals showed the same abstracting impulse. By the time fully fledged systems of divination were well described, the typical method was based, not on computing the relations of planets at the time of conception or birth, but on counting off the year, month, day and hour of birth within purely numerical cycles of 60. Since these cycles are constructed by pairing much older cycles of ten days and 12 hours, we have reason to suspect astral origins, although we do not yet have clear evidence for the connection.[6] But even if individual fates were originally rooted in the sky, the point is that those roots were pulled up.

Hellenistic horoscopes entered China in the third century AD and had a brief vogue there in the eighth and ninth centuries. But horoscopic astrology failed to influence the development of astronomy as it did so decisively in Europe, Islam and India (Nakayama, 1966). At the same time, we have recently uncovered a remarkable stratum in popular religion, and even in medical therapy, in which ancient men of knowledge, dancing out in ritual the shapes of constellations, launched themselves on spirit-journeys through the stars (Schafer, 1977; D. J. Harper, n.d., unpublished PhD dissertation, University of California at Berkeley, 1982). These are only a couple of examples of Chinese particularities, but they will perhaps make clear my point that many of the significances of Chinese astronomy and astrology are not at all universal.

6.4 Research issues

Given the richness of the Chinese record, we should be able to shed light on almost any archaeoastronomical question. Unfortunately the bulk of research has been devoted to questions of what astronomical feats the Chinese performed before anyone else. That is an understandable righting of the balance, considering how regularly Western histories of astronomy have ignored China, or devoted a few perfunctory pages to it using sources a generation out of date. This preoccupation with priorities does not help us to answer the sorts of questions that students of astronomy ask today. Let me remark briefly on what light the study of China might shed on a few of those issues:

(1) 'What are the *origins of astronomy?*' is a question that I doubt has one answer for all civilizations. I would go so far as to doubt that Chinese astronomy has a specifiable origin. Putting together the archaeological and historic evidence, it seems to me that a number of strands – the observational, the computational, the metaphysical, the mythological, the ritual – began to evolve very gradually, each at its own pace, and only came together to form the complex that we think of as Chinese astronomy about 100 BC, with the Grand Inception calendar reform as part of a great cosmological synthesis (Sivin, 1969). It is tempting to read that complex into earlier times, but, without strong evidence, anachronism is anachronism.

(2) I hope I have already made it clear that the *aims of astronomy* reflect the character of Chinese society, especially its monarchy. The *relations of public and private practice* similarly mirror the values and structures of the larger society. Chinese government from its creation has been totalitarian in principle, convinced that its mandate to provide public order necessitates regulating every aspect of private thought and behavior. In practice it has lacked the means to control more than the actions of officials and the education of the elite, and that only very imperfectly. Legislation to restrict astronomical research to specialist officials was seldom zealously, and never successfully, enforced. We can find a few instances over the last 2000 years in which talented scholars avoided astronomy, and private experts were harassed, but time and again important calendar reforms were initiated by private individuals. These experts were consistently co-opted, given official appointments after the fact. Intensive observation remained centered in the Imperial Bureau of Astronomy not because of an effective 'science policy', but because there was no other source of funds for large-scale research projects.

(3) The *relations between astronomy and cosmology* puzzle me perhaps even more than they do my colleagues. The conventional discussion of Chinese cosmology mentions three or more 'schools', without pausing over what sort of collectivity that word implies. The oldest of these traditions sees the earth and sky as two parallel cones or vaults; it is related to an early attempt to fix the basic dimensions of the cosmos using no tool but a gnomon. A second, related to the armillary sphere, views the earth as a disc or flat-topped hemisphere suspended like the yolk of an egg within a spherical sky. The third has the sun, moon and stars moving freely through boundless space; unfortunately the oldest account of it (early fourth century) speaks of it as a lost tradition.[7]

I find it difficult to understand what these conceptions have in common, aside from being conveniently grouped under a characteristically Western category that had no Chinese counterpart. The first comes from an exercise in mathematical cosmography; the second seems to represent the cosmic rationale of a new astronomical instrument; and the third is a fragment of speculation that we have not succeeded in connecting with any philosophical or practical trend. It is well to leave open the question of whether full-blown cosmologies are likely to appear, or are even useful, except in cultures where astronomy is a handmaiden of philosophy.[8]

Astronomy was not a handmaiden of philosophy in China. I have already stated my belief that astronomy must be based on assumptions about physical reality. Such beliefs in China originated among philosophers, or at least non-astronomers, by about 100 BC. But thereafter these ideas continued their evolution among astronomical specialists, who over the past two millennia have accepted remarkably little influence from the philosophical mainstream.

(4) Were there important *oral traditions* independent of the voluminous written archives? It would be difficult to maintain that there were not, in a civilization most of whose population was illiterate until less than 30 years ago. But the issue is how oral traditions interacted with advanced practice. In the West, the demise of Greek culture led to a separation of theory and practice that only began to be repaired in the Renaissance. In China, where there was no separation, the record in every sphere of endeavor from poetry to medicine reflects a continuing infusion of popular ideas that are anything but bookish, but that invigorated the thought of scholars who had more in mind than annotating the classics. In astronomy near its beginnings we find many allusions to images from song and myth (Major, 1978). It is about time that its mature phase was studied from this point of view.

But nothing in our general understanding of the relations between oral and written traditions in China justifies occasional assertions in archaeoastronomical writings that 'there must have been an oral tradition', or a relentlessly effective suppression of knowledge, or mass amnesia, behind some pet idea of the author for which he cannot find evidence. Such arguments admit a distaste for studying in a disciplined way the intellectual habits of the ancients.

(5) How was the *astronomers' reality* defined out of their collective work? How, in other words, did institutions and techniques interact to form and maintain conceptions of the sky embodied in tasks, usually bureaucratic tasks? To throw some light on that question, perhaps the biggest of all, we need a much better understanding of many complicated problems. Some are technical. Hashimoto Keizo, for instance, is now working on the evolution of Chinese notions about increasing accuracy.[9] Other are social. Yamada Keiji (1980), to take the outstanding example, has given us a detailed study of a major calendar reform, relating administrative organization, observation and mathematical technique.

None of these crucial problems, as these examples indicate, is purely technical or purely social. Accurate prediction of celestial phenomena, for instance, was an open sesame to appointment or promotion.

This list of issues could be prolonged indefinitely, but that is enough to suggest how the study of Chinese astronomy might converge with other explorations now afoot in the history of science.

6.5 Current research trends in China

Work in these relatively new directions will continue to draw on the publications of the 150 or so full-time historians of astronomy in Chinese research institutes. Their concerns are much more old-fashioned, primarily concerned with establishing Chinese priorities, finding precursors of present-day astronomical knowledge. This emphasis is now beginning to change, as new methodologies, and new disciplines such as archaeoastronomy, are introduced through correspondence and personal contact. But it takes more than enthusiasm to revolutionize a research establishment, and we can expect most work to continue along well-worn tracks. Let me summarize general characteristics of research results seen in recent Chinese publications.[10]

Chinese research tends to be much more philological than historical in approach. That is, its primary thrust is understanding ancient documents and explaining them in modern terms. Histories of astronomy and other sciences tend to be topical summaries of sources in chronological order. Determining patterns of change and the processes that underlie them is a matter of comparatively minor interest, except of course to demonstrate that the evolution of astronomy conforms to the laws of historical materialism.

This positivist approach is not a matter of fixation on an outmoded Western model. Chinese philology as applied to scientific documents draws on its own sophisticated traditions, which were well developed long before significant contact with Europe (Elman, 1984). This tradition is now reinforced by the expectation of the P.R.C. government, the employer of first and last resort, that students of early science contribute to the national pride of all Chinese.

The logical outcome of philological studies is not explanations but compilations, and that is very much the case for work on Chinese astronomy. Every astronomer is aware of, and many have used, catalogues in Western languages based on Chinese surveys of comets, novas and supernovas, sunspots, auroras, and so on, even detailed earthquake histories for every part of China. Most of these compilations go back 3500 years and are rather full for roughly half that time.[11]

These surveys draw mainly on the dynastic histories. They have been made obsolete by a search an order of magnitude broader, carried out by a team of roughly 100 scholars from 1975 to 1978. It resulted in a register of astronomical phenomena and other data roughly four-million characters long (or the equivalent of perhaps eight-million English words). China's very limited publication facilities give low priority to voluminous technical reference works of this sort, so the appearance of this book has been repeatedly delayed. A few publications that draw on it give some impression of its richness. It is now possible to see the fine detail in such gaps as the Maunder Minimum, a lull in the solar activity cycle previously considered devoid of sunspots.[12]

A most promising new area of exploration is the ethnoastronomy of the Chinese minority peoples, one area in which mythology and calendar computation are looked at together. This topic was almost completely neglected before the last decade, but government emphasis on ethnic minorities has drawn attention to it. We now have, in addition to scattered reports, a volume that draws together astronomical studies of a variety of peoples (Anon., 1981), and a substantial historical monograph on the astronomy of the Yi people of Yunnan province (Chen Jiujin, Lu Yang and Liu Yaohan, 1984). Such work has already thrown light on less local questions. The authors of the monograph, finding that the Yi used a solar calendar with ten 36-day months, realized that they could make sense for the first time of a Confucian text more than 2000 years old if it were describing a calendar of the same kind.[13]

The Cultural Revolution, which closed the historical institutes and left most historians of astronomy no option for research except to collaborate with archaeologists, accustomed them to studying images and artifacts. The endless succession of spectacular excavations made this work highly productive for astronomical studies (Xia, 1979, 1980; Wagner, 1980; and Xi, 1981).

Astronomical instrumentation has been particularly well covered in research publications, although a book that will tie these findings together is badly needed.[14] A team of scholars has been systematically going through the immense collections of the Palace Museum, studying both traditional instruments and those imported from the West. In 1977 I saw two of their discoveries, English heliocentric orreries made in 1730 that I had described four years earlier in a study of Copernicanism in China without suspecting that they still existed.[15] Recent studies of instruments have remained largely descriptive and antiquarian. Few have estimated the precision of the instruments they examine, just as the many publications on systems of computational astronomy have had very little to say about their accuracy. The same can be said of most writings about early European and Islamic instruments.

As for studies of site orientations, one can only say that it is time they began. What they will yield is impossible to predict. That major Chinese buildings were directionally oriented is universally known. Their positions have been adjusted for more than 1000 years – conceivably for more than 2000 years – by the peculiarly Chinese science of siting, or geomancy (Bennett, 1978). Students of archaeoastronomy who ignore it will spend a great deal of time reinventing the wheel. We know very little about the sources of this science in star lore. Surveying is the obvious way to find out. Chinese historians who have collaborated with archaeologists are prepared to do this work. Such skills were very much in evidence, for instance, in the recent survey and restoration of the great 40-foot gnomon observatory of 1280 at Dengfeng, Henan province (Zhang Jiatai, 1976). Analysis of aerial photographs and site plans of tombs from the second and first millennia BC by Vance Tiede (n.d.) suggest sorts of alignments to look for.

6.6 The next step

The challenge in the midst of this roiling activity is making it add up to something coherent. It ought to be possible to make equally productive use of the professional team working in the research institute on familiar tasks and the serendipitous archaeoastronomer who knows something about the sky-knowledge of several cultures.

The characteristics of Chinese astronomy that I have enumerated make it unlikely that sound contributions will come from anyone unwilling to study seriously both astronomy and China. Collaboration is an obvious way to maximize contributions, I mean collaboration as Norbert Wiener defined it: two or more people working together who have learned each other's fields. Still it is to everyone's benefit to ease the work of anyone who is willing to get his computations right and find out what the Chinese knew and did at a given time.

The simplest way to do that would be to compile a chronological catalogue of astronomical capabilities. It ought to be possible to set down in one place what we know about what types of observation were being made at a given time, with what instruments, and to what precision. We might interweave with that an account of what was being computed in each period, by what means, based on what data. We would want to specify what conceptual tools were guiding the design of Chinese models. It would not be wise to include only notions familiar in Europe. For instance, we have often missed the point that what we call the precession of the equinoxes was for the Chinese not a secular displacement of the vernal point but a numerical difference (not always considered constant) between the lengths of the tropical and sidereal years. That is why the Chinese term for it was 'annual difference' (*sui cha*).

A great deal of the basic work for such a catalogue has already been done. Much of the information is at hand in histories of Chinese astronomy published in Chinese and Japanese, although the data need to be reorganized for this purpose and brought up to date.[16]

A certain amount of the updating has even been done, but it tends to be lost in the flood of periodical literature. To the point is Maeyama Yasukatsu's brilliant error analysis which showed that the star catalogues usually dated to the fourth century BC were based on observations not earlier than 70 BC. Although this essay is a model for critical archaeoastronomy, it has hardly been cited since it was published.[17]

In the sort of compilation I am proposing there will undoubtedly be many gaps to fill. There is bound to be disagreement about dates and significances. It is time these divergences were faced and settled. Let the work begin!

Notes

1 In the introductory discussion of 'Limitations' in his *magnum opus*, Otto Neugebauer expresses the wish for a corpus of medieval Indian, Islamic and Western sources, but lack of early influence on the West somehow justifies leaving China off this list. The Western hemisphere is summarized in one sentence: 'No relation whatsoever exists between our study and Mayan astronomy' (Neugebauer, 1975, vol. I, p. 2). Nevertheless the title of the book remains unqualified: *A History of Ancient Mathematical Astronomy*.

2 Needham has emphasized repeatedly (e.g., Needham, 1954–, vol. I, pp. 5–6) that his *magnum opus* is a preliminary reconnaissance. The astronomical part is largely based on Chinese publications and is not very attentive to the important work published in Japan since about 1950. For an assessment see Cullen (1980a).

3 Needham's remains the only scholarly survey in a Western language. The secondary literature to about 1971 is surveyed in Ho (1977). There are several excellent general studies in Chinese and Japanese. The most detailed is Chen Zungui (1980–84). From Japan see the many publications of Yabuuchi Kiyoshi (especially 1969); there is a list of his works in Anon. (1982). The best technical analysis of the character

of Chinese computational astronomy is that in Nakayama (1969). A useful recent monograph on methods underlying early ephemerides is Teboul (1983).

4 The first point was made long ago in Burtt (1925). The second is being related to simple as well as elaborate cultures by anthropologists, e.g. in Horton and Finnegan (1973) and Horton (1983).

5 Sivin (1969); for a more general discussion see Sivin (1982).

6 Chao Wei-pang (1946) has yet to be superseded as a general description, but see also Doebereiner (1980). The best study of the relation between divination and astronomy is Nakayama (1966). On astronomical connections of early autochthonous divination see Kalinowski (1983).

7 Needham (1954– , vol. III, pp. 210–27) discusses the basic sources. Some additional material has been added to the summary in Needham (1975). Few colleagues familiar with the primary sources will agree with his attempts to relate teachings after the fourth century to the third 'school'. Needham's account of the first, the *kai t'ien* or 'sky as a cover' model, ignores two distinct strata; see Nakayama (1969, pp. 24–40). The question of whether the earth is spherical, hemispherical or discoidal in the second (*hun t'ien* or 'uranosphere' theory) remains contentious. In my view

the only adequately critical discussion is that of T'ang Ju-ch'uan (1962). T'ang concludes that in the *hun t'ien* world picture the earth is not spherical, but that the evidence does not allow a choice between the other two possibilities. What '*hsuan-yeh*' means is not clear; E. H. Schafer suggests 'unrestricted night' (Schafer, 1977, p. 38).

8 For an interesting, flexible approach to the definition of 'cosmology' see Henderson (1984).

9 K. Hashimoto (unpublished PhD dissertation, University of Cambridge, 1985) briefly takes up the theme of progress in accuracy.

10 Xi Zezong (1981, 1983) has discussed new tendencies. See also Sivin (1981).

11 Among publications of the last generation see: Imoto and Hasegawa (1958); Needham and Ho (1959); Ho (1962); Ho and Ang (1970); Kiang (1972); and the comprehensive tables in Chen Zungui (1980–84, vol. III). Earlier publications are fairly fully listed in Bibliography B of Needham (1954–, vol. III). The most detailed of three large recent compilations of earthquake records is Xie and Cai (1983). I have had access only to volumes I and V.

Like other ancient observational data, these materials must be used judiciously. Clark and Stephenson (1977) exemplifies critical exploitation for astronomical purposes.

12 For instance, Luo and Li (1978); Xu Zhentao (1981, unpublished papers jointly offered at the 6th International Summer College on Physics and Contemporary Needs, Pakistan); and a report on two earlier Chinese publications in Cullen (1980b). The Records of Phenomena project has been described in Zhuang Weifeng (1982). Its first fruit has finally appeared: a union catalogue of the more than 8200 local gazetteers from 190 collections used to compile the register of phenomena (Zhuang Weifeng *et al.*, 1985).

13 See also Chen Jiujin (1982). Chen's hypothesis does not dispose of the problem, since the book in question is organized according to a scheme of 12 months.

14 See, for instance, Pan Nai (1975) and Yi Shitong (1978). The latter article is concerned with the metrological standard used in the design of instruments. A splendid volume of photographs illustrating star maps, instruments and artifacts pertinent to early astronomy has been compiled by the Institute of Archeology, Chinese Academy of Social Sciences (1980).

15 Sivin (1973), especially pp. 105–7 and plates 10–11, which reproduce contemporary woodcuts of the instruments. A study published simultaneously in China, Hsi Tse-tsung *et al.* (1973), contained photographs. It was not possible in the early 1970s to establish communication with Chinese colleagues.

16 The closest thing to a pattern for such a project is the systematic outline of early cosmological capabilities in Nakayama (1969).

17 Maeyama (1975–76, summarized in 1977). Cullen (1982) throws new light on the earliest astronomical use of trigonometric functions. An equally important contribution that reassesses geometric proof in fifth-century mathematics is Wagner (1978).

References

Anon. ed. (1981). *Zhongguo tianwenxue shi wenji (Dierji).* (*Collected essays on the history of Chinese astronomy, 2nd collection*). Beijing: Kexue chubanshe.

Anon. ed. (1982). *Tōyō no kagaku to gijutsu. Yabuuchi Kiyoshi sensei shōju kinen rombunshū.* (*Science and skills in Asia. A Festschrift for the 77th birthday of Professor Yabuuchi Kiyoshi.*) Kyoto: Dōhōsha.

Bennett, J. (1978). Patterns of the sky and earth. A Chinese science of applied cosmology. *Chinese Science* 3, 1–26.

Burtt, E. A. (1925). *The Metaphysical Foundations of Modern Physical Science.* New York: Harcourt, Brace & Co.

Chao Wei-pang (1946). The Chinese science of fate-calculation. *Folklore Studies* 5, 279–315.

Chen Jiujin (1982). Lun *Xia xiao zheng* shi shi yue taiyangli. (On the 'Lesser annuary of the Xia dynasty' as a ten-mouth solar calendar.) *Ziran kexueshi yanjiu* 1 (4), 305–19.

Chen Jiujin, Lu Yang and Liu Yaohan (1984). *Yizu tianwenxue shi.* (*History of the astronomy of the Yi minority people.*) Kunming: Yunnan renmin chubanshe.

Chen Zungui (1980–84). *Zhongguo tianwenxue shi.* (*History of Chinese astronomy*), 3 vols. Shanghai remnin chubanshe.

Chung-kuo k'o-hsueh-yuan. K'ao-ku-hsueh yen-chiuso. *See* Institute of Archeology, Chinese Academy of Social Sciences (1980).

Clark, D. H. and Stephenson, F. R. (1977). *The Historical Supernovae.* Oxford: Pergamon Press.

Cullen, C. (1980a). Symposium: The work of Joseph Needham. *Past and Present* 87, 17–53.

Cullen, C. (1980b). Was There a Maunder Minimum? *Nature* 283, 427–8.

Cullen, C. (1982). An eighth century Chinese table of tangents. *Chinese Science* 5, 1–33.

Doebereiner, P. (1980). *Die chinesische und die abendlaendische Astrologie: Ein Vergleich zwischen den fernoestlichen und den Europaeischen Tierkreiszeichen.* Munich: Heyne.

Elman, B. (1984). *From Philosophy to Philology. Intellectual and Social Aspects of Change in Late Imperial China.* Harvard East Asian Monographs, 110. Cambridge, MA: Harvard University Press.

Hartner, W. (1965). The earliest history of the constellations in the Near East and the motif of the Lion-Bull Combat. *Journal of Near Eastern Studies* 24, pp. 1–2 – 1–16, plate 1–16. (See the additional remarks appended to the reprinted version in Hartner (1968–84, vol. I, pp. 227–59).)

Hartner, W. (1968–84). *Oriens-Occidens. Ausgewaehlte Schriften zur Wissenschafts- und Kulturgeschichte*, 2 vols. Hildesheim: Georg Olms Verlagbuchhandslung.

Hartner, W. (1969). *Die Goldhoerner von Gallehus*. Wiesbaden: Franz Steiner Verlag GmbH.

Hartner, W. (1973). Hellenistic science and ancient oriental tradition on North Germanic soil. *Rete* 2, 45–62.

Henderson, J. B. (1984). *The Development and Decline of Chinese Cosmology*. Neo-Confucian Studies. New York: Columbia University Press.

Ho Peng Yoke (1962). Ancient and medieval observations of comets and novae in Chinese sources. *Vistas in Astronomy* 5, 127–225.

Ho Peng Yoke (1977). *Modern Scholarship on the History of Chinese Astronomy*. Occasional Papers, Faculty of Asian Studies, The Australian National University, 16. Canberra.

Ho Peng Yoke and Ang Tian-se (1970). Chinese astronomical records on comets and 'guest stars' in the official histories of Ming and Ch'ing and other supplementary sources. *Oriens Extremus* 17, 63–99.

Horton, R. (1983). Tradition and modernity revisited. In *Rationality and Relativism*, eds. M. Hollis and S. Lukes, pp. 201–60. Oxford: Basil Blackwell.

Horton, R. and Finnegan, R., eds. (1973). *Modes of Thought. Essays on Thinking in Western and Non-Western Societies*. London.

Hsi Tse-tsung (Xi Zezong), Yen Tun-chieh, Po Shu-jen, Wang Chien-ming [Wang Chien-min], Chen Chiu-chin [Ch'en Chiu-chin] and Chen Mei-tung [Ch'en Mein-tung] (1973). Heliocentric theory in China—in commemoration of the quincentenary of the birth of Nicolaus Copernicus. *Scientia sinica* 16, 270–9.

Imoto Susumu and Hasegawa Ichiro (1958). Historical records of meteor showers in China, Korea, and Japan. *Smithsonian Contributions to Astrophysics* 2 (6), 131–44.

Institute of Archaeology, Chinese Academy of Social Sciences (1980). *Zhongguo gudai tianwen wenwu tuji. (Illustrations of Artifacts Pertaining to Ancient Chinese Astronomy.)* Archaeological Monographs, B17. Beijing: Wenwu chubanshe.

Kalinowski, M. (1983). Les instruments astro-calènderiques des Han et la méthode *liu jen*. *Bulletin de l'Ecole Française d'Extrême-orient* 73, 309–416.

Kiang, T. (1972). The past orbit of Halley's Comet. *Memoirs, Royal Astronomical Society* 76, 27–66.

Lockyer, J. N. (1906). *Stonehenge and Other British Stone Monuments Astronomically Considered*. London: Macmillan. (Second edition, 1909.)

Luo Baorong and Li Weibao (1978). Cong woguo gudai jiguang, dizhen jilu de zhouqi fenxi kan taiyangheizi zhouqi de wendingxing. (On the stability of sunspot periods as seen from the analysis of ancient Chinese aurora and earthquake records.) *Kexue tongbao* 23 (6), 362–6.

Major, J. S. (1978). Myth, cosmology, and the origins of Chinese science. *J. Chinese Phil.* 5, 1–20.

Maeyama Yasukatsu (1975–6). On the astronomical data of Ancient China (ca. −100 – +200): a numerical analysis. *Archives internationales d'histoire des sciences* 25, 247–76 (1975); 26, 27–58 (1976).

Maeyama Yasukatsu (1977). The oldest star catalogue of China, Shih Shen's *Hsing ching*. In *Prismata. Naturwissenschaftgeschichtliche Studien. Festschrift fuer Willy Hartner*, eds. Maeyama Yasukatsu and W. G. Salzer, pp. 211–45. Wiesbaden: Franz Steiner Verlag GmbH.

Nakayama, S. (1966). Characteristics of Chinese astrology. *Isis* 57, 442–54. (Reprinted in Sivin, 1977, pp. 94–106.)

Nakayama, S. (1969). *A History of Japanese Astronomy. Chinese Background and Western Impact*. Cambridge, MA: Harvard University Press.

Needham, J. (1954–). *Science and Civilisation in China*, 7 vols. projected, 13 fascicles to date. Cambridge University Press.

Needham, J. (1975). The cosmology of early China. In *Ancient Cosmologies*, eds. C. Blacker and M. Loewe, pp. 87–109. London: George Allen & Unwin.

Needham, J. and Ho Ping Yü [Ho Peng Yoke] (1959). Ancient Chinese observations of solar haloes and parhelia. *Weather* 14, 124–34.

Neugebauer, O. (1975). *A History of Ancient Mathematical Astronomy*, 3 vols. Studies in the History of Mathematics and Physical Sciences 1. New York: Springer.

Pan Nai (1975). Nanjing de liangtai gudai cetian yiqi – Ming zhi hunyi he jianyi. (Nanjing's two ancient instruments for astronomical observation: the Ming armillary sphere and the simplified instrument.) *Wenwu* 7, 84–9.

Schafer, E. H. (1977). *Pacing the Void. T'ang Approaches to the Stars*. Berkeley: University of California Press.

Sivin, N. (1969). *Cosmos and Computation in Early Chinese Mathematical Astronomy*. Leiden: E. J. Brill.

Sivin, N. (1973). Copernicus in China. *Studia Copernicana* 6, 63–122.

Sivin, N., ed. (1977). *Science and Technology in East Asia*. New York: Science History Publications.

Sivin, N. (1981). Some important publications on early Chinese astronomy from China and Japan, 1978–1980. *Archaeoastronomy* 4 (1), 26–31.

Sivin, N. (1982). Why the scientific revolution did not

take place in China – or didn't it? *Chinese Sci.* **5**, 45–66.

Tang Ruchuan (1962). Zhang Heng deng huntianjia de tian yuan di ping shuo. (On the theory of Zhang Heng and other uranosphere school cosmologists that the sky is spherical and the earth flat.) *Kexueshi jikan chik'an* **4**, 47–58.

Teboul, M. (1983). *Les Prémieres Théories Planetaires Chinoises*. Paris: Collège de France.

Tiede, V. (1988). Astronomical orientations of tombs in Ancient China. *Proceedings of the Third International Conference on the History of Chinese Science.* Beijing, August 1984. (In press.)

Wagner, D. B. (1978). Liu Hui and Tsu Keng-chih on the volume of a sphere. *Chinese Sci.* **3**, 59–79.

Wagner, D. B. (1980). Archeological sources for the history of science, technology, and medicine. Some supplementary references. *Chinese Sci.* **4**, 53–60. (Published with Xia Nai, 1980.)

Xi Zezong (1981). Chinese studies in the history of astronomy, 1949–1979. *Isis* **72**, 456–70; especially p. 470.

Xi Zezong (1983). Chinese researches in the history of science and technology, 1982. *Chinese Sci.* **6**, 59–83.

Nicolaus Copernicus. *Scientia sinica* **16**, 270–9.

Xia Nai (1979). *Kaoguxue he kejishi.* (Archeology and the History of Science and Technology.) Archeology Monographs, A1. Beijing: Wenwu chubanshe. (With English summaries.)

Xia Nai (1980). Bibliography of recent archeological discoveries bearing on the history of science and technology. *Chinese Sci.* **4**, 19–52. (Translation by D. B. Wagner of an article in Xia, 1979, with emendations by the author.)

Xie Yushou and Cai Meibiao (1983). *Zhongguo dizhen lishi huibian.* (Collected Materials for the History of Chinese Earthquakes.) 5 vols. Beijing: Kexue chubanshe.

Xu Zhentao. Ancient records of sunspot and aurora [*sic*] of China and investigations of history of solar activity *and* On the Sporer Minimum. Unpublished papers jointly offered at the 6th International Summer College on Physics and Contemporary Needs, Islamabad, Pakistan, June 1981.

Yabuuchi Kiyoshi [Yabuuti Kiyosi] (1969). *Chūgoku no temmon rekihō.* (Observational and Computational Astronomy in China.) Tokyo: Heibonsha.

Yamada Keiji (1980). *Jujireki no michi. Chūgoku chūsei no kagaku to kokka.* (The Road to the Season-Granting System. Science and the State in Medieval China.) Tokyo: Misuzu Shobō. (Concerned with the calender reform of 1280.)

Yi Shitong (1978). Liang tian chi kao. (On the astronomical foot.) *Wenwu* **2**, 10–17.

Zhang Jiatai (1976). Dengfeng guanxingtai he Yuan chu tianwen guance de chengjiu. (The Dengfeng observatory and the Achievements of early Yuan astronomical observation.) *Kaogu* **143**, 95–102.

Zhongguo kexueyuan. Kaoguxue yanjiusuo. *See* Institute of Archeology, Chinese Academy of Social Sciences (1980).

Zhuang Weifeng (1982). Woguo difangzhi zhong tianwen ziliao de pucha he zhengli. (Survey and compilation of astronomical materials in Chinese gazetteers.) *Zhongguo difangzhi jiu zhi zhengli* **6**, 52–7.

Zhuang Weifeng, Zhu Shijia, Feng Baolin, Wang Shuping, Jiang Huanwen, Li Bingqian, Shen Yingyuan, Xu Zhongkai, Chen Peirong, Zhang Shengze, Qi Shenqi, Jiang Guangtian and Liu Zhisheng (1985). *Zhongguo difangzhi lianhe mulu (Union Catalogue of Chinese Local Gazetteers)*. Beijing: Zhonghua shuju.

7

The cosmic temples of old Beijing

E. C. Krupp *Griffith Observatory, Los Angeles*

Beijing, the capital of Imperial China in the Ming (AD 1368–1644) and Qing (AD 1644–1911) dynasties, was the center of political power, and it was furnished with a set of symbolically located temples dedicated to cosmic gods. These temples and the ceremonies staged in them spotlight the relationship between monumental architecture, the emperor's authority to rule, and celestial order.

Even the fundamental plan of the capital carried a celestial connotation. Laid out on a north–south axis that was neither arbitrary nor routine, Beijing put cardinality on display in its streets and palaces. Cardinal directions originate with the recognition of the north celestial pole (Krupp, 1982a, pp. 9–10, 1982b, p. 11), the single stable spot in the skies that turned over China. For the Chinese, the sky's north pole was the unmoving hub of the universe. They called the circumpolar zone the 'forbidden polar palace' (Needham, 1959, p. 258), and the pole itself was the symbolic face of heaven, or *Shang di*, who ordered the world through cycles revealed in the motion of celestial objects.

Cardinality and celestial order were brought to a focus in the Imperial Palace, where the emperor resided, held audiences, announced the New Year, and eased the calendar along its way through the seasons by making timely appearances in various halls (Dorn, 1970, p. 87). His palace, the 'polar forbidden city' (Lin, 1961, p. 34), was the terrestrial counterpart of the circumpolar zone, and the emperor, the terrestrial counterpart of the pole, occupied what the Chinese regarded as the world's center. All of the world revolved about him as all of the sky turned about the pole (Yu, 1982, p. 18).

Beijing, established as the capital by the third Ming Emperor, Yong le, was designed to be a 'place where sky and earth meet' (Wheatley, 1971, p. 462).

Astronomical connotations emerged in the city plan of old Beijing, and astronomy in Imperial China was official business. Heaven presided over earth and harmonized its affairs. The emperor was the pivot of the world. As heaven's agent, he ruled with a celestial mandate. It was the foundation of his authority. His ability to establish and maintain order originated in the cyclical patterns of night and day, lunar phases, seasons, and planetary movements, all monitored by his own court astronomers.

Among the Chinese, then, there was the sense of an intimate connection between earth and sky. The sky's cyclical order governed earth. That's why the high sky god was accorded his exalted status. Order is the sky's real power, and the emperor was the conduit of celestial power (Krupp, 1983, pp. 196–8). Through appropriate ritual activity, he conveyed celestial power to earth to preserve world order, and the capital was designed to facilitate the harmonious convergence of influences.

Although the present government of the People's Republic of China deliberately broke and blocked the old central axis still visible in the historic gates, pavements and palaces of Beijing by building the Monument to the People's Heroes and the tomb of Mao Ze dong right on the north–south line in Tian an men Square, these intrusions on the imperial plan participate in the traditional cosmological layout and suggest that the old symbols are still at work on behalf of the contemporary government (Krupp, 1982a, p. 10).

65

Figure 7.1 At the Hall of Supreme Harmony (*Tai he dian*), the principal station on Beijing's cosmic axis, the emperor held his audiences. The central pavement defines the primary north– south line, and the view is north. (Photograph Robin Rector Krupp.)

Only the emperor and his court could occupy the 'polar forbidden city', but Beijing was the point of mediation between heaven and earth. The emperor interceded with heaven on behalf of his people. While it is clear that he exercised conventional power to carry out the ordinary business of empire, his real privilege and responsibility were the performance of sacrifices on behalf of the entire population and the promulgation of the calendar, by which all human activity would be organized and harmonized (Krupp, 1982c, p. 18). As a sacred king, he embodied the life, identity and place of the general population. So the calendar, then, would march through time, and the emperor would march through the rooms of the Imperial Palace to the appropriate doorway and announce an intercalation for the calendar. He would hold a grand audience in the Hall of Supreme Harmony on New Year's Day, and he would preside over seasonal ceremonies in Beijing's cosmic temples.

At the New Year – in early February by our calendar – the emperor went to the Hall of Prayer for Good Harvests (or for a Good Year) in the Temple of Heaven (*Tian tan*) to offer sacrifices for an abundant grain harvest in the coming year. Located about two-and-a-half miles south of the Imperial Palace and just east of the city's primary axis, the Temple of Heaven is a cardinally oriented complex of buildings, altars and walkways. Three of them – the Hall of Prayer for Good Harvests (*Qi nian dian*), the Imperial Heavenly Vault (*Huang qing yu*), and the Round Mound (*Huan qiu tan*) – are strung along a processional pavement, the primary north–south axis of the whole temple ensemble (see Figure 7.2). An outer enclosure wall defines the sacred precinct of the Temple of Heaven. Its north end is curved, to protect the temples from the pernicious supernatural influences that originate from the north (Bredon, 1931, p. 156). Rectilinear walls on the south stand for earth, whose shape is said to be square.

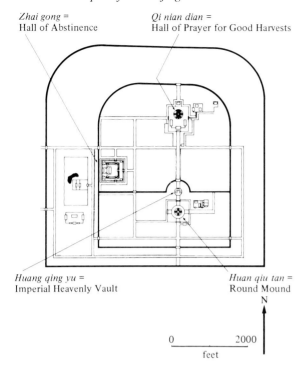

Zhai gong =
Hall of Abstinence

Qi nian dian =
Hall of Prayer for Good Harvests

Huang qing yu =
Imperial Heavenly Vault

Huan qiu tan =
Round Mound

N

0   2000

feet

Figure 7.2 Several structures comprised the Temple of Heaven, and the three principal monuments – the Hall of Prayer for Good Harvests, the Imperial Heavenly Vault and the Round Mound – established its primary north–south axis, an elevated walkway known as the Bridge of Cinnabar Steps (*Dan bi qiao*). (Griffith Observatory, Joseph Bieniasz.)

Round in shape, like heaven, the Hall of Prayer for Good Harvests incorporates numerological and calendrical symbolism in its architecture. Three rings of cylindrical pillars inside the building support its three fluted roofs, and the four tallest and most elaborately decorated pillars comprising the inner circle stand for the four seasons. The 12 columns in the middle ring refer to the 12 lunar months in the year, and the 12 'double-hours' in the day are symbolized by the pillars in the outer ring. Taken together, the 28 pillars also represent the 28 *hsiu*, or 'lunar mansions', reference zones in the Chinese system of mapping the sky (Zhou, 1984, p. 116).

After gaining heaven's confidence in the New Year sacrifice at the Hall of Prayer for Good Harvests, the emperor exited the main gate on the west side of the Temple of Heaven and crossed the primary axis of the city over to the Temple of Agriculture (*Xian nong tan*), another collection of altars and temples. With a curved wall on the north and a straight wall on the south, the Temple of Agriculture enclosure followed the same principles of design encountered at the Temple of Heaven. In area, the Temple of Agriculture was smaller than the Temple of Heaven. The grounds inside the Temple of Heaven's inner enclosure wall are comparable to the original size of the Temple of Agriculture.

There is a link between the seasonal cycle and the agricultural enterprise. And farming, naturally, sustained the economic and political vitality of the state. Dedicated to Shen nong, the second of China's five legendary emperors – the one who introduced farming to China – the Temple of Agriculture included a ceremonial field in front of its Altar of Agriculture, and calendric modulation of imperial ritual brought the emperor there at the New Year to cut the year's first furrows with a team of yellow oxen and a yellow plough (Favier, 1897, p. 356). At the same time, a clay ox was taken to the field, broken up, and ploughed under as a kind of sacrifice for a new season of fertile land. Yellow, in the Chinese system of correspondences, was the color of the center and the color of the earth. The emperor even wore yellow robes for the occasion.

Other platforms in the Temple of Agriculture include the Altar of the Gods of the Sky (*Tian shen tan*), the Altar of the Gods of the Earth (*Di qi tan*), and the Altar of the Year God (*Tai sui dian*). The 'Year God' is the planet Jupiter, which moves through approximately one zodiac constellation each year and completely around the sky in about 12 years. The Chinese 12-year calendar cycle derives from this planet's behavior.

Sacrifices at various temples and altars of Beijing made visible the emperor's personal relationship with the cosmic forces that ruled the universe and so legitimized his authority. His personal participation obviated any need for intermediaries. Neither his ancestors nor other spirits were invoked on his behalf. By performing sacred rituals to cosmic gods shared by all at shrines outside the privacy of the Imperial Palace, the emperor became the only licensed mediator with the universe. Sacrifices to heaven and earth, to sun and moon, were not just a responsibility, they were his prerogative. The right to conduct the imperial sacrifices in part defined what being an emperor meant (Bredon, 1931, p. 158). They were explicit evidence of his mandate (Wechsler, 1985, pp. 107–9).

Three classes of sacrifices – 'great', 'medium' and 'small' – required the emperor's presence or his delegated representative, and of the 'great sacrifices', the winter solstice observances at the Round Mound ranked highest. Its service was also the most complete. Elaborate offerings of food, beverages, incense, jade and silk were prepared and inspected in advance, while the final order of ceremony was worked out by the Board of Rites according to ancient tradition (Bredon and Mitrophanow, 1927, pp. 58–61).

Three days before the solstice, the emperor began a fast in which certain foods were excluded (Bredon and Mitrophanow, 1927, p. 56). On the third day of the fast, he left the Imperial Palace, traveled south to the Temple of Heaven, and moved into its Hall of Abstinence (*Zhai gong*). There, through isolation and self-discipline, he was to put himself into the proper mental and spiritual state the solemnity of the impending ceremonies required.

Before dawn the next morning, the day of winter solstice, he left the Hall of Abstinence in the company of a rich and colorful parade. The soldiers, courtiers, musicians, dancers, singers and high officials could number as high as a thousand, and they marched through the darkness with banners, flags, parasols and fans until they came to the south stairway of the Round Mound (Arlington and Lewisohn, 1935, p. 111; Wechsler, 1985, p. 119).

In the palace, the emperor was the north celestial pole. His subjects approached him from the south, and his top officials and family surrounded him as the circumpolar stars circle the pole. But at the Round Mound, on the winter solstice, the emperor was himself a subject, and in homage to Shang di, the Lord of Heaven, and the emperor faced him from the south. Although Shang di was unseen, the sky's north pole symbolized his presence (Bredon, 1931, p. 157; Wechsler, 1985, p. 115).

Heaven's jade tablet, usually stored in the Imperial Heavenly Vault north of the Round Mound, occupied a spot on the north side of the third and top platform of the altar for the winter solstice rites. As the emblem of heaven and the sky's hub, it faced the emperor as the north celestial pole faced the earth. Elsewhere on the circular shrine, inscribed tablets for the sun, the moon, the north star, the five planets, the 28 constellations, the host of stars, the clouds, the rain, the wind, the thunder, and the five emperors acknowledged the cosmic forces that orchestrated the world.

Only the emperor could presume to engage heaven, and to do so he had to climb from the ground level, that represented the earth, to the top of the Round Mound (see Figure 7.3). His ascent brought him physically closer to the sky, and the platform – open to the air – exposed him directly to heaven. In the same way, the architecture of the Round Mound made the emperor's ritual activity more visible and therefore more public than an offering made inside an enclosed and more private ancestral temple. Although the general populace would never witness the emperor's sacrifice to heaven, he had an audience. His elite entourage of officials, royalty, servants and guards were present, and what they saw, in the humbling of the emperor before heaven, was the true sign of his power and authority, his personal bond with the supreme ruler of the universe.

Numerous architectural details at the Round Mound symbolize heaven and its relation to earth. The outer enclosure wall is square. This is the shape of earth, according to Chinese tradition, and it originates in the Chinese emphasis on the significance of the four cardinal directions. At the square enclosure, then, one is still, in a sense, in the terrestrial realm, but each step closer to the Round Mound is a step closer to heaven. A second enclosure wall, circular to signify the sky, suggests that the north–south promenade really is a highway to heaven, and the Round Mound is itself circular to invoke the image of the horizon and the sky. Blue tiles capped the enclosure walls as a nod to the sky's blue color, and the tablet of heaven was blue jade. In a blue robe the emperor offered heaven rolls of blue silk.

Cardinal stairways provide access to the stack of three elevated decks, and each section of staircase has nine steps. Nine is heaven's special number at the Round Mound. The pavement on the upper terrace is laid out in concentric rings, and the number of stones in each ring increases as a multiple of nine around the central disk. These rings presumably symbolize the nine heavens, and the ninth and outermost ring has $9 \times 9 = 81$ stones. The scheme of multiplying nines continues on the middle and lower platform as well, and there were nine acts in the emperor's winter solstice sacrifice (Bredon and Mitrophanow, 1927, p. 58).

Figure 7.3 The emperor approached the Round Mound (*Huan qiu tan*) from the south along this paved walkway, climbed the three stairways to the top platform, and prostrated himself before heaven, the supreme ruler of the universe. In humbling himself before heaven, he exalted his unique status on earth. (Photograph by author.)

In the course of the ceremony, the emperor tendered portions of flesh, skin and blood of sacrificed animals to heaven and then a cup of wine. After all of the appropriate offerings had been made, they were collected and burned. Their smoke, rising to the sky, again alluded to heaven, their ultimate destination.

A complementary sacrifice performed six months later, at the summer solstice, directed the emperor's attention to earth (Bredon and Mitrophanow, 1927, p. 61). Here again, only the emperor was sanctioned to intercede on behalf of the world, and his privilege, exercised on another open-air altar, outside the privacy of the palace, put his power on 'public' display for the privileged spectators who attended him at the Temple of the Earth (*Di tan*) in the northern suburbs of Beijing.

All of the enclosure walls of the Temple of the Earth are square and not round (Figure 7.4). The altar is a two-level square platform, oriented cardinally with cardinal stairways. Because earth's color is yellow, tiles at the Temple of the Earth are yellow, and not blue (Figure 7.5). During the ceremony, the emperor wore yellow robes, faced a yellow tablet of the earth, and offered rolls of yellow silk and a yellow gem. Eight, rather than nine, was the operational number at the Altar of the Earth, and each set of steps numbers eight. A moat surrounds the platform, probably a reference to the deep watery trench imagined sometimes to surround the earth and connect with the vault of the sky. Tablets for the whole 'Ministry of Heaven' appeared on the Round Mound at the winter solstice. Here, in summer, the terrestrial divinities received their due in the tablets for the five sacred mountains (one for each cardinal direction and one for the center), for other important mountains, for the hills that protect imperial tombs, for the four seas, and for the four

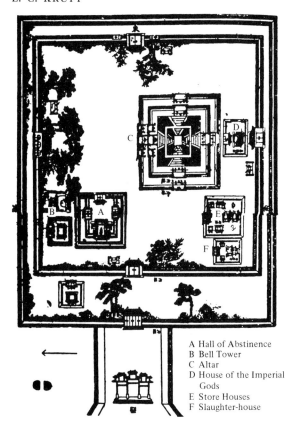

A Hall of Abstinence
B Bell Tower
C Altar
D House of the Imperial
   Gods
E Store Houses
F Slaughter-house

Figure 7.4 At the Temple of the Earth (*Di tan*), the enclosure walls and altar are all square because the shape of earth is square. Heaven, by contrast, is round. (Arlington and Lewisohn, 1935.)

sacred rivers. At the Temple of the Earth, the offerings were not burnt, but buried in the ground, the obvious domain of earth. There are four entrances to the inner precincts of the Temple of the Earth, and the main gate, with three portals instead of one, is on the north. It suggests that the emperor approached the altar from the north, in contrast to his southern approach at the Round Mound.

These opposite but complementary ceremonies – one at the winter solstice, the other at the summer solstice; one on the south side of the city, the other on the north – reflect the traditional Chinese concept of a world in balance between complementary principles in a dualistic cosmos. In Chinese terms, these two aspects of nature are known as *yin* and *yang*, and they represent the contrast between

    feminine and masculine,
    passive and active,
    dark and light,

    moon and sun,
    winter and summer,
    north and south,
    cold and warm,
    watery and fiery,
    soft and hard,
    still and moving,
    negative and positive,

and more. The Chinese saw the universe as a dynamic system in which these cosmic principles alternated in strength. Their oscillation was reflected in the diurnal cycle of day and night and in the annual cycle of the seasons. At the winter solstice when the sun was weak, *yang* was said to be diminished and heaven was on the downside. The emperor's sacrifice of precious materials and sanctified food on the *yang* side of town re-energized the celestial, male side of the cycle of undulating order. But when *yin* was at its ebb in the height of summer's heat, the emperor's efforts in the suburban *yin* territory on the summer solstice midwifed the rebirth of the terrestrial, female force.

By forging a balance in the cyclical pattern of nature through cosmo-magical ritual at crucial seasonal transitions, the emperor preserved the world's pattern and cohesion. And, in doing so, he reinforced his mandate to rule. There was then a political payoff to his solstice duties. With his authority upheld through cosmic ritual, the emperor really did stabilize society. Harmonizing the affairs of empire with nature involved a self-fulfilling process.

The original Round Mound was built in 1420, and the Altar of the Earth followed more than a century later in 1530. Both are relatively late as far as Chinese history is concerned. But an early book, the *Rites of Zhou* (*Zhou li*) verifies that a round altar winter solstice sacrifice to heaven and a square altar summer solstice sacrifice to earth were performed by at least the second century BC (Wechsler, 1985, p. 109). Another text of comparable antiquity, the *Book of Rites* (*Li chi*), refers to burnt offerings for heaven and buried offerings for earth at suburban altars (Wechsler, 1985, p. 109). In *Offerings of Jade and Silk*, Wechsler (1985), the American professor Howard J. Wechsler describes persistent confusion over the proper form of these ceremonies through the T'ang dynasty (AD 618–917). By then, however, they had become the most important state rituals. In them, the emperor confirmed his celestial

Figure 7.5 The altar walls at the Temple of the Earth are tiled in yellow, the color of earth. Summer solstice sacrifices here, on the city's north side, complemented the winter solstice sacri- fice at the Round Mound in the southern suburbs. (Photograph Robin Rector Krupp.)

lineage. He was the Son of Heaven. Centuries later, in the Ming and Qing dynasties, the emperor still consolidated his status at altars for heaven and earth.

Early T'ang dynasty sources indicate that imper- ial sacrifices took place at open-air altars on the east and west sides of town, too, in the appropriate seasons (Wechsler, 1985, p. 113). Ming and Qing emperors of Beijing seem to have preserved the tra- dition in their 'medium' sacrifices at the Temple of the Sun (*Ri tan*) and the Temple of the Moon (*Yue tan*) (Bredon and Mitrophanow, 1927, p. 65).

Enclosed by an outer wall that is rounded on the east and straight on the other three sides, the Temple of the Sun is situated in the eastern suburbs of Beijing (see Figure 7.6). It was built during the Ming dynasty, in 1530, and its square, cardinal altar and circular inner enclosure wall still survive. Ancillary structures include the house where the emperor changed into ceremonial robes, a bell

tower, a storage place for musical instruments and ceremonial paraphernalia, the room where the spirit tablets were housed, and a slaughter house and kit- chen for the preparation of sacrifices.

The sun's gold tablet inscribed with red charac- ters that said 'Spirit of Great Light', was set out on the altar for the vernal equinox sacrifice to the rising sun. A round, red gem symbolized the sun in this ritual because the sun's symbolic color was red (Arlington and Lewisohn, 1935, p. 265). Red, too, is the wall of the inner enclosure, and its blue– green tiles symbolize the east. Spring, sun and east are all three expressions of *yang*. The principal gate on the west side, conducted the emperor, to the east, to ascend the altar and face the sun's tablet and the direction of the rising sun. Upon its appear- ance, the sacrifice was carried out.

A complementary service for the rising moon was timed by the autumnal equinox and held in the Temple of the Evening Moon (*Yue tan*) on the west

Figure 7.6 The circular enclosure at the Temple of the Sun is painted red for the sun and tiled in blue–green for the direction east. This is the south portal. (Photograph Robin Rector Krupp.)

side of Beijing. Its outer enclosure is rectangular and its inner wall is square because the moon is associated with the earth and *yin* (Williams, 1941, p. 278). The round walls at the Temple of the Sun are a nod to *yang* and sky. Today, the old buildings are in use by other agencies, and a huge television antenna, erected in 1969, occupies the site of the moon's altar. This was said to have been a square sunken altar oriented cardinally, in accordance with the moon's female/*yin* character and in contrast with the elevated altar of the male/*yang* sun. High modern walls and restricted access prevented on-site confirmation of this description.

White robes, white silk and a white jade all invoked the moon, whose symbolic color was white, and the direction west, also associated with the color white.

Juliet Bredon, an Anglo–American resident of Beijing in the late 19th and early 20th centuries,

recorded the temple caretaker's eyewitness account of the last sacrifice performed on the moon's altar, in her book *Peking* (Bredon, 1931, p. 259):

It was on the evening of the autumn equinox that His Majesty came. When the Harvest Moon shone full on the altar spread with white offerings, white silks, white jades, milky pearls, The Lord of Ten Thousand Years bowed before the creamy tablet with the silvered characters meaning 'Place of the Spirit of the Light of Night'. Afterwards four animals were sacrificed, a pig, an ox, a sheep, and a deer, while the bell tolled from a near-by tower. Then the emperor changed his sacrificial robes in the pavilion yonder and returned to his palace while we 'humble folk,' he added with a chuckle, 'shared the meat offerings with the moon'.

The caretaker told her the moon had been so honored 20 years earlier, but it is not clear from

Bredon's text what year he meant. *Peking* was first published in Shanghai in 1919, and the year of the last sacrifice may have been 1899.

L. C. Arlington and William Lewisohn, also residents of Beijing, wrote *In Search of Old Peking* in 1935 and indicated that the sacrifice for the moon ordinarily took place at 10.00 p.m. on the 18th day of the eighth moon of the lunar year. They also confused this, however, with beginning of autumn, which usually occurs in the seventh month and in August (Arlington and Lewisohn, 1935, p. 250). Nathan Sivin has argued that, in general, the first day of the Chinese month was the day of conjunction (Sivin, 1969; O'Neil, 1975, pp. 110–14). It is an interesting coincidence that the moon's 'age' on the autumnal equinox in 1899 was 18 days.

Although some details of the emperor's equinox sacrifice at the Temple of the Moon are congruent with the overall pattern of use of these open-air altars dedicated to cosmic gods, some aspects of the Temple of the Moon are still puzzling. The main entrance is on the east side, and the moon's tablet was displayed on the west side of the altar. Emphasis on the west agrees with the *yin* character of the autumn and the moon, but the emperor would not have faced the rising moon as he faced the rising sun. Perhaps the moon's presence in the sky was an adequate constraint on the ceremony.

According to Chinese thought, the interaction of *yin* and *yang* creates the five forces (*Wu Hsing*) or 'elements' – water, fire, wood, metal and earth, and these active principles of nature provide a foundation for a system of correspondences that involves seasons, directions, mountains, colors, talismanic animals, planets, atmospheric conditions, grains, sounds, musical notes and many other components of experience (Burkhardt, 1982, pp. 204–5).

For example, color, direction, and season have these traditional associations (see Figure 7.7):

east	blue–green	spring
south	red	summer
center	yellow	late summer
west	white	autumn
north	black	winter

And some of these correspondences have shown up in the symbolic system of cosmic temples in Beijing. They are even more explicit in one more temple, the Temple of Land and Grain (*She ji tan*). It is slightly southwest of the *Wu men*, or Meridian Gate, of the Imperial Palace, and in a symbolic sense in

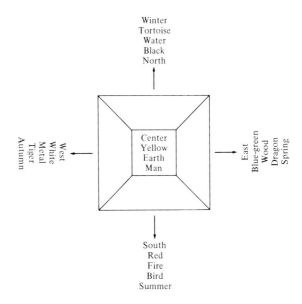

Figure 7.7 In China, a system of symbolic correspondences ordered the world with a pattern based on the principle of cardinal directions. This layout conforms, as well, to the layout of the five differently colored soils on top of the altar at the Temple of Land and Grain (*She ji tan*). (Griffith Observatory.)

the center of the city. Today it is part of Central Mountain Park, and although it did not serve as the center of imperial authority, as the abode of the gods of the soil, it stood for the central zone of the land. As the site of 'great' sacrifices linked to planting and harvest in the second (spring) and eighth (autumn) moons (Arlington and Lewisohn, 1935, p. 72), it brought the ritual calendar back to the center of the city. Emperor Yong le erected the altar here in 1420. It is a low pile of three marble platforms – square, cardinally oriented, and furnished with cardinal stairways. Its upper level transforms Chinese color–direction symbolism into a schematic picture of the world. Five colored soils for the five regions of China divide the top of the altar into four segments around a central square, and, although the soils have not been maintained, their color differences are still discernible. Each side of the square enclosure wall is properly coded with the correct color of tile: white on the west, black on the north, blue–green on the east, and yellow – in this context, an acceptable substitute for red – on the south. The present government does not conduct sacrifices here for favorable weather, but it has installed a protective iron fence

Figure 7.8 The white wall of the west meets the black wall of the north at the northwest corner of the enclosure of the Temple of Land and Grain. The idea of world quarters shows up in the paving as well. A diagonal line cuts through the square stones and restates the zones of the north and the west on the floor. (Photograph by author.)

space to time. The specific pattern of this activity was ordained by the fundamental cardinality of the Chinese cosmos and the junctions in their perception of cyclical time (see Figure 7.9).

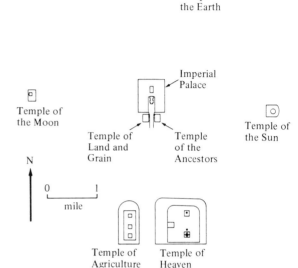

Figure 7.9 Seasonally timed ceremonies, dedicated to cosmic forces and celebrated in the cardinal quarters and center of Beijing, put the emperor's special relationship with heaven on display. The pattern of his sacrifices was intended to harmonize the forces of nature with human activity. He was a participant in the cycle of cosmic order and his participation took him around the cosmic temples of old Beijing. (Griffith Observatory, Joseph Bieniasz.)

around the altar. Each side has been painted in the color appropriate to the direction, this time with red on the south (see Figure 7.8).

Although this has hardly been a complete account of the imperial temples of Beijing and of the ceremonial responsibility of the emperor, even this sketchy report shows how the emperor's ritual activity was intended to enhance a salutary and congruent combination of influences of natural forces. His balancing act, in a world built on duality, allowed him to maintain a stable cosmos, climate and kingdom. His pilgrimages to cosmic altars sent him on an annual journey around the center, and back to it on several occasions. He moved counterclockwise around the city and, in doing so, married

Beijing is one of those pivots of the four quarters Paul Wheatley talks about – an urban center mobilized and stabilized by the cosmo-magical thought that brings the sky down to earth, where it is impressed into the temples, rituals, and layouts of the sacred capital of the divine king (Wheatley, 1971, p. 418). Beijing's temples turned Chinese cosmological thought into ritual architecture. As a sacred king, the emperor manipulated the symbols of cosmic order on ceremonial stages that made him a conduit for celestial power. He is, I believe, the natural descendant of the calendar-keeping, season-monitoring, resource-controlling shaman who knew an environmental impact report in the sky when he saw one.

References

Arlington, L. C. and Lewisohn, W. (1935). *In Search of Old Peking.* New York: Paragon Book Reprint Corp. (1967 reprint).

Bredon, J. (1931). *Peking.* Oxford University Press.

Bredon, J. and Mitrophanow, I. (1927). *The Moon Year.* Shanghai: Kelly & Walsh, Ltd.

Burkhardt, V. R. (1982). *Chinese Creeds and Customs.* Hong Kong: South China Morning Post Ltd.

Dorn, F. (1970). *The Forbidden City.* New York: Charles Scribner's Sons.

Favier, A. (1897). *Peking, Histoire et Description.* Peking: Imprimeries des Lazaristes au Pe-T'ang.

Krupp, E. C. (1982a). The mandate of Heaven. *Griffith Observer* **46**, (6), 8–17.

Krupp, E. C. (1982b). Tombs that touched the China sky. *Griffith Observer* **46**, (7), 9–17.

Krupp, E. C. (1982c). Shadows cast for the Son of Heaven. *Griffith Observer* **46**, (8), 8–18.

Krupp, E. C. (1983). *Echoes of the Ancient Skies.* New York: Harper & Row Publishers Inc.

Lin Yutang (1961). *Imperial Peking, Seven Centuries of China.* New York: Crown Publishers Inc.

Needham, J. (1959). *Science and Civilisation in China*, vol. 3. Cambridge University Press.

O'Neil, W. M. (1975). *Time and the Calendar.* Sydney University Press.

Sivin, N. (1969). *Cosmos and Computation in Early Chinese Mathematical Astronomy.* Leiden: Brill.

Wechsler, H. J. (1985). *Offerings of Jade and Silk.* Yale University Press.

Wheatley, P. (1971). *The Pivot of the Four Quarters.* Edinburgh University Press.

Williams, C. A. S. (1941). *Outlines of Chinese Symbolism and Art Motives.* New York: Dover Publications Inc. (1976 reprint).

Yu Zhuoyun (1982). *Palaces of the Forbidden City.* New York: The Viking Press.

Zhou, S. (1984). *Beijing Old and New.* Beijing: New World Press.

8

Ancient Maya ethnoastronomy: an overview of hieroglyphic sources

Dedicated to Floyd Lounsbury

John S. Justeson   *IBM T. J. Watson Research Center*

What remains of Lowland Maya astronomical theory and observation has come down to us in ethnohistoric records, mostly ethnographic accounts compiled by their Spanish conquerors; in the alignments of their surviving 'permanent' structures; and in native accounts, mainly in their hieroglyphic records. This chapter is an overview of data from sources of the third kind. The chief advantage of these sources is the precise temporal information they explicitly provide; patterns of dates in and between many hieroglyphic texts demonstrate their astronomical significance. These provide much material for investigating the formal means by which the Maya analyzed and anticipated celestial events.

Astronomical content of Maya hieroglyphic manuscripts was demonstrated almost from the beginning of serious research on Maya writing, in extensive computing tables. The extant copies date to within a few centuries of the Spanish Conquest, between AD 1200 and 1500 (M. D. Paxton. Stylistic and iconographic analysis of a Maya manuscript. Unpublished PhD dissertation, University of New Mexico, 1986) but internal evidence in the table indicates that they were adapted over the years from prototypes going back to *c.* AD 755.

Successes with the manuscript tables were soon extended to the carved inscriptions, almost all of which date to the Classic period of AD 300–900. Apart from the recognition of records in a lunar calendar, the results were generally not as reliable as those in the manuscripts; the reasons are sketched in Section 8.4. Since the discovery that most hiero-

glyphic texts record the activities and pedigrees of rulers, attention has focused on the construction of dynastic sequences and a skeleton of historical events involving the persons named in them. But, as Maya astronomy concerned the behavior of the sky gods, 'deities whose activities vitally influenced human affairs' (Kelley and Kerr, 1973, p. 180), astronomical correlates of historical events have begun to be recognized in the essentially historical narratives. Classic texts almost never mention these correlates; seldom do they make *any* explicitly astrological statements, referring instead to associated human events. As in the interpretation of structure alignments, these unstated correlates must be inferred from distinctive patterns, and demonstrated by statistical argumentation. While initially impeding recognition of celestial observations, this pattern of reference provides a means for recognizing the cultural significance of celestial phenomena in the events the Maya say occurred on these dates.

This chapter surveys hieroglyphic evidence for the means by which Maya astronomers attempted to develop formal models for the activities of the sky gods, and analyzes the intrinsic potential and consequences of these methods for the models they could develop. It sketches the relations of celestial to human affairs, so far as hieroglyphic sources reflect them; unfortunately, we are allowed only isolated glimpses into these relations, since the evidence comes from elite sources that treat relatively few topics, and these mainly as they involve the ruler himself.

8.1 Mesoamerican calendars and astronomy

The Maya calendar system

Although Maya predictive astronomy had both spatial and temporal dimensions, what survives in hieroglyphic texts is concerned almost entirely with predictions for the timing of events. The standard units of time measurement were not simply the tools by which the dates of celestial events were recorded, but also tools for recognizing the periodicities in those events via periodic repetition at or near the same point in one or more of these standardized measures of time. As the celestial objects were sacred deities, influencing the fates of human beings, the measure of sacred time in the ritual calendar was primary among these tools; but all the basic calendars were involved. Maya temporal astronomy must therefore be surveyed, as it developed, first of all in terms of Maya calendars.

Structure There were three basic Maya calendars. One was a ritual calendar of 260 days, used throughout Mesoamerica for prognostication; it is often called the sacred round (SR). There was also a so-called vague year of 365 days, and a civil year or tun of 360 days within a cycle called the long count; periods of 20 360-day years (the katun) and 400 360-day years (the baktun) were also commonly employed; higher powers of 20 were more rarely used to form multiples of the 360-day year. Most historical dates were specified in both the 260- and 365-day calendars, and in that order; the two commensurate in 73 sacred rounds, or 52 vague years, forming an 18 980-day cycle called the calendar round.

The names above are those used by scholars today. There is confusion in the literature over the referents of the Mayan words attached to these cycles. Names applied to the ritual calendar meant 'count/recitation of days' or 'series of days' (e.g., Cholan *čol k'in*, Yucatec *ţolkin*, Quichean *alah q'ix*), but these may refer instead or in addition to the almanacs for prognostication in terms of the ritual calendar days. A glyphic spelling that designates 260-day spans (but not the ritual calendar itself) has been identified by Mathews (Tonińa dates 1. A glyph for the period of 260 days? *Glyph Notes 8*, privately circulated, 1979); it can be transliterated [13]**SAK-HAB**, perhaps reading *sak ha ʔb'* (lit., 'white year', 'white' having connotations of 'diminished

substance' in some Mayan languages, 'resplendent, magnificent' in others); I suspect that the numeral 13 is a semantic determiner.

The pan-Mayan word for 'year' approximates *ha ʔb'* in almost all Mayan languages; this term was applied at Spanish contact to both the 365-day and 360-day years. A word apparently unique to the Lowland Mayan languages was *tu·n*, meaning the end of a year of either 360 or 365 days, e.g. a station in an absolute time count in the civil calendar or from another anchoring event whose anniversary was celebrated, especially installation into rulership (Fox and Justeson, 1984, p. 53, n.32). This word is conventionally used by Mayanists for the period of 360 days. This misuse is minimized but not entirely avoided here; when it occurs, it is distinguished by Roman typeface from the Mayan usage of *tu·n*. The 360-day years of the civil calendar were grouped into 20s; this grouping is referred to as the katun, after the Yucatec name for the period (**k'atun* < earlier **k'al tu·n* '20 year (endings)'), but it was evidently also referred to as **may* in Quichean languages, hieroglyphic texts, and colonial historical texts (Justeson and Campbell. The linguistic background of Maya hieroglyphic writing: arguments against a Highland Mayan role, unpublished manuscript, 1982). No Mayan name for the 400-year baktun is known (baktun is a scholarly term based on Yucatecan *b'á·k* '400' and *tù·n* 'year', on analogy with *k'atun*), or for higher multiples of the civil year. No Mayan term for the calendar round is known.[1]

Although these were the principal calendars, and the only ones appearing in the Postclassic hieroglyphic manuscripts, other calendars or temporal cycles were also recorded in the inscriptions. The most common was a nine-day cycle (designated Glyph G, and indexed G_1 through G_9). The hieroglyphs corresponding to the days in this cycle evidently named deities rather than days, deities generally equated with the nine Lords of the Night (a group of nine Aztec deities who ruled the fortunes of each night in succession); Kelley (1972), Kelley and Kerr (1973, p. 201) and Schlak (The gods of Palenque, unpublished manuscript, 1985) have suggested planetary identities for these deities. The glyphs referring to them seem to state that their authority has ended, suggesting that they are daytime records. Because 9 and 18 980 are relatively prime, a calendar round date referencing this nine-day cycle was thereby fixed historically in a cycle of nine calendar rounds or 468 years.

Next most commonly attested are records in a lunar calendar (see Section 8.2), about 170 cases being known. Rarely recorded are two other cycles. One is of 819 days, the least common multiple of three calendrically and ritually important numbers (9, 13 and 7); texts in which it occurs enumerate the days elapsed since the last station in the cycle, stations being 819 days apart. A final unknown cycle is Glyph Y, with coefficients 1–6; no current proposal successfully accounts for its coefficients. Formerly a Glyph Z was distinguished, always followed immediately by Glyph Y, but it is simply a numeral coefficient (5 + the grammatical suffix -b'iš) of Glyph Y.

The position of a date in the above cycles was presented almost invariably in the following order: long count – ritual calendar – nine-day cycle – Glyph Y – lunar series – vague year. The 819-day calendar usually occurs between ritual calendar and vague year dates, the intervening data being suppressed.

History   Calendars are essentially systems of cyclical numeration, specialized in content for application to days. The structures of the basic Mesoamerican calendars directly reflect the general Mesoamerican system of numeration: in all Mesoamerican languages the numeral systems were in base-20, and subdivision into cycles of 20 days is a common feature of the 260-, 360- and 365-day calendars. Thereby, they parallel the subdivision of lunar months into groups of ten days in the calendar sticks of the Cheyenne, Osage, Winnebago and Zuni (Marshack, Chapter 24, this volume; McCluskey, Chapter 28, this volume), numeral systems in the languages of these North American Indian groups being base-10. This is evidently a formal, numerical pattern, unrelated to natural cycles.

The calendar periods have been examined for astronomical bases. Interestingly, the Maya made no attempt to keep their 365-day vague year in line with the seasons or the sun, in spite of a clear seasonal basis for at least two of the Yucatec month names – *yáʔš=k'i·n*, the time of new growth, and *k'an=k'i·n*, the time of ripening or maturity. It was not subdivided into 'natural' units, e.g. approximating lunar months, but only into 18 named groups of 20 days each, followed by a year-ending period of five 'nameless', unlucky days.

The 260-day period has been taken as a deliberate approximation to various natural intervals, most significantly the time to birth since missed menses (e.g. Brotherston, 1983, and Chapter 21, this volume) and the time of zenith passage at *c.* 15° north latitude[2] (most recently, by Malmström, 1973, 1978; Coggins, 1982). Undoubtedly, such correlates were noted and used once the system was in place (cf. Kelley and Kerr, 1973, p. 180), but it seems unlikely that the system intentionally approximated *any* interval. Structurally, the ritual calendar is a permutation of two cycles, one of 20 named days and one of 13 numerals. Such a structure is unlikely to arise in a calendar whose essential rationale was its overall length; subdivision in such instances is usually into sequential units. Rather, it parallels the structure of the calendar round: there, two separate, coexisting cycles together formed a 52-year cycle; they came to be cited together since their permutation was useful for fixing dates in historical time, but no one doubts that the constituent cycles were independent. Most likely, the 260-day period was also the effect of combining two pre-existing ritual cycles, one of 20 named days and one of 13 numerals.

The separate origins of the constituent cycles are lost. The rationale for the 13 numerals designating days in the ritual calendar is uncertain (13 was an important mythic and ritual number, but these may be results rather than sources of its position in the structure of the ritual calendar). The cycle of 20 named days presumably reflects the base-20 system of numeration that was universal in Mesoamerica; it recurs in the subdivisions of the vague and civil years. This 20-day cycle can be reconstructed linguistically to a time before ritual calendar records are attested archaeologically (see below).

The period of 360 days was a later approximation to the 365-day period. It was used in a system of positional notation for recording arbitrary spans. One of its chief applications was to fix the cyclic dates of the ritual calendar and the vague year in a time count whose cycle was so long that it was effectively linear during the historical era; it is known among epigraphers as the long count. The long count was an enumeration of days from a mythological base some 3000–4500 years prior to the Maya historical era, expressed as a sum of years and days via multiples of 20; multiples of the higher powers were named before multiples of the lower powers in the long count and in everyday Mayan speech. Thus, a Maya date would be given effectively as, e.g., '9 baktuns, 15 katuns, 1 year, 6 uinals (a 20-day unit), and 3 days' (abbreviated 9.15.1.6.3), this sum being counted from an unstated but fixed historical date assignable to late August or early September of 3114 BC. Arbitrary spans of time counted from other, expressed dates were represented in the same system but, in the inscriptions, with the baktuns, katuns, years, uinals and days presented in reverse order.

The 360-day year clearly developed to accommodate the 365-day period to some other exigency; not only are these spans of almost equal length, but the same words, *ha?b'* and *tu·n*, were applied to both. Most likely, the accommodation was to the base-20 system of positional notation, for recording long count dates or other temporal spans: the 360-day unit is found only among those Mesoamerican groups that recorded the months of the vague year; and the place values in the base-20 system of positional notation were in series, not as $1, 20, 20^2, 20^3, 20^4, \ldots, 20^n$ (as in Mayan languages), but rather as $1, 20, 360, 20 \cdot 360, 20^2 \cdot 360, \ldots, 20^n \cdot 360$. However, it could have become the base for positional notation after having been devised for another purpose.[3]

The only calendrical cycles shared with other Mesoamerican systems were the ritual calendar and the vague year, along with their permutation as the calendar round. Both cycles are documented by *c.* 500 BC in Oaxaca, the ritual calendar (Marcus, 1976) evidently a century or two earlier; and ritual calendar dates may be recorded by *c.* 900–700 BC on Olmec-style artifacts and murals. Based on their seasonal correlates, the months of the vague year seem to have been named, among the Maya, by *c.* 500–400 BC, but the earliest documentation for the Maya or their neighbors is from the era 100 BC–AD 200. The named days of the Mayan ritual calendar existed by *c.* 600 BC, judging from linguistic data (by 1000 BC, if glottochronological dates are relied upon).[4] The lunar calendar and nine-day cycle are documented on a Preclassic Mayan text dated AD 199. The Glyph Y cycle is documented by AD 498, the other cycles not until after AD 600.

The astronomical content of the codices

Tables of numerals in the hieroglyphic manuscripts are invariably for counts of days between recorded positions in the ritual calendar; in some, the recorded time intervals had clear astronomical significance. The astronomical tables are now understood in fair detail, due especially to the trailblazing work of Förstemann (1904, 1906), Meinshausen (1913), Willson (1924) and Teeple (1925, 1930), but advances concerning essential features continue to the present day. Paxton (unpublished dissertation, 1986, pp. 35–50) provides a balanced assessment of controversial aspects, although neglecting Kelley (1983).

Planetary tables The Venus table spans five pages of the *Dresden Codex* (Figure 8.1), introduced by a one-page preface. 584 days accumulate on each of these five pages, via smaller intervals of 236, 90, 250 and 8 days. The full-page interval of 584 days approximates the 583.92-day average synodic period of Venus. The subintervals approximate the average times of visibility of Venus as morning star, invisibility around superior conjunction, visibility as evening star, and invisibility around inferior conjunction, respectively; but, as Aveni (1980, p. 187) comments, 'it is puzzling that the 90-day interval in the table is so different from the true disappearance interval (about 50 days) and that the morning and evening star intervals are represented as being unequal.' Another puzzle has been the use of a consistent value, 584, for the length of the synodic period. The synodic period varies annually, but has the same integer approximation every fifth year; the five-year cycle of Venus years repeats in a symmetrical sequence, 587, 583, 580, 583, 587. Thus, each of the pages could have been assigned a Venus year of 580, 583 or 587 days and thereby accurately reflected the variations in the Venus cycle. Likely solutions are noted in Section 8.3.

The chief controversy concerning the Venus table surrounds the Maya use of a set of four numerals in the preface to the table. They are generally seen as means of generating a new, viable starting point for the table from stations in the current cycle, on a day 1 Ahau (sacred to Venus) at or near heliacal rising, once calculations from the current 1 Ahau base began to accumulate observational error. Closs (1977) demonstrates that the numerals

Figure 8.1 — Scheme of the Venus cycle on *Dresden*, pp. 46–50 (restored and corrected). (From Thompson, 1972.)

Line	Page 46				Page 47				Page 48				Page 49				Page 50			
	Cib	Cimi	Cib	Kan	Ahau	Oc	Ahau	Lamat	Kan	Ix	Kan	Eb	Lamat	Etz'nab	Lamat	Cib	Eb	Ik	Eb	Ahau
1	3	2	5	13	2	1	4	12	1	13	3	11	13	12	2	10	12	11	1	9
2	11	10	13	8	10	9	12	7	9	8	11	6	8	7	10	5	7	6	9	4
3	6	5	8	3	5	4	7	2	4	3	6	1	3	2	5	13	2	1	4	12
4	1	13	3	11	13	12	2	10	12	11	1	9	11	10	13	8	10	9	12	7
5	9	8	11	6	8	7	10	5	7	6	9	4	6	5	8	3	5	4	7	2
6	4	3	6	1	3	2	5	13	2	1	4	12	1	13	3	11	13	12	2	10
7	12	11	1	9	11	10	13	8	10	9	12	7	9	8	11	6	8	7	10	5
8	7	6	9	4	6	5	8	3	5	4	7	2	4	3	6	1	3	2	5	13
9	2	1	4	12	1	13	3	11	13	12	2	10	12	11	1	9	11	10	13	8
10	10	9	12	7	9	8	11	6	8	7	10	5	7	6	9	4	6	5	8	3
11	5	4	7	2	4	3	6	1	3	2	5	13	2	1	4	12	1	13	3	11
12	13	12	2	10	12	11	1	9	11	10	13	8	10	9	12	7	9	8	11	6
13	8	7	10	5	7	6	9	4	6	5	8	3	5	4	7	2	4	3	6	1
14	4	14	19	7	3	8	18	6	17	7	12	0	11	1	6	14	10	0	5	13
16	Yaxkin	Zac	Zec	Xul	Cumku	Zotz'	Pax	Kayab	Yax	Muan	Ch'en	Yax	Zip	Mol	Uo	Uo	Kankin	Uayeb	Mac	Mac
17	N	W	S	E	E	W	S	E	N	W	S	E	N	W	N	E	W	W	R	E
18	A	B	C	D	E	F	G	H	I	J	K	L	M	N	O	P	Q	R	S	T
18	Red‡	Red‡	Red‡	Red‡	Red	Red	Red	Red	Red	Red	Red	Red	Red	Red	Red	Red	Red	Red	Red	Red
19	Venus	Venus	Venus	Venus	Venus	Venus	Venus	Venus	Venus	Venus	Venus	Venus	Venus	Venus	Venus	Venus	Venus	Venus	Venus	Venus
19	236	326	576	584	820	910	1160	1168	1404	1494	1744	1752	1988	2078	2328	2336	2572	2662	2912	2920
20	9	19	4	12	3	13	18	6	2	7	17	5	16	6	11	19	15	0	10	18
21	Zac	Muan	Yax	Yax	Zotz'	Mol	Uo	Zip	Muan	Pop	Mac	Kankin	Yaxkin	Ceh	Xul	Xul	Cumku	Zec	Kayab	Kayab
21	T	A			D	E			H	I			L	M			P	Q		
22	Winged Chuen	Winged Chuen	Winged Chuen	Winged Chuen	:	:	:	:	Winged Chuen	Winged Chuen	Winged Chuen	Winged Chuen	Winged Chuen	Winged Chuen	Winged Chuen	Winged Chuen	Winged Chuen	Winged Chuen	Winged Chuen	Winged Chuen
23	Red Venus	Red Venus	Red Venus	Red Venus	Red Venus	Red Venus	Red Venus	Red Venus	:	:	:	:	Red Venus	Red Venus	Red Venus	Red Venus	Red Venus	Red Venus	Red Venus	Red Venus
24	E	N	W	S	E	N	W	S	E	N	W	S	E	N	W	S	E	N	W	S
25	19	4	14	2	13	3	8	16	7	17	2	10	6	16	1	9	0	10	15	3
26	Kayab	Zotz'	Pax	Kayab	Yax	Muan	Ch'en	Ch'en	Zip	Yaxkin	Uo	Uo	Kankin	Cumku	Mac	Mac	Yaxkin	Zac	Zec	Xul
26	236	90	250	8	236	90	250	8	236	90	250	8	236	90	250	8	236	90	250	8

Figure 8.1 Scheme of the Venus cycle on *Dresden*, pp. 46–50 (restored and corrected). (From Thompson, 1972.)

can be used instead as devices for locating the recorded bases in the long count from the ostensible base in the table's preface at 9.9.9.16.0. Regardless of the nature of this mechanism, the location of the implied historical sequence of basedates in the long count is secure from 10.10.11.12.0 to 11.5.2.0.0 (*c.* AD 1038–1324), Teeple's full sequence, under the usual correlation of Maya with European chronology. Lounsbury (1983b) relates the 9.9.9.16.0 date to a historical base one cycle earlier than Teeple, at 10.5.6.4.0; Thompson (1950) interposed a full series of bases beginning 9.10.15.16.0. A clear presentation of the alternatives, and the basic arguments for each, are given by Paxton. Paxton and Closs prefer Teeple's sequence, while Aveni (n.d.) and I favor Lounsbury's; Thompson's sequence appears to lack current support.

The synodic periods of Mars, Jupiter, Mercury and Saturn have been identified in other tables in the manuscripts. Since the tables are structured as normal multiplication tables for time spans relevant to the major calendars, intervals recalling planetary periods or their subdivisions could be coincidental; because none shows deviations conforming more to planetary motions than to multiplication tables, no proposed table of planetary 'mean motion … has ever survived the stage of simple suggestion' (Aveni, 1980, p. 199). Thompson (1972, pp. 22–3, 107–8) provides a brief, scathing critique of these hypotheses.

In spite of Thompson's critique, Willson's (1924) view that the table on pp. 43b–45b of the *Dresden Codex* relates to Mars continues to be entertained (Aveni, 1980, pp. 195–8; Kelley, 1980, p. 30S, 1983, pp. 178–9). Based on Bricker and Bricker's (1986) results, the 3 Lamat base of the table in the ritual calendar can be interpreted as the canonical date for the heliacal rise of Mars, assuming a correlation in the Goodman family; and this parallels the use of canonical heliacal rise as the ritual calendar base of the Venus table. The table is 78 days long (subdivided into three rough quarters of 19 days and one of 21 days), re-entered ten times for a total of 40 stations spanning 780 days (= 3 · 260), just a couple of hours longer than the average synodic period of Mars. A table of multiples of 780 in the preface contains four numbers deviating by 260 from multiples of 780, recalling the corrective numerals in the preface to the Venus table;[5] Bricker and Bricker

(1986) and Section 8.3 show that they recover Martian heliacal rise at the 3 Lamat base of the table. The 78-day span evidently relates to the roughly 75-day period of retrograde motion. What I consider least securely established about this table is the function of the subdivisions (prior to the Brickers' work, it was whether the table related to Mars at all); I also differ with them on the placement of tabular retrograde motion in the long count, although their argument is in some respects more straightforward than my own.

Lunar tables The *Dresden* eclipse table is a sequence of 70 stations spanning 46 sacred rounds; the stations, occurring at intervals of six and, more rarely, five lunar months (177–8 or 148 days), are those on which solar eclipses might be visible. Details of the use of the tables are controversial, but their essential identity was realized on the basis of the six- and occasional five-month groups of lunar months. The positions of the nodes can be determined, to within a few days, from two lines of evidence. (1) A five-month period between eclipse dates joins an eclipse near the postnodal extreme to one near the prenodal extreme; the nodes fall about 14 days later. (2) The range of variation of the recorded dates from the 173.31 days between nodes can be calculated, from an arbitrary base; the position of the node will lie near the midpoint. The results of the two types of calculation are in agreement, placing a node at, or within a couple of days, of the first and last stations of the table.[6] Given these placements of the nodes, the number and internal arrangement of lunar groups is within a day of a true eclipse-possible date of new moon. Pictures are placed after the five-month groups, evidently referring to lunar eclipse possibilities or to actual lunar eclipses. Means of updating the table are provided, as in the Venus tables.

Calendrical commensuration of synodic cycles

The Venus table, Mars table, eclipse table, and prognostication tables are all entered via positions in the ritual calendar; the registered accumulation of time reaches specified stations in that calendar; and the full length of the tables is always an integral multiple of 260 days. The 405 lunations of the eclipse table amount on the average to 11 959.89 days, less than three hours short of the 11 960 days spanned by 46 ritual calendar periods. In the Venus table, five canonical Venus years (one pass through five pages) commensurate the vague year: $5 \cdot 584 = 2920 = 8 \cdot 365$; this commensuration was also exact or only a day short for the true Venus year, any five adding to 2919.6 ($= 5 \cdot 583.92$) days. 65 canonical Venus years commensurate the ritual calendar in two calendar rounds, and the Venus tables do cover this 104-year period; they move ahead of 65 true Venus years by only 5.09 days. The true Venus year commensurates 137 sacred rounds to within 0.78 days, four Venus years before the end of the table; provision is made in the preface to the Venus tables for a shift of tabular base at the point of this commensuration. The accumulation of 0.78 days positive error every 61 Venus years is balanced by provision for commensuration of 128 sacred rounds to 57 Venus years every fifth or sixth basedate shift, yielding an error of ± 0.4 days; the next (unrecorded) base would have been the occasion for such an adjustment.

In less than 16 years, the 260-day calendar commensurates the synodic period of the moon and of all the visible planets to within 4.31 days (see Tables 8.5 and 8.6 and Section 8.3). Three sacred rounds equal one Mars year, four are three days less than nine Mercury years, five are 16 hours less than 44 lunar months, 16 are a day more than 11 Saturn years, and 23 sacred rounds are three days less than 15 Jupiter years. Allowing spans as long as the Venus table, the synodic periods of Mercury, Venus, Mars and the moon commensurate with the sacred round to less than a day; the other calendars do not improve on the 260-day cycle in regard to Saturn and Jupiter.

Such commensurations were in use not only in the Postclassic manuscripts but also in the inscriptions of the Classic period. In both, the Maya joined simple numerological calendrics with their use of astronomical periodicities. Lounsbury (1976) established a pattern of backward projection of current historical events to a prior event of the same sort – a birth projected back to a birth, an accession to an accession – occurring on a date that fell in the same point in various calendrical cycles. Often, the projections are quite large, a common variant being a projection back to a sacred era, just before the base of the long count; this base was apparently itself a projection from the katun 7.6.0.0.0 backward by two cycles of $365 \cdot 1440 = 73 \cdot 7200$ days each, commensurating the vague year and the katun (Jus-

teson *et al.*, 1985, n.32). Other examples, recognized as such since Lounsbury identified the basic pattern, occupy small segments of real historical time, and refer back to genuine historical precursors on dates occupying similar positions in calendrical cycles.

This numerological use of planetary and lunar periodicities is frequent relative to other recognized astronomical references. Long-term cycles commensurating basic calendrical intervals with true astronomical synodic intervals, especially eclipse cycles, are amply attested in the Classic period. In the case of planetary cycles, Classic period usage is often essentially calendrical, involving the span of a canonical planetary year rather than approximate average planetary cycles: for long calculations, the whole-number values for synodic periods were presumably known to be inaccurate in fixing the date of a phenomenon, since the discrepancies become appreciable even during the career of an individual skywatcher; however, the synodic cycles of Mercury, Jupiter and Saturn were difficult to commensurate effectively within the calendrical framework (see Section 8.3). One text, Caracol Stela 3, records several intervals that do not closely approximate an average number of planetary synodic cycles or commensurate planetary with calendrical cycles, but do fall within the range of variation of synodic intervals of two planets at once (Kelley and Kerr, 1973, pp. 197–201; Kelley, 1975, 1977a, 1983, pp. 184–93); this suggests observational rather than canonical or predicted linkages with the cycles of the planets involved. There are many dates of planetary and lunar phenomena, explicitly marked as such in accompanying texts, that reflect the observations on which longer-term generalizations were based.

MAYA MODEL BUILDING

The basic approach to temporal predictive astronomy among the Maya was in long-term, cyclical commensuration of synodic cycles with basic calendrical cycles; such commensurations were codified by Postclassic times, some having Classic traces, with substantial observational records dating at least to the Classic period. By systematically applying this approach ourselves, its potential for developing useful predictive models can be determined; in some respects, the kind of astronomical knowledge codified in the manuscripts can be clarified. Section

8.2 treats these topics for lunar records, the most common celestial records explicitly designated as such in the texts; Section 8.3 treats planetary records. In both cases, the predictive knowledge exemplified by the Postclassic tables can be seen as emerging via attempts to commensurate the ritual calendar with other cycles.

8.2 Lunar models

Commensuration with the ritual calendar

Several multiples of 260 days commensurate the 29.530588 average synodic period of the moon (Table 8.1). All commensurations accurate to within $\pm 2\frac{1}{2}$ days constitute either an eclipse interval or an interval halfway between successive eclipse intervals; and the category alternates with successive commensurations.[7] Each off-phase commensuration is three months longer than one eclipse interval, and three months longer than another (when doubled, of course, each is a full eclipse cycle; half are to within $\pm 2\frac{1}{2}$ days, the rest deviating by $\pm 2\frac{1}{2}$–5 days). Thus, when the Maya attempted to rationalize the lunar month with the ritual calendar, that calendar highlighted eclipse recurrence intervals; the recovery of good eclipse cycles was *imposed* on the Maya astrologer, even if he did not immediately notice the eclipse relationship. The eclipse cycle of 405 lunar months in the *Dresden Codex* emerges as the shortest to rationalize the lunar synodic month with the sacred round to within less than a day; to improve the fit, a 4772-month cycle is required – almost 12 times longer.

These lunar/sacred round cycles could have been known quite early, especially if not only eclipses but also the monthly disappearance of the moon was an occasion for ritual prognostication. As soon as one of these cycles was applied to a lunar eclipse series – consecutive visible eclipses occurring every six months for two or three years – the dates reached would be a series in which lunar eclipses would again be visible, though with gaps in as many as half the positions. Thus, applying the cycles to lunar eclipse series provides immediate success in long-term eclipse prediction. Conversely, Mesoamerican astronomer–priests must soon have noticed that relatively small multiples of 260 days often separated pairs of visible eclipses, and that it is always the same multiples that do so.

Since eclipses were times of ritual danger,

Table 8.1. *Commensurations of ritual calendar with lunar months to within $\pm 2\frac{1}{2}$ days during a span of four calendar rounds.*

Filled circles mark commensurations used in the Classic period, the 405-month *Dresden*/Palenque cycle and the 1937-month Copan cycle; unfilled circles mark other commensurations that are likely to have been in use; and daggers follow multiples of the *Dresden*/Palenque cycle. Italics mark commensurations with the 360-day year. None of the cycles commensurate the 365-day year (and thus the calendar round), but a close approach occurs at four calendar rounds: $2571L = 4CR + 3.14d$, an eclipse interval. The clean alternation between commensurations falling into an eclipse cycle, and those three months out of phase with an eclipse cycle, was presumably known to the Maya; see also note 7.

Eclipse intervals $\pm$ three months	Eclipse intervals
$44L = \;\;\;5\cdot 260 - 0.65$	$88L = \;\;10\cdot 260 - 1.31$
$\circ 132L = \;15\cdot 260 - 1.96$	$229L = \;26\cdot 260 + 2.50$
$273L = \;31\cdot 260 + 1.85$	$317L = \;36\cdot 260 + 1.20$
$361L = \;41\cdot 260 + 0.54$	$\bullet\;405L = \;46\cdot 260 - 0.11$
$449L = \;51\cdot 260 - 0.77$	$493L = \;56\cdot 260 - 1.42$
$537L = \;61\cdot 260 - 2.07$	$634L = \;72\cdot 260 + 2.39$
$678L = \;77\cdot 260 + 1.74$	$722L = \;82\cdot 260 + 1.08$
$766L = \;87\cdot 260 + 0.43$	$810L = \;92\cdot 260 - 0.22\dagger$
$854L = \;97\cdot 260 - 0.88$	$898L = 102\cdot 260 - 1.53$
$942L = 107\cdot 260 - 2.19$	$1039L = 118\cdot 260 + 2.28$
$1083L = 123\cdot 260 + 1.63$	$\circ 1127L = 128\cdot 260 + 0.97$
$1171L = 133\cdot 260 + 0.32$	$1215L = 138\cdot 260 - 0.34\dagger$
$1259L = 143\cdot 260 - 0.99$	$1303L = 148\cdot 260 - 1.64$
$1347L = 153\cdot 260 - 2.30$	$1444L = 164\cdot 260 + 2.17$
$1488L = 169\cdot 260 + 1.52$	$1532L = 174\cdot 260 + 0.86$
$1576L = 179\cdot 260 + 0.21$	$1620L = 184\cdot 260 - 0.45\dagger$
$1664L = 189\cdot 260 - 1.10$	$1708L = 194\cdot 260 - 1.76$
$1752L = 199\cdot 260 - 2.41$	$1849L = 210\cdot 260 + 2.06$
$1893L = 215\cdot 260 + 1.40$	$\bullet 1937L = 220\cdot 260 + 0.75$
$1981L = 225\cdot 260 + 0.09$	$2025L = 230\cdot 260 - 0.56\dagger$
$2069L = 235\cdot 260 - 1.21$	$2113L = 240\cdot 260 - 1.87$
$(2157L = 245\cdot 260 - 2.52)$	$2254L = 256\cdot 260 + 1.95$
$2298L = 261\cdot 260 + 1.29$	$2342L = 266\cdot 260 + 0.64$
$2386L = 271\cdot 260 - 0.02$	$2430L = 276\cdot 260 - 0.67\dagger$
$2474L = 281\cdot 260 - 1.33$	$2518L = 286\cdot 260 - 1.98$

Mesoamerican astrologers would surely have used these ritual-eclipse cycles to project from a visible eclipse to its next few recapitulations in ritual time; after adjusting by up to $2\frac{1}{2}$ days to capture the correct date of full moon, these would reliably warn of certain dates as eclipse possibilities. Table 8.2a illustrates the results of such projections, applied to eclipses visible in the Yucatan peninsula in the fifth century AD; the figures in the table represent the distances between projected stations. In addition,

Table 8.2a. *An eclipse warning table based on eclipses visible during the fifth century* AD *in the Yucatan peninsula.*

Stations are situated at the dates of visible lunar eclipses, and at full moons occurring within ±2½ days of 10, 26, 36 or 46 sacred rounds after a visible lunar eclipse; figures are the distance from the last station to the current station. Brackets indicate that a station was the occasion of a visible lunar eclipse that was not projected; boldface indicates visible eclipses that were projected. Stations projected from eclipses, but on which no eclipse was visible, are in normal type. Superscripts indicate which ritual-eclipse cycles join the station to a prior visible eclipse, 1, 2, 3 and 4 corresponding, respectively, to 10, 26, 36 and 46 sacred rounds. Spans between eclipses are from Aveni (1980, Table 18), with columnar format keyed to his Table 19; selection among 176-, 177- and 178-day spans is also based on Aveni's Table 19, to facilitate comparison; they have not been checked against computed dates of full moons.

AD 400

[1920]	354^{1}	[326]	177^{24}	177^{3}	[176]	177^{2}
[502]	178^{1}	177^{1}	502^{1}	177^{234}	178^{13}	177^{2}
177^{1}	[679]	177^{1}	177^{23}	502^{1}	177^{124}	325^{14}
[354]	[176]	[177]	178^{23}	177^{24}	177^{12}	177^{1}
[178]	178^{2}	178^{1}	177^{123}	177^{34}	177^{23}	177^{12}
[1211]	177^{1}	177^{2}	177^{234}	177^{1234}	325^{4}	177^{14}
176^{1}	177^{1}	177^{23}	177^{34}	177^{14}	177^{13}	177^{24}
502^{1}	679^{1}	325^{1}	325^{1}	178^{12}	177^{24}	177^{234}
177^{1}	[178]	177^{13}	177^{2}	502^{4}	177^{12}	177^{2}
	[354]	177^{1234}	177^{1234}	177^{134}	178^{14}	148^{1}
	354^{2}	177^{1}	177^{3}	177^{1}	177^{1}	177^{13}
	325^{1}	177^{3}	178^{12}	177^{24}	502^{13}	177^{24}
	177^{12}	178^{3}	177^{23}	178^{34}	177^{1}	177^{124}
	177^{123}	325^{1}	[502]	[176]	178^{14}	177^{123}
	176^{1}	177^{1}	177^{34}	177^{2}	176^{13}	177^{23}
	177^{12}	177^{2}	177^{1234}	325^{3}	177^{13}	177^{23}
	177^{2}	177^{12}	178^{234}	178^{24}	178^{234}	[325]
		177^{12}	177^{34}	177^{34}	[325]	177^{1}
		177^{23}	177^{14}	[177]	177^{124}	177^{123}
		177^{234}	502^{13}	177^{13}	177^{3}	[177]
		[325]	177^{234}	[177]	177^{3}	177^{3}
		177^{14}	178^{4}	325^{1}	178^{2}	177^{34}
				177^{1234}	177^{24}	177^{3}
					177^{2}	325^{4}
					177^{4}	177^{23}
						177^{23}

AD 500

it is assumed that the ritual-eclipse cycles were unknown until detected in that century. Accordingly, the first few eclipses cannot be anticipated; 14 viable eclipse stations occur before the shortest of the ritual-eclipse cycles has elapsed, 38 before the next shortest. In fact, eclipses were seen 10, 26, 36 and 46 sacred rounds after the first visible eclipse in fifth-century Yucatan. Thereafter, 37 out of 46 visible eclipses are anticipated by projecting ritual-eclipse cycles from prior visible eclipses, a success rate that actually begins somewhat earlier, after the first 25 years of eclipse observation. Together with the warnings for dates on which eclipses did not occur, 113 of 148 eclipse stations are projected after this point; and the structure is clear enough that the 35 missing stations could be filled in, apart from some uncertainty in the placement of the 148-day spans. Two or three decades of eclipse observation and recording are necessary and sufficient to produce a model for the timing of eclipses so complete that a system for anticipating all eclipse-possible dates would be revealed – a model essentially identical in structure to that of the *Dresden Codex*. Since the lunar commensurations not forming eclipse cycles are spans three months larger and smaller than eclipse cycles, these intervals might also have been applied to the same problem; the smallest periods generated thereby, of ±3 months, are the 41- and 47-month divisions of the saros eclipse cycle (47 + 41 + 47 + 41 + 47 months). Use of the eclipse cycle ±3 month groupings would generate an eclipse warning table such as Table 8.2a more quickly and completely.

A similar sequence of eclipse warnings can be projected into any arbitary period by using the smallest few ritual-eclipse cycles, if records have been kept for at least 33 years (46 sacred rounds) before that period. I constructed such models for each katun from 8.13.0.0.0 through 10.0.0.0.0. Any given katun contains 41 or 42 eclipse stations; for comparability, projections were into a span, not of a katun, but of 42 eclipse stations. Each projection was forward from a visible eclipse, by the four smallest ritual-eclipse intervals (10, 26, 36 and 46 sacred rounds); projection therefore begins with eclipses visible 46 sacred rounds prior to the first day of the katun. About 80 per cent of the eclipse stations were projected by the ritual-eclipse cycles, most of them by more than one. By including also the off-phase commensurations (5, 15, 31 and 41 sacred rounds) ±3 lunar months, about 90 per cent of the stations are projected, most both by off-phase projections and by ritual-eclipse cycles.

Since the pattern of six-month groupings, with rarer five-month groups, emerges clearly in the last two-thirds of Table 8.2a, it could readily be reconstructed for the few cases of 11-, 12-, 17- and 18-month groupings, apart from some leeway in the placement of the five-month groups within the 11- and 17-month subdivisions. In fact, the century charted in the table is more than twice the length necessary to work out the pattern; forward projections into years 25–40 (column 3 and the lower part of column 2) would be quite adequate. Thus, a good model for predicting all eclipse-possible dates, applicable to both solar and lunar eclipse prediction, could have been generated solely via commensurations of the lunar synodic month with the ritual calendar given records of less than half a century; such tables could well have been known by the beginning of the Early Classic in southeastern Mesoamerica.

Even without extrapolating a complete eclipse warning table, successfully projecting so high a proportion of observed eclipses would have to have structured the astrologers' conceptions of the eclipses they had *not* anticipated. Referring again to our fifth-century Yucatan model, most of the eclipses that were not anticipated could be postdicted from subsequent eclipses, by *backward* projection of ritual-eclipse cycles (Table 8.2b); they were 'normal' eclipses, in that visible eclipses follow them at ritually appropriate intervals. In fact, 72 of the 75 lunar eclipses visible in Yucatan during the fifth century are separated from another visible in the same century by an eclipse interval of 10, 26, 36 or 46 sacred rounds, none longer than the Dresden's ritual-eclipse cycle. Given that they were considered normal eclipses, how might they be anticipated? Backward projection would immediately yield the answer: almost every one could be projected forward by a ritual-eclipse cycle from stations that they *had* projected, but on which no eclipse had been seen. In this way, they were indistinguishable from the eclipses projected from visible eclipses, and were captured by the same ritual constructs.

Such a synthesis of projected and unanticipated eclipses must have affected the astrologers' conceptions of the projected stations on which *no* eclipse was visible; since eclipses could be reliably projected forward from them, they were probably considered occasions of some sort of eclipse phenomena – not simply as mistaken predictions. Furthermore, by projecting forward from projected eclipse stations, *every* eclipse station occurring after the establishment of the two smallest ritual-eclipse cycles is captured, in the fifth century, developmental model and

in the mature, katun models (even without the off-phase projections).

The lunar calendar in the inscriptions

Accompanying many long count dates are records in a Maya lunar calendar. Teeple (1930) showed that lunar months were enumerated, from first to sixth moon numbers. The days within a lunar month were enumerated, roughly from new moon; moonages from 3 to 29 are attested. Finally, the total number of days in the lunar month was specified as 29 or 30. Other information, of an uncertain

Table 8.2b. *An eclipse warning table projected from lunar eclipses visible in fifth-century Yucatan.*

The table is a fuller version of Table 8.2a: all spans reach forward to eclipse stations; boldfaced spans reach visible eclipses projected, forward and/or backward, from visible eclipses, with daggers marking those reached only by backward projections; and italicized spans reach stations projected backward from visible lunar eclipses. Gaps are introduced to highlight positions at which eclipse stations are not projected.

AD 400

	177		177	177	*176*	177
354	177	326†		177	**178**	177
177	178	**177**		*177*	177	
		177	502		177	325
		177†	177	325	177	**177**
		178	178	177	*148*	177
	679†	177	177	177	177	177
	176†	177	177	177	177	177
	178	325	177	177	177	177
1211	177	**177**		178	177	177
177†	177	177	325	*177*	178	148
	502	177	**177**		177	177
		177	177	325		177
502†		178	177	177		177
177	177		178	177	502	177
	178†	325	177	177	177	177
354†	*177*	**177**		178	178	177
[178]	177†	177		176†	176	
	177	502†	177	177	[325]	
	354	177	177	*148*	178	177
	177	178	177		177	
679	325	177		178	[325]	[177]
177	177			177	**177**	**177**
177	177	325†		177†	177	177
177†	176	177		177	177	177
176	177	177		177†	178	
	177	177				325
	177			325	177	
502			177	177		
177						

AD 500

85

character but correlating with the moon number, was given in Glyph X. Occasionally, in place of the moonage a compound appears that was formerly treated as corresponding to a moon number of 0; along with an absence of any moonage record, it corresponds to the period of the moon's invisibility (F. G. Lounsbury, personal communication, 1978; L. D. Schele, personal communication, 1985).

Different records of the same date sometimes differ in moon number and/or moonage. Within a site, only one instance of a given date is normally contemporaneous; because the others are several years before the date of the monument on which they appear, discrepancies here may be due to back-calculation from contemporaneous records via a formal system. Between sites, the duplicate dates are always civil year endings, mostly contemporaneous; these discrepancies reflect different models or different parameters of shared models (for moonages, also different observational results). The systems can only be worked out in detail on a site-by-site basis.

There is scant opportunity to determine arithmetically whether moonages specify days elapsed within a lunar month, or the current day (see Lounsbury, 1978, p. 776). Readings of hieroglyphs, however, suggest that they were elapsed; I read the usual moonage compound as 'n (days) have passed'. The verbal suffix habitually associated with the count of days seems surely to represent either -$i(h)$ or -$i\check{s}$ (Fox and Justeson, 1984, pp. 58–62), either one of which indicates action already completed; the usual verb seems to be 'to pass' (Justeson, Norman and Hammond, 1988; cf. Schele, 1982, pp. 83–5; Stuart, 1984).

The starting point of this count is almost always within a day or two of new moon; disappearance, new moon, and reappearance all remain under consideration. Deviations of three to four days in the placement of moonages relative to each other suggested to Lounsbury (1978, p. 774) that moonages were calculated from different bases at different sites, and at different times within the same site. The count was from reappearance in texts explicitly stating that the moonage is the number of days since the moon was born (Lounsbury, 1978, p. 774); 'the birth of the moon' (e.g., colonial Yucatec *yí?x ù·h*) is a Mayan metaphor for the first appearance of the crescent after new moon. However, this starting point was probably not usual when it was not expli-

citly stated: moonages of one and two days are apparently never recorded, evidently falling in the period of invisibility;[8] new moon or the moon's disappearance must frequently have been the starting point. Lounsbury (1978) takes estimated new moon (conjunction) as the alternative to reappearance, while A. Schlak (unpublished manuscript, 1985) argues for disappearance; the differences involve a two-day discrepancy in the correlation constant each prefers. New moon ties better to the strongest candidate for a solar eclipse date; and, in the early system of moon numbering reconstructed below for Tikal, new moon was probably the first day of the lunar month.

That moon numbers range from 1 to 6 indicates that months in the lunar calendar were enumerated in groups of six or less; this, in turn, suggests that that calendar was at least loosely based on eclipse seasons, though not necessarily upon a formal system of eclipse cycles. The earliest evidence for calculations involving the cycle of the *Dresden* eclipse table is from AD 683 (Lounsbury, 1978, pp. 775, 811). Most of the Maya area was then using the Uniform System of numbering moons in groups of six, with no five-month groupings as in the eclipse cycle (Teeple, 1930); it was a purely formal system. Teeple hypothesized that this system replaced one keyed more faithfully to an eclipse cycle, although this has not been demonstrated in the intervening years.

Reviewing all lunar records prior to the ostensible introduction of the Uniform System, I find support for Teeple's hypothesis in the earliest records at Tikal. Records there, through 9.4.13.0.0, are consistent with an enumeration of elapsed lunar months, with the last month in an eclipse cycle beginning at new moon on the date of a solar eclipse possibility. Tentatively, the early Copan sequence appears to have been in an eclipse cycle, offset by three months so that the date of a solar eclipse would be assigned to the third month; a similar system may have been present after the end of the Uniform period at nearby Quirigua. If this is so, it may relate to the lunar month commensurations with the lunar calendar in a span three months offset from an eclipse cycle.

Further support for Teeple's hypothesis is provided by the deviant moon number in the earliest known lunar calendar record (J. S. Justeson, P. Mathews and F. G. Lounsbury. The chronological portion of the Seattle stela and the early history of the Maya eclipse calendar, unpublished manuscript, 1982). A miniature Preclassic stela in the Seattle Art Museum records a moon number of 17, with no compound for the moonage. The date, 8.8.0.7.0 (AD 199), occurs after 47 + 17 lunar months had just been completed since the last solar eclipse visible in the area. Apparently, the subdivisions of the saros cycle were recognized at the time (hence the cycle was begun after 47 months were completed), though not necessarily the cycle itself; the 47-month group could be a 44 + 3 month period, involving the shortest lunar/sacred round commensuration, with three months added to relate it to the eclipse cycle. This record suggests that the subdivisions of the saros cycle had been recognized, although the further subdivisions had either not been recognized or not incorporated into the lunar calendar, by AD 199 among the Lowland Maya.

Since no moonage is recorded on the Seattle stela, the reference is ostensibly to invisibility; given Thompson's correlation constant, 584283, one day had been completed since new moon, so the moon would not have been visible. If the month count was of elapsed months, then the first month of an eclipse cycle included a solar eclipse station at new moon; otherwise, the last month included one. If the moonage was counted from reappearance and the 584283 correlation is correct, the eclipse station fell near the end of the month, and the moon numbering system was essentially equivalent to Tikal's. If moonage was counted from disappearance or new moon, or if the 584285 correlation is correct, the eclipse station fell near the beginning of the month, and the Seattle system of moon numbering was distinct from Tikal's. Unfortunately, because the monument was recovered by looters its source is unknown.

The Uniform System replaced the eclipse cycle system at Tikal during the Early Classic. Satterthwaite (1958a, pp. 132–3) and Jones and Satterthwaite (1982, p. 57) postulated its adoption between 9.4.0.0.0 and 9.4.13.0.0, since the latter and all subsequent Tikal dates are in the Uniform System, while the former date is not.[9] Subsequently, John Graham (1972, pp. 106–14) established a different Uniform System as having been used at Altar de Sacrificios (it was also at neighboring El Pabellón) until the Uniform Period. This was a system of regular six-month groupings, as in the classical Uniform System, but the moon numbers were smaller by one. Both Satterthwaite's and Graham's observations are overlooked in the recent literature on the subject, which still maintains the late introduction of Uniform moon numbering. The only other sites with many pre-Uniform moon numbers are Pusilha and Piedras Negras; Pusilha also used Uniform moon numbering from the start. At all other sites near Tikal, which have only one or two pertinent texts, the Uniform System is found back to 8.16.0.0.0, and the Altar de Sacrificios variant seems to be found elsewhere along the Usumacinta (and at Uaxactun, which Mathews (1986) suggests had ties with Yaxchilan). The Uniform System was evidently in use throughout the Maya area from the beginning of the Early Classic, a conclusion reached independently by Schele (personal communication, 1985) in her examination of the Early Classic lunar records.

A few sites, with only one or two legible pre-Uniform lunar dates, do not agree with Uniform coefficients; they usually deviate by only one. These discrepancies could reflect different uniform bases, or eclipse cycles, but with only one or two dates at each site the discrepancies cannot be profitably investigated. Piedras Negras is the exception: it fits neither a uniform system nor an eclipse cycle, yet it has seven pre-Uniform moonages with known long count positions. Although there is still not enough evidence to work out the details, the Piedras Negras system seems to advance about one month per five civil years faster than the Uniform System, as though one five-month period was introduced during this period. There are enough dates at five-year intervals to show that the same moon number occurs after 59 lunar months (60 in the Uniform System); 59-month groups would have to consist of nine six-month periods and one five-month period. 59 months form an eclipse recurrence interval, one that occurs in each 135-month division of the *Dresden's* 405-month eclipse table, but discrepancies in assignments of moon numbers to true eclipse stations begin to appear late in the second group of 59. The system would normally provide the same moon number to all eclipse stations during a given hotun,[10] and thus may have served as a short-term eclipse calendar; perhaps the Uniform System itself began as an accommodation (12 months long) of an eclipse cycle to the civil year.

Implications of the early data are that eclipse cycles were being tracked, with accurate predictive models, during the Preclassic and in the first half of the Early Classic – at least in the interval from AD 199 to AD 514. From about AD 350 onwards, however, most sites adopted simpler systems of moon numbering. Most common was the Uniform System, in the Tikal orb, with a minority variant along the Usumacinta having moon numbers smaller by one; the majority variant may have enumerated current months, the minority elapsed months.[11] After c. AD 750, some sites are thought to have attempted to adapt the lunar calendar to the eclipse cycle once again. This hypothesis rests mainly on deviations from uniformity; moon numbers have not been shown to conform to stations in an eclipse cycle.

Although moon numbering does not generally fit into an eclipse-cycle system, eclipse cycles commensurating the sacred round were used in long backcalculations to compute moonages associated with mythological dates. At Palenque, the ratio of 405 lunar months to 46 sacred rounds was the basic cycle, implemented by a ratio of 43 30-day months to 38 29-day months (Lounsbury, 1978, p. 775) that produces a theoretical synodic month of 29.530864 days. The cycle at Copan was different, a ratio of 79 30-day to 70 29-day months corresponding to a theoretical month of 29.530201 days. Although, according to Lounsbury 'It has not been determined how the astronomers of Copan arrived at this ratio or on what theory it was based', the Copan system is simply a different implementation of the same theoretical apparatus as was in use at Palenque: a long eclipse cycle was commensurated with the ritual calendar.[12] The difference was in the commensuration chosen. The Copan cycle was one of 1937 lunar months, nearly 157 years; this span falls on the average just 18 hours short of 220 sacred rounds, and gives precisely the 79:70 ratio of 30-day to 29-day months.

Lounsbury (1978) notes that seven ratios involving small numbers of 30- and 29-day months give better results than the Copan system, and that five give better results than the Palenque system. Similarly, the *Dresden* and Copan ritual-eclipse cycles were not the shortest available. It appears that the lunar cycles used by the Maya were those for which the number of months in the ritual eclipse cycle could be subdivided evenly by one of the short and accurate month ratio systems. Table 8.3a displays the most accurate commensurations from Table 8.1, in order of accuracy, along with the 30- to 29-day month ratios they impose. Only one of the eclipse cycles (a doubled half-cycle) commensurates with the ritual calendar with greater accuracy than the Palenque and Copan systems. The difference was not significant; it would take over 8000 years for its average performance to better Palenque's system by a day, and over 200 to better Copan's. Further, the most accurate short-term behavior of that system requires a 30- to 29-day month cycle about ten times longer than the Palenque and Copan systems. So Copan's and Palenque's systems were not measurably inferior to any other commensuration of eclipse cycles with the ritual calendar; and, among these commensurations, they admitted the simplest systems of month-length alternations. Short and accurate, the Palenque/*Dresden* system was the best practical system available to the Maya for long-term use.

Table 8.3b shows that the commensurations of the ritual calendar with lunar positions three months (half a cycle) out of phase with an eclipse cycle almost all involve these long 30- to 29-day cycles; the exception, a 32-month system, is detectably less accurate only after about a century. Finally, Table 8.3c displays the many month-length alternation systems less than 200 lunar months long, together with the shortest eclipse cycle they fit into evenly, in order of the accuracy of the synodic period estimates they induce. The only ones that have been recognized in the texts are the Palenque and Copan systems, suggesting that commensuration with the ritual calendar was of prime importance in selecting among the possible alternation systems. Note, however, that such systems are represented so poorly that this restriction does not strongly support a preference for ritual-eclipse cycles over off-phase commensurations. Indeed, the 132-month off-phase commensuration fits the month alternation scheme of the *Dresden* eclipse table (as the 81-month Palenque system does not), perhaps because the related 132 + 3 month Metonic cycle is a major subdivision of the table (and the only eclipse cycle to subdivide the 405-month ritual-eclipse cycle).

The Copan and Palenque systems represent equally viable solutions to the formalization of eclipse seasons via the same essential model: both were extremely accurate in their arrangement of

Table 8.3. *Ratios of 30-day to 29-day lunar months and corresponding ritual calendar commensurations.*

Synodic month estimates differ depending on their derivation from ratios of the two month lengths vs. ritual calendar commensurations. (a) and (b) list all ritual calendar commensurations yielding synodic month estimates accurate to within ±0.001 day in a span of four calendar rounds or less; in (a), daggers mark cycles that are twice a commensuration of months with sacred rounds in an eclipse cycle ±3 months. (c) lists all month length ratios whose synodic month estimates are accurate to within ±0.001 day and that involve systems of alternation in a cycle of fewer than 200 lunar months. Six systems appear in both (a) or (b) and (c), simultaneously achieving an accurate month alternation system and a short and accurate commensuration of the ritual calendar with the lunar month; filled circles mark month ratios that correspond to ritual-eclipse cycles, while empty circles mark ratios that correspond to ritual half-eclipse cycles. At least two of these systems, both eclipse cycles, were used in the Classic period. Notations are the same as in Table 8.1.

Month ratio		Synodic estimates		Error
30:29	months:rounds	30:29	days in cycle months	SR − true
(a) Eclipse cycles				
621 550	2342 266	29.530316	29.530516	−.00027†
43 38	405 46	29.530864	29.530864	+.00028
79 70	937 220	29.530201	29.530201	−.00039
203 180	1532 174	29.530026	29.530026	−.00056†
668 591	2518 286	29.530580	29.530374	+.00079†
26 23	1127 128	29.530612	29.529725	−.00086
1121 992	2113 240	29.530525	29.531472	+.00088
(b) Eclipse cycles ±3 months				
633 560	2386 271	29.530595	29.530595	+.00001
150 133	1981 225	29.530540	29.530540	−.00005
209 185	1576 179	29.530457	29.530457	−.00013
656 1581	2474 281	29.530315	29.531124	+.00054
1098 971	2069 235	29.530691	29.531174	+.00059
17 15	1664 189	29.531250	29.531250	+.00066
335 296	1893 215	29.530903	29.529847	−.00074
(c) 30-day:29-day month-length alternation systems				30:29 − true
•26 23	1127 128	29.530612	29.529725	+.00002
87 77	9676 1099	29.530488	29.530798	−.00010
○95 84	537 61	29.530726	29.534451	+.00014
61 54	7475 849	29.530435	29.530435	−.00015
69 61	12740 1447	29.530769	29.530612	+.00018
96 85	9412 1069	29.530387	29.530387	−.00020
•43 38	405 46	29.530864	29.530864	+.00028
○35 31	132 15	29.530303	29.545455	−.00029
103 91	5626 639	29.530928	29.530750	+.00034
○60 53	678 77	29.530973	29.528024	+.00039
•79 70	1937 220	29.530201	29.530201	−.00039
77 68	4640 527	29.531034	29.530172	+.00045
44 39	7387 839	29.530120	29.530256	−.00047
94 83	5133 583	29.531073	29.530489	−.00049
97 86	17019 1933	29.530055	29.530525	−.00053
53 47	8100 920	29.530000	29.530864	−.00059
○17 15	1664 189	29.531250	29.531250	+.00066
62 55	5265 598	29.529915	29.530864	−.00067
71 63	5494 624	29.529851	29.530397	−.00074
80 71	3020 343	29.529801	29.529801	−.00079
89 79	6216 706	29.529762	29.530245	−.00083
93 82	15575 1769	29.531429	29.530658	+.00084
98 87	15170 1723	29.529730	29.530653	−.00086
76 67	8866 1007	29.531469	29.530792	+.00088
59 52	3108 353	29.531532	29.530245	+.00094
101 89	9500 1079	29.531579	29.530526	+.00099

lunar months over the short term while admitting full accuracy over the testable long term. The difference in the accuracy of the Copan and Palenque commensurations of the ritual calendar with the eclipse cycle amounted to only one day in over 800 years (i.e., it was not noticeable). The Palenque eclipse cycle of 46 sacred rounds has the smallest absolute error, in days, of any single cycle short of 542 sacred rounds; the Copan cycle has the smallest error of any other single cycle of its length or less. Copan's system was inferior to Palenque's: its commensuration took five times as long, and its 30- to 29-day month alternation schema was almost twice as long.

The most serious alternative to the Palenque and Copan systems would be a ritual/eclipse cycle of 128 sacred rounds, about a day less than 1127 lunar months. It is one of the more accurate of these cycles, the only one other than Copan's and Palenque's with a short enough month alternation system to be viable. Although a less accurate ritual/eclipse commensuration than Copan's, it takes almost 200 years for it to accumulate an extra day of error. Its advantages are that it is half the length of Copan's, and the 1127-month cycle subdivides into 23 segments of 49 months; the 49-month month alternation scheme is at once the most accurate and the shortest available. The 1127-month or 128-round cycles are important candidates to watch for as investigations into Classic period commensurations continue. No good evidence for the 49-month cycle has been published, but Lounsbury (personal communication) has discovered one indication of its use. On Palenque's Tablet XIV, Chan Bahlum's apotheosis on ritual day 9 Ahau is related to the birth of the god G-I of the Palenque Triad on the day 9 Ik 3000 years earlier, which is linked to a primordial event on the same day nearly a million years ago. The distance between the latter two dates is $2 \cdot 11 \cdot 41 \cdot 1447 \cdot 260$; in terms of Maya numerology, the equivalence of 1447 days with 49 lunar months ($= 1446.999$ days) is highly suggestive; $2 \cdot 41 \cdot 260$ is also a ritual-eclipse cycle, although not a particularly accurate one.

Some sites may not have used commensurations to the ritual calendar to fix their lunar ratios. At Uaxactun, the magnitude of error in the estimate by which a moonage of three days was projected back to 7.5.0.0.0 suggests that they used a ratio of $9:8$, $25:22$ or $100:89$ 30- to 29-day months in their cycles, unless their alternation system was much longer than elsewhere; the two shorter periods also form eclipse intervals, of 17 and 47 months, while the longer is an eclipse half-cycle. The 17-month alternation system fits within the 493-month eclipse cycle ($= 17 \cdot 29$ months), commensurating 56 sacred rounds. However, the approximation is poor by the standards of Copan and Palenque; a difference of at least four days would have accumulated since Uaxactun's lunar records began, while the 405-month Palenque cycle would still have shown no discrepancy. Similar results are found for the 47-month alternation system. A very long alternation system, using a month ratio of $620:551$, also agrees with the Uaxactun data and fits into the most accurate eclipse cycle in Table 8.3a, of 2342 months (though a $621:550$ ratio is more accurate); but Uaxactun's astrologers are unlikely to have projected moonages using this uniquely long and recognizably inaccurate system of month projection in place of the very accurate eclipse cycle system it presupposes. Finally, seven eclipse cycles of 405 months each can be subdivided into 15 groups of 189 months, but the 81-month cycle provides detectably better short-term results while fitting within the eclipse cycle ($5 \cdot 81 = 405$). Thus, Uaxactun was probably not making use of eclipse cycle commensuration with the ritual calendar.

An alternative in the spirit of the lunar month commensurations is that the projection involved commensuration of an eclipse half-cycle, transformed by adding ± 3 lunar months into an eclipse cycle – here, of $44 + 3$ months. If so, a 47-month cycle is suggestive of the Preclassic system of the Seattle stela. Otherwise, the projection may have been based solely on a short-term, formal arrangement of 30- and 29-day lunar months (more likely in 17 or 47 than in 189 months).

Finally, Classic inscriptions record another lunar cycle whose nature is undetermined. The Glyph X cycle consists of 12 known forms, half of which are *prima facie* variants of each other; they are designated 1, 1a; 2, 2a; 3, 3a; 4, 4a; 5, 5a, 5b; and 6a. They are not enumerated periods, since they lack numeral coefficients; they appear to function as names. Teeple (1930, pp. 61–2) observed that a given moon number normally admitted only two possible forms of Glyph X, and a given form of Glyph X normally occurs with only two moon numbers. The same correlations of the moon numbers and Glyph X forms is found, whether or not five-month groups are interposed among the more frequent six-month groups. Accordingly, the Glyph X cycle is keyed directly to the moon number cycle. Because there are at least five different systems for moon numbering before 9.12.15.0.0 – the Tikal eclipse calendar, two separate Uniform Systems, the Piedras Negras 59-month count, and the more obscure Copan system – Glyph X must also be treated on a site-by-site basis. Linden's (1986) attempt to predict the coefficient of Glyph X in a pan-Mayan system based simply upon the long count date while maintaining a strict correlation of Glyph X and Glyph C variants is therefore internally inconsistent. However, his proposal that Glyph X reflects an 18-lunar month cycle might be accommodated by supposing that it was reduced to a 17-month cycle when a five-month group occurred.

8.3 Predicting planetary motions

The success in anticipating visible eclipses, and the accuracy with which lunar phases could be anticipated over long spans of time, must have confirmed for the Mesoamerican astrologer the essential linkage of ritual time with sacred time – with the activities of the deified sun and moon. Its application to the motions of the planets was a logical extension. The much broader range of variation, in days, separating recurrences of the same events in the cycles of the planets would make their short-term motions more difficult to capture via projections of long-term commensurations; the only universally recognized planetary table includes no mechanism for dealing with such variation, nor, in my opinion, does the Mars table. Immediate success would again emerge with application of the ritual cycle to the motions of two of the planets, in fact within a single planetary cycle: three sacred rounds closely approximate the synodic period of Mars, and one approximates the 263 days of continuous visibility of Venus, as morning and evening star. While this augured well for the results to be obtained by the traditional commensuration, longer-term planetary cycles could not emerge so readily as did the lunar eclipse patterning; and, for Saturn and Jupiter, the commensuration would never be very successful. This may have been a serious cosmological problem for the Maya. But, applied to Venus, the brightest of the planets, the success of the standard approach

would be impressive: the canonical synodic period of 584 days commensurated the vague year in the same span as it did the ritual calendar, and it did so 13 times during that span.

The Venus table

Figure 8.1 is a schematic of the five pages of Venus table stations. Three lines of vague year dates are given, lines 14, 20 and 25. The last entry of each (13 Mac, 18 Kayab, and 3 Xul) is the starting point (basedate) of a Venus cycle; the three recorded bases correspond to successive historical bases near the heliacal rise of Venus, about a century apart. From a given station, one proceeds forward in time to the next station by the amount specified in line 26 below the station reached (e.g., by 90 days from 19 Kayab to 4 Zotz'). The ritual calendar dates of the stations are given in lines 1–13. The first pass through the five pages of the table uses the first line of ritual calendar dates, and each subsequent pass uses the next line below; the final entry of the 13th line is the basedate in the ritual calendar, 1 Ahau. The span of the table is thus 65 Venus years (= two calendar rounds).

Canonical heliacal rising The structure of the Venus table is such that its base is located at a date of predicted heliacal rising. Its content is such that this base occurs on a day 1 Ahau in the ritual calendar. 1 Ahau was the name of the god Venus; the day was probably sacred to the god, whenever it occurred, and was presumably of special significance when it occurred on the date of any special event in the life of the planet, such as heliacal risings and settings or stationary points and renewed motion. This combined structure and content imposes much of the remaining detail of form and content of the tables.

To have a regular system for anticipating heliacal rises on 1 Ahau, recourse might be had to records of such risings; but in any long series the pattern is quite erratic owing to variations in the time between successive inferior conjunctions, in periods of invisibility after inferior conjunction, and in viewing conditions (although the Maya could obviously have taken account of the latter in developing a model to handle the first two types of variation). However, it would have been sufficient to look instead at data concerning *all* ritual calendar dates and their recurrences *near* heliacal rising to get a good idea of when a new base could be anticipated.

Such recurrences can take place in very short order: four synodic periods average 4.31 days short of nine sacred rounds, and perfectly commensurate it after the 580-day year in the five-Venus year cycle $(587 + 583 + 583 + 587 = 2340 = 9 \cdot 260)$; depending on the position in the five-year cycle and with an unusually long disappearance interval, it can return very rarely after eight or 53 Venus years. However, it takes records of about a century to recognize when the recurrence of heliacal risings on roughly the same date in the ritual calendar would *regularly* occur, and another few decades to smooth out the variations into a regular system based on integer values. Apart from the near miss after four Venus years, the same point in the Venus year normally first returns on the same SR only after 57 or 61 Venus years, averaging, respectively, 3.53 days later or 0.78 days earlier. Any earlier pattern is off, on the average, by more than five days. So regular recurrences are noted only after a bit over 91 and 97 solar years – just under two calendar rounds (104 vague years).

These spans define the period during which the table *had* to be used from one 1 Ahau basedate before shifting to a new 1 Ahau basedate for heliacal rising – at least 57 Venus years – and the times (57 or 61 Venus years) at which corrections would have to be instituted. This period was convenient in terms of Maya analytic constructs: the least common multiple of the 584-day Venus year and the 260-day ritual calendar defines the length of the table, with no correction; this period was two calendar rounds, just four to eight Venus years longer than the timing of the base correction.[13] So the period of a 'complete' table was just long enough to accommodate the period after which the first basedate substitution would have to be made. A series of numerals permitting calculation of one basedate from another is provided in the preface to the table, p. 24.

The synodic periods of Venus go through a regular, symmetrical pattern of variation in length that repeats every five Venus years: 583, 587, 580, 587, 583. Thus, from a particular basedate, a given page of the Venus table always corresponds to the Venus year variant. The table, however, consistently registers the *average* figure, 584. Although this failure to register regular variation in the synodic period that so perfectly fits the structure of the table has occasioned comment and various types of explana-

tion, the Maya *could not* incorporate these variations and still be able to recycle the table by recapturing new 1 Ahau basedates. When basedate substitution recovers 1 Ahau as a date of heliacal rising, it also produces a phase shift of one or two Venus years within the cycle of five Venus years. Thus, if the five pages corresponded under the first base to Venus years with synodic periods 583, 587, 580, 587, 583, and the basedate is shifted after 57 Venus years, the pages would subsequently correspond to the series 580, 587, 583, 583, 587; a basedate shift after 61 years would give the series 587, 580, 587, 583, 583. So the regular 584-day intervals are simply the consequence of the reinstallation of 1 Ahau basedates for heliacal rising after two calendar rounds had almost run their course.

This regular interval, in turn, constrains the precise values of the adjustment numbers. Since any pair of computed heliacal risings are separated from 1 Ahau by some multiple of 584 $(= 20 \cdot 29 + 4)$, only five days of the 20-day veintena, separated by intervals of four days each, could serve as dates of any of the four stations of the Venus year. Since the basedate shift must preserve this feature when recapturing a heliacal rising on 1 Ahau, the numbers used to shift basedates have to be less than an even multiple of 584 by a small multiple of four days[14] (any multiple of 260 $(= 4 \cdot 65)$ will differ from a multiple of 584 by *some* multiple of four). The table of basedate adjustment factors consists of a series of four numbers; they are read from largest to smallest, as usual for multiplication tables).[15] All are multiples of 260 days, as necessary to lead from one day 1 Ahau to another. The three largest represent the numbers of days necessary to move from one date of heliacal rising on 1 Ahau to another, roughly correcting for accumulated errors; all three adjust by a multiple of four days less than a multiple of 584, as required. They could be used to calculate new bases from each other or from the canonical 1 Ahau 18 Kayab base, and thus are related directly to the structure of the table.

The first and smallest in the series of adjustment numbers is 9100 days, written 1.5.5.0. Like the other three, it is a multiple of 260 and thus leads from one day 1 Ahau to another. However, it disagrees with the other three intrinsic characteristics of the other three numbers in the adjustment table: (1) It does not approximate a multiple of 584 and thus cannot lead from one heliacal rising date to another. Either there is an error here, or the use of the number was markedly different. In moving from one 1 Ahau date to another, at least one of the dates involved must not be heliacal rising but presumably some other Venus station. (2) The rough magnitude of this number is such that about one day's error would have accumulated in the relation of the 584-day count to the average period of 583.92 days; if it is a correction factor of some sort, it corrects for an error of about one day. This difference is related to the first, since the deviations by multiples of four were a consequence of leading from one heliacal rising date to another within the table. (3) 9100 is not a small multiple of four days less than a multiple of 584; it factors as $15 \cdot 584 + 340$ or $16 \cdot 584 - 244$.[16]

Partly because 9100 does not lead from one date of heliacal rising to anywhere near another, and because it is unique among the correction factors in this regard, few discussions of the table's structure accommodate it; Thompson (1950, p. 225) posited scribal error and reconstructed another figure (also deviating in feature 3).[17] The recorded figure joins recorded stations of the table (canonical settings of morning and evening star); this is probably relevant, since only two other multiples of 260 can (along with their differences from $146 \cdot 260 = 65 \cdot 584$, the span of the table), and suggests the recorded number be accepted as written. Since it is in series with the basedate adjustment factors, its use was most likely related in some way to theirs, although it could not be used itself to effect such adjustments.[18] Its precise function remains enigmatic, but one likely class of hypotheses is developed on pp. 96–7. The Venus cycle was associated with eclipse cycles in a variety of ways (see *The Venus-eclipse cycle*, pp. 96–8). It is argued below that the number 1.5.5.0 was used with the 4.18.17.0 and 9.11.7.0 adjustments to locate eclipse intervals within the table, and that it may relate historically to the 1 Ahau 18 Uo basedate.

In summary, the attempt to regularly capture heliacal risings on the day 1 Ahau generates the regular 584-day lengths of the Venus years, and thereby the sequence of heliacal rising dates from a given basedate for two calendar rounds (after which the cycle, with no shift of base, would repeat). It also fixes the timing and magnitude of the correction factors. The epoch of these innovations can be secured via Lounsbury's reconstruction of the

rationale for the 1 Ahau 18 Kayab base of the table. Evidently no correction was applied during the first cycle through the table from that base, in AD 934, but a correction using the procedures registered in the preface to the tables was applied to institute the 1 Ahau 18 Uo basedate in AD 1129, and regularly thereafter near the end of the table, each 57 or 61 Venus years. So the uniform 584-day period was instituted with the 18 Uo base, if it was not in place earlier for other reasons.

The other stations The remaining structure of the table is defined by the internal subdivision of the 584-day periods into four portions. These subdivisions of the cycle are of the same length in each of the five pages of the planet's cycle. The lack of variation of the subdivision probably reflects the consistent 584-day length of the overall period rather than requiring separate explanation in terms of the basedate shift or some other factor; given the uniformity assigned to the Venus year, there is scant rationale for variation in the subdivisions. Because the invariance of the subintervals imposes or presupposes the uniformity of the 584-day period, the subintervals too were presumably defined by AD 1129. It is possible that the subdivisions were reached by different intervals at that time than they were when the next base, 13 Mac, was instituted, *c*. AD 1227; since that base was evidently the current one when this copy of the table was written down (see Lounsbury, 1983b, p. 7; Kelley 1983, p. 177), the present intervals were established no later than AD 1227.

The final station in a given Venus year, reached by a count of eight days, provides the base for the count to the first position of the next Venus year; the final position in the table gives the 1 Ahau base of the entire table. The eight-day interval is identified with the period of invisibility leading up to heliacal rising as morning star because it matches the average duration of the period; Aztec sources specify this canonical length for the period; and the associated iconography agrees with the ethnohistoric discussions of the activities of the gods on that day.[19] The subsequent intervals approximate the periods of visibility as morning star (*c*. 263 days, here 236); invisibility around superior conjunction (*c*. 50 days, here 90); and visibility as evening star (*c*. 263 days, here 250). Under this system, visibility as morning star is projected to end 17 to 29 days too soon (an average of 23 days early); visibility as evening star is projected to begin three to 14 days too late (an average of nine days late).

Since the approximations to the visibility and invisibility periods fall outside the range actually observed, either these periods are not what the subdivisions are intended to monitor, or they, like the 584-day units, were adopted to effect some kind of accommodation to another cycle.

Gibbs (1977) has proposed that the unusual units reflect ritual implications of certain days in the veintena for the stations selected; the chief evidence is that half of the days of the veintena do not occur in the table, while all would occur if the canonical series of intervals were the average values 263 + 50 + 263 + 8. She notes more specifically that there is a special structure to the placement of these veintena dates with respect to Venus stations: veintena days on which the evening star could set are the same as those on which the morning star could rise or set; and the veintena days on which the evening star could rise are midway between the veintena dates for the other stations. With various additional assumptions, she shows that the intervals recorded are the closest to the canonical values that could be used.

Although the schema developed by Gibbs is possible in principle, and has been favorably received (cf. Aveni, 1980, p. 190; n.d.), no positive data strongly supports or even suggests it. External support for the most basic assumption, that the days are ritually selected, is extremely tenuous: the absence of an assignment of certain veintena days to any station. Because the position of each Venus round is offset four days in the veintena from the position of the previous round, five different veintena days out of the 20 will occur with each station, one on each page; this gives at most four sets of names available for any station under any system. Given the eight-day period correctly attributed to invisibility around inferior conjunction, the set of veintena days for heliacal setting as evening star at the end of one Venus round is *forcibly* the same as for heliacal rising as morning star at the beginning of the previous Venus round; this agreement is not imposed by special ritual concerns, except that the specific set involved is imposed by assignment of helical rise to a 1 Ahau basedate. Any one of the four sets could occur with either of the other two stations, giving 16 possible set assignments. Six

of these assignments place one or the other of the stations at the same veintena sets as the stations around inferior conjunction, while another puts both in this set; and ten out of the 16 assignments restrict the possible veintena days in the table to at most half (one to a quarter) of the 20. So the assignment of sets found in the codex, using only half the days of the veintena, is typical of a chance assignment, i.e. one not based on concerns for days of the veintena. Given the actual assignment of veintena sets to Venus stations, there are 25 possible assignments of specific veintena days to a given Venus year; 20 per cent of these would place the specific day in the set for setting of morning star halfway between the day for the preceding and following stations, and of course this is only one of many 'special' formal patterns that are possible. So the assignment of sets is typical of their possible assignments, and the assignment of days within the actual sets is typical of their possible assignments. There is no evidence here for special selection of veintena dates having influenced the structure of the table.

There is positive evidence for another line of explanation. The tabular deviations from the reference points in the Venus cycle result in dates at which the planet was visible near the point of invisibility; the hieroglyphs above each day-count have been interpreted as referring to the visibility of the planet. Further, Teeple (1930) noted that the temporal distances were roughly equal to multiples of lunar months, or a half lunar month greater: $236 = 8 \cdot L$, $90 = 3 \cdot L + 1\frac{1}{2}$; $250 = \cdot 8\frac{1}{2} - 1L$. Since the last figure is imposed by the other two, given the eight days assigned to invisibility, with $1\frac{1}{2}$ days leeway around an exact lunar month the likelihood of getting such agreement by chance would be at most $\frac{4}{29}$ for the 90-day case (87–90 being within $1\frac{1}{2}$ of $88\frac{1}{2}$) and $\frac{3}{29}$ for the 236 day case (235–237 being within $1\frac{1}{2}$ of 236); the likelihood of getting both at lunar month intervals would be no more than $12/841$, less than 1.5 per cent. Teeple's observation therefore seems very likely to be relevant.

Putting these observations together, the essential structure of the table is determined: the first station specifies the last date on which the morning star is still visible when the moon is at the same phase as when it rose heliacally; the next specifies the first date on which the evening star is visible when the moon is again at that phase; the next station specifies the expected date of heliacal setting of the evening star, one half lunar phase offset from the previous three dates; and the last station specifies heliacal rising of morning star eight days later.

This system accounts for the difference between the *Dresden* intervals and the natural periods they approximate, and in particular for the fact that unequal intervals are used to approximate the 263-day morning star period and the 263-day evening period.

The apparent interest in placing the stations of a Venus year at a fixed position in the lunar month is tantamount to an association of the Venus stations with an eclipse cycle.[20] Although not part of astronomical reality, a relationship of Venus to eclipses is verified by the ethnohistoric association of Venus with eclipses, thought to be due to the planet's appearance during solar eclipses before any star or other planet. More to the point, there was a definite *calendrical* association of the Venus cycle with eclipse cycles: the calendric commensuration of the sacred round with the Venus cycle in 65 Venus years effectively commensurated it with eclipse cycles (see *The Venus-eclipse cycle*, pp. 96–8), and historic basedate corrections were made by means of shorter Venus-eclipse cycle commensurations with the sacred round. Kelley (1977a) has postulated such longer-term Venus-eclipse cycles as constructs in the inscriptions, and Aveni (n.d.; see also below) has shown that the historical basedates of the Venus table can be linked to a cycle of lunar eclipses visible in Yucatan.

Given the long-term calendrical Venus-eclipse cycles preserved in the Venus table, a shorter-term association of Venus stations with an eclipse cycle is plausible. It is most strongly supported by the specific number of lunar months specified in the Venus table: $8 + 3$ lunar months and $3 + 8\frac{1}{2}$ lunar months are spans frequently separating, respectively, pairs of visible lunar eclipses and a lunar–solar or solar–lunar eclipse sequence. The placement of Venus stations at consistent lunar phases would capture short-term repetitions of eclipses within a Venus year, if they happen to occur at the appropriate stations.[21]

The one oddity is that 88 or 89 days might be expected where 90 is consistently recorded, and the 250-day figure correspondingly increased by one or two to complete the sum of 584 days (alternatively, that the 250-day figure is one or two days short of what would be expected, and the 90-day figure being correspondingly increased from $88\frac{1}{2}$). The expectations are based on average lengths of lunar months, of course, and three synodic months can amount to 90 days observationally. This suggests that the model was based on a specific historical point at which the 90-day figure was correct. It could have been either an observed length of three lunar months, or the correct length within a 584-day grouping projected from a basedate. That the point was significant enough to incorporate an unusual lunar interval within a lunar phase recurrence system into the essential structure of the table suggests that it was the date of an eclipse associated with the corresponding Venus station – one that occurred 326 days after the canonical date of Venus' heliacal rising as morning star with a new or full moon, before the current base of AD 1227. It may be significant, then, that 326 days ($=236 + 90 = 178 + 148$) form an eclipse recurrence interval.

Eclipses visible in the Maya area rarely occur at the margins of Venus' visibility around superior conjunction. One of them is of special interest. On 11 January, AD 1126, a lunar eclipse occurred that was visible in the Maya area, *c.* $236 + 90$ days after Venus rose heliacally on a full moon (uncertainties concerning viewing conditions make it impossible to know the exact date of past heliacal risings); probably more to the point, it was exactly on the 326th day of the third 584-day Venus year before the 1 Ahau 18 Uo base.[22]

This association is the more significant because Lounsbury found that this date was one on which the correction mechanism was inaugurated. The previous basedate, 1 Ahau 18 Kayab, was one on which Venus and Mars rose heliacally and in conjunction as morning stars. The event led to the backward projection of six calendar rounds, the common multiple of the Venus and Mars canonical rounds, to a prior 'historical base', verifying the relevance of the Mars association. Although a new 1 Ahau basedate for heliacal rise can be instituted near the end of two calendar rounds, the Maya apparently passed by the opportunity to do this on 1 Ahau 8 Yax, after 61 Venus years; they permitted the Venus calendar to slip out of phase with the planet's cycle. Lounsbury suggests that the celestial display associated with the prior base was considered too special to permit human tampering.

A further reason may be suggested, based on Aveni's (n.d.) results: the 8 Yax date was not closely preceded by a lunar eclipse, whereas both the 18 Kayab basedates were preceded by a lunar eclipse a month before the basedate, as were the prior 18 Kayab basedates projected back through six calendar rounds.

The next feasible basedate occurred 57 Venus years after the second 1 Ahau 18 Kayab base, on 1 Ahau 18 Uo,[23] 61 Venus years since the missed 8 Yax base. 18 Uo was instituted as a base, recorded as such in the preface to the tables. Lounsbury argues that it did so because it recapitulated the display of 18 Kayab but with Jupiter in the role of Mars; it was also preceded by a lunar eclipse a month earlier, as the 18 Kayab bases had been. Whether these events provided a celestial mandate for the correction scheme, or simply rendered the original base a bit more prosaic, the corrections were regularly applied thereafter: 61 Venus years later a 13 Mac base was instituted, the current base whose vague year stations are recorded in the uppermost of the vague year sequences; 61 Venus years later, a 3 Xul base would be instituted, provided in the lowermost of the vague year sequences. These had neither the planetary nor eclipse correlates of the previous bases. The next expected base would be 1 Ahau 8 Ch'en, 57 Venus years later.[24]

Accordingly, the lunar eclipse on the 326th day of the Venus round just prior to the one ending on the new base, 18 Uo, heralded both the heliacal rising of Venus as evening star and the inauguration of the new base of the Venus calendar. To complete the parallels between the bases, the 18 Kayab base of AD 934 had itself been preceded by an eclipse, this one a solar eclipse, that occurred in the lunar month just before heliacal setting of morning star (i.e. at the first station of the Venus table); it was on the 238th day of the seventh Venus round before 1 Ahau 18 Kayab (on or near the 236th day since heliacal rising). The following basedate would be heralded by an eclipse at the following Venus station. All of this strengthens the parallelism of the 18 Uo base with the 18 Kayab base, and thereby supports Lounsbury's proposed rationale for the installation of the 18 Uo basedate. Evidently, it was the occasion for the values assigned to the intervals within the canonical Venus year. It was probably also the occasion for their approximation to lunar month groupings in eclipse cycle intervals. Just as

the Maya had been observing the occurrence of eclipses during the month just before the last station of the planet in the lunar month preceding the basedate of the Venus cycle (Aveni, n.d.), they had also observed eclipses during the month just before heliacal setting of the morning star before the original 18 Kayab base and in the month after heliacal rising of evening star before the 18 Uo base. The four significant stations in the planet's visibility were thereby provided the same kind of formal relation to eclipse recurrence, canonical rising as morning and evening star being separated by an 11-month eclipse interval, canonical setting as morning and evening star by an $11\frac{1}{2}$-month eclipse interval.

In summary, the subintervals of the Venus year, the ritual calendar dates associated with each, and the corrective mechanisms for reaching new base dates were all established with the installation of the 1 Ahau 18 Uo basedate. The essential difference of the form of the table present at that time is that a sequence of vague year positions would have been given that was calculated from the 18 Uo base, and the line of vague year dates calculated from the 3 Xul base would not have been present. The 18 Uo line would have been the uppermost line of vague year dates and the 13 Mac line the lowermost, if the association of the upper line with the current base and the lower line with the coming base was conventional.

The Venus-eclipse cycle   It is not only the short-term eclipse recurrences within a single Venus year that relate the Maya conception of Venus to eclipses. The Maya Venus cycle is intimately linked with an eclipse cycle by virtue of its length of two calendar rounds. This span is 13.19 days longer than an integral number of lunar months, and the position of the nodes advances by a little over five days in that time. Passing the two calendar rounds of the canonical Venus cycle can lead from a solar to a lunar eclipse, or conversely. Further, moving 13.19 days short of the Venus cycle, the nodes fall back about eight days; so, from postnodal eclipses, a Venus cycle less half a lunar month is an eclipse station, and from prenodal eclipses a Venus cycle plus half a lunar month is an eclipse station. The result is that the position of the eclipse station falls back by half a month for two Venus cycles, then moves forward a full month and thus reinstates an eclipse station within three or four days of the original position. Depending on the position of the nodes, there is a possibility of missing a lunar eclipse station every fourth cycle, when the node is roughly centered on new moon; in this case, a solar eclipse station appears about where the lunar eclipse station was two cycles earlier. Because the nodal limits are wider for solar eclipses, their stations are consistently reinstated. Finally, the fit to 65 true Venus years is even closer than to the canonical cycle; the node returns to virtually the same position, while the position in the lunar month advances by ′eight days.

Aveni (n.d.) has identified a historical sequence of lunar eclipses that not only fits this pattern, but does so in the lunar month just prior to the historic basedates of the *Dresden* Venus table. A lunar eclipse during the lunar month completed before the disappearance of Venus as morning star[25] is associated with each of the 18 Kayab basedates; the repetitions are a consequence of the Venus-eclipse cycle just noted. In addition, an eclipse occurred in the lunar month just before the 18 Uo basedate. However, this was not due to the Venus-eclipse cycle, for the distance to the corrected basedate was eight Venus years shorter than that cycle. An association of eclipses proximate to Venus basedates is linked in the previous section to the placement of the stations of the Venus cycle at eclipse recurrence intervals from each other within the Venus year.

Additional evidence of a linkage of eclipse cycles with the Venus cycle is preserved in the Venus table. The factors that serve to link basedates also function to link stations of the table in various eclipse cycles. The most obvious of these is the 4.12.8.0 factor; it joins the last historic 18 Kayab basedate to the 18 Uo basedate via a commensuration of 128 sacred rounds with an eclipse cycle of 1127 lunar months, the second or third best commensuration of the lunar and ritual calendars. It is because of this commensuration that the sequence of lunar eclipses occurring just before canonical disappearance of evening star just prior to each of the 18 Kayab basedates was also reflected in the same eclipse association for the 18 Uo basedate.

This correction is only associated with baseshifts after 57 Venus years. Among basedates registered in the table, only one involved such a shift, and this was a historical anomaly due to the failure to apply a baseshift after the first 18 Kayab base; 18 Uo could also have been reached as the second

of two basedate shifts of 61 years, if a shift had applied after the first 18 Kayab base. Corrections occurring after 61 years are exact multiples of the sacred round that do *not* commensurate the lunar month in an eclipse cycle. Thus, when a prior basedate occurred with the true heliacal rise of Venus, 128 sacred rounds overcorrects the accumulated error; its application is accurate only if a canonical basedate is maintained for the better part of two Venus cycles rather than one, or after four 61-year baseshifts have been applied. The former is the situation postulated by Lounsbury for the *Dresden*: the historical 1 Ahau 18 Kayab basedate at 10.5.6.4.0 was retained as the base of the next Venus cycle, at 10.10.11.12.0; the correction instituting the base at 18 Uo was applied to this now discrepant canonical base. Subsequent corrections were near the end of a single Venus cycle, after 61 Venus rounds; this interval does not approximate an eclipse recurrence interval, and the cycle noted by Aveni fails to continue just prior to the next two bases (13 Mac and 3 Xul).

In the table, the (current) basedate 13 Mac is reached by adding 9.11.7.0 to the 18 Kayab base; this factor is the sum of 4.12.8.0 (leading to 18 Uo) and a one-cycle correction of 4.18.17.0. It is here, I believe, that the significance of the problematic 1.5.5.0 correction factor can be discerned. Although the 9.11.7.0 correction is not an eclipse cycle, the 1.5.5.0 adjustment transforms it into one. 9.11.7.0 less 1.5.5.0 is 2025 lunar months – five lengths of the *Dresden* eclipse table; similarly, a one-cycle correction of 4.18.17.0 less 1.5.5.0 is an eclipse cycle of 898 lunar months commensurating 102 sacred rounds. The shortening links an eclipse station near the last 18 Kayab basedate to one in the 16th Venus year before the 13 Mac basedate. By adding rather than subtracting the interval, the nodal and lunar month positions both advance by ten days; thus, the interval leads from an eclipse station near the 18 Kayab basedate to ten days after an eclipse station in the 16th year of the new 13 Mac basedate.

The relevance of these positions is not only in their characterization as eclipse stations: 1.5.5.0 is also one of only six multiples of 260 that can lead from any station in the table to any other in less than one full Venus cycle. Specifically, it links canonical setting of morning star with a subsequent canonical setting of evening star. Being associated with the basedate correction factors linking the last 1 Ahau 18 Kayab base to the next two basedates of the table, the linkage between these setting dates is most plausibly of setting of evening star just before a basedate with a prior setting of morning star. If the linkage is of actual stations, the only historical position for the linkage would be between canonical setting of morning star on 1 Ahau 3 Zotz' in AD 1105 and canonical setting of evening star on 1 Ahau 18 Uo in AD 1129. The reason is that this was the only time that the new basedate fell on canonical setting of evening star, owing to the failure to institute a new basedate after the first full Venus cycle; normally, the new basedate fell four days after canonical setting of evening star and four days before canonical rise of morning star. However, this date-linking interval may have been more a formal device for indicating an important eclipse recurrence interval imbedded in the structure of the table, giving a longer-term eclipse relationship to the short-term eclipse cycles that defined the positioning of these stations in the first place; it may not be historically anchored at all.[26]

The 4.18.17.0 ± 1.5.5.0 interval links the original historical 1 Ahau 18 Kayab base to stations around superior conjunction during the same Venus cycle, the one ending on the new basedate of 18 Uo. Since the original basedate was preceded by a lunar eclipse a month earlier, they link these stations to prior eclipse stations, but neither was near a visible eclipse of the sun or moon. This suggests a formal relevance of long-term Venus-eclipse cycles to the canonical positions of the stations of the Venus table around superior conjunction in the years leading up to the 18 Uo basedate.

The pattern discovered by Aveni associates the Venus table basedates with visible lunar eclipses, highlighting the significance of the Venus stations as eclipse stations. In particular, the timing of the shift from an 18 Kayab to an 18 Uo basedate is such as to preserve an eclipse association of the basedates, or more precisely of the heliacal setting of evening star about eight days before the basedate, for this association was lost at the next 1 Ahau 18 Kayab; and the rationale for the placement of the Venus stations in the lunar month proximate to true setting of morning star and rising of evening star based on historic eclipses at those positions is clarified by the habitual association of visible lunar eclipses in the period of visibility during the lunar month prior to heliacal setting of evening star just before the basedates of each Venus cycle. If the 1.5.5.0 factor was used to derive longer-term eclipse cycles relating these same stations to heliacal rising basedates, that use connects with these interrelated eclipse recurrences near the significant stations of the Venus cycle during the years leading up to the establishment of the 18 Uo base.

Commensuration models for the other planets

The special structure of the Venus table is partly due to the special relation of its cycle with the 365-day year; its 5:8 ratio permitted recycling of the same vague year positions every fifth Venus year, with an adjustment for slippage by returning to the same position in the ritual calendar once that period had commensurated the true average synodic period. For the other planets, this kind of structure is not feasible.

Mars Willson (1924) proposed that the almanac on pp. 43–5b of the *Dresden Codex* functioned as a Mars table, and aspects of Bricker and Bricker's (1986) study of the table provide solid support for this interpretation. With good reason, a relationship to Mars has long been disputed. Apart from the table of multiples of 780 days in its preface, the most striking piece of evidence for this function was the ring number of 352 days, a period approximating the interval from conjunction to retrograde motion. Ring numbers are distances prior to the basedate of the long count, always reached by projecting backward from a significant historical date by a common multiple of 260 and another topically pertinent interval. This ring number is no exception, being the smallest multiple of 780 (= 3 · 260) that can reach to before the long count base; if the historical basedate was chosen for its intrinsic relation to the Mars cycle, the fact that the ring number achieved was 352 must be a coincidence. In fact, given that the basedate in the sacred round was 3 Lamat, *any* ring number had to be of the form 780 · n + 92, 780 · n + 352, or 780 · n − 78; the latter two correspond to canonical distances between significant points in the Mars cycle. Either the 3 Lamat base was contrived, or the connection of the ring number to the Mars cycle is coincidental, under Willson's hypothesis.

Willson's evidence that this table concerned Mars could therefore all be construed as coincidental.

Thompson (1950, 1972) also argued that the structure of the table provides positive evidence *against* a Mars interpretation. The most unnatural aspect of this hypothesis has been the fact that its basic length is 78 days, so that a subdivision into even tenths is being used; and each of these is further subdivided, about equally, into segments of 19, 19, 19 and 21 days in sequence. A table having 40 evenly spaced stations does not correspond to natural subdivisions of a planctary cycle. Secondly, even apart from the issue of the number of stations involved, the table's structure differs radically from that for Venus, the only unmistakable planetary table in the manuscripts, and accords with that of the non-astronomical almanacs.

Following the method of the previous subsection, it can be shown that the structure of the Venus table could not provide an appropriate model for Mars, and, based largely on work by Bricker and Bricker (1986), that the subdivision of the 780-day cycle neatly accommodates a ritual Mars cycle.

The 1:3 ratio of the Martian integral synodic period to the ritual calendar is so close to the true average synodic period that the error is just under one day per calendar round. This means that, once a station in the cycle departs from a given ritual calendar date, it cannot return in historical time. Thus, the standard Mesoamerican association of deities with a ritual calendar date was not feasible for Mars: corrections to a table of positions for the planet, keyed to dates within three calendar rounds, would have to return the same vague year position while shifting the ritual calendar position; and the essential structure of the table would give fixed ritual calendar dates for stations, which would cycle through diverse vague year positions. A fragment of such a model might look something like that of Table 8.4. Such a table would include 73 vague year positions for each station, each being a multiple of five days distance from every other position for that station; and a position 22 Mars years further along reaches a station exactly five days later in the vague year. A corrective mechanism therefore must balance five days of accumulated error. This requires about 79 (i.e., $5/(780 - 779.93651)$) Mars years, and thus could occur after approximately one pass through the entire table; it would be implemented by beginning the next pass on line 58 rather than line 7 (or perhaps on line 52 rather than line 1). After this length of time, the average position

Table 8.4. *A hypothetical Mars table, commensurating the vague year and calendar round.*

The stations are essentially arbitrary.

1	14 Pax	6 Zotz'	19 Kayab	11 Zec
2	4 Uayeb	1 Xul	9 Pop	6 Yaxkin
3	14 Uo	11 Mol	19 Zip ...	
		...		
72	...	11 Kayab	19 Ceh	16 Cumku
73	4 Kankin	16 Pop	9 Muan	1 Zip
	4 Imix	4 Etz'nab	4 Chuen	4 Lamat
	273	117	273	117

of Mars in the ritual cycle would have shifted five days in the ritual calendar; one line would have to be given for each ritual calendar base date under this shift, parallel to the vague year lines of the Venus table. However, it would take about 200 years before a single correction would be applied. Adjusting for accumulated average error of the Mars cycle within a reasonable time span therefore requires an approach that does not reuse expressed dates, if the ritual calendar is involved. The manuscripts express all dates in the ritual calendar, and their almanacs are always structured for reuse; this precludes construction of a Mars table adjusting for average accumulating error, the *raison d'être* of the Venus table. In a non-adjusting table of the Mars cycle, then, vague year stations are superfluous; a Mars year almanac would be constructed with its stations located in the ritual calendar only. The failure of Willson's proposed Mars table to accord with the structure of the Venus table is therefore unavoidable.

Because a system of simple ritual calendar stations is intrinsic to representation of the Mars cycle in the ritual calendar, the traditional structure of a divinatory almanac – among which astronomical almanacs must be numbered – was appropriate. One feature of these almanacs is that their stations span a divisor of their total span; the almanac is usually re-entered four, five, ten or 20 times before completion of a full round. The 78-day length of the repeating portion of the table accommodates subdivision into tenths; it also closely approximates the period of Mars' retrograde motion, as Willson noted. These two features are sufficient to make the 78-day structure quite a feasible one for a Mars table. The conformity of the proposed Mars table of pp. 43b–45b with the structure of non-astronomical almanacs is therefore to be expected; it does not argue against the validity of Willson's hypothesis. The exceptional appearance of the Venus table – presenting fixed stations in the vague year – arose because the Venus cycle does not accommodate ritual calendar stations; they would regularly repeat in the 260-day year only after prohibitively long intervals, while repeating reliably within the 365-day year at eight-year intervals for a relatively long time.

The features which have been seen as inimical to Willson's hypothesis are therefore predictable characteristics of a Maya Mars table. Furthermore, certain of Bricker and Bricker's (1986) results concerning pp. 43–45b strongly support the view that it represents the synodic cycle of Mars. The ostensible base of the table is 3 Lamat, for this is the day of the ritual calendar with which the prefatory table of multiples of 780 is associated, as well as that of the ostensible historical base of the table (9.19.7.15.8). The Brickers show that 3 Lamat was a significant anchor in the Mars cycle. In the Terminal Classic and Early Postclassic periods, the date of heliacal rising is restricted to the quarter of the ritual calendar starting three days before 3 Lamat, with the concentration at this end of the range being on 3 Lamat itself. Thus, the ostensible basedate of the table is evidently selected for its relationship to heliacal rise, just as the base of the Venus table was the date of heliacal rise. We can interpret 3 Lamat as a viable canonical date of Mars' heliacal rise during the era to which the manuscript pertains; it provides a *terminus ante quem* for predictions of heliacal rise.

The Brickers argue that the retrograde period of Mars is the emphasis of the almanac as a whole. This is supported by their observation that the historical base of the almanac precedes a retrograde period by only 50 days. This raises a question concerning the relevance of the 3 Lamat base, ostensibly related to heliacal rise, since the historical 3 Lamat base is the one 260 days *after* canonical heliacal rise. The answer, I believe, can be inferred from a consideration of this preceding 3 Lamat. If canonical heliacal rise was the computational base from which the almanac was entered, the beginning of observed retrograde occurred within a day or two of the beginning of the fifth period of 78 days, that running from 3 Ahau to 3 Etz'nab; so perfect an alignment of the retrograde period with the regular

tenths of the almanac is quite rare. This, I would argue, is the reason for the selection of this particular retrograde period as the one inaugurating the table, one that has the strength of accommodating both the treatment of heliacal rise as the tabular base and its apparent concern with retrograde motion. Further, just as 3 Lamat is a *terminus ante quem* for heliacal rise, the subsequent 3 Lamat provides a loose *terminus ante quem* for the retrograde period; this is because retrograde follows heliacal rise by an average of around 292 days.

This property suggests a function for 'aberrant' multiples of 260, those that are not multiples of the canonical planetary synodic period, that parallels the use of such numbers in the Venus table. In the latter, multiples of 260 that deviate from multiples of 584 are used to recapture appropriate dates of heliacal rise. In the case of Mars, multiples of 260 that deviate from being multiples of 780, by being added to or subtracted from the historical 3 Lamat base, *must* lead to dates of canonical heliacal rising: subtracting $780 \cdot n + 260$ leads backward, while adding $780 \cdot n - 260$ leads forward, to canonical heliacal rise on 3 Lamat.

Any such number is appropriate to capture canonical heliacal rising. In the Venus tables, the usage was more precisely to maintain the canonical date of heliacal rise on or near its observed date in the ritual calendar. In the case of Mars, the synodic period is quite variable. However, variations in planetary synodic periods are quite regular, defined by commensurations of the planet's sidereal period with the solar year. The difference $n - m$ between any two factors $780n - 260$ and $780m - 260$ must be such that $(n - m) \cdot 779.93651$ is within a few days of being a multiple of both the sidereal period of Mars (686.99576) and the solar year (365.2422); the Maya did note sidereal/synodic commensurations (see Section 8.5). The pattern for synodic Mars years is $(7+8+7+8+7) + (7+8+7) + (7+8+7+8+7)$, in which the 8's may be further subdivided as $1 + 7$. The aberrant multiples actually entered in the table are $17 \cdot 780 - 260$, $40 \cdot 780 - 260$, $4 \cdot 780 + 260$ and $93 \cdot 780 + 260$; interpreting them as reaching canonical heliacal rise from the basedate, they must be written $-4 \cdot 780 - 260$, $-93 \cdot 780 - 260$, $17 \cdot 780 - 260$ and $40 \cdot 780 - 260$. Their magnitudes do in fact differ by appropriate intervals, the separation of each from the largest $(40 \cdot 780 - 260)$ being $133 \cdot 780$, $44 \cdot 780$ and

$23 \cdot 780$, respectively. Thus, if any one of the aberrant multiples reaches an observed heliacal rise on 3 Lamat, all of them should be within a few days of it. Without detecting the pattern of intervals relating these aberrant multiples of 260, nor the systematic basis for such a relation given the position of the historical base with respect to canonical heliacal rise on 3 Lamat, Bricker and Bricker (1986, p. 74) discovered that forward counts of $17 \cdot 780 - 260$ and $40 \cdot 17 - 260$ do lead to dates on or near true heliacal rise; I calculate that $-4 \cdot 780 + 260$ and $-93 \cdot 780 + 260$ do as well.[27]

Accordingly, Willson appears to have been correct in proposing that pp. 43–45b of the *Dresden* treat the synodic cycle of Mars.

The remaining structure of this table is its subdivision into four segments of 19, 19, 19 and 21 days each. The Brickers' conclusions concerning this structure are at the very least controversial. They treat the almanac as a means for anticipating variations in the length of the synodic period, specifically with respect to the timing of retrograde. When observed heliacal rising falls into the first or last quarter of a 78-day span, a correction of 19 or 21 days is applied to the expected beginning of retrograde – otherwise no correction is applied – and addition or subtraction depends on whether heliacal rise is in the first or second half of the range of tzolkin dates on which it can occur. This model does place the widely varying positions of the retrograde period appropriately, but this is not a very compelling result. Details aside, some such model has to work: an extreme synodic interval is the cue, not anything structural in the table; and movement from an extreme quarter to a middle half can seldom fail to keep the retrograde period roughly centered in the 78-day span. The success of the model stems from the division of this span into roughly equal quarters, given that that span approximates the retrograde period. Accordingly, the table can be used in the way the Brickers propose, but there is no evidence that it was in fact so used.[28]

Moreover, such a model is far less accurate than models that would have been available to the Maya. With the amount of data that would have been used in constructing this scheme, they could not have failed to detect the regularity in the patterning of the variations. All visible planets undergo a regular, symmetric pattern of variation in their synodic periods that repeats when the sidereal and solar years

Table 8.5. *Mars table intervals (in days), in relation to the Mars cycle.*

Defining intervals	Average	Modelled as	Canonical
Helical rise to sp$_1$	292½	$3 \cdot 78 + 19 + 19 + 19$	292
			$10L - 3$
sp$_1$ to sp$_2$	75	$78(21 + 19 + 19 + 19)$	78
sp$_2$ to helical set	292½	$21 + 78 + 19 + 19$	294
			$10L - 1$
Helical set to helical rise	120	$19 + 21 + 78$	118 $4L$

sp = stationary point; L = lunar month.

commensurate; these commensurations perforce commensurate the average synodic year, and commensuration of this span with the solar year and/or the vague year is established for the Maya. Over a period of several hundred years, from the beginning of the Early Classic to near the end of the Late Postclassic, I find less than one day's variation in the length of a given variant for Mars; relatively short-term patterns of *c.* 15–22 Mars years give far more useful anticipatory results than the model proposed. Had the Mars table been divided so that seven or eight synodic periods were explicitly displayed, the pattern of variation could have been captured; reusing the same synodic chart for all variants, as in the Venus tables, prevents exploitation of the seasonal patterning of synodic variation.

Rather than providing a crude mechanism for tracking highly predictable variations, the subdivision of the 78-day unit into four segments of 19, 19, 19 and 21 days each – or some other subdivision into approximate 19½-day quarters of the 78-day span – is in fact required to accommodate the 78-day almanac to other stations of the Mars cycle. The points of observational interest are heliacal set and rise around the period of invisibility (lasting about $1\frac{1}{2} \cdot 78$ days), and the stationary points (about 78 days apart, and separated from the adjacent rise/set date by about $3\frac{3}{4} \cdot 78$ days). Thus, the sequence of 19, 19, 19 and 21 days, summing to 78, permit one to capture fairly closely the average positions of heliacal rising and setting and both stationary points. With heliacal rise after invisibility as the base, the sequence is as shown in Table 8.5. They can also accommodate the interposing of opposition between the stationary points, dividing the interval into rough halves of 38 and 40 days (the former more closely approximating $37\frac{1}{2}$); and the interposing of conjunction between heliacal set and rise, dividing 118 into 59 + 59 (19+21+19 + 19+19+21 in the scheme above). Finally, the period from stationary point 2 to stationary point 1 is roughly 10 + 10 + 4 lunations (295 + 118 + 295 days), a possible eclipse interval. In the four-station table, the essential results are independent of which point in the Mars cycle is chosen as the base, except that for two of the four possible bases the assignments of 292- and 294-day values are interchanged; other variations of two days would be possible in a six-station table.

Mercury, Jupiter and Saturn   The cycles of the other planets are less tractable, with respect to the calendar round, than are those of Venus and Mars. The pertinent data are summarized in Table 8.6. They show that it is not feasible to structure a table of positions for Mercury, Jupiter or Saturn in terms of the number of calendar rounds that commensurates the integral approximation to the synodic period: in all these cases, the accumulated error by the end of such a table would amount to several synodic periods, vitiating its usefulness; and the table would span a period far beyond historical time, making its construction an unverifiable extrapolation.

The ritual calendar does rationalize each true planetary synodic period within the span of the Venus table to within the accuracy of that table, i.e. to within less than 5.09 days in less than 104 years. The Venus table's corrective capabilities, to within 0.78 days in 137 sacred rounds (61 Venus rounds), can also be met for Mercury (0.70 days in 119 sacred rounds) and Saturn (0.99 days in 16 sacred rounds), and approximated for Jupiter (1·45 days in 112 sacred rounds). Finally, five calendar rounds rationalize with the average synodic periods of Saturn and Mercury, quite precisely for Saturn; recorded Saturn dates are ancient enough that at least the third such pentad was elapsing when the *Dresden Codex* was being compiled. The Jupiter period commensurates with 12 calendar rounds; one such had probably elapsed by the time of the *Dresden Codex*, although the Maya should have been able to project the commensuration via the rate of regression of the Jupiter cycle with respect to perhaps five or six calendar rounds, or by projecting forward from the separate commensurations of the vague year and the ritual calendar (571 = 3·161 + 88 Jupiter cycles = 3·247 + 135 sacred rounds; 571 Jupiter cycles = 2·237 + 97 = 2·259 + 106

Table 8.6. *Commensuration of ritual calendar and vague year with planetary cycles.*

Format is $n:m + r$, where n is the number of calendrical cycles, m is the number of average planetary synodic periods, and n = m periods + r days. Several commensurations are given in some cases, with increasing accuracy as higher numbers of units are involved in the commensurations. Boldface entries represent commensurations occurring in Maya planetary tables. In italicized (and some bold) entries, the specified number of calendar periods exactly commensurates the integral approximation to the planet's synodic period; for the true average synodic period, the accuracy in most of these cases is lower than for prior commensurations.

	Commensuration of true synodic period with		
	calendar round	ritual calendar	vague year
Mercury		4:9 − 2.90	20:63 − 0.28
		29:65 + 7.96	*116:365 + 44.70*
		37:83 + 2.16	
		41:92 − 0.73	
		119:267 + 0.70	
		160:359 − 0.04	
	5:819 − 3.70		
	29:4750 + 1.69		
Venus		9:4 + 4.31	8:5 + 0.4
		128:57 − 3.53	
		137:61 + 0.78	
	2:65 + 5.09	**146:65 + 5.09**	104:65 + 5.09
Mars		3:1 + 0.06	47:22 − 3.60
	3:73 + 4.63		*156:73 + 4.63*
			203:95 + 1.03
Jupiter		23:15 + 3.26	47:43 + 2.98
		112:73 − 1.45	106:97 − 1.77
		247:161 + 0.36	153:140 + 1.21
			259:237 + 0.56
		399:260 + 30.11	*399:365 + 42.26*
	12:571 − 2.88		
	399:18985 + 203.27		
	= 18980·399		
Saturn		16:11 + 0.99	29:28 − 1.58
		189:130 + 11.97	231:223 + 0.47
	5:251 − 1.11	365:251 − 1.11	260:251 − 1.11
		397:273 + 0.86	
	189:9487 + 260.44		*378:365 − 33.61*
	= 9480·378		

vague years; see Table 8.6).

So the important calendar cycles do commensurate in reasonable spans for tables, except for Jupiter in the case of the calendar round, and sometimes more precisely than did Venus. The kinds of structures that can be based on them are akin to the lunar tables, with a succession of synodic periods laid out end to end, not repeating until the entire sequence has been completed; the interlocking of the planetary year with the ritual calendar and vague year, seen in the Venus table and feasible for a Mars

Table 8.7. *Commensuration of the moon and planets with the 360-day civil year.*

Moon	Mercury	Venus	Mars	Jupiter	Saturn
$5:61 - 1.37$	$9:28 - 4.57$	$73:45 + 3.53$	$13:6 + 0.38$	$41:37 + 1.28$	$21:20 - 1.84$
$21:256 - 0.17$	$19:59 + 3.22$	$= 72$ vague years		$277:250 - 1.05$	$209:199 - 0.32$
	$28:87 - 1.35$				
	$75:233 + 0.53$	$133:82 - 1.58$			

table, was not available. Close commensurations to the 360-day year are also found, illustrated in Table 8.7. These were almost certainly noted and used, since all numerical records were kept in terms of it. At least the 13-tun Mars cycle, the 21-tun Saturn cycle, and the 73-tun Venus cycle would have been significant: a 13-tun period was given special importance by the Maya within the katun; the 21-tun period commensurates the lunar cycle to a fraction of a day; and the 73-tun cycle commensurates the tun with the vague year, while a 73-katun period appears to have been the basis upon which the basedate of the long count was fixed.

Possibly relating to planetary cycles is the compound epigraphers have dubbed Glyph Y, distinguished by its T739 mainsign and by its environment, between the glyphs of the nine-day count and the lunar series. In this environment, it takes a numeral prefix between 2 and 6, or takes no prefix. (In one instance, at Xcalumkin, it is preceded by a skull sign that represents the numeral '10', via a rebus on pCT, pYu *la·xu·n 'ten' vs. pCT, pYu *lax 'to end; to die'. In this case, it is evidently the 'end' reading that is appropriate, a conclusion reinforced by the death markings on the Glyph Y mainsign.) A Glyph Z, formerly distinguished as a 'companion' of Glyph Y, is simply an elaborate spelling of the numeral 5, as **5-bi-IH** (= either **5-(bi)-BIŠ** or **5-b(i)-IŠ**) for *ho-b'-iš* '5 (days) later'; its separation is due to the presence of the *-b'iš* suffix, restricted to use with numerals 5 and 7.

The mainsign of Glyph Y also occurs as one of the nominal hieroglyphs in the 819-day count passage, although the position of a date in the 819-day calendar appears to be uncorrelated with the coefficient of Glyph Y on that date. With the numerical prefix 9, it is G_6, indicating the sixth position in the nine-day count (J. Linden, 1981, Glyph Y of the supplementary series, unpublished manuscript, and 1982, The Supplementary Series,

unpublished MA thesis, Tulane University; F. G. Lounsbury, 1982, Glyph G_6 of the supplementary series, unpublished manuscript). These two contexts are related in that the basedate of the 819-day count falls on the last G_6 before the basedate of the long count; one usage is presumably a partial explanation for the other. Lounsbury and I have speculated that this count relates to the motions of Saturn and/or Jupiter, because of common factors (21 for Jupiter, $3 \cdot 21$ for Saturn) of their synodic periods with 819 and a high frequency of Saturn and Jupiter events associated with 819-day counts, and the factors of 819 (7, 9 and 13) show up in structurally parallel ways in canonical synodic periods of all the planets except Venus. This suggests that the 819-day count may have served as a formal tool for tracking the behavior of the planets, or perhaps the relative behavior of pairs of planets, in the short term, perhaps via seven-, nine-, 13- and/or 21-day counts.

If the 819-day count did have such a function, Glyph Y is both a likely candidate mechanism for tracking planetary motions and the most promising candidate for investigating this hypothesis. Its function has not been successfully identified.[29] It takes varying numeral prefixes, often occurs with suffixes indicating time counts, and occurs between various counts of days and lunar months; evidently it is a count of units of some sort. The day count suffix for *-b'iš*, specific to numerals 5 and 7 in Mayan languages and writing, is sometimes represented when the numeral prefix to Glyph Y is 5; otherwise, the distance number suffix for *-e·x* (or *-xe·y*), indicating counts of arbitrary units forward (or backward), is suffixed to the unit. Thus, it is not clear whether days within a unit or units are being counted, although probably it is uniformly one or the other. At this writing I have not successfully fit the data to any formal count of units having fixed length, and am still investigating the hypothesis via formal and naturalistic units of variable lengths reflecting subdivisions of planetary cycles and accommodations among cycles. Independently, Victoria Bricker (n.d.) suggests that it is a count of lunar months from or to Jupiter's period of retrograde motion, but this hypothesis fits only about two-thirds of the cases.

Useful constraints on a solution are provided by a series of relatively closely spaced records at Yaxchilan, from 9.16.0.0.0 through 9.16.10.0.0; also of special usefulness is the apparent 'end of Y' record at Xcalumkin, suggesting either the end of a unit of more than one day's duration or the possibility of more than six units in the count.

CULTURAL ASSOCIATIONS OF CELESTIAL EVENTS IN THE INSCRIPTIONS

Major celestial events were used to schedule human affairs, ostensibly as sacred mandates for elite decision-making. Postclassic astronomical tables provide temporal information concerning the timing of celestial events, showing that the elite could anticipate coming astronomical events; this would aid preparations for activities of their own selection in the guise of submission to the divine will. Although indicating the enhanced capabilities of the Maya elite for implementing their decisions about public action, the manuscripts say little if anything about those decisions. Thus, investigation of the astrological scheduling of activities depends primarily on Classic inscriptions, in which the events taking place on a given day are explicitly noted.

General content Maya historical texts are typically quite terse in their statement of events. There is usually a single main verb. Topics covered are few: birth, death, marriage, and genealogy; accession to office; performance of various rituals; and raiding and warfare. There are numerological exercises in which past, mythological events are projected backward by significant chronological units, evidently to manufacture 'precursors' for the current events, providing anachronistic 'precedents' for their timing. The only verbose parts of the texts are those naming the ruler: these may refer to him often by a series of epithets, e.g. a personal name, a series of titles of office, designations as captor of various individuals, ruler of his site, descendant of his royal predecessors, etc. In much of the area, each event was mentioned after the date of its occurrence, but

especially in the southern highlands many successive events might be mentioned without reference to the dates on which they occurred (a more typical pattern in the languages). Even in these texts, time anchors are given for the initial and final events, as well as for some intermediate events.

The most common types of event referred to in the hieroglyphic texts were the ceremonies conducted by the ruler at the end of the civil year. Usually, these events were recorded only at the end of the katun, a 20-year cycle, or more rarely at the quarter or half divisions of that cycle, or after 13 civil years had passed in the cycle. Other points, 'arbitrary' years in the katun, were seldom the occasion for the erection of monuments, although year-ending ceremonies were most likely performed at the end of every year. In the Postclassic, the year-ending (and year-opening) ceremonies were performed annually, but in relation to the vague rather than civil year.

Celestial events treated Few intrinsically astronomical signs or compounds are known: these include signs for 'sun', 'moon' and 'star' (including 'planet'; specific stars or planets are typically specified linguistically by compounds of the form noun + 'star', e.g. 'great star' (Venus), 'rattle star' (Pleiades)). As a result, the only firm evidence concerning the timing of human activity in relation to celestial events involves the sun, moon and planets. Zodiacal signs or constellations (see Section 8.6) are mentioned in a few texts. Research to date has neither focused upon nor established any pattern of scheduling according to stellar phenomena; however, events timed for fixed dates in the solar year could be connected with the position or visibility of a particular star on that date, and the year-ending rituals recorded on Stela 2 from Bejucal, Guatemala, may commemorate the appearance of a supernova that year. Similarly, dates of the appearances of comets are not securely tied to historical records; Halley's comet may be reflected in the taking of the royal name *k'uk'* by three rulers of Palenque who acceded four to six years after its appearance (A. Schlak, unpublished manuscript, 1985), but this pattern is probably coincidental.

The solar, lunar and planetary events are all, though not always, associated with rituals; apart from bloodletting, they do not involve offerings, but some are of an unknown character. So far as I am aware, eclipse dates are not recorded as dates of *other* activities, apart from such accidental associations as upon the birth of Chan Bahlum. Major stations in the cycles of the planets were occasions for initiating other activities, the prime examples being accessions and military campaigns. Most seem to have been anchored in observation rather than being scheduled according to formal or predictive cycles; this is likely, for instance, in Lounsbury's (1983a and this volume) examples of events occurring upon the first perceptible motion of Venus, Jupiter or Saturn from a stationary point. Such activities were presumably planned in advance: the occasions for their enactment would have been anticipated by rough prediction, and watched for; observation of the anticipated celestial event would then provide the signal for initiating human action.

8.4 Explicitly astronomical references

Almost all explicit lunar references in the inscriptions are records in the lunar calendar. Around 200 of these records are known, and a substantial minority are associated with non-year-ending dates. Although these records are calendric rather than astronomical, it would be useful to examine them for regularities in the time of month of the more frequent activity classes recorded in association with these dates, with particular attention to records near full or new moon. To my knowledge, this has not yet been done, and it is not attempted here; the extent to which phases of the moon may have structured the recorded activities of the Maya elite remains unknown. Explicit astronomical references to the moon are restricted to a few likely solar eclipse dates. One, at Poco Uinic, is marked by a solar eclipse sign quite similar to that of the codices (Teeple, 1930, p. 115); another, at Tres Zapotes, may be accompanied by solar iconography (J. A. Fox, 1986, The eclipse record on Tres Zapotes Stela C, unpublished paper presented at the Second Oxford International Conference on Archaeoastronomy, Merida, Yucatan; see p. 495 for abstract); a few seem to be marked by compounds whose root was a word for 'black(ness)' (Kelley, 1977b); and one or two are seemingly designated 'hidden lord' (L. Schele, personal communication).

The only recurrent astronomical references are those using the star sign; other events, referred to in text without the star sign, are depicted with it

in accompanying scenes. Since the star sign appears repeatedly in the Venus tables, usually in a compound read 'great star' (the Maya name of Venus) and rarely as the generic 'star', these events have been investigated for Venus correlates (Kelley, 1977a; Closs, 1979, 1981; Lounsbury, 1983a); as a result, stages in the cycle of Venus are the most securely documented planetary events. Other events linked to star events also show Venus correlations; Lounsbury (1983a) argues that many such dates at Bonampak are identifiable as Venus events, though not explicitly marked as such, both because of their relationship to explicitly indicated star events and because the sheer frequency of apparent Venus dates at that site seems too great to be due to chance (I have not tested this, however).

Venus events in the inscriptions

The events associated with star signs are quite limited in content. One, at Copan, is an heir designation, ostensibly of Venus, and its verb is the same one used, roughly, to indicate the planet's visibility in the *Dresden* Venus table. In the inscriptions, this verb is used only on occasions of accession, accession anniversaries, and heir designations. Although such events need not occur on Venus dates, associations of these events with Venus in some instances appear to be significant. At Bonampak and Piedras Negras, accessions and accompanying bloodletting rituals occurred on Venus dates within a few months of prior Venus dates, marked as such by the star sign; this association strengthens the case for the significance of the Venus dates of accession. At Palenque, the ruler dubbed Chan Bahlum acceded when Venus was at or near maximum elongation, and celebrated the fifth Venus-year anniversary of his accession when Venus was again at maximum elongation (A. Schlak, unpublished manuscript, 1985; Lounsbury, Chapter 19, this volume; Schele, 1982, had previously noted that this date was the eighth solar anniversary of his accession).

In almost all other events the star sign aids in the identification of what epigraphers have dubbed the 'shell star' and 'earth star' compounds. The essential referent of these compounds was first identified by Riese, in 1977 (cf. also Riese, 1982), as warfare; much hieroglyphic evidence has subsequently borne out Riese's conclusion, and the accompanying iconography is transparently war-related. V. Miller (1986, 'Star warriors at Chichen

Itza', unpublished expanded draft of a paper presented at the 1985 meeting of the Society for American Ethnohistory) shows that when star signs occur in iconographic context, the scenes are related to warfare or to human sacrifice in the ball game.

There are contrasts of content between the rituals and battles associated with Venus events, and those not so associated. Among rituals, only accessions, bloodletting and the ball game are clearly associated with Venus dates; offertory rituals, apart from bloodletting, are not, and bloodletting may be associated only or primarily in cases related to accessions and accession anniversaries. Contrasts among types of battles requires more extended discussion.

Soon after Riese pointed out the military implications of the star events, Mathews showed that these events refer to major intersite warfare: the 'shell star' events typically specify the site at which or against which war was waged, and at the site so designated hieroglyphic stairways were erected by the victors to commemorate the victory.[30] Mathews is probably correct in positing the T575 'shell' as a grammatical suffix on the verb, rather than contributing referential meaning, a view also held by Schele (1982, p. 101). The 'earth star' compounds are simple variations on the same structure, with 'earth' appearing where the place name would be expected. 'Earth' (comprising 'world', 'land', 'dirt' and 'territory' in Lowland Mayan languages) presumably functions as the place designation; it is not clear whether it refers to foreign territory, home territory, or both.

Mathews distinguished these cases of territorial conquest from the more common instances of armed conflict. These refer to the seizing of individual named captives (Proskouriakoff, 1960, 1963–4) by a verb generally read *čuk* 'seize'. Subsequent references to the ruler responsible would often designate him as 'captor of / who captured' these individuals (Proskouriakoff, 1960, 1963–4) – apparently via a word, **kan*, used in many Mayan languages for the hunting of game (Norman, 1984; cf. Houston, 1984) – or as 'he of *n* captives' (Stuart, 1985); rulers are not cited as conqueror or ruler of a conquered site.[31] Rulers and their associates are depicted in battle dress for these captures.

One battle or series of battles could be designated in both ways, as at Piedras Negras, Tortuguero and Yaxchilan. In none of these cases is a site name specified; Tortuguero specifies 'territory'. Other occasions of the taking of captives, though not designated as intersite war, were clearly not simple ambushes of isolated individuals, because several captives were taken and many of these episodes lasted for an extended period; and, in some cases, place names are cited in connection with the names of the captives. The events that are referred to via the ruler's taking of captives may be construed as raiding; and the taking of captives may have been more the propagandistic emphasis than the purpose of such raids, which probably involved the pillaging or despoiling of villages and lands. Especially since some cases are explicitly tied to warfare, and that at Tortuguero seemingly to 'earth', some of this raiding probably had a territorial dimension; since intersite warfare is not normally involved, this was probably aimed at wresting control over adjacent hinterlands, but they were not oriented to the actual capture of centres of power. Given the closer similarities to the raiding in chiefdoms than to war in state societies, and the propagandistic emphasis in the repeated references to the ruler as the captor of this or that individual, Webster's (1975) model for the role of territorial war at political boundaries in strengthening and maintaining a stable central authority appears apropos, although here in a primitive state rather than an advanced chiefdom.

Lounsbury (1983b; cf. Table 8.8, column b) showed that star events and, at Bonampak, raids were often scheduled for significant points in the Venus cycle. Now W. Nahm (unpublished, 1988; see also note 20) has demonstrated an even stronger tendency for both types of warfare to be scheduled for limited periods during the visibility of Venus. Mary Miller (1986) relates the Bonampak raids to the taking of captives in preparation for accession to office; since some level of success in warfare may have been prerequisite to the taking of office in this case at least – there are other examples of accession on the heels of captures or war, as at Piedras Negras and (Hassig, 1988) routinely among the Aztec – the scheduling may have been in effect a kind of anticipatory accession comparable to heir designation. However, since raiding is elsewhere connected with war, it seems more straightforward to interpret this as territorial warfare, whether at the boundaries or in an unsuccessful (thus unrecorded) attempt to take over a neighboring centre, or in a defense of their own territory (LaMar Stela 3 evidently records the capture of

Table 8.8. *Dates of events marked by the star glyph, and their planetary associations.*

All planetary dates are given in the 584 283 correlation.

a	b Venus	c Event	d Saturn	e Jupiter	f Mars	g Gregorian	h Mercury
War events:							
9.9.18.16.3	rE + 5	war				631 Dec 25	
9.10.17.2.14		war	op	op	1 : 0	649 Dec 21	me + 1
9.10.19.8.4		war	2 : −1			652 Mar 30	
9.11.9.8.12?		war	1 : −31	2 : −8		662 Feb 14	me − 1
9.11.16.11.6		war			2 : +6	669 Mar 3	
9.11.19.4.3		war	op	2 : −23		671 Sep 26	me − 11
9.12.0.8.3	Dme	war	2 : +17	op + 13		672 Dec 9	me − 11
9.12.5.9.14		war	op			677 Dec 14	
9.13.4.17.14		war		2 : +8		697 Feb 12	me − 2
9.13.13.7.2	sM	war			2 : +15	705 May 30	
9.14.2.0.14		war	1 : −21		op+ * 3	713 Dec 6	me − 7
9.15.3.8.8	ic − 2	battle	2 : +4		1 : −7	735 Jan 19	
9.15.3.8.12	rM − 2	battle	2 : +8		1 : −3	735 Jan 23	
9.15.4.6.4	rE − 1	war	op			735 Dec 1	me − 12
9.15.9.3.14	rE	battle		2 : −18		740 Sep 15	me − 2
9.15.12.2.2		war				743 Jul 30	me
9.15.12.11.13		war		2 : +17		744 Feb 6	me + 7
9.15.16.7.17	ic	ballgame				747 Nov 1	
9.16.4.1.1		war	1 : −5			755 May 7	
9.17.9.5.11		war	1 : +7		2 : +20	780 Mar 26	me − 3
9.17.10.6.1	me	capture	1 : 0	2 : +22		781 Mar 31	
9.17.15.3.13	me	faces captive				786 Jan 16	me − 12
9.18.1.15.5	ic − 2	battle	1 : −17			792 Aug 4	me − 13
9.18.9.4.4	rE + 4	war	1 : −13			799 Nov 15	
9.15.1.6.3		death (in war?)	2 : −3	op − 5	1 : −6	732 Dec 15	me − 10
Accession events:							
9.12.11.6.8	Dme	accession		1 : +10		683 Sep 8	
9.15.15.12.16	rE − 6	heir designation	1 : +6			747 Feb 13	
9.17.5.8.9	me	accession, autosacrifice	2 : −9	1 : −8		776 Jun 13	me + 2
9.17.10.9.4	rM	accession	op			781 Jun 2	
9.18.1.2.0	rE	dynastic ritual (heir designation?)				791 Nov 13	

Abbreviations. 1: stationary point preceding inferior conjunction, at which retrograde motion begins; 2: stationary point following inferior conjunction, at which retrograde motion ends; D: departure from; E: Venus as Evening Star; ic: inferior conjunction; M: Venus as Morning Star; me: maximum elongation; op: opposition; r: first appearance (rising), as morning or evening star; s: last visibility (setting), as morning or evening star. Deviations are given in days; − indicates that the recorded date precedes the event, + that it follows. Positive deviations should be reduced by two, negative deviations increased by two, if the 584 285 correlation favored by Lounsbury is used.

an individual from Bonampak about this time).

The timing of these raids in the solar year supports this territorial interpretation. Arguing from the records of Aztec warfare, Hassig (1988) has found that intersite war could not be sustained at long distances except when corn was standing in the fields of the intended victim; food carried in on the back would be consumed by its bearers within

a relatively short distance from their starting point. In the Maya area the relevant part of the year varies somewhat, but can be considered as concentrated between the autumnal and vernal equinoxes. Examination of all cases designated as war by the Venus/star verb (Table 8.8, column g) shows that they are indeed seasonally restricted and in the expected way. The restriction is on the borderline of statistical significance: six or seven of 23, and seven or eight of 24 dates fall outside the expected region, uncertainties being (a) the categorization of a date falling just four days before the autumn equinox, and (b) whether a Piedras Negras capture date associated with some of the dates where the star event glyph is removed should be allowed in the sample. These uncertainties leave a range of probabilities from a low of 1.73 to a high of 7.58 per cent for a chance concentration of dates in the expected region. The dates are actually more concentrated than this, occurring mainly in a 'war season' running from mid-November through mid-February, Gregorian.[32]

The Bonampak dates fall in the war season, and thus pattern with major territorial warfare. Nahm shows that the capture events are just as strongly correlated with this war season; evidently the Bonampak case is typical, raids often being long-term or long-distance campaigns. Accessions are also independent of season, except that some are apparently tied to success in war.

Several positions in the Venus cycle are associated with these events. They include inferior conjunction, as well as heliacal setting of the evening star before it and heliacal rising of the morning star after it; stationary points, or perceptible motion from them; maximum elongations; and heliacal rising as evening star. Not included are maximum brightness, heliacal setting as morning star, or superior conjunction. The number of agreements of star events with points in the Venus cycle is statistically significant: ten or 12 warfare dates out of 25 explicitly marked as star events have precise associations with points in the Venus cycle; given that none occur during invisibility around superior conjunction, there is less than a 0.15 per cent chance of getting as many as ten accidental agreements of this sort.[33] Military iconography is associated with the Venus gods depicted in the Venus tables, paralleling Aztec ethnohistoric accounts (Seler, 1904).

Although the cycle of Venus affected the scheduling of a limited class of events, there is no clear relationship between the position of a date in the Venus cycle and the type of event scheduled for that date. There is a certain asymmetry in that war events seem almost always to have been scheduled in connection with planetary cycles, while accessions were so scheduled in a minority of cases – perhaps only or disproportionately when war events led up to the accession. But by patterning similarly with respect to the Venus cycle, the events so scheduled form a single class that differs from other types of events, including the performance of at least those rituals not directly related to accession. The common denominator could in fact be warfare, if the only accessions so scheduled were those related to war, but this is unlikely since some occur at heliacal rising of evening star, long after any prior Venus event could have occasioned the scheduling of battle. Otherwise, the common denominator linking successful intersite warfare, accession and heir-designation, and what distinguishes accession-related rituals from offertory and other rituals that are not directly related to accession, is the taking or perhaps holding of power/authority. This alternative provides a link between the structure of the Venus table and the positions of the Venus cycle at which battles and accessions were located. In the tables, Venus is the subject of a verbal compound that, in Classic texts, describes only accession-related events. It consists of a hand holding a mirror – a symbol of authority, worn on the forehead (Schele and Miller, 1983). The stations nearest superior conjunction are placed so as to ensure that they never actually fall on a date of the planet's invisibility, whereas the dates of heliacal setting of the evening star before and of heliacal rising of the morning star after inferior conjunction are placed at the average positions, guaranteeing that they frequently fall during invisibility (and increasingly for heliacal setting as the stations move ahead with respect to the true Venus cycle by 0.09 days per Venus round). By interpreting the verb in the codex as referring to the regency or power of the Venus god, the deviation from canonical rising and setting near superior conjunction indicate that Venus was still powerful during invisibility around inferior but not superior conjunction. The latter, then, was a period during which accessions and wars should *not* be undertaken, at least under the auspices of Venus.

Along with the scheduling of events via temporal correlates in the Venus cycle, the position of Venus in the sky might correlate with the location or orientation of the events scheduled for those dates. There is little opportunity to know in the case of rituals, except those few whose architectural context is known. In the case of conquests, it can readily be tested, and there is no obvious directional connection in cases in which the conquered and conquering sites are known.[34] One reason may be the constraint on timing of warfare in relation to the agricultural year. For example, significant points in the cycle of Venus as morning star are distributed through about 70 days of the 584-day cycle; a battle against an eastern opponent could be delayed several years waiting for an eastern position for the planet. If political and military exigencies dictate battle during the proximate war season, the range of possible Venus dates will be restricted. Indeed, if superior conjunction falls in that season, dates near maximum elongations, stationary points, or inferior conjunction are not available, and if the season begins shortly after or ends shortly before invisibility around superior conjunction, maximum elongation is the only choice if even it is available.

The other planets

Less data are available concerning the cultural correlates of other planets. Only isolated examples of events involving Jupiter, Saturn and Mars have been noted in Classic inscriptions, almost all at stationary points or first detectable movement from them. The only Mars events in Classic texts that I know of and find plausible are ones coinciding with a stationary point of Saturn (Fox and Justeson, 1978; discussion of the Mars association was omitted from the published version) and one involved in triple conjunctions with Saturn and Jupiter (Lounsbury, Chapter 19, this volume), and possibly conjunctions of Mars on Caracol Stela 3 (Kelley, 1983, pp. 190–1); in addition, Lounsbury (1983b) has pointed out that Venus and Mars rose heliacally and in conjunction on what seems to have been the historical base of the *Dresden* Venus table. Stationary points of Saturn have also been recognized only in association with other planets, Mars (noted above) and Jupiter (Tate, 1988; Lounsbury, Chapter 19, this volume) being attested, while Caracol Stela 3 seems to relate the motions of Saturn and Jupiter (not necessarily by conjunctions, although this is often assumed).

Although these planetary dates are too few to permit secure conclusions concerning the kinds of events enacted on them, they appear to be of the same type as those occurring at all positions of the Venus cycle. The Saturn–Jupiter stationary points are identified as occasions for accession-related ritual by Lounsbury and Tate. The Mars–Saturn stationary point was probably the occasion for a

battle or raid: it is marked as the death date of an aged ruler of El Cayo, and occurs in the middle of the war season. Indeed, this was a war season in which there may have been no viable Venus position; the rising of the evening star would have been at or before its beginning, and it ended before maximum eastern elongation.

This leads to a hypothesis that significant points in the cycles of Saturn or Jupiter may have been selected as occasions for warfare when no Venus date is appropriate to the war season in a given year. I have reanalyzed the 25 war dates marked as star events in relation to the stationary points of Saturn and Jupiter, and their disappearance around inferior conjunction (Table 8.8, columns d and e). The suspected association was not confirmed, for the same proportion of Saturn and Jupiter events is found for dates with Venus associations as for those without. Unexpectedly, however, dates apparently associated with Saturn or Jupiter outnumber those associated with Venus: while half of the dates related to positions in the Venus cycle, fully three-quarters of them occur at the significant points in the cycle of Saturn or Jupiter. Eleven of them lie within ten days of a stationary point of Saturn and/or Jupiter, and five within ten days of opposition; eight of the Jupiter/Saturn dates are among the 13 with no Venus association, and eight are among the 12 having Venus associations. Turning to the five accessions marked as star events, and discussed by Lounsbury (1983a), the same associations appear: two prove to be at significant points in the Saturn cycle, one in the Jupiter cycle, and one in both cycles. These associations are statistically significant,[35] whether opposition dates are excluded or included. The Saturn–Jupiter dates discussed by Tate and Lounsbury are also dates of rituals – non-offertory, except for bloodletting rituals – that are related textually to prior accessions.

With this expanded sample of Saturn and Jupiter dates, it can be seen that the same types of events are being scheduled according to the same essential points of the planetary cycles as for Venus. They also show the same avoidance: there are no examples during Jupiter's or Saturn's disappearance around superior conjunction. However, this apparent avoidance is not statistically significant given the current sample size.

The association of not only Venus but also Jupiter and Saturn with dates of war and other events marked by the star sign suggests a check of Mars as well (Table 8.8, column f), particularly since it is brighter and faster than Saturn. Six of the 25 dates occur within 20 days of a stationary point in the 780-day synodic period of the planet, and another is within two or three days of opposition; these, too, are statistically significant numbers. The explicitly marked star events therefore relate to planets generally, not specifically to Venus as the star par excellence; correlatively, the prime referent of the star sign is in fact either planet or star, as Kelley has long argued, and not specifically Venus as is generally supposed. The rare use *e·k'* 'star' rather than the compound *čak e·k'* 'Venus', lit. 'great star', to refer specifically to Venus, is simply the use of a generic term in a context in which its specific referent has already been established.

It can be seen in the table that, although a few dates are within a few days of opposition of Saturn, Jupiter or Mars, only one occurs on a date for which no Venus event or stationary point of Jupiter, Mars or Saturn was also proximate. Thus, it is possible that only stationary points were relevant in the case of the superior planets. Alternatively, it may be their retrograde periods that are relevant (recall the axe event of *Dresden*'s 78-day Mars almanac); of the 13 war dates lacking Venus correlates, nine fall during and 12 within ten days of the retrograde period of Saturn, Jupiter or Mars.

Taking the four major planets together, the likelihood of a given date falling within ten or 20 days of a significant point (for the superior planets) or at the more precisely defined points in the Venus cycle, is quite high – between 36.02 per cent and 64.50 per cent. Nonetheless, the large number of positive cases renders the overall pattern statistically significant, with only a fraction of 1 per cent probability of attaining the observed number of dates showing a planetary association with ten- or 20-day leeway, including or excluding dates of opposition, without a genuine association. Given 20 days leeway, only one date (9.15.12.2.2) fails to have an association. Lounsbury notes that Venus and Mercury were in conjunction on this date; Mercury was also at maximum elongation.

In summary, almost every star date is correlated with one of the seven major parts of the Venus cycle or with stationary points (with or without oppositions) of Jupiter, Mars or Saturn. This need not indicate a rigorous planning of intersite warfare according with planetary behavior, however. Because the data suggest that ten or 20 days may separate an actual celestial event from the date recorded in association with it, it is usually possible to find an appropriate planetary event near a time chosen on the basis of political, economic or military opportunity: in individual cases, the probability of falling within 20 days of a significant date, is either 52.67 per cent or 64.50 per cent, depending on the inclusion of opposition dates; even restricting to cases within ten days, the likelihood is 36.02 per cent or 43.55 per cent. Although the Maya were clearly manipulating the dates of their enterprises, or ceremonies associated with them, in relation to the planetary cycles, this evidently imposed little constraint on more mundane considerations governing the strategic timing of events. It may have served more to supernaturally sanction strategic decisions of the elite, or to present them as the decisions of the god planets.

As a caveat, the leeway may seem broader in our data than it was in Maya practice, for it may have a number of different sources; for example, apart from questions of perceptibility of motion and conditions of observation, the activity most precisely associated with the planetary event may not be the one recorded There is no evidence for any leeway at all with respect to the retrograde period, from one perceived stationary point to the next.

The case of Mercury is not treated here. Its period is so short that, with any deviation from the exact date of a phenomenon, an extremely high proportion of dates must fit to be able to show a significant association. Because the planet's proximity to the sun makes for extreme difficulty in viewing, A. Schlak (unpublished manuscript, 1985) has suggested that proximity to the position of maximum elongation would be the prime point of interest in its cycle; this position was relevant to Venus as well so its pertinence is plausible. In fact, nine of the 24 star dates are within ten days of Mercury's maximum elongation (Table 8.8, column h). This is comparable to the numbers of Venus, Saturn and Mars dates, so an association of Mercury's maximum elongation with star events is feasible; indeed, the one star date not associated with any other planet occurs exactly on the date of Mercury's maximum elongation in Thompson's correlation (at 18.6°). Although the hypothesis is reasonable and may be correct, even this high a proportion of Mercury dates has a 52.4 per cent chance of occurring at random.

Thus, an association of star dates with Mercury's elongation from the sun cannot yet be substantiated. If substantiation is to occur, a substantially increased body of evidence will have to be gathered. Schlak has proposed a correlation of Mercury's maximum elongation with two classes of events, accessions and the holding of a symbol of authority representing God K (Bolon Dzacab), but his examples are selective rather than exhaustive. The selection criteria may be related systematically to maximum elongation, or to some associated characteristic such as a calendric pattern, since the sample appears to be biased in favor of such dates; and in a larger sample, including all accessions listed in Schele (1982), I find no statistically significant association of accession dates with large elongations of Mercury. The association with the holding of the God K scepter may turn out to be genuine, but I have not attempted to test this. If it does, this would provide circumstantial support for its relevance in the case of the star events.

8.5 Inferred astronomical dates

Although confirmation of the planetary associations of dates marked with the star sign is important for confidence that astrological correlates of dates are deliberate, and for establishing what types of planetary events were of interest, they are so rare that they provide only an introduction into the investigation of Classic Maya astrology. The leeway in the association of planetary event dates with the recorded dates of associated human activities, whatever the cause, increases the chance for an apparent celestial connection of an arbitrarily selected date to quite high levels (*c.* 30 per cent). The most widely used means of controlling for this is by focusing on inscriptions 'in which dates are separated by intervals of known astronomical meaning' (Kelley and Kerr, 1973, p. 181). Although this was the method by which manuscript tables were identified as astronomical in focus, the astronomical intervals there recurred in regular patterns, with subdivisions of major cycles registered, and with repeated examples sufficient to demonstrate the structure. In the inscriptions, comparable control is occasionally provided, when a series of events is explicitly linked by textual statements (not necessarily within a single inscription), and with several of the events separated

by intervals relating to the same bodies (see the cases discussed by Kelley, 1975, and by Lounsbury, Chapter 19, this volume).

Usually, however, we are dealing with a single interval, potentially of astronomical significance, between a single pair of dates, so it is difficult to know whether the interval was chosen for its astronomical relevance or not. This is particularly true when the intervals involved commensurate calendrical cycles: although this sort of commensuration is a hallmark of Maya astronomical inquiry, it is also a general characteristic of Maya numerology and thus does not intrinsically relate to the celestial domain; and the chances of a planetary interval being accidentally captured by a random multiple of a calendar cycle is quite high. Consider, for example, the timing of the accession-related rituals conducted by Chan Bahlum five Venus years and eight solar years after his accession. Being two days more than five Venus cycles or eight vague years, the interval ostensibly links the commemoration more closely to the solar year (or to stellar positions) than to the Venus year; however, the interval links a nearly maximum elongation of Venus upon accession to a genuine maximum elongation at the enactment of the ritual, while the dates are not associated with a significant point in the solar year. Either or both cycles could be significant, or the rituals may simply have been calendrical, planned for a similar point in the vague year.

When individual intervals are involved, the best chance of securing astronomical associations is via a recurrent class of events, referred to textually or iconographically, in which a non-random recurrence of comparable astronomical associations can be sought. Astronomically significant positions that simply chanced to coincide with historical events being discussed (cf. Thompson, 1935, p. 82) cannot be verified by this procedure, even if the celestial correlate was intentionally referenced via a telltale interval. Although this means that some genuine references are not always captured, a compensating advantage is that what gets recovered is a program by which certain classes of activity were regularly (not necessarily frequently) scheduled with respect to celestial events.

The solar/Venus/vague year accession rituals are a case point. The pertinent complex consists of accession-related rituals performed by rulers who had long since taken office. Most examples

Table 8.9. *Numbers of planetary synodic intervals separating synodic intervals of the same length.*

Commensuration accuracy	Mars	Jupiter	Saturn
$c.$ 24d	7, 8, 7, 8, 7		
	7, 8, 7		
	7, 8, 7, 8, 7		
$c.$ 12d	15, 7, 15, 22, 15, 7, 15	11, 11, 11, 10, 11, 11, 11	
$c.$ 6d	37, 22, 37	11, 54, 11	28, 29, 28, 29, 28, 29, 28

occur at a precise multiple of 360 days after a true accession date. Occasionally, a ruler will accede on such an anniversary of the accession of a predecessor; at El Cayo, a site subsidiary to Piedras Negras, accession anniversaries of Piedras Negras rulers were noted as such proximate to local accessions, and local accessions were scheduled at anniversaries of accessions at Piedras Negras. This recurrence suggests that there is a real recurrence involved in Chan Bahlum's accession. A parallel case from Tikal tends to confirm that the celebration was an anniversary intentionally commensurating the Venus year and the solar or vague year. One of the most ancient rulers whose inauguration is recorded in Maya texts, dubbed 'Curl Nose' by epigraphers, evidently acceded to the thrones of Tikal and another site (probably Uaxactun) eight years apart on the same day of the vague year (J. S. Justeson, 'Revisions in the earliest Maya dynastic sequence', unpublished manuscript, 1985), and not at a significant point in the Venus cycle, under the usual correlation. Evidently, solar or vague year commemorations within the real or canonical Venus year were a recurrent, if rarely recorded, class of events, going back to $c.$ AD 378.[36]

Planetary variational data

Although the variation in the synodic periods of the planets can be quite large, it is quite structured, repetitions occurring when the earth and the planet return to the same regions of their elliptical paths around the sun. The commensuration of the solar year with planetary sidereal periods thereby induces a commensuration with the average synodic period of the planet. After the induced number of synodic periods has passed, the series of synodic lengths exhibited during that span begins to repeat; the more accurate the commensuration, the more precisely the synodic period variants resemble one another.

For Venus, the only relevant commensuration is

five Venus years; for Mercury, the only relevant commensuration is seven Mercury years. These simple patterns result from the relation of $n_{sid} - n_{syn} = n_{sol}$; the error in the relation accumulates. In the outer planets, the corresponding relation is $n_{sid} + n_{syn} = n_{sol}$, the error gradually shifting from positive to negative. Thus, depending on the accuracy of the commensurations involved, the subgroupings of Table 8.9 define the periods within which repetitions occur for the outer planets.

Variations within the subgroups are approximately symmetrical, as are the variations among the subgroups, since the planets' velocities and paths are symmetrical about the axes of their elliptical orbits. Because the commensuration of planetary synodic intervals and the solar year is apparently a recurring Maya practice (see the following subsection), the associated rituals occur when the planet involved is in the same part of the sky at the same time of year. If, in such cases, the celestial location was a focus of the rituals as much or more than of any celestial event, or numerological relationship to prior events, then this should be the direction that can be observed from the rooms in which the rituals were performed; in many cases, these were clearly the structures that held the texts describing the rituals. However, the Maya seem not to have recorded sidereal spans, apart from those that are multiples also of the synodic period, so celestial location was probably not crucially involved in structuring the events discussed in these texts. Alignment studies have demonstrated that the Maya paid attention to seasonal variations at least of Venus, with observations particularly of extreme positions of rising and setting (Aveni, Gibbs and Hartung, 1975; Aveni 1980, pp. 92–4, 260–4, 277; Closs, Aveni and Crowley, 1984, pp. 222–3, 236–7). Sightings of returns to these extremes would aid in the discovery and manipulation of the temporal recurrences required to accurately assess the average synodic cycles, with which the manuscripts and numerological exercises are so unwaveringly concerned. But McCluskey's (1983) argument, that these assessments were sources of temporal recurrence constructs independent of the ritual calendar and that would subsequently be adjusted by various error equations to relate them to ritual time, is unconvincing.[37] Rather, the observations most likely were *made* in the framework of the 260-day cycle; this strategy lends itself to the description of mean motion, with which the surviving planetary tables are concerned, but it does not readily capture regular variation.

The solar year

The most problematic class of historical events with potential celestial correlates, because of susceptibility to accidental calendar commensurations, are those relating to the solar year. One reason is that scheduling of historical events in relation to the solar year is not usually done for essentially astronomical reasons, or even using celestial observation, but rather involves the organization of group activity with respect to seasonal variations in the cultural cycle. Cycles of ceremonies tied to hunting, agriculture and animal husbandary are given in ethnohistoric sources, and prognostications dated only in the 260-day ritual calendar are sometimes demonstrably keyed to the agricultural year via planting, harvesting and similar references (e.g. J. A. Fox and J. S. Justeson, 'A sixteenth century Cholti calendar', unpublished manuscript, 1982). In the Classic period, seasonal cycles of rituals are known for the ritual cave site of Nah Tunich, Guatemala (Stone, 1983): dates in a given portion of the cave are restricted to a corresponding quarter of the calendar year. An astronomical emphasis for such dates depends on their assignment to specific solar events, the primary candidates being eclipses and solstices, the rising of particular stars, and planetary sidereal periods.[38]

According to Landa, many ceremonies were specific to a given month in the vague year. Thus, apparent solar year events may well be calendrical events scheduled within the vague year; in some cases it is not possible to determine which was the relevant unit. Even longer spans, in which an exact solar interval is a few days more than a vague year interval, may not provide a clear resolution to the question. Consider again the solar/Venus accession

anniversary at Palenque: it occurs after exactly eight solar years, two days later than eight vague or five Venus years. Yet the solar/Venus commemoration at Tikal included an associated ritual two days *earlier* in the vague year. Thus, a precise solar year anniversary position cannot be insisted upon as part of the scheduling complex; it may have been deliberately selected at Palenque, but if so perhaps partly because this also captured Venus' maximum elongation more precisely. In any event, since planetary timing recorded for military activity deviates by a few days from the celestial events, such deviations cannot be excluded for solar or vague year events.

To separate the vague year and solar year possibilities, a linked series of intervals is necessary. If the relation is to the solar year, the deviations from the vague year should accumulate; otherwise, they should be randomly distributed. Two examples illustrate the two patterns.

Ian Graham (1967, pp. 97–8) found that monuments at the site of Machaquila recorded as their final date a day, in the year following the civil year ending, that fell in the month Cumku; the exact choice was regularly 13 Cumku (Table 8.10), except that 1 Ahau was chosen if it fell in that month. Because the series is so complete, erected every fifth civil year for 40 years with only one gap, we know that there is no gradual drift in dates in this month, such as would characterize a solar correlate. Several of the texts designate these dates by the 'end of the year' compound; this sign group is a verb that habitually accompanies the date on which the year ended, perhaps explicitly by stating that it was the date on which the year-ending monument was erected (*tu·n* meant both 'stone' and 'year' (ending)'). The dates so marked, and the year-ending ceremonies accompanying them and portrayed on the monuments, in the Classic period are almost uniformly for the end of the 360-day civil year; in the Postclassic they were for the vague year. The Machaquila monuments (along with Jimbal Stela 1), seem to represent a shift toward year-ending ceremonies of the Postclassic type, since dates in the last month of the vague year are being designated as year ending.

The selection of a 1 Ahau alternative, when available, may have been occasioned by its special ritual associations with the planet/god Venus. In the years of the 13 Cumku ceremonies, leading up to 1 Ahau

Table 8.10. *Machaquila vague year ending dates.*

		Vague year end		
	Civil year end	Long count	Vague year	
missing	9.18.10.0.0	9.18.10.7.5	13 Cumku	
missing	9.18.15.0.0	9.18.15.8.10	13 Cumku	
missing	9.19.0.0.0	9.19.0.10.0	18 Cumku	on 1 Ahau
	9.19.5.0.0	9.19.5.11.0	13 Cumku	on 1 Ahau
	9.19.10.0.0	9.19.10.12.0	8 Cumku	on 1 Ahau
	9.19.15.0.0	9.19.15.13.0	3 Cumku	on 1 Ahau
	10.0.0.0.0	10.0.0.14.15	13 Cumku	
	10.0.5.0.0	10.0.5.16.0	13 Cumku	
	10.0.10.0.0	10.0.10.17.5	13 Cumku	

dates in Cumku, the day 1 Ahau was associated with a Venus/lunar eclipse cycle. In the year ending 9.18.9.0.0, a year 1 Ahau and thus sacred to Venus, a lunar eclipse fell in the month just before this date (recalling the association of lunar eclipses with 1 Ahau Venus bases), shortly after the second stationary point (9.18.8.7.13), and followed 14 days later by a solar eclipse. Shortly thereafter a series of five lunar eclipses was seen, almost all associated with points in the Venus cycle, and ending with a solar eclipse:

*9.18.8.7.13		stationary point 2
(9.18.8.8.7		solar eclipse)
+7L =	9.18.9.0.0	year 1 Ahau, day 237 since inferior conjunction
? +11L =	9.18.9.16.3	stationary point 1
? +22L =	*9.18.11.14.3	maximum eastern elongation
+6L =	9.18.12.5.0	heliacal setting as morning start imminent
+6L =	9.18.12.13.18	(stationary point of Saturn)
+6L =	*9.18.13.4.14	stationary point 2
+5.5L =	(9.18.13.12.17	solar eclipse)

The day 1 Ahau occurred in the lunar month proximate to dates marked by an asterisk. Thereafter, eclipses and eclipse stations continued to occur in the month before alternate 1 Ahau dates, and in the lunar month after the intervening 1 Ahau dates, including those of the Machaquila year-endings; eclipse dates are in bold type.

The Machaquila series therefore reflects a ritual cycle of vague year ceremonies, possibly linked to

Table 8.11. *Yaxchilan dates of 'torch staff' ritual and Structure 23 'fire' events.*

Long count dates after Tate (1988). The passage from which the third entry is drawn is eroded; its long count assignment is a 'probable reconstruction'.

Long count	Vague year	Gregorian	Ritual
9.14.8.12.5	13 Yaxkin	25 June, AD 720	fire
9.14.14.13.17	15 Yaxkin	26 June, AD 726	fire
(9.15.3.16.6)	19 Yaxkin	(25 June, AD 732)	
9.15.9.17.16	19 Yaxkin	26 June, AD 741	torch staff
9.15.16.1.6	19 Yaxkin	25 June, AD 747	torch staff
9.16.17.6.12	end Yaxkin	18 June, AD 768	torch staff

Venus through associations in the ritual calendar and Venus years.

Tate (1988) and, independently, P. Mathews (personal communication, 1979) have isolated a set of rituals in the inscriptions of Yaxchilan that always occurred during the month Yaxkin (in fact, always in the last third of that month). Tate determined the Julian dates of these ceremonies; finding that they cluster around June 20, she suggested a connection between these ceremonies and the summer solstice. The ceremonies were conducted over a number of years, and generally show a gradual forward movement of a few days with respect to the vague year (Table 8.11). Recomputed in the Gregorian calendar, the table indicates that they fall consistently a few days from (usually later than) the summer solstice. Nonetheless, the textual and iconographic references to these ceremonies occur on the lintels of a building that C. Tate ('Astronomical and commemorative events of Yaxchilan Structure 23', unpublished manuscript, n.d., and Chapter 32, this volume) has shown was oriented to the point of summer solstice sunrise; its rooms are illuminated on that date and, because the sun is virtually at a standstill near the solstices, for a few days before and after. The ceremonies may therefore relate to the category of first perceptible departure from a stationary point (isolated by Lounsbury, 1983a, and Chapter 19, this volume), or perhaps to the stationary period. However, controlling for ritual type, the ceremonies appear quite fixed with respect to the vague year, while slipping with respect to the solar year. The setting of the ritual may have been selected opportunistically, in relation to solstice standstills, although its timing was fixed calendrically.

Civil year ceremonies

The most difficult class of dates to evaluate in relation to their astronomical correlates are those occurring near the end of the civil year of 360 days. Several have been suggested as having significant correlations of this sort. A solar eclipse was visible on or near 9.17.0.0.0, a katun ending. Venus was at significant points in its cycle on year-ending dates marked textually or iconographically with Venus or star signs (Kelley, 1977a, cf. 1977b; Closs, 1979, 1981; Lounsbury, 1983a).

The special difficulty posed by these dates is that the events most frequently recorded in the inscriptions were precisely the rituals performed at the end of the civil year (beginning in the Terminal Classic and increasingly in the Postclassic, at the end of the vague year). Since colonial accounts tell us that events of public importance were timed in part according to astrological and other omens, the types of rituals performed on year endings, or the details of their performance, may have been related to recent, current or imminent celestial events. If so, these effects are especially difficult to identify. Most such events are referred to tersely, with only generic statement of the performance of year-ending ceremonies rather than specification of which ceremonies were performed; variations in a specified type of ceremony seldom if ever enter the texts. So the features that may correlate with specific astronomical events generally are not reflected by direct textual statements; if their presence is marked at all, it is in accompanying scenes.

Monuments were erected at regular intervals at many Maya sites, at structural stations in the long count: characteristically recorded are katun endings, tenth-year endings (within the katun), fifth- and 15th-year endings, and 13th-year endings, in that order (with a roughly geometric rate of decline from class to class). Because the erection of monuments on these dates was scheduled mechanically via the civil calendar, not via the cosmos, and because such dates are almost always recorded as the dedicatory date of the monument, dates alone can provide no evidence of the pertinence to the Maya scribe or ruler of any celestial phenomena that may have occurred on or near them.

There are two ways of dealing with this problem. One is to look to the accompanying sculpture for themes whose presence and absence covaries with the celestial phenomenon on the major year endings, or to specific ritual events that do so. Exemplifying this approach is A. Schlak's (unpublished manuscript, 1985) attempt to correlate the holding of the so-called 'manikin scepter', an image of a snake-footed deity held as a symbol of authority, with the visibility and especially maximum elongations of Mercury; he identifies the deity with that planet.[39]

The other way to deal with this problem is simply to concentrate one's efforts only on dates of other types – those on which the dedication of monuments was unusual. These are the odd year-ending dates and especially non-year-ending dates; for them, reasonably strong cases might be made by simple calendrical argumentation. Records of these dates must indicate that something special happened on them – or, in the case of year endings, that something special happened during the year leading up to it or was anticipated for the coming year. Exemplifying this approach is my own examination of the texts for possible supernova records. Only one historically documented supernova took place during the Classic period – the supernova of AD 393. I undertook a systematic search to see if the date of its appearance, disappearance or maximum brightness was recorded, either to the day or by reference to the appropriate year ending. In fact, there was one, and only one, such date; it is the year-ending dedicatory date of Bejucal Stela 2. This monument, the only one referring to the time of the supernova, was dedicated at 8.17.17.0.0, the end of the year during which the supernova appeared. The only events mentioned are standard year-ending rituals. Given the number of dates recorded in the Early Classic, both odd-tun ending and non-tun-ending, there is about a 2 per cent chance of getting one that refers to the time of the supernova if the date was not mentioned in part *because* of the supernova.

Tempering this possibility is the circumstance that the only other monument from Bejucal was also an odd-tun-ending monument. Based on examination of unpublished photos in the archives of the Corpus of Maya Hieroglyphic Inscriptions, Mathews and I date it as one of the earliest in the Maya corpus (8.15.18.0.0). The monuments were erected by successive rulers, whose practice was apparently to schedule the erection of monuments by events or calendar stations other than major civil year endings. Since we do not know what that 'something' was, the calculated chance of hitting a supernova date is unreliable. For example, both monuments were erected at the end of a tun 11 Ahau, which was also the ending date of the katun on which the long count was inaugurated (because that katun began on the first day of the ritual calendar, 1 Imix). If this was the basic pattern at the site, then there would be a 16.7 per cent chance of hitting the supernova date. Another possible factor is the circumstance that intervals of 13 tuns bring back the same day of the Mars year, with a slippage of only one-third of a day.

8.6 Positional data

Architectural and archaeological studies have made positional patterns of celestial bodies a familiar source of information on the knowledge and practices of ancient astronomers. Epigraphic evidence concerning Maya astronomy is essentially complementary to that from architectural approaches, consisting almost entirely of temporal patterns. Because positional and temporal patterns correlate, some positional inferences might be made from patterns of dates in the epigraphic record, but attempts of this sort are rarely reported. The lack of apparent interest by epigraphers in relating temporal to positional patterns seems to reflect a lack of success in determining any non-trivial consequences for the interpretation of the epigraphic record.

Asterisms

The most significant possibility, one much studied by epigraphers, is a possible Maya zodiac (see Kelley, 1976, pp. 47–51, and Aveni, 1980, pp. 199–202 for summaries of published work). Colonial Yucatec

dictionaries provide the names for three asterisms – ‡ab' 'rattle of rattlesnake', for the Pleiades; á·k 'turtle' for certain stars 'in the sign of Gemini'; and sí·ná²n 'scorpion', evidently for Scorpio ('scorpion, and the constellation of that name'), though conceivably the reference means that sí·ná²n means 'scorpion', and that there is also a constellation named sí·ná²n. Depictions of a rattlesnake, a scorpion and a turtle are found in the *Paris Codex*,[40] suspended from a 'sky band', along with ten other animal depictions, three obliterated; and seven surviving figures, depicting a subset of the same animals as in the *Paris Codex*, and in the same order, occur in a sky band on the Monjas at Chichen Itza, surmounting or surmounted by a star sign. The Monjas stands across a plaza from the nearby Caracol, a building with confirmed alignments to stations of Venus (Aveni *et al.*, 1975).

The most important recent development in this area is Lounsbury's (1983, p. 166) recognition that the Turtle asterism is formed by the three stars in the belt of Orion. Miller (1986) noted the occurrence of a peccary and a turtle, three stars associated with each, in the vault of a room with a battle scene and bounded by sky bands; these recall the members of the Maya zodiac discussed above. The turtle, with three stars in a line on its back, clearly correlates with Colonial Yucatec ⟨ac⟩, which meant 'turtle' (<W m + Yu *ahk), and was also the name of the 'three adjacent stars that are in the sign of Gemini, which with others take the form of a turtle' (Motul Dictionary; my translation). Lounsbury noted that when the dictionary was compiled, the astrological *sign* of Gemini included the stars of Orion; and that Thompson (1950, p. 116) was told by a modern-day informant that the turtle *constellation* was Orion. Previously, the turtle asterism was associated with the constellation Gemini, partly because the printed version of the Motul omitted the word *tres* 'three' from the entry (Martínez Hernández, 1929, p. 66). Lounsbury noted that Orion would have been visible during early August in the eighth century, the date of the associated battle.

Lounsbury's demonstration that the Turtle was Orion's belt is the first major breakthrough providing solid confirmation that the Monjas and Paris series were in fact of asterisms; the association of a turtle and a peccary, the marking of both with star signs,

Table 8.12. *Possible physical arrangements of zodiacal series.*

(a) *Paris Codex:*

1 turtle	2 ×	3 death god	4 ×	5 frog
6 ×	7 serpent	8 scorpion	9 rattlesnake	
10 peccary	11 deer	12 bird	13 vulture	

(b) *Venus table directional deities:*

1: line 17 aligned with stations

corn g		hawk	Pauahtun	vulture/peccary			boa?		Imix g
Q	J	G	H	E	R	O	P	M	F
17.0	33.7	64.3	72.2	88.9	105.7	136.2	144.1	160.9	177.6
rabbit	turkey	death			Moon?	sun g	scorpion	deer	turtle??
C	D	A	N	K	L	I	B	S	T
−155.8	−143.9	−127.2	−110.4	−79.9	−72.0	−55.2	−38.5	−7.9	0.0

2: line 21/22 aligned with stations

boa?	sun g	Imix g	hawk	turkey	corn g	death		Moon?	vulutre/peccary
P	I	F	G	D	Q	N	O	L	E
17.0	33.7	64.3	72.2	88.9	105.7	136.2	144.1	160.9	177.6
scorpion	rabbit	turtle??			Pauahtun	death		deer	
B	C	T	M	J	K	H	A	R	S
−155.8	−143.9	−127.2	−110.4	−79.9	−72.0	−55.2	−38.5	−7.9	0.0

(a) The *Paris* zodiac under Kelley's model; (b) the directional deities from the *Dresden* Venus table, under the two alternative alignments to stations. Letter labels for deities in (b), are after Thompson (1950, p. 223, 1972, p. 66). *Abbreviation: g* = god.

and the clear depictive relation between the markings on the Turtle and the known referent of the colonial Turtle asterism, together leave no room for doubt. Lounsbury (personal communication) presumes that the Peccary would have been visible on the same night as the Turtle, and thus that they were in the same portion of the sky. Its three star signs appear to be arrayed in a non-linear arrangement.

Lounsbury's results provide the first firm basis for investigating the absolute positions of these asterisms, with two firm identifications – Rattle with Pleiades, Turtle with Orion's Belt – and a very likely identification of Scorpion with Scorpio. They are treated here, initially, via the zodiac of the *Paris Codex*.

The *Paris* zodiac names 13 asterisms. Assuming it is a complete cycle, the average distance between them is 27.67° in space, 28.07 days in time; in the body of the table, a series of 28-day intervals is accumulated in five groups of 13 (commensurating the ritual calendar), suggesting divisions of the year into 13ths. If it is assumed that the pictorial stations represent a physical sequence of successive asterisms, they should be separated by about 27.67°. This was the basis for Spinden's (1916, 1924, pp. 54–5) analysis of the table as a zodiac. Turtle is adjacent

to Rattlesnake, and Orion's Belt is indeed separated from the Pleiades by about 28°. However, Scorpion is then associated roughly with Gemini or the Little Dipper, not Scorpio.

Kelley (1976, p. 49) and F. G. Lounsbury (personal communication, 1986) believe instead that the asterisms occur at stations separated by 168 days, as that number is recorded between each pair in the upper register. 168 is 6 · 28, the nearest whole number approximation to $(\%_{13}) \cdot 365$, so passage through the entire table takes 6 · 364 days; the interval also imposes a spacing of the 13 zodiacal signs roughly at 13ths of the vague year, achieved only by multiples of 28 days. This provides a *prima facie* case for the positional sequence in Table 8.12a (according roughly with Kelley's Table 6), and the span between Scorpio and Orion's Belt does agree with that between the Scorpion and the Turtle in this scheme; however, the position of the Pleiades is discrepant from that of Rattlesnake, being assigned instead to Vulture.

There is support for each of the straightforward models in the spacings they impose, but Kelley's model accords better with the structure of the table. The discrepancy in either model can be accounted for if the asterisms were not evenly spaced, or if

Table 8.13. *Correspondences between the Dresden Venus table and other zodiacs.*

In the *Dresden* Venus table listing, the first line gives transliterations of spellings, the second gives linguistic readings of these forms, and the third gives translations.

Paris Codex:						
	turtle	scorpion	vulture	bird	deer	death god
Chichen Itza, Monjas:						
bird	turtle	scorpion	vulture			
Dresden Codex, Venus table:						
u-lu-m(u)	si$_2$-na$_3$-n(a)	VULTURE	TURTLE		⁷HOOF	DEATH GOD
ú·lum	*sí·ná ʔn*				*yù·k*	*kisin*
turkey	scorpion	vulture?	turtle?		brocket deer	death god

13 was not their total number. Rejection of Kelley's suggestion entails the assumption that the Spanish association of the Scorpion with Scorpio was a Spanish imposition, or that in the definition of *sí·ná ʔn* as scorpion, or the constellation 'of that name', 'that name' was Yucatec *sí·ná ʔn* rather than Spanish *scorpione*. Rejection of Spinden's sequential interpretation requires that Rattle (of Rattlesnake) and Rattlesnake were separate asterisms. Neither rejection is readily made; more evidence is needed.

Comparison with the Monjas is one source of evidence. The seven asterisms recorded, and verified as such by the association of a star sign with each – as at Bonampak – are all among the 13 in the *Paris Codex*. There is general agreement in sequence; the adjacent peccary and death god of the *Paris* are successive, but in opposite order, on the Monjas, while the Bird, if it is the same asterism in each, is displaced. The Monjas had eight or nine asterisms, ten or 11 if Imix and/or the moon are included; $(\frac{4}{9}) \cdot 365 = 162$, and $(\frac{5}{11}) \cdot 365 = 166$, so under Kelley's hypothesis both the agreement in asterisms, and the general agreement in their sequencing along with some discrepancy are accounted for; under a physical sequencing, the discrepancies are not readily explained.

Recently, another source for a Maya zodiac was proposed by Thompson; it is also relevant to the modelling of the locations.

A series of 20 'directional deities' are associated with the 20 stations of the Venus table. They are cited on two different lines, 17 and either 21 or 22; the sequence is the same on each line, but the series are offset by one position. A given deity in line 17 is associated with directions in line 16; named in line 21/22, the same deity is associated with the same direction, from line 24. Perhaps

because of this directional association, Thompson (1972, pp. 65, 67) suggested in passing that they refer to constellations. He did not attempt to draw any parallels with the zodiac, but there are several points of contact with the *Paris* and Monjas series (these series are reduced in Table 8.13 but remain in their original sequence). The Venus table series differs from the others in placing Turtle after Scorpion and Vulture instead of before them. W. Nahm and I identified the reference to Scorpion only in 1985, for readings of two constituent signs, na$_3$ by Lounsbury and si$_2$ by Fox, were not made until 1978 and 1979, respectively. Thompson had evidently not recognized the vulture sign or the deer compound as such, nor associated the carapace sign with turtles. So evidence for Thompson's hypothesis is growing, and supports the other glyphic and iconographic traces of the Maya zodiac.

Given this apparent relation to the proposed Maya zodiac, it is clear that the references to the same deity in lines 17 and 21/22 involve the same station; an interval of 90 days cannot separate two occurrences of one constellation, while 236 separates that for another, via a single mechanism (such as the passage of time) for offsetting them. Under the assumption that the Venus table does reference a set of asterisms, they can be associated roughly with points of the sky via the number of days separating them. Fixing the starting point, for convenience, on the Scorpion, and assuming the reference system of the first grouping is the more basic (it correlates the eastern deities with heliacal rise), the reference system of Table 8.12b1 can be reconstructed; that of Table 8.12b2 follows from the alternative assignment, of the second line of directional deities to the stations at which they are referenced.

The identifications of Turtle and Rattle(snake?)

are not secure enough to permit exploration of relationships between the directional deities and actual asterisms, but the results can be compared with the sequence reconstructed under Kelley's model, and with the actual sequence, of the *Paris Codex*. The Death God is mentioned twice (D and N), so there may only be 19 distinct entities. In Table 8.12b1, his two occurrences are separated by just 34 days, with the Turkey asterism intervening; this is close enough to arouse suspicion that the Death Gods are related, if not the same; possibly it was a long, gangly constellation, corresponding to the skeletal figure that represents him. Scorpion occurs 258 and 292 days from the Death God (hence also 197 and 73 days). In the *Paris* system they are separated by about 4 · 28, or 112 days; the range of uncertainty around these figures overlaps. However, Deer in the *Paris* is about 4 · 28 days from the Death God in the opposite direction, and 5 · 28 days from Death in the other; in the Venus tables, Deer and Scorpion would be adjacent, separated in the table by only 8°. In summary, there is sufficient correspondence among these three systems to be fairly sure they are related, but the nature of the relation and probably of the individual systems is not yet clear. There is no correspondence when the system of Table 8.12b2 is used.

Directional terms

Another possible avenue into positional correlates is via the directional compounds. The Yucatec Maya distinguished seven directions: east, west; north, south; zenith, nadir; and center. The *Madrid Codex* spells all of these on p. 78, but most often only east, west, north and south are given.[41] In the Venus tables, the expected correlation of east and west with the rising of morning and evening star are found on line 16 (see Figure 8.1), but the same directions are associated with stations offset one position in line 24. Thompson (1950, p. 223) interprets the second passage referring to the directional deities as associated with the intervals, in line 26, leading from the prior station, but if the directional deities are asterisms, or regions of the sky, the reference would apparently be back to the prior station. Support for the latter interpretation comes from the 'winged Chuen' verbal compound, which suggests that the second passage is a back reference; it says that the directional deity '*has/will have done*' what that compound indicates

on the date in question (the verb takes the -*ih* third person marker on verbs in the completive aspect). However, the rationale for attributing north and south to the setting of morning and evening star, respectively, is obscure. An alternative is that the associations are of different directions with the same stations in different basedate systems, evidently with the 13 Mac vs. 3 Xul or 18 Kayab basedates. In this case, the directional cycle would be ritualistic rather than terrestrial, as it evidently is elsewhere in the codex and in colonial ritual documents, often following a four-day cycle. Similarly, I know of no successful attempts to treat the standard directional compounds as providing directional information in Classic inscriptions; for example, the directional compounds associated with the emblem glyphs of Copan, Tikal, Seibal, and a 'Site Q' (Calakmul? El Peru?) do not correspond to their geographic relationships. A compound (T663:23 + EARTH) that I interpret as 'zenith', based on its context in a *Madrid Codex* directional sequence (see note 41), occurs fairly often in both the inscriptions and the codices, the latter at least sometimes in astronomical contexts; an investigation of its dated contexts might prove more fruitful, since zenith references are not strongly implicated in ethnohistoric ritual documents.

8.7 Conclusions

Cyclic commensuration in the ritual calendar was the basic approach used by ancient Mesoamerican astronomers to arrive at models for predicting the occurrence of celestial events in linear time. This paper has examined the potential and limitations of this method, applied to the moon and the visible planets; related the celestial events explicitly mentioned in texts to the activities scheduled in terms of them; and addressed methodological problems faced by epigraphers attempting to recognize unstated astrological content in hieroglyphic texts. Applied to the lunar cycle, Maya astronomical method was strikingly successful: it would lead to immediate success in long-term eclipse prediction, and to accurate placement of eclipse stations once records of roughly 50 years of eclipse observation were available. It was also successfully applied to Venus, and in a more limited way to Mars. For the other planets it was unsatisfactory. It meshed neatly with the synodic period of Venus and the 365-day vague year to create a long-term Venus calendar, one whose structure also incorporated lunar synodic month stations and eclipse recurrence intervals. The Mars cycle matched a tripled ritual calendar so closely that a single synodic period was presented in tabular form, reduced further to the subdivided ritual calendar spans typical of Maya ritual almanacs, but corrective mechanisms could not accommodate the very slow departure of the planet's motions from their ritual calendar stations. On the other hand, it did not accommodate long-term motions of Mercury, Jupiter or Saturn.

The Maya calendar priest's approach to the anticipation of celestial events did not simply permit the development of useful predictive models; it also constrained the essential form that these models could take. Included in this set of constraints was a failure to accommodate regular, symmetrical variations in the synodic periods of the planets. The Maya regularly made use of solar/sidereal/synodic commensurations that structure the recurrence of synodic variants in dynastic ritual (and of no other multiples of sidereal periods), but the tables' imposed structures required reuse of a given synodic chart with different synodic variants; an average value was the closest that could be used.

Such constraints, however, did not preclude the development of simple formal *calendars* related to the moon and perhaps to Venus that differed in structure from the predictive models of celestial activity. Stations in planetary cycles were occasions for sacred and secular war. The main verb of the Mars table of the *Dresden Codex* refers to warfare, and is so used consistently in the inscriptions; the Venus table depicts the god star Venus in warrior's garb. In the Classic period, events explicitly related to the cycles of the visible planets (save perhaps Mercury) were recorded as occasions for territorial warfare and for certain non-offertory rituals. The latter are mainly rituals of accession to supreme political power (perhaps especially when otherwise linked to warfare) and bloodletting rituals that were linked to warfare or accession, recalling references to Venus as the subject of an accession verb in the *Dresden Codex*. The planet involved is evidently not specified in the Classic records.

Demonstrating the pertinence of astronomical correlates of most Maya dates depends upon their recurrent association with iconographically or epigraphically definable complexes; even events explicitly marked as having celestial correlates refer essentially to human activities, and some activities with demonstrably pertinent celestial correlates lack any explicit indication of them. Statistical validation therefore proves as crucial in assessing epigraphic evidence for ancient astronomical practices as it is in architectural and archaeological studies. Such assessments are made here for events explicitly marked as referring to 'stars'. The dates of these events are verified as taking place at significant points in the cycles of Venus, Mars, Jupiter and Saturn. The test depends, however, on a correlation of Maya to Christian chronology in the Goodman family, which I accept along with most other epigraphers; Kelley (1983) presents a cogent critique on epigraphic and ethnohistoric grounds that has to be rebutted in detail before this hypothesis can be considered secure.

Positional astronomy is reflected only in the increasing evidence for reference to asterisms along the ecliptic ('the Maya zodiac'), but epigraphic evidence for Maya astronomy is otherwise exclusively in temporal patterns; this is a bias of the source rather than a reflection of the full range of Maya astronomical knowledge and inquiry.

Acknowledgments

Thanks are due to James Fox, Peter Mathews, Linda Schele, and especially Floyd Lounsbury for discussion of various astronomical aspects of Maya hieroglyphic texts. Anthony Aveni, Victoria Bricker, Werner Nahm, Merideth Paxton, Arthur Schlak and Gordon Whittaker provided access to unpublished work; this overview has been substantively improved by reference to these works, both in content and via the issues they have raised. Special thanks are due to Dennis Sinnott, who provided his computer program for computing planetary, solar, and lunar positions, and to Manfred Kudlek, for his 1978 tables of eclipses visible at Tikal.

Notes

Some of the results in this paper depend upon, and others are phrased in terms of, a correlation of Maya chronology with the Julian day count on or within a few days of that championed by Thompson, and originally proposed by J. T. Goodman. Most Mayanists are firmly convinced that this family of correlations is correct, but Kelley (1983) presents a strong case against it that will have to be refuted before this confidence is justified.

The orthography of this paper is that standard for linguistic transcriptions, when italic or bold script is used. Thus, *š* (not *x*) is for English *sh*, *x* is for Spanish *j* in opposition to *h* for English *h*, and *k*, *k'* are used rather than the colonial ⟨c, k⟩, for plain and glottalized velar stops.

1 Edmonson's (1982) definitions of the words *may* and *k'intun y-á?bil* ('drought') as '260-year period' and 'calendar round' are interpretive errors.

2 The pertinence of zenith passage in these latitudes was originally ascribed to the importance of Copan, more recently to Izapa, the ritual calendar then diffusing to the rest of Mesoamerica. However, the ritual calendar is attested in Mesoamerica well before either site attained cultural prominence, and in no other respect did Izapan civilization have such a pan-Mesoamerican effect.

3 Two examples: (1) A solar year amounts to $12\frac{1}{3}$ lunar months; 360 days could have been a formal span relating to both cycles via 12 equal units. The units would then have to be 30 days long; they would thereby approximate the lunar synodic month, as the Venus year was approximated, by a single whole-number value, and the excess of 5.63 days over 12 lunar months would roughly match the shortfall of 5.24 days in the solar year. As a possible parallel, dividing the year into 13 equal integer units yields a 28-day month in a 364-day cycle, a period used in calendrical computations although it was never an actual calendar period. (2) 360 days accommodates the veintena to the vague year, making it useful as a calculating device for relating the 260-day and 365-day calendars. Nothing strongly indicates these or any other alternative to the origin in positional notation.

4 Differences across languages in the pronunciation of several day names (e.g., WM *b'e?n*, EM *to·x*, GLM *manik'*) reflect very early changes in some branches of the Mayan stock. Especially indicative is *tiɲaš* for the day 'Flint' (Fox and Justeson, 1980, n.20), shifting to GKn *činaš*, EM *tixaš* only after an extremely ancient cleavage of the Mayan stock; see Justeson *et al.* (1985) for an absolute linguistic chronology independent of glottochronology.

5 All these deviations are considered errors by Thompson; if so, they are so numerous as to render the table useless for doing computations.

6 I do not see how it is possible to justify Bricker and Bricker's (1983) placement of the nodes 14 days earlier, in connection with their proposed recycling mechanism, with this data; cf. Lounsbury (1982).

7 The same pattern characterizes commensurations with smaller limits, though the alternation need not be as perfect; with a $\pm 1\frac{1}{4}$-day limit, the series in sacred rounds is 5,(36); 41,46; 51,(82); 87,92; 97,128; 133,138; 143,174; 179,184; (189),220; 225,230; (235),266; 271,276; parenthesized figures would be exceptional under a ± 1-day limit.

Few spans other than 260 have this property: among numbers less than 819 days long, 89, 260 and 449 behave essentially the same, giving a clean alternation between eclipse cycles and half-cycles; about half the commensurations of 119-, 164-, 224- and 656-day spans yield eclipse intervals, although with no special pattern of alternation nor a concentration among eclipse half-cycles; while commensurations of 178, 328 and 520 with the lunar synodic month yield over 85 per cent eclipse intervals. Many of these factors cease to capture eclipse intervals at multiples amounting to about a century, but the 328-day cycle remains viable about as long as the 260-day span. Some may wish to see in this viability of the 260-day span an explanation for the origin of this calendar. However, as the many other proposed rationales for this span make clear, some dramatic correspondence with some celestial or natural phenomenon is apt to attend the permutation of the 20-day cycle with any ritual cycle of a reasonably low number of days.

8 Moonages of 30 are also missing, but they are only half as likely to occur as any other; their absence is not statistically significant. If the gap *is* real, it indicates that the invisibility period was specified when current (there is no grammatical evidence on this point).

9 The 9.14.13.0.0 date also agrees with the eclipse-cycle system, so the introduction of Uniform moon numbering at Tikal may not have occurred until *c.* 9.6.0.0.0.

10 Hotuns are the quarter divisions of the civil year; they were major stations in the katun, the occasions for erection of most monuments.

11 The Tikal Uniform System makes the current era (since the base of the long count) begin with the first lunar month, while the Altar de Sacrificios Uniform System makes the previous era end on its last lunar month (these being the same lunar month). This is probably fortuitous; it is unlikely that the Maya could

have projected backward with such accuracy at this time, for they did not do so later. However, a comparable relation may have existed for a contemporaneous katun-ending basedate of the Uniform System of moon numbering (if so, it was probably inaugurated at 8.13.0.0.0 or 8.10.0.0.0).

12 Thompson (1950, p. 236) argued for the Copan cycle, formulating it as an attempt to correct the negative error of the 46-SR cycle by using two 41-SR cycles (just over 361 lunar months) for every three 46-SR cycles; he noted that 361 months does not make up an eclipse interval, but failed to note that two such cycles do form an eclipse interval.

13 Hence the last four pages of line 13 of the SR positions were never used; when corrections were made after 57 Venus years, the last three pages of the 12th line would also not be used. They are present because the last entry was the current basedate, and perhaps also for completeness in presenting the canonical correlation of the SR and Venus years in two calendar rounds, an elegant structure in its own right according to Maya constructs.

14 An error of somewhat over four days actually develops during the 57–61 Venus years of one basedate, so there is a slow accumulation of error of 0.47–0.78 days per pass through the table.

15 Discussions of the tables, normally proceeding in the opposite direction for clarity of exposition, have sometimes given the impression that the usual left-to-right, top-to-bottom reading order was reversed in these tables. For producing a given multiple, either order would be equally viable; the largest-to-smallest ordering is convenient for the reverse problem of removing the least multiple from a given number. The largest multiple smaller than the number to be reduced is subtracted repeatedly until the remainder is smaller than this factor; then the next largest factor is subtracted.

16 On this basis, Thompson revised it by the addition of 260, yielding 9360 or 1.6.0.0; this number is $16 \cdot 584 + 4 \cdot 4$ days. This still does not admit passage from one heliacal rising date to another, since it deviates from a multiple of 584 in the wrong direction; it would move approximately from the stationary point before inferior conjunction to heliacal setting. Thompson used it to move from a recorded historical date that was about 20 days before heliacal rising to one that occurred roughly at inferior conjunction, according to his correlation.

17 Closs (1977, p. 97) notes that, if 1 Ahau 18 Uo is taken as the astronomical base of the table, then 9100 days lead to a position 324 days into the Venus period, 260 days before its end. Although not occurring at a recorded station of Venus, this position approximates heliacal rising as evening star; it could be that morning star and evening star underwent heliacal rising on the same canonical date in the sacred round.

18 One possibility, then, is that it was used to aid in the recognition of a one-day error accumulation and perhaps for adjusting observational expectation as opposed to ritual timing of Venus phenomena; in fact, 9100 is the *only* multiple of 260 that can approximate a one-day adjustment while leading from one station of the table to another. Conversely, subtracting 9100 from the end of the 65-Venus year cycle would locate a date not present in the table at which the accumulated error had reached 3.83 days; this date would fall eight days after the station near first visibility as evening star, and indicates that the true position should be four days earlier. Such a shift could be assigned to the second station, and accommodated by shifting from the 5 Ik position, on the 11th line of p. 50 to the 1 Etz'nab position on the 12th line of p. 49. Such hypotheses seem unlikely, however, since this small *average* deviation would be swamped by short-term variations in the timing of planetary events.

19 The pertinence of Aztec sources is supported by the names of three Venus gods depicted in the table. Fox and Nahm in unpublished work and Whittaker (1986) show that three Nahua-sounding names are indeed Nahua. Whittaker and Nahm independently recognized in ta₃-wi-s(i₂)-ka-l(a₂) a name *tawiskal* for Tlahuizcalpantecuhtli, a Venus god. Whittaker identifies ČAK-T1048-wi-te₃-x with Tetzauhteotl, another Venus god, transcribing it te₃-sa-wi-te-ul; this depends on seeing Čak as a copying error from te₃ (Whittaker equates the signs) and on an improbable reading *sa* of T1048, for which most epigraphers would expect a Ci or Ce value. Nahm reads T1048 as *ši*, as in certain spellings for the month name Pax, yielding *čak šiwte..* for 'red Xiuhteotl' (or 'Xiuhtecuhtli', another name for the same god); the accompanying illustration depicts the god with Xiuhteotl's diagnostic turquoise bird (but on his shoulder, not his headdress). Finally, ka-ka-to-na₃-la₂, evidently for *kak tonal*, is taken by Fox as a Yucatecanism for ⟨tonallocactli⟩, the sun-sandal worn by Ixtilxochitl and his Venus-god brother Macuilxochitl; since the suffix *-al* normally occurs after the second root in a noun + noun compound in Yucatecan, adapting the word entailed an inversion of the structurally anomalous *tonal kak*. Whittaker takes *kak tonal* as a 'day name Sandal', but this is a purely hypothetical construct based mainly on the apparent day name Foot (Oc?) at Xochicalco.

20 W. Nahm ('The dates of Maya warfare', unpublished manuscript, 1988) provides evidence from the scheduling of war with respect to the Venus cycle that some such system existed in the Classic period. Work by Kelley, Closs and Lounsbury (see Section 8.4) had shown that about half of what Mathews characterizes as territorial wars were begun at special points in the Venus cycle. Nahm shows that almost all other warfare was scheduled according to a lunar count from the heliacal rise of Venus; even treating the special points as exceptions, the concentration of warfare in these spans is highly significant ($p < 0.0006$). In two three-month periods, warfare is favored; in the rest it is only sparsely attested. War was avoided during the first five months of the count, an eclipse interval. Then begins the first three-month span auspicious for war, which ends precisely with the opening of *Dresden*'s canonical three-month span for invisibility around superior conjunction. This period for avoidance of war lasted five months, again an eclipse interval, followed by a second three-month period auspicious for war. The span from the beginning of the first auspicious period to the end of the second is 17 months, another eclipse interval. (Section 8.4 shows that 12 of 13 wars not associated with a special point in the Venus cycle fall during periods assignable to retrograde motion of Saturn, Jupiter or Mars; the pattern is statistically significant, $p \approx 0.041$. In most Venus years, retrograde periods of both Saturn and Jupiter overlap at least one of Nahm's auspicious periods; in about half, Martian retrograde does.)

This demonstrates both the existence of a Classic period count of lunar months from the heliacal rise of Venus, and the organization of its months according to eclipse intervals. Its month grouping and that of *Dresden* are derivable from a more basic eclipse-structured sequence, for example of 2, 3, 3, 3, 2, 3, and $3\frac{1}{2}$ months.

21 They probably were not capturing recurrences outside a given Venus year. After one Venus year of just under 20 lunar months, the moon's phase is offset by about 23 days; the lunar phase does not return to the same position until eight more Venus years have been completed, after which the lunar cycle is offset by half a day. A station in the Venus cycle that had hosted an eclipse during one year cannot do so again nine Venus years later; $9 \cdot 584$ days $= 178$ lunar months, which cannot separate two eclipses (nor can its lower multiples). Accordingly, the lunar phase regularity appears to be a phenomenon of a single Venus year, permitting the association of eclipses at one Venus station with eclipses at another. The only possibilities are of two lunar or two solar eclipses, at heliacal risings of morning and evening star, or of one lunar and one solar eclipse, in either order, at disappearances of morning and evening stars. For the former possibilities, lunar eclipses are more likely than solar owing to the far great frequency with which they can be observed at 11-month intervals from a given location.

22 This presupposes the 584285 correlation, which Lounsbury used in providing the rationale for the 1 Ahau 18 Kayab and 1 Ahau 18 Uo basedates. Even under the 584283 correlation, Lounsbury's rationale is almost as powerful, the events of those days coinciding with heliacal rise of Venus only two days away from the sacred 1 Ahau Venus base; in this case, the intervals are correct projected from true rather than canonical heliacal rise.

23 The span from the 18 Kayab base was 128 sacred rounds, commensurating an eclipse cycle of 1127 months.

24 The best short ratio of replacements is two of 57 years to every nine of 61. The 18 Uo base was instituted by a 57-year shift after an 18 Kayab base; it could be the astronomical base of the table. In this case, the next shift after 3 Xul would occur after 61 Venus years, as usually posited. This, however, leaves no good rationale for the historical 9.9.9.16.0 1 Ahau 18 Kayab base of the Venus table. Using Lounsbury's (1983b) system, which provides this rationale, the 18 Uo base is simply the last basedate prior to the current one, instituted not after a 57-year shift but after two 61-year shifts. In this case, 10.5.6.4.0 1 Ahau 18 Kayab was the astronomical base of the table; error accumulation by the 3 Xul base would then have reached $+3.11$ days, and would be $+3.89$ if the basedate shift occurred after 61 years, as is usually posited. Replacement after 57 years produces a shift of -3.53 days, giving a total error accumulation of -0.42 since the original 18 Kayab base.

25 Thus, also in the synodic period from full moon to full moon completed just before canonical heliacal rising on the actual basedate.

26 The forward count may also have been relevant. It does not join recorded positions in the table. However, it links a position four days before a date of canonical heliacal rise of morning star with a position ten days after canonical heliacal rising of evening star. These figures match others just discussed: it was noted above that $1.5.5.0 + 4.18.17.0$ joins one eclipse station to a subsequent date ten days after another eclipse station, so canonical heliacal rise of evening star on a day 4 Oc is an eclipse station if 9.11.7.0 joins a prior eclipse station to a day 1 Ahau four days before canonical rise of morning star; the latter position is precisely where a new basedate is to be located

in the 61st year after the current basedate. For the eclipse station at rising of morning star to be a recorded position, the current basedate has to be retained rather than replaced. Historically, the only retained base was the 18 Kayab of AD 1034; if the relevance of a forward count of 4.18.17.0 + 1.5.5.0 is historical rather than formal, it therefore corresponds to an eclipse station at the canonical rising of morning star at 4 Oc 13 Mol in AD 1057, projected from an anticipated 1 Ahau 8 Yax basedate that was not actually adopted.

27 Bricker and Bricker devise admittedly *ad hoc* explanations for the other factors, attempting to relate them to their proposed mechanism for anticipating the onset of retrograde motion. Each receives a separate type of explanation in this regard, and neither is reflected in the structure of the table.

28 As is usual, they treat the 78-day unit and its subdivisions as applying throughout the Mars cycle; this is suggested by the table of multiples of 78 days, through 10 · 78, that immediately precedes it. However, they treat the accompanying scenes as referring only to one particular 78-day span out of 780, the period of retrograde motion to which they feel the almanac primarily pertains. The span to which the scenes pertain, they argue, has 3 Cimi as its base. The evidence is that, while the bases of the Venus and eclipse almanacs appear in a column adjacent to the almanac portion, in the table of multiples, 3 Lamat is not adjacent to the 78-day almanac. The absence of 3 Lamat, however, is simply a consequence of the reuse of the almanac ten times from the original base; *no one* 78-day base can properly be associated with the 78-day almanac if it is intended to apply throughout the 780 days implied by the table of multiples. Other almanacs with this structure are introduced by a series of distinct ritual calendar dates, usually having the same numerical coefficient since most almanacs span some multiple of 13 days before recycling; such an introduction is found in the prefatory table, which provides the multiples of the 78-day period in association with the dates to which these multiples lead from a day 3 Lamat. The 3 Cimi date adjacent to the almanac is simply the lowest multiple of 78, and it is in the position that is usual for the lowest multiple in the *Dresden* multiplication tables.

The Brickers (p. 54) take the 1 · 78 day multiple as a distance number to be counted to the 78-day span to which the pictures pertain; while this is presumably the function of these numbers, it is equally true of all the other $n \cdot 78$-day spans. Given the parallelism with the structure of other tables, 3 Cimi is not a uniquely significant implied entry date for the 78-day almanac. The only rationale for using this

particular entry is that for the greater part (about two-thirds) of that span, Mars is in retrograde. Since there is no intrinsic link of the pictures to any particular multiple, the association of the pictures with this particular period is purely hypothetical, not a point established by any line of evidence. Ethnohistoric parallels from the burner ceremonies of the Tizimin indicate that depictions in almanacs generally correspond to each multiple, not only one; and if only one multiple were involved, the *a priori* best position would be the last or possibly first entry, not the second. Finally, while it is possible to see the iconographic variations in the pictures as related to the relative visibility of the planet, such a connection is speculative at best. The only support is in a reading of the verb in the accompanying passages as *wak* 'protrude'. Without going into details, the phonetic justification for this reading seems weak, as is the semantic relation of 'protrude' to the visibility or brightness of a planet. It is shown in Section 8.4 that significant points in planetary cycles were occasions for warfare, and the hafted axe sign is frequently used to signal such events; the warlike associations of the Venus table's iconography support the view that it is overtly war that is being discussed.

29 J. Linden (1986; unpublished, 1982) has identified this compound as a count of days in a cycle of six days, with a one-day uncertainty in the coefficient (i.e. a given long count date can be assigned either of two adjacent Glyph Y coefficients unpredictably). This fits with the magnitude of the coefficients as well as with the use of the *-b'-iš* '(days) later' suffix on the numeral 5. However, this hypothesis fails to predict the correct form of Glyph Y about a third of the time.

30 This pattern has been verified in partly derivative work by Houston.

31 In the Dos Pilas area, however, a conquest state emerged in the Late Classic that involved the incorporation of several previously independent sites within the polity named by the Dos Pilas emblem (Houston and Mathews, 1985).

32 Nonetheless, the seasonal dispersion is greater than among the Aztec; it may be that the Maya zone provided better river transport, or greater seasonal availability of crops for the support of an invading army.

33 12 of 25 days explicitly marked as 'star events' involve a significant association with the cycle of Venus, while 13 do not; and, of the positive cases, two have to be treated as one for purposes of statistical analysis, since they reflect battles four days apart during the same war, and thus are not statistically independent trials. In addition, one of the dates (actually, the pair

separated by four days and here treated as one agreement) attains a Venus significance only if it is assigned a date one calendar round later than that assigned by Mathews (1980) using strictly historical evidence. So we have ten or 11 agreements in 24 dates marked as star or Venus events.

Statistical assessment of the significance of this rate of agreement requires a definition of a Venus association. There are two possibilities: they might correspond to a natural subdivision of the Venus year, or to significant points within it. The significant points in the cycle are the two periods of invisibility, the two stationary points, the two maximum elongations, and the first and last appearances of the morning and evening stars. The natural subdivisions would be the periods of morning and evening star, and the periods of invisibility separating them. To my knowledge, no one investigating the possible Venus correlates of Maya dates has considered the latter possibility, so the appropriate model for analysis can be restricted to the model of significant points in the Venus year.

Analysis also requires a decision upon how precisely the significant points of the Venus cycle are to be defined. For example, Lounsbury takes arrival at and departure from maximum elongations and stationary points as being especially significant; however, departure is distinguished from approximate maximum elongation by the positions of only three dates, and, given that at least three days would be suitable at the rate of relative motion and this over a span of only 10–12 days, such placement cannot be reliably distinguished from approximate maximum. A period roughly comparable to the period in which motion away from maximum cannot be perceived would precede the attainment of the maximum elongation, so a span of at least 20 days must be allotted to both maximum eastern and maximum western elongation in a given Venus year. The other periods are more straightforward. Invisibility lasts an average of eight days around inferior conjunction and 50 around superior conjunction; about three days should be assigned to the possible dates for first and last appearance of evening and morning star and on the location of stationary points, owing to variable viewing conditions and uncertainty in the precise determination of these days under optimal conditions. Thus, about 116 days out of 584 would provide rather precise agreement with points of the Venus cycle, increasing to 130 if two additional days are allotted to each span in view of a two-day uncertainty concerning the proper correlation coefficient. The probabilities of getting ten and 11 chance associations with the Venus cycle in 24 dates would then be 2.61 per cent and 0.89 per cent, respectively; the Venus associations are

therefore supported. (These probabilities should probably be lowered, since there are no instances in the 50-day period of invisibility around superior conjunction. The probability of ten or 11 agreements, given that no instance falls in this period, are 0.15 and 0.03 per cent, respectively.)

34 The chance of getting 15 or more dates within ten days of one of a stationary point or opposition of Jupiter and/or Saturn is only 1.1 per cent, so these associations are also probably genuine. They include eight of the 13 dates not associated with a significant point in the cycle of Venus. Invisibility around superior conjunction seems to be avoided, with only one instance (for Jupiter); this instance coincides with a stationary point of Saturn. However, this apparent avoidance may not be genuine, since there is a 7.0 per cent chance of getting one or fewer examples in 29 cases.

35 Kelley (1976, p. 50) suggests and Closs (1981) argues more strongly for a directional association for the star date on Tikal Temple IV, Lintel 3, B4-A5. The main evidence is the Turtle compound that immediately follows the star event compound, which they relate to the Turtle asterism. However, the parallels with the other star event compounds suggest that it should name the place conquered, and indeed the Turtle designates the nearby site of Yaxha. Closs misreads the *yaš* 'green' prefix as *čak* 'red, great' in making his argument; *yaš* is the canonical 'water group' prefix associated with the Yaxha site name.

36 They were in fact more broadly applied; Chan Bahlum also celebrated the 12th vague year anniversary of his accession; Lounsbury (Chapter 19, this volume) notes that this is within seven days of 11 average synodic periods of Jupiter. In the case of Venus and Jupiter, these commensurations are multiples of four vague years, and thus recover the same day in the veintena, but this may not have been a deliberate emphasis: the shortest approximate commensurations for the other planets would be seven vague years for Mercury, nine for Mars, and 29 for Saturn; recall also the deviation of two days in the solar/Venus anniversary celebrated by Chan Bahlum, thereby unnecessarily missing the veintena recurrence.

37 McCluskey suggests a departure from calendrical commensuration as a strategy for constructing the long-term Venus cycles, on the assumption that long-term seasonal Venus cycles were the focus of interest. He notes that good seasonal cycles recur after the 157th and 313th Venus years, spans that do not accommodate commensuration with the ritual calendar. He interprets the 301 Venus years of the reconstructed Venus table correction cycle as an

approximation to this span that also accommodates the ritual calendar. However, this interval is not actually a part of the structure of the table. What we glean from the Venus table's preface is the use of 61 and occasionally 57 Venus synodic periods to recover 1 Ahau at or near heliacal rise. The cycle of 301 Venus periods is simply the result of the best sequence of these elementary cycles; it is nowhere mentioned in the tables, and in practice selection of which correction to apply could have been based on the amount of error *being observed* in the projected dates of heliacal rise. The 'closeness' of the 301 to 313 Venus synodic periods is no sound basis for this proposal, since shorter and longer commensurations of the earth's and Venus' sidereal periods are available to produce seasonal recurrences at any general magnitude desired. In any event, the system evidenced by the tables requires about a century to devise and validate; McCluskey suggests that a cycle of 301 Venus years was accommodated to this 'Beginning with an observation of 314' Venus synodic periods – nearly 500 years! He supposes that this span, 13 days off from a commensuration to the sacred round, was linked on the one hand with the 313-period cycle he postulates and with a 317-period cycle, off by 17 days – quite crude commensurations by the standard of the Venus table. He then supposes that this last estimate was corrected to a 301-period cycle by knowledge that 16 Venus periods accumulate a four-day average difference from the sacred round, roughly cancelling the (*average!*) error in the 500 year span. The complexities and assumptions required by McCluskey's reconstruction are far more difficult than Teeple's model; a model of simple addition of shorter-term ritual commensurations produces the same result, and its elements are built into the structure of the table as the elements of McCluskey's model are not.

38 Solar year commensurations with a planetary synodic period simultaneously commensurate the sidereal period of the same planet, placing it in the same part of the sky relative to the stars. Sidereal commensurations with demonstrated astronomical significance are the Venus/solar commensurations discussed above, along with a parallel case for Jupiter (note 30; Lounsbury, Chapter 19, this volume), and long-term examples involving Jupiter and Saturn (Kelley, 1975).

39 It should be noted that Schlak's result is not definitive, since the sample of monuments depicting the holding of this scepter was quite small and since it was not contrasted with the results for the dedicatory dates of monuments that lack such depictions.

40 Sky bands are rectangular bands with enclosures containing signs whose referents are predominantly signs for sky, moon, sun and star; they seem to indicate celestial locations of illustrated deities.

41 Coggins (1980) has argued that the glyphic compounds identified with north and south by previous epigraphers were in fact for zenith and nadir. Based on these identifications, Bricker (1983) proposed phonetic readings of the directional compounds; these readings do not provide substantive support for this interpretation, however, for they hinge upon controversial and, in my view, unlikely readings of the main signs of each compound (T1016c as *ka?n*, T575 as *mak*), and in addition upon hypothetical rather than attested linguistic forms for 'nadir' and 'zenith'.

The attested forms for 'zenith' and 'nadir' contain the word for 'earth' in Classical Yucatec, with preposed qualifiers meaning 'within' and 'above'. Directional compounds customarily occur in a series of four; each directional compound in these four-part series is always unambiguously identifiable with one of the basic four directional compounds. There are cases in which series of six or seven directional compounds occur. In the series of six in the *Madrid Codex* on pp. 77–8, four are the traditional ones known from the four-direction series, while the other two never appear in series of four. Both of the extra directional compounds include the sign for *kab'* 'earth', along with the phonetic sign **ba₄** as a phonetic complement securing that reading. There is therefore a *prima facie* case that these compounds refer to zenith and nadir; thus, the compounds in the four-part series do not, since these spellings with the earth sign do not appear as alternatives to what have traditionally been interpreted as north and south compounds in the four-part directional series. Each of these directional earth compounds also includes a qualifier. In one compound, T96 and/or T283 is the modifying element. T96 evidently represents one or more words meaning 'in, within', in the codices, since it is used for locations of deities pictured in, not simply at, their named locations; T283 represents *c'* + an undetermined vowel, since it is a phonetic rendering of the final consonant of the month name *so·c'/suc'* in two Classic period spellings of that name. The spelling variations are best accounted for by supposing that T283 represented phonetic *c'u*, since *c'uh* also means 'in, within'; the compound would then represent *c'uh kab'*, the Classical Yucatec term for 'nadir'. I know of no compelling evidence, outside these directional compounds, for the meaning of T663:23, the modifier in what appears to be the 'zenith' compound.

Lounsbury (1984, pp. 179–80) discusses a phonetically explicit spelling of 'north', which is simply a variant of the usual spelling of what is traditionally identified as the 'north' compound. The usual main-sign, T1016, is followed by a sign pair that can be securely read **-ma-n(a)**, indicating that the word represented ended in *-man*; he also interprets the traditional prefix as representing *šam*, although this is at least debateable. Even restricting ourselves to the phonetically secure portion of the spelling, the traditional interpretation of the compound as a spelling for Yucatec *šaman* 'north' is confirmed. Finally, Lounsbury notes that an interpretation of T74 as meaning 'great' would account for its optional use in various titles as well as its use as a semantic determinative in the south compound, Mayan terms for this direction being typically based upon a word for 'big, great', as the direction when facing the rising sun of what in Mayan is the 'greater' (i.e. the right) hand.

References

Aveni, A. F. (1977) (ed.). *Native American Astronomy*. Austin: University of Texas Press.

Aveni, A. F. (1980). *Skywatchers of Ancient Mexico*. Austin: University of Texas Press.

Aveni, A. F. (n.d.). The moon and the Venus table in the *Dresden Codex:* an example of commensuration in the Maya calendar. Paper presented at the *Conference on Ethnoastronomy*, Washington, DC, September, 1983 (in press).

Aveni, A. F., Gibbs, S. L. and Hartung, H. (1975). The Caracol tower at Chichén Itzá: an ancient astronomical observatory? *Science* **188**, 977–85.

Bricker, H. M. and Bricker, V. R. (1983). Classic Maya prediction of solar eclipses. *Current Anthropology* **24**, 1–24.

Bricker, V. R. (1983). Directional glyphs in Maya inscriptions and codices. *American Antiquity* **48**, 347–53.

Bricker, V. R. (n.d.). Classic Maya observations of planetary retrograde motion. Paper presented at the *Conference on Ethnoastronomy*, Washington, DC, September, 1983 (in press).

Bricker, V. R. and Bricker, H. M. (1986). The Mars Table in the Dresden Codex. In *Research and Reflections in Archaeology and History: Essays in Honor of Doris Stone*, ed. E. W. Andrews V. New Orleans: Middle American Research Institute Publication 57.

Brotherston, G. (1983). The year 3113 BC and the fifth sun of Mesoamerica: an orthodox reading of the *Tepexic Annals*. In *Calendars in Mesoamerica and Peru. Native Computations of Time*, eds. A. Aveni and G. Brotherston, pp. 167–220. Oxford: British Archaeological Reports Series S174.

Closs, M. (1977). The date-reaching mechanism in the Venus table of the *Dresden Codex*. In *Native American Astronomy*, ed. A. F. Aveni, pp. 89–99. Austin: University of Texas Press.

Closs, M. (1979). Venus and the Maya world: glyphs, gods, and associated phenomena. In *Tercera Mesa Redonda de Palenque*, ed. M. G. Robertson, pp. 147–66. Pebble Beach: Robert Louis Stevenson School.

Closs, M. (1981). Venus dates revisited. *Archaeoastronomy* **4**, 38–41.

Closs, M., Aveni, A. and Crowley, B. (1984). The planet Venus and Temple 22 at Copan. *Indiana* **9**, 221–47.

Coggins, C. (1980). The shape of time: some political implications of a four-part figure. *American Antiquity* **45**, 727–39.

Coggins, C. (1982). The zenith, the mountain, the center, and the sea. In *Ethnoastronomy and Archaeoastronomy in the American Tropics*, eds. A. F. Aveni and G. Urton, pp. 111–23. Annals of the New York Academy of Sciences, vol. 385.

Edmonson, M. S., ed. and trans. (1982). *The Ancient Future of the Itza: The Book of Chilam Balam of Chumayel*. Austin: University of Texas Press.

Förstemann, E. (1904). Page 24 of the Dresden Maya manuscript. *BAE Bulletin* **28**, 431–43. Washington: Smithsonian Institution.

Förstemann, E. (1906). *Commentary on the Maya Manuscript in the Royal Public Library of Dresden*. Peabody Museum of Archaeology and Ethnology, Paper 4.2, Cambridge, Mass.

Fox, J. A. and Justeson, J. S. (1978). A Mayan planetary observation. *UCARF Contributions* **36**, 55–9.

Fox, J. A. and Justeson, J. S. (1980). Mayan hieroglyphs as linguistic evidence. In *Palenque Round Table, 1978, Part 2*, ed. M. Greene Robertson, Palenque Round Table Series, vol. 5, pp. 204–16. Austin: University of Texas Press.

Fox, J. A. and Justeson, J. S. (1984). Polyvalence in Mayan hieroglyphic writing. In *Phoneticism in Mayan Hieroglyphic Writing*, eds. J. S. Justeson and L. Campbell, pp. 17–76. Albany: IMS Publication 9.

Gibbs, S. L. (1977). Mesoamerican calendrics as evidence of astronomical activity. In *Native American Astronomy*, eds. A. F. Aveni, pp. 21–35. Austin: University of Texas Press.

Graham, I. (1967). *Archaeological Explorations in El Petén, Guatemala*. New Orleans: Middle American Institute Publication 33.

Graham, J. A. (1972). *The Monumental Art and Hieroglyphic Inscriptions of Altar de Sacrificios, Guatemala*. Peabody Museum of Archaeology and Ethnology, Paper 64.2. Cambridge, Mass.

Hassig, R. (1988). *Aztec Warfare: Imperial Expansion and Political Control*. Norman: University of Oklahoma Press (in press).

Houston, S. D. (1984). An example of homophony in Maya script. *American Antiquity* **49**, 790–805.

Houston, S. D. and Mathews, P. (1985). *The Dynastic Sequence of Dos Pilas, Guatemala*. Pre-Columbian Art Research Institute, Monograph 1.

Jones, C. and Satterthwaite, L. (1982). *The Monuments and Inscriptions of Tikal: the Carved Monuments*. Tikal Report 33, Part A. Philadelphia: University Museum.

Justeson, J. S. and Campbell, L. (1984) (eds.). *Phoneticism in Mayan Hieroglyphic Writing*. Albany: IMS Publication 9.

Justeson, J. S., Norman, W. M. and Hammond, N. (1988). The Pomona flare: a Preclassic Maya hieroglyphic text. In *Maya Iconography*, ed. E. P. Benson. Princeton University Press (in press).

Justeson, J. S., Norman, W. M., Campbell, L. and Kaufman, T. (1985). *The Foreign Impact on Lowland Mayan Language and Script*. New Orleans: MARI Publication 53.

Kelley, D. H. (1972). The nine lords of the night. In Studies in the Archaeology of Mexico and Guatemala, ed. J. A. Graham. *UCARF Contributions* 16, 58–68.

Kelley, D. H. (1975). Planetary data on Caracol Stela 3. In *Archaeoastronomy in Pre-Columbian America*, ed. A. F. Aveni, pp. 257–62. Austin: University of Texas Press.

Kelley, D. H. (1976). *Deciphering the Maya script*. Norman: University of Oklahoma Press.

Kelley, D. H. (1977a). Maya astronomical tables and inscriptions. In *Native American Astronomy*, ed. A. F. Aveni, pp. 57–73. Austin: University of Texas Press.

Kelley, D. H. (1977b). A possible Maya eclipse record. In *Social Processes in Maya Prehistory*, ed. N. Hammond, pp. 405–8. New York: Academic Press.

Kelley, D. H. (1980). Astronomical identities of Meso-american gods. *Journal for the History of Astronomy* 9, Suppl. 2, S1–54.

Kelley, D. H. (1983). The Maya calendar correlation problem. In *Civilization in the Ancient Americas: Essays in Honor of Gordon R. Willey*, eds. R. M. Leventhal and A. L. Kolata, pp. 157–208. Albuquerque: University of New Mexico Press.

Kelley, D. H. and Kerr, A. (1973). Maya astronomy and astronomical glyphs. In *Mesoamerican Writing Systems*, ed. E. P. Benson, pp. 179–215. Washington: Dumbarton Oaks.

Linden, J. (1986). Glyph X of the Maya Lunar Series: an eighteen month lunar synodic calendar. *American Antiquity* 51, 122–36.

Lounsbury, F. G. (1976). A rationale for the initial date of the Temple of the Cross at Palenque. In *Primera Mesa Redonda de Palenque, Part II*, ed. M. G. Robertson, pp. 5–19. Pebble Beach: Robert Louis Stevenson School.

Lounsbury, F. G. (1978). Maya numeration, computation, and calendrical astronomy. *Dictionary of Scientific Biography* 15, 759–818. New York: Scribner's.

Lounsbury, F. G. (1982). Comment on Bricker and Bricker (1982). *Current Anthropology* 24, 22.

Lounsbury, F. G. (1983a). Astronomical knowledge and its uses at Bonampak, Mexico. In *Archaeoastronomy in the New World*, ed. A. F. Aveni, pp. 143–68. Cambridge University Press.

Lounsbury, F. G. (1983b). The base of the Venus table of the Dresden Codex, and its significance for the calendar-correction problem. In *Calendars in Mesoamerica and Peru*, eds. A. F. Aveni and G. Brotherston, pp. 1–21. London: British Archaeological Reports S 174 (International series).

Lounsbury, F. G. (1984). Glyphic substitutions: homophonic and synonymic. In *Phoneticism in Mayan Hieroglyphic Writing*, eds. Justeson, J. S. and Campbell, L., pp. 167–84. Albany: IMS Publication 9.

McCluskey, S. C. (1983). Maya observations of very long periods of Venus. *Journal for the History of Astronomy* 14, 92–101.

Malmström, V. H. (1973). Origin of the Mesoamerican 260-day calendar. *Science* 181, 939–41.

Malmström, V. H. (1978). A reconstruction of the chronology of Mesoamerican calendrical systems. *Journal for the History of Astronomy* 9, 105–16.

Marcus, J. (1976). The origins of Mesoamerican writing. *Annual Review of Anthropology* 5, 35–67.

Martínez Hernández, J. (1929). *Diccionario de Motul, maya espanol, atribuido a fray Antonio de Ciudad Real y arte de lengua maya por fray Juan Coronel*. Merida: Tipográfica Yucateca.

Mathews, P. (1986). Early Classic Maya monuments and inscriptions. In *The Maya Early Classic*, eds. G. R. Willey and P. Mathews. Albany: IMS Publication 10.

Meinshausen, M. (1913). Über Sonnen- und Mondfinsternisse in der Dresdener Mayahandschrift. *Zeitschrift für Etnologie* 45, (2), 221–7.

Miller, M. (1986). *The Murals of Bonampak*. Princeton University Press.

Norman, W. M. (1984). Grammatical analysis of Mayan hieroglyphs. Paper presented at the *2nd Annual Workshop on Maya Hieroglyphs*, University Museum, University of Pennsylvania, Philadelphia.

Proskouriakoff, T. (1960). Historical implications of a pattern of dates at Piedras Negras, Guatemala. *American Antiquity* 25, 454–75.

Proskouriakoff, T. (1963). Historical data in the inscriptions of Yaxchilan. *Estudios de Cultura Maya* 3, 147–67.

Proskouriakoff, T. (1964). Historical data in the inscriptions of Yaxchilan. *Estudios de Cultura Maya* 4, 177–201.

Riese, B. (1982). Kriegsberichte der klassischen Maya. *Baessler-Archiv: Beiträge zur Völkerkunde* **30**, 255–321.

Satterthwaite, L. (1958a). Early 'Uniformity' Maya moon numbers at Tikal and elsewhere. *Thirty-third International Congress of Americanists* **2**, 200–10.

Satterthwaite, L. (1958b). Five newly discovered monuments at Tikal and new data on four others. *Tikal Reports* **4**. University Museum, Philadelphia.

Schele, L. D. (1982). *Maya Glyphs: the Verbs*. Austin: University of Texas Press.

Schele, L. D. and Miller, J. H. (1983). *The Mirror, the Rabbit and the Bundle: 'Accession' Expressions in the Classic Maya Inscriptions*. Studies in Pre-Columbian Art and Archaeology 25. Washington: Dumbarton Oaks.

Seler, E. (1904). Venus period in the picture writings of the Borgian Codex Group. *BAE Bulletin* **28**, 353–91. Washington: Smithsonian Institution.

Spinden, H. J. (1916). The question of the zodiac in America. *American Anthropologist* **18**, 53–80.

Spinden, H. J. (1924). *The Reduction of Maya Dates*. Peabody Museum of American Archaeology and Ethnology, Paper 6.4, Cambridge, Mass.

Stahlman, W. D. and Gingerich, O. (1963). *Solar and Planetary Longitudes for Years −2500 to +2000 by 10-day Intervals*. Madison: University of Wisconsin Press.

Stone, A. (1983). Epigraphic patterns in the inscriptions of Nah Tunich cave. In *Contributions to Maya Hieroglyphic Decipherment*, ed. S. D. Houston, vol. 1, pp. 88–103. New Haven: HRAF.

Stuart, D. (1984). A note on the 'hand-scattering' glyph. In *Phoneticism in Mayan Hieroglyphic Writing*, eds. J. S. Justeson and L. Campbell, pp. 307–10. Albany: IMS Publication 9.

Stuart, D. (1985). The 'count of captives' epithet in Classic maya writing. In *Fifth Palenque Round Table, 1983*, eds. M. Greene Robertson and V. M. Fields, Palenque Round Table Series, vol. 7, pp. 97–101. San Francisco: Pre-Columbian Art Research Institute.

Tate, C. (1988). Summer solstice ceremonies performed by Bird Jaguar III of Yaxchilan, Chiapas, Mexico. *Estudios de Cultura Maya* (in press).

Teeple, J. (1925). Maya inscriptions: Glyphs C, D, and E of the Supplementary Series. *American Anthropologist* **27**, 108–15.

Teeple, J. (1930). *Maya Astronomy*. Washington: CIW Contributions to American Archaeology 1.2.

Thompson, J. E. S. (1935). *Maya Chronology: the Correlation Question*. Washington: CIW Publication 456, Contribution 14.

Thompson, J. E. S. (1950). *Maya Hieroglyphic Writing: An Introduction*. Washington: CIW Publication 589. (Reprinted 1960, 1971, 1979 by University of Oklahoma Press.)

Thompson, J. E. S. (1972). *A Commentary of the Dresden Codex, a Maya Hieroglyphic Book*. Philadelphia: APS Memoir 93.

Webster, D. (1975). Warfare and the evolution of the state: a reconsideration. *American Antiquity* **40**, 464–70.

Whittaker, G. (1986). The Mexican names of three Venus gods in the *Dresden Codex*. *Mexikon* **8**, (3), 56–60.

Willson, R. W. (1924). *Astronomical Notes on the Maya Codices*. Peabody Museum of American Archaeology and Ethnology, Paper 6.3. Cambridge, Mass.

9

Cosmological, symbolic and cultural complexity among the contemporary Maya of Yucatan

John R. Sosa *State University of New York at Cortland*

The documentation of the cosmologies of the contemporary traditional societies of the world has reached a significant degree of thoroughness in the past three decades. As a result, the extent to which all humans, at all technological and social levels, express and rely upon considerably complex cultures and cosmologies in symbolic form, is becoming more and more apparent. Nevertheless, and more often than not, part of our own Western socialization process, and part of our own self-image as a people, is our classification of such societies, by both academics and lay people alike, as 'primitive'. It seems even more apparent then, that we, as so-called archaeo- and ethnoastronomers, should seize the opportunity to challenge this pejorative and unnecessarily limiting perspective that the West has on the rest of humanity. Because today, in the subject matter that we concern ourselves with, we have at our disposal evidence of the extraordinary capabilities that humans have for learning about, and finding cultural and social significance in, their total natural environment. Indeed, the only thing that is primitive is our understanding of this process. For my part, I have tried to begin to contribute towards this end by documenting and analyzing the relation between a cosmology, or a culturally specific theory of the universe, and social behavior, through ethnographic inquiry among the Maya of Yucatan. I undertook this field study in the community of Yalcobá and many of its surrounding villages and hamlets during 1982 and 1983. But before detailing my findings and suggesting some of their implications for the aforementioned problem, it is necessary to review a certain theoretical

and methodological rationale which structured my research.

A major point of consensus between the larger disciplines of sociology, anthropology, psychology, linguistics, folkloristics, poetics, philosophy, literary criticism and the history of religion, and this list is probably not complete, is that all humans live within a conceived 'world' that needs to be explained and interpreted continually. And this is not just considered to be a common human trait, which it surely is, but is borne out of the need of all people to survive through the exploitation of their 'world's' natural environment. The resulting interaction, though, between humans and nature, is also an inseparable dimension of *social* survival, since all humans live in groups.

This also being a characteristic of the human condition, and therefore having been one as long as humans have been physically and mentally human, a period of at least 50 000 years, we can expect that all groups, as social units of interaction, which compart the learned information necessary for both social and physical survival, will have cosmologies of comparable complexity. In short, this information, which gains significance in the dynamic and fluid medium of human social intercourse, is culture, and quite simply all people in all groups share some variant of it, and have done so for a long time.

Unfortunately though, we have not learned this soon enough, and the baggage of a Victorian age evolutionary paradigm still colors our perception of 'cultural otherness'. Words like 'primitive', 'civilization', 'developed', 'advanced', and even 'barbarian' and 'savage', still obscure our potential for appreciating the complexity of other cultures, irrespective of their technology or social complexity. While it is true that the larger the society, the larger the number of potential interpersonal relationships, and the more complex the resulting social system, our fatal flaw is that we still view other simpler societies from our assumed superior position, and further assume cultural simplicity. This is indeed the worst part of our version of ethnocentrism, and it stands as a major obstacle to any discipline that would seek to cross cultural boundaries and try to understand why other people do what they do, or did what they did.

I know I do not stand alone when I ask that whether we are trying to describe and analyze the cosmologies or the remnants thereof, of a particular complex or simple society, *and* its relevance to the social life of its members, that we consider the weight of the multidisciplinary evidence, which identifies a common human characteristic of having a complex cosmology and a complex culture, regardless of their numbers, their technology, or even whether or not they wear clothes.

If this premise is sound, and I believe that it is, then the task becomes to test for this complexity and interpret at least some of its significance. From my point of view, an appropriate methodological and theoretical foundation is provided by the anthropologists Victor Turner and Clifford Geertz, who approach the problem from the standpoint of the component symbols of a cosmology and the culture of which it is a part. Whether they be objects, actions, relationships or words, symbols have been found to be multivocalic, or having 'many voices' or many meanings at the same time, and if the interpretation of these meanings is to be a scientific and rigorous one, then some method must be employed which maximizes analytical objectivity.

Turner (1967, 1977) indeed provides this, and isolates three levels of significance and resulting interpretability that all symbols have. The first is called the exegetical level, which consists of descriptions of meanings by one's informants. The second is the operational level, or the documentation of behavior as observed by the social scientist, and the third or positional level of significance results from the scientist's comparison of the first two, through which the many meanings of a symbol may be interpreted, as they result from its position

in a total and larger cultural system of significance.

Of course, it follows that the symbolic interpretation of cosmologies is not absolute or provable, but instead substantiable by a rigorous comparison of different kinds of data, or what Geertz (1973) calls 'thick description'. The thicker the description then, the more substantiated an interpretation will be, and this process may in fact be a never-ending one, given the dynamic and changeable quality of human culture and symbolization.

What may be most enduring, however, is a cosmological structure, if it continues to be socially relevant. As a set of organizing principles, it represents a society's need to interface with a given environment, and its geological, topographical, floral, faunal, meteorological, climatic and celestial components. Moreover, this will be combined at a symbolic level with emotional and moral significance, and will simply make a society's dominant and most powerful symbols, those which its members express through spoken, written or behavioral activity, as being most important to *them*. This is what I believe I have found to be the case among the Yucatec Maya I worked and lived with, who after more than 400 years of attempts at forced acculturation still maintain a traditional cosmology, every bit as *important*, *vibrant* and *essential for life*, as that of their ancestors.

In the contemporary Maya of Yucatan,[1] we find people who are surrounded by reminders of their present social position. Their first names are Spanish, the names of many of the saints they worship are often Spanish, and many Spanish words have become part of the Maya language through changes in accentuation. In addition, they may also be reminded every day in their contacts with Spanish speakers that theirs is the lowest level of contemporary Yucatec society, and the overpowering nature of this total awareness results, for the most part, in their not actually calling themselves Maya, for this they claim to be the name of their language, but *mestizos*, or 'mixed people'. The complexity of their integration into larger Yucatec society might therefore threaten our potential for treating these people as a traditional non-Western society, but in the many isolated communities in which they live this is indeed possible.

The setting of my attempt to interpret the social relevance of a Yucatec Maya perspective on a total environment was the village of Yalcobá, in the eastern part of the state of Yucatan (Figure 9.1). In this community of 1500 people, I sought to test the basic premises of symbolic theory, through the documentation of the thick description necessary for responsible interpretations. Towards this end, I compiled and compared data sets on both belief and practice from the general public, as well as from the *hmèen*, or the Maya ritual specialist, who also acts as teacher, doctor, diviner, priest and advisor.

As the mediator between the realm of conceived deities and that of humans, his is a unique perspective, and while we might believe that his cosmological knowledge is more sophisticated than that of the public, we must also consider that his higher degree of confidence in the subject matter, given his belief that he is fulfilling a divine calling, will affect his exegetical responses. We might then consider that, given a greater level of insecurity among the general public to discuss their universe, they more easily express the social importance of their cosmology at the operational or behavioral level, and indeed I have found in Yalcobá a rather consistent correlation between the two.

The available historical documents, and the presence of archaeological remains in Yalcobá itself, indicate that this community has been the site of extended if not continuous Maya occupation for a very long time. Perhaps as a result it is an attractive location for testing whether traditional Maya beliefs in a culturally defined 'world' are still socially important. Redfield and Villa Rojas (1934), and Villa Rojas (1945) himself, have described this Maya world as flat and square, with its four corners and sides having directional significance. In Yalcobá, this world is actually called *u yóok'ol kàab'* (Figure 9.2), or 'upon the earth', which focuses our attention on the surface of a three part, but somewhat contiguous universe of sky, flat earth, and 'within the earth'.

The four corners, or *kan tu'uk'*, are described by the *hmèen* as being equivalent to the solstice extremes of rising and setting solar horizon positions, and this horizon is called *u šùul u yóok'ol kàab'*, 'the limit of the earth', or *čùun ka'an*, 'the source of the sky'. Both he and the general public easily discuss the importance of these positions for many facets of daily ritual, social and economic life, and make important distinctions between the sides of the quadrilateral and its corners.

The names of the directions themselves corres-

pond to the *sides*, each of which is called *u tàan*, 'its face', or 'its side', and is therefore not at all a 'cardinal point' as has so often been assumed in the Maya and the Mesoamerican literature. Our Western directional system is in fact based on a completely different set of assumptions and techniques, including a spherical earth concept, a telescopic astronomical technology, and an attention to magnetic attractions. The Maya directions, however, although roughly corresponding to the 'points of the compass' of east, north, west and south, are unquestionably considered by Maya *mestizos* to be 'sides' and not 'points', and reflect their cosmology as much as our Western directions reflect ours.

The solar determinant of these 'sides' and 'corners' is specifically described by both the *hmèen* and the general public, with the latter not making the equivalence with the solstices *per se*, but relating that these corners can be reckoned along the horizon as a calculation one makes, based upon where the sun is seen to rise or set, since *lak'in* or 'east' is always between two of the corners, while *čik'in* or 'west' is between the other two. *Šaman* or 'north' and *nohol* or 'south', are therefore the complements of the first and more primary set of directions. All four may be referred to on a daily basis, either in describing movements as important as those of the sun and moon, whether in their deified or light-emitting forms, or in discussing the mundane details of one's changing position as one walks within Yalcobá or outside of it, usually to one's *kòol*, or 'cornfield'.

This relationship between the *tàan* of *lak'in* and the rising sun, is further expressed by the *hmèen* as it changes during the course of a year. He claims that during June, the sun, or *Hahal Dios* 'the true God', as it can also be called, reaches the corner of *lak'in šaman*, or is comparable to our 'northeast', and that by December it has moved to *lak'in nohol*, or 'southeast'. And in actual public or specialized practice, this is only one way in which this or any other of the corners can be named, with the other possibilities for this one being, *nohol lak'in*, or 'southeast', *lak'in yéetel nohol*, or 'east with south', or *nohol yéetel lak'in*, 'south with east'.

A given corner, then, shares the names of both of the directional sides that it divides, and in this case can even be referred to also as *lak'in* or *nohol*, depending on the speaker's intent. Moreover, both the public and the *hmèen* can call this corner *lak'in*

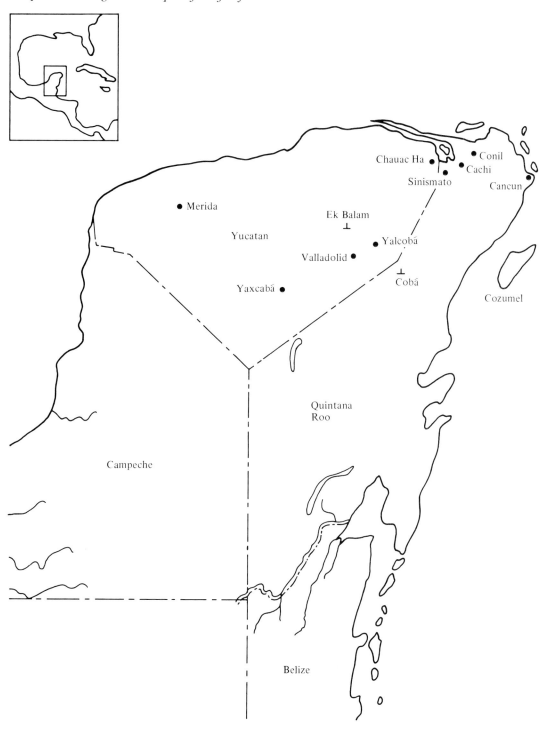

Figure 9.1 Map of Yucatan Peninsula.

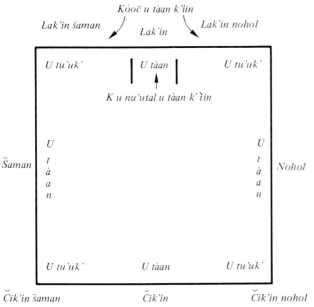

Figure 9.2 U *yóok'ol kàab'* – 'upon the Earth'.

nohol, *ti' lak'in*, for example, or 'towards east', in which case the direction is being 'sighted' as if the speaker were walking along the *nohol* 'side' of the perimeter of the earth, and 'towards' the 'side' of *lak'in*. A reference to the corner *ti' lak'in*, therefore, is the same as naming *u tu'uk' lak'in nohol*, 'the corner of east–south', since the direction of this naming practice is always *no'oha'an*, 'to the right' or 'counterclockwise'. Here, some clarification is necessary.

While this may be illogical for Westerners, who would consider a counterclockwise circuit to be 'to the left', the critical difference is that for the Maya, east, or *lak'in*, is the beginning or 'up' direction, and not the 12 o'clock or 'north' that would determine a Westerner's consideration of 'counterclockwise' as 'to the left'. So, for the Maya, *no'oha'an* is an expression of the direction of movement around a quadrilateral determined by *Hahal Dios'* conceived right hand side as he is rising, facing the observer, as evidenced in the *hmèen's* repeated reference to the 'right hand of *Hahal Dios*' in prayers. It is the rising position of the sun, therefore, with its 'right hand' being towards *šaman*, or north, which gives a 'to the right' meaning to any 'counterclockwise' circuit.

Using the same positional logic, the sun's daily motion in the sky is also described as *no'oha'an*,

because in reaching the western horizon it has moved 'to the right' of its original rising position, as if it had travelled around the perimeter of the quadrilateral earth. On the basis of this *no'oha'an* primacy in considering the directions, the other corners can then also be referred to as, *ti' šaman*, or 'towards north', for *lak'in šaman*, or 'east north', *ti' čik'in*, or 'towards west', for *čik'in šaman*, or 'west north', and *ti' nohol*, or 'towards south', for *čik'in nohol*, or 'west south'.

To continue with our specific discussion of the *hmèen's* reckonings, he describes the condition of the sun's position at either *lak'in nohol* or *lak'in šaman* as *kóoč u tàan k'ìin*, or 'the side of the sun is wide', since it has reached the limits of its annual motion along the horizon, and its rising position will now begin to move in the opposite direction towards the other eastern corner. During March and September however, the *hmèen* describes the sun as reaching the midpoint of this 'side', and calls its position *k u nu'utal u tàan k'ìin*, or 'the sun's side has become narrow'. In this way, the sun's rising position along the eastern horizon is described in terms of its relative 'narrow' position along the total possible extent of its annual motion, as bounded by two of the world corners.

This culturally specific Maya interpretation of the natural environment of their everyday experience is intertwined with notions of how it should and can be exploited for survival, and how this is even possible, through the belief in certain related deity symbols. The most dominant of these is the combined solar/Jesus Christ deity of *Hahal Dios*, who is believed to have created the earth and provided cosmic order. All of his other assistants, such as the *čàako'ob'*, or rain deities, nevertheless also compart with him the quality of being able to send evil as well as good, which plays a major role in daily considerations of morality and appropriate social and ritual behavior.

The seasonally essential rains for the successful production of life-giving corn or *'išì'im*, make the *čàak* deities particularly important as dominant symbols, and, while the general public again has many and varied beliefs about them, the *hmèen* offers greater exactitude in his descriptions. For him, the four primary *čàako'ob'* are the *babahtun* deities at the 'four corners of the limit of the earth'. Figure 9.3 depicts this arrangement in which the *hmèen* places *sak babahtun* at *lak'in šaman*, *ek' babahtun* at *čik'in*

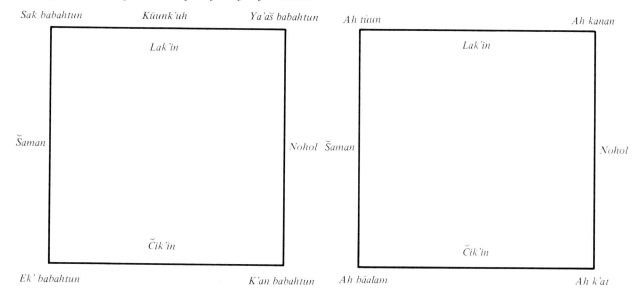

Figure 9.3 The *čàako'ob'* rain deities.

Figure 9.4 The *báalamo'ob'* guardians.

šaman, k'an babahtun at *čik'in nohol* and *ya'aš babah-tun* at *lak'in nohol*.

The prefixes of each are the names of colors, and the *hmèen* explains that the *sak*, 'white', *ek'*, 'black', *k'an*, 'yellow', and *ya'aš*, 'green', correspond to the colors of the clouds of these *čàako'ob'*. For both the *hmèen* and the public though, *Kúunk'uh* is the great thundering *čàak* who lives at the eastern horizon and is heard during July and August. The necessity for rain during a specific 15-day period in the corn plant's growth cycle, thereby makes these deities the focus of the year's most important *hmèen*-directed ritual, the *č'a čàak*, or 'take, *čàak*'. And just as the *čàako'ob'* are assigned a high value of importance due to their role in maintaining life, so are the *yum báalam* deities accorded this same status.

Both of these sets of symbols share the characteristics of being organized by the four corner structure, and both the *hmèen* and the public describe these 'guardians' as fulfilling their protective function from the four corners of either a town, a hamlet, a *kòol* cornfield, a cattle corral, a houseplot, a section of the forest, or *u yóok'ol kàab'*, 'the world' itself, as depicted in Figure 9.4.

In this, we see the *hmèen's* more specific distinctions of *yum báalam* positions, described as *lak'in šaman* for *ah tùun*, 'the stone', *čik'in šaman* for *ah báalam*, 'the guardian', *čik'in nohol* for *ah k'at*, 'the

clay dwarf', and *lak'in nohol* for *ah kanan*, 'the protector'. The multifaceted beliefs in these deities' capabilities are perhaps most evident in their role in *loh kàah*, or the ritual 'redemption of the village'.

Although this ceremony has not been performed in Yalcobá for about 12 years, *loh kàah* is described as an all night ritual, ideally performed by two *hmèeno'ob'*, as one remains at the church and other makes the ritual circuits along *u pàač kàah*, the path that surrounds Yalcobá. The intent is to purge an evil wind that is making the residents ill, or to prevent one from entering the community, and to accomplish either of these objectives ritual offerings are taken to *u pàač kàah*, and placed at all the *u hol kàah*, the 'holes of the town', or 'entrances to the town'. Specifically, these offerings, made to *Hahal Dios* and Yalcobá's guardian *báalamo'ob'* will be placed on the stone altars of the wooden crosses, which both represent these deities and mark these 'holes' in the 'edge' of the town. Particular attention is paid to *u tu'uk' kàah*, or 'the four corners of the town', marked by their crosses, since they are considered to be its primary guardians.

Figure 9.5 illustrates the extent to which *u pàač kàah* encompasses Yalcobá, and along with the position of the crosses serves to define the extent of the community, which in recent years has expanded to the east. When this happens, however, a new *pàač kàah* is cut, and, even though *loh kàah* has not

Figure 9.5 Map of Yalcobá. Wells are represented by ●.

been performed for a while, this footpath is still referred to on a daily basis for describing one's own or someone else's movements in the community. What this indicates is that this 'edge of Yalcobá' still has the delimiting meaning that it has in *loh kàah*.

So, in a number of different contexts, we can see how the quadrilateral structure of a conceived Maya world, which is itself defined by annual and daily solar motion, is duplicated and expressed through symbolic expression. In addition to the complexity of these relationships, and the inextricable role of cosmological beliefs, we can gain some insight into the cultural meanings that the Christian cross has for the Maya, and how it also occupies a logically ordered position in a cosmological and social scheme.

For the Maya, a cross can exist in two forms, with both being called *le sàantoho'* or 'that saint', but the most common kind is the standard Christian form, while the second version assumes the form of a tree trunk with two branches tilted upward (see Figure 9.6). This is possible because included in the cross symbol is the meaning of the tree, and the term *sàantoh de če'*, 'cross of wood', refers to these crosses, usually about three feet tall, as they are located at the 'holes', where roads lead out of Yalcobá, as well as to the standard wooden crucifixes of Maya households.

These meanings of tree and wood are actually inseparable in Maya, since the word *če'* is used for both. That is not to say that Maya is 'missing a word', but simply that in Maya the distinction is not necessary. So the crosses being made of wood

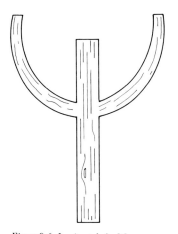

Figure 9.6 Le sàantoho' – Maya cross.

also allows them to share the 'tree' meaning. Also, virtually all household crucifixes, whether new or old, in 'cross' or 'tree' form, are painted green, and hence are called *ya' aš če'*, or 'green tree', 'green wood'. Not coincidentally, the *ya' aš če'* is a common species of tree, *ceiba pentandra* (Standley, 1977, p. 421) or simply *ceiba* in Spanish. Although beyond the bounds of our present discussion, this was the great 'first tree of the world' of the pre-Columbian Maya, and some future analysis would do well to compare the many pre-historic and historic references to this tree with contemporary meanings. For now though, we must question what the significance of the green coloring is for these crosses or *sàanto-ho'ob'*.

Exegetically, this is explained by the Maya as depicting the tree or cross as *kuša'an*, or 'alive', and is supplemented by the belief that one should not cut bush during Easter week or *semàanah sàan-tah* in March or April, even though this is the time when felling bush can be taking place, because of the fear that, by cutting down a tree at this time, one might be cutting the body of Jesus Christ. So the 'living cross' usage of *sàantoh* combines the meanings of tree, *ya' aš če'* and Jesus Christ's body in what for the Maya is a completely logical progression. There are still other meanings though, and to understand these we must return to the large crosses at the edge of Yalcobá.

The *sàantoh de če'* located at all the *u hol kàah*, or 'the entrances of the community', are where incoming roads perforate *u pàač kàah* (see Figure 9.5), or 'the back or the surface of the community', and where *k'ak'as íik'*, or 'bad winds' are believed to be able to enter and bring illnesses. For this reason, it is deemed necessary for the *sàantoh* crosses to be placed there as 'guardians' to protect the community, and indeed I have been told that every Maya town or village in Yucatan, whether old or new, has its guardian crosses at its entrances. In fact, a new *kàah* cannot be thought of as complete until its spatial extent is defined by the *hol č'ak*, or the path cut through the bush around the *kàah* and equivalent to *u pàač kàah*, and its guardian *sàanto-ho'ob'*.

Additional meanings for crosses are seen then, when we observe that many *mestizos*, as they leave Yalcobá and pass them, pick up a small pebble, touch it to their shoe, and place it on the stone slab 'altar' at the foot of the *sàantoh de če'*. These large stones are called the *méesah* or 'table' of the cross, and it is almost inconceivable to have a cross without it being on a 'table' of some kind. The exegetical explanation for this practice is that these crosses are the position of *yum báalam*, the guardian deity, and offering such a stone is intended to ask for his protection as one leaves the confines of the community under his care. *Yum báalam* is also described as resting at this location, and the 'table' can also be referred to as *u k'àan če'*, or his 'bench'. When we examine the components of this symbol's name, we find additional evidence of the basis for these symbolic relationships between its many meanings.

Yum translates as 'father' or 'owner', and today *báalam* is merely the name of this deity, 'the guardian'. But the Cordemex dictionary (Barrera Vásquez, 1980, p. 32) also offers 'guardian', 'priest' and 'jaguar' as definitions, and indeed the Colonial reference to *Chilam Balam* was translated as the 'jaguar priest'. The significance of all this is that, although the current term for jaguar is *čak mòol*, or 'red paw', it is rather common *mestizo* knowledge today that the few Yucatec jaguars that remain habitually rest in the large, shady *ya'aš če'* trees and wait for their prey. Moreover, the *hmèen* describes the *ya'aš če'* as *u k'àan če le yum báalamo'obo'*, or 'the bench of those guardians', and *te'elo' k u he'es kubàah*, 'there he rests himself'. Apparently then, although *bàalam* has lost the meaning of 'jaguar', comparing beliefs of the deity of the same name with practices of this animal allows us to be able to further interpret the related meanings of tree, *ya'aš če'*, Jesus Christ, *yum báalam* and guardian, all contained at the same time in the cross version of *sàantoh*.

What is perhaps most significant, though, is that as we have postulated beliefs and practices can be related by the unifying qualities of symbolic meanings, as social and natural phenomena are interpreted to share certain characteristics. In addition, it has been an attention to thick description which has enabled us to comprehend some of the many meanings of *sàantoh*, since we have not assumed that a Christian cross is just a Christian cross, and have probed for Maya cultural significance.

The expression of Maya cosmology at a behavioral level can also be seen in the procession that members of special social groups in Yalcobá make to the four external corners of the church, just as the *hmèeno'ob'* must visit the four corners of Yalcobá during *loh kàah*. The performance of this procession is the culminating event of the *fiesta*, or 'celebration' of the Sacred Sacrament, and during this the residents of Yalcobá make some of their strongest public statements. The focus of this *fiesta* is 40 hours of Christ's suffering before the crucifixion, with the host or sacrament being the major symbol, and with each of the four days of the event corresponding to ten hours of adoration, five in the morning and five in the afternoon.

Beginning with the first day, February 27th, or 28th if it is a leap year, the responsibilities of the day's social events and of the veneration to the *sàantoh* fall on that year's *presidente*, or head of one of the four groups, or *Gremios*. These four groups, to which one has affiliation by familial or social ties, correspond to their respective days, in the sequence of *Jovenes*, or 'young people', *Damas*, or 'women', *Deportistas*, or 'athletes', and *Agricultores*, or 'farmers', and each of these is also expressed as equivalent to ten of the 40 hours being celebrated.

In many Yucatec towns, especially the larger cities, *Gremios* are forms of trade guilds, and their number indicate the size and complexity of the community's economy. Because Yalcobá is small, however, there are only four, and their names reflect only the broadest categories of social or economic divisions, although they do not segregate membership on that basis. Rather, here as elsewhere in Yucatan, the *Gremios* have become the social unit for a particular *fiesta*, and they serve as vehicles for acquiring prestige in the fulfillment of the group's duties on such occasions.

These include (1) providing an open house at the *presidente's* residence on the corresponding day, in which invited as well as non-invited guests can partake of the most valued of all Maya meals, *relleno negro*, (2) making a procession late in the day to the church, at which there is mass, and then afterwards making another to next year's *presidente's* residence, and (3) making one of the four arches at the four corners of the church, through which all four *Gremios* will walk in procession on the culminating day of March 3rd. All three of these tasks are taken quite seriously, and the group's ability to do them well adds to their social status, and especially to that of the *presidente*.

The final obligation of each *Gremio* and its *presidente* merge with the other three groups as all make the final procession together. In Figure 9.7, we see how the motion of the *Gremios* (arrow) around the outside of the church and through the four arches is counterclockwise, as well as how the sequence of arch placement corresponds to the order of the *Gremio* days, in a *no'oha'an* direction. While *Yalcobail* just say that the procession is *hač k'a'abèet*, or 'very necessary', and that they only do it out of habit, our outsider's perspective and certain symbolic components allow us to offer interpretations of some of the significance it is expressing.

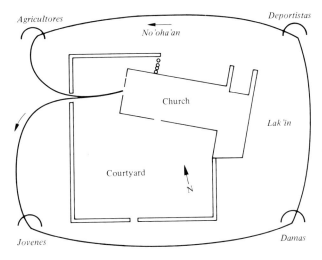

Figure 9.7 The procession of *Gremios*.

Of particular symbolic importance is the host symbol or *monstrance*, a large golden sunburst form (Figure 9.8) which is carried during the procession by the Catholic priest contracted by the community for just this purpose. The significant correlations included in this practice are (1) the priest's name of *yum k'iin*, or 'father sun', which is the exact same term used to refer to the 'sun', (2) the fact that the procession moves *no'oha'an*, 'to the right', which again is the exact same way the sun is described to move in its daily path, and (3) the time of day, which allows the *sacramento's* exit from the church's door to catch rays of the late afternoon sun and emit a brilliant light. Together, these three correspondences hint at a sun/host/Jesus Christ multivocalic set of meanings for this procession, and again indicate some of the logic of Maya culture and cosmology as it is expressed in social behavior.

Our continued appreciation of this logic, which joins and makes intelligible the shared quadrilateral

Figure 9.8 The *sacramento* monstrance.

structure of these practices and Maya cosmological order, is augmented even further by the ritual altar used by the *hmèen* in his many rituals which ask for 'evil' to be taken away, or 'good' to be brought. This altar, table, or *ka'an če'* as it is called (Figure 9.9), actually serves as a model of the quadrilateral earth, and is also referred to by the *hmèen* as *u yóok'ol kàab'*, 'the world', in prayer and in conversation.

As a dominant symbol, its complexity is multidimensional, but certain major features can be easily isolated. Of these, we can note its four domes, which are made of tree branches and create a matched set of arches on the *lak'in* and *čik'in* sides of the table, and are intended to represent the sky. Of central importance as well are *lùuč* gourd vessels which mark key deity positions on this 'earth'. But perhaps more important still are the components of *u yíič kib'* and *špeten ka'an*.

U yíič kib', or 'the running wax of the candle', is situated atop a stick piercing the center of the table and below *špeten ka'an*, 'the hanging platform of the sky'. Both together are intended by the *hmèen* to represent the foremost focus of his ritual attention, *u hol gloryah*, or the 'hole in the sky', which he believes to exist at *čumuk ka'an*, or 'the middle of the sky'. It is to this zenith position, especially when the sun/*Hahal Dios* is there, that his prayers

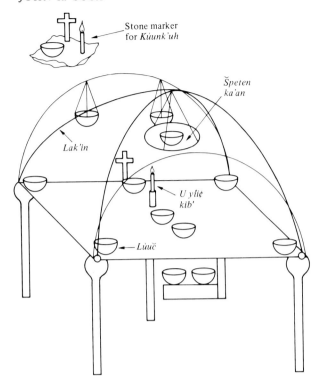

Figure 9.9 The *ka'an ĉe'* of *ĉ'a ĉaak*.

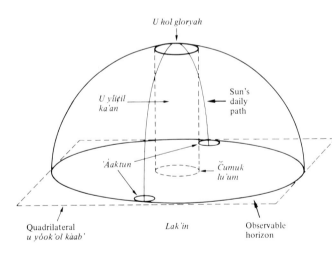

Figure 9.10 *U yíiçil ka'an* – the cosmological conduit.

are intended to arrive, and from which this same deity and his assistants are then believed to send the requested assistance. The result is a form of 'cosmological conduit', depicted in Figure 9.10.

The shared qualities of 'perforation' at the 'edge of Yalcobá' as well as at the 'edge' or 'center' of

the world, seem to be a function of the Maya belief in a flat world, since such a conception requires some sort of 'hole' to explain how the sun, moon and stars can seemingly pass through its plane at the horizons. At the center, the realm of humans and that of deities are joined at *ĉumuk lu'um*, 'the center of the earth', and *ĉumuk ka'an*, 'the center of the sky', where the *hmèen* believes *u hol gloryah*, 'the door or hole of heaven' to exist. The conduit itself, then, which achieves this joining is *u yíiçil ka'an*, 'the liquid substance of the sky', and corresponds to *u yíiç kib'* of the ritual table, while *u hol gloryah* is equivalent to the *špeten ka'an* hanging platform.

As for the horizon, the *hmèen* believes *Hahal Dios* provides a hole in the form of *'áaktun*, 'a cave', and that there is one for traversing the earth in *lak'in* and one in *ĉik'in*, *yo'o sal k u mèentik sùut*, 'so that he can go around'. He also says this as *Dios çáamile' u súutik u yóok'ol kàab'*, or 'God dives as he is making his circuit around the earth'. The *hmèen* then explains the same is necessary for *màa-mah lùunah*, or 'the mother moon' and *éek'*, 'stars', which again expresses his assumption that the earth is the dividing surface between the sky and the subterranean realm.

The symbolic expression of these cosmological assumptions in the structure of the ritual altar isolates some of the basic characteristics of spatial order that are also expressed in public rituals and *fiestas*. It may also be possible to interpret, however, a similar expression of Maya cosmology in the actual times at which these public functions begin and end, since the hours chosen seem to be due to more than just coincidence.

These beginning or end points can be either 3 a.m. or 3 p.m., and are depicted in Figure 9.11. Since the Maya often describe noon as '12 o'clock', setting sun or the Western horizon as '6 o'clock', and rising sun or the eastern horizon as '6 o'clock', even though most cannot afford to buy watches, we can interpret these two 3 o'clock hours as 'midpoints' of motion. Perhaps then, we can see the difference between 3 a.m. and 3 p.m. as the difference between the sun's movement from the earth to the sky for 3 a.m., since at this time it is believed to be within the earth and moving towards dawn, and its movement from the sky to the earth at 3 p.m., at which time it is 'diving to earth'. Significantly, I at no time documented any ritual or *fiesta* action that was prescribed to begin or end at 9.0 a.m. or 9.0 p.m., which would be the complementary 'mid-point' positions between either the 'top' at noon or the 'bottom' at midnight, and one of the horizons. Because of the absence of a 9 o'clock meaning, the 3.0 a.m. and 3.0 p.m. positions therefore seem to express the difference between the sun either moving towards the sky, or coming to the earth, at the in-between 'moments of motion' when these various activities begin or end.

In addition to the other meanings we might interpret as being expressed during these events, we can also consider the way in which this model of temporal and spatial equivalents depicted in Figure 9.11

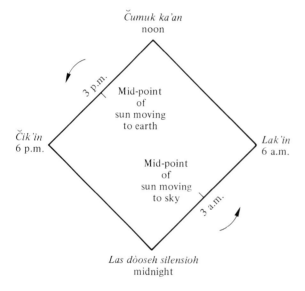

Figure 9.11 A spatio-temporal model of solar ritual significance.

helps us to interpret some of the significance of the beginning or end points of a ritual or *fiesta*. For whether 'good' is being asked to be brought by *Hahal Dios*, or 'evil' taken away, the assumed direction of motion of the intended offering, requested blessing, purged agent or the deity himself can be expressed as 'sky to earth' or 'earth to sky', and the difference between 3.0 a.m. and 3.0 p.m. may indeed represent this difference in time and space. And, if this is a valid representation of what the attention given to a 3.0 a.m. or 3.0 p.m. initiating or concluding point is expressing, then both the time and spatial location of the sun are the determining factors in the logic of this expression.

The symbolic structure which the various forms of Yalcobá ritual and social expression make manifest to us can therefore be interpreted to be a multi-dimensional set of temporal and spatial equivalences, organized by cultural interpretations of the visible properties of solar motion through and around a flat earth based cosmos. The very redundancy of the expression of this structure argues forcefully for considering it to be primary in providing a framework for the Maya symbolic system. And, since we have reviewed something of how the sharing of symbolic structure is so pervasive for these Maya, it has been necessary to give major attention to understanding how this is interpreted by these same Maya culture bearers, so that we might better comprehend how this society conceives of itself and its existence within the boundaries of a 'world'. Only with such analytic attention can we transcend the cultural boundary between ourselves and, in this case, the Maya, and begin to understand who they believe themselves to be.

In the case of the Maya of Yalcobá, the force with which they repeatedly express this cosmological structure apparently has everything to do with their reliance on corn agriculture, as the *kòol* cornfield itself is literally hacked out of a wild terrestrial environment, and can be productive only if uncontrollable weather conditions are favorable. The Maya expression of their cosmology, then, is continually re-created, shared, maintained, and also changed, in the inseparable realm of social interaction, very much in the same way as are the meanings of the cornfield upon which this society so critically depends. Indeed, the addition of electricity, a new road and new external economic attractions has brought some surface changes to Yalcobá and has accelerated its Western acculturation. But the vast majority of its *mestizo* inhabitants still hold to the most traditional beliefs and practices, which demonstrates, I believe, the true strength of the bond between a symbolic system, a cosmology and a means of subsistence.

Also in a surface form are the 'Catholic' identities of many of the system's component symbols, even though this is the name by which many *mestizos* now refer to themselves and their religion in response to the regional rise in Protestantism. For a multifaceted examination, encompassing many dimensions of expressive behavior and cultural information, has shown these forms to be multivocalic, interlocked sets of significance, which depend on the Maya interpretation of the relationship between the natural and social orders for their logical coherence, as they are made meaningful in a variety of socially interactive contexts. In the process, the members of Yalcobá society experience some of their strongest emotions and partake in a living, breathing cultural tradition that, far from being 'primitive', is tremendously complex and remarkably sophisticated.

Given the cultural, symbolic and cosmological complexity of the Yucatec Maya people that I have briefly described, and even though I consider my total effort at description to be 'thick', it could always be 'thicker', considerations of 'primitiveness' are exposed for the woefully inadequate and misinformed judgements that they are. For, whosoever might think of another society as 'primitive', or 'backwards', or 'underdeveloped', has surely never known any of its members on a profound first-hand basis, and tried to appreciate the unending complexity of their variant of human existence. A cross-cultural contact, resulting in a degree of cross-cultural understanding, perhaps above all else forces us to confront our quite natural and quite essential ethnocentrism, since human groupings could not exist without it, and informs us of the equitability with which we should treat humankind at all technological and social levels.

As for archaeostronomy, then, the contemporary traditional societies of the world need not only be treated as the degenerate 'leftovers' of ancient grandeur, to be used as the testing ground for revealing insights into the ways of their more 'complex' ancestors, but should be considered as comparable components to any other data set comprising any other culture, symbols and conceptions of world order. In fact, the value of seeing how a cosmology provides logically ordered, holistic systems which coordinate belief and practice is in the extent to which we might become aware of how its dynamics actually resulted in the archaeological or historical remains that concern the archaeoastronomer in the first place. It seems apparent, then, that the opportunity to better understand what these remains tell us about being *human*, and having to interact with and exploit an environment for survival, is at our disposal, but only if we as Westerners can overcome our learned biases towards much of the rest of our species, and try to appreciate the unique qualities and capabilities of conception, and action based on that conception, that we all share.

Acknowledgements

The author wishes to express his gratitude to Gilda Mason and Janet Dauley for the typing of this manuscript.

Notes

1 Yucatec Maya is a spoken language and therefore must be represented using a phonemic orthography. The system I and many other Maya ethnographers now use was first presented by Blair and Vermont Salas (1965) and includes the following signs for Maya sounds:

　　′ = glottal stop or 'throat click'
　　¢ = tz
　　¢′ = dz
　　č = ch
　　č′ = chh
　　š = sh
　　o′ob′ = plural

References

Barrera Vásquez, A. (1980). *Diccionario Maya Cordemex.* Ediciones Cordemex: Merida, Yucatan.

Blair, R. W. and Vermont Salas, R. (1965). *Spoken Yucatec Maya.* Department of Anthropology, University of Chicago.

Geertz, C. (1973). *The Interpretation of Cultures: Selected Essays.* New York: Basic Books Inc.

Redfield, R. and Villa Rojas, A. (1934). *Chan Kom. A Maya Village.* Revised in 1962. New York: Basic Books Inc.

Standley, P. C. (1977). La flora. In *Enciclopedia Yucatanense*, vol. 1, pp. 273–523. Merida, Yucatan.

Turner, V. W. (1967). *The Forest of Symbols.* Cornell University Press.

Turner, V. W. (1977). Symbols in African ritual. In *Symbolic Anthropology*, eds. J. L. Dolgin *et al.*, pp. 183–94. Columbia University Press.

Villa Rojas, A. (1945). *The Maya of East Central Quintana Roo.* Carnegie Institution of Washington publ. 559.

10

Keeping the sacred and planting calendar: archaeoastronomy in the Pueblo Southwest*

Michael Zeilik   *The University of New Mexico*

10.1 Introduction

The US Southwest defines a cultural area that is large in space and deep in time. During the 1000 years before the *entrada* of the Spanish, the Southwest harbored numerous, interacting cultures with possible connections to prehistoric Mesoamerica. Those cultures included the Patayan, Hohokam, Mogollon, and Anasazi. Archaeoastronomy in the Southwest aims to infer the content and the impact of astronomy in these prehistoric societies. This region has a great historical advantage – the people now living there have a cultural descent from prehistory. That cultural connection runs especially deep with the historic Pueblo people directly from their immediate protohistoric ancestors and on to their Anasazi (and Mogollon) ancestors who lived throughout the northern Southwest in the broad Anasazi region (now known as the Four Corners area).

I will focus in this paper on the Anasazi, in part because of the rich cultural heritage of the present-day Pueblos (revealed in the ethnographic record) and in part because of the rich archaeological tradition of the Anasazi region. Here the past comes alive when a Turtle Dance takes place at San Juan Pueblo – perhaps a displaced winter solstice dance (A. Ortiz, 1985, private communication). I can easily imagine a similar dance – different in detail but with the same ritual intentions – in the sunny plaza of a prehistoric pueblo. That image contains the essence of the oft-maligned argument by analogy that I will use in a conservative way: as a culturally appropriate source of hypotheses for testing in a

prehistoric context (Binford, 1972; Reyman, 1975). I will presume that those Puebloan concepts about astronomy that are most frequent within the diversity of the historic Pueblos mark cultural traditions that are the deepest in time. Specific cultural discontinuities certainly *did* occur between the past and the present – but aspects essential to survival and adaptation to the high desert stayed mostly intact. In particular, the sacred and planting calendars, kept by observations of the sun and the moon, control the life of the historic Pueblos – an essential integration of astronomy, agriculture, and ritual. The same was likely an astronomical practice of the Anasazi. As Cordell (1984, pp. 271–2) has put it:

> Agricultural people throughout the world are concerned with calendrical observations. These are crucial to the time of planting and harvesting outside of equatorial regions and to water management where supplemental watering is necessary to insure the success of crops. In Chaco Canyon, and in the San Juan as a whole, short growing seasons and inadequate rainfall are problems for agriculturists, so it is likely that *accurate* astronomical observations were important to the Chaco Anasazi; [author's italics].

Overall, I find ethnographic data to be a mind-opening, wondrous window on Pueblo life. Ethnographic analogy, properly used, can flesh out the old bones of prehistoric ruins. Ethnographic analogy does not straitjacket our development of testable hypotheses. To the contrary, it illuminates rather than retards our insights to Anasazi life.

We shall consider three points about methods for practising archaeoastronomers. First, we must respect the archaeological record and the interpretations derived from it. We should work with an archaeological awareness for every site and culture – and even generate hypotheses that the archaeologists can test for us! While we do so, we must remember that a major problem for the archaeologist is the *control of time* at a site; dating information might not be there or it has a large uncertainty. Second, I feel arguments for astronomical aspects of Anasazi life for which no ethnographic analogy exists have an additional burden of proof placed upon them. To demonstrate that a site 'works' astronomically is not enough; it should 'work' culturally, also. This means marshalling evidence from different aspects of the archaeological record so that any astronomical proposition makes cultural sense. Third, we need to involve expertise from diverse disciplines: art, art history, archaeology, anthropology, astronomy, ethnology, folklore, and history. This strategy requires that we form multiple hypotheses, involve multiple investigators (to avoid observer bias), and following multiple lines of evidence – for we know from the historic Pueblo example that cultural artifacts have multiple uses and meanings. This multivariate approach (Young, 1987b), however, must be conducted in the appropriate cultural context.

Finally, we must ask how *valid* our conclusions are, given our methods and ethnographic base from the historic Pueblos. Hedges has made some wise observations in this regard with reference to California archaeoastronomy. His points (Hedges, 1985, pp. 35–7) also hold true, in my opinion, for the Pueblo Southwest, especially his plea that 'we must develop a coherent, logically consistent line of reasoning in the interpretation of archaeoastronomical sites' (p. 35).

10.2 Astronomy in the historic Pueblos

We owe a dubious debt to ethnographers who visited the Pueblos before the economic encroachment of Anglo society in the 1930s. These people – Jesse W. Fewkes, Frank H. Cushing, John G. Bourke, Alexander M. Stephen, Matilda C. Stevenson, Elsie C. Parsons, and Edward S. Curtis – left us detailed records of many aspects of Pueblo life. So many and so detailed are they that the astronomical unity

of the calendar gets lost – in part, because it was so transparent to the Pueblo people. (See *Sun Chief* (Simmons, 1942) and *Pueblo Indian Journal* (Parsons, 1939), for wonderful examples of the 'right time' for religious and planting activities.)

Few attempts have been made to gather this material in one place and make sense of the often contradicting pieces. Reyman (1971, pp. 114–22) made a substantial effort for his pioneering study of kiva orientations at Mesa Verde and Chaco in the context of the influence of Mexican ceremonialism on the prehistoric southwest. (His work influenced that of Williamson *et al.*, 1975, for astronomical alignments at Chaco.) Ellis (1975) revealed her intimate association with Rio Grande ceremonialism in an attempt by ethnographic analogy to place the hypothetical 'AD 1054 supernova' pictograph at Chaco (Brandt *et al.*, 1975; Brandt and Williamson, 1977) in the context of the Pueblo calendar watch. For Hopi First Mesa, McCluskey (1977, 1981, 1982) broke new ground in a historical understanding of the synchronization of solar- and lunar-timed ceremonies. M. Jane Young (1987a) and Barbara Tedlock (1983) have given insights about Zuni from their own experiences there. (See also Young and Williamson, 1981, for information on Zuni constellations.) Feeling the need for coherent, comprehensive information, I have gathered material from 19 historic Pueblos on sun- and moon-watching for calendrical purposes (Zeilik, 1983b, 1985a, 1986a) and am continuing work on miscellaneous sky watching (planets, stars, comets, and meteors).

The prehistoric observation of and possible rock art record (Miller, 1955) about the 1054 supernova (which formed the supernova remnant now called the Crab Nebula in the constellation of Taurus the Bull) played a key role in sparking interest in Southwestern archaeoastronomy – at least among astronomers. The pictograph in Chaco on the canyon wall below Penasco Blanco (Figure 10.1) is one of the better known of the rock art sites (Brandt and Williamson, 1979; Koenig, 1979; Mayer, 1979). Most of the *astronomical* objections to this rock art site as representing the 1054 supernova can be eliminated simply by flipping the view (Gaustad, 1980); but the nagging doubt remains that this configuration represents Venus next to the crescent moon rather than the supernova (Ellis, 1975, pp. 86–7). Other sites seem to have to be carried

Figure 10.1 A portion of the rock art panel, below Penasco Blanco in Chaco Canyon, which may or may not represent the AD 1054 supernova in conjunction with a waning crescent moon (M. Zeilik).

along a very thin line of reasoning to be interpreted as the 1054 supernova (Hedges, 1985, p. 26). Workers now have moved away from the search for the supernova; yet we need to recall that it motivated much early work, especially at Chaco.

In general, we should distinguish between astronomical *purposes* and the astronomical *practices* (that derive from the purposes). In the Pueblo Southwest context, astronomy serves the purposes of establishing and validating (1) sacred directions and cosmic patterns, (2) cosmic mythology, (3) certain ritual sites and shrines, (4) the ritual and planting calendar, and (5) times for hunting and gathering. These desired ends prompted the development of horizon calendars, light and shadow markers, and lunar phase counts for tracking the calendar. The main task of the calendar watch centers on methods to *anticipate* festival dates (McCluskey, 1982; Zeilik, 1985a, 1986a). Pueblo ceremonies must be announced ahead of time so that ritual preparations

can be properly carried out, and the meshing of solar/lunar ceremonies worked out by the proper intercalation (McCluskey, 1977). The ceremonial cycle typically spans a year, and solar and lunar observations conducted by the appropriate religious officials set the timing for the rituals that are laid out in a sequence, where the end of one ceremony marks the start of the next.

Anticipation provides a culturally correct way to understand the terms 'precision' or 'accurate', when combined with historical information about festival dates as has been done by McCluskey (1977) for Hopi and Lyon (1985) for Shalako at Zuni. Sun priests historically at Hopi and Zuni were able to fix solstice ceremonies to within the astronomically correct day (or two) by making anticipatory observations about two weeks ahead of time (Zeilik, 1985a). In contrast, lunar ceremonies appear to be moveable festivals, with a scheduling decision under the power of the person responsible for setting the date (Frigout, 1979; Sekaquaptewa, 1983; Zeilik, 1986a). And the date of a particular lunar phase (such as full) may have been established to within a day (or two) of the astronomically correct phase (Bunzel, 1932b, pp. 702–3).

An emphasis on calendric anticipation switches our attention from directions to rate of change of direction (angular speed) of the sun. This mental realignment should not, however, overlook the importance of direction to the Pueblo sense of sacred space. The special directions are usually the four solstitial ones (winter and summer rising and setting), as well as the zenith and nadir, such as embodied in the placement of corn ears in the Chief kiva during the Hopi Powamu ceremony (Stephen, 1936, p. 228, fig. 140). Special colors are also linked with each of the sacred directions, often with several animal associations, too (Parsons, 1939). Such sacred directions may result in buildings with special orientations in historic and prehistoric times. The point here is that cycles such as the sun's seasonal swing along the horizon involve *motions* to achieve the extreme positions. The Pueblos embodied both in their astronomical practices: the directions tend to be cosmic; the motions, calendrical.

Finally, we need to consider how the calendrical observations were made. For the sun, a horizon calendar was most commonly used, and we have good maps for some of them (Stephen, 1936, maps 4 and 12). Astronomically, such horizons need to be fairly distant (a few kilometers at least) with some, but not necessarily dramatic, relief. The sun priest also goes to a special sunwatching station, which is typically not marked by rock art (Zeilik, 1985d; Young, 1987a). The ethnographic analogy then raises a key archaeological problem: What would be the material evidence for a sun-watching station? At Zuni, the historic sunwatching station at the old village of Matsakya was well marked (Figure 10.2), but it no longer exists today.

The second calendrical technique involves the use of light and shadow cast through windows/portals against a wall with markers (Lange, 1959, p. 56); this method has more hope for archaeological remains. At Zuni, the famous passage of Cushing (1979, p. 117) indicates markers of some kind on the wall. The diaries of John G. Bourke tell us what was going on (Bourke, 1881, entry for 19 November):

After breakfast, Cushing, Pedro Pino [the Governor of Zuni] and myself went to the upper story of one of the highest houses on the Eastern side of the Pueblo; here in the West wall was an old blue china plate fixed there, so the head of the house said, in the time of the Spaniards, to conceal a painting of the Sun, which faced a small rectangular aperture in the eastern wall. When the sun shone through the aperture farthest to the North, Spring had come and the season of planting had arrived; the more Southerly aperture allowed the rays of the Sun to fall upon the center of the plate (in ancient times upon the face of the sun picture) about the period of the Autumnal equinox – and when the light struck a certain point in the wall, it was the time of the Winter Solstice.

This statement reveals the practices of solar calendars at Zuni at a time before Anglo culture had much impact. Sunlight hits the sun face during harvesting season; planting and the winter solstice were also noted. And we also know that at the same time, Pekwin (Figure 10.3), the sun priest (or 'cacique of the Sun', as Bourke called him) did a horizon watch (Bourke, 1881, 19 November) as well as kept a wall calendar in his own house (according to his brother). The half of that calendar that marked the time from spring planting to the summer solstice consisted of a horizontal line of scratches onto which sunlight fell through an east-facing window. The half of the year to the winter solstice was marked by strings of abalone shells, hanging on the wall. From this information and the argument by analogy, we can tell archaeologists to look for rooms that have windows situated so as to cast sunlight (probably in the morning) against a wall. Any markers on these walls, however, would probably not have lasted 1000 years.

I find these Zuni examples especially curious because we are told that Pekwin keeps both a horizon watch (Stevenson, 1904, pp. 109, 117–18) and also one involving a wall calendar. What was the cultural reason for this redundancy? Did Pekwin use both to anticipate and confirm important times? For instance, one major sequence of the wall calendar includes the time of planting, winter solstice and summer solstice (Bourke, 1881, 3 September). Pekwin's horizon calendar involves similar time sequences. Did these two parallel calendars both support the degree of precision that Pekwin was

Figure 10.2 Sun-watching station at Matsakya, one of the old Zuni villages. Pekwin used this location for horizon observations prior to the summer solstice (Stevenson, 1904, pp. 148–9). Note the upright stone with the sun face. Photo by M. C. Stevenson (Smithsonian Institution photo no. 84–7550).

supposed to achieve? Bourke (1881, 25 November) states that

> Cushing took me to call upon the Priest of the Sun. This official, in the different Pueblos, is supreme during time of peace, that is to say his edicts can only be revoked by a council of all the Caciques. The present incumbent of the office in Zuni is a very young man and has but recently entered upon his duties. I have noted elsewhere [20 November 1881] that *by a mistake in his orientation* he ante-dated the festival of the Winter Solstice by some (15) or eighteen (18) days, a mistake very summarily corrected by the Council of the Caciques [author's italics].

Note that this passage tells us that other religious officials kept watch on Pekwin's work, and could track the calendar as well as he to insure that 'correct' times were announced. Also, note that Pekwin's mistake in anticipatory observations was far too large to be acceptable to the council.

I have discussed Zuni in some detail as a specific example of pan-Pueblo practices and as a case where we have substantial information – most of it consistent! – predating the 20th century. The practices then may reflect the survival of an astronomical tradition that traces back to prehistoric times.

In summary, the ethnographic evidence about the historic Pueblo calendar indicates:
(1) attention to a one-year *seasonal cycle* with the winter solstice usually as the 'middle';
(2) a *fixed sequence* of ceremonies, in which the completion of one sets the stage for the next;

Figure 10.3 Pekwin at Zuni in 1896. Photo by M. C. Stevenson (Smithsonian Institution photo no. 2250).

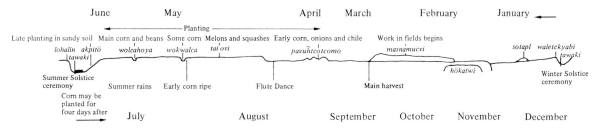

Figure 10.4 Planting and ritual horizon calendar, from winter to summer solstice, for the Hopi village of Shimopovi (after Forde, 1931, fig. 6B). The interval from December to June is read from right to left along the top; July to December from left to right along the bottom.

(3) an *anticipation* of ritual times so that proper preparations can make the ceremonies effective;

(4) *astronomical observations* that set the calendrical schedule.

In what follows, I will use the winter solstice, planting season (mid-April up to the summer solstice as for the Hopi – see Forde, 1931; Beaglehole, 1937; Titiev, 1938), and the summer solstice as the order of importance for seasonal dates (Figure 10.4). (The pueblos rarely mark the equinoxes as such; those they do correlate roughly with planting and harvesting. See Ortiz, 1969, p. 104, for explicit comments about the importance of the equinoxes to the Tewa, which marked the seasonal transfer of power between the summer and winter people.) I will also define *calendrical site* to mean one that could be used to achieve at least the same degree of accuracy of scheduling as attained by Puebloan calendar keepers in historic times.

Finally, my focus on the *Pueblo* Southwest should not result in the inference that only the Pueblo people paid attention to the sky. The Navajo and Apache did, also, in their own cultural context (see Farrer and Second, 1981, for examples from the Mescalero Apache). And the Indian peoples of North America in the greater Southwest (Spier, 1955) and all other regions had some type of calendrical and cosmological systems (see Cope, 1919, and Stirling, 1945, for general references; and see the journal *Earth and Sky* for California).

10.3 Horizon calendars

Because they are so prosaic, horizon calendars have received little attention – even though we have the most and best ethnographic information about

them! They are also hard to substantiate archaeologically, for it is difficult to ascertain where a sun priest might have stood to make his observations. On the other hand, if the horizon is distant – ten or so kilometers – the choice of the standing spot is not crucial if it is within or close to a pueblo. Among the historic Pueblos, the sun priest rarely travels very far to make his observations – from the roof at Walpi (Stephen, 1936, p. 62), near the edge of the mesa for Hano (Yava, 1978, p. 72), or a few kilometers east from Zuni (Halona) to the deserted village of Matsakya (Stevenson, 1904, p. 118). Hence, by ethnographic analogy, one can search for a spot in or close to the archaeological site of a village that provides a view of the horizon to see if a calendar can be kept in this fashion. If not, a search should then be made for light/shadow markers. That conclusion can be reached immediately if no good horizon line presents itself for the appropriate spans of time.

Archaeological sites that have been tested for horizon watching are (Table 10.1): Penasco Blanco (Williamson, Fisher and O'Flynn, 1977; M. Zeilik, unpublished field notes), Wijiji (Williamson, Fisher and O'Flynn, 1977; Williamson, 1983; Zeilik and Elston, 1983), Fajada Butte (Williamson, Fisher and O'Flynn, 1977), and Pueblo Bonito (Zeilik, 1985e) – all in Chaco Canyon; and Tsiping Pueblo in the Chama valley of New Mexico (Zeilik, 1984a), Tsankawi (M. Zeilik, unpublished field notes) on the Pajarito Plateau, and Kuaua along the Rio Grande (J. Lowther and M. Zeilik, unpublished field notes). The Chacoan Great House sites range in dates from AD 900 to 1100 (those at Fajada Butte date to the 1200s); Tsiping is a later occupation, from AD 1250 to 1450, and a probable prehistoric

Table 10.1. *Possible Anasazi horizon calendars*

Place	WS	VE	SS	AE	Anticipation
Chaco					
Wijiji	+	−	−	−	+ WS
Pueblo Bonito	−	+	+	+	+ except WS
Penasco Blanco	poor				
Fajada Butte	poor				
Rio Grande/Chama					
Tsiping Pueblo	−	+	−	+	+ equinoxes
Tsankawi	+	+	+	+	+ all
Kuaua	+	+	−	+	+ WS

WS = winter solstice; SS = summer solstice; VE = vernal equinox (planting); AE = autumnal equinox (harvesting).
Key: '+' means the site works for the time given; '−' that it does not; '?' that insufficient information is available.

Tewa site, as is Tsankawi (which is now part of the Bandelier National Monument), which may have been occupied up to the late 15th century. Kuaua was occupied into historic times and may have been visited by the Spanish (Dutton, 1963, pp. 3–18).

The view from Penasco Blanco, at the edge of the mesa where the prehistoric road leads to a point above the hypothetical '1054 supernova' site, spans the year but has little to offer for anticipatory markers or planting (Zeilik, unpublished field notes). The horizon from the Wijiji rock art site, which has a nearby rock pillar, can anticipate the winter solstice (Williamson, Fisher and O'Flynn, 1977; Williamson, 1983; Zeilik and Elston, 1983). The confirmatory alignment above the rock pillar on the winter solstice only is visible from a narrow range along the viewing ledge, because of the closeness of the pillar. The cultural context is confused between Anasazi and Navajo (Williamson, 1983); overall, a clear conclusion is hard to reach. The site does *not* work for a planting sequence or the summer solstice.

The horizon view from the east side of Pueblo Bonito provides a reasonable profile for summer solstice through planting, but it has only one possibility for an anticipatory marker for the winter solstice. The view from atop the canyon's rim, near the staircase, might be better and needs to be checked out. From a ceremonial area just outside Tsiping, the historically sacred mountain of Pedernal provides a profile (Figure 10.5) for anticipation of and tracking of a planting/harvesting calendar at sunset. For the solstices, the horizon from Tsiping offers no clear anticipatory or confirmatory pos-

sibilities either at sunrise or sunset (M. Zeilik, unpublished field notes).

Tsankawi sports a stunning view of the Sangre de Cristo mountains, which lie about 50 km to the east, and a panorama of the Rio Grande Valley and Jemez range (Hewett, 1906, pp. 20–1). From the tip of Tsankawi Mesa (see Hewett, 1906, pl. VIIa), the view encompasses a horizon profile that easily can be used for winter solstice, summer solstice, and a planting calendar (Zeilik, unpublished field notes). Other prehistoric pueblos in the same region have roughly the same view, so they can have similar horizon calendars. We also have enthnographic evidence from the Tewa pueblo of San Ildefonso (which is the historic Pueblo closest to Tsankawi) that the same mountain range was used to anticipate (by 12 days) and confirm the summer solstice (Stevenson, n.d.).

Kuaua contains, in its square kiva, the famous ceremonial murals (Dutton, 1963). Some of these sacred figures depict the Sun Father (based on ethnographic interpretations) and relate to ceremonies at the winter and summer solstice, as well as general symbolism related to fertility. At the site today (part of Coronado State Monument), the tree line of the Rio Grande makes observations difficult from the ground. We therefore observed from the roof of the Museum, which would mimic observations from the second story of the pueblo. The Sandia and Ortiz Mountains to the east provide a clear horizon profile at sunrise from the equinoxes to the winter solstice (Lowther and Zeilik, unpublished field notes). We know that at Santa Ana Pueblo, the Sandia Mountains are used to track the approach of the winter solstice (White, 1942, pp. 92, 103), for

Figure 10.5 Sunset on the vernal equinox behind the sacred mountain of Pedernal as viewed from the amphitheater area at Tsiping Pueblo (M. Zeilik).

which Masewi (the war priest and one of the Twin War Gods) begins the watch in September, just the time when the sun begins to climb up the slope of Sandia. Dutton (1963, p. 205) states that some of the Kuaua people may have joined those at Santa Ana.

I conclude that Anasazi sites need *far* more work in the search for potential horizon calendars. Our firmest ethnographic information about calendrical sun-watchers involves horizon profile sequences. By the ethnographic model – especially for the Hopi – we expect that each pueblo kept its own calendar by its own sun-watcher or watchers. Hence, each site should be checked carefully for a useful horizon track – or lack of one! If the horizon is not workable

for certain intervals, we may then infer that light and shadow techniques might have provided an alternative procedure.

10.4 Light and shadow

Light and shadow phenomena draw our attention by their visual appeal, especially through the medium of photography. The three-slab site and other sites at Fajada Butte (Sofaer and Sinclair, 1987) and the petroglyph sites in the Petrified Forest (Preston and Preston, 1985, 1987) have sparked public attention to this aspect of Southwestern archaeoastronomy. The petroglyph site near Holly House in Hovenweep is also visually stunning (Williamson and Young, 1979; Williamson, 1979, 1981,

1984), but has a lower visibility in the public eye. Other light and shadow sites associated with window/portals in buildings include the corner openings at Pueblo Bonito; the northeast opening in Casa Rinconada in Chaco; and openings in add-on rooms at Hovenweep Castle, Unit Type House, and Cajon Group Ruins at Hovenweep.

A general claim is sometimes made that such sites work as intentional and accurate calendrical markers. To make such a case requires, in my opinion, that the site serves to anticipate the solstices within an accuracy of a day or two and to keep track of a planting calendar. This requirement brings up the issue of *practical resolving power* – the ability of an observer to interact with a site to mark reliably one-day differences in shadow positions. Note that an emphasis on anticipation shifts our attention from the *positions* of the light-shadows to the rate of change of position on a surface, i.e. *linear speed*.

Two physical elements enter into this resolving power issue. The first is the linear speed along the surface per day, as the shadow mimics the seasonal motion of the sun. Imagine the shadow as a lever with a fulcrum at the point of the shadow-casting edge. Then, the longer this throwing arm, the larger the linear motion for each degree, say, of solar motion. So better resolving power requires longer shadow throws. The second is the sharpness of the shadow edges. When the light source has a finite size, such as the sun (which subtends $\frac{1}{2}°$ in the sky), a shadow has a fuzzy edge, even if the shadow-casting edge is smooth. An empirical examination of the degree of fuzziness gives, with a lever arm of 1 m, a width of about 1 to 2 mm. The fuzziness increases in direct proportion to the length of the shadow throw. So, shorter is sharper.

The two physical elements work against each other. A larger linear motion requires a longer throw; for sharper edges, a shorter one. Given the eye's perceptive ability for edge contrast, I suspect that an optimal length exists that balances these two considerations. From my experience, I surmise that this length is about 1 m (though this optimum length needs to be established by experiment). My subjective judgement is that a linear motion of at least 0.5 to 1 cm/day is needed for reliable observations with a one-day resolving power.

The final element for shadow-casting reliability is the actual visibility of the shadow edge and ability to note its position from a selected observing position. That is, what you can do 'being there' at the site, *not* what can be done after the fact with photographic records. (Photography increases the contrast along a shadow's edge.) To analyze the site correctly requires that we interact with it in the same way as a prehistoric observer might. So, the roughness of the shadow-casting edges (which will increase the fuzziness of the shadow), the texture and color of the surface on which the shadows fall, the sharpness of petroglyphs/pictographs on the surface, and the viewing position and angle all enter into the character of the observations. These are all specific to the site, so each site should be judged based on these details of the site's geometry. Claims for precision cannot exceed the practical resolving power that the site offers.

The above analysis is only *one* part of a site's interpretation. An essential piece is the archaeological context of the site – especially difficult for rock art images because they cannot be dated absolutely as can buildings by tree-ring or radiocarbon techniques. Again, the problem is the archaeological control of time. The archaeological context then illuminates the cultural one, an important consideration at sites that may have been used by different cultures at different times.

I will now examine light/shadow sites, dividing them into two groups: interior (buildings) and exterior (rock formations with art).

Interior sites

At Hovenweep, Williamson, Fisher and O'Flynn (1977) investigated a number of tower structures. Williamson (1981) elaborated on Hovenweep Castle and the Cajon Group ruins. These places and the Unit Type House have add-on rooms that contain portals apparently positioned to let in light for calendrical purposes. (Some portals are vents; others provide sights to other buildings in the canyons – not *every* one has an astronomical purpose.) Cajon lies about 10 km south of Hovenweep Castle and Unit Type house, which are within a kilometer of each other. Zeilik (1983a, 1987, and unpublished field notes) re-examined these sites and found that all could be used in an anticipatory fashion and that the average speeds of the light beams on the walls ranged from about one to a few centimeters per day (Table 10.2). Photos taken by Fewkes (1919) just prior to 1900 indicate that, for Unit Type House and Hovenweep Castle, the portals were intact

Table 10.2. *Anasazi interior sites*

Place	Average speed (cm/day)	WS	VE	SS	AE	Anticipation
Hovenweep						
Hovenweep Castle:						
WS port	1	+	−	−	−	+ WS
SS port	4	−	+	+	−	+ SS
Unit Type House:						
WS port	2	+	−	−	−	+ WS
E port	3	−	+	−	+	+ E
SS port	2	−	+	+	−	+ SS
Cajon Ruin 'Tower':						
WS port	?	?	?	?	?	?
E port	3	−	+	−	+	+ E
SS port	2	−	+	+	−	+ SS
Chaco						
Pueblo Bonito						
Room 228	3	+	−	−	−	+ WS
Casa Rinconada	6	−	−	+	−	+ SS

WS = winter solstice; SS = summer solstice; E = equinox.
Key: '+' means the site works for the time given; '−' that it does not; '?' that insufficient information is available.
Average speeds are used, rounded off to one significant figure, to allow direct comparison among sites.

before any stabilization. Hence, they were originally in the structures, although we are not sure to what extent the insides of the openings were plastered and so would restrict the view compared with today.

Cajon and Hovenweep Castle work calendrically at sunset; Unit Type House in early morning (about half an hour after local sunrise because of a nearby ridge that blocks the sunlight). In all three places, the horizon view does not work well for a calendar. The play of light within the rooms against the walls of each can anticipate the solstices, confirm them, and keep track of a planting calendar. So I judge them to be the best examples we have found to date of intentional calendrical structures in a large part because of the integrity of the structures. The next archaeological step would be to excavate these rooms to see if they contain any artifacts that might show a different ceremonial usage than other rooms also trenched for a controlled comparison.

In Chaco, Reyman (1976) noted that corner openings to rooms 225B and 228B at Pueblo Bonito are oriented to the winter solstice sunrise (Figure 10.6). The horizon profile from Bonito contains little relief from the end of October up to the winter solstice. The openings to room 228 and 225 will work to anticipate and confirm the winter solstice (Zeilik, 1985e). In room 228, the sunlight first enters the room at the end of October and moves horizontally at an average speed of about 3 cm/day. Hence, the beam's motion could be used to anticipate the winter solstice to within an accuracy of one day.

The main problem with these corner openings is that we do not know whether they were exterior windows or interior doors! I have examined pre-reconstruction photos of Pueblo Bonito taken in 1920; they clearly show the opening to room 228 and a barely visible outline of the one to room 225. but these photos provide no clear way to tell if the openings were corner doorways − of which Pueblo Bonito has a number of examples − or actually were windows. Archaeologists could help here by providing information as to the nature of the openings. Lekson (1985, private communication) suggests that they are most likely doorways, because of their size, location above the floors, and their identical construction to other interior openings at the first-story level of Bonito. In contrast, Reyman (1976, and 1986, private communication) has noted that photographic plates from the 1890s (by G. Pepper and R. Wetherill) show the windows; and that stabilization reports and present conditions at the site indicate that the exterior walls perpendicular to the south walls were buttressing walls added for structural support.

Upon first examination, the northeast opening of

Figure 10.6 The southeast corner of Pueblo Bonito today. The second storey corner opening just to the right of center is that to room 225B; the corner opening to the far left is that to room 228B. The stepped stonework in front of the wall at bottom right may have served as a shrine for offerings of effigies (M. Zeilik).

the great kiva of Casa Rinconada at Chaco appears to be aligned to the summer solstice sunrise (Williamson, Fisher and O'Flynn, 1977; Williamson, 1981, 1982). The beam of light strikes one of the large niches about half an hour after local sunrise and makes a visually appealing display. Sunlight first enters the opening at the beginning of May (Zeilik, 1984b), so the horizontal motion along the wall can easily be used to anticipate the summer solstice to the day – the average speed is some 6 cm/day (Table 10.2). But the archaeological evidence argues against this aspect of Casa Rinconada as calendric. Firstly, photos prior to the reconstruction clearly show that the opening was in shambles; in particular, the right-hand edge (as viewed from the inside) was non-existent. So the position of that edge and hence the width of the window are a result of the reconstruction rather than original. Second, a low wall probably cut the opening off from a view of the sky. Third, even if all else worked correctly,

154

the light beam would have struck the northwest roof support post rather than the niche, which was probably bricked in. All in all, I judge the evidence weights against an intentional calendrical phenomenon at Casa Rinconada on the part of the Chacoan Anasazi.

Table 10.2 summarizes the known properties of most of the Anasazi interior sites, which may have had an analogous function to those at Zuni. Note that for most sites, the average linear speed of the sunlight beam is 2^+ cm/day – enough speed to achieve one-day resolving powers. Clearly, this area will be a fruitful one for future work, if we can get around the reconstruction problem.

Exterior sites

The three-slab site atop Fajada Butte in Chaco has the widest reputation among the Anasazi exterior sites (Sofaer, Zinser and Sinclair, 1979a,b). Sofaer and Sinclair (1987) have improved the knowledge of the ethnographic context and also have expanded the view to other petroglyph sites on the butte, which they claim mark local solar noon and the seasons. Sofaer and colleagues find particularly attractive the purported integration of the sites for sun/moon and season/noon observations. Newman, Mark and Vivian (1982) infer that the three-slab site is likely a natural rockfall, rather than a 'construct'.

I have re-evaluated the three-slab site in some detail (Zeilik, 1985b,c) and will not repeat those comments here. (Carlson, 1987, has also analyzed the functions of the site.) But let me make four general remarks. One, any claim for 'precise' or 'accurate' must be made explicit and should be put in a cultural context. For example, the shafts of sunlight at the three-slab site do not, in my judgement, permit the anticipation of the solstices to within a day or two – they do not have sufficient practical resolving power. Two, the Fajada Butte sites must be placed in a firmer archaeological context. I note that the small site 29 SJ 1360, which was occupied during the 10th and early 11th centuries, is located at the base of the butte. Could it have housed the sun priest who might have monitored the astronomical observations on the butte (McKenna, 1984, p. 389)? Three, the ethnographic information shows that the historic Pueblo people clearly distinguished between sun shrines (see Fewkes, 1906, for examples) and sun-watching stations for calendrical purposes – a distinction often confused in the archaeoastronomical literature. Four, calendrical sites tend to be accessible, since at times the sun-watcher needs to make daily observations. The three-slab site is a hard climb (especially in winter) from the occupation sites within the canyon. The butte does contain two small habitation sites, which, from an analysis of ceramics found associated with them, appear to have been occupied late in Chaco's history (Toll, Windes and McKenna, 1980), around AD 1175–1225. We cannot be certain whether these habitation sites have any temporal connection to the rock art; if so, we may be seeing a development on the butte *after* the Bonito phase, which ended around AD 1130.

In my judgement, the three-slab site fits much better with the attributes of historic sun shrines rather than calendrical observational places. If so, the butte may have been a sacred – even cosmological – place, but I think that its astronomical use for calendrical keeping was much less likely its pur-

pose. That is, the sunlight motions at the three-slab site act as a *seasonal hierophany* that contains calendrical information but has limited practical resolving power (see Hedges, 1985, p. 27 for a similar conclusion). Put another way: I think that the site has more of a *commemorative* than a calendric function, if it were intentionally used by the Chacoan Anasazi.

Preston and Preston (1985, 1987) have worked on a number of so-called 'solar observatory' sites in Arizona – most of which are in the Petrified Forest National Park. They propose that petroglyphs were used prehistorically in two ways for calendrical observations: (1) by light/shadow interacting with the rock art in 'significant' ways, and (2) by using a glyph to position an observer's head for sighting a horizon feature at sunrise or set. (No ethnographic evidence has been found to date for the use of rock art as a head-positioning device for calendrical observations.) The Prestons define 'significant interactions' as tangent/center shadows striking spirals/circles (and in other interactions for anthropomorphic figures). The light/shadow phenomena occurring here strongly resemble those at Fajada Butte, but rather than being concentrated in one locale they are spread out over many locations.

I have visited the site close to the Puerco ruin at the time of the summer solstice. The horizon view from the ruins would not make a good calendar. To the east and south of the ruin lie many boulders with rock art. The one noted by the Prestons is a small (10 cm wide, 12 cm high) spiral-like figure, whose pecked line is about 1 cm wide and very irregular. At mid-morning, a shaft of light (formed by a natural slit in the boulders to the northeast some 1.6 m away and so has noticeably fuzzy edges) moves down along the rock and crosses the spiral in about half an hour. Park Rangers at the site told me that the light first comes through the slit (which is fairly narrow – about $\frac{1}{2}°$ on the sky as seen from the petroglyph) roughly a week before the solstice.

These observations raise a number of questions. One, given the density of petroglyphs and rock edges at the site, what was the probability that the interaction was accidental? Two, to what extent does the fuzziness of the shadow edge and the irregularity of the pecking limit the practical resolving power of the site? The Prestons' sites need to have such questions asked about them. Preston (1984) has made one stab at the probability problem; he finds that the Cave of Life site is the only one where the statistical chances that the interactions are accidental are very small.

Finally, the Holly site at Hovenweep presents a dramatic display at (and before) the summer solstice. Then, about a half an hour after local sunrise, two shafts of light move horizontally across a petroglyph panel that includes spiral and circle shapes. Only about seven minutes pass from the first appearance of the beams and their merging near a spiral. Williamson and Young (1979) have noted symbols on the panel that may relate to the Twin War gods (Venus as Morning and Evening star) and a snake/serpent that may be a water symbol. The petroglyphs here have many motifs similar to those around the '1054 supernova' site below Penasco Blanco at Chaco (M. Zeilik, unpublished field notes). Williamson (1984, pp. 93–103) has concluded that the site may have been used to keep a planting calendar similar to that at Hopi – especially pertinent to the farming area in which it is located. He also suggests that the change in the positions of the light shafts would allow a practiced observer to anticipate the summer solstice to '...within three or four days'. (In contrast, the summer solstice portal in Hovenweep Castle would allow a trained observer to predict the summer solstice to within a day or two.)

In summary about the exterior sites: I concur with Young (1986) and Schaafsma (1985, pp. 264–6) that some Anasazi rock art served to interact with light and shadow. However, I think that a better case needs to be made that these interactions work *calendrically* in the appropriate cultural context. Because it works today does not guarantee that it was used that way prehistorically.

No ethnographic analogy has been found yet for exterior sites whose light/shadow interacts with rock art for calendrical purposes. (Of course, the physical principle is the same as that for light entering the windows of buildings.) The closest process comes from Zuni, where Pekwin visited a 'small post of petrified or silicified wood' (Fewkes, 1891, p. 4). Fewkes states that this post '...is in certain particulars a gnomon' – an implication that the position of its shadows was used as a marker at sunrise. Bourke (1881) clarifies the situation a bit, when he writes that the stone pillar was a place for Pekwin

to make sacrifices and watch the sun. Bourke and Cushing go with the Zuni governor, Pedro Pino, to

> the Eastern boundary of the Pueblo to an open field in the center of which was a vertical, four sided piece of petrified wood, put there, Pedro said, a long, long while ago. In line with it away off to the East, six or seven miles on the crest of a mesa could be discerned another monument, pointed out by Pedro.

The implication to me is that Pekwin used the pillar in the field to position himself in the right alignment with the pillar on the mesa. This case does not much resemble prehistoric exterior light/shadow sites.

10.5 Lunar observations

Ethnographic information shows that some Pueblos paid close attention to the phases of the moon and used them as a timing device (among other functions, such as weather prediction) for some festival dates – most notably Shalako at Zuni (Zeilik, 1986a; and see McCluskey, Chapter 28) and Powamu at Hopi (McCluskey, 1977). The regular phase cycle serves as a nightly marker that can be used to anticipate events that coincide with a specific phase. For example, at Zuni, the head of the Long Horns had to anticipate full moons (for preparation and planting of prayer feathers by the impersonators of the Shalako) by two days (Bunzel, 1932c, p. 963). He could have done so by spotting the first visible crescent and then counting for ten days (Zeilik, 1986a).

Because the Pueblos had a reliable seasonal calendar (based on solar observations), the use of the moon to schedule ceremonies brings up the ancient problem of synchronizing the solar and lunar calendars. This intercalation issue clearly arises at Zuni in the coordination of Shalako, the winter solstice ceremony, and the New Fire ceremony that marks the start of a new ritual year. Pekwin sets the date for the winter solstice ceremony by observations of the sun; Sayatasha (the head of the Long Horn Kachina group) establishes the Shalako schedule by observing the phases of the moon and has the final decision for the date (Bunzel, 1932c). Sayatasha consults with Pekwin to ensure that Shalako does not conflict with the Winter Solstice festival.

Meanwhile, we are also informed that Pekwin is supposed to arrange for the winter solstice ceremony to occur at the full moon in December (Bunzel, 1932a, p. 354; see also Tedlock, 1983)! Reyman (1980) has emphasized how troublesome this scheduling requirement will be for Pekwin, for even given a three-day range, at best the full moon/winter solstice coincidence can happen only every three years. However, Pekwin does *not* control the date of Shalako, which *is* based on lunar phases. The 49-day countdown to Shalako begins on the day of a full moon (Bunzel, 1932a, p. 523), and that is directed by Sayatasha. Another articulation of the sun and moon occurs at the winter solstice ceremony, for the first full moon that follows begins the count of ten full moons that is the first stage of the timing for Shalako. Overall, I do not see how Sayatasha and Pekwin kept the timings in line.

One hint comes from the sequence of Zuni months, in which the first six are named and the next six are nameless or repeated in names (Stevenson, 1904, p. 108). Parsons (1917, pp. 300–1) suggested that the non-naming of the November moon, in particular, provides the elasticity to reconcile the Shalako and winter solstice ceremonies. Shalako dates (Lyon, 1985) need to be compared carefully to winter solstice celebratory dates to illuminate this issue (as McCluskey, 1977, has done for the Hopi).

The Pueblos with sun- and moon-watching invited the intercalation problem. The months began with the observation of the new (crescent) moon, the seasonal cycle by sun-watching. The months tended to be named after key ceremonies within them (especially at Hopi; see Ellis, 1975, pp. 65–8 for a summary) or seasonal events (see Harrington, 1916, pp. 63–6 for the Tewa month names and a comparison to those at Jemez and Zuni). The Pueblos that tally and name months usually have 12 or 13 in a seasonal year (see Cope, 1919, pp. 151–3 for a summary). I suspect those with a fixed 12 months have adopted, at least in part, a European calendar. Those that have 13 months (or more) probably are the closest to the aboriginal calendar, for those allow intercalation by a 'short month' technique. One Hopi, Abbott Sekaquaptewa (1983), has written in a Third Mesa newspaper that, for the case of the winter solstice ceremony and the next new moon (Pamuya), the adjustment involves 'the elimination of one or a couple of the four day cycles after the final rites of the winter solstice ceremony'. This procedure speeds up the arrival of the next lunar cycle and 'allows a continuity of the normal ceremonial and spring planting times'.

McCluskey (1977) has analyzed the actual dates for the winter bean ceremony (Powamu), where the full moon after the close of the winter solstice ceremony signals the planting of beans in kivas (Stephen, 1936, p. 162; Parsons, 1939, p. 505). Now, each year a given phase of the moon arrives 10.9 days earlier, so that, after three years, the moon phases are 32.7 days 'ahead' of the sun. A simple intercalation scheme would occur every 2.71 years. McCluskey (1977) finds this periodicity in the Powamu dates. The inference is that one 'month' is not counted every 2.71 years; it may be a very short interval. Unfortunately, similar evidence needs to be gathered for other Pueblos, and no general, consistent scheme of intercalation appears to have been used in historic times.

No firm information is yet available about an interest in or knowledge about the lunar 18.6-year standstill cycle. One hint comes from Curtis (1922, p. 156, n. 1) in his discussion of the Flute ceremony at Hopi: 'It is said that at villages other than Walpi, the rising of the moon at a certain landmark on the horizon determines the date; and the date is not the same at Walpi as at the other pueblos.' If this information is correct, then some Hopi religious officers observed the horizon position of the rising moon, which goes through a complete north–south azimuthal swing in a month. That swing is especially noticeable at major standstill, but it does not immediately follow that the standstill cycle was noted at Hopi. In fact, the context of the reference is the Flute/Snake ceremonies, which occur at the same time of year but in *alternate* years (Curtis, 1922, p. 156), rather than an 18.6-year (or 19-year) cycle.

Sofaer, Sinclair and Doggett (1982) have argued that the Anasazi marked the standstill cycle at the three-slab site in Chaco at moonrise. Such a cycle does not fit into the typical historic Pueblo ceremonial framework. That ritual round usually encompasses but one seasonal year in a fixed sequence of events. Longer cycles are typically not counted. For example, Dutton and Marmon (1936, p. 5) state that

> ... the Lagunans have had no very accurate
> system of 'year counts'. Those who have not
> adopted use of modern calendars, keep track of
> time as did their ancestors, by particular events
> or phenomena, or by saying that such-and-such

Figure 10.7 Zuni calendar stick, drawn by J. G. Bourke (1881, 21 November).

occurred when so-and-so was a boy. They have no eras, cycles, or periods of several or many years.

One exception to this framework of short cycles is given in an obscure reference in Stephen (1936, p. 1039, n. 1) in which, from early notebooks, he writes: 'The kachinas reckon their feasts by the year of thirteen moons, consequently their Keli only coincides with the secular Keli once in thirteen years, which period is called a glad year.' The context here is the Hopi way of reckoning the months. The 'Keli' probably refers to the initiation into the Hopi kachina societies. The general tribal initiation occurs in November, and every four to six years (Curtis, 1922, pp. 107–8) takes place in a long form (16-day long). No other source contains information on a Hopi 'glad year' cycle, nor does Stephen mention it elsewhere in *Hopi Journal*. It may have been bad information, or it may have come from a Hopi clan that died out during Stephen's stay there.

If the Fajada three-slab site did intentionally mark the lunar extremes, questions naturally follow based on ethnographic analogy. What ceremony did the major or minor standstill trigger? How were these ceremonies anticipated? Let me speculate that they were anticipated within an accuracy of a year. Can an observer reliably see this motion in moonlight, limited to northern declinations, with the given fuzziness of the shadow's edge? A practical test of lunar shadow watching needs to be made to ascertain if that site is usable for tracking the standstill cycle. (I note here that Zeilik 1985b and 1985c contain an error in the rate of motion of the moon's shadow over the standstill cycle; the numbers have been corrected – see Sofaer and Sinclair, 1986a,b. The corrections do not negate my point about the limited resolving power of the motion of the edge of the moon's shadow to anticipate the time of the standstills.) In other words, the moonlight shadows at the site may have provided a *cyclical hierophany* with a modest resolving power

for calendrical purposes (if the lunar standstill cycle did in fact matter to the Chacoan Anasazi).

Lunar calendars kept by phases (with the first visible crescent starting the month) present a serious archaeological problem: What would be the material remains? We have two hints. At San Ildefonso, Stevenson (n.d.) reported that the moon-watcher kept a tally for the year by notching the edge of a flat stone on which was drawn a moon face. These stones were then deposited in a ceremonial chamber. At Zuni and Hopi, we have evidence of the use of calendar sticks to track the months. Fewkes (1892, p. 151) described the one at Hopi, and Alexander Marshack, at my suggestion, found the same stick in the Smithsonian Institution. The Hopi stick has a total of 28 grooves, which could be used to count or record a month of phases if the times of invisibility are not counted. The Zuni stick (Figure 10.7) is described and drawn by Bourke (1881, 21 November), based on information acquired in a meeting with the Zunis Nanahe and Nayuchi (Figure 10.8), who was Head Bow Priest at the time), who implies that it was used to tally a year, starting with '. . . the moon of the Fire Festival, or winter solstice'. By Fire Festival, Bourke means the New Fire ceremony, which takes place right after the Zuni winter solstice ceremony. (I note that the Zuni stick more closely resembles the Winnebago one described by Marshack, 1985, than does the Hopi one.) Fewkes (1892, p. 151) states that he has seen calendar sticks at Walpi and Hano. He describes them as follows:

These sticks are about a foot or a foot and a half long, and are divided into two parts, one section being round, the other flattened on one side. The round section is girt by fifteen shallow parallel grooves, and occupies about a third of the whole length of the stick. The remaining two thirds of the stick has a number of parallel grooves or notches cut upon the flattened surface. Five of the latter grooves, which are situated at equal distances, are deeper than the

Figure 10.8 Zuni religious officials in the 1880s, photographed in Boston in 1882. From left to right: Nanahe, a Hopi who married in Zuni, Nayuchi, the Head Bow Priest; Layutsailunkia, foster father to Frank Cushing; Pahlowahtiwa, governor of Zuni; and Kiasi, Junior Bow Priest. (Courtesy Photo Archives, Museum of New Mexico, negative no. 89333, Benjamin W. Kilburn, photographer.)

159

remaining, and between each pair there are four smaller parallel grooves arranged at equal distances. The space in which these grooves are cut occupies about one half of the flat portion of the stick. The remaining half, or that more distant from the round section, is divided into two parts, which are separated by a rectangular space, in the centre of which there is a depression called the '*nā-tā'l-tcĭ*'. On one side of the depression there are three notches, on the other, seven.

The stick found by Marshack (Figure 10.9) matches this description quite well, and was probably collected by Fewkes at Hopi.

Sticks or rocks (even rock art?) with similar markings could be inferred to be Anasazi devices for tracking months, if found in the right archaeological context. More important, the mode of keeping such tallies would reveal the Anasazi solution to the problem of lunar calendar keeping.

10.6 The Mesoamerican connection?

Archaeologists have debated for many years the possible influences of Mesoamerican cultures on the prehistoric societies of the US Southwest. The interpretations tend to fall into two camps: those that argue for a strong (controlling?) Mexican influence, specifically from the Toltec–Chalchihuites cultures (Kelley and Kelley, 1975), and those that claim that the Southwest – specifically Chaco – developed in essentially a cultural isolation (McGuire, 1980). I do not want to be caught in the cross-fire between these two groups but do want to propose one possible solution to their feud. It lies in a careful comparison of prehistoric Southwestern and Mesoamerican astronomical practices (Reyman, 1971) – a comparison that I now feel we can begin to make with some reliability.

A general comparison of historic Pueblo practices shows that, when compared with calendrical activities from Mayan and Central Mexican cultures, the Southwest *lacked*: (1) written calendars, (2) a numerical system with long counts, (3) 260-day ritual 'years', (4) close attention to conjunctions of Venus, (5) a system of year bearers, and (6) zenith passages of the sun.

Some Southwestern investigators make the claim that the kachina cult is an outgrowth of Mexico that worked its way up the Rio Grande valley in the 14th century (Ellis and Hammack, 1968),

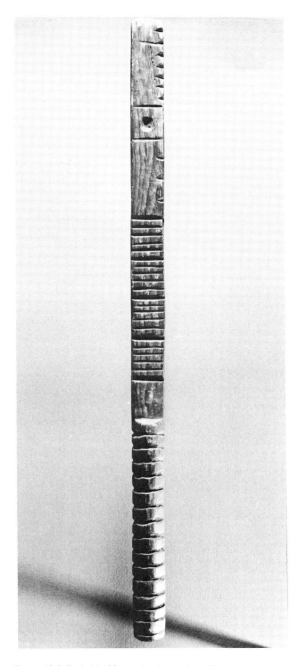

Figure 10.9 Probable Hopi calendar stick, collected by Fewkes. (Photo by Alexander Marshack.)

especially on the basis of a change in the rock art (Schaafsma and Schaafsma, 1974). Sometime after AD 1300, a new rock art style pervades the Rio Grande valley (and extends as far west as Hopi) – a style that includes masked figures similar to the kachina dancers. The motifs derive from the

Jornada Mogollon of southern New Mexico and include representations of the Mexican deity Quetzalcoatl. One of the most striking elements is that of Venus as Morning or Evening star (Schaafsma, 1980). Another Mexican influence relates to pecked-cross figures, which are abundant in the Mogollon region (Zeilik, 1980), but which are less common in Anasazi territory.

Reyman (1971), Ellis (1977) and Young (see Chapter 11, this volume) have taken steps in a cosmological context by comparing Hopi and Zuni kachinas with Mesoamerican deities – specifically for Ellis and Young through parallels in iconography that may derive from parallel functions of supernatural beings in different (but perhaps intertwined) cultural contexts. Fundamental to the Puebloan attitude towards the gods is the notion of *reciprocity* between human beings and supernaturals so that the cosmic order can be sustained. That reciprocity shows up strongest in the sun/water cult relating to fertility – and that lays the base for sacred directions in both Puebloan and Mesoamerican world views. But, we still need to reveal the connections to astronomy, if a cross fertilization has indeed occurred.

One striking area of similarity relates to sacred directions (Reyman, 1971; Broda, 1982; Vogt, 1985; and Young, Chapter 11 this volume). The important parallels here include the connection of specific animals and colors to each direction, as well as mountain shrines as the physical embodiments of the sacred space. But the most promising aspect to me is that of the *solstitial* points as cardinal points, rather than the European notions of north, south, east and west at right angles to each other. When seen in the Pueblo context, they may represent cultural borrowings or a misunderstanding by an investigator from a European background who 'expected' the 'proper' cardinal directions. Zenith passages do integrate particular directions and times in Mesoamerica (Broda, 1982, pp. 90–3); they do not occur in the Southwest. I do find it intriguing that the Isleta ceremonies of drawing down the sun (Parsons, 1932, pp. 290–3) and the moon (Parsons, 1932, p. 342) involve the use of sunlight and moonlight at meridian passage – perhaps an analogy to Mesoamerican zenith tubes (Broda, 1982, pp. 90–3) even though the sun does not transit the zenith at Southwestern latitudes.

I suggest that one way to clear up the cloud around the Southwestern–Mesoamerican connection requires a close look at Venus in the Southwestern context – a job yet to be done. We have many tales of the Pueblo Twin War Gods, who are related to Venus as Morning and Evening star (Young and Williamson, 1981). In the Rio Grande Pueblos, the two War Chiefs (or Outside Chiefs) are the two aspects of Venus. Curiously, only they can criticize the actions of the cacique – the religious head of the Pueblo – who does the sun-watching. As Bandelier noted in 1880:

> If the cacique shows unfavorable character, is not peaceable, quiet, or is ambitious, then the capitan de guerra [War Chief] calls a meeting at the estufa [kiva], which meeting is attended by all those who are regarded as capable of understanding the matter and worthy of participation in it ...

(Lange and Riley, 1966, p. 179). The cacique represents Earth Mother, who must know that the people of the Pueblo are doing the proper rites at the proper time. The War Chief, in contrast, represents Father Sun – a position that fell to him when the older position of War *Priest* disappeared after the Spanish contact with the Pueblos (F. H. Ellis, private communication). The War Priest and the cacique together headed the Pueblo, with equal but complementary duties as earth's representatives of Father Sun and Earth Mother. The War Priest had the responsibility of keeping the Pueblo safe from invaders and witches; his two assistants in this task were the War Chiefs, sons of the Sun. When the office of the War Priest died out, the 'elder' War Chief took over some of the War Priest's duties and carries them out today. Hence, one aspect of Venus has acquired some of the Sun's power in the Pueblo symbolic context. Needed here is a detailed investigation of Pueblo recordings of observations of Venus, a comparison to Venus configurations on those dates, and the search for a cycle (five or eight years, for example) in the historical record.

10.7 Conclusions and suggestions

Workers in the Southwest have come a long way since the concerted efforts to find rock art that may represent the 1054 supernova (Miller, 1955; Young, 1986). That field work drew attention to the issue of astronomy in the prehistoric Southwest. The next step involved possible kiva alignments (Reyman,

1971) and the search for sun-watching stations (Williamson *et al.*, 1975; Williamson, Fisher and O'Flynn, 1977) and the first instances of light and shadow interactions. That stage culminated in the recognition of the sunlight interactions at the three-slab site on Fajada Butte by Jay Crotty and Anna Sofaer and the subsequent long-term efforts there by Sofaer and co-workers, especially Rolf Sinclair at that site and others on the butte. We have recently begun to appreciate the possibilities of concrete expressions of cosmic structure by the Anasazi – such as with Casa Rinconada (Williamson, 1982, 1987) – based on available ethnographic information of Puebloan cosmology (such as described for Hopi Third Mesa by Titiev, 1944, and by Parsons, 1939, more generally). However, such possibilities will be very hard to demonstrate conclusively (Reyman, 1971, p. 321, rejected this notion based on his field work).

We are now at a new stage in the search for Anasazi astronomy – that of trying to place sites and phenomena into a specific cultural context. To do so successfully requires that we draw the archaeologists into the arena as active participants. They will tell us about the problems of the control of time with which we must now grapple in order to connect at the right level in the cultural matrix.

How to entice the archaeologists, most of whom have so far stood at the sidelines? I propose that we focus on the *material correlates of astronomical activities*. We need to demonstrate that the focus on the sky results in material remains that differ from those developed for other ritual activities. I have mentioned some so far: windows/portals that admit light onto walls with calendrical markings; calendar sticks or stones; light/shadow interactions on rock art at key seasonal times (solstices, planting); offerings from rituals that occur seasonally.

What can we do for ourselves? First, we need more *cooperative* work to make effective use of our energies that must span a wide geographic area. Second, in contrast, we also need more *independent* study of the same sites to avoid observer bias. Third, we must *share* our results, even in preliminary form, faster so that we can develop effective ventures in new directions. Fourth, we should develop a *broader* context with the Southwest and Mesoamerica. Fifth, we need *intensive* work in specific areas, such as a field school at Chaco Canyon. Sixth, light/shadow sites need an *objective* evaluation of their practical resolving power. Finally, we must hammer out appropriate methodologies so that we can trust our interpretations and conclusions, which we should view as tentative insights into Anasazi astronomy and cosmology.

Acknowledgments

I thank Rolf Sinclair, Anna Sofaer, M. Jane Young and Steve McCluskey for sending me copies of pertinent papers. Ray Williamson, Tony Aveni, Jonathan Reyman, Curt Schaafsma, Polly Schaafsma, John Carlson, Steve McCluskey and M. Jane Young read over and commented upon early drafts of this paper. Tom Windes and Joan Mathien made substantial comments on a later draft, as did Florence Ellis, whose knowledge of the Pueblos is deep and intimate.

References

Beaglehole, E. (1937). Notes on Hopi economic life. *Yale University Publications in Anthropology* 15.

Binford, L. R. (1972). Methodological considerations of the archaeological use of ethnographic data. In *An Archaeological Perspective*, pp. 59–67. New York: Seminar Press.

Bourke, J. G. (1881). *Diary*. West Point, New York: United States Military Academy Library, Special Collections and Archives. Photofacsimile in Special Collections, Coronado Room, Zimmerman Library, University of New Mexico. Entries for November, 1881.

Brandt, J. C., Maran, S. P., Williamson, R. A., Harrington, R. S., Cochran, C., Kennedy, M., Kennedy, W. J. and Chamberlain, V. D. (1975). Possible rock art records of the Crab Nebula supernova in the Western United States. In *Archaeoastronomy in Precolumbian America*, ed. A. F. Aveni, pp. 45–58. Austin: University of Texas Press.

Brandt, J. C. and Williamson, R. A. (1977). Rock art representations of the AD 1054 supernova: a progress report. In *Native American Astronomy*, ed. A. F. Aveni, pp. 171–7. Austin: University of Texas Press.

Brandt, J. C. and Williamson, R. A. (1979). 1054 Supernova and rock art. *Archaeoastronomy Supplement to the Journal for the History of Astronomy*, no. 1, pp. S1–S38.

Broda, J. (1982). Astronomy, *cosmovision*, and ideology in pre-Hispanic Mesoamerica. In *Ethnoastronomy and Archaeoastronomy in the American Tropics*, ed. A. F. Aveni and G. Urton. Annals of the New York Academy of Sciences, vol. 385.

Bunzel, R. L. (1932a). Introduction to Zuni ceremonialism. *Bureau of American Ethnology 47th Annual Report*, pp. 467–544.

Bunzel, R. L. (1932b). Zuni ritual poetry. *Bureau of American Ethnology 47th Annual Report*, pp. 611–835.

Bunzel, R. L. (1932c). Zuni katcinas. *Bureau of American Ethnology 47th Annual Report*, pp. 837–1086.

Carlson, J. B. (1987). Romancing the stone or moonshine on the Sun Dagger. In *Astronomy and Ceremony in the Prehistoric Southwest*, eds. J. B. Carlson and W. J. Judge. Albuquerque: Maxwell Museum Press. Papers of the Maxwell Museum Press, no. 2, pp. 71–88.

Cope, L. (1919). Calendars of the Indians north of Mexico. *University of California Publications in American Archaeology and Ethnology* 16, (4), 119–76.

Cordell, L. S. (1984). *Prehistory of the Southwest*. New York: Academic Press.

Curtis, E. S. (1922). The Hopi. *The North American Indian* 22.

Cushing, F. H. (1979). My adventures in Zuni. Reprinted in *Zuni*, ed. J. Green, pp. 46–134. Lincoln: University of Nebraska Press.

Dutton, D. P. (1963). *Sun Father's Way: The Kiva Murals of Kuaua*. Albuquerque: University of New Mexico Press.

Dutton, B. and Marmon, M. A. (1936). The Laguna calendar. *The University of New Mexico Bulletin, Anthropological Series*, whole number 283, 1, (2).

Ellis, F. H. (1975). A thousand years of the Pueblo sun–moon–star calendar. In *Archaeoastronomy in Precolumbian America*, ed. A. F. Aveni, pp. 59–87. Austin: University of Texas Press.

Ellis, F. H. (1977). Distinctive parallels between Mesoamerican and Pueblo iconography and deities. Paper delivered at Guanajuato, Mexico.

Ellis, F. H. and Hammack, L. (1968). The inner sanctum of Feather Cave, a Mongollon sun and earth shrine linking Mexico and the Southwest. *American Antiquity* 30, 25–44.

Farrer, C. R. and Second, B. (1981). Living the sky: aspects of Mescalero Apache ethnoastronomy. In *Archaeoastronomy in the Americas*, ed. R. A. Williamson, pp. 137–51. Los Altos, CA: Ballena Press/Center for Archaeoastronomy.

Fewkes, J. W. (1891). A few summer ceremonials at Zuni Pueblo. *Journal of American Ethnology and Archaeology* i, 1–61.

Fewkes, J. W. (1892). A few summer ceremonials at the Tusayan Pueblos. *Journal of American Ethnology and Archaeology* ii, 1–161.

Fewkes, J. W. (1906). Hopi shrines near East Mesa, Arizona. *American Anthropologist* (NS) 8, 346–75.

Fewkes, J. W. (1919). Prehistoric villages, castles, and towers of Southwestern Colorado. *Bureau of American Ethnology Bulletin* no. 79, plates 14a, b, c; 19c and 32b.

Forde, C. D. (1931). Hopi agriculture and land ownership. *Journal of the Royal Anthropological Institute of Great Britain and Ireland* 61, 357–405.

Frigout, E. (1979). Hopi ceremonialism. *Handbook of North American Indians*, ed. A. Ortiz, vol. 9, pp. 564–76. Washington: Smithsonian Institution.

Gaustad, J. E. (1980). The Chaco Canyon supernova pictograph – a reorientation. *Archaeoastronomy* III, (4), 33–4.

Harrington, J. P. (1916). *Bureau of American Ethnology, 29th Annual Report*, pp. 29–618.

Hedges, K. (1985). Methodology and validity in California archaeoastronomy. In *Earth and Sky*, eds. A. Benson and M. Hoskinson. pp. 25–39. Thousand Oaks: Slo'w Press.

Hewett, E. L. (1906). Antiquities of the Jemez Plateau. *Bureau of American Ethnology Bulletin* no. 32.

Kelley, J. C. and Kelley, E. A. (1975). An alternative hypothesis for the explanation of Anasazi cultural history. In *Collected Papers in Honor of Florence Hawley Ellis*, ed. T. R. Frisbie. Papers of the Archaeological Society of New Mexico, no. 2, pp. 178–223.

Koenig, S. H. (1979). Stars, crescents, and supernove in Southwestern Indian art. *Archaeoastronomy Supplement to the Journal for the History of Astronomy*, no. 1, pp. 539–50.

Lange, C. H. (1959). *Cochiti: A New Mexico Pueblo Past and Present*. Austin: University of Texas Press.

Lange, C. H. and Riley, C. L. (1966). *The Southwestern Journals of Adolph F. Bandelier: 1880–1882*. Albuquerque: University of New Mexico Press.

Lyon, L. (1985). Chronology of the Zuni Sha'lak'o Ceremony. In *Southwestern Culture History: Collected Papers in Honor of Albert H. Schroeder*, ed. C. H. Lange. Papers of the Archaeological Society of New Mexico, no. 10, pp. 233–49.

McCluskey, S. C. (1977). The astronomy of the Hopi Indians. *Journal for the History of Astronomy* viii, 174–95.

McCluskey, S. C. (1981). Transformations of the Hopi calendar. In *Archaeoastronomy in the Americas*, ed. R. A. Williamson, pp. 173–82. Los Altos, CA: Ballena Press/Center for Archaeoastronomy.

McCluskey, S. C. (1982). Historical astronomy: the Hopi example. In *Archaeoastronomy in the New World*, ed. A. Aveni, pp. 31–57. Cambridge University Press.

McGuire, R. H. (1980). The Mesoamerican connection in the Southwest. *The Kiva* **46**, 3–38.

McKenna, P. J. (1984). The architecture and material culture of Chaco Canyon, New Mexico. In *Reports of the Chaco Center*, ed. W. J. Judge. Albuquerque: Division of Cultural Resources, National Park Service, no. 7.

Marshack, A. (1985). A lunar–solar year calendar stick from North America. *American Antiquity* **50**, 27–51.

Mayer, D. (1979). Miller's hypothesis. *Archaeoastronomy Supplement to the Journal for the History for Astronomy*, no. 1, pp. S51–S74.

Miller, W. C. (1955). Two prehistoric drawings of possible astronomical significance. *Astronomical Society of the Pacific Leaflet* no. 314.

Newman, E. B., Mark, R. K. and Vivian, R. G. (1982). Anasazi solar marker: the use of a natural rockfall. *Science* **217**, 1036–8.

Ortiz, A. (1969). *The Tewa World*. University of Chicago Press.

Parsons, E. C. (1917). Notes on Zuni. *Memoirs of the American Anthropology Association* **4**, (3).

Parsons, E. C. (1925). A Pueblo Indian journal 1920–1921. *Memoirs of the American Anthropological Society*, no. 32.

Parsons, E. C. (1932). Isleta, New Mexico. *Bureau of American Ethnology 47th Annual Report*, pp. 193–466.

Parsons, E. C. (1939). *Pueblo Indian Religion*. University of Chicago Press.

Preston, A. L. and Preston, R. A. (1985). The discovery of 19 prehistoric calendric petroglyph sites in Arizona. In *Earth and Sky*, eds. A. Benson and M. Hoskinson, pp. 123–33. Thousand Oaks: Slo'w Press.

Preston, R. A. (1984). Calendrical petroglyph sites in Arizona: new evidence and statistical studies. Paper presented at the 1984 International Conference on Prehistoric Rock Art and Archaeoastronomy, Little Rock, Arkansas.

Preston, R. A. and Preston, A. L. (1987). Evidence for the calendric function at 19 prehistoric petroglyph sites in Arizona. In *Astronomy and Ceremony in the Prehistoric Southwest*, eds. J. B. Carlson and W. J. Judge. Albuquerque: Maxwell Museum Press. Papers of the Maxwell Museum of Anthropology, no. 2, pp. 191–204.

Reyman, J. E. (1971). Mexican Influence on Southwestern Ceremonialism. PhD dissertation, Southern Illinois University, pp. 114–22.

Reyman, J. E. (1975). The nature and nurture of archaeoastronomical studies. In *Archaeoastronomy in Precolumbian America*, ed. A. F. Aveni, pp. 205–15. Austin: University of Texas Press.

Reyman, J. E. (1976). Astronomy, architecture, and adaptation at Pueblo Bonito. *Science* **193**, 957–62.

Reyman, J. E. (1980). The predictive dimension of priestly power. In *New Frontiers in Archaeology and Ethnohistory of the Greater Southwest*, eds. C. L. Riley and B. C. Hendrick. Transactions of the Illinois Academy of Science **72**, (4), 40–59.

Schaafsma, P. (1980). *Indian Rock Art of the Southwest*. Albuquerque: University of New Mexico Press.

Schaafsma, P. (1985). Form, content, and function: theory and method in North American rock art studies. *Advances in Archaeological Method and Theory* **8**, 237–77.

Schaafsma, P. and Schaafsma, C. F. (1974). Evidence for the origins of the Katchina Cult as suggested by Southwestern rock art. *American Antiquity* **33**, 535–45.

Sekaquaptewa, A. (1983). Out of phase with the moon phase. In *Qua'toqi*. Arizona: Oraibi.

Simmons, L W. (1942) *Sun Chief: The Autobiography of a Hopi Indian*. New Haven: Yale University Press.

Sofaer, A. and Sinclair, R. (1987). Astronomical markings at three sites on Fajada Butte. In *Astronomy and Ceremony in the Prehistoric Southwest*, eds. J. B. Carlson and W. J. Judge. Albuquerque: Maxwell Museum Press. Papers of the Maxwell Museum of Anthropology, no. 2, pp. 43–67.

Sofaer, A. and Sinclair, R. M. (1986a). Letter. *Science* **231**, 1057–8.

Sofaer, A. and Sinclair, R. M. (1986b). Appraisal. *Archaeoastronomy Supplement to the Journal for the History of Astronomy*, no. 10, pp. S59–S66.

Sofaer, A., Sinclair, R. M. and Doggett, L. E. (1982). Lunar markings on Fajada Butte, Chaco Canyon, New Mexico. In *Archaeoastronomy in the New World*, ed. A. F. Aveni, pp. 169–81. Cambridge University Press.

Sofaer, A., Zinser, V. and Sinclair, R. M. (1979a). A unique solar marking construct. *Science* **206**, 283–91.

Sofaer, A., Zinser, V. and Sinclair, R. M. (1979b). A unique solar marking construct of the ancient Pueblo Indians. In *American Indian Rock Art*, eds. F. G. Bock, K. Hedges, G. Lee and H. Michaelis, vol. V, pp. 117–25. El Toro, CA: American Rock Art Association.

Spier, L. (1955). Mohave culture items. *Museum of Northern Arizona Bulletin* no. 28.

Stephen, A. M. (1936). *Hopi Journal*, ed. E. C. Parsons. New York: Columbia University Press.

Stevenson, M. C. (1904). The Zuni Indians: their mythology, esoteric societies, and ceremonies. *Bureau of American Ethnology 23rd Annual Report*.

Stevenson, M. C. (n.d.). Material on the Tewa, Harrington Papers. Washington: Smithsonian Anthropological Archives.

Stirling, M. W. (1945). Concepts of the sun among American Indians. *Annual Report of the Smithsonian Institution*, pp. 387–400.

Tedlock, B. (1983). Zuni sacred theater. *Native American Quarterly* 7, (4), 93–110.

Titiev, M. (1938). Dates of planting at the Hopi Pueblo of Oraibi. *Museum Notes of the Museum of Northern Arizona* xi, (5), 39–42.

Titiev, M. (1944). Old Oraibi: a study of the Hopi Indians of Third Mesa. *Papers of the Peabody Museum, Harvard University* 22, (1).

Toll, H. W., Windes, T. C. and McKenna, P. J. (1980). Late ceramic patterns in Chaco Canyon: the pragmatics of modeling ceramic exchange. In *Models and Methods in Regional Exchange*, ed. R. E. Fry. Washington: Society for American Archaeology.

Vogt, E. Z. (1985). Cardinal directions and ceremonial circuits in Mayan and Southwestern cosmology. *National Geographic Society Research Reports* 21, 487–96.

White, L. A. (1942). The Pueblo of Santa Ana, New Mexico. *Memoirs of the American Anthropological Society*, no. 60.

Williamson, R. A. (1979). Field report Hovenweep National Monument. *Archaeoastronomy* 2, (3), 11–12.

Williamson, R. A. (1981). North America: a multiplicity of astronomies. In *Archaeology in the Americas*, ed. R. A. Williamson, pp. 61–80. Los Altos, CA: Ballena Press/Center for Archaeoastronomy.

Williamson, R. A. (1982). Casa Rinconada, twelfth century Anasazi kiva. In *Archaeoastronomy in the New World*, ed. A. Aveni, pp. 205–18. Cambridge University Press.

Williamson, R. A. (1983). Sky symbolism in a Navajo rock art site, Chaco Canyon. *Archaeoastronomy* VI, (1–4), 59–65.

Williamson, R. A. (1984). *Living the Sky: The Cosmos of the American Indian*. Boston: Houghton Mifflin.

Williamson, R. A. (1987). Light and shadow, ritual, and astronomy in Anasazi structures. In *Astronomy and Ceremony in the Prehistoric Southwest*, eds. J. B. Carlson and W. J. Judge. Albuquerque: Maxwell Museum Press. Papers of the Maxwell Museum of Anthropology, no. 2, pp. 99–119.

Williamson, R. A., Fisher, H. J., Williamson, A. F. and Cochran, C. (1975). The astronomical record in Chaco Canyon, New Mexico. In *Archaeoastronomy in Precolumbian America*, ed. A. F. Aveni, pp. 33–43. Austin: University of Texas Press.

Williamson, R. A., Fisher, H. J. and O'Flynn, D. (1977). Anasazi solar observatories. In *Native American Astronomy*, ed. A. Aveni, pp. 203–17. Austin: University of Texas Press.

Williamson, R. A. and Young, M. J. (1979). An equinox sun petroglyph panel at Hovenweep National Monument. In *American Indian Rock Art*, eds. F. G. Bock, K. Hedges, G. Lee and H. Michaelis, vol. V, pp. 70–80. El Toro, CA: American Rock Art Research Association.

Yava, A. (1978). *Big Falling Snow*. Albuquerque: University of New Mexico Press.

Young, M. J. (1986). The interrelationship of rock art and astronomical practice in the American Southwest. *Archaeoastronomy Supplement to the Journal for the History of Astronomy*, no. 10, pp. 543–58.

Young, M. J. (1987a). The nature of the evidence: archaeoastronomy in the prehistoric Southwest. In *Astronomy and Ceremony in the Prehistoric Southwest*, eds. J. B. Carlson and W. J. Judge. Albuquerque: Maxwell Museum Press. Papers of the Maxwell Museum of Anthropology, no. 2, pp. 169–90.

Young, M. J. (1987b). Issues in the archaeoastronomical endeavor in the American Southwest. In *Astronomy and Ceremony in the Prehistoric Southwest*, eds. J. B. Carlson and W. J. Judge. Albuquerque: Maxwell Museum Press. Papers of the Maxwell Museum of Anthropology, no. 2, pp. 219–32.

Young, M. J. and Williamson, R. A. (1981). Ethnoastronomy: the Zuni case. In *Archaeoastronomy in the Americas*, ed. R. A. Williamson, pp. 183–91. Los Altos, CA: Ballena Press/Center for Archaeoastronomy.

Zeilik, M. (1980). Pecked-cross-like petroglyphs in New Mexico. *Archaeoastronomy* III, (1), 21.

Zeilik, M. (1983a). Anticipation in Anasazi astronomy. Paper presented at the *56th Annual Pecos Conference*, August, 1983.

Zeilik, M. (1983b). Historic Puebloan sun watching. Paper presented at the *First International Ethnoastronomy Conference*, Washington, DC, September, 1983.

Zeilik, M. (1984a). A possible equinoctial sun-sighting station at Tsiping, New Mexico. *Archaeoastronomy* 7, (1–4), 70–5.

Zeilik, M. (1984b). Summer solstice at Casa Rinconada: calendar, hierophany, or nothing? *Archaeoastronomy* 7, (1–4), 76–81.

Zeilik, M. (1985a). The ethnoastronomy of the historic Pueblos. I. Calendrical sun watching. *Archaeoastronomy Supplement to the Journal for the History of Astronomy*, no. 8, pp. S1–S25.

Zeilik, M. (1985b). The Fajada Butte solar marker: a reevaluation. *Science* 228, 1311–13.

Zeilik, M. (1985c). A reassessment of the Fajada Butte solar marker. *Archaeoastronomy Supplement to the Journal for the History of Astronomy*, no. 9, pp. S69–S85.

Zeilik, M. (1985d). Sun shrines and sun symbols in the US Southwest. *Archaeoastronomy Supplement to the Journal for the History of Astronomy*, no. 9, pp. S86–S96.

Zeilik, M. (1985e). Keeping a seasonal calendar at Pueblo Bonito. *Archaeoastronomy* 8.

Zeilik, M. (1986a). The ethnoastronomy of the historic Pueblos. II. Moon watching. *Archaeoastronomy Supplement to the Journal for the History of Astronomy*, no. 10, pp. S1–S22.

Zeilik, M. (1986b). Reply. *Science* **231**, 1058.

Zeilik, M. (1986c). Response. *Archaeoastronomy Supplement to the Journal for the History of Astronomy*, no. 10, pp. S66–S69.

Zeilik, M. (1987). Anticipation in ceremony: the readiness is all. In *Astronomy and Ceremony in the Prehistoric Southwest*, eds. J. B. Carlson and W. J. Judge. Albuquerque: Maxwell Museum Press. Papers of the Maxwell Museum of Anthropology, no. 2, pp. 25–41.

Zeilik, M. and Elston, R. (1983). Wijiji at Chaco Canyon: a winter solstice sunrise and sunset station. *Archaeoastronomy* **VI**, (1–4), 66–73.

11

The Southwest connection: similarities between Western Puebloan and Mesoamerican cosmology

M. Jane Young   *University of New Mexico*

11.1 Introduction

Various scholars have suggested that Western Puebloan ideological systems – such as those manifested in the kachina cults of the Hopis and the Zunis – are derived largely from concepts that originated in Mesoamerica (Parsons, 1933; Anderson, 1955; Ellis and Hammack, 1968; Kelley, 1966; Schaafsma and Schaafsma, 1974; Schaafsma, 1975; Ellis, 1977). Whether one posits that the kachina cult came to the Southwest only relatively recently, for instance, around AD 1325, or much earlier, there is no doubt that the extensive although intermittent contact between the peoples of Mesoamerica and the American Southwest has resulted in a number of striking parallels in world view and religious practice, as well as in the more practical domains of agriculture and textile and pottery production. It is now known that extensive trade networks existed between the two groups and that cultural items and ideas were exchanged as well as trade goods over a long period of time: the precursors to the modern-day Puebloans adopted maize and squash – later to become central foods in their diet – along with the techniques for cultivating them, from Mesoamerican peoples at an early date. Maize domesticated in central Mexico was being grown, at least sporadically, by the ancestral Puebloans at around 1000 BC (Ford, 1981). It seems unlikely, then, that the foodstuffs were adopted entirely apart from the socio-religious complex to which they were central (Kelley, 1966). As Florence Ellis puts it, 'we cannot imagine the seed and the techniques for growing corn having been transmitted without added instructions as to gods and rituals believed necessary for success' (Ellis, 1977, p. 4).

Nevertheless, at present there are scant archaeological data to support a northward transmission of ideology from Mesoamerica to the American Southwest during the middle to late Archaic period (ca 1500 BC to AD 100); perhaps this is due to the fact that ritual paraphernalia and other items of material culture from this early date have not survived the ravages of time.[1] There is some evidence that prayer sticks were deposited in caves by the Archaic cultivators of maize in the Southwest (Martin *et al.*, 1952), but there is little else that relates to the Mesoamerican religious complex that may have come with the maize. I do not suggest that one may therefore conclude that there was no ideological impact, but rather that we must look towards the less easily quantifiable and less tangible areas of cosmology and world view to begin to formulate an idea of the extent of Mesoamerican ideological influence on ancestral Western Puebloan cosmology and, hence, astronomical practice. Although this influence was certainly pan-Puebloan, I focus on the Western Pueblo groups of Hopi and Zuni because they have been subject to less Spanish and Anglo influence than the Eastern Puebloans.

One means by which to explore this Southwest–Mesoamerican connection is to look for parallels in iconography; I do not mean to suggest, however, that I advocate comparisons based strictly on stylistic criteria such as one would find in a study of correspondences between masks of Western Puebloan kachinas and the codex depictions of various

167

Mesoamerican deities. I am more concerned with similarities between religious concepts – such as those illustrated by the parallel functions and attributes of the supernatural beings – than with pictorial/stylistic correspondences between representations of deities (these parallels are depicted in Figure 11.1). Furthermore, it must be stressed that it is highly unlikely that the ancestral Puebloans adopted Mesoamerican concepts without subjecting them to change and variation. Certainly differences in climate and ecology would have necessitated adjustments, but it is also significant that the character of these Southwestern peoples tended to be dynamic: they borrowed much from surrounding groups, but tended to mold what they adopted to their own particular cultural style (Ellis, 1977).

To keep this paper to a manageable length, I will limit my discussion of Mesoamerican ideology to the Valley of Mexico, focusing particularly on the Aztec pantheon of the late Post Classic Period. The Aztecs did not, however, exist in isolation from the rest of Mesoamerica, and their culture was subject to many influences both from peoples in nearby geographical areas and from the pervasive ideologies of earlier cultures. This influence through time and over space is integral to the hypothesis that the Aztec religion was dominated by three Gods who appear to have persisted since Olmec times, their forms only slightly changed (Covarrubias, 1946; Joralemon, 1976; Nicholson, 1976; von Winning, 1976; Ellis, 1977). These deities are: *Huehuetéotl*, the Old Fire God, descended from the Olmec serpent–jaguar with flame eyebrows; *Tláloc*, the representative of water and growth who was derived from the Olmec dwarf or infant symbolic of maize; and *Xipe Tótec*, the patron of spring and the annual rebirth of plant life whose progenitor was the Olmec masked god.

Of these three, perhaps the most important was *Tláloc*, the rain god. He was frequently pictured with his eyes encircled by raised rings, originally in the form of two snakes that represented rain clouds – this has been described as a 'goggle-eyed' effect (Furst, 1974, pp. 69–71); further water symbolism is seen in *Tláloc's* down-turned and cavern-like mouth, pointing to his association with caves and underground springs (Grove, 1970, pp. 11, 32; Joralemon, 1976, pp. 37–40; Ellis, 1977). In certain depictions he is clutching lightning bolts, rendered in other portrayals as serpents (Nicholson, 1976,

p. 168; Ellis, 1977), and in some representations he wears a fringed kilt signifying rain (Peterson, 1961, fig. 31; Anton, 1969, pl. 101).

Paralleling its importance in Mesoamerica, a rain–water cult, seemingly derived from the *Tláloc* cult, took hold rapidly and at an early date in the Southwestern United States. Polly Schaafsma has pointed to Mesoamerican prototypes for many of the iconographic features of the Jornada rock art style which first appeared in the Southwest sometime around AD 1050. Of particular note in this respect are the goggle-eyed *Tláloc*-type figures and possible representations of *Quetzalcóatl* found in this rock art style complex. Schaafsma concludes that there is a logical historical and cultural connection between cults such as that of *Tláloc* and *Quetzalcóatl* in Mesoamerica and the later Pueblo Kachina cult (Schaafsma, 1975). Of course, it is not surprising that the cults of Tláloc and Quetzalcóatl – deities of the rain and wind that often brings rain – spread so rapidly, for both the Mesoamericans and the ancestral Puebloans faced the contingencies of sustaining an existence dependent on agriculture in a frequently dry and somewhat capricious climate.

I will turn now to a comparison of some aspects of the Western Puebloan pantheon and cosmological concepts with those of the Mesoamericans (see Figure 11.1 for a diagram of the parallels drawn throughout this discussion). I regard this sort of comparison as a necessary first step in delineating the range and extent of the interactions between the two groups that had a major effect on their perceptions of astronomical phenomena as well.

To begin with ideas of the beginning, the Aztecs and their predecessors posited a dual creative principle: *Ometecuhtli* and *Omecíhuatl* – a male god and female goddess who are never represented pictorially. According to one legend, this pair had four sons 'to whom they entrusted the creation of the other gods, the world, and man' (Caso, 1958, p. 10). These four sons, associated with the cardinal directions and their corresponding colors, were the *Red Tezcatlipoca*, the *Black Tezcatlipoca*, the *White Tezcatlipoca* or *Quetzalcóatl*, and *Huitzilopochtli*, the *Blue Tezcatlipoca*. Although there are diverse opinions about this, there appears to be some similarity between the Mesoamerican generative pair and the Zuni primary principle of light and life: *'Awonaawi-l'ona*. This term is used as an epithet both for the Sun Father and the Moon Mother, translating in

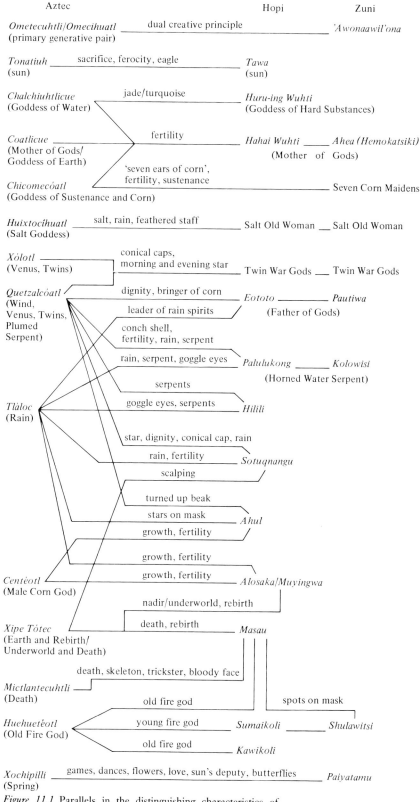

Aztec   Hopi   Zuni

Ometecuhtli/Omecihuatl ———— dual creative principle ———— *'Awonaawil'ona*
(primary generative pair)

Tonatiuh ———— sacrifice, ferocity, eagle ———— *Tawa*
(sun)   (sun)

Chalchiuhtlicue ———— jade/turquoise ———— *Huru-ing Wuhti*
(Goddess of Water)   (Goddess of Hard Substances)

Coatlicue ———— fertility ———— *Hahai Wuhti* ——— *Ahea (Hemokatsiki)*
(Mother of Gods/   (Mother of Gods)
Goddess of Earth)

Chicomecóatl ——— 'seven ears of corn', fertility, sustenance ——— Seven Corn Maidens
(Goddess of Sustenance and Corn)

Huixtocíhuatl ——— salt, rain, feathered staff ——— Salt Old Woman ——— Salt Old Woman
(Salt Goddess)

Xólotl ——— conical caps, morning and evening star ——— Twin War Gods ——— Twin War Gods
(Venus, Twins)

Quetzalcóatl ——— dignity, bringer of corn ——— *Eototo* ———— *Pautiwa*
(Wind,   leader of rain spirits   (Father of Gods)
Venus, Twins,   conch shell, fertility, rain, serpent
Plumed
Serpent)   rain, serpent, goggle eyes ——— *Palulukong* ——— *Kolowisi*
                                                  (Horned Water Serpent)
                 serpents
                 goggle eyes, serpents ——— *Hilili*

                 star, dignity, conical cap, rain
Tláloc   rain, fertility ——— *Sotuqnangu*
(Rain)   scalping

                 turned up beak
                 stars on mask ——— *Ahul*
                 growth, fertility

                 growth, fertility
                 growth, fertility ——— *Alosaka/Muyingwa*

                 nadir/underworld, rebirth
Centéotl   death, rebirth ——— *Masau*
(Male Corn God)

Xipe Tótec
(Earth and Rebirth/
Underworld and Death)

                 death, skeleton, trickster, bloody face
Mictlantecuhtli ———
(Death)   old fire god   spots on mask

Huehuetéotl ——— young fire god ——— *Sumaikoli* ——— *Shulawitsi*
(Old Fire God)   old fire god ——— *Kawikoli*

Xochipilli ——— games, dances, flowers, love, sun's deputy, butterflies ——— *Paiyatamu*
(Spring)

Figure 11.1 Parallels in the distinguishing characteristics of
Aztec, Hopi and Zuni deities.

169

the plural as 'The Ones Who Hold the Roads' (Tedlock, 1979, p. 499). A major difference, however, is that, in addition to this male/female pair, the Zuni term refers to an entire class of supernaturals – 'the sun, the earth, the corn, prey animals and the gods of war' (Bunzel, 1932, p. 486). The Hopi also ascribe subsequent generations to a male–female creator pair (Parsons, 1939, pp. 212–13).

Both the Western Puebloans and the Mesoamericans believe that the sun travels from its eastern to its western house during the course of the day – the Aztecs considered these houses to be paradises to which certain people went upon death (Parsons, 1939, p. 212; Caso, 1958, pp. 58–60). Just as the Aztecs and Mayans perceived the universe to be layered, so do the Zuni, although not in a nine underworlds/13 heavens pattern (Caso, 1958, pp. 60–5; Henderson, 1981, p. 83). Instead, for the Zuni there are four underworlds, each associated with a tree and a direction; four upperworlds, each associated with a particular kind of bird; and in the middle, the familiar world, adding up to nine levels in all (Tedlock, 1979, p. 499).

The six points of orientation emphasized by the Hopis and Zunis correspond in many respects to those of the Aztecs and Maya, although the colors they assign to those directions are somewhat different, and the Western Puebloans, at least in some cases, refer to semi-cardinal, or solstitial, directions rather than the cardinal directions of the Aztecs and Maya.[2] For all four groups, the ritual direction is generally counterclockwise (Parsons, 1933, pp. 618–19; Henderson, 1981, p. 83; Young, 1988). Both Mesoamericans and Puebloans include two points which refer to an up–down dimension, or zenith and nadir, in their directional scheme. Thus, while four is an important ritual number for these peoples, so is five, signifying four plus the center, which also includes the vertical dimension: up, down, and center. (Frequently, the zenith and nadir are included as discrete points: the ritual number then becomes seven.) The Hopi and Zuni associate separate colors with the zenith and nadir, but the Maya used green to symbolize the entire up–down dimension, including the center (Henderson, 1981, p. 83). The concept of the center has great significance for the Western Puebloans as it did for the Mesoamericans. Just as the Aztecs were induced by *Huitzilopochtli*, their tribal god, to leave their mythical homeland and undertake long wanderings

in search of their ideal abode, so, after they left the four underworlds, the Zunis travelled for a period described by some as four days and others as four years until they reached their ideal home, 'the center' (Caso, 1958, pp. 34–5; Young, 1985, p. 14). The Western Puebloans and the Mesoamericans do not limit their directional associations to colors, but extend them to include almost everything in the world – birds, mountains, trees, animals, and so on – in a vast network of symbolic associations (Caso, 1958, p. 11; Young, 1985, pp. 16–18). In the Aztec pantheon, certain of the gods, including the rain gods, were related to the directions, as were human beings whose directional associations were determined by the day on which they were born (Caso, 1958, p. 11). Similarly, the Zuni assign the rain priests, or *'uwanammi*, and the Hopi, the Cloudyouths, to the six directions. These Puebloan Chiefs of the Directions are also intimately associated with mountain tops, as were the Aztec rain gods, or *Tlálocos*: thus both groups perceive a fundamental relationship between mountain ranges and rains (Parsons, 1933, p. 612; Caso, 1958, p. 42).

Western Puebloan and Aztec beliefs about reincarnation also reveal certain similarities. The Aztecs held that there had been four previous creations which were destroyed in cataclysms and that the fifth and final existence would also succumb to destruction (Caso, 1958, pp. 14–17). The Zunis, too, foresee the end of the current world: whereas the Aztecs predicted destruction by a great earthquake, the Zunis say that 'at the end ... our tools and utensils and everything we have will rise against us; the stars will fall and we will all be boiled by a hot rain' (Tedlock, 1975, p. 270).

Similarly, for both the Zunis and the Aztecs there are different afterlives, depending on the position one held in life. The Zunis believe that deceased rain priests join the deified rain priests of the six directions, just as the bow priests and members of the beast god societies join their deified counterparts. Members of kachina societies become kachinas upon death and may return among the living as clouds. When a person goes through four reincarnations, that person can choose to return among the living as an animal, depending on the knowledge acquired in life. Young girls and uninitiated boys, for example, become turtles or watersnakes (Tedlock, 1979, p. 507). In contrast, the Aztecs believed that where a person's soul went after death was determined, not by that person's conduct in life, but rather by the manner of death and occupation in life. The souls of those who fell in combat or died on the sacrificial stone went to the eastern paradise of the sun; from thence they would return to earth after four years, transformed into hummingbirds and other exotic birds (Caso, 1958, pp. 58–64). To the western paradise went women who died in childbirth; they were believed to have possessed great magical power. Those who died by drowning, by the strike of a lightning bolt, and from other illnesses that were thought to be related to the water gods, went to the paradise of the rain god *Tláloc* in the south – a place of fertility, where all kinds of trees and foodstuffs existed in abundance (Caso, 1958, p. 60). Those not selected either by the sun or *Tláloc* went to *Mictlan* in the north and underwent a series of magical trials, passing through nine hells until, after a period of four years, they reached their final rest. In a somewhat related vein, the Zunis believe that after one has died four times – descending death by death through each of the four underworlds – one returns to the place where the people originated (Tedlock, 1979, p. 508). Finally, the Aztec believed that children who died before they reached the age of reason joined the dual creator gods in the 13th and highest heaven – perhaps to become the new human race when the present one is destroyed in the great earthquake (Caso, 1958, p. 64).

Lack of space precludes a detailed comparison of Western Puebloan and Mesoamerican ritual practice but I will briefly summarize some of the most important similarities. The ritual life of both groups included or includes: a period of fasting and continence for four days before the enactment of some ceremonies; the performance of certain rituals that could be described as 'killing the god'; rites of running and kindling new fire with a wooden drill; various sorts of divination, including perceiving omens in the appearance of certain birds or certain types of phenomena such as water and fire; burning offerings of food to the deities; using ritual shields and crooked staffs; and observing a period of idle days at the end of the ceremonial–calendrical year (Parsons, 1933). Of special note is the Zuni practice of dividing the months into three groups of ten, as well as that of dividing the 49-day Shalako count into four periods of ten days each with nine remaining: Parsons describes this

as an Aztec, not a Puebloan count (Stevenson, 1904, p. 108; Parsons, 1917, p. 187; Parsons, 1933, p. 626).[3]

Having given a brief introduction to the Mesoamerican and Western Puebloan world views in general, I will turn to an examination of the similarity in function and attributes of the various deities of both groups, beginning with the sun (see Figure 11.1 for a graphic depiction of these similarities).

11.2 Sun

Although different deities took on the role of the sun during the four previous creations, the Aztecs seem to have perceived the sun as an entity distinct from these other deities. This Sun, *Tonatiuh*, was invoked by various names, including 'shining one', 'beautiful child' and 'eagle that soars', and was generally represented by a disk – such as the one that makes up the core of the Aztec calendar stone (Caso, 1958, p. 32). In the center of the disk *Tonatiuh* is depicted with his tongue hanging out; his hands, at the side of the disk, are tipped with eagle claws clutching human hearts (Caso, 1958, p. 33). It is of note that one aspect of this Aztec Sun is eagle, for the Zuni beast god of the zenith is sometimes eagle, and sometimes knifewing – the mythical creature with wings and tail of knives – although this being is not synonymous with the Sun Father (Young, 1985, p. 18). Furthermore, the Aztec equation of the sun with sacrifice and a measure of ferocity, as illustrated by his eagle claws clutching human hearts, is reiterated in a Hopi myth in which 'Sun had to be helped to move on across the sky by killing a child...', and his daily movement depended on some form of dying at various times, and 'on ritual racing by the town youths' (Parsons, 1939, p. 212). The sacrifices that accompanied Aztec sun worship were thought to provide sustenance to the sun who could be kept alive only by life itself; this underscores the reciprocal relationship that existed between the Aztecs and their deities – a characteristic of Puebloan relationships with the supernaturals as well (Caso, 1958, p. 12). Despite Hopi and Zuni legends hinting at human sacrifice to the solar deity, however, the most prominent feature of the sun is kindness and helpfulness. The Hopi describe their sun god, *Tawa*, as 'young, handsome, gentle, kind, and helpful' (Colton, 1959, p. 80). When impersonated, *Tawa* wears eagle and parrot feathers inserted in a plaited corn husk in a circle around his face like the rays of the sun (Fewkes, 1903, pp. 138–41).

11.3 Goddesses of water

According to some Hopi myths, the wife of the sun god is *Huru-ing Wuhti*, the goddess of hard substances, that is turquoise in particular or wealth in general (Voth, 1905, pp. 1–9). It seems likely that this personage is a parallel to the Mesoamerican *Chalchiuhtlicue*, 'the lady of the jade skirts', who was the goddess of water and, according to different legends, the wife or sister of *Tláloc*, the rain god (Caso, 1958, pp. 42–4). *Chalchiuhtlicue* was the special patroness of the sea and much revered by fishermen. However, for those who traded in salt, there was another patroness, *Huixtocíhuatl* – either the sister or daughter of *Tláloc* and *Chalchiuhtlicue* (Caso, 1958, pp. 44–5). This goddess may have been a precursor to Salt Old Woman at Hopi and Zuni – an extremely important personage who brings rain as well as salt (Bunzel, 1932, p. 1035). The Zuni kachina of Salt Old Woman carries a feathered staff with which she pulls down the rain clouds – a feature quite similar to the befeathered stick of the Aztec salt goddess (Bunzel, 1932, p. 1032 and pl. 42). Furthermore, the Aztec story of the banishment or departure of the salt goddess seems to be a variant of the Western Puebloan tale in which Salt Old Woman is offended by the way she is treated and goes away to the south (Bunzel, 1932, pp. 1031–5; Sahagún cited in Parsons, 1933, p. 628).

11.4 Wind and rain

The Aztecs assigned *Quetzalcóatl* to the cardinal direction west and the color white, and regarded him as a creator god along with the *Black Tezcatlipoca*. Their colors are opposite, reflecting the fact that so are their personalities: the struggle between these two gods seems to symbolize the struggle between good and evil in the universe. Whereas *Quetzalcóatl* was a beneficent god and a hero – the discoverer of corn and the founder of agriculture and industry – the *Black Tezcatlipoca* was a god of darkness, the patron of sorcerers and evil ones; yet he was also the patron of warriors and the discoverer of fire. *Tezcatlipoca* means 'mirror that smokes', and he is generally depicted wearing such a mirror at his temple and another in place of his right foot which was torn off by an earth monster (Caso, 1958, pp. 10, 14, 25, 27–30).

The image of *Quetzalcóatl* in the Codex Borbonicus reveals certain important distinguishing attributes (Caso, 1958, pp. 19, 21–33). For instance, he carries in one hand an incense pot with its handle in the form of a serpent. Covering his mouth is a red mask in the form of a bird's beak; in some representations this beak is also set with the fangs of a serpent. This particular mask identifies *Quetzalcóatl* as the god of wind, in which form he was worshipped under the name of *Ehécatl*, 'wind' (Caso, 1958, p. 22). On his head is a conical cap made of ocelot skin. His breastplate, described as the breastplate of the wind, is made from the transverse cut of a large seashell; perhaps it is this sort of association which links a spiral form with wind as well as with water, an association which seems to hold in the Southwest as well as in Mesoamerica (Fewkes, 1892, p. 20; Ellis, 1977; Young, 1985, p. 16).

Like many other Mesoamerican deities – and, as will be shown, Puebloan deities as well – *Quetzalcóatl* comprised a number of different and seemingly unrelated aspects synthesized in a single god. Furthermore, he was an extremely ancient god, known among the Maya as *Kukulkán* and *Gucumatz*. He was *Quetzalcóatl*, the god of many entities: the wind, life and morning, the sun during one of the ages of creation, and the planet Venus – the god of twins and monsters. His name literally means 'the plumed serpent' but also translates somewhat esoterically as 'the precious twin'. The second meaning refers to *Quetzalcóatl* as representing Venus, or the morning star; while his twin brother, *Xólotl*, represented Venus as the evening star. It has been suggested that the apparent motion of Venus – its appearance as evening star, its disappearance, and its reappearance as morning star – is symbolized in an Aztec myth about *Quetzalcóatl* and his twin brother *Xólotl*. According to this myth, the two descended to the world of the dead where they underwent various trials, finally asking *Mictlantecuhtli*, the god of the underworld, for the bones of the dead so that they could bring about a new creation. After escaping from this god they reappeared and recreated man. *Quetzalcóatl* then discovered corn, showed the people how to weave and do mosaic work and taught them science; in fact, it is he who endowed humans with 'the means to measure time and study the movements of the stars'. *Quetzalcóatl* not only invented the calendar, but he designated specific days for the performance of ritual activities. In his benevolent aspect *Quetzalcóatl* is the essence of saintliness (Caso, 1958, pp. 23–6): his life of fasting and penitence and his priestly character resemble the qualities of the Zuni *Pautiwa*, chief of the kachinas, who 'displays the most honored of Zuni virtues, dignity, kindliness, and generosity, and also beauty'. *Pautiwa* brings the corn maidens to Zuni after Shalako inaugurates the winter solstice ceremony, and comes at the new year with the crooks of appointment for the principal kachina impersonators of the coming year (Bunzel, 1932, p. 909).[4] Furthermore, like *Quetzalcóatl*, *Pautiwa* makes up the yearly ceremonial calendar.

In his manifestation as Venus, as well as in his aspect as Sun, *Quetzalcóatl* is strikingly parallel to the Zuni and Hopi Twin War Gods (Caso, 1958, pp. 15, 23–7): at Zuni, these twins are sons of the sun who represent the morning and evening stars and are sent to the fourth underworld to bring the people to the surface of this earth (Parsons, 1939, pp. 236n, 239–42; Young and Williamson, 1981, p. 184; Young, 1985, p. 12). In a Hopi kiva mural at Awatovi painted during the late 15th or early 16th century, a figure, strikingly similar to the Hopi and Zuni War God images, is depicted wearing a conical cap very much like those worn by *Quetzalcóatl* in various codex depictions (Stevenson, 1904, plates 137, 138; Smith, 1952, pp. 302–3, 316–18, figs. 28i, 65a, color plate E).[5] Schaafsma (1975) has documented similar *Quetzalcóatl*-like figures with conical caps in Jornada style rock art at Hueco Tanks State Park, Texas. These figures are associated with cloud and water symbolism as are the Hopi and Zuni Twin War Gods and the Hopi deity *Sotuqnangu*. According to Schaafsma and Schaafmsa (1974), the Jornada complex was Mexican-derived and appeared in the Southwest at about AD 1050. A further link with Mesoamerica is demonstrated by the fact that the Western Puebloans make miniature weapons, such as bows and arrows, as well as rain-associated shields and lighting sticks for the Twin War Gods (Parsons, 1939, pp. 305–7) – an offering paralleling that which the Mesoamericans made to their own war gods (Parsons, 1933, p. 617).

Quetzalcóatl may also be linked to the Hopi deity *Sotuqnangu*, 'god of the sky, the clouds, and the rain ... good, dignified, and powerful' (Ellis and Hammack, 1968, p. 41). He is impersonated only occasionally and then may be masked or unmasked (Colton, 1959, p. 78). His white mask is topped with one vertical horn colored blue and slightly curved at the tip. In some depictions his face is marked with cloud designs. In his left hand he carries a netted gourd of water and in his right hand a lightning stick (Fewkes, 1903, p. 178, pl. 58; Stephen, 1936, pls. V, VI). When not masked, the impersonator of *Sotuqnangu* wears a star-shaped hat from which feathers hang over his face (Colton, 1959, p. 78). This deity is described as 'both Star and Lightning, the god who kills and renders fertile and who initiated the practice of scalping...' (Parsons, 1939, p. 178). Thus we see in this Hopi god one version of *Quetzalcóatl*, with traces of *Tláloc* (rain, fertility), and also *Xipe Tótec*, who was ritually honored by the taking of scalps (Parsons, 1933, p. 617).

While strongly resembling the Hopi and Zuni Twin War Gods, *Quetzalcóatl* is also akin to the Zuni and Hopi Horned or Plumed Water Serpent, although *Tláloc* also figures in this association. The Hopi Horned Water Serpent is called *Palulukong* and plays a central role in a dramatic ceremony in late February or early March when puppet effigies of this serpent knock down miniature corn fields and fight noisily with the mudheads (patterned after Zuni clowns) amid loud roars made by actors blowing through empty gourds (Fewkes, 1900; Stephen, 1936, pp. 287–349; Young, 1987). The Zuni Horned Water Serpent, *Kolowisi*, figures prominently in the initiation of boys into the kachina society, and the performer who creates the serpent's roar blows through a conch shell (Stevenson, 1904, pp. 94–102; Bunzel, 1932, pp. 975–80). As mentioned earlier, *Quetzalcóatl's* breastplate is also made from a conch shell, perhaps symbolizing a wind/water/spiral relationship which is also integral to this Zuni drama. Ellis notes that, at both Hopi and Zuni, the mudheads 'are fertility figures and, interestingly, have the thick ringed goggle eyes and thick ringed mouth reminiscent of *Tláloc*' (Ellis, 1977, p. 10).[6] The Hopi and Zuni Horned Water Serpent puppets also have goggle eyes, which are filled with corn kernels and the seeds of important plants: this points to a water/fertility association which is characteristic of both *Quetzalcóatl* and *Tláloc* (Caso, 1958, pp. 15, 24–7, 41; Young, 1988). There is an apparent sun–serpent antagonism in the Hopi performance: the puppets of *Palulukong* push aside the sun symbols in the curtain as they dart at the imitation hills of corn; the Hopis believe that if the Horned Water Serpent isn't properly propitiated, it might hinder the journey of the sun along the horizon (Fewkes, 1897, pp. 270–1). Yet, the Horned Water Serpent is also connected with agricultural fertility, as indicated by the seed-filled eyes of the puppets. Similarly, the puppet figure at Zuni spurts forth water from a sacred spring as well as grasses which symbolize the long life of the initiate into the kachina society (Young, 1988). Finally, both the Hopis and Zunis believe that the Horned Water Serpent has the power to send floods and that he controls the underwater springs – in this respect he is much like the Mesoamerican *Tláloc*. In one Zuni tale, the sacrifice of a boy and girl to the Horned Water Serpent causes the flood waters to abate – a practice similar to the Mesoamerican tradition of child sacrifice to the mountain rain gods (Parsons, 1933, p. 616; Caso, 1958, p. 42; Young, 1988).

11.5 Mothers of the gods/goddesses of fertility

The Hopi and Zuni puppet performances described above reveal an important link between the mother of the gods, also associated with fertility, and the Horned Water Serpent: this female deity, called *Hahai Wuhti* at Hopi and *Ahea* or *Hemokatsiki* at Zuni, suckles the Horned Water Serpent puppets and offers them sacred corn meal – an act of propitiation that serves to quell the antagonism of the serpents toward the young corn plants (Young, 1987). The Puebloan mother of the gods is similar to the Mesoamerican *Chalchihuitlicue*, goddess of water, described earlier. She also resembles the Aztec mother of the gods and old goddess of the earth, *Coatlicue*, who is sometimes depicted carrying a child in her arms (Caso, 1958, p. 12; Ellis, 1977). The Hopi *Hahai Wuhti* is the first image presented to newborn babies (Wright, 1977, p. 56). Her mouth is turned up in a perpetual smile and her cheeks are red disks. In her right hand she carries a gourd of water and frequently in her left an ear of corn or a tray of corn meal – thus she may also be connected with the Mesoamerican goddess of sustenance *Chicomecóatl* who is called 'seven ears of corn' (Fewkes, 1903, p. 76; Stephen, 1936, pp. 297–8; Caso, 1958, p. 45).

11.6 Gods of death and rebirth

Xipe Tótec was mentioned earlier as a deity who had his origins in Olmec ideology: the god of spring and the annual rebirth of plant life. Yet the Aztec added a terrifying aspect to this god who became the *Red Tezcatlipoca*. As 'our lord the flayed one', his cult consisted of flaying a slave and covering a priest with the victim's skin, symbolizing the arrival of spring when the earth covers herself with a new coat of vegetation (Caso, 1958, p. 47). *Xipe* thus reveals characteristics similar to those of the Hopi *Masau*, deity of earth and rebirth as seen in spring growth and renewal, but also god of the underworld and death (Parsons, 1939, pp. 183, 789–90; Ellis, 1977): it is perhaps because of this dual and contrasting role that the Hopi describe him as one 'who does things by opposites' (Titiev, 1944, p. 72n). The patron of travelers as well as the Hopi Old Fire God, *Masau* is impersonated with or without a mask, his face painted red or blotched with blood (Fewkes, 1903, p. 90; Parsons, 1939, p. 184; Colton, 1959, p. 78). According to the Hopi origin myth, the people first met *Masau* in his unmasked form as they emerged from the underworld. Although they were terrified of him, he spoke kindly to them and in his earth deity guise gave them land to plant (Voth, 1905, pp. 12–13). In his gruesome aspect as god of the underworld – especially his appellation of 'skeleton' – *Masau* resembles the Aztec god of the underworld, *Mictlantecuhtli*, who is frequently depicted with his body covered with bones and wearing a human skull for a face mask (Voth, 1905, pp. 12–13; Caso, 1958, p. 56). Although he is most often characterized as kindly though fearsome, *Masau* is infrequently described as a thief, liar, and practical joker; these qualities further link him with *Mictlantecuhtli* who, according to some accounts, was 'double-dealing and mistrustful' (Stephen, 1929, pp. 55–7; Caso, 1958, p. 24).

11.7 Father of the gods

Masau is related to, and in some respects almost antithetical to, one of the most important Hopi personages – *Eototo*. Called the 'father' or chief of the kachinas, as is the Zuni *Pautiwa*, *Eototo* controls the seasons and is sometimes referred to as the husband of *Hahai Wuhti*, mentioned earlier as the mother of the kachinas who suckles the Horned

Water Serpent during the Palulukonti ceremony (Parsons, 1939, pp. 175, 205; Wright, 1977, p. 34).[7] According to both Ellis and Colton, 'Eototo knows all the ceremonies ... and leads in the spring Bean dance at which kachinas of all types appear, though emphasis is on those wearing the flowers and butterflies of spring' – for the Hopi this comprises a 'return of the gods ceremony' (Colton, 1959, p. 22; Ellis, 1977). Just as *Masau's* mask is of the helmet-type with circles for mouth and eyes, so is *Eototo's*, although *Eototo* lacks *Masau's* many-colored spots; while *Masau* carries yucca whips in his hands, *Eototo* carries a pouch of sacred meal in one hand and a gourd of sacred water in the other (Fewkes, 1903, pl. 14, pp. 90–3). According to Fewkes, the similarity in symbolic designs on the masks and costumes of these two kachinas shows that these two gods are 'virtually dual appellations of the same mythological conception' (Fewkes, 1903, p. 93). Interestingly, the Aztec gods who correspond somewhat to these Hopi gods are also symbolically linked to one another. As just mentioned, *Masau* is related to *Xipe Tótec* who as god of seed time and planting is in turn linked to *Tláloc* and *Tláloc* resembles *Eototo* to a great degree. As chief of the kachinas or rain-fertility spirits, *Eototo* holds a position similar to *Tláloc* who is leader of the Aztec *Tlálocos* or rain spirits. Furthermore, *Eototo's* prominence in the Hopi 'return of the gods ceremony', which is part of the pre-planting ritual of Powamu or the Bean dance, corresponds to *Tláloc's* role in the Aztec preplanting celebration (Parsons, 1933, p. 625; Ellis, 1977). Like the Powamu, the Aztec festival was also known as 'when they eat bean food' (Spence, 1912, pp. 48, 56–8).

11.8 Gods of germination

Another Hopi deity who, like the Mesoamerican *Tláloc*, could be described as 'he who makes things grow', is *Alosaka* or *Muyingwa*, the Hopi patron of reproduction for man, animals and plants (Caso, 1958, p. 41; Colton, 1959, p. 79). Predominant features of his mask are two mountain sheep horns curving backwards while the lower section of his mask is black (Fewkes, 1903, pl. 59, and p. 182; Stephen, 1936, pl. VII). Interestingly, although a god of earth and growth, *Muyingwa* is also called 'Chief of the Nadir' and 'father of the underworld', a description which calls to mind *Xipe Tótec's* dual function as a god of death and rebirth (Parsons,

1939, p. 178). This Hopi god of growth in a slightly different form appears as the Hopi Chief Kachina *Ahul*, who is also the lieutenant or companion of *Eototo*; both *Ahul* and *Muyingwa* are described as germ god kachinas (Colton, 1959, p. 20; Wright, 1977, p. 35). On the back of *Ahul's* many-colored cloak is painted a likeness of *Alosaka*, who is responsible for the germination of seeds (Colton, 1959, p. 32); *Alosaka* is also depicted in the center of the shield carried by the impersonator of the sun during the Walpi winter solstice ceremony (Fewkes, 1918, p. 502, pl. 2). Of course, in this aspect of enabling the germination of seeds, both *Alosaka/Muyingwa* and *Ahul* resemble the Aztec *Centéotl*, the male corn god (Caso, 1958, p. 46). *Ahul* is further characterized by a turned up beak and a mask painted over with stars; in some depictions, he carries a gourd full of sacred water and, in others, a bundle of bean sprouts (Fewkes, 1903, pl. VII, p. 74; Colton, 1959, p. 20). One can see in certain of these characteristics of *Ahul* a similarity to *Tláloc* who also has stars scattered over his mask (Caso, 1958, p. 40). But *Ahul* fulfils a function parallel to *Quetzalcóatl*, too, in his guise as the *White Tezcatlipoca*, water and fertility deity (Caso, 1958, pp. 9, 23–6). According to Ellis, the turned up beak of *Ahul* is related to some Mesoamerican depictions of *Quetzalcóatl* as *Ehécatl*, the wind god (Ellis, 1977).

11.9 Fire gods

Both the Hopis and the Zunis have a dual counterpart to the Mesoamerican Old Fire God, *Huehue-téotl* – in Western Puebloan ritual drama these roles are played by a youth and an older man. At Hopi the older fire god is *Kawikoli* and the younger is *Sumaikoli*, a possessor of powerful magic who kindles new fire during a spring festival and carries it to the shrines of the fire god at the four solstitial points. *Kawikoli* and *Sumaikoli* live on the edge of a mesa and are regarded by the Hopis as kind and helpful (Fewkes, 1903, p. 131; Colton, 1959, p. 85). Of course, *Masau* is also a version of the Hopi Old Fire God and the shrines at the four solstitial points are dedicated to him – evidence perhaps of a symbolic association between sun and fire. The new fire ceremony itself may be a ritual enactment of the rekindling of the sun's warmth in the spring which causes the growth of young plants; on the other hand, it is also a dramatization of that part of the emergence myth in which *Masau* teaches the Hopi the use of fire (Titiev, 1944, p. 135). Furthermore, Stephen suggests that the Hopis regard the speed with which this new fire is kindled as an omen of the good or ill that the coming year will bring (Stephen, 1936, p. 964n). At Zuni the kindling of the new fire is a crucial part of the winter solstice ceremony – throughout the ten days of the solstice observances, the fire is tended and watched over by an older man of the Badger clan appointed by the Zuni priests. This man, in turn, selects the impersonator of *Shulawitsi*, the Young Fire God, from among the young boys of his own family. The young fire god lights signal fires prior to Shalako, 'the coming of the gods ceremony', and leads the Zuni 'council of the gods' into the village (Bunzel, 1932, pp. 637, 659, 958–61). The colored spots on *Shulawitsi's* black-painted body and mask suggest gleaming sparks and resemble the multi-colored disks on the mask of the Hopi *Masau* (Tyler, 1964, p. 125). The Western Puebloans create new fire by the friction produced from a wooden drill and they carry it out from the kiva; the Hopis perform a ritual in that kiva for *Masau*. Similarly, the Aztec produced new fire with a wooden drill and carried it out from the temple of the Aztec fire god (Sahagún, cited in Parsons, 1933, p. 621). In both Mesoamerica and the Southwest, the fire gods and their societies are extremely old and important aspects of ceremonial life.

11.10 Gods of music, flowers and love

The attributes of the Mesoamerican and Western Puebloan gods of love, music and flowers point to another sun/fertility parallel. The Aztec god of summer, *Xochipilli*, is called the 'prince of flowers' and symbolizes dances, games and love. His symbol is formed by four points signifying the heat of the sun. He is adorned with flowers and butterflies and carries a staff on which a human heart is impaled. As a god of summer and fertility he is also connected with *Centéotl*, the Aztec corn god (Caso, 1958, p. 47). *Xochipilli* seems to be a prototype for the Zuni *Paiyatamu*, the gay and youthful flute player associated with music, poetry, flowers and butterflies, who is also described as the Sun's deputy (Bunzel, 1932, p. 530). According to Zuni mythology, *Paiyatamu* became enamored of the chaste Corn Maidens who brought corn to the Zunis: at first his passion frightened them and they fled, but he was also instrumental in causing their return (Stevenson, 1904, pp.

48, 56). It is of note that the Aztec goddess of sustenance, *Chicomecóatl*, was also called 'seven ears of corn' – a striking parallel to the seven Corn Maidens associated with *Paiyatamu* at Zuni.[8] These seven maidens represent corn in the colors of the six directions plus the center which is symbolized by sweet corn (Cushing, 1896, p. 433). There is also some indication that the Zuni Corn Maidens symbolize the seven stars of the Big Dipper. The Zuni word for this constellation means 'seven ones', and Cushing explains that the Zunis first planted corn in the spring 'by the light of the seven great stars which were at that time rising bright above them' (Cushing, 1896, p. 392; Young and Williamson, 1981, pp. 183–4, 191).

11.11 Conclusion

Although there are many other parallels to be explored in establishing the connection between Mesoamericans and Western Puebloans, what I have outlined points to a number of significant similarities in the cosmological outlook of both groups. Not only do many of the Puebloan deities and kachinas seem to be based on Mesoamerican prototypes, but the entire Puebloan religious perspective is shaped by a belief in the reciprocity between human beings and supernaturals – the interdependence of the people and their gods – which harkens back to the Aztec premise that sacrifices provided necessary nourishment for the gods. The Zuni Sun Father, for instance, sent his sons, the Twin War Gods, to bring the ancestral Zunis out from the fourth underworld, not only because he took pity on their miserable semi-human existence in the darkness, but also because he wanted their offerings and prayers; and both the Hopi and the Zuni perform ritual activities which are aimed at helping the sun to complete his daily and annual travels. Furthermore, as we have seen, many of the Western Puebloan kachinas and other deities – such as *Masau*, the Hopi earth and death god, and the Horned Water Serpent – have the same sort of dual or multiple personalities, sometimes demonstrating conflicting aspects as well, that characterized the Aztec gods. Interestingly, the Hopis say that the Horned Water Serpent came 'from the Red Land of the South', and that he is a patron of the Water–Corn clan which 'also came from that Red Land' (Parsons, 1939, pp. 184–5).

A pervasive theme in both the cosmology of the

Western Puebloans and that of the Mesoamericans is the necessary interrelationship of sun and water as the basis for human and agricultural increase or fertility. This significant association is exemplified in the Zuni myth that ascribes the birth of the Twin War Gods to the union of the sun's rays with the foam of a waterfall (Cushing, 1896, p. 381; Stevenson, 1904, p. 24). Sometimes this sun/water relationship is antagonistic, however, as when the Horned Water Serpent who controls the underground springs causes a flood which destroys the benefits wrought by the sun in the Hopi and Zuni fields; or when the Aztec gods who took on the role of Sun destroyed four consecutive creations by invoking water, wind, rain and jaguars (Caso, 1958, p. 15).

Finally, this sun/water cult is the basis of the all-pervasive directional scheme, with its accompanying color symbolism and emphasis on ritual numbers such as four and five, employed by both the Mesoamericans and the Western Puebloans. The vast network of symbolic associations generated by these two principles is derived not only from the apparent motion of the sun, but also from the belief in its interconnection with rain and water more generally. Although the Western Puebloan semi-cardinal directions refer to the solstice positions, they also refer to four oceans and to the areas in which the rain priests dwell; similarly, the deities the Aztecs associated with directionality were related to both the sun and water.

This discussion has been necessarily general, but I hope that it will provide an impetus for more detailed contextual analyses of the range and extent of the interactions between the Mesoamericans and Western Puebloans – interactions that had a major effect on their perceptions of astronomical phenomena as well. Certainly we see a common focus on observing the ordered motions of the sun, the moon, Venus, and various constellations which are central to the interlocking spheres of religion, ceremonial practice, and world view. Nevertheless, the Southwest–Mesoamerican connection is a complex and complicated scheme to unravel, and many questions concerning its nature and impact are yet to be answered.

Acknowledgments

I am grateful to Polly Schaafsma for reading and commenting on this paper. I also thank Robert H. Leibman for his help in devising Figure 11.1.

Notes

1 The archaeological evidence that we do have of a Mesoamerican–Southwest connection is summarized in Kelley (1966) and McGuire (1980).

2 The Hopis and Zunis allocate the same colors to the semi-cardinal directions: northwest/yellow, southwest/blue, southeast/red, northeast/white, zenith/all-colored, and nadir/black (Stephen, 1936, p. 961, Fig. 483; Young, 1985, p. 18; 1988). The Maya and Aztec add no new colors, except, perhaps, for green, although blue for the Hopi and Zuni is often described as blue–green (Stephen, 1936, p. 51). The Maya color-directional system is: east/red, north/white, west/black, south/yellow (Henderson, 1981, p. 83). For the Aztecs the scheme is: east/red, north/black, west/white, south/blue (Caso, 1958, pp. 10–11).

3 The ten-day count *is* unusual in the Southwest, but it is not clear why Parsons attributes it to the Aztecs.

4 For similar reasons Ellis and Hammack (1968, p. 41) link *Tláloc* with the Zuni *Pautiwa*, the 'kindly, generous, and dignified head of the katcina village, leader of the gods, and, like *Tláloc*, associated with deer'.

5 The comparison to *Quetzalcóatl* is suggested by Ellis (1977). She further suggests that this cap is 'like the one of the four rays usually shown for such stars', adding that somewhat similar figures are found in the Pottery Mound paintings (Hibben, 1975, Figs. 24, 26, 105; Ellis, 1977).

6 A further Puebloan link with *Tláloc* may be evidenced by the fact that, in one sculpture of *Tláloc*, the circles around his eyes are formed by two serpents so that this face resembles the mask of the Hopi kachina *Hilili*, sometimes described as the Witch Kachina (Caso, 1958, pp. 42–3; Switzer, n.d.; Wright, 1977, pp. 40, 43–4, pl. 4).

7 According to Fewkes, however, *Hahai Wuhti* is the wife of the Sun. Thus, her performance in the Palulukonti drama symbolizes a prayer on the Sun's behalf that the angry serpents cease their destructive activities (Fewkes, 1902, p. 29).

8 Both the Western Puebloans and the Aztecs connected corn (sustenance) and sacrifice. The Aztecs tinted certain offerings to *Chicomecóatl* with blood and sacrificed a young maiden (whose flayed skin one of the priests

wore) to *Chicomecóatl* and her son *Centéotl* (Sahagún,
1932, pp. 80, 114). Of course, the flayed skin and association with death and sustenance relates to the worship
of *Xipe Tótec* as well. The Zunis also linked corn with
death and rebirth. According to the Zuni origin myth,
after the people emerged from the four underworlds,
a witch also emerged, bringing corn. The witch
demanded the sacrifice of a chief's child in return for
the corn. The child died and returned to the place
from which the people had emerged, carrying on in
death activities similar to those undertaken in life
(Bunzel, 1932, p. 593; Tedlock, 1972, pp. 258–63).

References

Anderson, F. G. (1955). The Pueblo kachina cult: a historical reconstruction. *Southwestern Journal of Anthropology* **11**, 404–19.

Anton, F. (1969). *Ancient Mexican Art.* London: Thames
and Hudson.

Bunzel, R. L. (1932). Introduction to Zuñi ceremonialism. Zuñi origin myths. Zuñi ritual poetry. Zuñi kat-
chinas: an analytical study. *Bureau of American
Ethnology, Annual Report* **47**, 467–1086.

Caso, A. (1958). *The Aztecs: People of the Sun.* Norman:
University of Oklahoma Press.

Colton, H. S. (1959). *Hopi Kachina Dolls,* revised edn.
Albuquerque: University of New Mexico Press.

Covarrubias, M. (1946). El arte 'Olmeca' o de La Venta.
Cuadernos Americanos **28**, 153–79.

Cushing, F. H. (1896). Outlines of Zuñi creation myths.
Bureau of American Ethnology, Annual Report **13**, 321–
447.

Ellis, F. H. (1977). Distinctive parallels between Mesoamerican and Pueblo iconography and deities. Paper
delivered at Guanajuato, Mexico, August, 1977.

Ellis, F. H. and Hammack, L. (1968). The inner sanctum
of Feather Cave, a Mogollon sun and earth shrine linking Mexico and the Southwest. *American Antiquity* **33**,
25–44.

Fewkes, J. W. (1892). A few Tusayan pictographs. *American Anthropologist* **5**, 9–26.

Fewkes, J. W. (1897). Tusayan katcinas. *Bureau of American Ethnology, Annual Report* **15**, 245–313.

Fewkes, J. W. (1900). A theatrical performance at Walpi.
Proceedings of the Washington Academy of Sciences **2**,
605–29.

Fewkes, J. W. (1902). Sky-God personations in Hopi
worship. *Journal of American Folk-Lore* **15**, 14–32.

Fewkes, J. W. (1903). Hopi katcinas drawn by native artists. *Bureau of American Ethnology, Annual Report* **21**,
13–126.

Fewkes, J. W. (1918). *Sun worship of the Hopi Indians.*
Smithsonian Annual Report for 1918, pp. 493–526.
Washington.

Ford, R. I. (1981). Gardening and farming before AD
1000: patterns of prehistoric cultivation north of Mexico. *Journal of Ethnobiology* **1**, 6–27.

Furst, P. T. (1974). Hallucinogens in Precolumbian art.
In *Art and Environment in Native America,* eds. M. E.
King and I. R. Traylor, pp. 55–101 (sp. publication
no. 7). Lubbock, Texas: The Museum, Texas Technical University.

Grove, D. C. (1970). *The Olmec paintings of Oxtotitlan
Cave, Guerrero, Mexico.* Studies in Pre-Columbian Art
and Archaeology, 6. Washington: Dumbarton Oaks.

Henderson, J. S. (1981). *The World of the Ancient Maya.*
Ithaca: Cornell University Press.

Hibben, F. C. (1975). *Kiva Art of the Anasazi at Pottery
Mound.* Las Vegas: K.C. Publications.

Joralemon, P. D. (1976). The Olmec dragon: a study
in Pre-Columbian iconography. In *Origins of Religious
Art and Iconography in Preclassic Mesoamerica,* ed. H.
B. Nicholson, pp. 27–71. UCLA Latin American Series
31. Los Angeles: UCLA Latin American Center Publications.

Kelley, J. C. (1966). Mesoamerica and the Southwestern
United States. In *Handbook of Middle American Indians,*
Vol. 4, ed. E. A. R. Wauchope, pp. 95–110. Austin:
University of Texas Press.

Martin, P. S., Rinaldo, J. B., Bluhm, E. A., Cutler, H.
C. and Grange, R., Jr. (1952). *Mogollon Cultural Continuity and Change: The Stratigraphic Analysis of Tularosa
and Cordova Caves. Fieldiana Anthropology,* **40**.

McGuire, R. H. (1980). The Mesoamerican connection
in the Southwest. *The Kiva* **46**, 3–38.

Nicholson, H. B. (1976). Preclassic Mesoamerican iconography from the perspective of the Postclassic: problems in interpretational analysis. In *Origins of Religious
Art and Iconography in Preclassic Mesoamerica,* ed. H.
B. Nicholson, pp. 157–75. UCLA Latin American Studies Series **31**. Los Angeles: UCLA Latin American
Center Publications.

Parsons, E. C. (1917). Notes on Zuñi. Part I. *Memoirs
of the American Anthropological Association* **4**, (3).
Menasha, Wisconsin.

Parsons, E. C. (1933). Some Aztec and Pueblo parallels.
American Anthropologist **35**, 611–31.

Parsons, E. C. (1939). *Pueblo Indian Religion.* University
of Chicago Press.

Peterson, Frederick. (1961). *Ancient Mexico.* London:
George Allen and Unwin, Ltd.

Sahagún, B. (1932). *A History of Ancient Mexico: 1547–
1577* (translated by F. Bandelier). Reprinted in 1976,
Glorieta, New Mexico: The Rio Grande Press, Inc.

Schaafsma, P. (1975). Rock art and ideology of the
Mimbres and Jornada Mogollon. *The Artifact* **13**, 2–14.

Schaafsma, P. and Schaafsma, C. (1974). Evidence for
the origins of the Pueblo katchina cult as suggested

by Southwestern rock art. *American Antiquity* **39**, 353–545.

Smith, W. (1952). Kiva mural decorations at Awatovi and Kawaika-a. *Papers of the Peabody Museum of American Archaeology and Technology* 37. Cambridge, Mass.

Spence, L. (1912). *The Civilization of Ancient Mexico.* New York: G. P. Putnam's Sons.

Stephen, A. M. (1929). Hopi tales. *Journal of American Folk-Lore* **42**, 2–72.

Stephen, A. M. (1936). *Hopi Journal of Alexander M. Stephen*, ed. E. C. Parsons. New York: Columbia University Press.

Stevenson, M. C. (1904). The Zuñi Indians: their mythology, esoteric fraternities, and ceremonies. *Bureau of American Ethnology, Annual Report* **23**, 3–634.

Switzer, R. (n.d.). The origin and significance of snake-lightning cults in the Pueblo Southwest. Special Report no. 2. El Paso Archaeological Society, Inc.

Tedlock, D. (1972). *Finding the Center: Narrative Poetry of the Zuni Indians, by A. Peynetsa and W. Sanchez.* New York: Dial Press.

Tedlock, D. (1975). An American Indian view of death. In *Teachings from the American Earth*, eds. D. Tedlock and B. Tedlock, pp. 248–71. New York: Liveright.

Tedlock, D. (1979). Zuni religion and world view. In *Handbook of North American Indians, Vol. 9, Southwest*, ed. A. Ortiz, pp. 499–508. Washington, DC: Smithsonian Institution.

Titiev, M. (1944). Old Oraibi: a study of the Hopi Indians of the Third Mesa. *Papers of the Peabody Museum of American Archaeology and Ethnology* **22**, 1. Cambridge, Mass.

Tyler, H. A. (1964). *Pueblo Gods and Myths.* Norman: University of Oklahoma Press.

von Winning, H. (1976). Late and terminal Preclassic: the emergence of Teotihuacán. In *Origins of Religious Art and Iconography in Preclassic Mesoamerica*, ed. H. B. Nicholson, pp. 141–56. UCLA Latin American Studies Series **31**. Los Angeles: UCLA Latin American Center Publications.

Voth, H. R. (1905). The traditions of the Hopi. *Field Columbian Museum Publication 96, Anthropological Series* 8. Chicago.

Wright, B. (1977). *Hopi Kachinas.* Flagstaff: Northland Press.

Young, M. J. (1985). Images of power and the power of images: the significance of rock art for contemporary Zunis. *Journal of American Folklore* **98**, 3–48.

Young, M. J. (1987). Humor and anti-humor in Western Puebloan puppetry performances. In *Humor and Comedy in Puppeting*, eds. D. Sherzer and J. Sherzer, pp. 127–50. Bowling Green, OH: The Popular Press.

Young, M. J. (1988). Astronomy in Pueblo and Navajo world views. In *Ethnoastronomy: Indigenous Astronomical and Cosmological Traditions of the World*, eds. V. D. Chamberlain, M. J. Young and J. B. Carlson. Los Altos and Thousand Oaks, CA: Ballena and Slo'w Press.

Young, M. J. and Williamson, R. A. (1981). Ethnoastronomy: the Zuni case. In *Archaeostronomy in the Americas*, ed. R. A. Williamson, pp. 183–91. Los Altos, California: Ballena Press.

II
Archaeoastronomy: the textual basis

12

Antares year in ancient China

I. Ecsedy *Orientalist Research Centre, Budapest*
K. Barlai *Konkoly Observatory, Budapest*
R. Dvorak *Universitäts-Sternwarte, Vienna*
R. Schult *Zentralinstitut für Astrophysik, Sonneberg*

The centre of the official Chinese world of stars was the Northern Pole Star, a ruler of the sky reflecting the desire and later the reality of the ancient Chinese centralized government. Polaris as Pole Star appeared in the last centuries of the Shang (-Yin) dynasty, thus giving an objective date for the beginning of the official Chinese picture of the sky (Ecsedy, 1981).

An earlier popular tradition survived, however, based on the Star of Fire – Antares – which had been marking the time of popular feasts for village people after the establishment of the imperial calendar (Ecsedy, 1981, 1984a). Based on recent archaeological results, it is not unrealistic to suppose an ancient agricultural tradition which refers to ancient times when Antares was still able to indicate the beginning of the agricultural year, as shown by the pictogram of 農, *nung*, agriculture; the mistaking of that time must have been a ritual crime; 辱, *ju*, shame. Today, both of these characters contain a part of the constellation Scorpio in the form 辰, *ch'en*.

At that time the Sun must have stood just in the 'centre' of the anticipated constellation *jih-chung*; i.e. in Scorpio. In short, all this must refer to a time when the vernal equinox was connected with the appearance of Scorpio, i.e. with a heliacal rise of Antares. The role of Antares in ancient Chinese popular time reckoning is uncertain. Recently, however, a Chinese author summarized the records referring to a possible ancient Chinese year of Antares (Pang Pu, 1984). The meeting of the Fire Star and the Sun in the early morning at the beginning of spring marked the beginning of field work

for the peasants, while its disappearance meant the end of the agricultural year.

For the chronological (and local) orientation of ancient Chinese agricultural civilization we have to know the yearly movement of Antares from its heliacal rise at spring equinox to its disappearance in the sunset at different geographic latitudes.

In the southwest provinces of China a 12-month year has been preserved, beginning at various times apparently connected with a cult of Antares (Ecsedy, 1978, 1984b). It could determine the beginning of the year among village people, even at the time of a different official calendar. Naturally, the division of a year could be borrowed, e.g. from China, without changing the seasons' natural order, including the beginning of the year. (The two-week long ceremony, which since ancient times has traditionally begun the year in China, would make it possible for the 'first month' to be the 'first month' of the agricultural year.)

When the spring equinox became the fixing point of time, an attempt was probably made to harmonize this with the return of Antares. However, as a consequence of the apparent movement of the zodiacal constellations – among them Scorpio, including Antares – the star began to return later than the spring equinox, and continued to appear later and later. Half a month's delay would be incurred in about 1000 years; and the regular astronomical observations actually began a millennium later, in the middle of the first millennium BC.

Finally, Chinese sources show records, concerning certain 'countries' of the Southern Barbarians, to the effect that their year began in this or that

183

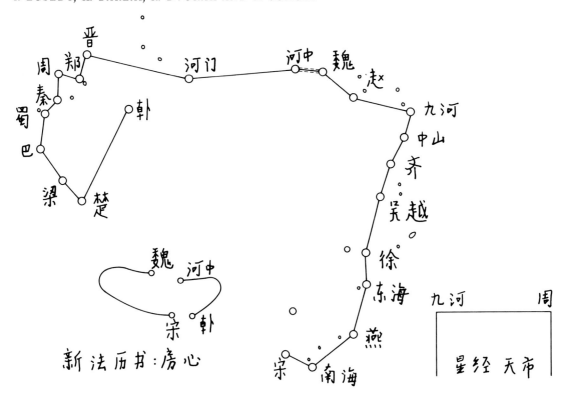

Figure 12.1 'Celestial Market': the outlines of ancient Chinese civilization as seen in the field of the constellation Scorpio. It also reflects a possible ancient change of orientation from the ritual southerly position based on the sighting of Antares to a northern orientation turning towards the Pole Star. The reconstruction changed the spoiled order of stars (seen in certain records, cf. right corner).

month, i.e. not simultaneously with the Chinese official year. In the literary records a celestial map can be concluded from the names of stars ruled by the Star of Shang (or Wei, i.e. the state of Wei, inheritor of the territory of Shang). The third celestial 'empire', called Celestial Market Fortress 天市垣 *T'ien-shih yuan*) is 'walled' by a series of stars having the names of Chou-time states and geographic regions of archaic times (Figure 12.1). And on the peripheries of Chinese culture, among popular beliefs the cult of Antares survived until modern times (Yünnan).

Attempts have been made to find the place of the astronomical events of the remote historical past described above. Data on the heliacal rising and setting of the Fire Star Antares have been determined for the equinoxes at different grographic latitudes. Calculations have been carried out concerning Aldebaran as well (Tables 12.1 and

Table 12.1. *Heliacal rising and setting for spring equinox*

Geographical latitude (degrees)	Antares		Aldebaran	
	hel. rising	hel. setting	hel. rising	hel. setting
20	−17 150	−13 750	−4450	−1000
25	−17 300	−13 350	−4600	−650
30	−17 500	−12 850	−4850	−150
35	−17 750	−12 250	−5100	450
40	−18 150	−11 500	−5500	1200

12.2). This bright red star in the zodiacal constellation Taurus could also have been the Fire Star (different form – according to some commentaries – and maybe prior to that called Ch'en).

The precessed coordinates of the stars were used in the calculations (Wislicenus, 1892, 1895, p. 35; Ginzel, 1906, p. 18; Berger, 1977). Proper motions,

Table 12.2. *Heliacal setting for autumn equinox*

Geographical latitude (degrees)	Antares hel. setting	Aldebaran hel. setting
20	−2000	−15 000
25	−2100	−15 100
30	−2200	−15 200
35	−2250	−15 300
40	−2300	−15 400

however, have not been taken into account because of the uncertainty of the time intervals involved. At any rate, it would not modify the results significantly. Local climate and the horizon also influence the visibility of the phenomenon (Schaefer, 1985), but *in situ* observations were inaccessible to us. Therefore, only the accessible 100 years of the phenomenon are given in the tables.

References

Berger, A. (1977). Long-term variations of the Earth's orbital elements. *Celestial Mechanics* **15**, 53.

Ecsedy, I. (1978). On a few traces of ancient Sino–Tibetan contacts in the early Chinese mythic tradition. *Proceedings of Csoma de Körös Memorial Symposium, Bibliotheca Orientalis Hungarica, XII*, p. 89. Akadémiai Kiadó, Budapest, 1978.

Ecsedy, I. (1981). Far Eastern sources on the history of the steppe region. *Bulletin l'École Française d'Extrême-Orient* **LXIX**, 263.

Ecsedy, I. (1984a). Capitals and village communities at the beginning of China's history. *Acta Orientalia Hungarica* **XXXVIII**, 7.

Ecsedy, I. (1984b). Nanchao: an archaic state between China and Tibet. *Tibetan and Buddhist Studies, Bibliotheca Orientalis Hungarica* **XXIX**, 165.

Ginzel, F. K. (1906). *Handbuch der Mathematischen und Technischen Chronologie.* Leipzig: J. C. Hinrichs.

Pang, Pu 庞 朴 (1984). 火 历 三 探, Huoli san tan. *Wen shi zhe* **1**, 21.

Schaefer, B. (1985). Predicting heliacal risings and settings. *Sky and Telescope* **XXX**, 261.

Wislicenus, W. F. (1892). Tafeln zur Bestimmung der järlicher Auf- und Untergänge der Gestirne. *Publ. der Astronomischer Gesellschaft* **XX**.

Wislicenus, W. F. (1895). *Astronomische Chronologie.* Leipzig: B. G. Teubner.

13

Comets and meteors in the last Assyrian Empire

Robert Chadwick   *John Abbot College, Quebec*

13.1 Introduction

Assyriologists have known for over a century that the Assyrians and Babylonians developed a great civilization in the Tigris–Euphrates River valleys, in what is now modern day Iraq. The Assyrians and Babylonians were originators of scientific astronomy (Schaumberger, 1935; Neugebauer, 1957b; van der Waerden, 1974) as well as divination, or 'fortune-telling' by the medium of celestial objects (Oppenheim, 1969; Thompson, 1900). This celestial divination, under additional Greek influence (Oppenheim, 1978; Rochberg-Halton, 1984), would later become horoscopic astrology (Sachs, 1952).

The Babylonians and Assyrians were semitic speaking peoples who shared many of the same cultural traits such as language, architecture, religion and mythology. For most of the first and second millennia BC, first the Assyrians and then the Babylonians would dominate the Tigris–Euphrates valley, as well as much of the rest of the Near East, until the last Babylonian king, Nabû-naid (Nabonidus) was overthrown by the Persians in 539 BC (Lewy, 1946).

In the 18th century BC the first evidence of divination by celestial objects comes to use from Babylon, as can be seen in the following passage, which reads (Bauer, 1936; van der Waerden, 1974, p. 48):

> If on the day of the crescent [moon], god does not disappear quickly enough from the sky, the quaking disease [a form of epilepsy?] will come upon the land.

This type of divination by celestial objects is referred to as 'omen astrology' (van der Waerden, 1974, p. 48), and had as its principal object to predict the welfare of the king (Labat, 1939, pp. 352–60; Frankfort, 1948, pp. 231–48, 262–74; Oppenheim, 1964, p. 226) (who was the embodiment of the nation), price fluctuations in the market place, the condition of the harvest, or any military threats to the nation. This type of celestial divination existed throughout most of the first and second millennia BC before being replaced by another form of celestial divination, known as horoscopic astrology. The first evidence of this type of astrology emerges late in the fifth century BC in Babylonia (Sachs, 1952, p. 50), but is only fully developed later in Greco–Roman times.

In the late Assyrian period (the seventh and eighth centires BC), celestial observers from around the Tigris and Euphrates river valleys (see Figure 13.1) sent dispatches[1] and letters to the court of the king concerning the activities of many kinds of celestial phenomena (Oppenhein, 1969, p. 98; Parpola, 1973, 1983). This constant flow of celestial reports and their accompanying omina kept the king and his advisors aware of all possible manner of good as well as evil events that were foretold by the stars, constellations and planets. The largest number of such documents, numbering over 500,[2] comes mainly from the period between 680 and 660 BC, and give a detailed account of the nature, extent and sophistication of celestial observation during the reigns of the kings Esarhaddon and Assurbanipal. These texts always contained some

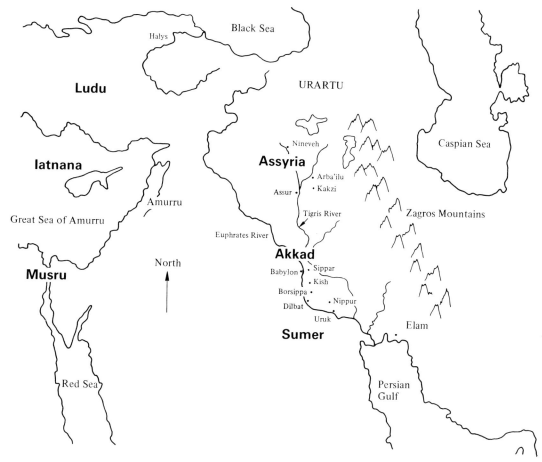

Figure 13.1 Centers of celestial observation in the Tigries–Euphrates region in the seventh and eighth centuries BC. Approximate (straight line) distances between major centers:

Nineveh to Assur = 100 km; Nineveh to Arba'ilu = 90 km; Nineveh to Uruk = 600 km; Nineveh to Babylon = 470 km; Nineveh to Nippur = 500 km.

kind of record of a celestial observation such as (Thompson, 1900, 183, 1):

Šumma MUL ᵈMAR.TU ina REŠ MU IGI-ir

If the star of the god Marduk [Jupiter] is seen
at the beginning of the year,

which is then followed by one of the standard omens[3] that were used by the celestial observers of the time such as (Thompson, 1900, 184, 2):

MU BI AB.SÍN-šu SI.SÁ

In that year the crops will be prosperous.

Sometimes more than one planet was mentioned (Thompson, 1900, 195, r. 5):

Šumma MUL SAG.ME.GAR MUL MAN TE-šu

If Jupiter approaches the 'strange star' [Mars], and the resulting omens were not always favourable (Thompson, 1900, 195, r. 6):

ina MU.BI LUGAL URI KI Uš-ma EBUR KUR SI.DÁ-si

In that year the king of Akkad [Babylon] will die [but], the crops of the land will prosper.[4]

There were hundreds of such omen texts which mentioned many different celestial objects both with favorable and unfavorable omens resulting from them.

Besides the planets, moon and stars, a small number of reports, as well as other forms of literature from the Late Assyrian period (seventh and eighth centuries BC), mention both meteors and comets.

One of the principal reasons why meteors and comets have been so poorly understood in the past by Assyriologists is because of the lack of precise terms that were used with any consistency by the celestial observers of the time to refer to such objects. Often the observers referred to heavenly objects using only the words *MUL* or *kakkabu*[5] which means 'star'. Thus, a constellation, a star, planets and even meteors and comets were referred to simply as 'stars', usually with little or no attempt being made by the celestial observers to clarify what kind of 'star' was meant.

13.2 Meteors

Despite the lack of clarification mentioned above, we can be certain that the appearance of at least meteors, and probably comets as well, were recorded. Both the Assyrians and the Babylonians accorded great importance, for calendrical and especially divinatory purposes, to the observation and recording of celestial events. Any civilization that gave such importance to the recorded observations of the sun, moon and planets could not have helped noticing and recording the flashing brilliance of the many meteors that shoot across the night sky.[6] Furthermore, it is known that there was a large and well organized network[7] of celestial observation stations scattered throughout the Tigris–Euphrates river valleys in the Late Assyrian period and that these stations, or observation posts, were manned by teams of professional observers who 'kept watch' for the king, looking for any ominous signs, both in the heavens and on earth, that would have served to warn palace leaders of danger.

Even though there were no precise terms that were used with any consistency to refer to meteors, one word, *kiṣru*, which also means 'lump or knot'[8] was sometimes used to indicate meteors or shooting stars. For example (Thompson, 1900, 28, 2–3):

Šumma MUL SUR-ma ki-ma ki-iṣ-ri
TA ᵈUD.ŠÚ.A ana ᵈUD.È

If a star flashes like a meteor from West to East. This word is also used to refer to meteorites in several versions of the *Epic of Gilgamesh*,[9] one dating from the old Babylonian period (2000–1600 BC), as well as another from around the beginning of the first millennium BC (Tigay, 1982, pp. 242, 256ff.). The later version (known as the GEI version) reads as follows (Tigay, 1982, p. 270, 27–8):

ib-šu-mim-ma kakkabani (MUL.MEŠ) šamê
(ANe)
kima (GIM) ki-iṣ-ru ša ᵈA-nim im-ta-qu-ut e-li
ṣēri-(EDIN)ia

there were stars in the heavens, like a meteor from the sky god Anu it fell down on top of me.[10] The exact meaning of this text is somewhat unclear since this meteor reference occurs within the context of a dream that predicts the arrival of Enkidu, the friend and companion of Gilgamesh and which the mother of Gilgamesh, Ninsun, is about to interpret for her hero son.[11] Since the meteor incident occurs in rather strange circumstances it would be difficult to believe that this passage is an accurate record of a real event. Nevertheless, it does indicate that meteorites were known already in the second millennium BC and that they, like other celestial phenomena, were the harbingers of important or ominous events.[12]

Another word that may be referring to meteors, at least in some texts, is *ṣallammu/zallammu* (Thompson, 1900, 183, 1).[13]

Šumma MUL ṣal-lum-mu-u ina šūt ᵈAnu innamir

If a *ṣallummu* star [meteor] is seen in the path of the god Anu.[14]

In one instance the sign *DINGIR*, signifying a deity, is placed before the word *zallummu* indicating that it was thought to be a god.[15] It is probable that the word *zallummu* is used in a small number of texts to refer to comets. This will be discussed below in the section on comets.

Two other terms used to refer to meteors in the *Enuma Anu Enlil* are *MUL. AN.TA.SUR.RA* (*eliš ṣariru*) 'star flashing from above'[16] and *MUL. AN.-TA.ŠUB.ŠUB BA)* (*eliš maqatu*).[17] These 'stars' sometimes foretell meteorological changes such as the wind:

MUL.AN.TA.SUR.RA ana IM.ZI.GA.

The flashing star is for the rising of Wind.[18]
or it could presage the opposite:

MUL.AN.TA.ŠUB.ŠUB.BA ⟨ana⟩ IM.ŠUB.BA

The Star falling from above is for the abating[?] of wind.[19]

Even though both of these examples, and others like them, refer to meteors there is a better method that can be used for identifying meteors and separating them from comets and other celestial phenomena.

Instead of looking for words or terms that ambiguously refer to such objects it will be useful to examine the way celestial objects are described by the authors of celestial observation literature. For example (Thompson, 1900, 201, 1–7):

KASKAL.GID GE it-ta-lak

MUL.GAL TA.IM⁶SI.-SÁ
ana IM.GAL.LU
iṣ-ṣa-ru-ur

When the night had advanced one double hour
a great star flashed from North to South.[20]

In the above report it is clear that some kind of
celestial object, most likely a meteor, 'flashed' or
moved rapidly across the sky from north to south.
A number of other reports refer to meteors simply
as a 'great star', but the texts do give enough information about them to make classification possible
in a number of instances (Thompson, 1900, 202,
1–3):

Šumma MUL.GAL TA ᵈUTU.È
ana ᵈUTU.ŠÚ.A SUR-ma ir-bi
u mi-ši-šu

If a great star flashes from the east to the west
and its glow disappears [sets].

In this report we have two terms which refer to
meteors. The first, *MUL.GAL*, or 'great star', and
the second, *mišhu*, which refers to some kind of
luminous phenomena, such as a mirage, a comet
or a meteor.[21] The term 'great star' is used in a
number of texts to refer to the planet Jupiter.[22] However, in the above report, because of the description
of the object, and because of the use of the
additional term *mišhu*, which is rarely used to refer
to planets, we can be fairly certain that this passage
is referring to the observation of a large meteor.

The same report continues with an additional
reference to meteors (Thompson, 1900, 202, r. 1–
4):

II MUL.GAL.MEŠ
ina EN.NUN.MURUB₄.BA
EGIR a-ha-meš
iṣ-ṣar-ru

Two great stars [meteors] flashed in the middle
watch one after the other.

This passage is probably referring to two meteors
(perhaps one meteor that split up on contact with
the earth's atmosphere)[23] that were seen shooting
across the sky, or it may be referring to one of the
numerous meteor showers that occur at different
times during the year.[24]

In these and a number of other texts, mostly
reports, the 'star' or celestial object referred to is
said to 'flash' or move quickly across the sky. The
writers of these texts make it clear that this is not
something that lingers for any long period of time
in the heavens. The ideas of 'flashing' and 'rapid

movement' are both inherent in the verb *ṣararu*[25]
which is used in these texts when referring to
meteors. This verb is also used when referring to
other fast entities such as the infamous headache
demons:

ana ameli lemniš kima kakkabu i-ṣar-ru- ru

They [the demons] flash evily like stars for the
man.[26]

A similar 'demon' example reads as follows:

kima kakkab samê i-ṣar-ru-ur kima me muši

the headache demon flashed like the stars in the
sky [and] moves [swiftly] like the waters at night.[27]

'Flashing' and 'rapidity' are important for determining whether our texts are referring to meteors
or other celestial objects. When a reference is made
to a 'shooting' or 'flashing' object (i.e. *MUL iṣ-ṣar-
ru-ur*), provided that the text contains no additional
commentaries to the contrary, we can be reasonably
certain that the celestial observers were in fact
recording the passage of a meteor.

The key to understanding and identifying
meteors is that they move very fast, and generally
burn up in a few seconds after entering the earth's
atmosphere, causing a bright steak of light to move
across a portion of the sky. Meteors become visible
80–100 km above the earth, and are seldom visible
for more than a few seconds (Abell, 1964, p. 109).
They enter the earth's atmosphere at speeds up to
75 km/s, and plunging through the earth's atmosphere they are heated to incandescence, becoming
visible to observers on earth.[28]

In summing up so far, the two most important
criteria for identifying references to meteors in
cuneiform sources which enable us to distinguish
them from comets, planets and other celestial
objects are, first, that meteors are flashing, rapidly
moving objects, whose description, in most
instances, incorporates the verb *ṣararu* 'flashing'
into its description, and second, these same texts
often give the impression that the object traverses
a large portion of the sky. References to meteors
usually mention that the object flashed from one
horizon to the other or from one cardinal point
to another.

An interesting and important meteor reference
comes from a celestial observer named Nabu-mušiṣi
who wrote celestial observation reports during the
early part of the seventh century BC. One of these,
RMA 200, contains the following passage (Thompson, 1900, 200, 1–2):

Šumma MUL.SUR-ma ṣi-ri-ir-šu GIM ṣeti na-mir

ina ṣa-ra-ri šu GIM nam-aš-ti GÍR.TAB KUN GAR-in

If a star flashed and its flashing is as bright as daylight, and it has a tail like the scorpion [i.e.] the animal [not the constellation] while it is falling.[29]

Even though the text occurs in a broken context it does include references to the north and south cardinal points. This text is somewhat difficult since it incorporates two criteria that apply to meteors, flashing and directions and yet the observer had enough time to see a tail in the shape of a scorpion. Seeing the tail of a meteor in the shape of a scorpion's tail would indeed be rare but certainly possible (Brown, 1974, p. 219ff., Figure 3b) and the author of the text may have in fact been using the curved tail motif as a metaphor for a great 'bolide'[30] or meteoric fireball in the shape of a *zuqaqipu*[31] or scorpion. It is still puzzling why the 'animal' scorpion is specifically pointed out instead of the constellation. But I do not believe that this text records the passage of a comet.

13.3 Comets

In the ancient Near East, comets, like meteors, have not been studied systematically in recent years. Apart from one article by F. Boll (1921, columns 1153–8) and another by Largement (1957) most assyriologists have remained silent on the subject, probably due to the lack of specific terminology or solid references that refer to such things.[32]

Recently interest in comets have been rekindled with the appearance of a very important article written jointly by F. R. Stephenson, K. K. C. Yau and H. Hunger (n.d.) entitled 'Records of Halley's Comet on Babylonian tablets' (see also Walker, 1985). This article contains several solid references to the passage of Comet Halley that were seen and recorded by celestial observers in Babylon in November of 164 BC and in late July to mid-August of 87 BC. Until the appearance of the above article, precise, datable descriptions of comets, which were accompanied by unmistakable terms used to refer to them, had been lacking, and it has been practically impossible to relate such terms as, for example, *zallummu* to any references to comets in the first millennium BC.[33] It is now clear that the term used

to refer to comets in the first and second centuries BC was *ṣallummu*.[34] In the appearance of 87 BC, the comet was even referred to as 'the god *ṣallummu*. The use of this substantive lends much support to the supposition of Reiner and Pingree that *zallummu* was used already in the celestial omen series *Enuma Anu Enlil*, which dates back to at least the seventh century BC, but which is probably older,[35] to refer to comets.

However, even though the word *zallummu* was used to refer to comets in the first and second centuries BC, this does not necessarily mean that it was used in the same way or with the same precision in earlier periods. Despite this it is still tempting to consider this word as a term for comets even though the references in the *Enuma Anu Enlil* are based only on the substantive *zallummu* and not on any kind of actual description of any celestial object.

Besides the references in the *Enuma Anu Enlil*, other sources from the late Assyrian period also contain several possible references to comets. As with meteors I am quite certain that, given the nature of the celestial observation posts, the education and professional status of the observers, there is good reason to believe that comets were observed and recorded in the literature of the time. An example of a text that probably makes reference to the appearance of a comet, based on the description of the object, reads as follows:

Kakkabu ša ina panišu šip-ra ina arkišu ṣibbata šanknu innamir ma

A star [i.e. comet] was seen which had a coma [beak] in front, and a tail in back.[36]

This is a fairly good description of a comet because the observer had enough time to observe more than just a flash of light in the sky and he could actually see a 'coma' or 'beak' in front and a tail in back. Observing a tail behind a comet is not unusual; naked eye comets always show tails because of their unique physical makeup and interaction with the sun. But to see a 'beak' on a comet is unusual, although totally possible. Comets do indeed have beaks, or as astronomers refer to them 'anti-tails' (Abell, 1964, p. 296, Figure 14.4),[37] and because our observer recorded this information we can be reasonably certain that the object in question is a comet. Another comet reference, based on both the description of the object and the use of the word *ṣallummu*, reads as follows:

*Šumma kakkabu ša ina panišu ṣipru ina arkišu
zibatta šaknu innamir-ma šamê ZALAG-ir ki-ma
ṣal-lum-mu-u kima me-ših MUL.MEŠ ṣal-lum-
mu-u meš-hu-sa MUL X ia-a-nu*

If a star, which has a beak in front and a tail
in back is seen and illuminates the sky like a
comet, [variant] like the glow of the stars
[explanation] *zallummu* [the comet] equals the
glow of a star.[38]

Again we have an object with a tail in both the front
and the back. But further importance is given to
this text because it includes the word *ṣallummu*
which, as we saw above, was used as the substantive
for comet.[39] There are several other passages that
make use of the word *ṣallummu* when referring to
what appear to be comets. However, a number of
these also incorporate the use of another substantive, *bibbu*, 'wild sheep'[40] and thus their nature is
somewhat more complex and cannot be discussed
here, but will be the subject of a separate article
that will make up the second part of this study of
meteors and comets.

Another group of texts known as *A Babylonian
Diviner's Manual* (Oppenheim, 1974, p. 187ff.), a
kind of guide, or handbook for diviners and for-
tune-tellers of the period[41] (seventh century BC) also
makes reference to a comet (Oppenheim, 1974, p.
209, line 36):

*kakkabu ša ina pani-šu ṣipra ina arki-šu ṣibbatu
šaknu innamirma šamê namir*

If a star that has a beak in front and a tail placed
in back is seen and [it] lights up the sky.[42]

It would appear that this is a reference to a comet
that was so bright that it actually caused the sky
to light up, something that has been reported in
more recent times with the appearance of Halley's

and other large, sun grazing comets. If our understanding of comet is correct then *A Babylonian
Diviners Manual* makes an even more intriguing
reference to the existence of whole series of texts
devoted exclusively to comets. The title of this series
is the same as the above extract, 'If a star that has
a beak in front and a tail in back is seen and [it]
lights up the sky' (Oppenheim, 1974, p. 209). However, up until the present, this series has escaped
detection by assyriologists.[43]

Even though this series has not been located we
still have reasonably good evidence which points
to the existence of at least five instances in the Late
Assyrian period when references were made to comets. With the refinement of techniques and further
research I am quite certain that more comets will
be found.

One final point concerns the dating of these
comet references. In any discussion which
addresses the problem concerning the identification
of ancient references to comets, particularly in this
the year of the return of our solar system's most
famous comets, one must ask the question whether
or not any of the above comet references might possibly be referring to Halley's Comet? However, all
comet references previously mentioned in this article cannot be dated with the kind of accuracy
necessary to pinpoint the return of any major naked
eye comet. It is known only that they were written
sometime in the late eighth or in the first half of
the seventh centuries BC.

Nevertheless, the fact that comets can be identified with reasonable certainty at this early date
should serve as a useful tool for further research
in the history of astrology and astronomy in the
ancient Near East.

Notes

1 Referred to as 'Reports' which are 'mostly short, con-
cise, and very formalized – addressed to the King.
They are written on small, oblong, and often quite
thick clay tablets' (Oppenheim, 1969, pp. 97–8).

2 Oppenheim (1969, p. 127, note 2) gives the number
as 600. Parpola (1983, pp. 498–503) gives 532 celestial
observation Reports. If we add to this several-
hundred 'Letters', also concerning celestial matters,
the number is well over 700.

3 Such as *Enuma Anu Enlil* which contained about

70 tablets and 7000 omens relating to celestial objects.
See Weidner (1941–5, 1954–6, 1968–9), and also
Reiner and Pingree (1981).

4 For a similar passage see Reiner and Pingree (1981,
Text III, line 13b). Strangely enough, even though
the king's life is threatened by the 'stranger', the crops
of the land would have prospered. Thompson (1900,
vol. II) in his transliteration inserts the negation *la*.
This is not likely to be incorrect since it does not
occur in the autograph copy of the text in volume
I. *The Assyrian Dictionary of The Oriental Institute of
the University of Chicago.*

5 (*CAD*) sub voce *kakkabu* s.; 'star symbol, star-shaped ornament, star-shaped brand, written syllabically or *MUL*; and Gössman (1950, p. 105) 'Auch *MUL₂*; *UL*, *UL₈*: das sum. Wort fur Gestirn, Sternbild; Stern; akk. *kakkabu*'.

6 On a typical moonless night an alert observer can see a dozen or more meteors per hour (Abell, 1964, p. 302).

7 'A state-wide network of observation stations charged with dispatching such Reports to the capital was quite likely in existence' (Oppenheim, 1969, p. 114).

8 Cf. *CAD* sub voce *kiṣru* meanings 7 and 11a and b, and von Soden (1965–) sub voce *kiṣru(m)* 'knoten, zusammenballung; Miete' and meaning 6a. 'sternhaufen'. E. A. Speiser (1958) reads (incorrectly I believe) 'essence' for *kiṣru*.

9 Tigay (1982) is the most up to date treatment of this story. Heidel (1942) gives a less accurate translation of this passage.

10 Lines 6 and 7 of the Old Babylonian version (Gilg. P.) reads 'The stars of the heavens appeared [and] A "meteor"? of Anu descended upon me.' In the Late Version (GEI) the same author reads in lines 27 and 28

> The stars of the heavens appeared [and]
> [something] like a meteor [?] of Anu keeps descending toward me.

11 In the dream the meteorite is so heavy that Gilgamesh cannot even lift it:

> I raised it, but it was too mighty for me!
> I tried to move it away, but I could not dislodge it.

Tigay (1982, p. 271, 29–30). The meteorite turns out to be a metaphor for the savage man-companion of Gilgamesh whose name was Enkidu.

12 This is one of the few incidents in recorded history of someone actually being struck by a meteorite. See *Sky and Telescope*, March 1985, p. 222, and specifically 'Killer Meteorite', *Sky and Telescope*, December 1985, p. 553, for other accounts of people being struck by meteorites.

13 Depending on the author, *ṣallummu* or *zallummu* are both accepted ways of writing this word. The reading of comet for *ṣallummu* is questionable, cf. *CAD* sub voce *ṣallammu* s.; '(meteoric) fireball, meteor'; Gössmann (1950), #361 apud Schaumberger (1935) 'Glanzender Stern, Meteorschwarm'; Reiner and Pingree (1981, p. 19, section 2.2.5.1 and 2) are less sure of the meaning of this word and refer to it as a term which is ambiguous, but may from time to time refer to a mirage', or again 'may refer to the passage of a comet before *AŠ.GAN*' [*iku* the field or the square of Pegasus]. Text XV, line 23 in Reiner and Pingree (1981, pp. 72–3) reads:

UL sal-lum-mu-u ana IGI MULAŠ.GÁN

If a comet crosses towards the Field (The Square of Pegasus). Some of the texts cited in the *CAD* sub voce *ṣallammu* may be comets instead of meteors. See von Soden (1965–) sub voce *ṣalamu(m)*.

14 The 'path of the god Anu' was a section of the sky lying on or near the celestial equator. See van der Waerden (1949, p. 10).

15 Thureau-Dangin (1922) *ᵈṣal-lum-mu-u*.

16 'Probably a term for shooting star or meteorite' (Reiner and Pingree, 1981, p. 10, Text II 2; III 6, 6b; XIX 1–2). In Akkadian *eliš ṣariru* 'flashing from above', from *AN.TA = eliš*, cf. *CAD* sub voce *eliš* 'above', see lexical section, and *SUR.RA = ṣariru*, *CAD* sub voce *ṣararu* B 1. 'to flash (said mainly of shooting stars)' and lexical section.

17 Reiner and Pingree (1981, p. 10) 'probably a term for meteorite' and *CAD* sub voce *maqatu* 'to fall' written syllabically and *ŠUB*.

18 Reiner and Pingree (1981, Text III, line 6, pp. 40–1). See also line 6b:

> *MUL.AN.TA.SUR.RA ma-diš SA₅*
> The flashing star is very red

See also Text XIX, lines 1–2.

19 Reiner and Pingree (1981, Text I 20; III 20; IV 14) read 'The *AN.TA.ŠUB.ŠUB.BA* is for the abating of wind'.

20 Cf. *CAD* sub voce *beru* meaning A2, p. 210 for a slightly different translation and other examples.

21 Reiner and Pingree (1981, section 2.2.5.1 and 2) and cf. *CAD* sub voce *mišḫu* and von Soden (1965–, p. 660) sub voce *mišḫu* 'Aufleuchten, Meteor'.

22 Cf. Gössmann (1950, no 62, p. 18). This term is also used to refer to the Moon, Sirius, Saturn and meteors. However, the majority of the references to *MUL.GAL* in the Reports refer to Jupiter.

23 For a photograph of this phenomenon see Chapman and Brandt (1984, p. 109) and Olson (1985, p. 4, Figure 4, and p. 72, Figure 66).

24 For a list of the major meteor showers see Muirden and Robinson (1979, pp. 170–1).

25 Cf. *CAD* sub voce *ṣararu* v.; meaning B 1. 'to flash (said mainly of shooting stars)'. Written syllabically and *SUR*.

26 Cf. *CAD* sub voce *ṣararu* meaning B 2. 'to flit (said of demons), and lexical section.

27 Cf. *CAD* sub voce *ṣararu* p. 106, especially *CT* 1625 i 52f.

28 Meteors vary in weight from less than one gram to the largest known meteorite, found at Grootfontein, South Africa, which weighs 64 tons. See Wallenqvist (1968).

29 Cf. *CAD* sub voce *ṣararu* B meaning 1, p. 107. In lines 9ff. a similar passage occurs in a broken context.

From the autography copy of Thompson's edition of the *RMA*, vol. 1, the text appears to read:

Šumma MUL.GAL ul-tu ti-ib IM SI SÁ
a-na ti-ib IM.GAL.LU (Šuti) SUR-ma
mi-ši-hu ŠU GIM nam-maš- ti GÍR.TAB KUN
GAR-in

If a great star from the south to the north flashes and it takes a luminous glow like the tail of a scorpion.

30 'While the flight of a bolide lasts only a few seconds the dust trail remains visible sometimes for many minutes and even for over an hour' (Krinov, 1960, p. 66ff.).

31 CF. *CAD* sub voce *zuqaqipu* la, 'scorpion'.

32 O. Neugebauer (1957a, pp. 211–15) refers to comets only in the Islamic period. The subject is briefly mentioned by Parpola (1983, p. 96): 'There do not seem to be any certain references to comets in the Sargonid [Late Assyrian] astrological letters and reports'.

33 However, see Stephenson, Yau and Hunger (n.d.), Walker (1985) and also the references given in note 13.

34 Stephenson, Yau and Hunger (n.d.) from *BM* (British Museum) 41462, line 16 ... *ṣal-la m-mu-u.* ᵈ*ṣal-lam-mu-u*, Stephenson, Yau and Hunger (n.d.) from BM 41018 rev. 8′–10′ (unpublished).

35 'The actual tablets were found mainly in Assurbanipal's library at Kuyunjik [the Ancient Nineveh] and were inscribed in the seventh century BC.' However, the whole series may be based on an earlier source that 'dates back considerably earlier than *c.* 1000, possibly to the Old Babylonian period at the beginning of the second millenium BC.' (Reiner and Pingree, 1981, p. 1.)

36 Cf. *CAD* sub voce *ṣipru* 'crest coma (of a comet), summit (of a triangle) exoresence on an animal's head'. All of these contain the basic idea of something in front or on top of an object and fit the idea of a 'beak'. Marcel Leibovici has shown that the idea of a beak or something pointed is germain to *ṣipru* as is shown in Leibovici (1956, p. 145, line 4):

kakkabu ša ina pani-šu ṣipra

L'étoile qui par devant [est] pointue.

37 See also Chapman and Brandt (1984, p. 89), 'anomalous tails'. 'Such tails are detected only during short intervals around the time when the earth passes through the place of the comet's orbit.'

38 Cf. *CAD* sub voce *zallummu*, p. 75.

39 Leibovici published a similar text over 30 years ago in *Syria* (Leibovici, 1956). He translated *zallummu* as 'lumière brillante' and not as comet since, at that time, no one was certain of the exact meaning of *zallummu*.

40 Cf. *CAD* sub voce *bibbu*, p. 218, 'wild sheep', Largement (1957) and Gössman (1950) sub voce *bibbu*.

41 The original publication of these texts (Virolleaud, 1911, pp. 109–13) were entitled 'Tables des Matières de deux traités de divination: l'un terrestre l'autre astrologique, accompagnée d'instructions du mage à son élève' (Oppenheim, 1974, p. 197).

42 The transliterations on pages 199–200, lines 32, 36, 37 and 48, have slight variations and are not normalized in Akkadian.

43 This series 'is as yet unaccounted for in the thousands of "astrological" fragments found in Assurbanipals' library.' See Oppenheim (1974, p. 209).

References

Abell, G. (1964). *Exploration of the Universe*. NY: Holt, Rinehart and Winston.

Bauer, T. (1936). Eine Sammlung von Himmels-Vorzeichen. *Zeitschrift für Assyrologie* **43**, 308–14.

Boll, F. (1921). 'Kometen': *Pauly's Real-Encyclopädie der Klassichen Altertum Wissenschaft*, ed. G. Wissowa, half vol. 11 i. Stuttgart: Metzlersche Bucherei.

Borger, R. (1963). *Babylonisch–Assyrische Lesestucke*, cahier 1. Rome: Pontificum Institutum Biblicum.

Brown, P. L. (1974). *Comets, Meteors and Men*. NY: Taplinger Publishing.

Chapman, R. and Brandt, J. (1984). *The Comet Book*. Boston: Jones and Bartlett.

Frankfort, H. (1948). *Kingship and the Gods*. University of Chicago Press.

Gardner, J. and Maier, J. (1984). *Gilgamesh: The Sin-Leqi-Unninni Version*. NY: A. Knopf.

Gössmann, P. (1950). *Planetarium Babylonicum*, [[teil]] 1, vols. 1, 2, no. 274. Rome: Verlag des Päpstliches Bibelinstituts.

Heidel, A. (1942). *The Epic of Gilgamesh and Old Testament Parallels*. University of Chicago Press.

Krinov, E. L. (1960). *Principles of Meteoritics*. London: Pergamon Press.

Labat, R. (1939). *Le Caractère Religieux de la Royauté Assyro–Babylonienne*. Paris: Librairie d'Amerique et d'Orient.

Largement, R. (1957). Contribution a l'etude des astres errants dans l'astrologie chaldéene. *Zeitschrift für Assyrologie*, new series, **18**, (52), 235–63.

Leibovici, M. (1956). Les présages Hittites traduit de l'Akkadien. *Syria* **33**, 142–6.

Lewy, J. (1946). The Assyro–Babylonian cult of the moon and its culmination at the time of Nabonidus. *Hebrew Union College Annual* **19**, 405–89.

Muirden, J. and Robinson, J. H. (1979). *Astronomy Data Book*, 2nd edn. NY: John Wiley and Sons.

Neugebauer, O. (1957a). Notes on Al-Kaid. *Journal of the American Oriental Society* **33**, 211–14.

Neugebauer, O. (1957b). *The Exact Sciences in Antiquity*, 2nd edn. NY: Dover.

Olson, R. (1985). *Fire and Ice*. NY: Walker and Co.

Oppenheim, A. L. (1964). *Ancient Mesopotamia: A Portrait of a Dead Civilization*. University of Chicago Press.

Oppenheim, A. L. (1969). Celestial observation and celestial divination in the Last Assyrian Empire. *Centaurus* **14**, 97–135.

Oppenheim, A. L. (1974). A Babylonian diviner's manual. *Journal of Near Eastern Studies* **33**, 197–220.

Oppenheim, A. L. (1978). Man and nature in Mesopotamian civilization. In *Dictionary of Scientific Biography*, ed. C. C. Gillespie, 634–66. NY: Charles Scribner's Sons.

Parpola, S. (1973) *Letters from Assyrian Scholars to the Kings Esarhaddon and Assurbanipal*, vol. 1. Neukirchen-Vluyn: Verlag Butzon und Bercker Kevelaer, Neukirchener Verlag.

Parpola, S. (1983). *Letters from Assyrian Scholars to the Kings Esarhaddon and Assurbanipal*, vol. 2 Neukirchen-Vluyn: Verlag Butzon und Bercker Kevelaer, Neukirchener Verlag.

Reiner, E. and Pingree, D. (1981). *Babylonian Planetary Omens: Part Two Enuma Anu Enlil, Tablets 50–51*. Bibliotheca Mesopotamia 2, fascicle 2. Malibu, CA: Undena Publications.

Rochberg-Halton, F. (1984). New evidence for the history of astrology. *Journal of Near Eastern Studies* **43**, 115–40.

Sachs, A. (1952). Babylonian Horoscopes. *Journal of Cuneiform Studies* **6**, 49–75.

Schaumberger, J. (1935). *Sternkunde und Sterndienst in Babel Ergänzungen III*. Münster: Verlag der Aschendorffschen Buchhandlung.

Speiser, E. A. (1958). Myths and epics from Mesopotamia. In *The Ancient Near East: An Anthology of Texts and Pictures*, ed. J. B. Pritchard. Princeton University Press.

Stephenson, F. R., Yau, K. K. C. and Hunger, H. (n.d.) Records of Halley's Comet on Bablyonian tablets.

Thompson, R. C. (1900). *The Reports of the Magicians and Astrologers at Nineveh in the British Museum*, 2 vols. London: Lucas. Reprinted. 1977 by AMS Press, NY.

Thureau-Dangin, F. (1922). *Tablette d'Uruk a l'usage des Pretres du Temple d'Anu au Temps des Seleucids*, text AO 6455, column 11, line 4. Paris: Geuthner.

Tigay, J. H. (1982). *The Evolution of the Gilgamesh Epic*. Philadelphia: University of Pennsylvania Press.

van der Waerden, B. L. (1949). Babylonian astronomy II. The thirty-six stars. *Journal of Near Eastern Studies* **8**, 6–26.

van der Waerden, B. L. (1974). *Science Awakening II*. Leiden: Noordhoff International.

Virolleaud, C. (1911). Tables des Matières de deax traités de divination: l'un terrestre l'autre astrologique, accompagnée d'instructions du mage à son élève. *Babylonica* **4**, 109–13.

von Soden, W. (1965–). *Akkadisches Handwörterbuch*. Wiesbaden: Otto Harrassowitz.

Walker, C. F. (1985). *Nature* **314**, 576–7.

Wallenqvist, A. (1968). *The Penguin Dictionary of Astronomy*. Baltimore: Penguin Books.

Weidner, E. (1941–5). Die Astrologische Serie Enuma Anu Enlil. *Archiv für Orient Forschung* **14**, 308–18.

Weidner, E. (1954–6). Die Astrologische Serie Enuma Anu Enlil. *Archiv für Orient Forschung* **17**, 71–89.

Weidner, E. (1968–9). Die Astrologische Serie Enuma Anu Enlil. *Archiv für Orient Forschung* **22**, 65–75.

14

The ancient Vedic dice game and the names of the four world ages in Hinduism*

Luis González-Reimann   *El Colégio de Mexico*

According to classical Hinduism the world goes through four world Ages, called Yugas, which are repeated again and again throughout every day of the creator god Brahmā. The day of Brahmā is the basic time unit of creation, whereas his night brings about the destruction of the world.

The names of these four Yugas are Kṛta, Tretā, Dvāpara and Kali, and a clear numerical pattern is associated with them. The Kṛta Yuga is said to last 4000 divine years, the Tretā 3000, the Dvāpara 2000, and the Kali 1000, for a total of 12 000 divine years in each Mahāyuga or Great Yuga, as the cycle of four Yugas is called.[1] This 4 : 3 : 2 : 1 ratio is to be found also in other characteristics of the Yugas; the duration of human life, for instance, decreases according to this proportion and so does the observance of *dharma*, the Eternal Law.

The names of the four Yugas, however, go back a long way and they pre-date the theory of four world Ages. These names, Kṛta, Tretā, Dvāpara and Kali, are associated in Vedic literature with the game of dice. The game was well known during the Vedic period (c. 1200–600 BC), and it had both a secular and a ritual aspect. Already in the *Ṛg Veda* (*RV* 10.34; O'Flaherty, 1981, pp. 240–1), the earliest Vedic text, we hear a gambler grieve over the loss of his possessions. He has been abandoned by his wife and his friends because of his compulsive need to play dice, which have an intoxicating effect on him. And in the *Atharva Veda* (*AV* 7.109) there is a petition to the gods for success at the game. In other Vedic texts, such as the Brāhmaṇas and

the Śrauta Sūtras, the game of dice also plays a part in the *rājasūya*, the royal consecration ritual. (See, for instance, *ŚB* 5.4.4.6; for further references see Macdonell and Keith, 1912, vol. 1, p. 4, n. 28.)

Later, in the *Mahābhārata*, dice occupy an important place, as it is as a consequence of king Yudhiṣṭhira's defeat in a dice game against Duryodhana that the Pāṇḍavas must live in exile in the forest for 12 years, in addition to a 13th year during which they may appear in public but without being recognized.[2] Once this period is completed, the Pāṇḍavas return and claim what belongs to them. Their claim is rejected and the two armies meet at the battle of Kurukṣetra. These events are central to the Epic,[3] and this battle forms the setting of the famous religious text, the *Bhagavad Gītā*, which contains the teachings of Lord Kṛṣṇa to Arjuna. Elsewhere in the Epic, in the episode of king Nala, dice play an important part with thematic elements similar to those in the story of Yudhiṣṭhira (*Mbh* 3.50.78).[4]

The connection between the names of the Yugas and the game of dice lies in the fact that in some Vedic texts these names are given to the different throws of the game or, according to a different interpretation, to the different dice used in the game. The names appear in the *Taittirīya Saṃhitā* (*TS* 4.3.3; Keith, 1914, p. 328), in an invocation to the five world directions (cardinal points plus zenith) which is used during the building of an altar. Each direction is associated with different elements such as a season, a deity, a sage (*ṛṣi*) and a certain age in the life of a calf. The dice throws are among these elements. We also find them in the *Vājasaneyi Saṃhitā* (*VS* 30.18; translated in Eggeling, 1882–1900, vol. 5, p. 416) and in the *Taittirīya Brāhmaṇa*

* A slightly longer version of this paper is part of chapter 3 of L. González-Reimann, Tiempo Cíclico y eras del Mundo en la India. Mexico: El Colégio de Mexico (forthcoming).

195

Table 14.1. *The names of the dice throws in Vedic texts*

Taittirīya Saṃhitā	*Vājasaneyi Saṃhitā*	*Taittirīya Brāhmaṇa*
–	Akṣarāja	Akṣarāja
Kṛta	Kṛta	Kṛta
Tretā	Tretā	Tetrā
Dvāpara	Dvāpara	Dvāpara
Āskanda	Āskanda	Kali
Abhibhū	–	–

(*TB* 3.4.16; text and translation in Dumont, 1963, p. 181) in the section that deals with the *puruṣamedha* sacrifice. There are slight differences in the names as they are used in these texts, and this can be better appreciated by comparing them side by side as in Table 14.1.

The first thing we should point out is that here five throws are mentioned instead of the four of later times (at least only four are connected with the Yuga theory). It is also possible that Akṣarāja may not be the name of a throw. This seems to be the opinion of Eggeling (1882–1900, vol. 5, p. 416) and Dumont (1963, p. 181) since they both translate the word as 'dice-king' (*akṣa*, die; *rāja*, king), whereas they leave the other names untranslated and keep the original Sanskrit terms.

Another thing worth noting is that the lists of the *Taittirīya* and the *Vājasaneyi Saṃhitās* seem to indicate, as has been suggested by MacDonell and Keith (1912, vol. 1, p. 3), that Abhibhū and Akṣarāja are identical, and this is further confirmed by the fact that *abhibhū* (*abhi* plus the root *bhū*) means something that predominates, conquers or surpasses. We might add that the comparison of these two lists with that of the *Taittirīya Brāhmaṇa* points to an equivalence between Āskanda and Kali.

We do not know for sure how the game was played; it is even possible that in time the rules may have changed or that the ritual game may have been different from the popular one. But, if many details are obscure, it seems clear that the 4–3–2–1 sequence was an integral part of the game (Mac-Donell and Keith, 1912, vol. 1, pp. 3–4). The number four was related to Kṛta,[5] three to Tretā, two to Dvāpara and one to Kali. Of these, Kṛta was the winning throw[6] and it was followed by the others in a descending order down to Kali, the worst throw of all.

In the passage of the *Śatapatha Brāhmaṇa* that deals with the royal consecration ceremony (*ŚB* 5.4.4.6), Kali is associated with the idea of predominance, but this is infrequent.[7] The word also appears at the beginning of a hymn of the *Atharva Veda* (*AV* 7.109.1) in which help is asked of the gods in order to obtain good results at the game. MacDonell and Keith (1912, vol. 1, p. 4) believe this could be an allusion to Kali as the winning throw. In this hymn, Kali is invoked as the main die, and is propitiated with clarified butter as a means of helping the gambler. But this does not necessarily imply that Kali is the winning throw, and it could even mean the reverse. In other words, Kali could, as usual, stand for the much feared losing throw. As pointed out by Griffith (n.d., vol. 1, p. 380), this invocation could respond to the need to pacify Kali to keep it from harming the gambler. In a verse from the *Ṛg Veda* (10.34.12) the player addresses the general of the army of dice while promising to gamble everything, and O'Flaherty (1981, p. 241, n. 6) sees here a possible allusion to Kali as the losing throw. Both situations could be similar. In a more recent text, the *Mahābhārata*, Kali is personified as the god of gambling and discord.[8]

The equivalence between numbers four, three, two and one, and the names of the throws is confirmed, at least in two cases, by the etymology of the words. Dvāpara is related to *dva*, two, and Tretā to *tri*, meaning three. These words are cognates of those used in other Indo–European languages to signify the same numbers.

Kṛta is the past passive participle of the Sanskrit root *kṛ*, to do, and means 'done' or 'prepared'. By extension, it indicates something well done or good. This meaning fits well with the word being used as the name of the winning throw, the best of all. It also fits with its later use as the name of the best of the four Yugas, in which truth and understanding predominate. The Kṛta Yuga was also to be known as Satya Yuga, the Age of Truth (*satya*). In the *Mahābhārata* the monkey god Hanūmān ingeniously explains the origin of the name of the Kṛta Yuga. According to him, the best of the Ages is called Kṛta because in it 'things are done [*kṛta*], not left to be done [*kartavya*]' (*Mbh* 3.148.10; van Buitenen, 1973–8). The play on words of the original is lost in the translation.

But *kṛta* has another possible etymology

suggested by Pisani. It appears to be the one pre-ferred by Mayrhofer (1956–78, vol. 1, p. 258) in his Sanskrit etymological dictionary. In Pisani's view, *kṛta* comes from an Indo–European root meaning 'four', a root from which the names for this number in other Indo–European languages are also derived.[9] If this interpretation is correct it merely confirms the relationship between *kṛta* and the number four,[10] without setting aside the idea of completeness, since the number four can also represent totality.

The etymology of Kali is more complicated. Kali means discord or fight, and this agrees with it being the worst throw. Still more obvious is the fact that it is the name of the worst of the four Ages, the Kali Yuga, where discord and egotism prevail. What is not too clear is whether *kali* came to mean anything that produces quarrels and conflicts because it was the name of the losing throw, or whether, on the contrary, the worst throw was called Kali because this term referred to something negative, conflict producing. Mayrhofer (1956–78, vol. 1, p. 182) thinks that the first explanation (which we consider the most acceptable) could be correct but does not take a definite stand. On the other hand, different etymologies have been suggested for the word *kāla* (time), sometimes connecting it with *kali*. Theories fluctuate between the possible Indo–European origin of these words and the possibility that they may be derived from Dravidian roots belonging to Southern India.[11]

The word for die most often used in Vedic literature is *akṣa*, which is part of the term *akṣarāja*. As we have seen, this term is the name of one of the throws. But *akṣa* also has another meaning; it designates the axis of a chariot wheel. In this case, the word is related to the latin *axis*. It is very likely, however, that the etymological origin of *akṣa* may be different for these two meanings.[12]

Now that we have investigated the origin of the names of the dice throws, we may ask ourselves why they came to designate four time periods. What do dice and the Yugas have in common? The answer to this question, or at least a hint of its possible explanation, could lie in the *Ṛg Veda* itself. In the hymn that describes the grievances of the gambler who has lost everything it is said that dice function 'by rules as immutable as those of the god Savitṛ,' and that, 'they do not bow even to the wrath of

those whose power is terrifying; the king himself bows down before them' (*RV* 10.34.8; O'Flaherty, 1981, p. 240).

Now, the rules or laws (*dharma*) of the god Savitṛ, who is very likely a representation of the morning and evening sun or of its vitalizing power,[13] are surely the laws which cause the sun to rise and set, and to fulfil its cycles. In any case, this no doubt refers to natural or cosmic laws; and what this verse of the *Ṛg Veda* is saying is that dice also obey immutable laws which are beyond the control of humans and nearer to the gods. The forces that control the movement of dice are, then, just another expression of natural laws, to which time with its different cycles is also subject.[14] We could say that just as dice, while moving, function as agents of fortune and destiny, likewise the Yugas – and all time cycles – endlessly turn determining the destiny of the world.[15]

Another aspect of the relationship between gods and dice is expressed in a verse from the tenth *maṇḍala* of the *Ṛg Veda* which says that:

The gods move about like dice, who give us wealth and take it away.[16]

This confirms the cyclic character of the coming and going of fortune.

The interrelationship between the gods, dice and destiny, can be seen clearly in the similarity between certain words. The word for god is *deva*, whereas *devana* (or *adhi-devana*) is, in the *Ṛg Veda*, the place on which dice were thrown (MacDonell and Keith, 1912, vol. 1, pp. 5, 375). In time *devana* came to be a synonym of die. Destiny, on the other hand, is called (among other things) *daiva*, that which is of, or is related to, the gods. This term appears in the *Mahābhārata*, where destiny is closely linked with time.[17] It is interesting that the verb used to signify dice playing should be *dīvyati* (for instance, in *RV* 10.34.13 and *Mbh* 3.56.4). According to traditional grammarians, all these words are derived from one same root, *div*, which on the one hand is associated with the idea of brightness,[18] and on the other with playing dice or being happy. The common etymological origin of these two meanings of the root *div* is uncertain (Mayrhofer, 1956–78, vol. 2, p. 47–8), but nevertheless it is an interesting coincidence.

Another possible connection between dice and the Yugas is derived from the close relationship there is between concepts of time and space. Time and space constitute the coordinate system by means of which things and actions are located. Vedic literature goes into much detail when it comes to defining the appropriate time and place for the celebration of rituals. The interconnection between these two systems of reference is embedded in Sanskrit grammar itself, as the locative case is employed for defining position in space as well as in time.[19]

We have seen that in the *Taittirīya Saṃhitā* (*TS* 4.3.3; Keith, 1914, p. 328), each dice throw is associated with one of the five world directions. Starting with the east, and ending at the zenith, the sequence is as shown in Figure 14.1.

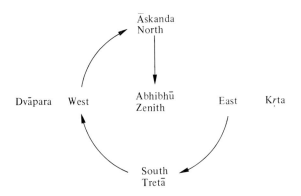

Figure 14.1 Cardinal directions and dice throws.

In the *Śatapatha Brāhmaṇa*, when dice are used during the *rājasūya*, the royal consecration ceremony, the priest places five dice in the king's hands while reciting a verse from the *Vājasaneyi Saṃhitā* that reads:

> . . . you are dominant; may these five directions (*diśaḥ*) of yours prosper![20]

It seems clear that, on receiving the dice, the king establishes his sovereignty over the directions and assures for himself full dominion over the territory of the kingdom. All this implies that dice could have been used to symbolize the cardinal points, and, therefore, it is not surprising that they were later used to designate the Yugas. From representing space, they went on to represent time, a very natural transition for a cosmovisión that establishes

equivalences between the microcosm and the macrocosm, and which accepts correspondences between the different elements of the universe.

A passage from the *Mahābhārata* (*Mbh* 3.121.19) states that the transition or union (*saṃdhi*) between the Tretā and the Dvāpara Yugas is the Narmadā River. In this way, time is identified with space, and more precisely a period of transition in time is identified with a dividing line in space. The choice of the river Narmadā is surely not accidental, since classical texts consider it (together with the Vindhya range) as the dividing line between North and South India; that is, between Uttarāpatha (the northern path) or Āryāvarta (the land of the *aryas*), and Dakṣiṇāpatha (the southern path).[21]

In a different section of the Epic, the relationship between time and space is made evident when Vaiśravaṇa instructs king Yudhiṣthira, explaining that:

> The persevering hero who knows time and place [*deśa-kāla*] and is informed of the rules of all the Laws, he . . . governs the earth.

Whoever acts in this manner, he goes on to say, obtains fame in this world and a good journey after death. And it was due to his knowledge of the right time and place to act that the god Indra was enthroned in heaven (*Mbh* 3.159.3–5; van Buitenen, 1973–8, vol. 2, p. 531).

A third way of explaining the use of the names of the dice throws for naming the Yugas could be the simple fact that such names indicated a quantitative and qualitative graduation. When the idea of the existence of a series of successively decaying Ages was taking shape, the terms Kṛta, Tretā, Dvāpara and Kali became obvious candidates for designating such periods. On the other hand, it is possible that the numerical precision implicit in the names of the throws, that is the 4–3–2–1 sequence, may have influenced some of the descriptions of the Yugas. This can be seen, for instance, by the fact that the duration of human life is considered to be 400 years in the Kṛta Yuga, and 300, 200 and 100 years in the following Yugas;[22] or by the idea that the Law (*dharma*) stands on four feet in the Kṛta Yuga, on three during the Tretā, on two in Dvāpara, and it wobbles on only one leg in the dark Age of Kali (*MDhŚ*, 1.81–2).

We have suggested three possible explanations of why these terms, originally associated with the game of dice, were later used to designate the Yugas of the classical period. Most likely, none of these explanations should be considered as the only one, and it is rather the combination of all three, together with other factors, that allows us to have a better idea of how this transition came about. We must say here that this transition probably took place during a long period of time. In the *Mahābhārata* itself, a text which in its final form belongs to the classical period and in which the Yuga theory is already well defined, we still find Kṛta and Dvāpara used for dice throws (*Mbh*, 4.45.23). In Kane's opinion,

Kṛta, Tretā, Dvāpara and Kali began to be used as names for the Yugas sometime around the fourth century BC (Kane, 1958, p. 687). This coincides, in a general way, with the transition from Vedism to Hinduism.[23]

Acknowledgment

My attendance at the Conference was made possible by a grant from the National Council on Science and Technology (CONACYT), Mexico.

Notes

1 One divine year equals 360 human years. The Mahāyuga includes 2000 years of transition periods called *saṃdhis*.

2 We cannot help noting that the duration of the exile of the Pāṇḍavas conforms to the numerical pattern set by the 12-month year with the addition of a 13th intercalary month. In a different section of the Epic (*Mbh* 1.55.31–2), king Yudhiṣṭhira sends his brother Dhanaṃjaya (Arjuna) to the forest for a period of a year plus one month; see van Buitenen's note to this passage (van Buitenen, 1973–8, vol. 1, p. 446). On the other hand, 12 years is also the length of Jupiter's cycle, which plays an important role in various astronomical and astrological texts, and which became the clock for setting the date for the celebration of the Kumbha Melā, one of Hinduism's main religious festivities.

3 The episode of Yudhiṣṭhira's dice game appears in the *Sabhā Parvan*, the Epic's second book (*Mbh* 2.43–72). van Buitenen (1973–8, vol. 1, p. xv) thinks that Yudhiṣṭhira could not avoid participating in the game of dice, even at the risk of losing his kingdom, because the game was part of the ceremony by which he was consecrated as king. Actually, to van Buitenen (1973–8, vol. 2, p. 6), the importance of the *rājasūya* in this book of the *Mahābhārata* is such that he suggests that the *Sabhā Parvan* is, in its structure, an epic dramatization of the Vedic ritual. In this respect, see his introduction to the *Sabhā Parvan* (*The Book of the Assembly Hall*) in vol. 2 of his translation. On pages 27–30 he deals with Yudhiṣṭhira's dice game.

4 According to van Buitenen (1973–8, vol. 2, p. 183), the game of dice is not such an integral part of the story of Nala as it is of Yudhiṣṭhira's.

5 Keith (1908, pp. 826–7) points to the comparison between Kṛta and the *catuṣṭoma* ritual (from *catur*, four) in *TB* 1.5.11.1 and in *ŚB* 13.3.2.1, as examples of the equivalence of Kṛta and the number four. But

he thinks that Kṛta could formerly have stood also for the number five, and that this tradition survived in the ritual game. In any case, common usage turned Kṛta into a synonym of a throw associated with the number four.

6 The *Chāndogya Upaniṣad* (4.1.4) states that all other throws belong to Kṛta, the winning one. MacDonell and Keith (1912, vol. 1, p. 4) point to *RV* 1.41.9 as evidence that Kṛta was the winning throw. In this verse it is said that he who holds 'the four' must be feared. The general opinion, including that of Geldner (1951–7, vol. 1, pp. 51–2) and Renou (1959, p. 110) is that this verse refers to the game of dice, although Ludwig and Bergaigne (according to Griffith, 1889, p. 28) interpret it differently.

7 Goldstücker (according to Eggeling, 1882–1900, vol. 3, p. 107) is of the opinion that in this case the predominant role of Kali symbolizes the triumph of the Kali Yuga, the present Age.

8 In the story of Nala, he is possessed by Kali (with help from Dvāpara, also personified), who makes him play dice and lose everything (*Mbh* 3.55–6). It is only when Nala has learned the secret of the dice that he is able to get rid of Kali's curse. Kali then takes refuge in the *vibhītaka* tree (*Mbh* 3.70.25–38). It is precisely the nuts from this tree that were used as dice since Ṛg-Vedic times (MacDonell and Keith, 1912, vol. 1, p. 2). In Yudhiṣṭhira's episode (*Mbh* 2.45.40) Kali, the spirit of dice, is also mentioned.

9 According to Pisani (Mayrhofer, 1956–78, vol. 1, p. 258), *kṛtam* comes from Indo–European *$q^w t(w)rto$-m*, through the intermediate stage *ktṛtam*. In the system used by *The Heritage Illustrated Dictionary of the English Language* (1975, p. 1525) the Indo–European root is *kwetwer*, whose zero-grade form is *kwt(w)r*. From this root are derived the Greek *tetra*, Latin *quattuor*, as well as Sanskrit *catur* or *catvāra*, all of which mean 'four'.

10 See note 5.

11 We find Kuiper's suggested etymology for *kāla* (Mayrhofer, 1956–78, vol. 1, pp. 202–3) particularly interesting. He derives the term from the Indo–European root *q^wel- (*kwel-[1], in the *Heritage Dictionary*, p. 1524), which implies movement and, above all, circular movement. That is to say that the notion that time is a cyclical process is implied here. And it is precisely from the reduplicated form of this root that the Greek *kuklos* (circle, wheel), English *wheel* and Sanskrit *cakra* (same meaning), are derived. Nevertheless, Parpola does not agree with this etymology nor with any of the other Indo–European etymologies cited by Mayrhofer, and suggests that *kāla* comes from Dravidian *kāl*, which means leg or foot. The idea is that just as a leg or a foot can represent (by association with four-legged animals) the fourth part of something – and this image is frequently found in India – it can also refer to a season, that is a period of time. The problem here is that this interpretation presupposes a four-fold division of the year into four seasons, something which is incorrect for the Vedic period. And Parpola's explanation (1975–6, p. 371) is not convincing when he writes that, 'even though the Vedic year did not consist of exactly four seasons, the length of one season, be it 1/3 or 1/5 or 1/6 of the year, very nearly agrees with that of 1/4 of the year.' It is true, as he states immediately afterwards (pp. 371–2) that the solstices and equinoxes establish a natural division of the year in four parts, and it is also possible that this division may have been used in Ṛg-Vedic times, but there is no clear evidence of such a division in Vedic literature, and it is particularly difficult to determine the role played by the equinoxes in Vedic times. The matter is however still open to discussion, and Parpola's etymological explanation remains interesting. For other proposals as to the possible Dravidian origin of *kāla* (and *kali*) see Przyluski (1938).

12 In its sense of 'axis', *akṣa* is derived from the Indo–European root *aks- (*Heritage Dictionary*, p. 1505), which is also the root of the Latin *axis*. In its sense of 'die' it could be related to Sanskrit *akṣi*, eye. See Mayrhofer (1956–78, vol. 1, p. 16), where it is suggested, following Pokorny, that this connection stems from the fact that the die 'has eyes' in reference to the markings indicating the numerical value of each side of the die. *Akṣi*, in turn, comes from the Indo–European root *okw-, to see, (p. 1531 of the *Heritage Dictionary*), which gives rise to Latin *oculus*, eye. Both these origins for *akṣa* are accepted by Pokorny (1949–69, pp. 6, 775–6).

13 This is Sāyaṇa's (14th century) interpretation of the meaning of Savitṛ (Monier-Williams, 1899, p. 1190) and it is accepted by O'Flaherty (1981, p. 339). Savitṛ

means he who impels, stimulates or vivifies, and it is associated with the idea of setting in movement. The Sanskrit root is *sū*, which carries the same meaning.

14 This concept, by the way, is in sharp contrast to Einstein's well known phrase stating that he did not believe God played dice with the universe.

15 In the *Nirṇayasindhu* of Kamalākara (17th century), Skanda relates how Nārada traveled to the sacred mountain Kailasa in the Himalayas, where he found the god Śiva playing dice with his consort Pārvatī. Nārada describes their game by saying that the whole universe is the ground on which the game takes place. He compares the 12 months of the year, the days of the lunar month, and the halves of the year (marked by the solstices or the equinoxes) with different elements of the game, and says that the two possible outcomes of a game – winning and losing – are creation and destruction. Thus, the whole world is nothing more than the dice game between Śiva and Pārvatī: when the goddess wins, there is creation; when her husband wins, the world is destroyed. Neither one, however, definitely conquers the other, therefore maintaining the eternal balance of the cycle of creation–destruction–creation. This episode has been translated by Shamasastry (1923–4, pp. 117–18).

16 *Ayā iva pari caranti devā ye asmabhyam dhanadā udbhidaś ca*; *RV* 10.116.9. This is Keith's translation (Keith, 1908, p. 827), which seems the most adequate. The second part of this passage is translated differently by Geldner (1951–7, vol. 3, p. 342), who, however, agrees with Keith in the general meaning of the verse. The difference in translation is due to a disagreement over the meaning of *udbhidas* (see Keith, 1908).

17 'With the greatest wisdom, who can ward off fate [*daiva*]? No one steps beyond the path the Ordainer has ordained. All this is rooted in Time, to be or not to be, to be happy or not to be happy.' *Mbh* 1.1.186–7, as translated by van Buitenen (1973–8, vol. 1, p. 30.)

18 From the root *div*, meaning brightness, come the words *deva*, god; *divā*, 'during the day'; *dyaus*, the sky; all of which are obviously associated with light and brightness. Some interesting cognates are: Greek Zeus, Germanic Tīwaz (the sky god), as well as Latin *deus*, god, and *diēs*, day.

19 This is a characteristic of Indo–European, from which Sanskrit is derived.

20 *ŚB* 5.4.4.6, taken from *VS* 10.28; my translation. The Sanskrit word I have translated as 'dominant' is *abhibhū*. For Eggeling's translation see Eggeling

(1882–1900, vol. 3, p. 107), where he also translates *TB* 1.7.10 in which the same idea is expressed.

21 See Fleet (1888, p. 13, n. 5, 6) (translations), where he points to *MDhŚ* (2.22), according to which Āryāvarta goes from the Himalayas to the Vindhya range (next to which flows the Narmadā). Fleet is of the opinion that the division is more accurately expressed by the poet Rājaśekhara (9th or 10th century), who speaks of the Narmadā as the dividing line between Āryāvarta and Dakṣiṇāpatha.

22 As in *MDhŚ* (1.83). It is common for these durations to be given in hundreds of years as in this case; but in *Mbh* (6.11.5–7) (see translation in Roy, n.d., vol. 5, pp. 25–6) they are given in thousands, so that human life lasts 4000 years in Kṛta, 3000 in Tretā, and 2000 in Dvāpara. In the degraded Kali Age (here called Puṣya) a fixed duration cannot be set as men can die even when still in the womb or shortly after birth. We find a further variation in the *Yuga Purāṇa*, which states that life lasts 100 000 years in Kṛta, 10 000 in Tretā, and 1000 in Dvāpara. Although not expressly stated, it is understood that during Kali life lasts 100 years (*YP*, 15, 42, 47–8; text and translation in Mitchiner, 1976, pp. 918–19).

23 It is interesting to note that in one of the late Vedic Brāhmaṇas, the *Ṣaḍviṃśa*, we find the names Kṛta, Khārvā, Dvāpara and Puṣya (later used as a synonym for Kali). The text makes use of them for referring to the new moon (Kṛta) and the full moon (Khārvā) as well as to the day before the new moon (Dvāpara) and the day before the full moon (Puṣya). No clear connection, however, can be found between this use of the terms and their use as the names of the four Yugas. The verse is *ṢB* (5.6), according to MacDonell and Keith, (1912, vol. 2, p. 193); according to C. Dimmitt Church ('The Yuga story: a myth of the four ages of the world as found in the Purāṇas'. PhD thesis, Syracuse University, 1970, pp. 80, 135) it is *ṢB* (4.6.5).

Abbreviations

AV:	*Atharva Veda*
Mbh:	*Mahābhārata*
MDhŚ:	*Mānava Dharma Śāstra*
RV:	*Ṛg Veda*
ŚB:	*Śatapatha Brāhmaṇa*
ṢB:	*Ṣaḍviṃśa Brāhmaṇa*
TB:	*Taittirīya Brāhmaṇa*
TS:	*Taittirīya Saṃhitā*
VS:	*Vājasaneyi Saṃhitā*
YP:	*Yuga Purāṇa*

References

Bühler, G. (1886), trans. *The Laws of Manu* [*Mānava Dharma Śāstra*]. Sacred Books of the East, vol. 25. Oxford University Press. Reprinted in 1982 by Motilal Banarsidass, Delhi.

Buitenen, J. A. B. van (1973–8), trans. *The Mahābhārata*, 3 vols. Vol. 1: book 1; vol. 2: books 2–3; vol. 3: books 4–5. University of Chicago Press.

Caland, W., ed. (1926). *The Śatapatha Brāhmaṇa in the Kāṇvīya Recension*. Revised by Raghu Vira. Panjab Sanskrit Series 10. Lahore (in Sanskrit). Reprinted in 1983 by Motilal Banarsidass, Delhi (in Sanskrit).

Dumont, P-E. (1963). The human sacrifice in the Taittirīya Brāhmaṇa. *Proceedings of the American Philosophical Society* **107**, 177–82.

Eggeling, J. (1882–1900), trans. *The Śatapatha Brāhmaṇa*. Sacred Books of the East, vols. 12, 26, 41, 43, 44, 5 vols. Oxford: Clarendon Press. Reprinted in 1972 by Motilal Banarsidass, Delhi.

Fleet, J. F. (1888) *Inscriptions of the Early Gupta Kings and their Successors*. Corpus Inscriptionum Indicarum, vol. 3. Varanasi: Indological Book House. Reprinted in 1963.

Geldner, K. F. (1951–7), trans. *Der Rig Veda*. Harvard Oriental Series, vols. 33–6. Harvard University Press.

Griffith, R. T. H. (n.d.), trans. *The Hymns of the Atharva-Veda*, 2 vols. Reprint (3rd. edn.) edited by M. L. Abhimanvu, 1962. Varanasi: Kherali Lal & Sons.

Griffith, R. T. H. (1889). *The Hymns of the Ṛg Veda*. Reprint (rev. edn.) edited by J. L. Shastri, 1973. Delhi: Motilal Banarsidass.

Heritage Illustrated Dictionary of the English Language. (1975). Appendix: Indo–European Roots, pp. 1505–50. Boston: American Heritage Publishing, and Houghton Mifflin Co.

Kane, P. V. (1958). *History of Dharmaśāstra*, vol. 5, pt. 1. Government Oriental Series, class B, no. 6. Poona: Bhandarkar Oriental Research Institute.

Keith, A. B. (1908). The game of dice. *Journal of the Royal Asiatic Society*, pp. 823–8.

Keith, A. B. (1914), trans. *The Veda of the Black Yajus School entitled Taittirīya Sanhita*. Harvard Oriental Series, vols. 18–19. Harvard University Press. Reprinted in 1967 by Motilal Banarsidass, Delhi.

MacDonell, A. A. and Keith, A. B. (1912). *Vedic Index of Names and Subjects*, 2 vols. London: John Murray. Reprinted in 1982 by Motilal Banarsidass, Delhi.

The Mahābhārata. (1933–59). Critical edition, eds. Vishnu S. Sukthankar and S. K. Belvalkar. Poona: Bhandarkar Oriental Research Institute. (In Sanskrit.)

Mayrhofer, M. (1956–78) *Kurzgefasstes Etymologisches Wörterbuch des Altindischen*, 4 vols. Heidelberg: Carl Winter, Universitätsverlag.

Mitchiner, J. (1976), trans. The Yuga Purāṇa. In *Indo–Greek and Indo–Scythian Coinage*, ed. M. Mitchiner, vol. 9, appendix 6, pp. 918–24. London: Hawkins Publications.

Monier-Williams, M. (1899). *A Sanskrit–English Dictionary*. Oxford University Press. Reprinted in 1976 by Motilal Banarsidass, Delhi.

O'Flaherty, W. D. (1981), trans. *The Rig Veda, an Anthology*. Penguin Classics. Middlesex: Penguin.

Parpola, A. (1975–6). Sanskrit *Kāla* 'Time', Dravidian *Kāl* 'leg', and the mythical cow of the Four Yugas. *Indologica Taurinensia* **3–4**, 361–78.

Pokorny, J. (1949–69). *Indogermanisches Etymologisches Wörterbuch*, 18 vols. Bern: Francke Verlag.

Przyluski, J. (1938). From the Great Goddess to Kāla. *Indian Historical Quarterly* **14**, 267–74.

Renou, L. (1959). *Études Védiques et Pāṇinéennes*, vol. 5. Publications de l'Institut de Civilisation Indienne, fascicule 9. Paris: E. de Broccard.

Roy, P. C. (n.d.), trans. *The Mahābhārata*, 12 vols. Calcutta: Oriental Publishing Co.

Shamasastry, R. (1923–4). Dice-play on the first day of the white half of the month, Kartika. *Quarterly Journal of the Mythic Society* **14**, 117–19.

15

The origin of the Chinese lunar lodge system

David S. Nivison   *Stanford University*

Many early calendar astronomies are what we would call 'zodiacs', of 27 or 28 spaces, conceived as successive 'houses' or 'lodges' of the moon as it circles the heavens each sidereal month (of approximately $27\frac{1}{3}$ days). In China there have been two such systems of 28 'lunar lodges' (*xiu*), with variations. Only one of these is described by Needham (1959, pp. 231, 234–7); it is the standard lunar lodge system of the past 2000 years, and is ascribed to an astrologer named Shi Shen, of the late fourth century BC. Until very recently, it has been the only one to be carefully studied by scholars (Xi, 1981). But as an eighth century AD encyclopedia of portents, the *Kaiyuan zhanjing* (ch. 60–3), makes clear, Shi Shen's system (if it is his) is actually the later of the two. The other is described, almost completely, in notes in that encyclopedia, quoting a lost first century BC book by Liu Xiang. Liu's system differs from Shi Shen's only in the widths of the individual lodges, and in the fact that, when the two systems are mapped, the beginning points – the first point of the lodge Jiao – are a few degrees apart in longitude.

Liu's lodge system, called the 'old degree' (*gudu*) system in the *Kaiyuan zhanjing*, has now been validated by an object found in 1977 in the tomb of Xiahou Zao of the Western Han Dynasty, who was buried in 165 BC (see Figure 15.1). This object is a pair of discs that probably had some kind of use in divination; one of the discs has the names of the lodges on the rim, spaced approximately in proportion to their equatorial extensions, and the extension of each is given beside the name, in Chinese degrees (*du*, 365 to the circle). The system

of extensions on this disc is nearly the same as Liu's 'old degree' system. A recent article by Pan Nai (1979, p. 159) uses the disc and the information from the *Kaiyuan zhanjing* to establish the exact widths in *du* for all the lodges in Liu's system.

I accept these figures of Pan. But I reject a basic assumption that he makes elsewhere in his study. Each of the two systems has lodges of very unequal width. In the Shi Shen system, the extensions range from 2 *du* to 33 *du*; in the old system, the smallest lodge is 5 *du*, the largest 29 *du*. The later system is a drastic reform of the earlier one. Each system has the same sequence of 28 lodge names, each name being the name of an asterism located wholly or partly within the longitudinal boundaries of that lodge. In the later system, each boundary is marked by a 'determinative star', which is a star in the asterism and at or near the western edge of it, i.e. at the imaginary line of longitude that the moon (or the sun) crosses as it enters the longitudinal space of that asterism. There is thus a ready explanation for the striking inequalities in the widths of the lodges: the asterisms associated with each are unequal in width, and are more or less randomly distributed around the zodiac.

Pan assumes that we can explain the unequal widths of the lodges in the old system in the same way: all we need to do is find the right determinative stars – different ones, of course – that must have marked the old system boundaries. But his attempt to work this out (Pan, 1979, pp. 160–5) forces him, frequently, to pick stars that are implausibly located. In nine cases, they are not at the western (right)

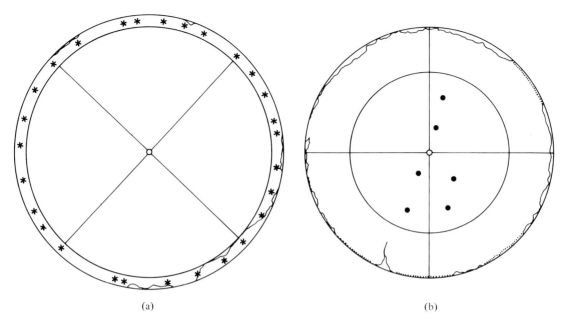

Figure 15.1 The Xiahou Zao Discs.[1] (a) Lower disc with lodges around rim. (b) Dipper handle forms 15° angle with vertical.

edge of the asterism (as one looks south), but at or near the eastern (left) edge.

And, in one case, Pan selects a star that is not in the given asterism at all, but is outside and to the left of it. This is his choice of a determinative star for lodge #21, Shen. (Figures 15.2a–c shows the problem, and also illustrates the concept of determinative star; Pan Nai's and Shi Shen's solutions are given, together with the one I will argue for.) The name Shen, cognate with *san*, 'three', means 'triad', and refers to the 'Belt' of Orion. Pan takes it to be the whole of the asterism Orion, and chooses Alpha Orionis, at the end of the left upper arm. Perhaps here he follows Sima Qian (see MH, 3.352), who takes Shen to be the whole figure, seen as a tiger, with Zi as its mouth; but if Shen is distinguished from Zi, as in the lodge system, this interpretation cannot be right. I conclude that Pan Nai is wrong in supposing that the old system's boundaries were determined by 'determinative stars'. (I add that his working out the correct lodge widths in the old system, which I do accept, is in no way dependent on his belief in determinative stars.)

Thus, as I see it, Pan Nai and Shi Shen made complementary mistakes: Pan looked at the old system's lodge widths, figured out about where (he thought) their boundaries would have to be in the zodiac, and tried to find stars to match, twisting the concept of a determinative star in order to do this. Shi Shen, 23 centuries earlier, likewise just assumed that the boundaries must be marked by stars, and, failing to find proper determinative stars where the boundaries were supposed to be, he moved the boundaries and altered the widths, not hesitating to reduce one already small lodge – #20, Zi, 6 *du* – to a mere 2 *du*. Each man let the determinative star concept lead him into absurdity. Rejecting that concept, my problem now is to find the true explanation for the unequal widths of old system lodges.

If the boundaries were not determined by where certain stars happened to be, I think that there is only one other reasonable possibility: somehow, the boundaries were fixed according to a concept of dividing the zodiac mathematically, most likely into parts that were equal in width; and, since the old system lodges are usually unequal in width, such a division into equal parts must for some reason have gotten distorted with the passing of time. So I have two problems: what was the principle, or principles, of division? And what was the cause, or causes, of distortion?

As to the first problem, is there any evidence for what we should expect early astronomers to have done? There are several pieces of evidence.

(1) Any division of the solar year is implicitly a division of the sun's annual course through the stars. In Shang oracle inscriptions dating from *c*. 1200 BC, a graph probably identifiable with the graph for *chun*, 'spring', is used, apparently for the half year beginning with what we would call winter. (Some of these '*chun*' inscriptions refer to numbered months, as early as the 11th, the first month of winter, and as late as the fifth – ordinarily the first month of summer, but the last month of spring if the calendar for the year started with the solstice month, as it sometimes did; see Shima, 1977, p. 187). If the solar year was divided into 12 'months' matching the (normal) 12 lunar months in a year, then one way of dividing the solar year, and likewise the circle of stars through which the sun moves, would have been simply to divide it in two, at the sun's locations at the beginning of winter and the beginning of summer. (Sinologists will recall that the term *chun-qiu*, 'spring and autumn', means 'year', in old Chinese.)

(2) This use of the word *chun*, 'spring', need not mean that the early Chinese did not know of four seasons. They could have used 'spring' in two senses, just as we use 'day' in two senses (for the 24-hour period from midnight to midnight, and also for the daylight part of that period). The 'Yao dian' chapter of the *Shang shu* (translated in Legge, 1865) is probably a fourth century BC text, containing older material; in it the legendary emperor Yao instructs his calendar astronomers in their art. Of the spring equinox (e.g.) he says 'the day has reached medium length, and the star [culminating at 6.00 p.m.] is the Bird, and thus you can fix mid-spring' (and similarly for the other season mid-points; Legge, 1865, pp. 19–21). This text has been much discussed (see Needham, 1959, pp. 245–6), especially on the question whether the culminations are observed at a varying 'dusk', or calculated for a fixed time of day and observed when observable (my own view). But in any case the concept is of a solar year, and so of a solar 'zodiac', divided into four seasons, with the solstices and equinoxes in the middles of their respective seasonal fourths.

Of the many early evidences of this standard concept, I will cite one more. The lower disc in the Xiahou Zao pair, with the lodges marked on its rim, also has diameter lines crossing at right angles (see Figure 15.1). The line from top right to bottom left connects the beginning points of Dou (#8), beginning of winter, and Jing (#22), beginning of summer. The other line shows that the concept employed here is the concept of four seasons of equal zodiacal extension. But, interestingly, this line does not cut the circle in the traditional way; indeed it cannot. I will have to return to this.

(3) In various places and ways, the early Chinese divided the solar year and zodiac into 12 equal parts (30 or 31 days or *du*). For example, the 'Yue-ling' chapter of the *Liji* (a Han text) uses this division of the year. The scheme is found also in a '*rishu*' (book of [lucky] days) found in the recently excavated tomb M11 at Shuihudi northwest of Wuchang, dated to 217 BC (see Rao and Zeng, 1982; Kalinowski, 1986, pp. 211–12).

Another very ancient scheme used the 12 'branches' *zi*, *chou*, *yin*, ... (one of the two short cycles that are combined to make the 60-day dating cycle) as names of 12 equal divisions of the horizon counted off clockwise, with due north at the center of *zi*; the handle of the Big Dipper at an ideal 6.00 p.m. pointed north at the winter solstice, and thereafter pointed at each 12th in turn, in successive months. The lunar lodge system was then thought of as mapped onto this system with the (traditional) winter solstice point (in the 11th lodge, Xu) at north, and with two, three and two lodges to each season-space of three 12ths. (See Needham, 1959, pp. 240, 243, 250. A curious result of this 'Dipper–lodge–dial' system is that the quarter of the lodge system that is the sun's abode in autumn is assigned to spring and east; and conversely.) The concept of 12ths, without the 'branches' as names, is used on the hamper lid found in 1978 in the tomb of Marquis Yi of Zeng in Suixian, Hubei Province; the distribution of the lodges into a scheme of 12ths has to be assumed in interpreting the date on the lid, which dates the tomb to 433 BC.

Jing   Shen   Zi   Bi   Mao

(a)

Jing   Shen   Zi   Bi   Mao

(b)

Jing   Shen   Zi   Bi   Mao

(c)

The lodge names circle a representation of the Big Dipper in the center, the whole design being flanked by the celestial Dragon, for spring, and the Tiger, for autumn. (See Figure 15.3; and see Thorp, 1979, p. 31; Wang, Liang and Wang, 1979, pp. 44–5; Harper, 1980–1, p. 49; also Legge, 1865, Prolegomena p. 94 (appendix by John Chalmers).)

(4) The middle Han scholar Liu Xin (son of Liu Xiang), in his *Shi jing* (Canon of generations), incorporated into the *Han shu* by Ban Gu (see *Han shu* 21B, 'Lüli zhi' 37a–39a), describes both a 12-fold equal division of the zodiac into named stations of the planet Jupiter, and also a further division into named 24ths. This scheme survives to this day in popular use as a division of the solar year, six 24ths to a season, solstice or equinox in the middle of its season (see Needham, 1959, p. 405). Modern versions are adjusted to respect the actual inequality of the seasons; in Liu's account of it, however, the seasons are assumed to be equal, each 24th being a division of the zodiac, defined precisely in terms of the lunar lodges in Shi Shen's revised system. Liu appears to be simply recording a common concept, which is probably very old. Inscriptional evidence (below) appears to indicate that a system of equal 24ths was already in use in the 11th century BC, more than 1000 years before Liu Xin.

(5) The fact that there are 28 lodges, in a zodiac

Figure 15.2 The Zi–Shen problem. (a) Pan Nai's solution. The diagram shows lodges #18 through #22, with asterisms:

#18 Mao (Pleiades) – meaning uncertain; 15 *du*; Pan: 13.37 *du* (13.19°);
#19 Bi (Hyades) – 'Bird Net'; 15 *du*; Pan: 14.85 *du* (14.65°);
#20 Zi (Lambda, Phi 1, Phi 2 Orionis) – 'Mouth'; 6 *du*; Pan: 6.34 *du* (6.25°);
#21 Shen (Delta, Epsilon, Zeta Orionis) – 'Triad'; 9 *du*; Pan: 8.11 *du* (8.00°);
#22 Jing (part of Gemini) – 'Well'; 29 *du*; Pan: 27.6 *du* (27.22°).

Pan's determinative stars are circled. He assumes an equator for 500 BC, represented here by the rising curve. (The horizontal line is the present equator.) Notice that Pan picks the last star in Bi (Alpha Tauri) as determinative, with the result that his Bi lodge contains none of the Bi asterism except this star. In Jing, he moves from the first star (Mu Geminorum – Shi Shen's choice) to a star (Gamma Geminorum) in the middle. (And obviously, for Pan the exact number of *du* in a lodge has no significance.) His most unfortunate choice, however, is Alpha Orionis as determinative for Shen. He appears to assume that Shen is simply Orion, as usually understood, and as drawn here. Sima Qian (in *Shiji*, 'Tian guan shu') does use the name Shen for the whole of Orion together with Zi, conceiving it as a tiger with Zi as the tiger's mouth. But Shen as distinct from Zi has to be just the 'Triad' – the literal meaning of the name. Thus Pan's 'Shen' lodge does not contain a single star of the Shen asterism.

(b) Shi Shen's solution. As in (a), determinative stars are circled. Shi Shen means them to mark boundaries; but he is obviously having trouble with Zi. The rising curve represents the equator as of 300 BC. Shi's widths for these lodges are as follows: Mao, 11 *du*; Bi, 16 *du*; Zi, 2 *du*; Shen, 9 *du*; Jing, 33 *du*. Shi Shen's problem, to reconcile the determinative star principle with the received system of lodges, confronts him here with its most intractable case. He knew that 'Shen' means 'Triad'. If Zi were to be allowed 6 *du*, it was obvious that the right boundary of the lodge would have nothing near it as a 'determinative'. So he picked a determinative star in the tiny Zi asterism, and assigned the lodge 2 *du* (perhaps thinking that '1 *du*' would be just too ridiculous – but the *Kaiyuan zhanjing* does in fact give it only 1 *du*). The concept of a 'lodge' has lost all meaning. Shi's choice of Epsilon Tauri as determinative for Bi needs comment. The first star ought to be Lambda Tauri, at the end of the handle of the 'Bird Net'. The *Kaiyuan zhanjing* (62.6a), however, identifies the determinative star as 'the first star of the left leg'. So another way to see Bi is as a pair of legs, with a stick torso, viewed from above left. From this perspective, Epsilon Tauri is the first star.

(c) Recommended solution. There are no determinative stars here, beyond the requirement that the Zi–Shen boundary passes between the Zi and Shen asterisms. The rising curve is the equator of 1700 BC; the dotted curve is the ecliptic. The boundary between lodges #17 and 18 (Wei and Mao) is set at the spring equinox point, and other boundaries are computed accordingly. (If an equator for 1750 BC is chosen, all boundaries will be shifted slightly left, so that the Mao–Bi boundary would be about at Gamma Tauri.) The longer boundary line (Shen–Jing) is the division between spring and summer. The dotted boundary is an early spring–summer dividing line, shifted 16 *du* to the right after about 1120 years of precession, so that the lodge Jing becomes abnormally large. The terminal bound of Jing is a residue of this earlier system; the space from it to the dotted line is 13 *du*, i.e. one-28th of the zodiac. The Zi–Shen boundary was not set until after 1000 BC, however. And the same must be said of the left boundary of Gui, the small 5-*du* lodge next after Jing; for at an earlier date it would not have passed between Gui (#23) and Liu (#24). Various other constraints (e.g. that Xing continues to begin at Alpha Hydrae and Xin at Alpha Scorpii) combine with the Zi–Shen and Gui–Liu problems to push the date when the Liu Xiang system was complete down into Western Zhou – my estimate being about 800 BC.

Figure 15.3 The Zeng Hou Yi hamper lid. The hamper is 82.8 cm in length, 44.8 cm high. The design, in lacquer, is red on a black base. The lodge names, circling the Dipper-graph, begin after the gap at the top, reading clockwise. There is a day-date below the second name (Kang). The handle portion of the Dipper-graph points at the fifth lodge name, Xin, approximately as a line drawn across the top of the Dipper and along the handle would point in the sky. (Tomb date = 433 BC.) Photograph (reduced) from Xi (1984). (Courtesy of the Center for Archaeoastronomy).

circle in Liu Xiang of 365 *du*, suggests that the lodges ought to be about equal in extension, 27 of them 13 *du* and one of them 14 *du*. That they were in fact originally conceived as equal, i.e. each of them corresponding to one day of lunar movement, is confirmed by the '*rishu*' from Shuihudi tomb M11; for the calendar system in this text uses the 28 lodge names as a 28-day cycle. And the '*rishu*' usage enables us to see that the 'Yue-ling' chapter of the *Liji* does the same thing, implicitly (see again Rao and Zeng, 1982, pl. 36–7; Kalinowski, 1986).

Thus we have evidence that a regular partitioning of the zodiac, at some early time, would have divided it into some or all of the following: halves (winter–spring and summer–autumn), seasonal fourths (the midpoint of each fourth being a solstice or equinox point), 12ths (tropical month spaces), 24ths (six to a season space), and 28ths (seven to a season space). And some of this evidence is centuries earlier than Shi Shen or Liu Xiang. The whole system is shown, together with Liu's, in Figure 15.4.

Now let us look at Liu Xiang's system (Figure 15.4, second circle in): here, as in the standard system, the seven lodges beginning (counter-clockwise, as the sun and moon move) with lodge #8, Dou, are assigned to winter, and the seven beginning with lodge #22, Jing, are assigned to summer.

Notice that in Liu's 'old degree' system the beginning points of Dou and Jing divide the circle as closely as possible in whole numbers of *du*, 183 *du* to summer–autumn and 182 *du* to winter–spring; so the upper right to lower left diameter on the Xiahou disc is geometrically correct. Further, this division of the year in two is exactly as should be expected for a Shang Dynasty system (middle 16th to middle 11th centuries BC).

Given these beginning points for winter and summer, the winter–spring break ought to be in the middle of Bi, lodge #14; we find no residual boundary there (see below). But the summer–autumn boundary ought to be exactly at the beginning of Zhen, lodge #28 – since Liu's first six summer lodges total 91 *du*, and Zhen plus his seven autumn lodges total 92 *du*. So three season boundaries out of four survive, though one, autumn (as the sun moves), is unrecognised in Liu's system, which takes Jiao as starting this season; but notice in Figure 15.1 that the upper left to lower right diameter does pick out what must have been the beginning of Zhen (though the rim is damaged at this point).

There appear to be three residual tropical month spaces, of 30 or 31 *du*, which would preserve five boundaries out of 12 in a system of regular 12ths: Xin–Wei–Ji (#5–7); Dou–Niu (#8–9); and Bi–Zi–Shen (#19–21). The Yi–Zhen (#27–8) boundary is a sixth tropical month boundary.

There appear to be four residual 24ths of 15 or 16 *du*: Mao (#18), Bi (#19), Zi–Shen (#20–1), and Zhen (#28); but, if we mark off ideal 24ths against Liu's boundaries, nine of his boundaries are 24th boundaries.

Finally, Xing, Zhang and Yi (#25, #26 and #27), each being 13 *du*, seem to be residual 28ths, since the end point of Yi is the residual summer–autumn season boundary.

Since evidence indicates that a regular system divided in these ways probably existed at an early date, let me tentatively suppose that these features of Liu's system are in fact residues of that system. Why should the early Chinese have wanted such a system? And why does it survive in Liu only in scattered residues?

The divisions into halves, fourths, 12ths and 24ths are simply finer and finer subdividings, and there is an obvious possible motive for this: a solar 12th is about $30\frac{1}{2}$ days; a synodic lunar month is

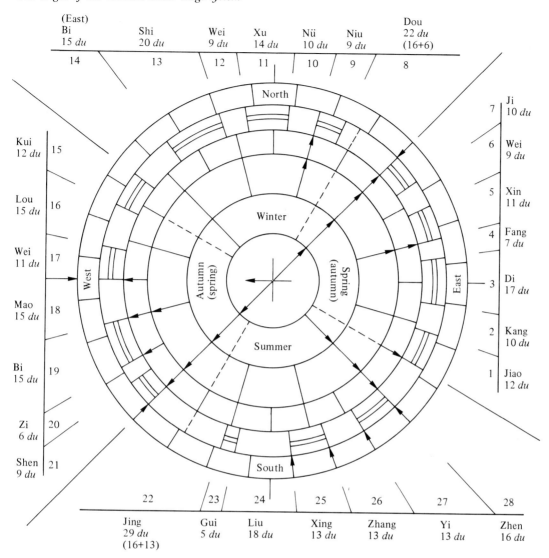

Figure 15.4 Liu's lodges, and the regular system behind it.[2] The Big Dipper's handle viewed at 6.00 p.m. was seen as pointing north in midwinter, east in midspring, etc., moving clockwise around the lodge system interpreted as a partitioning of the horizon, with Xu (winter solstice lodge) at north. The sun, however, moves around the circle of lodges in the sky counter-clockwise; hence the spring–autumn ambiguity.

Liu Xiang's system is represented by the second circle in, of spaces of varying width (odd numbered spaces are banded). Arrows indicate where Liu's boundaries are residues of structure in the antecedent regular system of half years (center circle), seasonal quarters (second circle out), tropical month 12ths (third circle out), 24ths (fourth circle out), and 28ths (outer circle). The cross at center represents the solstices (vertical) and equinoxes (horizontal).

The results are as follows:

solstices and equinoxes	1 of 4 (spring equinox), 25%;
half-year boundaries	2 of 2, 100%;
season boundaries	3 of 4, 75%
tropical month boundaries	6 of 12, 50%;
24th boundaries	9 of 24, 37.5%
28th boundaries	6 of 28, 21.4%

12 of 28 (42.9%) preserve structure in this underlying system.

In the scale of this diagram, differences in division of 1 *du* are disregarded – e.g. one-24th is a uniform width, whether it be 15 or 16 *du*. Broken lines represent the original season boundaries, before the clockwise shift of one-24th that created the abnormally large lodges Jing and Dou, at the beginnings of the two half years. All 24th boundaries are residues of this original system, including notably the beginning point of the whole system – the beginning of Jiao – and the original equinoxes – the Fang–Xin and Mao–Bi boundaries.

209

about $29\frac{1}{2}$ days. Once a day in the year is identified – e.g. 15 days before the winter solstice day, in some year when this day was the first of a lunar month – then by simple counting (and occasional checking) one can tell when an extra lunar month must be added, in order to make each lunar month contain a solar month midpoint (*qi*-center), as later calendar theory required (see Kong Yingda, commentary to *Zuo zhuan*, Xiang 27.11). This system would have enabled the early Chinese to add the intercalary month at any point in the year.

But was there in-year intercalation in very early times? This matter is debated. A set of 77 short oracle inscriptions recording a Shang royal campaign to the east in 1077–6 BC (see Chen, 1956, pp. 301–4; Nivison, 1983, pp. 501–2) seems to me to prove that in-year intercalation was being used. Each inscription bears a partial date. Together they appear to require a ninth lunar 'month' in 1077 running two lunar cycles; and, as it happens, the ninth lunar month of 1077 would have had to be extended by another cycle, to contain (as it should) the actual autumn equinox. But the actual equinox fell on 2 October in 1077, whereas the nominal equinox (*qiufen*) in Liu Xin's system would have been 29 September, if we count from the winter solstice for that year. As it happens, the first inscription in the set, inaugurating the campaign, is dated 29 September; so it would appear that already in the 11th century BC there was a system of 24ths, and that important first days in the system were lucky days. More evidence is found in the dates of events at the time of the Zhou Conquest in 1046–5 BC (see Nivison, 1983, p. 566): the Zhou campaign against Shang began just 15 days before the solstice, and the conquest battle was fought on the 16th day after the solstice. A system of 24ths, devised as the basis of the calendar but later used to select lucky days, would explain this.

But why divide the zodiac into 28ths? Whatever the reason, as the system was used by calendar astronomers the 28-fold division was apparently nonfunctional, for it eventually disappeared, and must now be reconstructed inferentially. My own assumption is that it had to be the first way a zodiac was conceived, because the first thing that early astronomers had to do, if they were to keep track of the sun's position in the zodiac, was to have a map accurate enough to serve as a basis for deciding where the sun was. We know how they did it, at

least by late antiquity: they observed meridian crossings at dusk and at dawn and deduced the sun's position as half-way between the two crossing points; see Needham (1959, p. 247). With the aid of a zodiac map, use of this method would also have enabled them to decide (more or less accurately) what celestial points must be crossing the meridian at midnight, and so also to decide where the Dipper's Handle must be pointing at 6.00 p.m., without using a clock, even if the stars were unobservable at that time.

To make such a map, the natural procedure would have been to notice the length of the sidereal month, and then to use the moon each night as a mapping tool, so to speak. The result, after some years of patient observation, would be a zodiac of 27 or 28 lunar 'lodges' of equal width (and in China it happened to be 28), with prominent stars marked in. Such a map could be used in decision procedures, even if the map, and observation techniques, were quite crude, so long as the concepts were clear. The lodges may have gotten their names only after this was done.

Liu Xiang's old degree system, of 28 lodges of mostly irregular width, could not have had a calendar use; I assume that it was a popular system used by astrologers. But if it derived from the regular system I have reconstructed, how did that system get distorted so much that only careful analysis reveals what Liu's system came from? Part of the answer is that a lodge system was supposed to have seven lodges to a season space; but some of the lodges come from a system of 28ths, seven to a season, and some from a system of 24ths, six to a season; these two kinds had to be unequal in width, and accommodating both must have forced distortions elsewhere. The requirement that each season have seven lodges, grouped 2–3–2, forced lodge-status on some spaces that were abnormally small or large. (Thus Shi Shen's Zi, although only 2 *du*, had to be a lodge, to make up the count.) And once the source of the old degree system in an earlier regular system had been forgotten, there may have been some attempt, already before Shi Shen, to adjust boundaries to fit observed asterisms. But the two most striking irregularities in Liu cannot be explained in any of these ways.

Working out this problem leads to an extension of my hypothesis, which will need fuller attention than I can give it here. I will sketch the argument, without noting all possible objections. The matter is especially important, because it provides a precise 'mechanical' connection between evidence (in this case certain features of Liu's system) and hypothesis, i.e. that there was at an earlier time a regular system such as I have described, with the uses I ascribe to it. The two features to be explained are the following:

(1) In Liu's system, the beginning point, and the beginning of the autumn season space (as the sun moves), is the beginning of Jiao. In the reconstructed system, the season space has to begin with Zhen.

(2) In Liu's system (and even more strikingly in Shi Shen's reform) the two largest lodges are directly opposite each other, and are very large – twice the average width. These are Dou, 22 *du*, beginning winter, and Jing, 29 *du*, beginning summer. (They are 26 *du* and 33 *du* in Shi Shen.)

The first feature has not been studied before (since I am the first to apply the regular reconstruction that brings it out). The second has been given much attention by 19th-century Western scholars. A favorite theory was that certain lodge boundaries were keyed to meridian crossings of certain prominent circumpolar stars (Needham, 1959, p. 233), and that there happened not to be any circumpolars to justify boundaries that would have created smaller lodges where Dou and Jing are located. This theory was worked out to apply to Shi Shen's system, not Liu's (and it doesn't work anyway), so I shall ignore it.

My own view is that one explanation accounts for both of these anomalous features. At some time, a new regular system was adopted, superseding a still older system, and the new one moved all boundaries east or west exactly 16 *du* – the width of the Zhen lodge; but only some, not all, of these changes were taken up in the evolving popular system that became Liu Xiang's. Did the shift move boundaries east, so that the beginning point moved from Zhen to Jiao, with the beginning points of Jing and Dou remaining untouched in the popular system? Or did it move boundaries west, with the popular system picking up the new beginnings of Jing and Dou, but sticking to the old beginning for the whole series of lodges?

If the shift were east, so that an original beginning at the beginning of Zhen was shifted 16 *du* to what is in Liu the beginning of Jiao, while the beginning

points of Dou and Jing remained unchanged, then the abnormally large size of Dou and Jing would remain unexplained. So what must have happened was a reform, by calendar astronomers, that shifted the boundaries west 16 *du*. This moved the beginning points of Dou and Jing 16 *du* west, enlarging those lodges by 16 *du* in the popular system; but the popular system resisted a change in the beginning point of the whole system, which in popular astrology had been and remained the beginning of Jiao. The result was that in Liu's (the popular) system the autumn space was reduced by 16 *du*, probably from an original 92 *du* to 76 *du*, and the summer space was enlarged by as much, from 91 *du* to 107 *du*.

This hypothesis can immediately be confirmed. Let us suppose that the original beginning point was the beginning of Jiao; the proper beginning points for winter and summer would then have been as shown by the broken lines in Figure 15.4, 16 *du* east of the later boundaries. This shows (see also Figure 15.2c) that Jing, or whatever it was called – 13 + 16 = 29 *du* – was originally just the eastern 13 *du* of Liu's Jing. This phantom 13 *du* space must be a residue of an earlier regular system; for, if we point off 13-*du* 28ths from its boundaries, we get as one of these 28ths lodge #11, Xu – of 14 *du*, not 13, because it was the most important: traditionally, and as implied by the 'Yao dian', it was supposed to contain the winter solstice point. (There has to be exactly one lodge of 14 *du*: 28 × 13 = 364, not 365.)

Analysis of what happens to Dou (22 *du*) is further confirmation. In Dou (see Figure 15.4) what remains when the western 16 *du* are deleted is a 6-*du* space. Combined with the next lodge Niu, 9 *du*, it makes a 15-*du* 24th – just as do another 6 *du* plus 9 *du* combination, Zi and Shen, on the other side of the zodiac. So the division of a 24th into the small lodges Zi and Shen can now be explained: this resulted from a further shift, in another reform, of all boundaries another 9 *du* west. Two important ones were kept in the evolving popular system: the new spring–summer boundary, Zi–Shen, was one. (Though in the end this boundary could not be recognized as the summer boundary; the requirement that there be seven lodges to a season prevented this, and also forced lodge status on Zi, in spite of its small size and its implausibly small asterism at its eastern edge.) And in a system that had

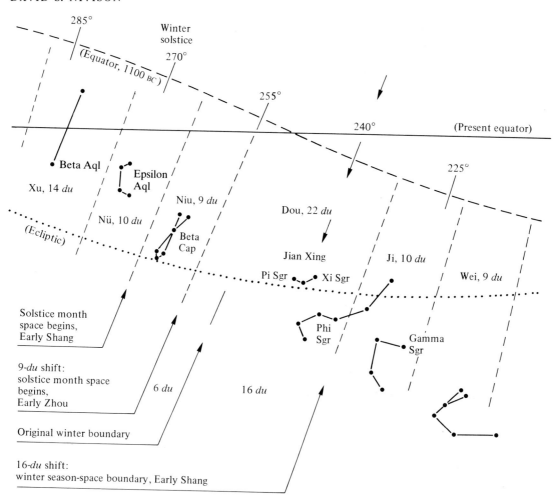

Figure 15.5 The Dou–Niu boundary and Jian Xing: the Early Zhou reform. The midpoint of the first solar month space of winter has to be two-24ths of the zodiac west of the winter solstice point. The Dou–Niu boundary in the old system is halfway between the probable Early Zhou locations of these two points; so it was probably set to fix the Early Zhou beginning point of the solstice month space. The name of the small asterism Jian Xing (the 'establishing stars', Xi–Pi Sgr) tends to confirm this. The midpoint of the first solar month space of winter was,

as the sun moves, approximately at the beginning of this asterism (marked with arrows); thus, when the sun had reached Jian Xing, one could know that the next lunar month must contain the solstice. In this way, the calendar experts could 'establish the *zi* month' (*jian zi yue*) as the first month of the official Zhou calendar (see MH 3.326, 320). The word *jian* is a technical term for the calendarist. For the identity of Jian Xing, see MH 3.355.

Zi–Shen as summer boundary, Dou–Niu was the beginning point of the winter solstice solar month space, the most important 12th in the zodiac (see Figures 15.5 and 15.2c).

But why this shifting in repeated reforms? Always, the shift is west. A regular system must be keyed precisely to the solstices and equinoxes, which must be the midpoints of their respective season spaces. West is the direction of the precession

of the equinoxes, at the rate of about one Chinese *du* every 70 years. The Chinese of these early times had not yet discovered precession; they were merely aware that, after many centuries, what they had thought was a secure basis for their calendar no longer worked, and had to be reformed.

Let me suppose that my readers are satisfied that a regular system once existed that leaves the most traces in Liu's old degree system, with the beginnings of Dou and Jing as half-year boundaries. As soon as we see this regular system as perhaps having been itself reformed, and as having been itself a reform of something still earlier, it then becomes reasonable to ask whether there may have been more reforms than the ones I have yet pointed out. Two matters need attention:

(1) Although Xu is part of a system that includes the 13-*du* remainder of Jing, Xu cannot have been the winter solstice lodge in the same regular system that had the beginning of Jiao as its beginning point. It is exactly one 13-*du* 28th too far west for that. I conclude that Xu as winter solstice lodge probably belongs to a reform on an archaic original system, that was carried out after precession had accumulated to just one 28th of the zodiac.

(2) Liu's old degree system preserves three out of four of the season boundaries in the regular system that I first reconstructed (Figure 15.4). The winter–spring boundary in Liu's system, between lodge #14, (East) Bi, and #15, Kui, is just 8 *du* too far east to have been the season boundary in my reconstructed system, and just 8 *du* too far west to have been the season boundary in an archaic system starting with the beginning of Jiao. I conclude cautiously that this season boundary is probably the residue of a reform done when precession had built up to approximately one-half of one-24th of the zodiac – enough to throw off astronomers' calculations by about a lunar quarter. The East Bi lodge, of 15 *du*, would be the residue of a 24th in this reform.

I have postponed to the end – not wishing to spend my credibility too rapidly – the related matters of mapping and dating. Once a regular system is mapped on the stars, we can date it from the way it locates the solstices and equinoxes. And since all of my reconstructions have been derived from analysing Liu's system, once that system is mapped, the reconstructions can be mapped too.

Liu's system must be mapped as follows. The asterism Zi is very small, and was very close to Shen (the belt of Orion) in longitude (though not in declination; therefore later precession causes the two asterisms to overlap in longitude). If we slide a boundary line between the two (as of about 1000 to 800 BC) and count *du*, it turns out that a boundary fits between the asterisms of Gui (#23) and Liu

(#24), which are also very close together in longitude (see explanation for Figure 15.2c). More counting shows that the boundary of Xing (#25) must be approximately Alpha Hydrae, the 'Bird Star', and the boundary of Xin (#5) must be approximately Alpha Scorpii, i.e. Antares, the 'Fire Star'. (Interestingly, both these stars are determinatives for these lodges in the Shi Shen system, according to the *Kaiyuan zhanjing*, and the distances in *du* from the Zi–Shen boundary to these two boundaries, 61 and 162 *du*, respectively, are exactly the same in Shi and Liu; these are almost the only coincidences in mapping between the two systems.)

The dates that result are shown in Figure 15.6. The archaic original system had the Fire Star as autumn equinox point – as is implied by the 'Yao dian' – correct for *c.* 2880 BC. The 8-*du* reform on that system had the Bird Star as summer solstice point, and approximately the Pleiades, Mao, as spring equinox point – as is implied by the 'Yao dian' – correct for *c.* 2310 BC. The 13-*du* reform, again reckoning from the archaic 'Fire Star' system, had, approximately, asterism Xu (at Beta Aquarii) as winter solstice point – as is implied by the 'Yao dian' – correct for *c.* 1960 BC. The reform I first reconstructed, made 16 *du* after the archaic system, was correct for *c.* 1750 or 1700 BC.

The surprisingly early dates for these systems must cause one to pause. (I of course must follow my argument where it leads.) Some comfort may be had from a recent discovery, at Dahecun on the outskirts of Zhengzhou in Henan, of some fragments of pottery (see Figure 15.7) that appear to have markings of astronomical significance. Certain pieces, fitting together, allow reconstruction of two pots, each with 12 sun designs equally spaced around the outside (Xi, 1984). Might these represent the concept of a 'zodiac' divided into 12 solar month spaces? Archaeologists argue about it. Carbon dating of the site suggests that the objects were made a little before 3000 BC (Li, 1983).

Since each of the magnitudes of the reforms – 8, 13 and 16 *du*, i.e. one-half of one-24th, one-28th, and one-24th of the zodiac – is theoretically important, I assume that the archaic system was part of the general culture, without political identity (which would be too early for us to know in any case), and that repeated attempts were made to go back to it and straighten it out at politically significant times. Traditionally, it was the task of the founder

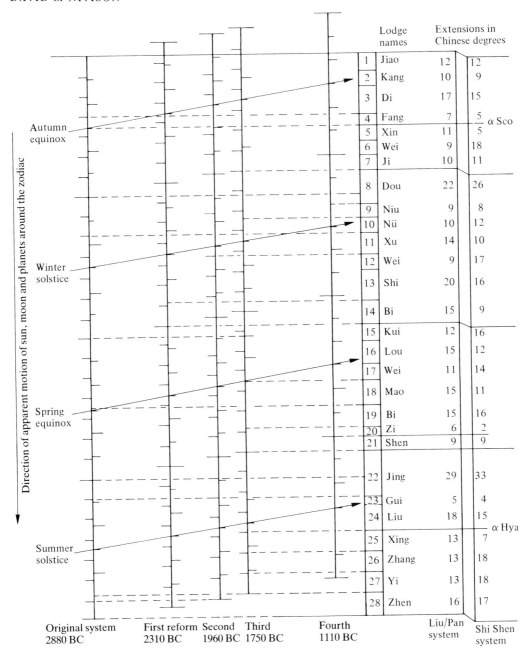

Figure 15.6 Evolution of Chinese lunar lodge systems. (Reconstructed early regular systems, showing probable major residues in Liu Xiang's 'old degree' system; more are possible). The dates at the bottom of the figure are the earliest astronomically possible ones; the actual date of adoption could be much later. The third and fourth reforms were probably made soon after the beginnings of the Shang Dynasty (*de jure* first year 1575 BC) and the Zhou Dynasty (*de jure* first year 1058 BC).

(a)

(b)

Figure 15.7 Pottery fragments from Dahecun, and reconstructions. (a) Two fragments fitted together, apparently with flaring sun designs. The curvature shows that the complete pot had 12 of these, if equally spaced, as shown in the reconstructions, (b). (From Xi, 1984.) Another fragment from the same site (not shown) has a design some take to be a representation of an asterism (see Li, 1983). The site is carbon-14 dated to 3075 BC ± 100 years and to 2550 BC ± 140 years.

of a new political order in the world to 'rectify the calendar'. Since the Chinese did not know what was happening – being still unaware of precession – we can expect the dates when these systems would have been astronomically correct to be earlier than the dates when the reforms were made. Such was the case in the Han 'Tai-chu' (great beginning) reform of 104 BC, which used astronomy that was two-and-a-half centuries out of date (Needham, 1959, p. 247; *Han shu* 21B).

The 16-*du* reform brings us to what we can with some assurance call historical time. The *de jure* first year claimed by Tang, the founder of the Shang Dynasty, must have been 1575. This is the year given as his first year in his local domain by the *Bamboo Annals* (*Zhushu jinian*, translated in Legge, 1865, Prolegomena, p. 125; this book is still supposed by many to be a faked text, but is not as is now proved; see Nivison, 1983; and especially Shaughnessy, 1986). And it is the year after a conjunction of the planets (*Bamboo Annals* distorts the date of the conjunction to 1580; the correct date, late 1576, is established in Pankenier, 1983); it is known that the Zhou founder Wen Wang claimed a conjunction (of 1059) as the sign of the divine transfer of authority to himself, and took the next year as the first year of his royal calendar; the Shang founder probably used the 1576 conjunction in the same way. It would seem, then, that the 16-*du* reform was done in early Shang; and, as we saw, it appears to have bisected the year as Shang calendarists would have done.

I see no theoretical significance in 9 *du*, or in (9 + 16 =) 25 *du*. I assume then that the last reform, correct for *c.* 1100 BC and so probably adopted *c.* 1000 BC or later, in early Zhou, was a reform of the Shang system. Perhaps the archaic Fire Star system had by then been forgotten (though hints of a simpler prehistoric Fire Star calendar of seasonal activities remain in early literature; see Pang, 1978). The one really important effect of this reform on later astrological systems was to create the small lodges Zi and Shen.

One remaining question puzzles me. If my analysis of the old degree system of Liu Xiang is right, why didn't Shi Shen see it, in the fourth century BC? Of course, I knew about the precession of the equinoxes, and he did not. But as I have presented my analysis, I turned to precession for explanation only after I had generated enough detail about earlier systems to require precession as an explanation of them. Shi Shen could have done this reasoning as well as I, if he hadn't been blinded by the concept of determinative stars. If he had just seen that he had the option of discarding the determinative star idea, and of asking himself why, e.g., the old system had a series of 15-*du* spaces (Mao, Bi, Zi–Shen) just before the summer boundary, he might have been the first astronomer in the world to discover precession – two centuries before Hipparchus in the West.

Acknowledgments

In the research and analysis that led to this article, I owe thanks especially to Alvar Ellegard for thought-provoking comments at the Merida Conference, and to a number of people there and elsewhere for advice or fruitful exchange of ideas at various times: Brad Schafer, Don Harper, Jeff Riegel, David Keightley, Wilbur Knorr, Dagfinn Follesdal, Nancy Cartwright, Lincoln Moses, Michel Teboul, Robin Yates, David Pankenier and Pang Pu. Anthony Aveni, David Keightley and Owen Gingerich have patiently given more than one of my manuscripts a critical reading. Nathan Sivin started me thinking seriously about Chinese astronomy in a long letter in 1980. Unpublished papers of Edward Shaughnessy have guided my interpretation of the campaign inscriptions of 1077–6 BC. I am also very grateful to Li Xueqin, Bo Shuren and Yin Lujun for confirming for me that the drawings of the Xiahou Zao discs as published in *Wenwu* (1978, 8, p. 19) are correct, and that the drawings in *Kaogu* (1978, 5, p. 342) are only concept sketches. And I am indebted to Shi Xingbang and to Xi Zezong, for inviting me in 1985 to present earlier versions of this article (in Chinese) at meetings of the Shanxi Institute of Archaeology in Xi'an, and of the Beijing Astronomical Society. I represent none of the people who have assisted me as either agreeing or disagreeing with me; and any mistakes of fact or reasoning in my argument are of course my own responsibility.

Notes

1 The discs were discovered in 1977 in Fuyang, Anhui Province, in the tomb of Xiahou Zao, who died in 165 BC (Pan, 1979, p. 158). They were evidently intended to be used together, one placed on top of the other, mounted on a spindle. They are about 25 cm in diameter, the upper disc being slightly smaller than the lower. They were found under water, and are kept under water, in the Anhui Provincial Museum, to prevent deterioration; therefore they cannot be photographed. Shown above are drawings, orientation modified, from *Wenwu* (1978, 8, 19). (See Cullen, 1980–1, p. 35, for the illustration as it appears in *Wenwu*.) A copy of the pair of discs has been made, and is kept in the institute for the History of Science of the Chinese Academy of Sciences in Beijing. The *Wenwu* drawings appear to conform exactly to this copy, and my analysis assumes that the copy itself is a faithful one. (For a different drawing of the discs, one on top of the other, as published in *Kaogu* (1978, 5, 342), see Harper, 1978–9, p. 5. I judge that the *Kaogu* drawing is only a concept sketch; it lacks details that enter into my arguments.) As copied, the discs are of wood, finished in dark brown lacquer, with white markings.

The original suffered some damage at the edges in the process of excavation in 1977. This damage includes the location and width of Zhen (#28) (lower disc, lower right); however, if we assume that the upper left to lower right diameter marks the beginning of Zhen at lower right, then the beginnings of Liu (#24), Xing (#25), Zhang (#26), Yi (#27, barely visible), and of Jiao (#1) are placed correctly. There is distortion elsewhere, especially at upper left; apparently a geometrically exact model was used by the original maker, who then made his lower disc conform (1) to his idea that the seasons must be equal, and (2) to the standard idea that Kui (#15) must begin the spring season. But if he knew that Jiao (#1) was supposed to begin autumn, he apparently saw that this would force too much distortion if summer and autumn (lower right) were to be equal; therefore he accepted Zhen (#28) as the beginning of autumn, leaving summer with only six lodges and autumn with eight. Except for the distortion, the original maker's season beginnings are the same as the ones in my reconstruction of the Shang equal-space system in Figure 15.4.

Moreover, when the upper disc is oriented as in my illustration, the vertical diameter can be interpreted as picking out the first month of the Shang calendar, namely the month following the winter solstice month (see MH 3.326). The lower disc's diameter lines taken as season boundaries imply an autumn equinox one-24th of the zodiac before the beginning of Xin (#5), i.e. in the middle of Di (#3). The Dipper handle, when conceived as continuing a straight line across the top of the Dipper, as it does in the upper disc, must have been thought really to point at the beginning of Xin (as it does in the Zeng Hou Yi hamper lid in Figure 15.3). The tradition was that the Dipper's handle points north at 6.00 p.m. in midwinter; by the beginning of Shang, however, this was only approximately true; at that time, the situation would have been thought to be as follows. Let the day of the year be when the Dipper handle in the sky is observed to point 15° right of north at 6.00 p.m.; and let the upper disc be oriented accordingly, so that the disc's diameter (rather than the handle) will point due north (straight up), as shown. At this time the sun, due west, must be 30° beyond the solstice, at the center of the solar month space corresponding to the Shang first lunar month. Both of the two discs therefore seem to conform to the regular system constructed in Figure 15.4 as the primary basis of Liu Xiang's 'old degree' system.

(Here I sketch a complex argument only in part; it will require a separate study. The orientations of the two discs are, of course, not fixed in relation to each other: the 6.00 p.m. pointing of the Dipper's handle moves clockwise 30° each solar month. I have shown the lower disc with the *zi*-sector (the winter solstice sector, centered on Xu, #11, as it was traditionally, and as it still was actually in early Shang) at north, i.e. up, conforming to Figure 15.4, and to the traditional horizon-sector interpretation of the zodiac as representing the natural year. But if the upper disc is oriented as shown (making the sun at 6.00 p.m. 90° + 15° from the handle, and so 90° + 30° from the equinox), then the lower disc must be rotated 30° counter-clockwise, so that the *chou*-sector (second solar month sector in the – reversed – horizon interpretation) is at the top, appropriately for a representation of the Shang civil year.)

2 Figure 15.4 (and Figure 15.6 also) are not to be interpreted as reconstructed maps of the zodiac stars. They are reconstructions of concepts: of different ways to divide up 365 *du* in a diagram, and of different sidereal identifications for some key point in the diagram. The hypothesis is that, when a reform occurred, the structure was kept intact, but shifted to fit the currently assumed location of a key point, such as the winter solstice. The reformer might then check his new diagram to see where the familiar stars belonged – prepared for the possibility that the authors of the system he was reforming had made other mistakes as well as mistaking the location of the solstice. Neither he nor they knew that the equator too was changing, and with it the distance (in some small degree) in right ascension between many of the stars he wanted to place in his diagram. Perhaps he made a correction of a star's

location relative to his diagram structure, perhaps not. To the extent that he did, we cannot say, e.g., that the first point of Jiao is through all the reforms a longitude line having exactly the same distance from Alpha Virginis. Also, he might fail to see the need for a correction of a star's location and therefore fail to make it, but still think of the relative distance between the boundary lines on his diagram as remaining the same.

References

Chavannes, E. (1895–1905). *Les Mémoires Historiques de Se-ma Ts'ien*. 5 vols. Paris: Leroux. Reprinted in 1967 by Adrien-Maisonneuve, Paris.

Chen, Mengjia (1956). *Yin-xu buci zongshu. (Compendium on Shang Oracle Inscriptions.)* Beijing Kexue Chubanshe.

Cullen, C. (1980–1). Some further points of the *shih. Early China* 6, 31–46.

Gautama, Siddhartha (Qutan Xida) (*fl.* 713–41). *Tang Kaiyuan zhanjing. (Kaiyuan era (713–41) encyclopedia of portents.)* Reprint (1973) of Wenyuange manuscript *Siku quanshu zhenben* collection #4, vols. 172–81. Taipei: Commercial Press.

Han shu: Wang, Xianqian (1900, date of preface). *Qian Han shu buzhu. (History of the former Han dynasty* (by Ban Gu and others, first century AD), with supplementary notes.) Taibei: Yiwen yinshuguan. Photographic reprint of Wang's original Changsha edition.

Harper, D. J. (1978–9). The Han cosmic board (*shih*). *Early China* 4, 1–10.

Harper, D. J. (1980–1). The Han cosmic board: a response to Christopher Cullen. *Early China* 6, 47–56.

Kalinowski, M. (1986). Les traités de Shuihudi et l'hémérologie chinoise a la fin des royaumes-combattants. *T'oung Pao* LXXII, 175–228.

Kaiyuan zhanjing: see Guatama, Siddhartha.

Kong, Yingda (574–648). *Chunqiu Zuozhuan zhengyi. (Commentary to the Spring and Autumn with Zuo commentary.)* Any edition of *Shisan jing zhushu (the Thirteen Classics with commentaries and subcommentaries)*.

Legge, J. (1865). *The Chinese Classics, vol. III: The Shoo King, or Book of Historical Documents*. London: Henry Frowde. Reprinted in China in 1939.

Li, Changtao (1983). Dahecun xinshiqi shidai caitaoshang de tianwen tuxiang. (Astronomical designs on neolithic painted pottery from Dahecun.) *Wenwu* 8, 52–4.

MH: See Chavannes, E. (1895–1905).

Needham, J. (1959). *Science and Civilization in China, vol. 3: Mathematics and the Sciences of the Heavens and the Earth*, collaborator L. Wang. Cambridge University Press.

Nivison, D. S. (1983). The dates of Western Chou. *Harvard Journal of Asiatic Studies* 43, 481–580.

Pang, Pu (1978). 'Huo li' chu tan. (A preliminary investigation of the 'fire calendar'.) *Shehui kexue zhanxian* **4**, 131–7.

Pankenier, D. W. (1983). Astronomical dates in Shang and Western Zhou. *Early China* **7**, 2–37.

Pan, Nai (1979). Wo guo zaoqi de ershiba xiu guance ji qi shidai kao. (A study of ancient determinants and dates thereof for China's 28 lunar lodges.) *Zhonghua wenshi luncong* **3**, 137–82.

Rao, Zongyi and Zeng, Xiantong (1982). *Yunmeng Qin jian rishu yanjiu. (A Study of the 'Book of Days' in the Qin Dynasty Bamboo Slips from Yunmeng.)* Hong Kong: Chinese University Press.

Shaughnessy, E. L. (1986). On the authenticity of the *Bamboo Annals. Harvard Journal of Asiatic Studies* **46**, 149–80.

Shima, Kunio (1977). *In-kyo bokuji sōrui. (Concordance of Shang Oracle Inscriptions.)* Revised reprint. Tokyo: Kyuko Shoin. Originally published in 1967.

Thorp, R. L. (1979), trans. Brief excavation report of the tomb of Marquis Yi [of] Zeng at Sui Xian, Hubei. Sui Xian Leigudun tomb I archaeological excavation team. (*Wenwu* **7**, 1–24.) *Chinese Studies in Archaeology*, winter 1979/80 1.3, pp. 3–45.

Wang, Jianmin, Liang, Zhu and Wang, Shengli (1979). Zeng hou Yi mu chutu de ershiba xiu qing long bai hu tuxiang. (The diagram of the 28 lunar lodges with the green dragon and the white tiger, excavated from the tomb of Marquis Yi of Zeng.) *Wenwu* **7**, 40–5.

Xi, Zezong (1981). Chinese studies in the history of astronomy, 1949–1979. *Isis* **72**, 456–70.

Xi, Zezong (1984). New archaeoastronomical discoveries in China. *Archaeoastronomy: Journal of the Center for Archaeoastronomy* **7**, (1–4), 34–45.

16

A reflection on the ancient Mesoamerican ethos

Miguel León-Portilla   *Instituto de Investigaciónes Antropologicas de la UNAM*

The organizers of the Second Oxford International Conference on Archaeoastronomy asked me to reflect on those traits that can be described as essential to the old Mesoamerican ethos. I have dared to accept the invitation in full awareness of the complexity of the subject.

To begin, there are two concepts included in the title I have given to this paper that require at least an introductory comment. One is the idea of *ethos*. The other is related to the possibility of applying that concept to the ancient Mesoamericans as if they had lived in basically one universe of spiritual traditions and forms of behavior.

A number of years ago a so-called 'ethos anthropology' developed and exerted various kinds of influence on the disciplines related to the study of culture. While not necessarily adhering to the conceptual implications of the ethos anthropology, I consider it adequate to guide this reflection from the perspective of the concept of *ethos*.

16.1 What do we mean by ethos and how can we apply this concept to the ancient Mesoamericans?

Let us turn to the two best dictionaries of the English language you all so often consult. The *Webster New Twentieth Century Dictionary* (the *Webster*) tells us that ethos is 'the characteristic and distinguishing attitudes, habits, etcetera, of a racial, political, occupational or other group' (*Webster*, 1978, p. 678). Members of the 'Oxford Conference on Archaeoastronomy' will find it necessary also to consult with the classical *Oxford English Dictionary* (the *OED*). Its answer is that ethos is 'the characteristic spirit, prevalent tone of sentiment of a people

or community, the genius of an institution or system' (*OED*, 1979, p. 314).

In sum, it being our aim to reflect on the ancient Mesoamerican ethos, we have to take into account the available sources, archaeological and documentary, to search for what they can offer about the characteristic attitudes, habits, world view, prevalent tone of sentiment, characteristic spirit, forms of behavior, institutions and, employing the *OED's* term, 'the genius' of the ancient Mesoamericans.

In terms of this description of the concept of ethos, we can ask ourselves if the ancient Mesoamericans, since the classic period and more intensely in the postclassic, shared attitudes and key concepts so basically similar that their world views, institutions and main forms of behavior will be correctly considered as variations derived from one and the same source.

I know this question has been posed many times. In my search for some key traits of the Mesoamerican ethos, I will not go back to the old arguments. Instead, I will present to you some converging evidence derived from the testimonies mainly of the Nahuatl speaking peoples of Central Mexico and of several members of the great Maya family. The evidence derived from testimonies of these peoples living so far away, as well as from other groups like the Mixtec of Oaxaca, will help us to appreciate whether we can speak of an ancient Mesoamerican ethos.

16.2 A sense of belonging

What we know from the texts of the pre-Hispanic tradition of the Nahua, Yucatec Maya, Quiché and other groups, and also from the familiarity with their contemporary descendants, reveals that among the Mesoamericans there is a deeply rooted sense of belonging. This encompasses the feeling of belonging to a family, nuclear and extended, often enlarged by new spiritual forms of relationship. The members of a family are named in Nahuatl *cencaltin*, 'those of the same house', and also those who share a same essence and form of existence, *cenyeliztli*. Each human being is essentially linked to a community that lives in a land, ancestral to all its members, and rich in a variety of symbols. The many levels of belonging include, in addition to the primary one of the family, those of the *calpulli* or ward, the town, province and chiefdom. To a Mesoamerican, whether Maya, Mixtec, Otomí or Nahua, it would have been unthinkable to consider himself a loose entity, kinless or in any form isolated.

The cycles of the feasts and religious ceremonies during the solar year helped to provide the Mesoamericans with an understanding and vital feeling of one's belonging to a sacred space and time. According to the ancient world view, about which there are texts in Nahua, Yucatec Maya, Quiché and Mixtec, man exists in a universe horizontally distributed into four cosmic quadrants, each one full of symbols. This universe encompasses also a series of superior and inferior levels, and is the scene where the gods perform their actions. Those divine performances, although mysterious, determine man's existence and fate on the earth. Everything happens in accordance with the rhythms of time, and of divine forces, those of the Sun in its daily and nocturnal paths, and of the many other celestial bodies which at determined moments enter and abandon the scene of man's visible universe.

Each family, *calpulli*, town, province and chiefdom exists in this universe, distributed into four quadrants. The gods had established the primeval measurements of the four quadrants. Also Huitzilopochtli, the tutelary god of the Mexica, gave the command to his people to distribute their city, Mexico–Tenochtitlan into four great *calpullis* or wards (Alvarado Tezozomoc, 1975, pp. 74–5).

I will add here a reference I dedicate to those who recently have proposed the idea that, at least to the Maya, the four-fold cosmic horizontal distribution was of little or no meaning. I will quote from the account of archaeologist Richard E. W. Adams about his recent findings in the zone of Río Azul in northeastern Petén of Guatemala (Adams, 1986, pp. 441–2):

> In 1985 we cleared another looted burial, Tomb 12, and found additional beautiful hieroglyphic paintings. On the smooth plaster of the tomb's four walls, in thick red line, are the signs for east, south, west and north. Although the meanings of these hieroglyphs have been known for more than a century, our discovery marked the first time they were found in actual natural context, for their appearance in Tomb 12 correctly matches the real directions.

In Río Azul, evidence was unearthed of the actual and 'natural' meaning of the hieroglyphs of the four cosmic quadrants. There, their corresponding spatial references are associated not only with man's existence in this quatripartite universe but also to the remains of those who have already departed to 'the Region of Mystery'. It is of interest to quote once more from the account of Adams, who points to another association of the hieroglyphs of the cosmic quadrants with the symbols of some of the celestial bodies, those who establish the divisions of time and carry the meanings and 'burdens' of it (Adams, 1986, p. 442):

> Moreover each directional glyph was supplemented by another representing its mythical cosmic association with, respectively, the Sun, Venus, darkness and the Moon.

The presence there of the hieroglyphs of these deities–celestial bodies is indeed significant. Their paths in the heavens had to be observed all year round, because as a function of them the destinies (*k'in* in Maya, *pije* in Zapotec and *Tonalli* in Nahuatl) had to be discerned and propitiated.

To man's consubstantial adherence to his family, kinship, town, province and chiefdom, as well as his essential relation to his own sacred space, one has to add his appurtenance to the realm of time, at once cyclic and divine, endowed with a large number of faces, those of the portentous carriers of the universe's and man's destinies. Yes, indeed, Mesoamerican man, far from considering himself in any form loose, insulated or deprived of ties, human and divine, had a deep sense of belonging. Two brief texts from the Nahua tradition give eloquent testimony to this. One, a part of a *Huehuehtlahtolli*, 'The Ancient Word', will help us to understand from the inside the meaning the Mesoamericans attached to the arrival of a new human being to the family. Among other things these words of great tenderness – in the mood of the Mesoamerican tone of sentiment – were addressed to the recently born baby (*Florentine Codex*, 1969, Book VI, p. 196):

The Lord of the Near and by [*Tloqueh Nahuaqueh*], the Giver of Life, has inclined his heart. Here a precious necklace is flaked off, here a precious feather is spread out. Here we dream, we see in dreams. In your hand [the parents' hands], on your neck, He placed a precious necklace, a precious feather, a jade, precious bracelet. Here we see your face. You have been formed; you have been born. . . .

Since the day of his birth the new human being *belonged* to a family. He was, to his parents and to the rest of his relatives, a precious necklace, a precious feather, a jade, a bracelet. . . .

From a different perspective, the other text describes the intrinsic link of man, from the moment of his conception, to the realm of the divine, from which his destiny descends (*Florentine Codex*, 1969, Book VI, ch. XXII):

It was said that in the thirteen heaven
our destiny [*tonalli*] was determined.
When the child is conceived,
when he is placed in the womb,
his *tonalli*, destiny, comes to him there;
it is sent to him by [the supreme god] the Lord
of Duality.

The overall Mesoamerican belief in the *tonalli*, with its rich complexity of meanings and interrelations within the ancient world view, is another element that intrinsically colored the ethos of these peoples.

16.3 The *tonalli*, destinies of man on the earth

The universe in which the Mesoamericans lived had been established and re-established several times, four according to the Maya, five in the Nahua tradition. There are more than 20 testimonies, in the indigenous texts and the codices, that speak of those 'Suns' or ages which have existed. They came into being and were later destroyed, not by any kind of chance but by a predetermined divine act, the result of a destiny, *tonalli*, *k'in* or *pije*).

An identical and very complex constellation of meanings was attached to these three words, from three different Mesoamerican languages: Nahua, Maya and Zapotec. 'Sun, day, feastday, time and destiny' are their basic connotations. That of des-

tiny permeated the others so deeply that for the Mesoamericans the reckoning of any period of time led them always to investigate the *tonalli* or destinies associated with it. Thus each of the 'Suns' or ages had been re-established as determined by its corresponding destiny. That is why each 'Sun' had its calendric name, precisely the name of the *day-tonalli* in which it began to exist: 4-Water, 4-Ocelot, 4-Rain-of-Fire, 4-Wind and 4-Movement. Those *day-tonalli* names were not mere words, they announced the destiny of each one of the 'Suns'.

Everything happening on the earth and in man's life from his birth to his death, is the outcome of a *destiny-tonalli*. And every one of the gods also has an intrinsic relation to the realm of the *tonalli-s* of time. Thus each god has likewise a calendric name, expression of the *destiny-tonalli* that belongs to him. Examples of this are 5-Flower, *Macuilxochitl*, He-She protector of the singers, dancers, and other artists, and 7-Serpent, *Chicome-coatl*, She (and also He), the deity of maize and of our maintenance in general.

Time penetrates the cosmic quadrants, endowed with preordained rhythms, as the *Book of Chilam Balam* declares it (Solís Alcalá, 1949, pp. 340–1):

When [the year] has gone round
the four quadrants,
east, north, west, south,
it is said to be *1-Katun*.
The same thing is said
when *1-Cauac* begins,
and each of the years
goes through this. . . .

A very similar Nahuatl statement is included in *Madrid Codex* (vol. 8, fo. 269r) 'Time penetrating space, the quadrants of the earth's surface, the heavens above us and the underworld, disseminates the *tonalli-destinies*, bringing about the realization everywhere of what has been determined by the gods and, above all, by Him–Her who is like the Night and the Wind.'

A distinguishing attitude of the Mesoamericans, an extremely significant element of their ethos, was their concern for time as a conveyor of destinies. Several texts from the pre-Hispanic tradition in Maya and Nahuatl proclaim that the *tonalli*, day-time-destiny of a given person, is precisely *i-macehual*, 'that which has been granted to him, what he deserves'.

The knowledge about the *tonalli-destinies*, still alive in some isolated spots within contemporary Mesoamerica (Villa Rojas, 1968, pp. 151–9), is an attribute of some of the priests and sages. In the old days these experts were called *ah k'inob* by the Yucatec Maya and *tonalpouhqueh* by the Nanua. *Ah k'inob* means, 'those of the days–destinies' and *tonalpouhqueh*, 'the ones who read and tell the days–destinies'.

The knowledge of the *tonal-pohualli* or *tzol-kin*, 'count of the destinies', implied very complex calculations and a great familiarity with the sacred beliefs and rites. The *tonalpouhqueh*, *ah k'inob* and other experts in different places within Mesoamerica, were in possession of various calendric counts whose remote origins must be traced back to the days of the Olmecs, to at least the middle of the first millennium BC (Coe, 1977, pp. 70, 79).

All periods of time, the 13 hours of the day, and the nine of the night, the days, the 13-day groups, the spatial orientations of them, the 20-day 'months', the years, and their different positions in the various cycles of 52, 104 or in the Maya Long Count, including their interrelations with the cycles of the 'Great Star' (Venus), all this and much more had to be taken into account in order 'to read and tell' a *tonalli-destiny*.

The *tonalli-day*, the carrier par excellence of the *tonalli-destiny* belonged to the 260 days *tonalpohualli*, the count of the days-destinies. This count was basically formed by four divisions of 65 days, each one broken into five 'weeks' of 13 days each. The names of the 260 days-destinies were structured by means of 20 day-signs, preceded by a numeral which ran from 1 to 13.

The 260 possible names of the day-destinies were incorporated in the *xihuitl* or *haab*, the solar-year-count of the Nahua and Maya, and colored it with the richness of their sacred connotations, like the mystic presence of one or several gods with their influences, good or adverse, the revelation of a time propitious for the performance of a determined form of action, a religious ceremony, the initiation of a war, the enthroning of a ruler.

Nothing in man's life would be set apart as being unrelated to the realm of the destinies of time. The religious celebrations with their many rites, prayers, hymns and dances; the sacrifices and re-enactments of the primeval happenings – everything had to be attuned to the rhythms of the days and destinies.

The gods brought down the destinies. They were the carriers of those *tonalli-s* that belong to them and which the *tonalpouhqueh* or *ah k'inob* could anticipate by knowing, among many other things, the calendric name of the 'arriving' deity.

There are some extant *tonal-amatl*, 'books of the days-destinies', such as *Codex Borbonicus* and *Tonalamatl of Aubin*, as well as some parts of the Maya *Dresden*, *Madrid* and *Paris* codices. By means of them one can try to penetrate into the subtleties of the computations of the day-destinies and their constellations of interrelations within the universe of divine realities.

Those books were consulted by the priests asked to advise about the *tonalli-s* which had to do with the birth of a baby, the giving of the most adequate name to the infant, the choosing of the most favorable date for his/her dedication to a school, the date for a wedding or for a declaration of the sexual trangressions to the Goddess who will cleanse them.

Those concerned with agriculture, the arts and crafts, commerce, war and the collection of tributes, the lords and high rulers, i.e. all the Mesoamericans, constantly had to consult with the experts about the meaning of their *tonalli-s* and about the remedies to escape as best one could the bad consequences of an ominous destiny.

16.4 Gods, destinies and acts of deserving performed by Mesoamerican man

When there was still night, in a primeval time, the gods had deserved, by their own sacrifice and death, the restoration of the Sun, Moon, earth and man. These gods, who so often appear acting by pairs are the same who everywhere, in space and time, make themselves present in what seems to be a preordained sequence. He–She, *Ometeotl*, the Dual God in the religion of the Nahua, the Grand Father–Grand Mother of the *Popol Vuh*, the divine Pair who resides in the uppermost of the heavens, as depicted in the Mixtec codices known as the *Selden Roll* and *Gómez Orozco*, is the Ultimate Reality, Begetter–Conceiver of all that exists.

He–She has a variety of names often expressed by a pair of interrelated words. Examples of this are *Tloqueh*, *Nahuaqueh*, 'The One that Is Near', 'The One that Is Close'; *Ometeuctli*, *Omechihuatl*, 'Lord of Duality', 'Lady of Duality'; *Yohualli*, *Ehecatl*, 'Night', 'Wind'; *Tezcatlanextia*, *Tezcatlipoca*, 'Mirror of Day', 'Mirror of Night'; *Citlallatonac*,

Citlalinicueh, 'Star Which Illumines Things', 'Skirt of Stars'.

An easily noticeable trait is the apparently androgynous character of many gods. Thus, for instance, *Tlalteuctli* is at once Lord and Lady of the Earth, and *Cinteotl* is He–She God of Maize. There are, besides, other gods who appear and act by pairs: Our Lord and our Lady of the Place of the Dead (*Mictlanteuctli* and *Mictlancihuatl*); The Lord of Rain and the Goddess of the Terrestrial Waters (*Tlaloc, Chalchiuhtlicueh*); and, in a very especial form, *Quetzalcoatl*, and *Cihuacoatl*, understood as 'The Precious Twin' and the 'Feminine Twin' – the word *coatl* meaning both 'serpent' and 'twin'.

On other occasions there are two pairs of divine beings who present themselves acting in what seems to be a preordained sequence. This is the case of the four *Tezcatlipocas*, 'Smoking Mirrors', who preside over the four quadrants of the world, or rule, in succession, the four previous cosmic ages (*Borgia Codex*, 1980, vol. I, p. 21). In cases like this, it is often said that 'other gods' can play the role of those who integrated the original quartet.

A number of ancient texts also register that one or another of the gods, or a pair of them, is invoked as 'Our Mother, Our Father' (*in Tonantzin, in Totahtzin*). Such is the case, for instance, in the following prayer from the *Florentine Codex*, 1969, book VI, ch. IX):

Mother of the gods, Father of the gods, God of Fire, the Old God, *Xiuhtecuhtli, Huehuehteotl*, reclined on the navel of the earth, within the circle of turquoise....

Several testimonies that pertain to 'the ancient word', the *Huehuehtlahtolli*, and some sacred hymns and other chants unveil for us what seems to be at the core of this Nahuatl pantheon, so rich in divine pairs and quartets. Indeed the Ultimate Reality, the divine source referred to as 'Giver of Life' (*Ipalnemoani*: literally 'Thanks To Whom One Lives'), is thought of as a dual entity.

Quetzalcoatl, the Feathered-Serpent god, symbol of the divine wisdom, was asked by the other gods to take care of restoring the human beings. *Quetzalcoatl* went to *Mictlan*, 'Place of the Dead', in search of the precious bones of men who had lived in previous ages. (*Leyenda de los Soles*, 1975, p. 120). In this manner he would restore the human beings to inhabit the re-established earth. In the Region of the Dead, *Quetzalcoatl* had to overcome many

obstacles. Once he could gather the precious bones, he took them to *Tamoanchan*, the place of origin, the abode of the supreme Dual God. There the Mother Goddess 'took them to grind and put them in a precious vessel'.

Quetzalcoatl had to transmit new life to the bones. 'He bled his virile member. He and the other gods at once did pennance, deserved it (*tlamacehuayah*).' And they said: the human beings have been born, i.e. the *macehualtin*, 'the deserved ones, because for our sake the gods did pennance, deserved it' (*topan otlamaceuhqueh*) (*Florentine Codex*, 1953, Book VII, ch. II). In fact, the word *macehualtin*, 'the deserved by the gods' pennance', became synonymous with human beings not only in Nahuatl but also, as a loan, in several other Mesoamerican languages. It is true that later a differentiation was introduced between *macehualtin* 'human beings' (understood as the common people), and *pipiltin*, 'those of lineage', members of nobility, the ruling class. But this did not alter the idea that all, men and women, whether of lineage or commoners, essentially were *macehualtin*, 'deserved by the gods' pennance'.

The key concept of *tlamacehua* denotes the primary and essential relation human beings have with their gods. These, with their own penance and sacrifice, deserved – brought into existence – the human beings. The gods did it because they were in need of some beings who would be their worshippers, the providers of sustenance to foster life on earth. Man also had to perform *tlamacehualiztli*, 'penance, the act of deserving through sacrifice', including the bloody one of human beings. If the gods *topan otlamaceuhqueh*, 'for us did penance', we ought to follow their example, to deserve our own being on the earth with our blood and life.

The often described as so 'utterly detestable human sacrifices', the consuming of small pieces of flesh of the human victim, the blood smeared effigies of the gods, were elements the Mesoamericans thought essential to act and respond in terms of their *tlamacehualiztli*. If the gods had sacrificed themselves when it was still night there in Teotihuacan, and if only thus, with their blood, they had deserved our being, to re-enact that primeval action was indeed to give in return, to pay and also restore. The victims were thus named *teomicqueh*, 'the divine dead'. With these *teomicqueh* man repayed and did his part in maintaining the flow of life on the earth, in the heavens and in the shadows of the underworld. There is a discourse of the *tlahtoani*, 'high ruler', in which he advises his sons to take firm hold upon whatever is related to this (*Florentine Codex*, 1969, Book VI, ch. XVII):

> In this manner there is entry near and close into *Tloqueh*, *Nahuaqueh*, 'The One Who Is Near', 'The One Who Is Close', where there is removing of the secrets from His lap, from His bosom, and where He recognized one, shows his mercy to one, takes pity upon one, causes one to deserve things [*macehualtia*] ... Perhaps He causes one to merit, to deserve [*quitemacehualtia*] virility, the eagle warriorhood, the tiger warriorhood. There He takes, he recognizes as His friend the one who addresses Him well, the one who prays well to Him ... In his hands He places the eagle vessel, the eagle tube [instruments for the sacrifice].

> This one becomes father and mother of the Sun. He provides drink, He makes offerings to those who are above us [*Topan*] and in the Region of the Dead [*Mictlan*]. And the eagle warriors, the tiger warriors revere Him.

With these words to his sons, the high ruler unveiled the meaning of the forms of action they should take, including that of 'providing drink and sustenance' to the gods through human sacrifice. To keep near and close to *Tloqueh*, *Nahuaqueh*, the Dual God, was difficult but if rulership, government, were to be alive, were to be deserved, *tlamacehualiztli*, the act of deserving it should be re-enacted.

Several codices, e.g. the *Borbonicus* and the *Tellerianus*, and some texts written in Nahuatl with the Roman alphabet, describe the great variety of performances that in perfect order were rendered in the public celebrations among the 18 groups of 20 days within the solar calendar. There, and also in private life at home, many rites were carried out to win the divine benevolence, to deserve a good *tonalli*.

The *tonalli*, destiny, depended on what Our Mother, Our Father had deserved and conceded to a person. Because of this, whatever possibility might exist to modify the *tonalli* had to be sought in the same book where one could decipher the mysteries inherent in the divine predestinations. There one would find the most adequate action, in a determined date, to give in return, so as to foster in the best possible way the divine flow of life. This, of course, presupposed the *tlamacehualiztli*, the act of meriting and deserving what is 'appropriate and righteous', what Our Mother, Our Father had determined and disposed for us when we were engendered and placed in the womb of our own terrestrial mother.

Perhaps as a projection of the deeply rooted belief in the divine Duality, one can trace a tendency in the Mesoamericans to conceive things and speak about them in pairs. Such a psychological leaning appears in a large number of concepts and expressions. Here are a few examples: the universe is composed by the earth's surface as the visible reality, and 'that which is above us, and the Land of the Dead', as an invisible and mysterious realm; the concept of person as 'the owner of a face and of a heart'; the basic two social strata the *pipiltin*, 'those of lineage', and the *macehualtin*, a term understood here not so much in its primary connotation but as meaning 'the ordinary people, the commoners'.

To concepts like these one must add a different kind of dualism, present in a great number of expressions in the Nahuatl and Maya languages. It consists of uniting two words which complement each other, either because they are almost synonyms or because they evoke a third idea, usually a metaphor. This linguistic stratagem can be compared to the kennings, compound metaphorical expressions used in ancient Anglo-Saxon. A few examples from Nahuatl are the following: 'flower, song' (meaning poetry, art); 'water, fire' (war); 'seat, mat' (authority), 'blouse, skirt' (womanhood).... From Yucatec Maya one can quote: 'eye, ear' (that which permits to know); 'father, mother' (protection); 'quetzal and blue bird' (a precious reality); 'stick, stone' (punishment)....

To feel and cherish the belonging to a family, clan, town, chiefdom ...; to believe and act in a world where time as an atmosphere permeates everything and introduces the *tonalli-destinies* ultimately determined by Our Mother, Our Father; to think and speak with a dualistic orientation, are traits essentially related to the Mesoamerican ethos.

16.5 Time, *tonalli* and stargazing

To live attuned to the rhythms of time, performing adequate *tlamacehualiztli*, 'acts of meriting and deserving', thus propitiating the good realization of a *tonalli*, one had always to consult the *tonalpohualli* (the 260-day system), and the *xihuitl* (365-day count). But there were other realities, related also to the cycles of time, which presupposed more complex forms of correlation and adjustment based on astronomical observations. Stargazing was an occupation of some of the priests and other members of the ruling group. Centuries of observation of the movements of the celestial bodies had actually led to the structuring of the basic calendrical systems. And stargazing as well had made it possible to introduce the required corrections in the calendar. As a Nahuatl text (*Coloquios ... apud*, Léon-Portilla, 1979, pp. 19–20) expresses it, one had often to consult with:

> those who guide us ... and instruct us how our gods must be worshipped ... Those who see, who dedicate themselves to observing and measuring with their hand the running and the crossing of the stars in the sky.

Thanks to those observers and measurers of the stars' movements a number of other cycles of primary importance to the Mesoamericans were charted. Mention can be made of the cycles of the Great Star (Venus); of the Moon and its eclipses; of the Pleiades (so closely related to the 52-year cycle), and of several other constellations and celestial phenomena.

This, of course, helps us to appreciate the importance attached to astronomical observations as an activity so intrinsic to the accuracy of the calendar and to whatever was ruled by it: the cult of the gods, the wisdom of the destinies, the basic duties of man in his belonging to his family, group, town, chiefdom, his activities as a farmer, warrior, artist, merchant or in any other profession. In brief, to exist for the Mesoamericans one had to observe the sky. Without skywatchers the ethos of this people, its distinguishing spirit, its own genius would not have developed.

At this point, I believe it is time for a warning. This paper is addressed to my friends, the archaeoastronomers. The *tonalli, k'in* or destiny of our moment and concern is bringing about the need for the warning. At least this is the way I see it. If skywatching – 'the observing and measuring of the running of the stars' – was an essential part of Mesoamerican culture, modern 'discoveries' of any of its ancient achievements will have always to do with celestial phenomena intrinsically related to the world view, religion and practices of the Mesoamericans. 'Discoveries' of archaeoastronomers dealing with celestial bodies or cosmic cycles, about whose meaning in the culture itself nothing is known, have to be held suspicious and put in parentheses. Those who claim that the cycles of planets like Mars or Neptune were well known to the Mesoamericans will have to tell us where in the Mesoamerican sources, archaeological or documentary, reference is made to the meanings of such cycles and the ultimate nature of those celestial bodies. There are many accounts about Venus and its cycles, a good number related to Quetzalcoatl–Kukulcan, but have we parallel texts or other references to gods associated to the cycles of Mars or Neptune?

This is just an example of what I want to declare as a 'reading' of the *tonalli* which belongs to my friends, the archaeoastronomers. A last word: if you are persuaded, as I am, that stargazing, or if you prefer astronomy, was an essential ingredient of Mesoamerican culture, do your best to study the achievements of its skywatchers, not from the perspective and previous knowledge of western astronomy, but from the point of view of those who, contemplating the sky with their naked eyes, unveiled many of its secrets, discovering at once a universe of celestial meanings they could understand only in terms of their own culture.

References

Adams, R. E. W. (1986). Archaeologists explore Guatemala's lost city of the Maya, Río Azul. *National Geographical Magazine* **169**, (4), 420–60.

Alvarado Tezozomoc, F. (1975). *Crónica Mexicayotl*. Translated from the Nahuatl by A. León. Mexico: Instituto de Investigaciones Históricas, Universidad Nacional.

Borgia Codex (*Códice Borgia*) (1980). Comentarios de Eduard Seler, 3 vols. Mexico: Fondo de Cultura Económica. First reprint of the 1963 edn.

Coe, D. (1977). *Mexico*, 2nd edn. New York: Praeger Publishers.

Florentine Codex, General history of the Things of New Spain [by] Bernardino de Sahagún, in 13 Parts, 12 vols. Book VII, 1953; Book VI, 1969. Translated from Aztec into English, with notes and illustrations, by Arthur J. O.

Anderson and Charles E. Dibble. Santa Fe, New Mexico, The School of American Research and the University of Utah.

Léon-Portilla, M. (1979). *Aztec Thought and Culture.* Norman: Oklahoma University Press.

Leyenda de los Soles (1975). In *Códice Chimalpopoca*, ed. and trans. P. F. Velázquez. Mexico: Instituto de Investigaciones Históricas, UNAM.

Madrid Codex (*Tro-Cortesianus Troano*) Museo de América, Madrid. Reprinted, 1967, Codices Selecti, vol. VIII.

Graz: Akademische Druck und Verlagsanstalt.

Oxford English Dictionary (1979). 2 vols. Oxford University Press.

Solís Alcalá, E. (1949). *Códice Pérez.* Mérida, Yucatán.

Villa Rojas, A. (1968). In *Time and Reality in the Thought of the Maya*, ed. M. León-Portilla. Boston: Beacon Press.

Webster New Twentieth Century Dictionary (1978). Unabridged, 2nd edn. New York: Collins World.

17

The Mexica leap year once again

Carmen Aguilera *Biblioteca del Instituto Nacional de Antropología e Historia, Mexico*

The problem of knowing if the Mexicas made some kind of adjustment to their solar calendar of 365 days in order to coordinate it with the astronomical events is still a matter of controversy. In this chapter I am not going to discuss the arguments sustained by the different positions. I want only to point out that if the Mexicas ruled their lives by a solar calendar of 365 days, and if this calendar was in accordance with solstices and equinoxes when the Spaniards arrived in Mexico in AD 1519, it is highly probable that they made some kind of adjustment to their calendar.

The way in which the Mexicas could have adjusted their calendar is also a matter of debate among researchers who admit this possibility. In a paper about calendrical correlation (Aguilera, 1982), I adopted the scheme of adjustment postulated by Castillo (1971). He studied the texts of Chapter 37 in the *Florentine Codex* in the Nahuatl and Spanish versions (1979, Book 2, fos. 96r–102r), and concluded that the Mexicas adjusted their calendar by extending the last day of Izcalli, called 20 Izcalli, to two days or 48 hours, every four years. With this very simple device, the Mesoamerican system of 18 months of 20 days plus five days called *nemontemi*, remained forever unchanged and coordinated to solstices and equinoxes.

In the present chapter I restudy this subject because of new data found in Chapters 37 and 38, Book 2, of the *Florentine Codex* (fos. 102r–106v). These reinforce the hypothesis of an extension every four years. Chapter 37 is entitled: 'De la fiesta y ceremonias que se hazian en las calendas del 18 mes que se llama Izcalli' ('Of the feast and ceremo-

nies that took place in the calends of the 18 months whose name is Izcalli'), and Chapter 38: 'De la fiesta llamada Oauhquiltamalqualiztli: quese hazian a los diez dias, del mes arriba dicho, quese hazian, a honrra del dios llamado Ixcocauhqui.' 'Of the feast called Oauhquiltamalqualiztli: that was done at ten days of the month mentioned above that were done in honor of the god called Ixcocauhqui.'

The chapters also have different folio headings. For Chapter 37: 'De la fiesta de Izcalli' ('Of the feast of Izcalli') and for Chapter 38: 'De la fiesta de Ixcoçauhqui' ('Of the feast of Ixcoçauhqui'). In spite of these differences in titles and headings, both chapters deal with the same four subjects: (1) the regular feast of 20 Izcalli; (2) the feast of 10 Izcalli called Huauhquiltamalcualiztli; (3) the feast of 20 Izcalli every four years; (4) the five days called *nemontemi*.

This duplication of chapters was probably due to the fact that the second chapter slipped in the arrangement of materials, done a little after 1575, when the first version of Sahagún's work, with the Spanish text on the left and the Nahuatl text on the right of the page, was integrated. In *c.* 1578, when the *Florentine Codex* was written from the same materials, the duplication persisted, probably because Sahagún's team was pressed for time in order to send the document to Spain.

Nevertheless, the copyists of Chapters 37 and 38 in the *Florentine Codex*, and probably those of the first version, knew the material was repeated. In the Spanish text of Chapter 37 it is said: 'This is the relation of this feast, although there is another one more extensive that will be put ahead'. In

Chapter 38 it is stated in the Spanish column on the left: 'The extensive relation is to follow...', but then the text says only a little about the feast of Huauhquitlamalcualiztli and ceremonies with captives that are not killed. The Spanish text was cut short in order to leave space for the 15 illustrations concerning this feast. On fo. 104r it is written: 'The rest done in this feast is contained in the text in the Mexican language... has been said before'.

Castillo found that the feast of 20 Izcalli, in the first part of Chapter 37, lasted two days. If in Chapter 38 there is another description of the same feast, not analysed by this author, it was possible that this feast also lasted two days. If so, it could reinforce the hypothesis that the Mexicas adjusted their solar calendar by extending 20 Izcalli to two days. The *Florentine Codex* says that something very special was done on the feast of 20 Izcalli every four years. Nahuatl text: 'And during Izcalli, the first, second and third years, nothing was done; but on the fourth years, it is made big', and the Spanish text: 'Three years in a row, they did what it has been said, in this month and in this feast; but the fourth year they performed many other things, as follows...'. The Nahuatl term for 'it is made big' is ambiguous and could mean big in splendor and/or big in time. Examination of the feast shows the word covers both meanings; on the other hand, the analysis of the feast in both chapters of the *Florentine Codex* disclosed that there are five relations of the feast, four on Chapter 37 and one in Chapter 38.

17.1 Relations Nahuatl 1 and Spanish 2

The Spanish and Nahuatl texts in Chapter 37 are extensive, have no illustrations and are very similar in their versions. These are the *Relations* that Castillo examined in their part of the feast of 20 Izcalli every four years to deduce his 'Nahuatl leap year' (Castillo, 1971). He notes in both texts the time at which the ceremonies take place and discovers they last two days. Other feasts of the solar year last two or more days; the significant fact about these texts is that they state clearly that the celebrations began at dawn of 20 Izcalli every four years.

228

First day

Auh in ye huallathui Izcalli, 'and when [the feast] Izcalli comes to clear' (dawns).

Captives to be sacrificed are taken up to a temple called Tzonmolco, dedicated to Xiuhtecuhtli, 'Lord of the year' and patron deity of Izcalli.

Auh in huacic yohualnepantla, 'and when midnight comes out'.

Captors cut hair from captives' heads.

Second day

Auh in otlahuizcalli moquetz in ye tlatlalchipahua, 'and when dawn stood up and the earth begins to be seen'.

Captives burn or give away clothes. Shortly afterwards dawn at a little later time is mentioned, only in the Nahuatl text, once more:

Auh in otlahuic, 'and when it cleared'.

Captives are dressed in ceremonial attire and are taken up to the place where they were to be sacrificed.

Auh in ye ommotzcaloa tonatiuh, 'and when the sun grew strong', which, the Spanish text says, is after midday, the impersonator of the Mexica god, Painal, descended from the temple and was the first one to be sacrificed, followed by impersonators of Xiuhtecuhtli. Then Motecuhzoma and his nobles dance. The name of the ruler is not mentioned in the Spanish Relation.

17.2 Relations Nahuatl 3 and Spanish 4

After the preceding two days' festivities some children's ceremonies are described in an apparent third day. Nevertheless the Spanish text asserts: 'This very same day', that is, the second day of 20 Izcalli, because the ceremonies narrated cover only one day, and after their description the Nahuatl text reads: 'Here ends the day Izcalli'.

First and only day

Oc huehcayohuan, 'very far or deep the night', or 'before dawn' (*Florentine Codex*, 1979, Book 1, fo. 11v), they pierced children's ears, pasted on their heads *toztli*, a young yellow parrot, and down feathers. The Spanish text explains that they pierced the ears of children three years old or less, which alludes again to the fact that this ceremony took place only every four years, and this way all children

Table 17.1

Nahuatl Relation 1	Spanish Relation 2	Nahuatl Relation 3	Spanish Relation 4	Nahuatl Relation 5
Auh in ye huallathui Izcalli And when Izcalli comes to clear	*En amaneciendo* At dawn			Seek godparents Give them gifts
Take captives up to Tzonmolco temple	Take captives up to temple			
Auh in huacic yohualnepantla And when midnight came Cut captive's hair	*Y llegada la media noche* And when midnight came Cut captive's hair			*Yohualnepantla* At midnight Godparents take children to Tezcacoac Pierce children's ears, paste them with feathers
Auh in otlahuizcalli moquetz in ye tlatlalchipahua And when dawn stood up and the earth began to be seen		*Oc huehcayohuan* Very far or deep the night	*Ese mismo dia, muy de manana, antes que amaneciese* That very same day, very early, before dawn	*Auh in otlahuic* And when it cleared Eat, drink go to *calmecac* Children and godparents dance
Captives burn clothes	Captives burn clothes	They pierce children's ears and paste them with feathers	They pierce children's ears and paste them with feathers	
Auh in otlahuic And when it cleared	*Y luego en amaneciendo* And then at dawn	Lustration Look for godparents Give them gifts	Look for godparents Give them gifts	
Captives bedight	Captives bedight			
Auh in ye ommotzcaloa Tonatiuh And when already the sun strong	*Despues del mediodia* After midday	*Nepatla tonatiuh* The sun in middle	*Al mediodia* At midday	
Captives sacrificed Motecuhzoma dance	Captives sacrificed Lord's dance	Back to temple Singing, dancing Children drink	Back to temple Singing, dancing Children drink	
		Auh in teohtlac And in the afternoon They go home In courtyard feast, old people drink	*En la tarde* In the afternoon They go home At feast, owners and neighbors drink	*Auh in ye achitlahca in ommopilo tonatiuh* And when a little late, when already hangs the sun In Tezcacoac children drink; everybody intoxicated

had their ears pierced before they were four years old. Mothers sought godparents for their children, gave them gifts, and at the children's ceremonies of piercing of ears and lustration they were appointed to their respective child by the priest; then all went home.

Nepantla tonatiuh, 'the sun in the middle', they went back to the temple where there was dancing and singing and then the ceremony of *pillahuano* or 'drinking by children', started. Adults also drank small amounts of pulque from very small vessels.

Auh in teohtlac, 'and in the afternoon', everybody went home where, in the courtyard, dancing and singing proceeded, and the relatives who were old were allowed to drink all the pulque they wished.

We have to note that the 'piercing of ears' ceremony took place very early in the morning, before dawn. Godparents had to be with their respective child to take them to the lustration ceremony and their appointment by the priest. In *Relation 5*, as we will see below, the choosing of godparents is done the day before the 'drinking of children' or *pillahuano* ceremony. This alteration of order was done probably because the Tenochcas had to accommodate all ceremonies, traditionally of two days, in only one.

17.3 Nahuatl Relation 5

The *Nahuatl Relation 5* in Chapter 38 is longer than the two preceding ones. The text lists two ceremonies as a pair: *nepilquixtiloya* and *pillahuano*, and then mentions a dance of lords called *netecuitotilo* that is never mentioned again, but that according to the text took place after *pillahuano*.

First day

The text does not state the precise moment at which *nepilquixtiloya* or 'taking out of the children' began, but it was in 20 Izcalli because the text says so immediately above. On that day, godparents were chosen and gifts were offered to them. The ceremony was so-called because at an undetermined time godparents took the children out to a *calmecac* or priestly house called Tezcacoac, and at

Yohualnepantla, 'midnight', the priests pierced the ears of children, pasted them with yellow feathers of the *toztli* parrot and down, and godparents took them home where they held vigil all night.

Second day

Auh in otlahuic, 'and when it cleared', they ate and drank. In the courtyard of the Tezcacoac, the singers and old men of the *calmecac* sang for the children and they danced with their godparents, the little ones on the back of their godparents, the older ones taken by the hand.

Auh in ye achitlahca in ommopilo tonatiuh, 'and when already a little late, when already hangs the sun', they returned to the Tezcacoac, where they danced, sang and proceeded to give wine to children. Everybody drank until intoxicated, from vessels called *tzicuiltecomatl* that had three legs and handles on four sides. They said: 'The wine feast has been observed, for the children have drunk wine'. Thus ended the feast of Izcalli every four years.

In the descriptions of the feast of 20 Izcalli every four years there are differences of structure and content that are meaningful. We have already noticed the different names of chapters and their headings. Another difference is that in *Relations 1* and *2* the *nemontemi* are discussed after the feast of Izcalli, and in *Relation 5* they are discussed after 20 Izcalli every four years. This alteration of structure indicates a possibly different author or provenance but neither alters the position of the *nemontemi* at the end of 20 Izcalli.

Relations 1 and *2* extensively describe war ceremonials, preparation for human sacrifices and then the sacrifice's ritual. Sacrifices took place at Tzonmolco's temple that was located in the ceremonial center of Mexico-Tenochtitlan (*Florentine Codex*, 1979, Book 2, fo. 177v). The first captive to die was Painal, Huitzilopochtli's vicar, and at a dance of lords appears Motecuhzoma (probably the second). All these data point to the fact that *Relations 1* and *2* were written by Tenochcas.

Relation 5 is Tlatelolca because it is transcribed verbatim from a text that appears in the *Memoriales en tres columnas* written around 1563–5 in that city (Cline and D'Olwer, 1973, pp. 190–1) and published in the *Códice Matritense del Real Palacio* (Paso y Troncoso, 1906, fo. 45r). This seems to be the traditional celebration of 20 Izcalli every four years. This feast was dedicated to children's ceremonies to propitiate their growth and they initiated the fertility cycle under Xiuhtecuhtli, 'Lord of the year'.

Relations Nahuatl 3 and *Spanish 4* are a condensed form of the one above. Apparently the Mexicas in their hegemony introduced and gave pre-eminence to war ceremonial and human sacrifices, in accordance with the militaristic orientation of their culture. Children's ceremonies in the text, which were traditional, are relegated to the end in a sort of appendix. Children's ceremonies take place at an unidentified temple and are confined to only one day, which is the second of 20 Izcalli every four years.

17.4 Conclusion

In conclusion, instead of only a single *Tenocha Relation*, a double day celebration on 20 Izcalli every four years, we have (a) Tenochca *Relations 3* and *4*, in which we detected that the two day ceremonies are confined to one day, (b) *Relation 5* from Tlatelolco and (c) the explicit statement by Sahagún: 'In this feast [of Izcalli] on the common years they did not kill anybody, but on the leap year, that was every four years, they killed in this feast captives and slaves' (*Florentine Codex*, 1979, Book 2, fo. 11v). With these new data Castillo's hypothesis is reinforced, and although more thorough research has to be pursued there is no reason to doubt that the Mexicas and Tlatelolcas used this adjustment and that either they or another pre-Hispanic cultures devised this very simple, efficient way to adjust their solar year every four years.

References

Aguilera, C. (1982). Xoppan y Tonalco. Una Hipótesis acerca de la correlación astronómica del calendario mexica, *Estudios de Cultura Náhuatl* **15**, 185–207.

Castillo, V. (1971). El bisiesto náhuatl, *Estudios de Cultura Náhuatl* **9**, 75–104.

Cline, F. and d'Olwer, L. N. (1973). Fray Bernardino de Sahagún, 1499–1590. In *Handbook of Middle American Indians*, ed. R. Wanchape, vol. 13 (Guide to Ethnohistorical Sources 2), pp. 186–239. Austin: University of Texas Press.

Florentine Codex (*Códice Florentino*) (1979). Manuscrito 218–20 de la Colección Palatina de la Biblioteca Medicea Laurenciana, Gobierno de la República Mexicana, Florencia, Italia, 3 vols.

Paso y Troncoso, F. del (1906). *Códice Matritense del Real Palacio*, Hauser y Menet, vol. VII. Madrid.

18

Astronomical references in the table on pages 61–9 of the *Dresden Codex*

Victoria R. Bricker and Harvey M. Bricker   *Tulane University, New Orleans*

18.1 Introduction

Pages 61–9 of the Precolumbian Maya hieroglyphic book known as the *Dresden Codex* (Thompson, 1972; Villacorta C. and Villacorta, 1976) contain two almanac-like tables preceded by a table of multiples and 15 base dates (one shown twice) in *pictun*, serpent number, and ring number format (Figures 18.1 and 18.2). Eleven of these dates can be used to enter the table at the top of pages 65–9, and in every case they lead to other dates that are associated epigraphically, iconographically, and in astronomical fact with eclipse seasons, solstices, and equinoxes. This upper table, which we refer to as 'the seasonal table', is the subject of this chapter. The three remaining base dates refer to the table at the bottom of pages 65–9, which contains no identifiable astronomical glyphs and which we do not deal with further here.

The seasonal table differs from other, better known astronomical tables in the *Dresden Codex* (e.g., the Venus table on pages 24 and 46–50 (Closs, 1977; Lounsbury, 1983), the eclipse table on pages 51–8 (Bricker and Bricker, 1983), and the Mars table on pages 43b–5b (Bricker and Bricker, 1986)) in two respects: it has 11 base dates for entering the table (more than twice as many as found in any other table), and it makes use of *pictun* and serpent number notation, as well as long rounds and ring numbers, for calculating those dates (see Willson, 1924, pp. 24–5, Beyer, 1943 for explanations of the relationship between the Maya long count and serpent numbers and ring numbers, respectively). In spite of the variation in their format, all the base

dates can be used for correlating eclipse seasons with solstices and equinoxes in the seasonal table. We believe that the Precolumbian Maya shifted to a new base date whenever the accumulated recession of the table through the tropical year compromised its usefulness for synchronizing these astronomical events, and we regard the many base dates in the introduction to the table as evidence that such changes were frequently invoked.

The 11 base dates for entering the seasonal table are distributed over *baktuns* 1, 8, 9 and 10 of the Maya long count as shown in Table 18.1. Seven of them are entwined in the coils of the serpents pictured on pages 61 and 62, three are expressed in terms of long rounds and ring numbers on pages 62 and 63, and one is given in terms of a *pictun* calculation on page 61 (Figure 18.1). Here, we limit our discussion to the five *baktun* 10 dates, and we concentrate particularly on the black ring number base date closest to the table of multiples, 10.6.1.1.5 3 Chicchan (8 Zac). As will become apparent, the text and the pictures in the table proper seem to refer most exactly to the period of time immediately following this date.

The column adjacent to the table of multiples on page 63 contains two black bar-and-dot distance numbers: a long round (10.8.3.16.4) and a ring number (7.2.14.19). One bar is missing from the second highest term in the long round, which should represent '13', not '8' (Thompson, 1972, p. 21). The long count position of this date is calculated by first subtracting the ring number from 13.0.0.0.0 4 Ahau 8 Cumku, the base date for the current calendrical era, and then adding the long

232

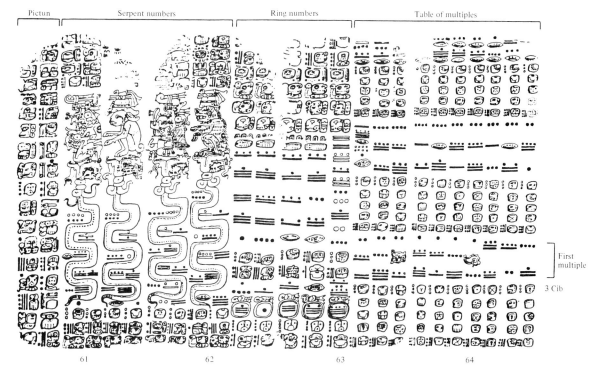

Figure 18.1 The introduction to the tables shown on pages 65–9 of the *Dresden Codex* (after Villacorta C. and Villacorta, 1976, pp. 132, 134, 136, 138).

round to the remainder, as shown in Table 18.2. Note that the *tzolkin* date, 3 Chicchan, which appears between the long round and the ring number in the column, is reached only if the *katun* term in the long round is changed from '8' to '13' (the uncorrected long round arrives at a day *13 Chicchan* in the Maya *tzolkin*).

The base date, 10.6.1.1.5, is mentioned only once in the introduction on pages 61–4. It is not repeated after the table of multiples, at the beginning of the table proper. In our previous work with the Mars table on pages 43b–5b of the same Codex (Bricker and Bricker, 1986), we demonstrated that this means that the base date cannot be used as it is to enter the table proper. Some multiple of the table must first be added to the base date, in this case the first multiple of 91 days (4.11 in Maya notation), which appears in the lower right-hand corner of page 64 (Figure 18.1). The entry date for the table proper is 10.6.1.5.16 3 Cib 19 Muan (= 10.6.1.1.5 3 Chicchan 8 Zac + 4.11). The *tzolkin* date, 3 Cib, appears directly below the distance number repre-

senting the first multiple of 91 days at the beginning of the table of multiples (Figure 18.1).

18.2 The table proper

Each of the 13 hieroglyphic clauses at the top of pages 65–9 has a pair of bar-and-dot numbers above it and another pair below it (Figure 18.2). In each of the 26 pairs, a black distance number records an interval in days, and a red number (shown outlined in black) specifies the coefficient of the *tzolkin* date reached by adding the distance number in question to the preceding date. As shown in the schematic sketch of this part of the table (Figure 18.3), the distance numbers, which range from 1 to 13 days, occur in an irregular order. Nine days added to the 3 Cib entry date of the table lead to a day 12 Chicchan; the addition of five more days leads to a day 4 Oc, and so on across the top row of numbers. The last day in the top row is 3 Manik, 91 days after the entry date. The further addition of 11 days, the left-most distance number in the

233

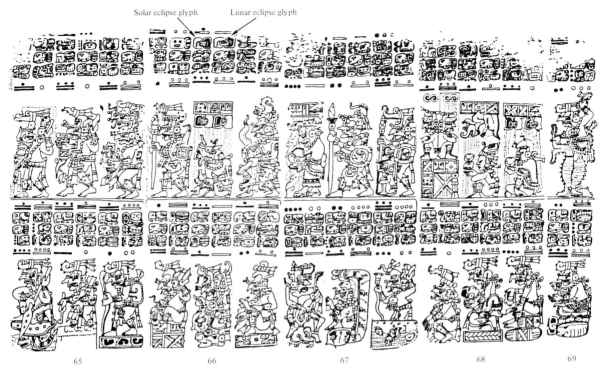

Figure 18.2 The tables on pages 65–9 of the *Dresden Codex* (after
Villacorta C. and Villacorta, 1976, pp. 140, 142, 144, 146, 148).

bottom row, leads to a day 1 Edznab, and so on
across. The right-most distance number in the bottom row, a '2', leads to a day 3 Edznab, which
would be the starting date for the next run of the
table.

The basic structure of the seasonal table may then
be described as follows. Although it is constructed
of explicit modules of 91 days, one full run or multiples of the table has a total length of 182 days
(2 × 91). The 13 pictures in the table and the clauses
associated with them may be relevant for the first
91-day component of the table or for the second.
Furthermore, two multiples of the table comprise
what has been called the 'computing year' of 364
days (Thompson, 1960, p. 256), one day shorter
than the Maya *haab* and one-and-one-quarter days
shorter than the tropical year.

The most obvious astronomical references in the
seasonal table are the solar and lunar eclipses that
appear in the second clause and picture on page
66a and in the third picture on page 68a (Figure
18.2). Depending on which of the four occurrences
one chooses as a starting point (two defined by the

top row of distance numbers, two defined by the
bottom row), the eclipse glyphs on these pages are
separated by intervals or distances of 49, 53, 45
or 35 days, none of which is an eclipse interval.
If the date associated with the second clause and
picture on page 66a falls within an eclipse season,
then the date associated with the third clause and
picture on page 68a (either 49 or 45 days later)
cannot fall within an eclipse season. The same is
true of the 53-day interval between the eclipse reference on page 68a and the immediately following
one on page 66a within any given 182-day multiple
of the table. It is theoretically possible for both the
page 68a reference near the end of one multiple
of the table and the page 66a reference near the
beginning of the following multiple to fall within
the same eclipse season, but *only* if a date of nodal
passage is precisely centered within the 35-day
interval that separates them. This would be a very
rare circumstance, and, as shown by data presented
below in our discussion of recycling the table, the
table's structure is predicated on its non-occur-
rence. In practical terms, only one of the two eclipse

234

Table 18.1. *Base dates of the seasonal table*[a]

P = *pictun*, R = ring number, S = serpent number

	Maya date		Gregorian date	Note
P	1.4.2.15.2	3 Ik 5 Zodz	9 Jul 2638 BC	b
R	8.11.7.13.5	3 Chicchan 8 Kankin	23 Feb AD 266	c
R	8.16.14.11.5	3 Chicchan 18 Zip	3 Jul AD 371	d
S	9.17.8.8.5	3 Chicchan 18 Xul	25 May AD 779	e
S	9.18.4.8.4	3 Kan 17 Uo	1 Mar AD 795	f
S	9.18.5.16.4	3 Kan 12 Yax	2 Aug AD 796	g
S	10.4.6.15.14	3 Ix 7 Pax	28 Oct AD 915	h
R	10.6.1.1.5	3 Chicchan (8 Zac)	12 Jul AD 949	i
S	10.7.4.3.5	3 Chicchan 13 Yaxkin	22 Apr AD 972	j
S	10.8.5.0.6	3 Cimi 14 Kayab	4 Nov AD 992	k
S	10.11.5.14.5	3 Chicchan 13 Pax	30 Sep AD 1052	l

(a) In every case, the base date must be calculated from a date in one of the other three formats. In making these calculations, we have accepted corrections to several of the serpent numbers originally determined by Beyer (1943, p. 404) as well as to all three of the ring numbers. These and other corrections affecting pp. 61–9 are listed by Thompson (1972, pp. 21–2).

(b) Inferred from information given in the first two columns of the table, on p. 61.

(c) Calculated from ring number (RN), 11.15, and long round (LR), 8.11.8.7.0 3 Chicchan 13 Kankin, in the first column of p. 63; date in *haab* corrected to **8** Kankin.

(d) Calculated from RN, 1.4.16, and LR, 8.16.15.16.1 3 Chicchan 13 Zip, in next-to-last column of p. 62; date in *haab* corrected to **18** Zip.

(e) Calculated from black serpent number (SN), 4.6.14.13.15.1 3 Chicchan 18 Xul, in left serpent on p. 61; SN corrected to 4.6.**0**.13.15.1.

(f) Calculated from red SN, 4.6.1.9.15.0 3 Kan 16 Uo, in right serpent on p. 62; date in *haab* corrected to **17** Uo.

(g) Calculated from red SN, 4.6.1.9.15.0 3 Kan 12 Yax, in right serpent on p. 61; SN corrected to 4.6.1.**11.5.0**.

(h) Calculated from black SN, 4.6.7.12.4.10 3 Ix 7 Zec, in left serpent on p. 62; date in *haab* corrected to 7 **Pax**.

(i) Calculated from black RN, 7.2.14.19, and LR, 10.8.3.16.4 3 Chicchan, in fifth column of p. 63; LR corrected to 10.**13**.3.16.4.

(j) Calculated from black SN, 4.6.9.16.10.1 3 Chicchan 13 Yaxkin, in right serpent on p. 61; SN corrected to 4.6.**10.9.10.1**.

(k) Calculated from red SN, 4.6.11.10.7.2 3 Cimi 14 Kayab, in left serpent on p. 62.

(l) Calculated from red SN, 4.6.0.11.3.(?) 3 Chicchan 13 Pax, in left serpent on p. 61; SN corrected to 4.6.**14.11.3.1**.

Table 18.2. *Calculation of the long count equivalent of the black ring number base date next to the table of multiples.*

(13.0.0.0.0)	4 Ahau 8 Cumku
− 7.2.14.19	ring number
(12.12.17.3.1)	13 Imix 9 Uo
+ 10.13.3.16.4	long round
10.6.1.1.5	3 Chicchan (8 Zac)

references is relevant during any set of several multiples of the table. For the 10.6 base date under discussion here, only page 68a is astronomically significant during the first two runs through the table. Page 66a becomes important for recording eclipse seasons when the table is recycled and when it is

used with other base dates.

Let us look next at the second picture and clause on pages 68a (Figure 18.4). If one uses the previously mentioned 3 Cib entry date for the table, it is the case that during the first 91-day half-run of the table this clause and picture refer to an eight-day period of time beginning on a calendar-round date of 4 Ik 0 Pop. The *haab* portion of the date (0 Pop) indicates that it refers to a New Year's day. The *tzolkin* portion (4 Ik) indicates that Ik is the year bearer for that year. The picture below that clause depicts the maize god sitting beneath a sky band from which what has been called the 'Mars beast' (Willson, 1924, p. 34) is dangling. The caption above the picture contains a reference to the maize god (in column 1, row 2 of the caption), and page 27a of the same codex (Figure 18.5) implies

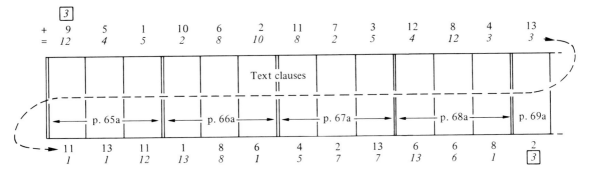

	3												
+	9	5	1	10	6	2	11	7	3	12	8	4	13
=	*12*	*4*	*5*	*2*	*8*	*10*	*8*	*2*	*5*	*4*	*12*	*3*	*3*

Text clauses

| ← p. 65a → | ← p. 66a → | ← p. 67a → | ← p. 68a → | p. 69a |

| 11 | 13 | 11 | 1 | 8 | 6 | 4 | 2 | 13 | 6 | 6 | 8 | 2 |
| *1* | *1* | *12* | *13* | *8* | *1* | *5* | *7* | *7* | *13* | *6* | *1* | *3* |

Figure 18.3 Schematic diagram of the structure of the Seasonal Table in the *Dresden Codex*, showing double sequence of distance numbers and, in italics, *tzolkin* coefficients.

caan-ah

Figure 18.4 Page 68a in the *Dresden Codex* (after Villacorta C. and Villacorta, 1976, p. 146).

that the maize god was the yearbearer for Ik years. All this suggests that the second clause and picture on page 68a are concerned with the New Year's ceremonies for an Ik year, which is corroborated by the fact that the calendar round date associated with them is 4 *Ik* 0 Pop. The long count date is 10.6.1.9.2, which corresponds to 16 December AD 949 in the Gregorian calendar according to the

Modified Thompson 2 correlation constant of 584,283 (Thompson, 1960, p. 305).

There are also references to Mars in the second clause and picture on page 68a (Figure 18.4). The clause begins with a verb which we read as *caan-ah*. The main sign of that verb resembles the sky glyph (Figure 18.6), which would have been read as *caan* in Classical Yucatec; the phonetic complement, *na*, confirms that reading. Juan Pío Pérez (1866–77, p. 37) glosses *caan-ah*, an inchoative verb derived from *caan* 'sky, above', as 'to rise'. The verb is followed by the glyph for Mars (Figure 18.4), which represents its subject. The sense of the clause is 'Mars rose', and, in fact, Mars was a morning star on 16 December AD 949 (Gregorian), rising at 5.01 a.m. (Table 18.3), about an hour and a half before sunrise (Kluepfel, 1980). Venus was also a morning star on that date, rising about 35 minutes after Mars. The sky band in the picture below the second clause on page 68a contains as its middle element the glyph for Venus, from which the 'Mars beast' is suspended. While not technically in conjunction, the two planets were very close to each other in the sky (Table 18.3), and this fact seems to be shown iconographically in the second picture.

The third clause and picture on page 68a refer to a four-day period starting on 24 December AD 949 (Gregorian). The beginning of the clause is missing, but solar and lunar eclipse glyphs are shown hanging down from a sky band. The left-most element of the sky band is a Sun glyph with a darkened area around it, a very probable representation of a solar eclipse. Column 64 of the

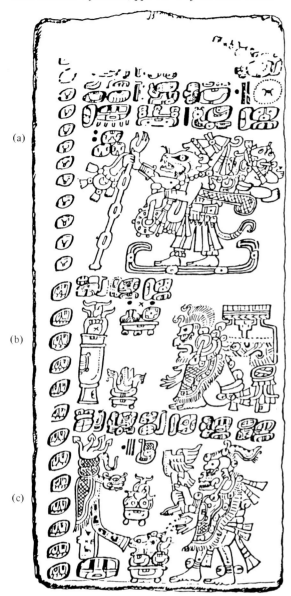

Figure 18.5 Page 27 in the *Dresden Codex* (after Villacorta C. and Villacorta, 1976, p. 64).

Figure 18.6 An inchoative verb derived from *caan* 'sky'. (a) D. 56a. (b) D. 68a. (After Villacorta C. and Villacorta, 1976, pp. 122, 146.)

Table 18.3. *Position of Mars, Venus and the Sun on 16 December* AD *949 (Gregorian), Julian Day Number 2,068,025, Maya day 10.6.1.9.2 4 Ik 0 Pop.*[a]

(a) *Positions as of 5.55 a.m. in the Yucatan Peninsula (20° 30' north latitude, 88° 30' west longitude).*

	Ecliptic longitude	Ecliptic latitude	Azimuth	Altitude
Mars	254.02°	0.21°	118.23°	10.93°
Venus	245.80°	−0.24°	115.37°	3.20°
Sun	265.05°	0°	112.35°	−7.42°

(b) *Times of rising (altitude = 0°), assuming a flat horizon.*

	Time	Azimuth
Mars	5.01 a.m.	112.97°
Venus	5.37 a.m.	113.75°
Sun	6.27 a.m.	114.95°

[a] Astronomical data obtained from Kluepfel (1980).

fifth multiple of the eclipse table on pages 51–8 of the same codex predicts a solar eclipse for the following day, 25 December (Bricker and Bricker, 1983). An eclipse occurred two days later but was not visible in the Maya area (Oppolzer, 1887, p. 206).

There is, then, a close fit between the second and third clauses and pictures on page 68a and the dates associated with them. They correlate three events of apparent significance to the Classic-period Maya: (1) New Year's day, as shown by the references to the maize god in the second picture and its caption; (2) the near conjunction of Mars and Venus shortly before sunrise; (3) a solar eclipse predicted by the fifth multiple of the eclipse table. Although 11 base dates for the seasonal table are given in the codex, it is only during the original multiple of the table using the 10.6.1.1.5 base date that a Maya New Year can be correlated with the astronomical events described on page 68a. We can, therefore, state with some confidence that the text and pictures of the seasonal table refer very specifically to the 182-day period between 11 October 949 and 11 April 950.

We turn now to the first clause and picture on

page 68a (Figure 18.4). It has been suggested by Herbert Spinden (1924, p. 184) that the picture refers to the reversal between the dry and the wet seasons. We believe that the 91-day length of the modules of the table also has a bearing on this question because it so closely approximates the interval between solstices and equinoxes. The so-called reversal signs (each resembling a recumbent letter 'S') directly above the picture and in the caption could also refer to solstices, when the day reaches its maximum or minimum length with respect to the night, or to equinoxes, when the day and the night are of equal length.

The clause over the first picture on page 68a refers during the first half-run of the table to a 12-day period starting on 4 December AD 949, which falls in the heart of the dry season and more than two weeks short of a winter solstice. However, in the second 91-day half-run, the relevant date is 20 March AD 950, coinciding exactly with the spring equinox of that year (Kluepfel, 1980). It is possible that the spring equinox was regarded by the Classic-period Maya as the symbolic dividing line between the dry and the rainy seasons. The picture shows the rain god, Chac, sitting on a sky band and facing in two directions. Rain falls from the 'reversal' sign onto the Chac facing right. Because the reading order of this table is from left to right, the dry season must precede the rainy season, as would be the case at the time of the spring equinox on 20 March AD 950. Another 'reversal' glyph appears in the caption over the picture in the eclipse table on page 52b (Figure 18.7). The *baktun* 10 date associated with that caption falls within one day of a summer solstice. This implies that the 'reversal' glyph was used for recording solstices as well as equinoxes.

Still other examples of the 'reversal' glyph are associated with Venus glyphs in the facade of the interior doorway of Temple 22 at Copan (Closs, Aveni, and Crowley 1984, p. 245, fig. 1). It is interesting that the earliest dates of Venus visibility in the window of that structure, using a diagonal sighting scheme (Closs, Aveni, and Crowley, 1984, pp. 239–41, table 1), always fell less than two months (and often less than one month) after the vernal equinox during the eighth-century AD period in which the structure was most likely built (as explained by Closs, Aveni, and Crowley, 1984, p. 228, 9.17.0.0.0 is a probable *terminus ante quem*

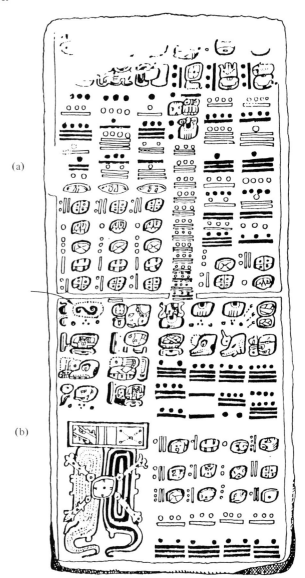

(a)

(b)

Figure 18.7 Page 52 in the *Dresden Codex* (after Villacorta C. and Villacorta, 1976, p. 114).

for the construction of Temple 22). This probable association of Venus, the vernal equinox, and the 'reversal' glyph at Copan draws attention to the fact that the sky band on which Chac sits on page 68a of the seasonal table of the *Dresden Codex* includes a Venus glyph under Chac's feet in the 'rainy' half of the picture, and, indeed, Venus was an 'evening star' on the relevant date, 20 March AD 950, setting about one hour after sunset (Kluepfel, 1980).

18.3 Recycling the table

That the seasonal table was intended for use over longer spans of time is implied by the table of multiples on pages 63 and 64, which contains 31 multiples of the 91-day modular value (Figure 18.1). The following even multiples are given: the 1st through the 20th and the 40th, 60th, 80th, 160th, 240th, 320th, 400th, 480th, 1200th, and 1600th. (Of course, all stated multiples above the 19th are also even multiples of 182 and 364.) There is also an 'aberrant' 1504th multiple with 20 extra days (19.0.4.4), which was clearly squeezed into the leftmost column in violation of the layout of the rest of the table (Figure 18.1).

Over the longer spans of time that result from recycling the table, it is relevant to examine its interaction with two other cycles of astronomical significance. First, as already noted, an interval of 91 days is the closest integral approximation of the mean distance between solstices and equinoxes, and four such intervals (364 days or two multiples of the table) fall 1.2422 days *short* of the true length of the tropical year. Second, a period of 182 days is 8.69 days *longer* than what is called the 'eclipse half-year' of 173.31 days (Teeple, 1931, p. 90; Aveni, 1980, p. 79), the periodicity of the arrival of the Sun at successive *nodes* of the Moon's orbit, which are the points of intersection of the plane of the Moon's orbit around the Earth and the plane of the Earth's orbit around the Sun (the plane of the ecliptic) (cf. Aveni, 1980, pp. 72–3). Eclipses occur only during an 'eclipse season', centered on such a node, if dates of new moon (for solar eclipses) or full moon (for lunar eclipses) fall sufficiently close to the date of nodal passage. Solar eclipses may occur within *c.* 18 days of a node (Teeple, 1931, p. 90) and lunar eclipses within *c.* 12 days of a node (Aveni, 1980, p. 77). The maximum limits of an eclipse season are, therefore, node day ± 18 days. Because of the interactions between a 182-day cycle and the two others just mentioned, the table *precesses* through the eclipse half-year at the same time that it *recesses* through the tropical year. The Classic-period Maya dealt with these problems in several different ways.

The first problem is solved by referring to eclipse seasons *four* times in a 182-day span rather than only twice, as one might expect. Textual and/or iconographic references to eclipses occur for the

intervals beginning on days 26, 75, 128 and 173 of the 182-day span; the first and third references relate to the second picture on page 66a, the second and fourth references to the third picture on page 68a (Figure 18.2). The long-term effects of this scheme are most easily seen by expressing each of the four relevant dates as a deviation score, in days, from the nearest eclipse node and plotting the results. Figure 18.8 shows such a plot for the 3 Cib original version of the table and the next 20 multiples. The second of the first four points on the graph is the date of the first occurrence of the third picture on page 68a; it is one day before a node day, with a deviation score of −1. The occurrence of a solar eclipse three days later has already been mentioned. The other three of the first four points have deviation scores greater than +18 or −18; therefore, for the 3 Cib original version, only the second of the four iconographic references correctly identifies an eclipse season. This is true as well for the second and third runs of the table. By the time of the fifth run through the 182-day table, the eclipse nodes move from page 68a to page 66a, where the second clause and picture become relevant. Three runs later, the nodes have returned to the third clause and picture on page 68a. The nodes continue to oscillate between these pages as recycling of the table continues. Because the shift from one pattern to another is, in fact, a transitional one, a given 182-day multiple of the table may occasionally contain two valid eclipse-season references (for example, as shown by Figure 18.8, in the eighth run both the first and the fourth references fall within 18 days of the node), but – as explained previously – these can never be immediately sequential references. Other multiples may contain no valid references, as is the case in the fourth run of the table. Which, if any, of the four possibilities was correct would have been perfectly obvious to the Maya, in advance of the event, because of the way the eclipse cycle interacts with the *tzolkin* cycle, as we have explained in detail elsewhere (Bricker and Bricker, 1983). Thus, the unusual spacing of references to eclipse seasons in the seasonal table serves as a compensatory mechanism for the discrepancy between the length of the table and the eclipse half-year.

The second problem, the recession of the table through the tropical year, is handled by correcting the table every few decades – adding extra days

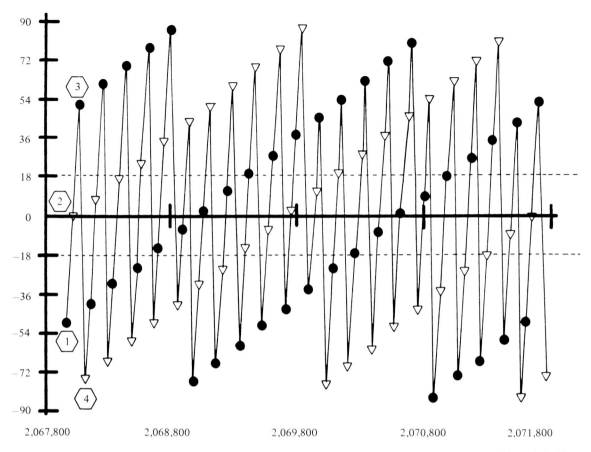

Figure 18.8 Relationship between node dates and the four dates associated with eclipse glyphs in the 3 Cib original version of the seasonal table and the following 20 multiples. The *Y*-axis shows the deviation, in days, of each tabulated date from the node date. The maximum duration of an eclipse season, node ±18 days, is indicated by dashed horizontal lines. The *X*-axis shows Julian Day Numbers. Tabulated dates associated with the second picture on page 66a are plotted as solid circles; dates associated with the third picture on page 68a are plotted as open triangles.

in order to compensate for the recession. In some of the other tables in the *Dresden Codex* – for example, the eclipse table and the Mars table – long-term corrections are signalled by 'aberrant' multiples; because such a multiple incorporates the extra days as well as the modular value, it allows the user to produce a new and corrected starting date for subsequent recycling. In the seasonal table, however, it is the *result* of correction that is supplied to the user, not the formula for producing it. Frequent correction of the seasonal table is indicated by the presence of the many base dates that have already been mentioned.

The mechanism and effect of correction for recession through the tropical year may be briefly illustrated by the shift from the 10.6.1.1.5 base date

to the 10.7.4.3.5 base date. These dates are separated by a total of 8320 days, 39 days or three *trecenas* more than 91 modules of 91 days. Because the correction value, 39 days, is an even multiple of 13 (as, indeed, is the modular value of 91 days), and because the total distance between base dates, 8320 days, is an even multiple of 260, the *tzolkin* date of the entry day is preserved – 3 Cib 19 Muan for the 10.6 version and 3 Cib 4 Ceh for the 10.7 version.

The effect on recession is not quite what one might expect. Because the intervals between columns in the first half of the 182-day table are not the same as those in the second half (although they both sum to 91), the addition of 39 days to the table's count effects a somewhat smaller correction

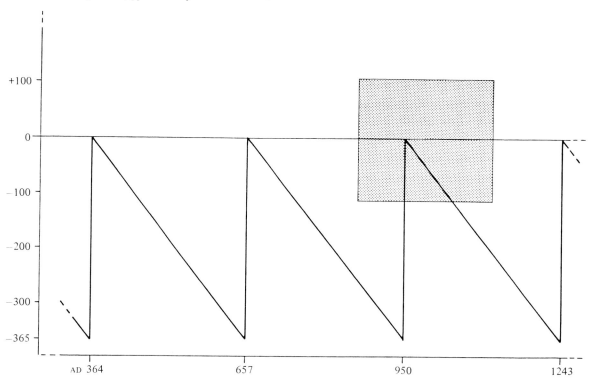

Figure 18.9 Rate of recession of the seasonal table through the tropical year, assuming perfect alignment with the vernal equinox in AD 950. Shaded area is the temporal range shown in Figure 18.10.

to the interaction of the table's iconography with the passage of the tropical year (a step-by-step presentation of the calculation details is given in the appendix to this chapter). Figure 18.9 shows the rate of recession of the table. If, for example, the iconography of the table were perfectly aligned with the vernal equinox in AD 950 and if the table remained uncorrected, the table would fall further and further behind the vernal equinox in subsequent years until, in AD 1243, the recession having reached 365.2422 days, the alignment would once again be perfect. Figure 18.10 shows what actually happened to the table in the 10th and 11th centuries AD (*baktun* 10 of the Maya). As we have already mentioned, the first clause and picture on page 68a fall exactly on the vernal equinox during the 10.6 3 Cib original, 20 March AD 950. In the following years, the recession accumulates to a total of 29 days by AD 973. However, the 39-day correction made as of July 972 has the effect of associating the first picture on page 68a with 15 March 973,

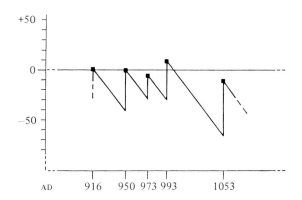

Figure 18.10 Adjustment of the seasonal table during the 10th and 11th centuries AD to correct for its recession through the tropical year. The *Y*-axis shows the deviation in days, from the actual date of the vernal equinox, of the corresponding tabulated dates in multiples based on the 10.4, 10.6, 10.7, 10.8, and 10.11 starting dates. The diagonal lines show the rate of recession, as in Figure 18.9.

241

Table 18.4. *Detailed calculation of the shift from the 10.6.1.1.5 to the 10.7.4.3.5 base date*

	Date (J.D.#)	Date (Gregorian and Maya)	Multiple of table	Starts column number	Day number of multiple	Remarks
(1)	2,067,868	12 Jul 949 10.6.1.1.5 3 Chicchan 8 Zac	–	–	–	10.6 base date
(2)	+ 91					add 91 to reach starting date
(3)	2,067,959	11 Oct 949 10.6.1.5.16 3 Cib 19 Muan	10.6 Orig.	col. 1	1	10.6 starting date
(4)	+ 91					sum of DNs in first half of table
(5)	2,068,050	10 Jan 950 10.6.1.10.7 3 Manik 5 Uo	10.6 Orig.	col. 14	92	begin second half of table
(6)	+ 69					sum of DNs in cols. 14 through 22
(7)	2,068,119	20 Mar 950 10.6.1.13.16 7 Cib 14 Zec	10.6 Orig.	col. 23	161	**vernal equinox**
(8)	+ 22					sum of DNs in cols. 23 through 26
(9)	2,068,141	11 Apr 950 10.6.1.14.18 3 Edznab 16 Xul	10.6 1st	col. 1	1	182 days later than 10.6 orig. starting date; component 'a' accounted for
(10)	+ 8008					$= 182 \times 44$
(11)	2,076,149	14 Mar 972 10.7.4.1.6 3 Cimi 14 Zec	10.6 45th	col. 1	1	component 'b' accounted for; begin first half of table
(12)	+ 91					
(13)	2,076,240	13 Jun 972 10.7.4.5.17 3 Caban 5 Yax	10.6 45th	col. 14	92	component (c) accounted for; begin second half of table
(14)	+ 39					**correction**; $= 13 \times 3$
(15)	2,076,279	22 Jul 972 10.7.4.7.16 3 Cib 4 Ceh	10.6 45th or 10.7 Orig.	– col. 1	131 1	component 'd' accounted for 10.7 starting date (= 10.7 base date + 91); *tzolkin* coeff. = 3 as required
(16)	+ 54					sum of DNs in cols. 1 through 9
(17)	2,076,333	14 Sep 972 10.7.4.10.10 5 Oc 18 Kankin	10.7 Orig.	col. 10	55	eight days before autumnal equinox
(18)	+ 8					

Table 18.4 (*contd*)

	Date (J.D.#)	Date (Gregorian and Maya)	Multiple of table	Starts column number	Day number of multiple	Remarks
(19)	2,076,341	22 Sep 972 10.7.4.10.18 13 Edznab 6 Muan	10.7 Orig.	–	63	autumnal equinox
(20)	+ 120					
(21)	2,076,461	20 Jan 973 10.7.4.16.18 3 Edznab 1 Zip	10.7 1st	col. 1	1	begin first half of table
(22)	+ 54					sum of DNs in cols. 1 through 9
(23)	2,076,515	15 Mar 973 10.7.5.1.12 5 Eb 15 Zec	10.7 1st	col. 10	55	five days before vernal equinox
(24)	+ 5					
(25)	2,076,520	20 Mar 973 10.7.5.1.17 10 Caban 0 Xul	10.7 1st	–	60	**vernal equinox**

DN = distance number

five days before the vernal equinox. Twenty-four of the 29 days of recession have been recovered. Thus, the 39-day correction may be seen as the best compromise between the competing demands of the *tzolkin* and the tropical year. With the shift to the new base date, the seasonal table becomes useful once again for correlating eclipse seasons with equinoxes.

It will be recalled that the introduction to the table gives a total of five *baktun* 10 base dates. Figure 18.10 makes it clear that *all* of them can be considered in the same way; the figure is a graphic record of Maya efforts at commensuration of eclipse cycles and the tropical year throughout a century and a half of their history. We consider this commensuration to be the principal function of the seasonal table of the *Dresden Codex*.

Acknowledgments

We acknowledge with gratitude the help of our colleagues Anthony F. Aveni (Colgate University) and Munro S. Edmonson (Tulane University). Their critical readings of an earlier version of this paper provided very useful suggestions, which we have attempted to implement.

Appendix to Chapter 18

The difference between the 10.6 base date (10.6.1.1.5; Julian Day Number 2,067,868) and the 10.7 base date (10.7.4.3.5; Julian Day Number 2,076,188) is 8320 days. This is also the difference between the starting dates of the tables based on these base dates (in each case, starting date = base date + 91 days). The explanation of how the 10.7 table represents an adjustment for the recession that had accumulated during the period of use of the 10.6 table may be couched in terms of either difference, but for ease of presentation we have chosen to treat the 8320-day quantity as the difference in starting dates.

As explained in the text, a period of 8320 days is equivalent to 91 modules of 91 days plus an additional 39 days. In order to facilitate the explanation presented below, it is, however, more useful

243

to think of 8320 days as being composed of four components, as follows:

(a) 182 days (182 × 1)

(b) 8008 days (182 × 44)

(c) 91 days (182 × 0.5)

(d) 39 days (13 × 3)

In going from the 10.6 starting date to the 10.7 starting date, one must account for all four components, the total of which is the 8320-day difference between starting dates.

Another terminological convention that will facilitate explanation is to refer to captions, distance numbers, and pictures as occurring in columns 1 through 13 if they apply to the first 91-day half of the table and in columns 14 through 26 if they apply to the second half. Thus, for example, the seated double-Chac figure associated with the vernal equinox and the shift from the dry to the rainy season may occur in either column 10 (first half of the table) or column 23 (second half).

Each quantity used in the calculation presented in Table 18.4 is numbered serially from '(1)' to '(25)', and these numbers are referred to in the discussion.

Three aspects of the calculation presented in Table 18.4 benefit from further clarification. First, it should be noted that quantity 6 is not equal to quantity 22. The sum of the distance numbers between the beginning of the *first* half of the table and the first occurrence of the tenth picture (column 10) is less than the sum of the distance numbers between the beginning of the the *second* half of the table and the second occurrence of that picture (column 23). Whereas the tabulated reference to the vernal equinox of AD 950 (quantity 7; = Julian Day Number 2,068,119) is 69 days after the beginning of the second 91-day half of the table, the reference

to the vernal equinox of AD 973 (quantity 23; = Julian Day Number 2,076,515) is only 54 days after the start of the first half of the table. With respect to these two vernal equinoxes, the structure of the table has resulted in the apparent 'loss' of 15 days (69 minus 54). However, what we may call a 'gross correction' of +39 days (quantity 14) has been made in AD 972. The 'net correction' is then:

$$
\begin{array}{r}
+39 \\
-15 \\
\hline
= +24
\end{array}
$$

Because of the different arrangement of distance numbers in the two halves of the table, only 24 days of the 29 days of accumulated recession are recovered by the 39-day correction.

The second point to be clarified is that, although the Julian Day Number shown as quantity 17 in the 10.7 Original multiple is in part the structural homologue of quantity 7 in the 10.6 Original (in the sense that both are starting dates of the interval associated with the tenth picture), it is not *seasonally* homologous. The quantity 17 date is clearly in the wrong half-year, near the autumnal rather than the vernal equinox. If, therefore, calculation of the correction for recession is to be made accurately, we must go one additional half-year (half of a computing year = 182 days = one full multiple of the table) forward in time to column 10 of the 10.7 1st multiple of the table, arriving at the Julian Day Number shown as quantity 23, a date near the *vernal* equinox of AD 973.

The third clarification is an alternative statement of the first. Quantity 23 minus quantity 7 is a period of 8396 days, which is equal to 46 multiples of the full table (46 × 182) plus an additional 24 days. Adjusting for recession by switching to the new 10.7 table has effected a net correction of +24 days.

References

Aveni, A. F. (1980). *Skywatchers of Ancient Mexico*. Austin: University of Texas Press.

Beyer, H. (1943). The long count position of the serpent number dates. *International Congress of Americanists, Proceedings of the 27th Congress*, vol. i, pp. 401–5. Mexico City, 1939.

Bricker, H. M. and Bricker, V. R. (1983). Classic Maya prediction of solar eclipses. *Current Anthropology* **24**, 1–23.

Bricker, V. R. and Bricker, H. M. (1986). *The Mars Table in the Dresden Codex*. Tulane University, Middle American Research Institute, Publication 57, pp. 51–80.

Closs, M. P. (1977). The date-reaching mechanism in the Venus table of the Dresden Codex. In *Native American Astronomy*, ed. A. F. Aveni, pp. 89–99. Austin: University of Texas Press.

Closs, M. P., Aveni, A. F. and Crowley, B. (1984). The planet Venus and Temple 22 at Copán. *Indiana* **9**, 221–47.

Kluepfel, C. (1980). *Planets.* (An astronomical software package available from the author.)

Lounsbury, F. G. (1983). The base of the Venus table of the Dresden Codex, and its significance for the calendar-correlation problem. In *Calendars in Mesoamerica and Peru: Native American Computations of Time*, eds. A. F. Aveni and G. Brotherston, pp. 1–26. Oxford: BAR International Series 174.

Oppolzer, T. (1887). *Canon der Finsternisse.* Wien: Kaiserlichen Akademie der Wissenschaften, Mathematisch-Naturwissenschaftliche Classe, Denkschriften 52.

Pío Pérez, J. (1866–77). *Diccionario de la Lengua Maya.* Merida: Imprenta Literaria de Juan F. Molina Solís.

Spinden, H. J. (1924). *The Reduction of Mayan Dates.* Harvard University, Peabody Museum of American Archaeology and Ethnology, Papers, vol. vi, no. 4.

Teeple, J. E. (1931). Maya astronomy. *Carnegie Institution of Washington, Contributions to American Archaeology* 1, 29–115.

Thompson, J. E. S. (1960). *Maya Hieroglyphic Writing: An Introduction.* Norman: University of Oklahoma Press.

Thompson, J. E. S. (1972). A Commentary on the Dresden Codex. *American Philsophical Society, Memoirs*, vol. 93. Philadelphia.

Villacorta C., J. A. and Villacorta, C. A. (1976). *Códices Mayas*, 2nd edn. Guatemala: Tipografía Nacional.

Willson, R. W. (1924). *Astronomical Notes on the Maya Codices.* Harvard University, Peabody Museum of American Archaeology and Ethnology, Paper, vol. vi, no. 3.

19

A Palenque king and the planet Jupiter

Floyd G. Lounsbury *Yale University*

The introductory passages of the inscription of the Temple of the Sun at Palenque[1] are concerned with mythological matters, in particular with the birth of the second of the deities of the 'Palenque Triad', an event that is declared to have taken place on a day which was number 1.18.5.3.6 (275 466) in the Maya day count and 13 Cimi 19 Ceh in the Maya calendar-round, this being equivalent, in European chronology, to October 28 (retroactive Gregorian) of the year 2360 BC.[2] The presentation of this date in the 'initial series' of the inscription is followed by its further orientation in what are known as 'supplementary series', which specify – amongst other things – its position in the lunar calendar (26th day since first visibility of the current moon, in this, a 30-day lunar month) and its position in an 819-day cycle.[3] But the passage which purports to place it in this latter cycle contains some glaring and puzzling discrepancies. These are well known to students of Maya epigraphy but have yet to be offered reasonable explanation.

An '819-day passage' is one which orients an initial-series date according to its position within a Mayan cycle of that length, doing so by stating the number of days since the last 'station' or epochal day of the cycle, and then – along with other details – naming that station. In the present case, the inscription states that 13 Cimi 19 Ceh, 1.18.5.3.6, the date of the initial series, was 1.2.11 (411 days) after the 819-day station of 1 Ik 10 Tzec. But this is an arithmetical impossibility. The interval from 13 Cimi 19 Ceh back to the most recent 1 Ik 10 Tzec is 10.10.8 (3808 days); while that between 13 Cimi 19 Ceh and the most recent 1 Ik 10 Tzec

that was a station in the 819-day cycle is 11.10.10.4 (83 004 days).[4] This was the 1 Ik 10 Tzec of 1.6.14.11.2. If we translate the record of the inscription, it amounts to saying that the day November 14 of the year 2360 BC was 411 days after August 14 of 2587 BC, which is a bit short of the mark: it was more than two-and-a-quarter centuries. This is one of the problems of the inscription. And, apart from the arithmetical problem of the 1 Ik 10 Tzec date, there is another, namely that the proper 819-day station for the initial date of this inscription, according to all existing precedents, should be 1 Imix 19 Pax (1.18.4.7.1), at a removal of 285 days (14.5) from the initial date; for this is the next prior station of that cycle. A second problem, then, is in the bizarre choice of station for the 819-day passage in this inscription: instead of the next prior (as are all other recorded examples), it is the 101st-prior.

After the mythological prologue, the text goes on to record what appears to have been one of the most important events in the life of the then ruler of Palenque, a king to whom we can refer confidently by his proper name, 'Serpent-Jaguar' in translation, or 'Chan-Bahlum' in a Chol–Mayan rendering, for both the sense and the reading of his nominal hieroglyph are now quite secure. This historical event was a three-day affair, obviously concerned with the gods, and with more than just the Triad. It began on the day 2 Cib 14 Mol, 9.12.18.5.16, which was July 20, AD 690; it continued on to the next day, 3 Caban 15 Mol; and it was concluded with an offering of some sort on the third day. The human protagonist who dealt with the gods on this occasion was Chan-Bahlum. The events are recorded not only in the Temple of the Sun, but also in the Temple of the Cross and in the Temple of the Foliated Cross (these accounts complementing one another in a certain respect); and the occasion was commemorated, along with its 12th-year anniversary, also in the Medallion Series that was placed across the piers and the lintels of the face of the Temple of the Inscriptions. The nature of the events and of the occasion that prompted them have been one of the many intriguing questions that are posed by the inscriptions of Palenque.

The remaining passages of the inscription from the Temple of the Sun are concerned with other chronological matters relating to Chan-Bahlum, viz the date on which he, as a boy of six, was named as heir-designate and future successor of his father Pacal; the numerical relation of this date to that of a similar event for a predecessor of similar age a century and a half earlier; the five-day proximity of the date to the summer solstice; the date of Chan-Bahlum's birth; his precise age on the day of his heir-designation; and the interval from this event to the half-katun date. These constitute the subject matter of the main text, apart from the passages that are inserted into the iconographic part of the central panel. Of those latter, one of them is again concerned with the date of his heir-designation, and the other with that of his inauguration as ruler 45 years later (his father lived to be 80 years old). The date of this last event will be one of those to concern us later, but at this point we return to the problem of the apparent error in the 819-day passage.

One suspects that an error so gross is unlikely to have been due to any ordinary sort of mistake in arithmetic. There is, in fact, only one other instance in Mayan inscriptions – from any site – of an error in an 819-day passage. That, curiously, is also from Palenque, and from the same 'Cross' group of temples, namely, from the temple of the Foliated Cross. In that passage the interval that is recorded (the 'distance number' in Mayanist jargon) was computed correctly, but it was mistakenly added to the initial-series date instead of being subtracted from it as it should have been. It is as if a competent calendar priest had done the initial calculation – the hard part – and then turned the result over to an assistant to take it from there, but to an assistant, unfortunately, who was still too

much of a novice to know that in this context (in contradistinction to most others) it should be subtracted rather than added. The error in the Temple of the Foliated Cross is thus explicable; but the discrepancy in the Temple of the Sun is of an order that calls for some other kind of explanation. In any case, its properties merit inquiry before dismissing it as hopeless; for the possibility of numerological manipulation is something which experience has taught us not to overlook.

Calendrical primes, such as are employed in certain other instances of Mayan numerology, fail to yield anything of significance in application to this problem. That being so, planetary constants are the next most plausible to be tried. The number 584, for Venus, also is unproductive; but 399, for Jupiter, offers a more promising result. The actual interval from the 1 Ik 10 Tzec 819-day station (at 1.6.14.11.2) to the initial-series date of 13 Cimi 19 Ceh (1.18.5.3.6), as previously noted, is equal to 83 004 days (11.10.10.4). The recorded interval, as also previously noted, is equal to 411 days (1.2.11). The difference between the true interval and the recorded one – i.e. the amount of the apparent computational 'error' – is thus equal to 82 593 days (11.9.7.13). This is an integral multiple of 399, the whole-number approximation to the mean synodic period of Jupiter (398.88 days, when averaged over very long time spans). It has thus the earmark of a Mayan numerological projection, which is typically in terms of integral values without application of a correction formula.[5]

This result, though suggestive of an interest in Jupiter, could yet be due simply to chance. But there is another facet to the puzzle. A day 1 Ik 10 Tzec is also recorded in a pair of medallions over one of the piers of the Temple of the Inscriptions. This is the same calendar-round day as the one recorded for the 819-day station accompanying the mythological initial date of the Temple of the Sun. An analysis of the Medallion Series by Peter Mathews (1980) showed that the role of 1 Ik 10 Tzec in the text of the medallions was that of the 819-day station proper to the initial series date of the text (the fourth of those commemorating the 2 Cib 14 Mol event of 9.12.18.5.16), and that it was recorded there properly distanced from that initial date (by an interval of 2.3.14, i.e. 794 days prior).

An 819-day station can occur on the same calendar-round day only once every 63 calendar rounds

(3276 years, omitting leap-year days). Or, to put it another way, an appropriate calendar-round day – such as 1 Ik 10 Tzec – can recur again as an 819–day station only after a lapse of 63 × 52 Maya years. That is the interval between the 1 Ik 10 Tzec of the 819-day station of the Temple of the Sun (necessarily 1.6.14.11.2) and the 1 Ik 10 Tzec of the 819-day station of the Medallion Series (9.12.16.2.2). Since the choice of the station to which to relate the initial date of the Temple of the Sun is entirely counter to the normal rule and to all precedents (being the 101st-prior such station rather than the first-prior), one is led to ask whether there could have been some overriding desideratum in this case, giving cause for replication in mythological time of the station that is appropriate in the historical context. And, since the numerical property of the 'error' in the Temple of the Sun is suggestive of Jupiter, one is led to inquire into the circumstances of Jupiter on the historical date initiating Chan-Bahlum's important rites for the gods (2 Cib 14 Mol, 9.12.18.5.16) and on the date of the 819-day station prior to those rites (1 Ik 10 Tzec, 9.12.16.2.2).

For the 2 Cib 14 Mol date – equal to AD 690 July 20 (Julian) by the 584285 correlation – data from the Tuckerman planetary tables may be seen as favourable to a hypothesis involving not only Jupiter, but also Saturn and Mars. During a period of some six months prior to that date, Jupiter, Saturn and Mars were in close proximity, Jupiter and Saturn having barely missed one of their rare 'triple conjunctions' with each other (see Figure 19.1).[6] And at intervals following this date there occurred first a conjunction of Mars and Jupiter, then one of Mars and Saturn, and then, somewhat later, one of Jupiter and Saturn (what would have been their third conjunction, had their paths actually crossed and recrossed on their first close encounter). This suggests that it might have been the unusual and impressive gathering of these three celestial figures – conceived to be gods – that occasioned the special Palenque observances.[7]

But this is only a vague description of the situation. When put into precise terms, it is inadequate to account for the choice of the day 2 Cib 14 Mol (July 20, AD 690) for the observances. The closest approach of Jupiter and Saturn to each other, and of Mars to them, prior to this date was some four months earlier. After a trend of separation for a

248

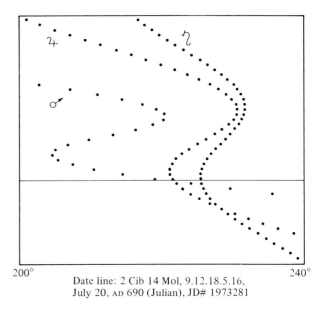

Date line: 2 Cib 14 Mol, 9.12.18.5.16,
July 20, AD 690 (Julian), JD# 1973281

Figure 19.1 Celestial longitudes of Jupiter, Saturn and Mars before, on, and after the date of Chan-Bahlum's rites for the gods at Palenque, July 20, AD 690.

few months, and then of reversal of that trend, they were now closing in on each other again, Mars moving rapidly and the other two more slowly, heading – no doubt obviously – for conjunctions. But the conjunction of Mars and Jupiter was not until several days after the Palenque observances had been concluded, that of Mars and Saturn only after another several days, and that of Jupiter and Saturn not until another month and a half after that. Thus, although suggestive, these seemingly notable aspects of the situation fall short of accounting for the particular date.

On the other problematic date, that of the historical 819–day station 1 Ik 10 Tzec (9.12.16.2.2), May 17, AD 688 (Julian), the situation of Jupiter was in a certain respect similar. Mars was, and had been, in proximity (these two also having barely missed a triple conjunction – see Figure 19.2), though Saturn was more distant. But there is a more exact similarity in the situations of Jupiter on the two dates of interest: in each case the planet had moved about one-third of a degree ($0°.32$ and $0°.30$, respectively) in departure from dead center of the second stationary point. This – very likely a 'just noticeable difference' for naked-eye observation – suggests another hypothesis, viz that it could have been Jupiter's

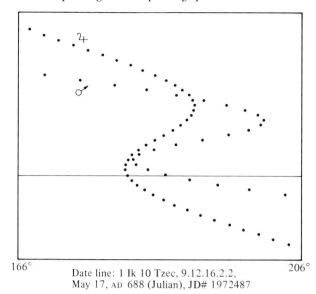

Date line: 1 Ik 10 Tzec, 9.12.16.2.2,
May 17, AD 688 (Julian), JD# 1972487

Figure 19.2 Celestial longitudes of Jupiter and Mars before, on, and after the date of the 819-day station prior to the rites, May 17, AD 688.

departure from the second stationary point that was the critical detail of interest in these situations. It appears to be the only one that allows specification of a precise similarity on these two dates.

Two instances, however, are insufficient to make a case for the hypothesis suggested by them; their similarity could have been due to chance. But if Chan-Bahlum or his court astrologers had really been interested in Jupiter's departures from the second stationary point, then some other dates might be expected also to give evidence of it. If so, the ruler's accession date is one that should be looked into.

The date of Chan-Bahlum's accession to rule is recorded at least seven times at Palenque, in six different inscriptions: it was on 8 Oc 3 Kayab, 9.12.11.12.10, which translates as January 7, AD 684 (Julian). On this date, it turns out, Jupiter was indeed at virtually the same point in its synodic period as it was on the two dates previously reviewed; its position this time was 0°.42 past dead center of the second stationary point (see Figure 19.3).

The Medallion Series from the Temple of the Inscriptions had already been mentioned, and two of its dates have concerned us. The first was that of its initial series, which was one of the four recordings of 2 Cib 14 Mol, 9.12.18.5.16, the date of Chan-Bahlum's special observances for the gods (first noted here in connection with its record in the Temple of the Sun). The second was the date of its 819-day series, the historical 1 Ik 10 Tzec station of that cycle, which has provided a possible rationale for the exceptional treatment of the 819-day passage in the Temple of the Sun. The concluding text of the Medallion Series made reference to yet a third date, which was the 12-year anniversary of the 2 Cib 14 Mol event. It included the 12-year distance number 12.3.0 (12(360) + 3(20) + 0(1) = 12(365)) and the calendar-round day to which this leads when applied to 2 Cib 14 Mol. That day is 1 Cib 14 Mol (9.13.10.8.16 = July 17, AD 702), for the 12 × 365-day increment is one which has the property of reducing the trecena component of a calendar-round date by an amount of one, while returning the veintena and the haab components unaltered. It must be supposed that this calendrical property provided a prime rationale for the selection of that particular date for an anniversary commemoration of the 2 Cib 14 Mol event. But was it the only criterion? Can there have been others?

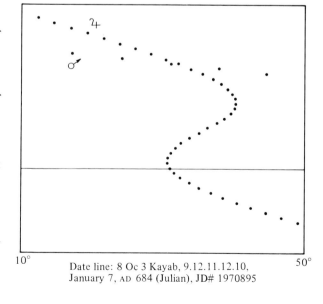

Date line: 8 Oc 3 Kayab, 9.12.11.12.10,
January 7, AD 684 (Julian), JD# 1970895

Figure 19.3 Celestial longitudes of Jupiter before, on, and after the date of the royal accession of Chan-Bahlum at Palenque, January 7. AD 684. (Also of Mars, top portion of chart.)

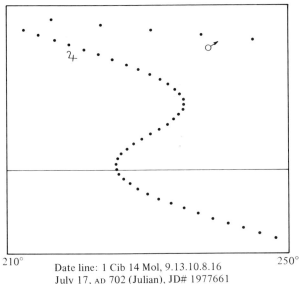

210°   Date line: 1 Cib 14 Mol, 9.13.10.8.16   250°
July 17, AD 702 (Julian), JD# 1977661

Figure 19.4 Celestial longitudes of Jupiter before, on, and after the date of the 12-year anniversary of the rites for the gods, July 17, AD 702. (Also of Mars, at top of chart.)

The 12-year interval (4380 days) is just seven or eight days short of the 11th multiple of the mean synodic period of Jupiter (11 × 398.88 = 4387.68). The planet would necessarily have been again on point of departure from its second stationary point, though its movement would probably have been still too slight for naked-eye detection. The amount was actually 0°.08 (see Figure 19.4). The 12-year period is also one which brings Jupiter back to approximately the same region of the zodiac. The mean sidereal period of the planet is 4332.59 days. Its celestial longitude on 2 Cib 14 Mol was 221°.43; on 1 Cib 14 Mol, 12 years later, it was 225°.73, just 4°.3 beyond its earlier position. This, then, was Jupiter's first departure from a second stationary point in the same constellation of the zodiac as it has been in on the 2 Cib 14 Mol date 12 years earlier. On both occasions it was in the region of Beta and Delta Scorpii.[8]

Chan-Bahlum died on 6 Chicchan 3 Pop, 9.13.10.1.5 (February 16, AD 702), only half a year before the anniversary date of 1 Cib 14 Mol; and his next-younger brother, who had already been designated to succeed him, was installed in office just 48 days before this date. Thus, although the planning for the 12-year anniversary observance

and for the commemorative Medallion Series could well have been Chan-Bahlum's, their execution must have fallen to his brother and successor (whom we call Kan-Xul, or Hok, though not with the degree of confidence that we have in the names of Chan-Bahlum and Pacal).

Another monument to Chan-Bahlum, with yet another postumous date, is in Temple XIV at Palenque. Its tablet shows him executing the dance that symbolizes emergence from the underworld, while he is offered the symbol of rulership by his long deceased mother.[9] (The dance step and the arm movement are known from other examples at Palenque and elsewhere, as well as from vase paintings which relate it to its mythological precedents.) The inscription of the panel begins with dates of events in an antiquity of deep cosmological proportions. It moves then from one of these by an interval of 5.18.4.7.8.13.18 (nearly a million years!) to the day 9 Ahau 3 Kankin (9.13.13.15.0 = November 2, AD 705). On this day there occurs an event of which Chan-Bahlum is the protagonist and which involves again some of the deities. The iconography and the posthumous timing implicate it as the date of the ruler's apotheosis, although the hieroglyphs designating the event are not yet sufficiently well understood to secure the interpretation from that line of evidence. The four preceding examples which we have seen of Jovian second-station dates lead us to inquire now into the position of Jupiter on this date. It turns out again to be the same: Jupiter has moved 0°.29 off dead center of the second stationary point (see Figure 19.5).

We have now, in all, five examples that appear to be indicative of an interest in the movement of the planet Jupiter. The critical datum would seem to be the planet's first perceptible move in departure from the second stationary point – the deity's embarkation, so to speak, on the next leg of his journey in his nearly 12-year circuit of the zodiac. Of these five examples, three are the dates of events whose timings were subject to human control: Chan-Bahlum's inauguration as ruler, his special rites for the gods, and his apotheosis, all of these coinciding with the same critical point in the synodic period of Jupiter. The other two are dates whose determinations followed independent arithmetical or calendrical criteria; but they happened to exemplify the same astronomical phenomenon, and the exploitation of the date in each case was clearly deli-

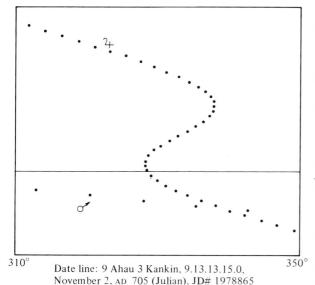

310°   350°

Date line: 9 Ahau 3 Kankin, 9.13.13.15.0,
November 2, AD 705 (Julian), JD# 1978865

Figure 19.5 Celestial longitudes of Jupiter before, on, and after the date of the apotheosis of Chan-Bahlum, November 2, AD 705. (Also of Mars, after this date.)

the long count for fixing dates given in the calendar-round. Two others are of natural events, Chan-Bahlum's birth, 9.10.2.6.6 (May 20, 635), and his death, 9.13.10.1.5 (February 16, 702), whose timings presumably were not subject to prior astrological calculation. (His birth, however, was the day before a predictable eclipse date,[10] a coincidence exploited in other ways in the inscriptions and mythological inventions of Palenque.) With these four removed from consideration in relation to the Jupiter phenomenon, the remainder is down to three.

One of these is the date of the rite that we call 'heir designation', this being the only interpretation that survives comparison of all of its hieroglyphic records and their contexts. For Chan-Bahlum, as for a ruler a century and a half earlier with whose similar rite his is compared, it took place when he was six years old – 6.2.17, to be exact. (There is wide variation in the ages of the rulers for whom it is recorded: the youngest were those designated in childhood or youth to succeed lineally, with actual succession coming many years or decades later; the oldest were those designated in adulthood to succeed laterally, with succession following after a relatively short interval.) An heir-designation, surely, is an occasion susceptible to astrological timing.

The heir-designation rite for Chan-Bahlum took place on 9 Akbal 6 Xul, 9.10.8.9.3, which was June 14 (Julian Calendar) or June 17 (retroactive Gregorian) of the year 641. In respect to the situation of Jupiter, this date, strictly speaking, ought not be eligible for inclusion in the same category as the five 'second-station' dates that have already been discussed, because the amount of the planet's movement away from the second stationary point was a bit more, really too much to be considered a 'just noticeable' difference or a 'first' indication of departure from the rest station. It had moved 1°.18 by that date. As compared with the others (0°.32, 0°.30, 0°.42, 0°.29 and 0°.08), this is of a different order, though of course still miniscule in relation to the range of possibilities. In a different respect, however, the situation bore an aspect of similarity to those of the 1 Ik 10 Tzec and 2 Cib 14 Mol dates (of 47 and 49 years later), this being in an involvement of Jupiter with Mars. In fact, it was the most special of all of these, and surely the most striking to observe. The rite for Chan-Bahlum fol-

berate and in response to the astronomical coincidence. One of these involved the extraordinary projection, employing Jupiter numerology, of a historical 819-day station back 63 calendar-rounds to create a mythic-time precedent; and the other gave recognition to an approximate sidereal, synodic, and calendrical anniversary of a Jupiter date and an event that were climactic in the life of this ruler.

For the sake of perspective on the statistical significance of these coincidences in timing, we should see them in relation to the entire corpus of recorded dates pertaining to this ruler. There are a total of 13 or of 14 if the third day of the rites for the gods is included. (Reference is made to the third day in the inscriptions, but it is not listed by name as are the first two days of the rites.) Of the 13 named days, five are those discussed above, associated with the second stationary point of Jupiter, and another of the 13 must be included with these, since it is that of the second day of the rites. This leaves a reminder of seven. Of the seven, two are period-ending dates in the long count: 9.10.10.0.0 (December 3, 642), and 9.13.0.0.0 (March 15, 692), strictly numerical in determination and with no significance in relation to critical points of Jupiter's period. They serve as round-number anchors in

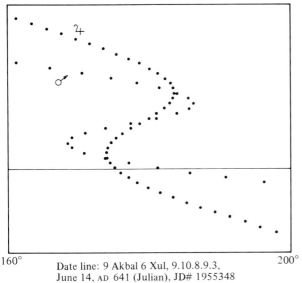

160° Date line: 9 Akbal 6 Xul, 9.10.8.9.3, 200°
June 14, AD 641 (Julian), JD# 1955348

Figure 19.6 Celestial longitudes of Jupiter and Mars before, on, and after the date of the rite of heir-designation for Chan-Bahlum, at age six, June 14, AD 641.

lowed upon a remarkable triple conjunction of those two planets, about the most symmetrical and 'perfect' example that might ever be witnessed, the conjunctions coinciding, respectively, with (1) Jupiter's first stationary point, (2) the virtually simultaneous oppositions of Mars and Jupiter to the sun, and (3) Jupiter's second stationary point. (See Figure 19.6.) It was a celestial drama which the astrologers must surely have been monitoring. And having long since personified and supernaturalized these zodiacal travelers, they could hardly have failed to attach significance to their encounters. This is what immediately preceded the rite that was performed before the gods for the boy Chan-Bahlum by his father Pacal, dedicating him and designating him (if our interpretation is correct) as his future heir and successor. But the date of that rite, while similar to the others reviewed here, was 13 or 14 days later (about 0°.75 further) in respect to Jupiter's departure from the second stationary point.

Another attribute of this date, however, deserves mention: it was the date of new moon (lunar–solar conjunction) just prior to the summer solstice – 'five days prior', as is specially noted in the inscription of the Temple of the Sun.[11] The fact that attention is directed to this in the inscription suggests that

the lunar and seasonal timings may also have been a desideratum, in which case the slightly greater measure of Jupiter's removal from the second stationary point could be seen as a minor accommodation to the lunar schedule. Also to be considered is that this was the first of these Jupiter timings at Palenque, preceding the others by from 43 to 64 years. And it is only with Chan-Bahlum that they are associated; the dates for no other ruler at Palenque conform to this pattern.

There is one other important date in Chan-Bahlum's official curriculum vitae: it is the eight-year anniversary of his accession to rule. This was observed on 5 Eb 5 Kayab, 9.12.19.14.12, which equates with January 7 (Julian) or 10 (Gregorian), AD 692. Although the accession date (v. supra) is one of those exhibiting the Jupiter timing discussed here, this eight-year anniversary date is not of that category. (One at 12 years would have done better.) For present purposes, i.e. for the hypothesis of deliberate timing of important events to coincide with Jupiter's departure from the second stationary point, this counts as a definite negative instance. It had, however, other attributes of potential astronomical interest. Eight years is the Venus recurrence period – five synodic periods, 13 sidereal periods (to within a day and a half), with completion of these at the same time of the year. The Mayan Venus cycle, as known from the later *Dresden Codex*, was of eight civil years, 2920 days. This anniversary, however, was after eight tropical years, 2922 days; it was 5 Kayab, not 3 Kayab. (Both the accession day and its anniversary equate with January 7 in the Julian calendar, with leap-year intercalations.) On both occasions the morning star Venus was close to its greatest elongation from the Sun (a little over 45°), though still about 1° from the maximum. And for whatever it may be worth in this context, it can be noted that on the morning of this anniversary date, January 7, 692, there was very likely (i.e. as nearly as the writer can determine) a heliacal rising of Jupiter, a first appearance at the eastern horizon after conjunction.[12] Thus the 5 Eb 5 Kayab date appears to have had its own qualifications suitable for such an occasion. But it pertains to an entirely different category from that which has emerged from the other cases here considered.

These are the dates connected with the name of Chan-Bahlum. The relevant data are summarized in Table 19.1. One would like to have had

252

Table 19.1. *Recorded dates pertaining to the ruler Chan-Bahlum, with positions of the planet Jupiter, in degrees of celestial longitude, and in days, relative to a preceding second stationary point.*

For significance of asterisk markings, see text.

Date (Julian)	Event		Degrees	Days
May 20, 635	Birth (father's age 32)		37.83	204
Jun 14, 641	Heir-designation (age 6)	?*	1.18	28
Jan 7, 684	Accession to rule (age 49)	**	0.42	15
May 17, 688	'1 Ik 10 Tzec' 819d station	*	0.30	14
Jul 20, 690	'2 Cib 14 Mol' rites	**	0.32	15
Jan 7, 692	Eight-year anniversary of accession		25.14	154
Feb 16, 702	Death (age 67)		40.18	247
Jul 17, 702	'1 Cib 14 Mol' commemoration	*	0.08	8
Nov 2, 705	Apotheosis of deceased ruler	**	0.29	12

(**) marks instances in which Jupiter's departure from the second stationary point appears to have been a criterion for the selection of the date.

(*) marks those instances in which the date was otherwise determined, but where its special observance appears to have depended on this position of Jupiter. The one marked (?*) may have had a similar connotation.

Unstarred items are on dates when Jupiter was not in immediate departure from the second stationary point.

more dates to test; but even with this number it would be difficult to suppose that the degree of association of controllable timings with a particular phenomenon of Jupiter had been a matter of chance, or that the emphasis given to the chance associations had been otherwise motivated. It appears that the ruler had associated his identity and his fortunes in some way with that planet or deity.

There survive no Mayan Jupiter tables, such as we have for Venus and for the eclipse cycle in the Dresden hieroglyphic codex. The skywatchers at Palenque, however, must have been in possession of a sufficient corpus of records of prior observations of Jupiter to have arrived at the correct integral value for the mean of the synodic periods; for this was employed numerologically in the reckonings or the 819-day station which accompanies the initial date in the inscription of the Temple of the Sun. And with such records, and in the process of determining the mean, they would have acquired some knowledge also of the extent and the patterns of deviation from the mean. A comparison of the

intervals between the Palenque dates shows that this deviation was taken into account. If we consider the dates of the three events whose timing was entirely free of calendrical, numerical, or other astronomical constraints (those marked with two asterisks in Table 19.1), we have the following:

9.12.11.12.10   8 Oc 3 Kayab, accession (January 7, 684)
+ 6.11. 6   = 2386 days = 6(398.88) −7.28 days

9.12.18. 5.16   2 Cib 14 Mol, rites to the gods (20 July, 690)
+ 15. 9. 4   = 5584 days = 14(398.88) −0.32 days

9.13.13.15. 0   9 Ahau 3 Kankin, apotheosis (November 2, 705).

It can be seen that the second interval, the longer one, is as close as possible to an integral multiple of the true mean synodic period of Jupiter (398.88 days), while the first, the lesser of the two, falls more than seven days short of an integral multiple of that period. When first noted, this was disconcerting, but it turns out to be in accord with proper expectations, supporting rather than contradicting the interpretation. The Jupiter period is a little more than a month longer than the solar year. Consequently each succeeding Jupiter period 'begins' approximately a month or so later in the calendar year than did the one preceding. If we count 'beginnings' from the second stationary point – as it appears the Maya at Palenque may have done – then we can say that the Jupiter periods that 'begin' in the spring months of the year are the shorter ones, including those that are as short as about 395 days; those that 'begin' in the fall months of the year are the longer ones, including those that are as long as about 402 days; and those that 'begin' in the summer and in the winter months run closer to the mean. In the above tabulation it can be seen that the six Jupiter periods that intervene between the first and the second dates have winter 'beginnings' (more or less average in length) and spring 'beginnings' (short); hence the accumulated shortfall from an integral multiple of the mean is to be anticipated. In contrast, the 14 periods that intervene between the second and third dates start with summer 'beginnings' (more or less average) and then include beginnings that range over an entire year thereafter (wherein the short balance the long); hence the optimum approximation to an integral multiple of the mean period in their total duration. (This is a drastic oversimplification, though true enough to illustrate the point in question. A more suitable locus for 'beginnings' would have been the planet's opposition to the Sun.)

Two hypotheses, related but distinct, have been advanced. The first is that a departure of Jupiter from its second stationary point marked a preferred time for dynastic rituals for a particular ruler at Palenque. Support for it derives from the associations discussed above and summarized in Table 19.1, and further, also from the intervals separating the doubly starred items of Table 19.1, whose deviations from mean-value multiples correspond to the actual deviations at those times. The second is that the coming together of Mars or of Saturn with Jupiter provided a favorable or motivating context in some cases, though it was not in general a necessary condition. This rests primarily on the fact that the close congregation of Jupiter, Saturn and Mars during their simultaneous retrograde periods prior and up to the 2 Cib 14 Mol event furnishes the only discoverable motivation for that event and for the manner in which it is recorded in the inscriptions. The triple conjunction of Mars with Jupiter prior to the 9 Akbal 6 Xul heir-designation event is also suspect as one of the determining components in its timing; and the approximation to a repetition of that relationship prior to the 1 Ik 10 Tzec 819-day station, although a near miss so far as a triple conjunction is concerned, is suspect as one of the reasons for the important attached to that date. Such assemblages of the planets during simultaneous retrogressions are rare enough to warrant attaching some significance to these, and much too rare to have been made a general requisite for the timing of dynastic rituals.[13]

This chapter can be concluded with a note returning to the matter with which it began, viz the anomalous 819-day passage in the Temple of the Sun. Its chronology can be reconstructed now as follows:

1.18. 5. 3. 6 13 Cimi 19 Ceh, birth of deity G3 (November 14, 2360 BC)

  − 1. 2.11 = 411 days, recorded interval

1.18. 4. 0.15 5 men 13 Yax, posited Jupiter phenomenon (September 29, 2361 BC)

−11. 9. 7.13 = 82 593 days (= 207 × 399 days)

1. 6.14.11. 2 1 Ik 10 Tzec, recorded station of the 819-day cycle and hypothetic Jupiter phenomenon (August 14, 2587 BC).

An 819–day passage normally specifies only a single interval and a single date. The date is that of the next-prior 'station' or starting point of the 819-day cycle, and the interval is that which mediates between it and the date of the initial series. The chronology of such a passage, if presented in the manner above, would require only three lines instead of five. And so it is also in the Temple of the Sun, where the format is normal though the content is not; the information contained in the third and fourth lines of the above reconstruction has been omitted, leaving the passage arithmetically absurd if interpreted literally. Whether this was by design, through an oversight, or due to an uncomprehending editor's deletions for preservation of conventional format, remains an unanswered question. The intent of the person who made the original computations appears to have been to set down a distance number representing the interval from the initial-series date back to a next-prior supposed Jupiter station, and then to go from there by successive mean Jupiter periods to a date that was both a hypothetic Jupiter station and an 819-day station with the same calendar-round name as the one that preceded the historical 2 Cib 14 Mol event, thus arriving at an ancient precedent or mythological 'charter' for the latter – a manipulation itself with ample precedents in the Mayan legacy. That the 63-calendar-round cycle necessary to this end must doom the result to failure (not being commensurate with the mean Jupiter period) appears not to have been an important consideration. In any case, pertaining to a mythological era, it was exempt from empirical verification.

A survey of the dates from other Mayan sites, with a view to testing for similar observations of this or of other phenomena of Jupiter, has not yet been made. A few instances, however, have been noted. An interesting one is the date of Lintel 25 of Yaxchilan: 5 Imix 4 Mac, 9.12.9.8.1, or October 20, AD 681 (Julian). This is the date of the blood offering and serpent vision that is depicted on that famous lintel (housed in the British Museum),[14] and it is the date of the inauguration of Shield-Jaguar as ruler of Yaxchilan. As of that date, 17 days past the date of dead-center of the second stationary point, Jupiter had moved 0°.18 away from its position at that time. Whether this is indicative of a pattern for that principality, or for that particular ruler, or whether it was an isolated chance occurrence, remains to be seen. It was 809 days before the inauguration of Chan-Bahlum at Palenque, an interval that is equal to two mean synodic periods of Jupiter plus 11 days, or to six days more than the two actual synodic periods that intervened (deviating on the long side from the mean).

Another one of interest is the date 6 Eb 0 Pop, 9.15.12.11.12, which is on Lintel 2 from Temple 4 at Tikal (one of the spectacular wooden lintels in the Ethnographic Museum in Basel).[15] As of that date, February 3, AD 744 (Julian), Jupiter had moved 0°.50, in 18 days, away from dead-center of the second stationary point. Mars also was in the scene on that date. The next day, 7 Ben 1 Pop, is recorded in the same inscription as the day of a 'star-timed' war expedition against an enemy principality. 'Star' timings of such undertakings in synchrony with critical points in the cycle of Venus have been previously noted (Kelley, 1977; Lounsbury, 1982). This one from Tikal Lintel 2, lacking any such Venus credentials but with a significant one from Jupiter, offers confirmation of a point that Kelley has made about the 'star' sign in these hieroglyphs, namely that its reference is not inherently restricted to the planet Venus, unless a specific prefix or an implicating context indicates it as such.

Acknowledgments

This paper is an abridgment (in its first parts) and an expansion (in its later parts) of one begun but not completed in the spring of 1977 while a Senior Fellow in Pre-columbian Studies at Dumbarton Oaks. I wish to acknowledge my indebtedness to Dumbarton Oaks, and to Elizabeth P. Benson, then Director of Pre-columbian Studies, for the opportunities made available to me there to work on problems of Palenque epigraphy.

I owe thanks also to John S. Justeson for helpful criticisms received during the revision and preparation of the final manuscript of this paper, and to Roger W. Sinnott for the use of his computer program, *Planet Positions*, which has facilitated computation, and which, with an appended graphing program, produced the figures which appear here.

The subject matter of this paper was presented publicly on a number of occasions prior to its presentation in Merida at the 2nd Oxford Conference viz at University of California at Los Angeles, Anthropology Graduate Student Association, April 13, 1982; Wellesley College, Henry R. Luce Lecture, April 26, 1983; Yale University, Anthropology Colloquium, September 15, 1983; State University of New York at Albany, Department of Anthropology and Institute of Mesoamerican Studies, October 7, 1983; University of Pennsylvania, University Museum Maya Symposium, April 7, 1984. It was also the topic which I first proposed as a contribution to the then forthcoming International Conference on Archaeoastronomy at Oxford (September, 1981), but which I later changed, when the exciting results concerning Venus phenomena at Bonampak and other sites began to accumulate while pursuing the problem to which Mary Miller introduced me in the fall of 1980 (cf. Lounsbury, 1982).

Notes

1 The archaeological site of Palenque is in lowland Chiapas, Mexico, at 92° 5′ 20″ west longitude and 17° 29′ 30″ north latitude (Maudslay, 1889–1902, vol. V: 'Text [for] vol. IV', p. 8). The inscriptions with which this paper is concerned date from the last two decades of the seventh century and the first of the eighth century of our era. The king to whom reference is made in the title lived from AD 635 to 702, and he reigned from 684 to 702. For reproductions of the panel from the Temple of the Sun the reader is referred to Maudslay (1899–1902, plates 87 and 88) (A. P. Maudslay's photographs and Annie Hunter's line drawings), and to Robertson (1980, p. 105) (Linda Schele's more recent drawing of the text).

But for a single exception (note 9), discussion of problems of hieroglyphic and iconographic interpretation of the Palenque tablets is not included in this present article.

2 The Maya day count, or 'long count' as it is generally known, was a native chronological system wherein the days, starting from a far-distant prehistoric epoch, were assigned numbers in sequence. In principle it was analogous to the Julian Day count introduced by Joseph Justus Scaliger in 1583 as a scheme for unifying Old World historical chronology. The institution of the day count in Middle America, however, antedated the invention of the Julian Day count in Europe by at least 16 centuries, its earliest surviving

examples (carved on stone monuments) dating from the first century BC. Its epoch, or day zero, was August 13 (retroactive Gregorian) of the year 3114 BC (-3113 in astronomical chronology). Like the epoch of the Julian Day count (January 1, retroactive Julian, 4713 BC), this was a backward projection, via cyclical reckonings, from its date of invention sometime prior to the mentioned earliest attestations.

The Mayan numeral system is vigesimal. Its application to the day count, however, entails a departure from base 20 in one position, viz a unit in the third position is equal to 18 (rather than 20) of the second. This can be viewed in either of two ways, depending on whether the 'tun' (360 days, unit in third position) or the 'kin' (day, unit in first position) is taken to be the fundamental unit of the count.

The Maya calendar-round was a native calendrical system wherein the days were designated by their coordinates in three modular dimensions: a cycle of 13 days, represented by the numerals 1 to 13; a simultaneous cycle of 20 days, represented by day names; and another simultaneous cycle of 365 days, numbered within 18 named 20-day periods and an additional five-day residue period. The first two of these cycles taken together define the 260-day sacred almanac, or 'tzolkin', wherein the days receive binomial designations (e.g., 13 Cimi, as in the date just cited in the text). The third cycle is that of the civil year, or 'haab', wherein the days are designated by numerals prefixed to the names of the subperiods (e.g., 19 Ceh, of the date cited in the text). The tzolkin and the haab, running simultaneously, together define the cycle of the 'calendar-round', 18 980 days in length ($=52 \times 365$). This system is older in Middle America than the chronological system described in the paragraph above, and with even earlier attestations.

The equation of Mayan dates with dates of the European calendar and Christian chronology presumes the correctness of the '584285' correlation (Maya Day Number $+$ 584285 $+$ Julian Day Number). This is the first of the values proposed by J. Eric S. Thompson (1927, 1935) for what is generally known as the Thompson (or Goodman–Martinez–Thompson) correlation. Although Thompson later (1950) proposed a two-day reduction of its value, and although there are a few scholars who are not yet convinced of the correctness of either of Thompson's two values, the present writer has no doubts whatever concerning the validity of the first of these for the Lowland Classic Maya inscriptions (restricting the second to Central Mexican and Highland Mayan dates, and admitting a minor asynchrony within the Mesoamerican area). Proofs will be presented in a future publication.

3 The term 'initial series' commonly designates the series of hieroglyphs standing at the beginning of an inscription, which introduce and specify the chronological and the calendrical attributes of its opening date. The term 'supplementary series' has been applied to the hieroglyphs which follow immediately after the initial series and specify the attributes of the date in the lunar calendar; and it has been extended by some writers to cover also the further specifications given in the 819-day orientation of a date. The term for this series is a relic of a time prior to the discovery of the significations of its glyphs. Usage has varied, some writers employing the dichotomy between 'initial' and 'supplementary' strictly on the basis of positions in an inscription, while others have employed it on a conceptual basis, with reference to content. In the positional usage, which was prior, the 'initial series' included also the glyphs known as 'G' and 'F' (placing the date in a nine-day cycle), and it might or might not include those that place the date in the haab (the civil-year component of the calendar-round). In the conceptual usage, on the other hand, it has come to include the day number (long-count number) and the almanac or calendar-round specifications regardless of where they are found, even if not initial to an inscription, and even to the inclusion of long-count dates recorded in the *Dresden Codex*.

4 The interval between any two given calendar-round days (cr) can be determined from the following relationships:

$$\Delta tz = 40 \cdot \Delta tr - 39 \cdot \Delta v, \bmod 260;$$
$$n(H) = \Delta tz - \Delta h, \bmod 52;$$
$$\Delta cr = 365 \cdot n(H) + \Delta h;$$

where the independent variables Δtr, Δv and Δh are defined as follows:

$\Delta tr =$ the interval between the two trecena coordinates (in the cycle of the repeating series of numbers 1 to 13);

$\Delta v =$ the interval between the two veintena coordinates (in the cycle of the repeating series of 20 day-names: Imix, Ik, Akbal, Kan, Chicchan, Cimi, Manik, Lamat, Muluc, Oc, Chuen, Eb, Ben, Ix, Men, Cib, Caban, Etznab, Cauac, Ahau);

$\Delta h =$ the interval between the two haab coordinates (in the cycle of 365 days, numbered 0 to 19 in the 18 named 20-day periods 'Pop, Uo, Zip, Zotz, Tzec, Xul, Yaxkin, Mol, Chen, Yax, Zac, Ceh, Mac, Kankin, Muan, Pax, Kayab, Cumhu', and 0 to 4 in the five-day residue period 'Uayeb');

and where the dependent variables *tz*, *n*(*H*) and *cr* are as follows:

$\Delta tz =$ the interval between the two tzolkin days (in the cycle of 260);

$n(H) =$ the number of whole haabs (365-day years) contained in the interval between the two calendar-round days; and

$\Delta cr =$ the interval between the two calendar-round days.

5 For examples of Mayan numerology involving calendrical, lunar–solar and planetary periods, see Lounsbury (1976, 1978, pp. 804–8; 1983).

6 Note concerning the figures. Celestial longitudes are plotted at ten-day intervals. Longitude increases from left to right, and the passage of time from top to bottom. The date of the noted event in each figure is that of the horizontal axis (designated the 'date line') one-third of the distance up from the bottom of the chart. Each figure displays the longitudes of Jupiter at ten-day intervals within a 40° sector of the zodiacal circle which includes its opposition to the Sun and its stationary points before and after opposition. Also dispayed are the longitudes of Saturn and Mars when either or both of these is (or are) within the same 40° sector. Any of these at opposition to the Sun is visible in the sky throughout most of the night. At first stationary point visibility is more in the morning sky, and at second stationary point more in the evening sky. The sequence of the figures represents the order of their consideration in the text. Their chronological sequence is 20.6, 20.3, 20.2, 20.1, 20.4, 20.5.

7 Dieter Dütting has also proposed this as a possible explanation of the significance of the 2 Cib 14 Mol date (Dütting, 1982). I came to it by way of a solution to the problem of the discrepancy in the 819-day passage in the Temple of the Sun, and I take that route again in the presenting the arguments here. The original hypothesis has been modified, however, in the light of the additional data here introduced.

The months during the spring and summer of 1981, just prior to the (First) Oxford Conference, presented a rare opportunity for the naked-eye observer to gain an appreciation of the events that led up to the 2 Cib 14 Mol observance at Palenque – a phenomenon still fascinating for a 'primitive' astronomer, as the writer can attest. The subject of the frequency of such occurrences received attention in several publications during the year preceding. See especially Meeus (1980).

8 Equatorial coordinates for Jupiter on the two dates of interest, July 20, 690 (2 Cib 14 Mol) and July 17, 702 (1 Cib 14 Mol), both at 6.00 p.m. CST (90° west), and for Delta, Beta and Alpha Scorpii during that period, were as follows (coordinates in degrees with decimal fractions; data for the stellar coordinates, AD 700, from Hawkins, 1968):

	RA	Decl.
Jupiter, July 20, AD 690 (Julian)	219.23	−14.57
Jupiter, July 17, AD 702 (Julian)	223.47	−15.92
Delta Scorpii, AD 700 (+/−)	221.40	−18.03
Beta Scorpii, AD 700 (+/−)	222.94	−15.36
Alpha Scorpii, AD 700 (+/−)	227.96	−22.55

9 Illustrations of the carved and inscribed panel from Temple XIV may be seen in Robertson (1974, p. 173; photographed by Merle Green Robertson), and Robertson (1979, p. 204; line drawing by Linda Schele). The 'dance' iconography and its significance are discussed in Lounsbury (1985, pp. 53–6).

10 The eclipse of May 21, AD 635 -the day after Chan-Bahlum's birth- was not observable in the Americas. The supposition that it would have been anticipated as a possibility by the Maya at Palenque is based on the fact that there are several attestations to knowledge of the 11 960-day eclipse cycle (405 lunations, 69 internodal intervals) at Palenque during Chan-Bahlum's lifetime and after, and that the mythological birth date of the Sun god, recorded with accompanying eclipse symbolism, is related to the birth date of Chan-Bahlum by means of that cycle and one of its subdivisions. This was before the institution of the 12 Lamat base of the table in AD 755 (November 8), and was some three to four centuries before the base of the update of the table which we have in the *Dresden Codex*.

11 That this is the import of the hieroglyphs at Q7 to Q8 in the inscription of the Temple of the Sun is premised on the following considerations: (1) the inscription states that on the fifth day after the heir-designation there was another event, involving [as protagonist(?)] the Sun; (2) that day, by the 584285 correlation, was June 19, AD 641 (Julian), or June 22 (retroactive Gregorian); (3) the summer solstice of that year was exactly on that day, at 14.35 UT, or 8.35 CST; (4) although the event phrase is without other attestation in this value, there is not much else that can be said about the Sun that would be of relevance on that particular date. (Another interpretation that has been placed on this event phrase would have the Sun as a predicate noun and the heir-designate as the subject, to the effect that the latter (metaphorically) 'became the Sun' on the fifth day after his initial designation; (Schele, 1984). Should that turn out to be the correct interpretation, one may ask 'Why on the fifth day?'; in which case, there being no other reason or precedent, the answer could only be that the summer solstice was a symbolically appropriate

time for such a status transformation.

12 The situation of Jupiter in respect to the Sun on the morning of the eight-year anniversary date of the accession, January 7, 692 (Julian), at 6.00 a.m. CST (12.00 UT), was as follows:

	Longitude	Latitude	Altitude	Azimuth	Magnitude
Sun	290.20	0.00	−10.8	110.00	−26.8
Jupiter	278,50	−0.18	0.4	114.7	−1.9

In ecliptic coordinates the difference in longitude was 11°.7 and in latitude 0°.18. In horizontal coordinates (at Palenque's geographic longitude and latitude) the difference in altitude (the 'arcus visionis') was 11°.2 and the difference in azimuth 4°.7.

13 The frequency of triple conjunctions of various pairs of planets, and of planets with stars, has been discussed by Meeus (1980). He notes that a theoretical frequency of triple conjunctions of Jupiter with Saturn, assuming circular orbits, had been calculated (by another writer) as an average of one every 125 years, and of Mars with Jupiter an average of one every 110 years. Their actual frequencies are much less, however, due to factors such as orbital eccentricities, perturbations, and other irregularities; and their occurrences cannot be predicted from any simple periodicities. He counted 20 of Jupiter and Saturn in the tables computed for the 3100 years from 101 BC to AD 3000, the shortest interval between them being of 40 years and the longest of 377 years. The 2 Cib 14 Mol event at Palenque, however, gives witness to the fact that near misses may also excite interest. Allowing a tolerance slightly greater than that exemplified on that occasion, and assuming circular orbits, it appears that approximations of that degree might be expected every 91 or 109 years, with occasional misses; but the sources of irregularity destroy also that periodicity and lessen the frequency of the phenomena.

14 For excellent photographs, and for an illuminating glimpse into the current state of our understanding of the ancient Maya, see the related publication by Schele and Miller (1986). For excellent drawings, see Graham (1977). Lintel 25 is on Plate 63 in Schele and Miller (1986, p. 199) and on p. 3:55 in Graham (1977).

15 For illustrations of Lintel 2 of Temple 4 at Tikal, and drawings of its inscription, see Maudslay (1889–1902, vol. 3, Plates 72–4).

References

Dütting, D. (1982). The 2 Cib 14 Mol event in the inscriptions of Palenque, Chiapas, Mexico. *Zeitschrift für Ethnologie* **107**, 233–58.

Graham, I. (1977). *Corpus of Maya Hieroglyphic Inscriptions*, vol. 3 [Yaxchilan], part 1. Cambridge, MA: Peabody Museum, Harvard Univerity.

Hawkins, G. S. (1968). Astro-archaeology. In *Vistas in Astronomy*, ed. A. Beer, vol. 10, pp. 45–88. New York: Pergamon Press.

Kelley, D. H. (1977). Maya astronomical tables and inscriptions. In *Native American Astronomy*, ed. A. F. Aveni, pp. 57–73. Austin: University of Texas Press.

Lounsbury, F. G. (1976). A rationale for the Initial Date of the Temple of the Cross at Palenque. The art, iconography, and dynastic history of Palenque, Part 3. In *Proceedings of the Segunda Mesa Redonda de Palenque*, ed. M. Greene Robertson, pp. 211–24. Pebble Beach, CA: Pre-Columbian Art Research, The Robert Stevenson School.

Lounsbury, F. G. (1978). Maya numeration, computation, and calendrical astronomy. In *Dictionary of Scientific Biography*, ed. Charles Coulston Gillispie, vol. 15 (Supplement 1), pp. 759–818. New York: Charles Scribner's Sons.

Lounsbury, F. G. (1982). Astronomical knowledge and its uses at Bonampak, Mexico. In *Archaeoastronomy in the New World*, ed. A. F. Aveni, pp. 143–68. Cambridge University Press.

Lounsbury, F. G. (1983). The base of the Venus table of the Dresden Codex, and its significance for the calendar-correlation problem. In *Calendars in Mesoamerica and Peru: Native American Computations of Time*, eds. A. F. Aveni and G. Brotherston, pp. 1–26. Oxford: British Archaeological Reports (Intl Series) no. 174.

Lounsbury, F. G. (1985). The identities of the mythological figures in the cross group inscriptions of Palenque. In *The Palenque Round Table Series*, vol. 6, eds. M. Green Robertson and E. P. Benson, Fourth Palenque Round Table, 1980, pp. 45–58. San Francisco: The Pre-Columbian Art Research Institute.

Mathews, P. (1980). The stucco texts above the piers of the Temple of the Inscriptions at Palenque. *Glyph Notes no. 10*, 6 pp. Cambridge, MA.

Maudslay, A. P. (1889–1902). *Archaeology, vol. IV: Palenque*. London: R. H. Porter and Dulau. Facsimile reprint 1974, prepared by F. Robicsek. New York: Arte Primitivo, Inc.

Meeus, J. (1980). Les conjonctions triples Jupiter–Saturne. *L'Astronomie et Bulletin de la Société Astronomique de France* **94**, 27–36.

Robertson, M. Greene, (1974) (ed.). *Primera Mesa Redonda de Palenque, [1973 Part 1]*. The Palenque Round Table Series, vol. 1. Pebble Beach, CA: Pre-Columbian Art Research, The Robert Louis Stevenson School.

Robertson, M. Greene, (1979) (ed.). *Tercera Mesa Redonda de Palenque [1978, Part 1]*. The Palenque Round Table Series, vol. 4. Monterey, CA: Herald Printers.

Robertson, M. Greene, (1980) (ed.). *Third Palenque Round Table [1978, Part 2]*. The Palenque Round Table Series, vol. 5. Austin: University of Texas Press.

Schele, L. (1984). Some suggested readings of the event and office of heir-designate at Palenque. In *Phoneticism in Mayan Hieroglyphic Writing*, ed. J. S. Justeson and L. Campbell, Publication no. 9, Institute for Mesoamerican Studies, pp. 287–305. Albany: State University of New York.

Schele, L. and Miller, M. E. (1986). *The Blood of Kings: Dynasty and Ritual in Maya Art*. Fort Worth: Kimbell Art Museum.

Thompson, J. E. (1927). A correlation of the Mayan and European calendars. *Publications of the Field Museum of Natural History, Anthropological Series* 17, (1), 22 pp. Chicago.

Thompson, J. E. (1935). Maya chronology: the correlation question. *Contributions to American Archaeology*, (14), 104 pp. Washington, DC: Carnegie Institution of Washington.

Thompson, J. E. (1950). *Maya Hieroglyphic Writing: An Introduction*. CIW Publ. 589. Washington, DC: Carnegie Institution of Washington. (2nd edn., 1960; 3rd edn., 1972. Norman: University of Oklahoma Press.)

Tuckerman, B. (1964). Planetary, lunar, and solar positions, AD 2 to AD 1649, at five-day and ten-day intervals. *Memoirs of the American Philosophical Society* 59.

20

A new Sun at Chichen Itza

Clemency Chase Coggins   *Peabody Museum, Harvard University*

20.1 Introduction

At Chichen Itza two or more ancient cultural traditions are manifest in the architecture and pictorial record, and may be detected in the time-keeping and the hieroglyphic inscriptions as well. These traditions are Mexican (often called 'Toltec'), deriving ultimately from the central highland regions of modern Mexico, and Maya as found throughout the Yucatan Peninsula and south into the Guatemalan and Honduran highlands. Recent work on Chichen Itza has emphasized the likelihood of the contemporaneity of these apparently distinct traditions (summarized in Lincoln, 1983), and in a recent paper I have suggested that the term 'Toltec', generally thought to apply only to tenth-century expansionistic warriors from the central Mexican site of Tula, Hidalgo, many instead have referred to all those peoples, whether superficially Mexican or Maya, who traced their ancestry and cultural traditions to Teotihuacan – the vast city that dominated Mesoamerica for about 600 years until its collapse in the eighth century AD (Coggins, 1987a).

Chichen Itza, it was argued, was founded toward the end of the eighth century by Mexicanized Maya who deliberately chose the location of Chichen Itza for strategic, symbolic and sky-viewing reasons, as one of several post-Teotihuacan religious and political centers that were all called Tula after the ancient name of Teotihuacan. (The Thompson correlation between the Maya and Christian calendars is used in this paper; as reconfirmed in Lounsbury, 1983 – but using the day equivalent 584283.)

Thus the founders of Chichen Itza and all of the rulers who eventually journeyed there to express

their politico-religious loyalty were Toltec by virtue of their ancestral ties to Teotihuacan, no matter what language they actually spoke.

It was also argued that the date of the founding of Chichen Itza was determined by the approaching end of the tenth baktun in the Maya Long Count (at 10.0.0.0.0). This event was to occur in significant proximity to the Toltec New Fire Ceremony for a doubled Calendar Round of 104 years (also 65 canonical Venus years) that was in turn immediately followed by a heliacal rise of Venus – all potentially cataclysmic events within the year AD 830.

In anticipation of these inevitable dates it is postulated that first, toward the end of the eighth century, the lower Caracol was built to check on the timing and location of the relevant astronomical events, particularly of Venus as Evening Star, and of Pleiades and solsticial settings. The tower of the Caracol may have been built as long as a century later, to a slightly different alignment that had been obtained by fine-tuning observations made from the earlier structure. It is further postulated that the outer Castillo was built between the construction of the early and later parts of the Caracol in comemmoration of the completion of the baktun, at 10.0.0.0.0 7 Ahau 18 Zip (March 13, 830) (Figure 20.1). The plan of the outer Castillo is the same shape as the Maya completion sign and, like dated structures at Tikal and Seibal to the south, this pyramidal structure was apparently built as an objective symbol of cyclic completion within the Long Count (Coggins, 1980, 1983, pp. 55–7).

The Castillo has traditionally been dated a century or more later than AD 830 (Tozzer, 1957, p.

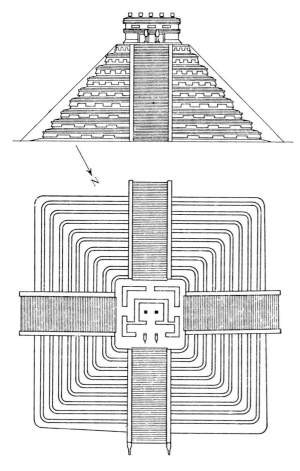

Figure 20.1 Plan and elevation of the Castillo, Chichen Itza. (From Kubler, 1982, Figure 60).

table 4).

In the previous paper (Coggins, 1987a) it was argued that a Mexican style, or Toltec (meaning Teotihuacan), New Fire Ceremony was celebrated at Chichen Itza, perhaps beginning in AD 830 at the Castillo where the fire was drilled on a pyrite 'back shield' that was set upon the chest of a Chac Mool statue; the fire drilling was performed at the moment when the Pleiades crossed the zenith at midnight in November (Milbrath, 1980; Broda, 1982). This scenario substitutes a stone Chac Mool for the noble victim on, or in, whose chest the Aztec drilled the New Fire 700 years later. Furthermore, this reconstruction suggests that, as on the Aztec Calendar stone, fire serpents descended from the celestial zenith at the moment of the Pleiades transit and of the fire drilling. These are represented by the descending feathered serpents that flank the portals of the Castillo, and of several other temples at Chichen Itza (Kubler, 1982, pp. 93–100). One of the most important elements in this postulated ceremony was the pyrite (fool's gold) mosaic back disc, or shield, that identified the Toltec elite (Figure 20.2b). These discs, which like the quadripartite plan of the Castillo are made in the form of the Maya completion sign, are prominently shown in the reliefs and paintings of Chichen Itza and burned examples were excavated from the Caracol, the Castillo, The Temple of the Warriors and the Sacred Cenote (Coggins, 1987a). These discs are identified in Central Mexico and Oaxaca by a cross that the Maya call a 'Kan Cross'. In Yucatec Mayan *K'an*, which means both yellow and precious stone, applies to the golden pyrite that, as its name 'pyrite' implies, was also used (world-wide) for making fire; whereas in the Maya regions its presence on discs (and in some inscriptions) reflects the presence of Toltec New Fire drilling ritual. (All translations and Mayan orthography are from Yucatecan; Barrera Vasquez, 1980.)

This chapter is concerned with styles of time-keeping and period commemoration, and with the associated choice of personal names and titles as they were symptomatic within ancient Mesoamerica of different world views and cultural identity; such variations in Mesoamerican style were partially reconciled at Chichen Itza in the ninth century AD, where differences in calendric emphasis led to contrasting meanings for the completion of the tenth baktun for the Maya, and for the inauguration of

41) but there is archaeological evidence for the earlier date. Many of the jades cached before the construction of the outer Castillo can be dated stylistically to late in the Late Classic period (Marquina, 1964, photo 428), and the majority of the dateable jades that were thrown into the Sacred Cenote are Terminal Classic in style (Proskouriakoff, 1974, plates 42–9, 53b–78a) and were probably offered in conjunction with the baktun completion ceremony. A ninth-century date for construction is also supported by dates in hieroglyphic inscriptions elsewhere at Chichen Itza that cluster between AD 869 and 884 (Kelley, 1982, table 1) not long after the baktun end which probably marked the formal foundation of Tula/Chichen Itza. Finally a carbon-14 date from the lintel of the Castillo is AD 810, ± 70 years (Andrews and Andrews, 1980,

(a)

(b)

Figure 20.2 Cache from foot of stairway, inner Castillo, Chichen Itza. (a) Two sacrificial knives and remains of two pyrite/mosaic discs: *k'ab k'u/pakal* (sacrificial knife/shield). (b) One of the pyrite/mosaic discs, reconstructed: *K'ak' pakal* (fire shield). (Photographs, Peabody Museum, Harvard University.)

the 11th Calendar Round for the Toltec.

At Chichen Itza this reconciliation had, by the tenth century, adopted more Toltec characteristics, whereas at eighth century Copan similar Toltec calendric ritual had remained within the Classic Maya long count idiom. Elite naming practices provide clues to the identity of some early Toltec, or Mexicanized Maya, at Copan, at two western Maya sites and at Chichen Itza – as does the importance to them of the ceremony that involved the use of the Toltec back disc or Fire Shield for the drilling of fire.

Finally I will tentatively outline the sacred drama of the solar and ritual year that began with the completion of the tenth baktun at Chichen Itza, signifying the end of one cycle, while also signifying the beginning of a new Toltec one.

20.2 The end of an age or Sun

One assumption of this analysis is that the well-known architecture and pictorial records at Chichen Itza document a deliberate juxtaposition and reconciliation of Toltec and Maya cultures as has long been supposed (Tozzer, 1930, 1957; Kubler, 1961; Ruz Lhuillier, 1962). It is postulated that this amalgamation probably began at Chichen Itza in the 52-year Mexican cycle that followed the end of Teotihuacan and that preceded the end of the Maya baktun, or AD 778–830. It is well known that Mexican peoples feared that the cataclysmic end of the age might come every 52 years. When the New Fire was successfully rekindled at midnight they insured that a new cycle would follow, and that the New Sun would emerge from the nadir to which it had descended just as the Pleiades reached their nocturnal zenith and signaled the moment of fire-drilling (Milbrath, 1980, pp. 292–4).

Since the Mexican ages, or Suns, had all lasted multiples of 52 years (Valazquez, 1945, *Leyenda de los Soles*, p. 119), it is clear that the people were at peril at the completion of each Calendar Round. What may have been the 11th New Fire Ceremony was performed in November, AD 830, the 13th (?) in November 934. These hypothetical durations are based on the work of David Drucker (n.d.) who has reconstructed a Teotihuacan calendar that began in AD 311/12 as recorded by one of the Pecked Crosses ('Teo 2') on the Street of the Dead at Teotihuacan. That critical Two Reed year would

have been inaugurated, as were later Mexican years, by a November New Fire Ceremony in the preceding year of One Rabbit (AD 310).

For the Classic Maya evidence of the 52-year Calendar Round ceremony is not clear; while they used this combined ritual and solar basic count, it apparently did not have the same astrological and apocalyptic meaning as for the Mexicans, or if it did this information did not regularly appear in the official Maya Long Count inscriptions. Instead the Maya thought more in terms of baktuns, or cycles of 400 tuns (360-day years), and of multiples of them. Although there were only two of these 400-year baktun endings during the Classic period, it is apparent that these were of extraordinary symbolic significance, both by virtue of the structure of the Long Count, and by the negative evidence of how these 'millenial' dates were celebrated. They were not commemorated overtly on stelae, as were katun (20 tun) and lahuntun (10 tun) endings. Only one monument is known to have celebrated the end of the ninth baktun, at 9.0.0.0.0 (El Zapote, Stela 5), and this may portray a foreign woman who was involved in correlating that critical Maya date with a Oaxacan one (Coggins, 1984a, Fig. 8). The solitary stela to record the end of the tenth baktun, at 10.0.0.0.0 (Machaquila, Stela 7; Ian Graham, 1967, pp. 78, 79) does so almost in passing, and only by Calendar Round reference. Such apparent avoidance of the most important Long Count dates suggests they had a very particular meaning and may have inspired terror about the continuation of the Maya age.

Baktun ten was the first 'skeletal' baktun; it was represented glyphically by the death's head number ten, which was the first of the succeeding nine skeletal numbers (Thompson, 1960, Figs. 24, 25). Thus the end of baktun nine signified the completion of the first ten Maya baktuns, and all subsequent time was in a new megacycle, as inscriptions at Chichen Itza seem to state. In fact baktun ten at Chichen is referred to directly only once – in the single Long Count inscription at the site. Instead, a New Cycle is noted (Kelley, 1982, pp. 13, 14), although record of this New Cycle was not made until the third katun of the new baktun (after 10.2.0.0.0). This New Cycle apparently replaces the Maya baktun as the largest time period, but it may not have begun until the katun One Ahau (between 10.2.0.0.0 and 10.3.0.0.0), thus leaving a two katun 'nameless period', before what was perhaps viewed as a New Age, or Sun, that would begin at Chichen Itza. An analogous situation is noted in a recent study of Mexican dates and cosmology by Emily Umberger (1987); she notes that the first 26 years of the fifth age (or Sun) passed in darkness, before the sun was born. This implies a period of time before days were counted (since the sun defines a day) just as the lack of Chichen dates implies a period of time when the days were not counted before the starting katun 'One Ahau'.

This new Maya hieroglyphic terminology generally avoids the forms of the Long Count, while still using its structure, and it is one of many examples of cultural synthesis achieved at Terminal Classic Chichen Itza. At this time the more traditional Maya katun cycle beginning date of Eleven Ahau was eclipsed by One Ahau for the beginning of the redefined 'baktun' or New Cycle. This was analogous to the Mexican emphasis on the beginning of the cycle and on One Tochtli, or One Rabbit, when the New Fire was drilled and the New Sun and Toltec history traditionally began (Velasquez, 1945, p. 4); such a restructuring also put new emphasis on the day One Ahau which was so important at Chichen, and in the later *Dresden Codex*, as the prime day for the heliacal rise of Venus.

20.3 Names and titles of Toltec Maya rulers in the south

Yax T'ul in the southwest

Another example of this cultural synthesis may be found in the names and titles of a, perhaps the, ruler of Chichen at this period. His name has been read as *Yax T'ul*, First (or New) Rabbit (see Figure 20.3a: A8, E6; Davoust, 1980; Kelley, 1982, p. 6).

This name may be understood as a Mexican calendar day and year bearer name referring to the new hybrid calendar that began with the drilling of New Fire in the year One Rabbit (AD 830). However, I do not believe the name was read only in this way. This Chichen ruler may have come from the Chiapas Maya regions where a similar name (?) is found on Stela 2 at Sacchana just before it appears at Chichen Itza (Figure 20.4a: B3).

While the name is rare, an early example of *Yax T'ul* is found on the Palace Tablet at Palenque where it is a secondary name, or title, of Kan Xul

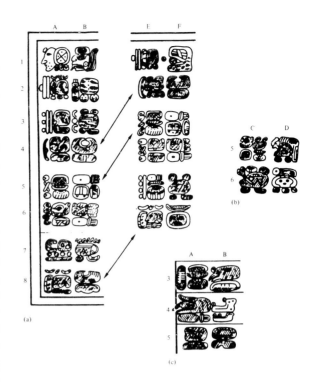

Figure 20.3 Lintel inscriptions, Temple of the Four Lintels, Chichen Itza. From Anderson (n.d.).

(a) Lintel 1. A1–B1 9 *Lamat*; *k'in* (day). A2–B2 11 Yax; *winal* (?) (month). E1–F2 11 Yax; One Sun at Horizon. A3–B3 13 tun; One *Ahau*. A4–B4, E2–F2 *U-ts-ol-a'n k'al* (count of the katun, 20). A5–B5, E3, F3 *U Pa-ka-b(a) ti-(i)-k(i)* (his lintel is revealed; Kelley, 1982, p. 5). A6, E4 *Ah K'ak'(a)* (Lord Fire, Burner; Kelley, 1982, p.5). B6, F4 *K'aba' ti'* (noble lineage). A7 *Bolton hel*, F5 *Bolton ti'k'ab (a')* (Many generations). B7, F5 *Ahaw*, Lord. A8, E6 *Ah (?) Yax To(h)k'(a)* (Lord (?) New fire/flint). B8, F6 *Ka-n(a)* (serpent or sky title).

(b) Lintel 1. C5 *K'a-ku-pa-ka-l(a)/K'ab ku pakal(a)* (Fire-its-shield/sacrificial knife-shield). D5 *Chuhil* (sacrifice by fire). C6 *Ahaw* (Lord). D6 *Hax* (fire-drilling; Kelley, 1976, pp. 144–8).

(c) Lintel 3. A3 *Ah Pa-ka-b(a)* (Lord (?) lintel; Kelley, 1982, p. 5). B3 *Chi'na* (door of the house). A4 *Ch'i-bal ti'* (the lineage; *ch'ich*, bird, *bal*, hidden). B4 (2 = *ka*) *Kabal chi'* (Two jawbone). A5 *K'aba'(ch?)* (noble lineage). B5 *Ka'an-(na)* (sky title).

II early in the eighth century AD (Robertson, 1985, Figs. 257–9: P3). Another occurrence is *Ix Yax T'ul* of Yaxchilan, a name of the wife (?) of the ruler who is portrayed about 50 years later on Stela 2 at Bonampak (Mathews, 1980, Fig. 2: H1–4). Both of these designations consist of a *yax* sign prefixing the head of an indeterminate mammal with a long drooping ear that is marked with the sign for *ets'nab*, or flint knife (T16.759; T designates glyphs in

Thompson, 1962; phonetic readings that are not proposed here may be found in Mathews, 1984, and Justeson, 1984).

A third example of *T'ul*, with an associated jaw-bone glyph like the ones at Chichen, is found on the Terminal Classic Stela 2 at Sacchana (Figure 20.3c: B4; 20.4a: B2) which is also a southwestern Maya site like Palenque, Yaxchilan and Bonampak. By the late ninth century when hieroglyphic inscriptions were first carved at Chichen Itza a group of traits may have been adopted from that southern region, since *Yax T'ul*, long-beaded hair and possibly bat designations all occur at Chichen. Important men with long beaded hair are shown in the ball court and in the Upper Temple of the Jaguars there (Coggins, 1984b, p. 160), while John Graham (1967, pp. 213–16) has noted the importance of long-haired people at Terminal Classic Seibal, another western site.

XVIII 'Jog', Yax T'ul, *and the Fire Shield at Copan*

Copan was designated by a bat emblem glyph (Figure 20.4b: E10), and the name of the 13th ruler of Copan is now usually interpreted as '18 Rabbit' (Schele and Miller, 1983, part 2). This name generally consists of T XVIII.122.757 (Figure 20.4b: C10; Schele and Miller, 1983, Fig. 12d): a mammal head with an infixed K'an Cross, preceded by the number 18 and often by a fire/smoke prefix. Schele and Miller argue that this animal head is both a generic animal, and may more specifically be a rabbit read *umul* or *umal*.

J. Eric S. Thompson suggested that T757 and T759 were composite beasts comprising jaguar and dog, and nicknamed it 'Jog', while he described T758 as the dog-derived (Mexican) name of the month *Xul* (Thompson, 1962, pp. 350–61). David Kelley also identifies as a dog the T758 animal with protruding tongue that accompanies fire-drilling phrases (Kelley, 1968a, pp. 151, 152, 1976, pp. 144, 148, Fig. 54). Schele and Miller's recent analysis of these important and puzzling glyphs does not pursue Thompson's or Kelley's dog, or Thompson's jaguar interpretations, even though the *mal* reading advanced is the phonetic equivalent of *bal* which is the main part of *balam*, jaguar. In the case of this Copan ruler I suggest that the name may have been read in more than one way, including rabbit, but that the principal *meaning* of this name was both fire and shield.

XVIII K'abal On Stela A, in a four glyph phrase referring to the ruler (Figure 20.4b: C9–F9), the first glyph, which is read *umul* (rabbit) by Schele and Miller, was more likely read *u k'abal*; this reading assumes this head with a jaguar ear is a jaguar, and that an initial *k'a* is indicated (in this instance) by the protruding ('lolling') tongue *ak'*, and reinforced by the infixed K'an cross. Finally, the *bal* reading is reiterated by the twisted cloth with a spot on it below the animal's head, since *bal* means to twist; this usage is found elsewhere to convey both twisted and jaguar meanings as Schele has demonstrated (Schele, 1985, pp. 62, 63). In Colonial times the *k'abal* was a name for the wheel that potters turned by foot; however, during the Classic period, it probably referred to the hand (*k'ab*) turned (*bal*) fire-drill, or to the object on which it was turned.

The infixed K'an cross refers directly to the pyrite inlaid Fire Shield upon which the drill was turned (Coggins, 1987a), while the next glyph in this inscription illustrates the Fire Shield itself (Figure 20.4b: D9a*; see also Figure 20.2b). The Fire Shield is followed by a serpent/sky, *Kan-na* title (D9b*). This fanged serpent head has the nose fret that signifies Venus, so it is the *Yax* serpent, patron of the month *Yax*.

Reading across this row, the glyph immediately following the 'Fire Shield, Sky' phrase is the familiar *Yax T'ul* that is found at the western sites and at Chichen Itza. I propose that this was read here as *Yax Tok*, New Fire, and that the animal head is a dog's; *toh* means woolly or fleecy dog in Yucatecan, and with the *ka* (T103) postfix *to(h)-k(a)* becomes *tok*, fire. Since, however, there is an *ets'nab*, flint knife, marking on the dog's ear it may also be read as the punning *tok'*, flint. Thus this *yax tok* glyph means New Fire, while it also includes reference to the flint used in making fire. The next glyph block is composed of a sequence of affixes that might be read *kahok lemchahal* (k'a-ha-k(a) le-m(o)-ka'-hel), meaning the thunder and lightning grow and exceed themselves (*kahoq*, Kaufman and Norman, 1984, p. 89). This apparently gradiloquent observation would be entirely appropriate for the accession statement of this ruler who identified himself with the storm god, Tohil (God K). His name and titles follow across the next row: 'XVIII

---

* a is the top and b is the bottom half of the compound glyph in position D9.

(a)

(c)

Figure 20.4 (a) Stela 2, back; Sacchana, Chiapas. A1–B1 Completion of the tun. A2–B2 *K'a-k'(a)-ma(l)* (to wither or fade); *Kabal chi' (chay) (k'amach*, to grow, wax). A3–B3 *U bal-k'a (u k'abal*, his fire-drilling); *U toh-?k'(a)* (*Tok'*, fire/flint). A4– Glyph D. Moon at conjunction (?).

(b) Copan Stela A (from Maudslay, 1889–1902, 1, pl. 30). C9 *U k'abal-bal* (His fire-drilling). D9a *K'ak' pakal* (Fire Shield; Kelley, 1968b, p. 257). D9b *kan-(na)* (Venus sky-serpent). E9 *Yax Toh-k(a)* (New Fire/flint). F9 *Ka-ho-k(a) le-m(o)-ka-hel* (thunder and lightning increase). C10–D10 18 *K'abal* (18 fire-drilling); *Tohil* (God K). E10–F10 *K'anhel Ahaw Sots'* (Copan 'emblem glyph'); *Bakab* (bearer, supporter title).

(c) 'First-Sun-at-Horizon', name of 16th ruler of Copan. Altar Q. (From Maudslay, 1889–1902 1, pl. 83.)

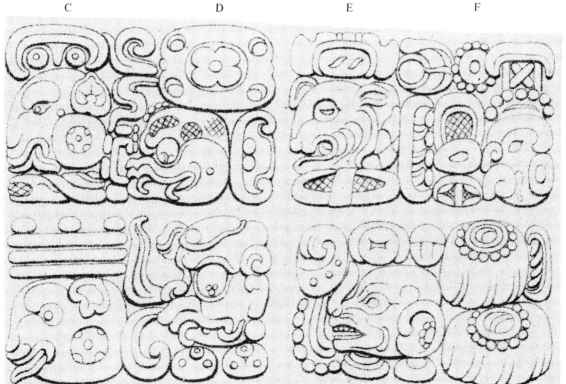

(b)

K'abal, Tohil, Kanhel Ahaw Sots' [Copan], *Bakab'* (for emblem glyph prefixes, and the reading of *hel* see Coggins, 1987b; in press).

This record of the accession of XVIII *K'abal* and of the associated drilling of New Fire took place at the dark of the moon on 4 *ahaw* 18 *Muan*, 9.14.19.5.0 (AD 730). *U k'abal*, his fire drilling, may also be read, in a typically Maya play on words, as *u k'amal*, his ascent, or accession, so that for a Toltec Maya ruler the drilling of the fire to create the New Sun, might also have the metaphorical purpose of creating a new ruler who was identified with the sun.

Glyph B This New Fire ritual was performed at night, and at the dark of the moon, as is confirmed by the presence of Glyph B of the Lunar Series accompanying this same accession date on this ruler's Stela H (Maudslay, 1889–1902, plate 61: A3). Glyph B also provides a confirmation of the *k'abal* and *k'amal* readings presented here in a substitution the Maya made for the *k'abal* animal head. Schele and Miller (1983, pp. 40, 42) emphasize that *bal* and *ma(l)* are interchangeable, and show that this head (T757, T758) can be the main element in one form of Glyph B (Schele and Miller, 1983, Fig. 10). I proposed that Glyph B, which refers to a phase of the moon, is to be read *k'amach*, meaning to grow stronger, T1.187:757:110; this interpretation reads the T187 'sky elbow' as *k'a* (for *k'at*, crossed bands, or *ka* for *ka'an*, sky) and the T110 'bone' as *ch'(ay)* ('bony'). Astronomers may be interested to find Glyph B reads *k'amach* (growing, waxing?) yet apparently refers to the dark phase of the moon (as checked by Floyd Lounsbury using a 584285 correlation; Schele and Miller, 1983, p. 51). This may be by analogy with the sun and Venus which were believed to grow inside the womb of the earth before they were reborn (Thompson, 1960, p. 236).

Glyph B can be read *k'amach*, growing, whether it has the animal head main element or two substitute glyphs, and whether or not it has the 'bone', *ch'* affix (although this is confirmed in its role by an *ich*, T287 substitute); *k'amach* is also virtually homophonous with *kama'ach* meaning lower jaw. *Kama'ach* in turn, is both synonymous and loosely homophonous with *kabal chi'*, lower jaw, which might be written phonetically (as in Glyph B), or logographically as at Sacchana and Chichen Itza

(Figure 20.3c: B4; Figure 20.4a: B2). There is an analogous iconographic use of *kabal chi'* on the sarcophagus lid of Pacal at Palenque. He is shown emerging from between the two sides of the lower jaw of the earth, like the newly rising moon or sun (Marquina, 1964, p. 954).

For the Copan inscriptions the most important aspect of these puns and substitutions is that *bal* and *mal* may be equivalent, so that the name of this Copan ruler, XVIII *K'abal*, fire-drilling, made a pun with *k'amal*, ascent – which is one of the most important glyphs in the phrase describing accession to office, as Schele and Miller demonstrate glyphically (1983, pp. 36–60).

XVIII Tok *and XVIII* K'abal On Stela D at Copan, erected by this same XVIII *K'abal*, the fire-drilling meaning of his name is confirmed, although the reading is different. There a large round chert (flint) nodule was deliberately left in the stone in order to substitute for his name which as a consequence read 'XVIII flint'.

Schele and Miller suggest this is to be read as 'ball' since ball-game balls and rabbits are closely associated with each other (Schele and Miller, 1983, pp. 49, 50, Fig. 19), but it seems more likely it is to be read as what it is, namely flint, thus giving the name, in this context for the ruler, as 'XVIII *Tok'/tok*' (flint/fire). Like XVIII *K'abal*, this refers to the Fire Shield, so that while this ruler's name was XVIII *K'abal*, he could be called XVIII *Tok* since they meant the same thing: fire-drilling. The names may possibly refer to his age, 18 annual fire-drillings (*k'abal*) at the time of his accession (*k'amal*).

My own recent work on emblem glyphs has shown that a constant sign (as for instance the Tikal emblem) may be read in a variety of closely related but different ways, while different signs (as with the Palenque emblems) may sometimes all be read the same (Coggins, 1987b), hence it would be entirely consistent with Maya practise to call this ruler XVIII Rabbit under certain conditions, and XVIII *K'abal* under others.

New-Sun-at-Horizon at Copan

Stela N, perhaps erected by the 15th ruler of Copan, '*k'ak'mo*', is dedicated to a 'One Ahau' date that celebrates the heliacal rise of Venus, among other significant astronomical, calendric and historical events (Coggins, 1988). Although no accession is recorded here, a Fire Shield, with a *kan*, serpent/sky title, begins the second half of the inscription (Maudslay, 1889–1902, plate 79). These, and the next three glyphs, are parallel to the Stela A statement; however, they commemorate a fire-drilling by this ruler that was perhaps not a New Fire Ceremony, since the following dog/fire glyph may lack a *yax* prefix, and the *hel* glyph lacks the '*kahok*', thunder and lightning and sky elements; here too, however, the lunar phase is described by Glyph B.

The 16th and last known ruler of Copan, who associated himself with Stela N by adding a carved base to the stela, had two different names. The first of these was *Yax Pak-(al?)* (W. L. Fash, Jr., 'Maya state formation: a case study and its implications'. PhD thesis, 1983, Harvard University, appendix C, p. 12), and the second a glyph that is read as 'First (or New) Sun-at-Horizon' (Figure 20.4c). Since this glyphic compound apparently shows the sun between earth and sky, it has been translated as 'dawn', or *madrugada* in Spanish. This glyph may refer to the rebirth of the sun at a new Calendar Round, since one would have occurred at AD 778 during this man's reign, and to the heliacal rise of Venus as well, since this was celebrated or commemorated on the associated Stela N.

Sun-at-Horizon, the *Yax Tok* name, and the Fire Shield are all commonly found in the inscriptions of the katun One Ahau three Calendar Rounds later at Chichen Itza, where they are perhaps directly associated with Calendar Round ritual, while at Copan they pertained more to the ruling Toltec Maya elite and its dynastic ceremony within the Maya Long Count, as exemplified by the stelae themselves that do not occur at Chichen Itza.

Hieroglyphic writing in the inscriptions at Chichen Itza is composed of more phonetic than logographic signs, but a Cholan/Yucatecan vocabulary probably served as the base (Campbell, 1984) for these late hieroglyphic inscriptions as well as most earlier ones everywhere; glyphic forms are most often translatable in Yucatecan or Cholan (Kaufman and Norman, 1984, p. 87), and they were probably understandable in both (S. R. Witkowski and C. H. Brown, 1986, 'Loan words and lowland Maya prehistory' unpublished manuscript).

The inscriptions and the educated ritual language of Copan may thus have been bilingually Cholan and/or Yucatecan, in distinction to another Mayan language that was spoken there but was expressed only selectively in the inscriptions.

The erection of a stela must have been preceded by a conclave of priests, scholars, and elders of the ruling family, many of whom would have been trained as scribes and artists; at such a council each glyph must have been the subject of debate since there were choices to be made about every constituent element in order to convey as much meaning as possible (logographic and phonetic) within a very limited format, while also achieving a work of art. This process was exceptionally successful at Copan, as can be seen in its beautiful inscriptions and symbol-laden sculpture, whereas 150 years later at Chichen Itza, with the loss of the richness and complexity of Classic Maya culture, the hieroglyphic writing had become more phonetic than logographic, and more like typing than calligraphy.

20.4 Sun-at-Horizon at Chichen Itza

The glyph Sun-at-Horizon has been interpreted as signifying the passage of one day (Thompson, 1960, pp. 168–9, Fig. 31: 41–51), although it may describe a horizon event that would have occurred periodically and thus be the equivalent of a particular day – whereas at Copan it is usually (also) a name or title. At Chichen Itza the glyph is used 13 times in what remains of the inscriptions at the site (Beyer, 1937, Figs. 337–45). At the Temple of the Four Lintels this event is noted four times: three times at 9 Lamat 11 Yax, 10.2.12.1.8, Katun One Ahau; once on the interior eastern Lintel 1 (Figure 20.3a: F1), once on the western lintel 4, both west-facing (Thompson, 1960, Fig. 38: 1–3), and once on the north-facing Lintel 3 (Ruppert, 1952, Fig. 109). This date which is filled with Venus (both Lamat and Yax) symbolism was July 11, 881, a date that preceded the One Rabbit New Fire ceremony in November 882 by one-and-a half years, but one that occurred when Venus was Evening Star, about 40 days after the first appearance (D. Drucker, personal communication). The date on the north lintel, 12 Kan 7 Zac is fourdays earlier (Thompson, 1960, Fig. 38.4) The sun-at-Horizon glyph may thus involve either morning or evening phases of Venus. At Copan the Venus association for the ruler Sun-at-Horizon is probably implied by the name's *yax* prefix, since the month *Yax* has the tutelary Venus monster (Thompson, 1960, Fig. 22: 50–2), and *yax* means first or new, just as Venus's heliacal (first) rise is also implied for this name by the particular importance for this Copan ruler of the One Ahau dedicatory date of Stela N. At Chichen, however, it is the associated Lamat and Yax dates that make the Venus association, while the west-facing lintels and the date itself indicate Venus was Evening Star on a historic date in the Katun One Ahau.

20.5 Toltec ritual at Chichen Itza

Yax Tok *and jawbone*

The Chichen ruler usually called *Yax T'ul*, or First Rabbit, was probably also named *Yax Tok'*, New Fire/flint, although the Rabbit name may have been used under certain circumstances. This name phrase appears many times at Chichen, four of them on the lintels of the Temple of the Four Lintels (Figure 20.3a: A8, E6). *Kabal chi'*, expressed as two (*ka*) jawbones (Figure 20.3c: B4) also appears three times on lintel 3, and as a jawbone and a 'longbone' in six other Chichen structures associated with other names (Beyer, 1937, pp. 55–7). Michel Davoust (1980) has treated these as different from each other, and often as names of rulers, but I suggest they are all the same and that they mean lower jaw, *kama'ach/kabal chi'*, and may be read as *k'amach*, growing, as in Glyph B. On lintel 3, in the Temple of the Four Lintels, the double jaw-bone is followed, in its three appearances, by the hand with crossed bands that means lineage (*k'aba'*), with a bone (*ch'*) postfix (Figure 20.3c: A5), further suggesting a reading of *k'abal chi'*, for the preceding two lower jaws, a punning reference to *k'amach*.

Lunar data at Sacchana The significance of the homonymous use of the *k'abal ch'* animal head and lower jaw substitutions become clearer in reference to Glyph B, when the Sacchana inscription is considered (Figure 20.4a). This continues from an Initial Series on the front of the stela to the notation of the completion of the *tun* at the top of the back which should be followed by a lunar series. The next two glyphs may indicate the condition of the moon and be read *k'ak'mal*, which means to wither or fade at A2, and *k'amach*, to grow or augment at B2, which serves as Glyph B. An explanation for these conflicting terms may be found in modern Chiapas, where the last day of the old moon is counted as immediately preceding the first day of the new moon 'even though the moon is not visible' (Thompson, 1960, p. 236). The ambivalent glyphic statement on this Chiapas monument is followed by *u k'abal* (actually *u-bal-k'a*), 'its fire-drilling', and by *u tok'* (*u toh-k'a?*), 'fire/flint' (Figure 20.4a: A3–B3). The inversion of phonetic elements in glyphic compounds, as in these words, is called 'sign order reversal'; it is not uncommon and may signal alternate readings (see Fox and Justeson, 1984, p. 22). After these two fire-drilling words there is a moon glyph (A4) which is probably Glyph D of the Lunar Series. Thompson describes this form as designating the dead, or invisible, moon between disappearance and reappearance – as is also stated by the contradictory preceding terms (Thompson, 1960, p. 238). Thus, this lunar report apparently records fire-drilling on a moonless night that occurred at the completion of the lahuntun 10.2.10.0.0, on 2 Ahau, 13 Ch'en, June 24 or 26, AD 879.

This inscription, with its ritual and lunar statement located on the back of the monument, suggests the importance of night-time ceremony, particularly fire-drilling, and shows that the timing of such Calendar Round activities might have been regulated by the phases of the moon, separated as they are from the solar Long Count on the other side. Such a lunar context is one in which it would be appropriate to read the all-purpose animal head as a rabbit. Schele and Miller (1983, Fig. 18) point out the identification of the rabbit with the moon, and Thompson (1960, pp. 231–2) notes that flint is an important, probably Mexican, component in this same symbolism, so it is not possible to rule out rabbit, or dog, or jaguar for this animal head; each of them may have been signified, according to the context of the reading.

The Temple of the Four Lintels, Chichen Itza The Temple of the Four Lintels' inscriptions make numerous references to fire ceremonies, and it is possible that, as on the Sacchana stela, a record of lunar position was made in order to document the character of the sky at the time of the fire-drilling, but the first 'lunar' glyph (T683c) in the inscription refers to the count of 20, or to the katun cycle (Figure 20.3a: B4). The characteristic Long Count dating system of Chichen Itza identified each katun, or 20-year period, by the *Ahaw* day on which it ended. In this inscription One *Ahaw* (Figure

20.3a: B3) is immediately followed by a bat-head 'count of the katun' phrase '*u tsola'n k'al*', 'the ordering of twenty' (A4–B4); this phrase is analogous to the name of the ritual calendar '*tsolk'in*', count or order of the days).

The next two glyphs on lintel 1 may refer to the lintel itself, as Kelley has suggested (Kelley, 1982, p. 5); '*U Pakab, tik*', 'his lintel is revealed' (Figure 20.3a: A5–B5). This is followed by the possessor of the lintel *Ah K'ak' K'aba'-ti*', 'Lord Fire of the noble lineage' (Figure 20.3a: A6–B6).

In the next block two more titles, '*Bolon Hel, Ahaw*', 'Many Generations, Lord' (Figure 20.3a: A7–B7) are followed by the *Ah Yax Tok*', New Fire/ Flint, name and *Ka-n(a)*, the serpent or sky title. *Yax Tok'* is identifiable, as in its other occurrences, by the *ets'nab* marked ears that also give a *tok'* (flint) reading to the *toh* (woolly dog) name (see Beyer, 1937, Figs. 131–7 for better examples); this reading is further reinforced by the T60 (knot) *k'a*, which is invariably postfixed. *K'ax* means to tie or knot, and this knot T60, when used in accession phrases with the T757 animal head, reads *k'amal*, or ascent, a close homonymn, certainly a pun, with *k'abal*, fire-drilling, as noted earlier (see Coggins, 1987b, for the T60, *k'a* prefix in the *Ti k'al* emblem glyph).

The Fire Shield at Chichen Itza

The ceremonies associated with this man and commemorated on these (and other Chichen) lintels are described as involving the use of the Fire Shield that has been read by David Kelley as the phonetically rendered *k'ak'-u-pakal* (Figure 20.3b: C5; Kelley, 1968a, p. 151, 1968b). Here the ceremony in question involves the Sun-at-Horizon event. Sun-at-Horizon was thus celebrated with burning ceremonies, and probably with sacrifice and fire-drilling since those glyphs are grouped together in these inscriptions (Figure 20.3b). It is clear from the date, however, these were not November New Fire Ceremonies; ritual fire-making probably also accompanied, or perhaps 'generated', Venus appearances as well as symbolically rekindling the sun annually, and more important every 52 years.

New Fire was also drilled, *k'abal*, to inaugurate the reign of the new ruler whose role and person were identified with the sun, and whose *k'amal*, ascent, may have been timed for an heliacal rise

of Venus. Fire-making with the Fire Shield was Toltec ritual. It was concerned with the beginnings of cycles, in distinction to Maya calendric ritual which traditionally emphasized the completion of cycles. These Toltec fire-drillers at Chichen Itza may have been the Itza, polyglot Maya from the south (Kelley, 1968b, pp. 260–2).

David Kelley has read both the logographic flaming shield on Stela A at Copan, and the phonetically designated Fire Shield at Chichen Itza as *k'ak'u-pakal*, translating it as 'Fire-his-Shield'. He identified this as the name of a Chichen ruler who, since the name occurred elsewhere and earlier, may have had a descendant at Chichen Itza (1968b, 1982, p. 10, Table 2); and since the name *kakupakal* is found in colonial Maya history, probably in reference to an Itza warrior, it is thus also demonstrated as having existed as a name. In view of the temporal spread and textual associations of this compound, it seems more likely to me that in its prehispanic role this glyphic compound was descriptive of a ritual act, or referred to a royal function rather than to a person.

In the previous paper on this topic I traced the occurrence and significance of pyrite 'mirrors' or back shields at Teotihuacan, in Maya regions, at Xochicalco and finally at Chichen Itza and Tula where they are prominent in the symbolic accoutrements of the sites (Coggins, 1987a).

At Chichen Itza these discs or shields were worn by warriors, and they, or their central pyrite mosaic 'mirrors', were cached, often burned, at the pivotal center of the Caracol (Ruppert, 1935, p. 86), and elsewhere (Figure 20.2a,b); the cache at the foot of the stairway of the inner Castillo included two of these fire shields that were wrapped up with two sacrificial knives (Figure 20.2a). In the earlier paper it was proposed that these back discs, emblematic of Teotihuacan and later 'Toltec' warriors, were placed on the epigastrium of the Chac Mools and that New Fire was drilled upon them. These discs upon which fire was drilled were Fire Shields and, following Kelley, I suggested *k'ak'upakal* ('Fire-its-Shield') was their Maya name (Figure 20.3b: C5).

In his study of Chichen inscriptions Thomas Barthel read this compound a little differently, since the first element is a hand (Barthel, 1964, pp. 224–5). He read the first two elements (Figure 20.3b: C5a,b) as *k'ab ku*, 'hand of God' – which is the euphemistic name of the Maya sacrificial knife. In view of the association of sacrificial knives with these shields in the Castillo cache, and of the sacrifice glyph that immediately follows this compound (Figure 20.3b: D5), I think Barthel was right. Thus this glyphic compound could be read *k'ab-ku-pakal*, in referring to the two closely associated ritual objects: the sacrificial knife and the fire shield, and thus to the sacrifice that accompanied ceremonial fire-making. *K'ab-ku-pakal* (sacrificial knife/shield) and *k'ak'-u-pakal* (fire-its-shield) are, however, virtually homophonous and they are equally accurate in describing the associated attributes of the shield and its ritual, so both were probably implied.

Conclusions

It is postulated that Chichen Itza was founded by Toltec or Mexicanized Maya not long after the fall of Teotihuacan in anticipation of the end of the tenth baktun, and the beginning of the 11th(?) Calendar Round, which were to be followed by the heliacal rise of Venus. The completion of the tenth baktun, on March 13, AD 830, was commemorated by the radial, quadrilateral form of the newly constructed outer Castillo. This great cyclic completion ceremony culminated at the equinox when the serpents appeared to descend the balustrades of the Castillo and head for the Sacred Cenote (Rivard, 1971), probably in their first performance. Then, 100 days, or five mesomonths, after the baktun ending came the summer solstice celebration, bracketed equally by the two solar zeniths.

There must have been great relief and jubilation over the continuation of the world, of time and of the usual ceremonial cycle after the end of the tenth baktun, but these Mexicanized people may have been equally anxious over the success of the impending 11th(?) New Fire Ceremony in November. For the Colonial Maya this *Xul* (November) festival was dedicated to Kukulcan or Quetzalcoatl. At this time the annual New Fire was drilled at the Castillo and feathered serpents descended to earth (Tozzer, 1941, p. 158). This fire-drilling was followed by the month *Yax Kin* or New (first) Sun, when the sun was presumed to ascend from its nadir position.

In the year of the end of the Baktun, AD 830, the sacrificial New Fire Ceremony that generated the sun was, however, followed by the heliacal rise of Venus on the long-awaited day One Ahau, December 18. This appearance was the ultimate guarantee of Toltec continuity, and signaled that the New Age or Sun at Chichen Itza was truly inaugurated.

Twelve of the inscriptions at Chichen Itza recorded dates within the katun One Ahau, 40 years later (AD 869–89). This katun included the 12th(?) New Fire Ceremony, the first celebrated since the apocalyptic year of AD 830. Keeping public counts, or recording them, appears to have commenced only when the people of Chichen Itza had arrived safely in the katun One Ahau. This katun may also have marked the departure of *Yax Tok'* from the western Maya regions and his arrival at Chichen Itza. Even though these ninth-century inscriptions are written in a more phonetic hieroglyphic writing than found on Classic monuments, they documented Toltec Maya names or titles, New Fire and Sun-at-Horizon, and rabbit and dog glyphs associated with fire-drilling rituals that had been recorded in sites on the Maya frontiers for more than a century.

There are very few later inscriptions at Chichen Itza, but the four-stairway platform known as the Venus Platform in front of the Castillo has an ideographic relief that commemorates the principal events of the year One Rabbit two Calendar Rounds (104 years or 65 Canonical Venus cycles) after AD 830 (Figure 20.5)–or AD 934.

However, in AD 934 the 13th(?) New Fire Ceremony and the heliacal rise of Venus, occurred in close succession. The Venus Platform tableros depict a Maya Pop sign, a Venus emerging-from-the-earth monster, and a composition of Toltec calendric signs. This latter includes a Year Sign, a *Xiuhmolpilli* (or year bundle), a Venus sign with a fire-drill and cord, and the numbers five (a bar) and eight which add up to 13 (possibly the number of the Calendar Round). The numbers also refer to the combined cycle of five Venus and eight solar cycles that culminated in the New Fire Ceremony and the heliacal rise of Venus that it is postulated were celebrated consecutively, culminating on One Ahau, 18 Kayab (November 22, 934). Evidence for continuation of the New Fire drilling and of its Maya terminology may be seen in the Venus, or star, element next to the year bundle. Star, *ek'*, if

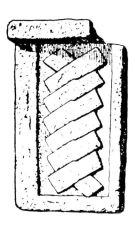

Figure 20.5 Venus Platform reliefs, Chichen Itza. (From Seler, 1915, Figures 241–3.)

prefixed to the twisted cord of the fire drill, *bal*, may be read *(e)k'-bal*, or *k'abal*, fire-drilling. The unusual importance of this date is discussed by Floyd Lounsbury who notes that it served as a calculating base for the Venus Table in the *Dresden Codex*, and that it had added significance, since Mars was in conjunction with Venus at the same time (Lounsbury, 1983, Table 4, Figs. 1, 2, pp. 8–10).

The Venus Platform was constructed during the period of dominance by the highland Toltec of Tula, Hidalgo, and represents a victory of the Toltec Calendar Round and the associated Venus cycles over the millennial, and historical Long Count of the lowland Maya – a process that was underway at Copan ten katuns before in the reign of XVIII *K'abal*.

Acknowledgments

This paper grew out of an earlier one on the New Fire Ceremony at Teotihuacan and Chichen Itza (Coggins, 1987a), but it would not have taken this direction without the work of several people. I am particularly grateful to Patricia Anderson (n.d.) for making available her catalogue of the inscriptions of Chichen Itza, and to Cornelia Kurbjuhn and John Justeson for comments on an early draft of this paper. It is also a pleasure to acknowledge the work of David Drucker, whose reconstruction of a Teotihuacan calendar is innovative and has been an inspiration (Drucker, n.d.). Finally David Kelley's analyses of the inscriptions of Chichen Itza (Kelley 1968a,b, 1976, 1982) have provided a foundation for these attempts at interpretation.

References

Anderson, P. K. (n.d.). Some preliminary observations on the Maya hieroglyphic inscriptions of Chichen Itza.

Andrews, E. W., IV and Andrews, E. W., V (1980). *Excavations at Dzibilchaltun, Yucatan, Mexico.* Tulane University, Middle American Research Institute, publication no. 48.

Barrera Vasquez, A. (1980). *Diccionario Maya Cordemex.* Merida: Ediciones Cordemex.

Barthel, T. S. (1964). Comentarios a las inscripciones clasicas tardias de Chichen Itza. *Estudios de Cultura Maya* **4**, 223–4.

Beyer, H. (1937). Studies on the inscriptions of Chichen Itza. *Carnegie Institution of Washington, Contribution to American Archaeology* no. 21.

Broda, J. (1982). La fiesta Azteca del fuego nuevo y el culto de las Pleyades. In *Space and Time in the Cosmovision of Mesoamerica*, ed. F. Tichy, pp. 129–57, Lateinamerikan Studien no. 10. Munch: Universitaet Erlangen.

Campbell, L. (1984). The implications of Mayan historical linguistics. In *Phoneticism in Mayan Hieroglyphic Writing*, eds. J. Justeson and L. Campbell, pp. 1–16, pub. 9. Institute of Mesoamerican Studies, State University of New York at Albany.

Coggins, C. C. (1980). The shape of time: some political implications of a four-part figure. *American Antiquity* **45**, (4), 727–39.

Coggins, C. C. (1983). *The Stucco Decoration and Architectural Assemblage of Str. 1-sub, Dzibilchaltun, Yucatan, Mexico.* Tulane Univesity, Middle American Research Institute, publication no. 49.

Coggins, C. C. (1984a). An instrument of expansion: Monte Alban, Teotihuacan and Tikal. In *Highland–Lowland Interaction in Mesoamerica: Interdisciplinary Approaches*, ed. A. G. Miller, pp. 49–68. Washington, DC: Dumbarton Oaks.

Coggins, C. C. (1984b). Catalogue. In *Cenote of Sacrifice: Maya Treasures from the Sacred Well at Chichen Itza*, eds. C. C. Coggins and O. C. Shane III, pp. 23–166. Austin: University of Texas Press.

Coggins, C. C. (1987a). New fire at Chichen Itza. *Memorias del Primer Coloquio Internacional de Mayistas*, pp. 427–84. Universidad Nacional Autonoma de Mexico, 5–10 August, 1985.

Coggins, C. C. (1987b). the names of Tikal. *Primer Simposio Mundial sobre Epigrafia Maya*, pp. 23–45. Instituto de Antropologia e Historia, Guatemala.

Coggins, C. C. (1988). On the historical significance of decorated ceramics at Copan and Quirigua and related Classic Maya sites. In *The Southeastern Classic Maya Zone*, eds. E. H. Boone and G. R. Willey, pp. 95–123. Washington, DC: Dumbarton Oaks.

Coggins, C. C. (in press). the manikin scepter; emblem of lineage. *Estudios de Cultura Maya* (1985).

Davoust, M. (1980). Les premiers chefs Mayas de Chichen Itza. *Mexicon* **2**, (2), 25–9.

Drucker, D. R. (n.d.). The Teotihuacan pecked crosses: models and meanings. Paper presented at *The 51st Meeting of the Society for American Archaeology*. New Orleans, 1986.

Fox, J. A. and Justeson, J. S. (1984). Polyvalence in Mayan hieroglyphic writing. In *Phoneticism in Mayan Hieroglyphic Writing*, eds. J. S. Justeson and L. Campbell, Institute for Mesoamerican Studies, pub. 9, pp. 17–76. State University of New York, Albany.

Graham, I. (1967). *Archaeological Explorations in El Peten, Guatemala.* Tulane University, Middle American Research Institute, publication no. 33.

Graham, J. A. (1967). Aspects of non-classic presences in the inscriptions and sculptural art of Seibal. In *The Classic Maya Collapse*, ed. T. P. Culbert, pp. 207–220. Albuquerque: School of American Research, University of New Mexico Press.

Justeson, J. S. (1984). Appendix B: Interpretation of Mayan hieroglyphs. In *Phoneticism in Mayan Hieroglyphic Writing*, eds. J. S. Justeson and L. Campbell,

Institute of Mesoamerican Studies, pub. 9, pp. 315–62. State University of New York, Albany.

Kaufman, T. C. and Norman, W. M. (1984). An outline of Proto-Cholan phonology, morphology, and vocabulary. In *Phoneticism in Mayan Hieroglyphic Writing*, eds. J. S. Justeson and L. Campbell, Institute of Mesoamerican Studies, pub. 9, pp. 77–166. State University of New York, Albany.

Kelley, D. H. (1968a). Mayan fire glyphs. *Estudios de Cultura Maya* **7**, 142–57.

Kelley, D. H. (1968b). Kakupacal and the Itzaes. *Estudios de Cultura Maya* **7**, 255–68.

Kelley, D. H. (1976). *Deciphering the Maya Script.* Austin: University of Texas Press.

Kelley, D. H. (1982). Notes on Puuc inscriptions and history. In *The Puuc: New Perspectives*, ed. L. Mills, Scholarly Studies in the Liberal Arts, publication no. 1, supplement. Pella, Iowa: Central College.

Kubler, G. (1961). Chichen Itza y Tula. *Estudios de Cultura Maya* **I**, 47–79.

Kubler, G. (1982). Serpent and atlantean columns: symbols of Maya–Toltec polity. *Journal of the Society of Architectural Historians* **51**, (2), 93–115.

Lincoln, C. E. (1983). Chichen Itza: clasico terminal o postclassico temprano? Boletin de la Escuela de Ciencias Antropologicas de la Univ. de Yucatan **10**, (59), 3–29.

Lounsbury, F. G. (1983). The base of the Venus Table of the Dresden Codex and its significance for the correlation problem. In *Calendars in Mesoamerica and Peru: Native American Computations of Time*, eds. A. F. Aveni and G. Brotherston, pp. 1–26. Oxford: British Archaeological Reports, International Series no. 174.

Marquina, I. (1964). *Arquitectura Prehispanica. Memoria* **1**, 2nd edn. Mexico: Instituto Nacional de Antropologia e Historia.

Mathews, P. (1980). Notes on the dynastic sequence of Bonampak, part 1. In *Third Palenque Round Table, 1978, part 2*, ed. M. G. Robertson, pp. 60–73. Austin: University of Texas Press.

Mathews, P. (1984). Appendix A: a Maya hieroglyphic syllabary. In *Phoneticism in Mayan Hieroglyphic Writing*, eds. J. S. Justeson and L. Campbell, Institute of Mesoamerican Studies, pub. 9, pp. 311–14. State University of New York, Albany.

Maudslay, A. P. (1889–1902). *Archaeology, vol. IV: Palenque.* London: R. H. Porter and Dulau. Facsimile reprint 1974, prepared by F. Robicsek. New York: Arte Primitivio, Inc.

Milbrath, S. (1980). Star gods and astronomy of the Aztecs. In *La Antropologia Americanista en la Actualidad: Homenaje a Raphael Girard*, vol. 1, pp. 289–303. Mexico.

Proskouriakoff, T. (1974). *Jades from the Cenote of Sacrifice, Chichen Itza, Yucatan*. Cambridge, MA: Memoirs of the Peabody Museum **10: 1**, Harvard University.

Rivard, J. J. (1971). *A hierophany at Chichen Itza*. University of Northern Colorado, Miscellaneous Series no. 26.

Robertson, M. G. (1985). *The Sculpture of Palenque, vol. 3. The Late Buildings of the Palace*. Princeton University Press.

Ruppert, K. (1935). *The Caracol at Chichen Itza, Yucatan, Mexico*. Carnegie Institution of Washington, publication no. 454.

Ruppert, K. (1952). *Chichen Itza: Architectural Notes and Plans*. Carnegie Institution of Washington, publication no. 595.

Ruz Lhuillier, A. (1962). Chichen Itza y Tula comentarios a un ensayo. *Estudios de Cultura Maya* **2**, 205–20.

Schele, L. (1985). *Bolon-ahau*: a possible reading of the Tikal emblem glyph and a title at Palenque. In *Fourth Palenque Round Table, vol. 6*, eds. M. G. Robertson and E. P. Benson, pp. 59–66. San Francisco: Pre-Columbian Art Research Institute.

Schele, L. and Miller, J. H. (1983). *The Mirror, The Rabbit, and The Bundle: Accession Expressions from the Classic Maya Inscriptions*. Washington, DC: Dumbarton Oaks.

Studies in pre-Columbian Art and Archaeology, no. 25.

Seler, E. (1915). Die ruinen von Chichen Itza in Yucatan. In *Gesammelte Abhandlungen zur Amerikanischen Sprach- und Alterumskiunde*, vol. 5, pp. 197–388. Berlin.

Thompson, J. E. S. (1960). *Maya Hieroglyphic Writing: An Introduction*, 2nd edn. Norman: University of Oklahoma Press.

Thompson, J. E. S. (1962). *A Catalogue of Maya Hieroglyphs*. Norman: University of Oklahoma Press.

Tozzer, A. M. (1930). Maya and Toltec figures at Chichen Itza. In *Acta*, 23rd International Congress of Americanists, New York, 1928, pp. 155–64.

Tozzer, A. M., tr. and ed. (1941). *Landa's Relacion de las Cosas de Yucatan*. Cambridge, MA: Papers of the Peabody Museum, Harvard University, **18**.

Tozzer, A. M. (1957). *Chichen Itza and its Cenote of Sacrifice: a Comparative Study of Contemporaneous Maya and Toltec*. Memoirs of the Peabody Museum, Harvard University **11**, **12**, Cambridge, MA.

Umberger, E. (1987). Events commemorated by date plaques at the Templo Mayor: further thoughts on the solar metaphor. In *The Aztec Templo Mayor*, ed. E. H. Boone, pp. 411–50. Washington, DC: Dumbarton Oaks.

Velazquez, P. F. (1945). *Codice Chimalpopoca: Anales de Cuauhtitlan y Leyenda de los Soles*. Mexico: Instituto de Investigaciones Historicas.

21

Zodiac signs, number sets, and astronomical cycles in Mesoamerica

Gordon Brotherston   *University of Essex*

What is surely known about Mesoamerican astronomy consists in the main of numerical data relevant to cycles of sun, moon, and the five visible planets. As spatial coordinates, this order of measurement requires only one or other of the horizons east and west, or one or other moment along the zodiac path that runs at about 18° either side of the ecliptic and hence to north and south of the celestial equator. In these circumstances the zodiac, which everywhere in tropical Mesoamerica runs through the zenith if not actually into the northern hemisphere, could be expected to have been a major object of attention; and just this is suggested by previous studies of possible zodiac signs in the corpus of Maya hieroglyphic texts, signs which by their inclusion of zoomorphs etymologically justify the term 'zodiac'.

The identity of these Maya hieroglyphic signs, in relation to each other and to the constellations of the sky, can be usefully explored through comparison with counterparts to them, much neglected, that exist in the corpus of texts written in the script sometimes called 'Mixtec–Aztec' and here referred to as iconographic. Indeed, such an exercise reveals a zodiac paradigm overlooked in the one comprehensive catalogue so far made of ritual texts in iconographic script, Nowotny's indispensable *Tlacuilolli* (1961). And it opens the way to a reading of allied numerical data found in these same iconographic sources, and to an awareness of the Mesoamerican distinction in principle between the synodic time of the horizon and the sidereal time of the zodiac.

21.1 Signs of the zodiac

Had the Mesoamerican zodiac simply consisted of a fixed sequence of never-varying signs, of the type our duodecimal version eventually became, then no doubt it would have been fully explained long ago. Major differences certainly exist between the sequences examined here. Yet their internal consistency may be demonstrated on the basis of two complementary factors: the correspondence in detail of particular signs and sequences of signs; and the existence of a zodiac paradigm involving the number 11 (Table 21.1).

In *Deciphering the Maya Script* (Kelley, 1976, pp. 49–51), David Kelley reviewed the evidence, as it then was, for a Maya hieroglyphic sequence, paying special attention to the 'sky-band' signs found at Chichen Itza, above the east door of Las Monjas, and in the *Paris* screenfold, in a table of sidereal months (pp. 23–4). Since then, still within the Maya hieroglyphic domain, further analogies have been demonstrated with the gods 'A' to 'T', as they have been called, which accompany the Venus tables in the *Dresden* screenfold; and with the signs in the top register of the murals in Room 2 at Bonampak, which likewise accompany Venus data (Lounsbury, 1982). This last sequence, like that at Chichen, appears to total 11 and clearly includes the 'turtle', *ac ek* in Maya, which the Motul dictionary defines as Gemini or Orion. In his 'Postscript' on astronomy at Bonampak, Lounsbury presses the case for Orion (1982, pp. 166–7); yet for our purposes it does not matter which it is (and it can be only one or

Table 21.1. *Sequences of zodiacal signs in various codices*

	e/w 1	2	3	s 4	5	6	w/e 7	8	9	n 10	11
(a)	Patecatl	4	3	2	1	11	Mayauel	9	8	7	6
(b)											
(c)											
(d)											
(e)											
(f)											
(g)											
(h)											

(a) *Magliabecchi;* (b) *Cospi;* (c) *P. Diaz;* (d) *Madrid;* (e) *Dresden;* (f) Chichen; (g) *Paris;* (h) Bonampak.

the other) since, as Kelley elegantly put it, the difference between them in right ascension, either side of 6 hours in the broad ecliptic band, is no greater than the range of error of the method. For its part, *Dresden* offers a visual plus phonetic rendering of the Scorpion constellation, *si'nan* in Maya, that is prominent at Chichen and in *Paris* (see Table 21.1e, f, g, h, position 1, and Chapter 8, this book, by J. Justeson).

Introducing evidence from the iconographic corpus both confirms these parallels and suggests further ones, along with a numerical paradigm. On the 11 pages of the *Cospi* screenfold (reverse), 11 zoomorphs appear, in three groups of 'stingers', 'snail-shells' and 'hearts', which so closely correspond to those in the hieroglyphic texts that there can be little doubt about their common identity (Table 21.1b). The Scorpion, whose stellar nature in Nahuatl is assured by the *Florentine Codex* (book VII; see Figure 21.1), is followed by a flying creature, another smaller 'stinger' and then by another

Figure 21.1 Scorpion constellation (*colotl*); *Florentine Codex.* (Schultze Jena, 1950, p. 402.)

'flier', in a pattern that is exactly matched in *Paris*, and is three-quarters present in *Dresden* and at Chichen. Structurally, the following pair of shell signs, which are accompanied by water and jade emblems, is just as exactly matched by the Chichen pair of moon and water-lily (*naab*) which have no sky band between them and uniquely in the sequence as a whole are devoid of the 'bright-star' infix. Also at appropriate positions are Snake, a rattle-snake in *Paris* and the pure bright-star Venus at Chichen; the Feathered Serpent who was 'always a rattlesnake' (Kelley, 1976, p. 47); Deer, alluded to as a constellation 'sign' in the *Popul Vuh* (Edmonson, 1971, p. 211), and shown in *Madrid* and *Dresden*, where it is given its number-place in the Twenty Signs (seven); and an unmistakable Turtle. Work in Guerrero by Schultze Jena in the 1930s (Schultze Jena, 1938) has convincingly shown a link between this chapter of the *Cospi* screenfold and native hunting rituals still practised today, a function to which such a zodiacal night sky would clearly be germane.

In the 11-page *Cospi* chapter, the 11 zoomorphs appear with 11 armed figures and with bar-and-dot arithmetic whose optimal number base is 11, which of itself cannot but draw attention to that total as a cipher or paradigm. It is one which on inspection proves to be generally valid where the zodiac is concerned. For 11 signs, ending with Turtle, succeed each other with all clarity at Bonampak; 11 is also the likely and 'implied' total at Chichen, as Kelley noted. It is at least consonant with the double five-and-a-half page arrangement found in *Madrid* (Kelley, 1976, pp. 13–18); and with what is still visible in *Paris*, with possibly six constellations in the upper register and five in the lower, though here some have suggested seven plus six, making the ritual lunar cipher 13. Also, the pair of scorpion and snail signs in the eight-page opening chapter common to *Borgia*, *Cospi* (r) and *Vaticanus* (r) appear to be set into a frame of twice-11 columns (Figure 21.2).[1] As for the cases where the total of units of signs is definitely *not* 11, they may be explained as consequences of the zodiac being thematically subordinate to other ritual patterns. In *Diaz* (Table 21.1c), but four zodiac signs attach themselves, via the *tonalamatl*, to a map of four quarters, with north right and south to left, a geographical model also implicit in the linking of zodiac and 'emblem' glyphs in Classic Maya inscriptions. In *Dresden*, as 20 the zodiac 'gods' are regularly allotted in fours to the five synodic years of Venus and aptly combine, in *tonalamatl* logic, with the 13 sidereal cycles traversed by that planet over that period ($13 \times 224 \cdot 7 = 5 \times 584 = 8 \times 365$). Wherever the zodiac remains a prime reference, however, it may be said to have 11 as its diagnostic, just as it governs today such rituals as that of the '11 virgins of the moon' honoured in Puebla (T. Knab, personal communication). Also, in the set of 13 Heroes (Nowotny, 1961, p. 217), the 11th place is reserved for Yoaltecutin, lord of the starry night sky.

If it is asked why 11 rather than another number, then the question must be extended beyond Mesoamerica, to the Kogi of Colombia for example, whose zodiac has 11 names and divisions (Mayr, 1985), and for that matter beyond the New World to the Old, where 'eleven went to heaven'. As Clerke reports, 'the earlier Greek writers – Eudoxus, Eratosthenes, Hipparchus – knew of only eleven zodiacal symbols', Geminus and Varro being the first to insert Libra (Clerke, 1911, p. 994). In proposing the 2383-year Great Trigon of Jupiter and Saturn as an hour-division on the clock of equinoctial precession, Santillana and Dechend (1969, p. 268) have *de facto* given an arithmetical reason for this total, 11 such Trigons being approximate to 26 000 years, much more so than a standardized 12. At a far simpler level, 11 of course represents the nights of the epact. Of the significance of the cipher 11 as such, full confirmation is given by the corpus of ritual texts examined below.

Given this degree of correspondence in both detail and numerical structure, and allowing for conventional differences of reading directions between left and right and of linear start and end points, the sequences of signs recorded in *Cospi* and the Maya hieroglyphic texts must be held to refer to the same phenomenon. That this phenomenon is indeed the zodiac can be shown by further comparisons with the constellations of the sky. When Kelley came to make this link he appears to have been guided by two considerations: the desire, reasonable enough, to maintain the well-attested Turtle–Orion/Gemini link; and respect for the mechanism of the sidereal month table found in *Paris*. The effect was to make him see alternate not contiguous signs as successive; and though this yielded a fit between Scorpion and our Scorpio, he conceded his scheme to be tentative, as indeed

Figure 21.2 Scorpion and snail signs (*Cospi*, r. pp. 1–2).

it would absolutely have to be if the obvious sequence of all the other zodiacs, from which this particular sidereal table is absent, is to be at all respected. This last point was apparently not lost on Spinden in his pioneer account of the *Paris* text (Spinden, 1924; quoted by Kelley, 1976).

Retaining the Turtle–Orion/Gemini link as an anchor, we may rather see the contiguous Scorpion as Leo and the other 'stinger' as Scorpio, these being the two patently 'tailed' clusters of the ecliptic, and the latter being in fact redder in *Cospi*, as Antares is (Table 21.1, positions 1 and 3). The corresponding pair of winged signs would then coincide with Virgo and Sagittarius, Libra not being a concept, as once was the case in the Old World (Table 21.1, positions 2 and 4; in *Mexicanus*, p. 36, Libra is said to deserve 'annihilation': *niman nimitz-mictin peso*). In this arrangement, the four Mesoamerican signs may be structurally matched with the four brightest stars of the Old World constellations, all of them lying very close to the ecliptic at appropriate intervals of right ascension: Regulus (Leo) 10 hours, Spica (Virgo) 13 hours, Antares (Scorpio) 16 hours, and Kaus Australis (Sagittarius) 18 hours. Also, with claws extended just over the celestial equator (12 hours) into the western part of Virgo, Scorpion touches on that sector of the ecliptic indicated to Schultze Jena (1950, p. 64) in his search for the constellation *colotl* ('scorpion').

Next in the sequence, the watery Snail-Shell and Water-Lily constellations, uniquely devoid as they are of the bright-star infix at Chichen (Table 21.1f, positions 6 and 7), would correspond to Aquarius and parts of Capricorn and Pisces at either side, there being in fact no star brighter than second magnitude all the way from Sagittarius to Aries. In *Cospi*, this part of the sky is further marked by the only two women in the team of 11 armed figures, and by the use of even, or 'female', bar-and-dot numbers (all the male ones are odd), a practice echoed at the appropriate place in *Dresden*. That birth from the uterus resembles emergence from the snail-shell is attested explicitly in many sources, including *Telleriano* (*Tecciztecatl*, 6 Monkey) and Nuttall (birth of Lady Three Flint, p. 16); and the motif as such, whose presence at Bonampak (Table 21.1h, position 7) is confirmed through comparison with the Palenque shell design (Figure 21.3), may perhaps even be read in the spiral of dim points that links Aquarius with Pisces. In the adjacent sign

Figure 21.3 Shell of growth; Palenque 'Maize' panel.

Death (position 8), the acute angle of this figure's foreleg in *Paris* recalls the eastern part of Pisces which overlaps with Aries (Alrisha, Shertana, Hamad); and finally coming back up to the Turtle (position 11), the Snake is left in an ideal position to represent the Pleiades, with which it is identified as the Maya rattle-snake *dzab* (Coe, 1966, p. 162).

It should be stressed, however, that possible signs for this part of the ecliptic, positions 5 to 11 and right ascensions 19 through 9 hours, exhibit neither the correspondence of number-item nor the clarity of detail characteristic of the initial fourfold sequence, positions 1 to 4 and right ascensions 10 through 18 hours; and in no circumstances could an absolute one-to-one relationship be claimed between the various textual sequences and between them and the sky.

Rather, the main weight of proof must lie with the numerical paradigm as such, of which more below; with its inner intervals, especially the alternating fourfold pattern between right ascensions 10 and 18 hours, and the ensuing 'non-bright' star area around Aquarius and Pisces; and with its structural fit with the course of the ecliptic to north and south of the equator. For wherever a spatial dimension is explicit or may be deduced, the relative positions of those signs that have been identified are perfectly apt. To the north lie Death, Turtle and Scorpion whose claws reach back down to the equator again; to the south lie the other stinger (Scorpio), and the watery Snail of Aquarius–Pisces. Read from left to right and starting physically in the south, the purely linear sequence of the Chichen facade begins appropriately with the southernmost constellation Sagittarius, which finally displaced Capricorn from the solstice and the tropic over two millennia ago, thanks to precession. As for the Scorpion, its attainment of the equator about 500 years ago could perhaps explain its initial prominence in

Figure 21.4 Scorpion and flier at the start of the *Fejervary* zodiac sequence, p. 5.

Table 21.2. *Ritual texts.*

Codex	Tonalamatl: Nine figures	Year: sidereal 11	solar 18
Borbonicus (32)	1–20[a]; 21–2		23–38
Tonalamatl Aubin (15)	1–20		
Borgia (33)	9–12	[1–2]	29–46[b]
Vaticanus (384)	r19–23	[r1–2]	
Cospi (79)	r1–8	vl–11[a,b] [r1–2]	
Culte rendu (14)	toponyms 1–5[a,b]		
Díaz (255)	1–2[b]	[10]	
Laud (185)	31–8[a,b]	39–44[b]	21–2[b]
Fejervary (118)	1; 2–4	5–14[a,b]	5–22[a,b]
.			
.			
.			
Telleriano (308)	8–24		1–7
Rios (270)	13–37		42–51
Magliabecchi (188)		37–47[b]	17–35
Mexicanus (207)		[1–11]	13–4

( ) Census number in Glass (1975); [a] contains astronomical data; [b] paradigm ignored by Nowotny 1961; [ ] zodiac signs made subject to the *tonalamatl*.

what are known to be more recent sources like *Fejervary* (Figure 21.4) and *Cospi*, and its absence from the eighth-century picture at Bonampak. Further, unlike north and south, the 'east' and 'west' equatorial positions in between, picked out in the pair of signs in *Borgia* (Figure 21.2), are purely conventional and invert depending on which way the zodiac is joined up two dimensionally; so that while Bonampak follows the common association of the female shell with west, in *Díaz*, from which the shell is absent, this position is occupied by Scorpion, north being a 'pointer' in both.

21.2 Ritual number sets

To assess further the significance of this Mesoamerican zodiac, and in particular its link with the number 11, involves focusing more closely on the whole 'ritual' corpus in iconographic script, to which *Cospi* and the other non-hieroglyphic texts referred to so far belong, and which relies on certain highly specific ciphers and number sets to establish reading sequence. (In hieroglyphics such ciphers are reduced almost always to just the 13 and the 20 of the *tonalamatl*.) This in turn involves a critique of the one catalogue published so far of the iconographic ritual corpus, Nowotny's *Tlacuilolli*. This work gathers together the nine original texts that are extant, arranging them into subgroups and comparing them with other genres like annals, and to later texts on European paper like *Rios* and *Telleriano* of Tenochtitlan (Table 21.2), so as to suggest provenance and the fact that though mantic they deal in part with material history; and it accounts for every page of every text. For all these reasons, because it is exhaustive and has no published rival, this catalogue must be held indispensable for any thorough enquiry into the problem of possible zodiac and related data in the iconographic script tradition, and by extension in Mesoamerican texts as a whole.

Throughout, Nowotny makes the principle of his analysis formal rather than thematic, rightly point-

ing to the excesses that the latter path has led to (e.g. with Seler). Respecting the constants inherent in the genre and which determine the extent and internal reading sequence of each chapter or other subdivision of the text, he concentrates on the 18 20-day 'Feasts' of the 365-day year and, above all, the 260-unit *tonalamatl* with its constituent sets, which he calls 'weeks' (*wochen*), of Nine Night Figures (Yoalitecutin), Thirteen Numbers, and Twenty Signs. Alone, and in combination, these are shown to determine not just reading sequence, chapter by chapter, but frequency and rhythm, as in a musical score. In other words, by the very formality and thoroughness of his analysis, Nowotny vindicates an overriding principle of numerology for the entire genre, one that relies on the three ciphers of the *tonalamatl* (9, 13, 20) and the cipher of the tropical year (18), the few passages he fails to explain in these terms providing our main focus below.

In the analysis of the *tonalamatl* chapters, there is little to object to, except perhaps a certain neglect of the Nine Night Figures announcing what proves to be a general anti-astronomical bias; he has nothing to say about the origin of these figures in midwifery, as representatives of the nine moons of human gestation, from the Fire of coition (1) to the full water-bag of the Rain-god (9), and from the first missed menses to birth, the exact span of the *tonalamatl* as a whole.[2] Overlooking their presence in *Laud* (pp. 33–8), *Diaz* (pp. 1–2), and the toponyms of *Culte rendu map*, he comments neither on the time-shift effected by the Nine between nights and moons, the Nine serving to identify both; nor on the sevenfold year-guardian sequence of the Nine, the midwives' gods (*dioses de las parteras*) illustrated in *Borbonicus* (pp. 21–2). To pick up these formal omissions in Nowotny's catalogue on the subject of the moon, along with its general lack of interest in astronomy, is essential to restoring to the *tonalamatl* its link with the sky, one which has always been accepted in principle in the case of the year and its 18 seasonal Feasts.

When it comes to that other chief domain of ritual organization, the year, the same anti-astronomical bias causes Nowotny seriously to underestimate its function and complexity, though of course he fully endorses the link between the cipher 18 and the sun's annual cycle. Again, at the most basic level he overlooks major examples of chapters whose paradigm is the 18 Feasts: *Borgia* (pp. 29–46), where these Feasts occupy a dramatic 18-page central section running from Tecuilhuitl to Etzalcualiztli; *Fejervary* (pp. 5–22) where 18 pages range from Tititl to Atemoztli/Tititl, and characterize the span from Micailhuitl with the 'heart' emblem found with the Death to Turtle constellations in *Cospi* (Table 21.1c, positions 8 to 11), which in fact dominate the night sky at that time of the year; and *Laud* (pp. 21–2) which highlights a similar span in its concentration on Micailhuitl (with its domestic dog and turkey) and Atemoztli ('water-falling'). In practice this omission precludes from the start any discussion on his part of the variant forms and names of the 18 Feasts, not least those found in Maya hieroglyphics, or those in the post-Cortesian transcriptions of Tenochtitlan where they are connected to dates in the Christian year; of their structure, arched as they are in pairs between the equinoxes in the Aztec tribute system; or of the support they lend, as a farmer's, hunter's and tribute calendar for the claim that the year in question is indeed agricultural and seasonal and therefore over the four-year span of its sign-names can and must include a leap-day (Broda, 1983), except of course in the special case of Maya hieroglyphics with its standardized 360-day unit, to which even the Nine Night Figures of the *tonalamatl* are made subject.

A yet more serious failure of *Tlacuilolli* is that it does not acknowledge even the existence of what turns out to be the missing paradigm of Mesoamerican ritual: the 11. For the only chapters and sequences extant that are not governed by the *tonalamatl* or, within the domain of the year, by the 18 Feasts, adhere, *without exception*, to the cipher 11, leaving not a single page now unaccounted for anywhere in the nine works that constitute the genre. It should be recalled that this is so at the most basic level of reading principle, of formal logic of sequence and periodicity; and this in turn must clearly affect our estimate of the significance and identity of the cipher 11.

Lumped together by Nowotny with unrecognized Feast chapters under the all-purpose rubrics 'Tempelkult' and 'Rituale mit Bündeln abgezählter Gegenstände', the cases of the missing 11 are: *Cospi* (reverse pp. 1–11), where the 11-fold total of the zodiac signs, already examined, governs the accompanying row of seated and armed figures, the number base of the bar-and-dot numbers, and the pages in the chapter; *Laud* (pp. 39–44), which opens with the same sacrifice scene as *Cospi* and where an upper row of 11 figures is introduced by a star-sky emblem; and *Fejervary* (pp. 5–14) which opens with the same zodiac stinger-and-flier as *Cospi* and where the row of seated figures and the bar-and-dot numbers likewise adhere to the cipher 11 (Figure 21.5).

This complete numerical consistency and the corresponding iconography of previously identified zodiac signs (*Cospi*, *Fejervary*) and stars (*Laud*) cannot but create the strong supposition that just as the nine of the *tonalamatl* relate to the moon, and just as the 18 Feasts relate to the sun, so the 11 relate to the night sky. In this reading, as complementary models of the year, solar and sidereal, the 18 and the 11 can be shown to enjoy a special and complementary relationship: for in the *Fejervary* sequence (pp. 5–14) their annual intimacy is spelled out in the fact that the former sequence is actually set into the first part of the latter's 18 pages, from Tititl to Tecuilhuitl, before the change in page format and the doubling of the number base to 22 which occur at Micailhuitl (p. 11). As yearly sets they are further similar in that like those of the 18 the identities of the 11 individual figures (as opposed to actual constellation signs) display more regional variation, within the surely fixed total of

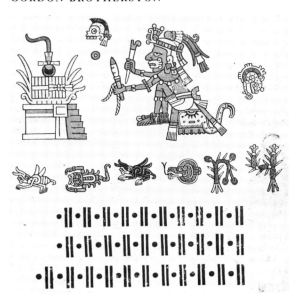

Figure 21.5 Fejervary (p. 5).

the set, than is the case with the *tonalamatl*. Finally, affirming this identity for the 11 makes plain that all the ciphers according to which the entire ritual corpus is regulated derive, when not from the human body (20 digits), then from the sky (13 moons, 18 Feasts, 11 constellations), or from a combination of both (nine orifices, moons).

Within this framework, further evidence for the idea of a zodiac 11 can be found in all that concerns pulque, the drink fermented from maguey, and the pulque bringers whose nose-adornment is the moon crescent and whose principals are Patecatl and his consort Mayauel. The cosmogonical link between pulque and the zodiac is stated by the *Popol Vuh*, where after the Flood when humankind reverted to fish, the 'four-hundred sons' who built the first house and brewed the first pulque got drunk and ascended to become 'The Four-Hundred', that is the zodiac asterism the Pleiades (*Motz*, in Maya, Edmonson, 1971, p. 48; cf. the Nahuatl Miec, 'The Many'). In the treatment of the same cosmic history in *Borbonicus* and elsewhere, the pulque-bringer Patecatl glossed by *Telleriano* as 'the one who survived the Flood' (*el que salio del diluvio*), presides over the 11th 'week' of the *tonalamatl*, whose initial Sign is XI, Monkey, also a character from that part of the *Popol Vuh* and the offspring of the 11th Bird, Macaw; and he does so under a zodiac sky of exactly 11 stars, a total repeated in *Telleriano* and *Rios* in

his 11 vertebral disks (Figures 21.6 and 21.7). Again in *Borbonicus* the same total is picked out in the stars over his consort Mayauel $(3 + 1 + 11)$, on a page where fermenting pulque is indicated by dots numbering about 400 in all (see Figure 21.12), the number of the Pleiades-drunks in the *Popol Vuh*. The 11 stars occur again with Mayauel in the *Aubin tonalamatl*, all of which must be deemed significant since otherwise such star totals vary considerably due in part to their being capable of representing calendrical time-units.

In any case, in the *Magliabecchiano* group of texts from Tenochtitlan these pulque gods actually appear as 11, 'ten local pulque gods, and one goddess of the same category' as Zelia Nuttall put it (Boone, 1983, p. xix), leaving no doubt on the question. In this sequence, symmetrically placed at positions 5 and 10, Patecatl and Mayauel (the one 'goddess') hold their fellows between them (Table 21.1a) these being explicitly identified with toponyms, beginning with 'he of Tepoztlan' and the Tlahuica pulque shrines south of Tenochtitlan cogently discussed by Seler (1904; on 'Tula', see

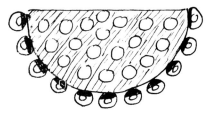

Figure 21.6 Patecatl's 11 stars (*Borbonicus*, p. 11).

Figure 21.7 Patecatl's 11 vertebral disks and axe (*Telleriano*, 15v).

Brotherston, 1985, p. 228). Hence, the *Magliabecchi* sequence overall may be seen to follow the 'drunken' swaying of the zodiac 11 to south and north of celestial equator positions occupied by Patecatl and Mayauel, a scheme which in principle echoes that of *Diaz* and which is physically represented by the frieze at Bonampak, in exactly the same numerical proportions (1:5:1:4) in and between the 'east' and 'west' equatorial positions. A yet further link between the 11 and the zodiac is provided by the axes they carry: used by the Pleiades 'Four-Hundred' in the *Popol Vuh* to cut their house-beams, this instrument is also carried by the Death and Rain-god figures among the *Cospi* 11 and in *Madrid*, as well as by Patecatl in *Borbonicus*.

In the opening comparison between native-script sequences known to refer to the night sky, the number 11 emerged as significant in at least three cases. In the subsequent analysis of the ritual corpus this same number emerged as the cipher essential to a complete and formally consistent reading of the texts in question. Between the two orders of evidence the cross-references were multiple and tied in thematically to maguey and the pulque bringers Patecatl and Mayauel. These circumstances would appear to justify the recognition of a Mesoamerican 'zodiac 11'.

21.3 Numerical data and astronomical cycles

In thus corroborating and illuminating the idea of the Mesoamerican zodiac, whose existence was first suspected on Maya hieroglyphic evidence, the corpus of iconographic texts helps to vindicate in turn the notion, often dismissed, that they carry numerical data of astronomical significance, solar, lunar, and planetary, counterparts for which are well established in Maya hieroglyphic. The problem here has been one both of assumed context – typified in Nowotny's anti-astronomism – and of the sheer nature of the recording medium. For, unlike Maya hieroglyphics, the iconographic script of Mesoamerica is not obliged formally to separate out 'writing' from 'arithmetic' or either from 'picture', or for that matter from symbolic logic or music. On the contrary, the ritual genre exults in the capacity of this script to propose multiple levels of reading, which may be conceptual, or exist as literal registers on the page, a capacity which the alphabet is ill-equipped to transcribe.

Hence, since the question has been raised by

others, the bar-and-dot numbers attached to the zodiac sequences in *Cospi* and *Fejervary* may of course represent 'Bündeln' on actual altars or 'Opfertische', as Nowotny, and Van der Loo (1982), following Schultze Jena, state they do; for they indeed visually resemble such bundles and the layouts of them used by Tlapanec and other shamans today. Yet it is one thing to press the anthropological claim that these shamans know nothing of astronomy or mathematics (something questionable enough in itself); and it is quite another implicitly to extend this denial back to the whole civilization of pre-Columbian Mexico, given what is surely known about its urban architecture, its calendars, and its written records, and given the patent complexity of a document like *Fejervary*. Whatever else may be said of it, *Fejervary* is certainly not in the business of *just* depicting shamanic altars: not least in both *Fejervary* and *Cospi* the layouts in question provide but a part, confined to the lower register, of the whole complex page-design which also involves the 11 figures and a highly technical use of the *tonalamatl* (Figure 21.5), a fact which Van der Loo and others choose to ignore. Further, while several of the subtotals recorded in *Fejervary* and *Cospi* (bars for 5, dots for 1) defy as yet any obvious astronomical or mathematical reading, and to that remain susceptible to the anthropological charge of being magic or merely ritualistic, the grand totals in both cases, as well as the optimal number base on which they are constituted (11), readily lend themselves to such a reading. Indeed, in sheer arithmetic, defined as they are by the zodiac 11, these bars and dots never fail to yield sets of interrelated formulae and totals relevant to the synodic and the sidereal cycles of the moon and Venus.

In the case of *Cospi* the focus appears to be the sidereal moon over a ritually significant period of nine, or 246 nights; the 2460 units given correspond to ten times that period (Figure 21.8). The wider echo, in ritual and in economics, of this sidereal-moon total emerges from its presence in formulae

	Spike male			Shell female		Heart male				
8×11	11	11	11	8	6	9	9	11	9	7	
297	341	189	189	72	54	2×9^2	9^2	11^2	9^2	7^2	
385	429	277	277	136	102	234	153	209	153	105	= 2460

Figure 21.8 Bar-and-dot totals and page distribution in *Cospi* (v. pp. 1–11). 2460 = 10 × 9 sidereal moons.

relevant to Mesoamerican tribute geography. In the case of Tenochtitlan, the towns of the Aztec empire are introduced, in the *Mendoza Codex*, by two 'zodiac' sets of garrisons, 11 in the highland Valley and 11 beyond, beginning with Citlaltepec, 'star-mountain'; and in the four provinces of the empire these towns total the sidereal 246, in an ingenious arithmogram which also highlights the synodic moon's 29 and the year's 365 (Figure 21.14). In the case of Coixtlahuaca, Tenochtitlan's major prize in conquest and the gateway to the east, the same total of 246 appears as dots subjected to Tepexic, Coixtlahuaca's northwestern ally (*Culte rendu map*, Figure 21.13).[3] As for the *Fejervary* total of 2914, its divergence from the standardized octaeteris of 2920 days found with the *Dresden* zodiac has previously been accounted for in terms of an elegant, and exact, harmonizing of the actual synodic periods of moon, sun and Venus, one which moreover invokes the time-shift between nights and moons, days and years (utas and octaeteris), preeminently identified in ritual with the ennead and the 4 + 4 structure inherent in this span (Brotherston, 1984; see Figure 21.9).

the order and indeed identity of the information given is the same as that in the zodiac bar-and-dot totals. The main celestial body referred to is again

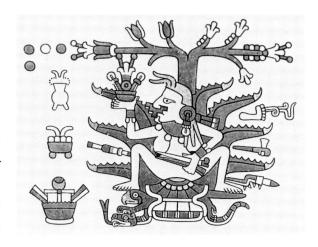

Figure 21.10 Mayauel (substituting for Jade-skirt, sixth of the Nine), on a turtle, and with a pulque pot set before her (*Laud*, p. 38).

Figure 21.9 Bar-and-dot totals and page distribution in *Fejervary* (pp. 5–14).

2914 = eight solar years less eight days at 365.24 days per year
= five Venus years less five days at 583.92 days per year
= 99 moons less nine days at 29.53 days per moon.

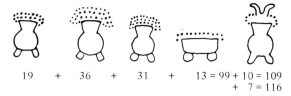

19 + 36 + 31 + 13 = 99 + 10 = 109
+ 7 = 116

Figure 21.11 Dots in pulque pots set before the Nine. 109 = four sidereal moons; 116 = one synodic Mercury cycle. (*Laud*, pp. 32, 34, 36–8.)

Analogous to these bars and dots is the case of dots alone, which in principle are likewise acknowledged to have an arithmetical potential, as when recording the name '20 Jaguars' in Mixtec annals, for instance. Three examples of possible astronomical interest occur in *Laud* (pp. 33–38, the Nine Figures chapter), *Borbonicus* (p. 8, Mayauel), and the *Culte rendu map* (the maguey with the upper-left toponym, Tepexic; Figures 21.10–21.13). In each case the dot totals are directly connected with the zodiac via the maguey plant sacred to Mayauel which is the source of pulque, whose fermentation is represented by the dots, and of paper, where dots are recorded on a page, on the page. In all cases

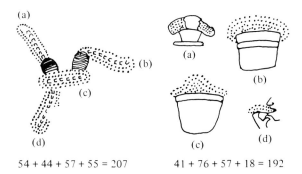

54 + 44 + 57 + 55 = 207   41 + 76 + 57 + 18 = 192

Figure 21.12 Dots in Mayauel's headdress (left) and pulque pots set before her (right). 207 = seven synodic moons; 192 = seven sidereal moons; 399 = almost 400. (*Borbonicus*, p. 8.)

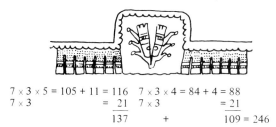

$7 \times 3 \times 5 = 105 + 11 = 116 \quad 7 \times 3 \times 4 = 84 + 4 = 88$
$7 \times 3 \qquad\qquad = \underline{21} \quad 7 \times 3 \qquad\qquad = \underline{21}$
$\qquad\qquad\qquad\quad \overline{137} \qquad + \qquad\qquad \overline{109} = 246$

Figure 21.13 Dots on maguey paper subject to the tribute center Tepexic. 116 = one synodic Mercury cycle; 88 = one sidereal Mercury cycle; 137 = five sidereal moons; 109 = four sidereal moons; 246 = nine sidereal moons. *Culte rendu map.*

the moon, plus the other inner planet Mercury, i.e. with Venus the three that pass between the earth and the sun, a course which gives them a more dramatic behaviour pattern on the east and west horizons than is the case with the outer planets. And again the moon totals have a known ritual resonance, as 9 (3×3, $4 + 5$), and 7 (orifices of the head, Caves of tribal origin, Night guardians of the year, and so on). That an absolutely sure iconography has yet to be agreed for Mercury need stand as no more of an objection than it has effectively done in the case of hieroglyphic data on that planet.

For a first example, the approximate 400 'Pleiades' dots found with Mayauel in *Borbonicus* when carefully counted prove to contain within themselves an opposition between the sidereal and the synodic moon: seven of the former occur as 191–2 dots in the fermented pulque itself while seven of the latter as 207 dots occur separately in the goddess's headdress (Figure 21.12). The sidereal moon, apparently a constant numerical concern in all maguey-related data in the genre, features again in *Laud*, this time to a total of 109 or four such moons, in the fermented pulque of the lunar Nine Figures; summing up this chapter and preparing for the next, which features the zodiac 11 (pp. 35–44), the sixth Figure or moon, Jade Skirt, actually becomes Mayauel and sits on the Turtle constellation (Figure 21.10).

In the more complex case of the *Culte rendu* text, the examples cited so far are strongly corroborated by being brought together and related to further planetary data. On the maguey paper beneath Tepexic 246-dots appear, the number of provincial tribute towns actually subject to Tenochtitlan (Figures 21.13 and 21.14), and the nine sidereal-moon cipher noted in *Cospi*. It is divided vertically,

left and right, into 137 (five moons) and the 109 of *Laud* (four moons); then a well-signposted horizontal subtraction of 21 from each yields, respectively, the subtotals 116 and 88 (Figure 21.13). These correspond exactly to synodic and the sidereal periods of Mercury. The latter, 88, exists as the dominant subtotal of the sidereal moon in *Cospi* (8×11, the optimal number base). The former, 116, also present as an approximate fifth of Venus's fifth (584) of the octaeteris in *Fejervary* (on page 15, thematically dedicated to the number 5); and it could be seen to offer a solution to the dilemma posed literally by the deer horns set into the last pulque-dot subtotal in *Laud* (Figure 21.11); for as 7, this additional item raises the already established lunar 109 up to 116. Deer is Sign VII; and as a zodiac constellation in *Dresden* it has both these horns and the coefficient 7 (Table 21.1e, position 10). For its part, the zodiac facade at Chichen similarly relates moon to Mercury in its totals of 'step' units, which read: $4(13 + 16) = 116$ (Figure 21.15), a formula distinguished in Van der Loo's present-day Tlapanec data, as '4 manojos de 29 zacates' (Van der Loo, 1982, pp. 223–4).[4]

That the astronomical significance of these totals is not illusory is made the more likely by the degree to which they reinforce each other between sources within the corpus, and by the fact that in so doing they all refer to such a highly defined and reduced range of celestial phenomena. This argument by negative definition may be taken further in the sense that throughout the corpus astronomical relevance seems impossible to detect in the absence of the zodiac pointers of number (11) and theme (maguey); for example, the bars in the Feasts chapter of *Laud* (pp. 21–22) yield totals that are quite opaque in these terms. Similarly, throughout *Borbonicus*, and for that matter the *Tonalamatl Aubin*, it is only Mayauel's maguey dots in the former that emerge clearly enough as units for counting to be physically possible.

Further analysis of numbers given in the iconographic ritual corpus, their subtotals, inner proportionality, and shifts between levels of time, is possible and has been attempted elsewhere (Brotherston, 1982a). Notable is the Metonic sequel to the octaeteris in *Fejervary* where, according to the principle testified to by Old World logarithms that

Gobernadores		Head towns		All towns									
Petlacalcatl	+123	centre	9	13	10	26	16	26	7	10	7	9	= 124
Atotonilco	+46	west	7	6	7	13	12	6	2	1			= 47
Tlachco	+69	south	7	10	14	12	14	8	6	6			= 70
Chalco	+82	east	7	6	22	11	11	3	22	8			= 83
Cuauhtochco	+45	north	8	7	6	7	11	7	2	5	1		= 46
	365		29										246

Figure 21.14 Toponyms subject to the tribute center Tenochtitlan. 365 = one year; 29 = one synodic moon; 246 = nine sidereal moons. (*Mendoza*, part 2.)

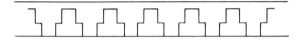

Figure 21.15 'Stepped' units from a total of 116 or 4 × 29; Chichen zodiac façade.

mathematical sophistication follows astronomical advance, play is made with the cycles of the outer planets through such devices as: sigma counting, which equates the Eagle of the 13 Birds (5) with the Eagle of the Twenty Signs (15); numbers squared, cubed and to other powers, likewise evident in Tlapanec knot totals of 169 or 13^2; and angle-value notation analogous to that found with kin- and uinal-units in hieroglyphic dates. Also, in the case of the sun, the constant and principled distinction between synodic and sidereal, horizon and zodiac time, can be shown to have yielded the precessional cycle of which the present calendrical age as the fifth is one-fifth (5200 years), and which serves as a time-unit in the further reaches of the cosmogony recorded in such texts as the *Popul Vuh* (Brotherston, 1982b, pp. 39–47). But that is another story and beyond the scope of the present enquiry, which is concerned exclusively with lunar and inner-planetary totals represented as simple additions of dots and of bars and dots.

21.4 In sum

Essential to the argument set out above is the notion that the hieroglyphic texts of the lowland Maya do not have a monopoly on sophistication in astronomy and mathematics, and that texts in the more wide-ranging and international iconographic script of Mesoamerica deserve more attention than they have received as evidence of astronomy that is both competent and formative of ritual, economic, and other types of social patterning. From a historical point of view this much could even be expected of the iconographic texts, since the script and calendar tradition to which they belong, while being intimately related to Maya hieroglyphics, is clearly older than it, as evidence from Monte Alban testifies. At any event, being closely interdependent the three stages of the argument surely present at least the beginnings of a case, of which a sum is: that the zodiac signs in certain hieroglyphic sources match those found in *Cospi*, especially in positions one to four within the set of 11, Scorpion and Turtle having a wider echo, also within the set; that at the same time 11 unequivocally reveals itself as the missing cipher of the ritual corpus, which can now be read entirely according to just the ciphers of the *tonalamatl* (9, 13, 20) and of the year (11, 18); and that data there given in bars and dots and just dots in the numerical and thematic context of the zodiac (11, maguey) correspond to cycles of the moon and the two inner planets.

Notes

1 This limit is given by the sign *cipactli* in the 22nd of the 52 columns; the zodiac signs appear, respectively, in the sources cited in columns 9 and 21, 8 and 21, 9 and 22. In line with its greater appeal to numeracy and logic, *Cospi tonalamatl* offers a footprint formula which highlights the lunar sevens and nines discussed

below: $2(7^2 + 9^2) = 260$. Note too the four sky scorpions on *Vaticanus* v47, which are subject to a four-way division of the *tonalamatl* like that found in the *Diaz* map.

2 This idea is set out more fully in Brotherston (1982b, p. 9), and a recent independent testimony to it is provided from a Quiche-Maya source by Tedlock (1985, p. 232). It is further supported by the role of the Nine in the 'Beget and bear, sever [the umbilical cord] and suckle' chapter in the ritual corpus (*Laud*, pp. 1–8, *Fejervary*, pp. 23–9, *Borgia*, pp. 15–7, *Vaticanus*, pp. 33–42, *Diaz*, p. 3). As the midwives' goddess, Oxomoco/Xmucane gives the new human form a ninefold blessing in the *Popul Vuh* (line 4801); on the grouping of the moons into threes, the fourth and sixth being those of major growth centred on the bone-former Hell-Lord (fifth), see *Florentine Codex*, vol. 6, pp. 24–9 and Cortez (1986). Set out in arabic numbers the sevenfold guardian sequence in *Borbonicus* reads as follows:

```
1 7 5 2 9 6 3
1 7 5 2 8 6 3
1 7 4 2 8 6 3
9 7 4 2 8 5 3
9 7 4 1 8 5 3
9 6 4 1 8 5 3
9 6 4 1 8 5 2
9 6 4
```

3 For a full discussion of this map and the relation of its toponyms to those in *Diaz* (of Cuicatlan) and *Laud*, see Brotherston (1985). Together with *Fejervary* these texts form a regional Upper-Papaloapan group distinct from the *Borgia* group probably from the Puebla plain, and from the *Tonalamatl Aubin* (Tlaxcala) and *Borbonicus* (Tenochtitlan). The 246 total in Mendoza is highlighted by both the format of the text and the sheer geography of the toponyms in question: Atotonilco, west, is distinguished by being the sole chief cabecera to be repeated in the entire sequence; Tlachco, south, and Chalco, east, stand out after the condensed 'end' pages of west and south, respectively; and Cuauhtochco, north, follows the special far-flung case of Xoconochco, the last in the eastern set. For their part, the zodiac sets of 11 garrisons in Mendoza find a clear counterpart in the 'Pintura de Mexico' transcribed by the Texcocan historian Ixtlilxochitl, for there towns said to be subject to Texcoco are listed in 11 columns and total 11^2 or 121 ('Sumaria relación', *Obras históricas*, Mexico, 1975, **1**, 383–4).

4 Like the zodiac 11 and the sidereal-lunar 246 of both the sky and tribute geography, this Mercury cipher also appears to find a practical application in economics: a plan of the tianquiz or market of Mexico shows an arrangement of 10×12 rectangular lots, in which the central four lots are replaced by a single octagonal structure; hence, $120 - 4 = 116$ lots (*Aubin* copy, Bibliothèque National de Paris 106; Glass, 1975, p. 332).

References

Boone, E. (1983). *The Codex Magliabecchiano*. 2 vols. Berkeley: University of California Press.

Broda, J. (1983). Ciclos agricolas en el culto. In *Calendars in Mesoamerica and Peru. Native American Computations of Time*, eds. A. F. Aveni and G. Brotherston, pp. 145–66. Oxford: BAR S174.

Brotherston, G. (1982a). Astronomical norms in Mesoamerican ritual and time-reckoning. In *New World Archaeoastronomy*, ed. A. Aveni, pp. 109–42. Cambridge University Press.

Brotherston, G. (1982b). *A Key to the Mesoamerican reckoning of time*, British Museum Occasional Paper 38. London.

Brotherston, G. (1984). 'Far as the Solar Walk': the path of the North American shaman. *Indiana* 9, 15–29.

Brotherston, G. (1985). The sign Tepexic in its textual landscape. *Iberoamerikanisches Archiv* 11, 209–51.

Clerke, A. M. (1911). The Zodiac. *Encyclopaedia Britannica*. 11th edn, vol. 28, pp. 993–8. NY: J. J. Little and Ives.

Coe, M. D. (1966). *The Maya*. London: Praeger.

Cortez, C. (1986). Aztec images of woman: the myth and ritual of pregnancy. Paper given at the IV LAILA Symposium, Merida, January 8th, 1986.

Edmonson, M. (1971). *The Book of Counsel: the Popol Vuh of the Quiche Maya*. Tulane: New Orleans Middle America Research Institute Publication no. 35.

Glass, J. B. (1975). A Census of Native Middle American Pictorial Manuscripts. In *Handbook of Middle American Indians*, vol. 13, pp. 81–252. Austin: University of Texas Press.

Kelley, D. H. (1976). *Deciphering the Maya Script*. Austin: University of Texas Press.

Lounsbury, F. G. (1982). Astronomical knowledge and its uses at Bonampak. In *Archaeoastronomy in the New World*, ed. A. Aveni, pp. 143–68. Cambridge University Texas.

Mayr, J. (1985). Astronomía de los Kogi. Paper given at the XLV Congress of Americanists, Bogota, 1985.

Nowotny, K. (1961). *Tlacuilolli. Die mexikanischen Bilderhandschriften*. Berlin: Gebr. Mann.

Santillana, G. de and Dechend, H. von (1969). *Hamlet's Mill. An essay on myth and the frame of time*. New York: Macmillan.

Schultze Jena, L. (1938). Bei den Azteken, Mixteken und Tlapaneken der Sierra Madre. *Indiana* (1st series), Jena, 3.

Schultze Jena, L. (1950). *Wahrsagerei, Himmelskunde und Kalender der alten Azteken.* Stuttgart: Gebr. Mann.

Seler, E. (1904). Die Tempelpyramide von Tepotztlan. *Gesammelte Abhandlungen* 2, 200–14.

Spinden, H. (1924). *The Reduction of Mayan dates.* Papers of the Peabody Museum 6, no. 4. Cambridge, MA.

Tedlock, D. (1985). *Popul Vuh.* New York: Simon and Shuster.

Van der Loo, P. L. (1982). Rituales con manojos contados en el grupo Borgia y entre los tlalpanecos de hoy. Coloquio internacional: *Los indígenas de Mexico*, ed. M. Jansen and Th. Leyenaar, pp. 232–43. Leiden: University of Amsterdam Press.

22

Comets and falling stars in the perception of Mesoamerican Indians

Ulrich Köhler   *University of Freiburg*

22.1 Introduction

This chapter summarizes the ideas of Mesoamerican Indians about comets, meteors and other celestial phenomena which are occasionally termed 'comets'. There are two reasons for treating comets and falling stars together: first, the latter are sometimes a result of the former; and, second, both in colonial and recent Spanish, the word *cometa* is not only used for comets, but for meteors as well. Therefore, in each instance where this word occurs, its respective meaning has to be determined from additional information. With *aerolito*, *meteoro* and *estrella fugaz* there are enough words in Spanish to designate meteors, but for one reason or another – at least for the present rural population of Mexico – *cometa* seems to be the preferred term.

The chapter focuses on two time levels: the 16th century, with a few notions from the preceding and the following ones, and then on present day Indians. Most of the historical sources refer to the Valley of Mexico, while the recent data originate from many parts of Mesoamerica. There have been attempts to interpret certain Mayan inscriptions from the classic period as comets (Hochleitner, 1974, 1979, 1985); the data presented so far do not seem, however, to be conclusive.

22.2 Comets

The typical image Mesoamerican Indians associate with a comet is that of a smoking star. This is the meaning both of the Aztec term *citlalin popoca* (Sahagún, 1979, book VII, fo. 8v)[1] or Mayan *budz ek*, *budzil ek*, or similar words as encountered in early

Yucatecan dictionaries (Álvarez, 1980, p. 106). This picture is supported by several sources from central Mexico which show a star, followed by waves of little clouds, as in Sahagún (1979) (Figure 22.1),

Figure 22.1 Sahagún (1950–69, book 7, fig. 21).

Codex en Cruz (Figure 22.2), as well as *Codex Telleriano Remensis* and *Vaticanus A* (see Figures 22.4 and 22.5). In Yucatan also the terms *ikom ne* (*cometa caudata*) and *kak noh ek*, (*cometa grande*) are encoun-

Figure 22.2 Codex en Cruz (1981, 13 Acatl).

tered, which render the image of a celestial phenomenon with a tail or that of a star with a huge fire (Álvarez, 1980, pp. 106f).

Sahagún (1979) describes the Aztec fear of comets, which were considered as omens for the death of a prince or king, or incipient war or famine. Very similar was the opinion of the Tlaxcaltec according to Muñoz Camargo (1892, p. 133), and Torquemada concludes that Mexican Indians feared comets as bad omens just as much as the Spaniards did (Torquemada, 1969, vol. I, p. 85).

With regard to the observation of specific comets, most unequivocal data refer to the early colonial period. Torquemada (1969, vol. I) mentions the appearance of a comet in the context of early Aztec history in the valley but gives no information about the exact year. *Codex Mexicanus* (1952, pl. 51) shows two comets for the year 1 Acatl (1363), and according to Baldet one was seen that year in China (Baldet, 1950, p. 41). For the year 10 Calli (1489) the *Codex Telleriano Remensis* informs us that *corrio una cometa muy grande*, called *xihuitli*, and it presents the picture of a snakelike being. The attribute of running, however, clearly speaks against an identification as comet and suggests that it was a meteor or a shower of meteors. This interpretation would also hold for a running *xiujtl* at the eve of the Conquest, which is translated as *cometa* by Sahagún (1979, book VIII, fo. 12r).

According to Torquemada (1969, vol. I, p. 235) a *cometa grande* appeared in 1519, the year of arrival of the Spaniards. In the list of Baldet there is, however, no entry for a comet, as observed in the Old World, including Chinese sources (Baldet, 1950, p. 45). From Torquemada's description of the phenomenon as *de grande resplandor*, that it was fixed in the air without moving, and that it lasted many days, one could imagine that it might possibly have been a nova or supernova. With regard to the latter, none is mentioned in the record of Clark and Stephenson (1977). Since this alleged comet is mentioned in the context of several bad omens preceding the conquest, one is tempted to conclude that it was added later to give the account a more dramatic appearance.

The first clearly discernible comet, attested to by several Mexican sources is one in 13 Acatl (1531), i.e. Halley's comet. References are found in Chimalpahin (1963–5, vol. II, p. 8), *Codex en Cruz* (1981), the *Anales de Tecamachalco* (1981, p. 8), and possibly also in *Codex Mexicanus* (1952, pl. 79).[2] *Codex Telleriano Remensis* (1899, p. 44r) and the related *Codex Vaticanus A* (1979, p. 88r), which were probably both copied from an earlier pictographical manuscript, show an interesting picture (Figure 22.3a, b). There is the night sky with stars, above it clouds of smoke – as in usual presentations of comets – and under it the eclipted sun. Surprisingly, the text of *Codex Telleriano Remensis* (*Vaticanus A* has none) only mentions the eclipse for that year, and indeed there was one which could be seen in Mexico (Oppolzer, 1962, p. 262, chart 131). The picture in the two codices differs from depictions

of comets of later years, which are also described as such, in one detail: the star in the middle, which elsewhere represents the head of the comet (Figure 22.5a, b), is not bigger than the others. This could explain the reason why there is no reference to a smoking star in the written comment, which was most probably added later.

The picture could possibly convey the idea that the eclipse of the sun of that year and the appearance of the comet were considered as one related phenomenon. This would have been logical, if both occurred at the same time. However, the eclipse was on March 18, while the comet passed perihelion on August 25, so that there could not have been any overlap. We may conclude that an imprecise drawing in the original manuscript was probably aggravated in the two later copies, thus obscuring the complete meaning of the pictograph.

The two codices *Telleriano Remensis* and *Vaticanus A* show five clearly discernible comets during the 1530s (Figures 22.4 and 22.5a, b).[3] The first

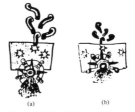

Figure 22.3 (a) *Codex Telleriano Remensis* (1899, fo. 44r). (b) *Codex Vaticanus A* (1979, fo. 88r).

Figure 22.4 (a) *Codex Telleriano Remensis* (1899, fo. 44v). (b) *Codex Vaticanus A* (1979, fo. 88v).

Figure 22.5 (a) *Codex Telleriano Remensis* (1899, fo. 44v). (b) *Codex Vaticanus A* (1979, fo. 88v).

in 2 Calli (1533) has the written comment that Venus was smoking (*Telleriano Remensis*, 44v). The idea that Venus may be smoking is also reported for the recent Pipil of El Salvador by Schultze Jena (1933–8, vol. II, pp. 100f). In that year of 1533, however, a real comet was generally observed in the Old World (Baldet, 1950, p. 46), and it is probably this one which is depicted.

The four other entries in these two codices also appear to be accurate. They refer to the years of 3 Tochtli (1534), 4 Acatl (1535), 6 Calli (1537) and 8 Acatl (1539). According to Baldet (1950, p. 46) a comet was seen in 1934 from China, in 1535 one from Milan, and for 1539 another one is reported from different parts of the Old World. This one is also mentioned in the chronicles of Chimalpahin (1963–5, vol. II, p. 12). For 1537 Baldet has no entry, but there is a comet listed which was first observed in Europe on January 17 of the following year. This one could possibly have been observed in Mexico towards the end of 1537.

For the following decade no information about comets could be found, beginning with the middle of the 1550s there are, however, several references. The principal sources are now the *Anales de Tecamachalco* (1981), a 16th-century chronicle in Nahuatl from a Popoloca community in the state of Puebla. Its information is of special interest, because in most cases the exact date of observation is given. For each of the years 12 Tecpatl (1556) and 13 Calli (1557) a comet is reported, and with March 5 and October 14 the entries in the *Anales* (1981, pp. 18, 24) are just four days later than first observation in the Old World according to Baldet (1950, pp. 46f). In 12 Calli (1569) the *Anales* report a comet for October (1981, p. 64) – without giving the exact date – while the first observation in the Old World was as late as November 9 (Baldet, 1950, p. 47). The *Anales* (1981, p. 66) also mention a comet for July 12 of 6 Tecpatl (1576), but there is no corresponding entry in Baldet's list.

The next comet, reported by several sources, is that of 7 Calli (1577), which was widely commented at its time in Europe, a period when ideas about the universe were changing from the Aristotelian point of view to the Copernican doctrine (Hellmann, 1944). In Mexico this comet is mentioned by Chimalpahin (1963–5, vol. II, p. 27), *Codex Aubin* (1981, pp. 49, 84), *Codex Mexicanus* (1952, pl. 86) and Torquemada (1969, vol. I, p. 643) for the area

around the capital, by the *Anales of Tecamachalco* for southern Puebla (1981, p. 70) and by the dictionaries of Motul and Vienna for Yucatan (Álvarez, 1980, p. 106). The *Anales* and *Codex Aubin*, respectively, give the dates of November 4 and 6, while the first observation in the Old World was already on November 1 (Baldet, 1950, p. 47).

For the year 10 Tecpatl (1580) the *Anales* are very specific and report a comet for the period between September 14 and November 20 (1981, p. 76), Chimalpahin mentions it for October 9 (1963–5, vol. II, p. 29), and with October 1 Baldet also gives a later date (1950, p. 47). In 2 Calli (1585) the *Anales* refer to a comet on October 20 (1981, p. 88) which had already been seen on the third day of that month in the Old World (Baldet, 1950, p. 47). Finally, the reappearance of Halley's comet in 1607 is attested to by the Indian chronicler Chimalpahin (1963–5, vol. II, p. 63), as well as by his Spanish counterpart Torquemada (1969, vol. I, p. 85). Both observed it personally from September 21 onward, that is ten days after its first spotting in the Old World (Baldet, 1950, p. 48).

As a result of this review, we have to conclude that the first worldwide observation of the comets of 1569 and 1580 is recorded in the chronicles of the Indian village of Tecamachalco in the highlands of central Mexico. With regard to the credibility of the *Anales* it may be useful to crosscheck the data on the two solar eclipses mentioned in it, that is for January 25, 1571, and November 2, 1575 (1981, pp. 50, 62). Both are exact to the very day (Oppolzer, 1962, pp. 264–7, chart 132, 133).

Since most concrete descriptions of comets refer to years after the conquest, we may wonder to what extent these reports contain European influence. After all, since the beginning of the Renaissance, and especially during the 16th century, celestial phenomena, including comets, played an important role in the minds of Europeans, and most of the sources quoted were written by Spaniards or by Indians influenced by European thought. I do not think the interest in reporting the appearance of comets was entirely due to the influence of Europeans, but possibly it was fostered by them.

The concepts of present day Indians about comets do not seem to have changed since the conquest. Among Nahua in the Sierra de Zongolica, Veracruz, I heard the term *citlalin popoca*, just as reported for the Aztec, and the variant I collected in the

Sierra de Puebla, conveys with *citalpol popocaya*, also the idea of a smoking star. This concept is also reported for the Nahua of Tetelcingo (Brewer and Jean, 1971, p. 24), the Totonac of Papantla (Aschmann, 1973, pp. 27, 178), the Zapotec of Mitla (Parsons, 1936, pp. 319, 340), the Zoque of Chapultenango (Báez-Jorge, 1983, p. 391), the Kekchi in Guatemala (Sedat S., 1955, p. 198) and in a slight change, as burning star, among the Huave (Lupo, 1981, p. 285; Stairs Kreger and Scharfe de Stairs, 1981, p. 122). Quite akin is still the visualization of a comet as *xohom' k'anal*, 'lightbeam star', as I was informed by Tzotzil of San Pablo. Only among Nahua of the Huaxteca I heard the term *citlal cuitlapil*, 'star-tail', which could possibly be a result of European influence.

Regarding omens related to the appearance of a comet, these are just as negative as back in the 16th century. Among the Nahua of the Sierra de Puebla I was informed that it announced war, Parsons noted that for the Zapotec of Mitla it was a sign of revolution, war, pestilence and famine (Parsons, 1936, p. 319), and the reports collected among the Maya of Chan Kom (Redfield and Villa Rojas, 1934, p. 332), the Huave (Lupo, 1981, p. 285) and Zoque (Báez-Jorge, 1983, p. 391) are just as pessimistic. My own data from the Tzotzil of San Pablo also fit well into this general picture.

The last appearance of Halley's comet in 1910 has been reported in a few ethnographies, and even today some older people still remember having seen it. A Nahua from Huahuaxtla in the Sierra de Puebla, whom I met in 1982 when he was about 80 years old, told me that he had seen a big comet when he was around ten years old (in *c.* 1910), and he pointed out that indeed war began soon after (the Mexican Revolution). Such direct references to the comet of 1910 are also known for Chan Kom in Yucatan (Redfield and Villa Rojas, 1934, p. 332), the Zoque of Chapultenango (Báez-Jorge, 1983, p. 391) and probably also for the Zapotec of Mitla (Parsons, 1936, p. 340). At least in Mexico, the last appearance of Halley's comet was indeed followed by calamities which helped to revive ancient fears!

22.3 Other spectacular phenomena, occasionally supposed to be comets

Between the years 1508 and 1510 a strange phenomenon appeared in the nocturnal sky which caused great fear and was later interpreted as a bad omen,

foreshadowing the conquest. There is almost no major chronicle which does not refer to it.

Durán, who wrote in the second half of the 16th century, describes it as a comet (Durán, 1967, vol. II, p. 467), and as such it appears also – in European style – on his Lám. 48 (Figure 22.6). From the con-

Figure 22.6 Durán (1967, vol. II, Lám. 48).

text, we can conclude, however, that he was referring to the same phenomenon as the other sources, and that seems to have been something quite different. Surprisingly, Sahagún also uses the term '*cometa*' in one of his accounts of the celestial appearance (Sahagún, 1979, book VIII, fo. 11r), although his description illustrates particularly that it was not a comet.

The earliest observation in the year 3 Tecpatl (1508) is reported by the *Historia de los Reynos*,[4] and according to this source it was a *mixpamitl*, 'cloud-flag' which appeared early in the morning on the eastern sky (Lehmann, 1938, pp. 284f). It is mentioned again for 4 Calli (1509), this time stating that it began then (Lehmann, 1938, p. 287). For this year there is a picture in *Codices Telleriano Remensis* and *Vaticanus A* which shows a line with several branches at both sides, connecting the earth with the sky (Figure 22.7a, b). The former terms

Figure 22.7 (a) *Codex Telleriano Remensis* (1899, fo. 42r). (b) *Codex Vaticanus A* (1979, fo. 85r).

it a *claridad de noche* which lasted 40 days[5] and reached from the earth to the sky (*Codex Telleriano Remensis*, 1899, 42r). *Codex Aubin* has just a short entry about a *tetzavitl*, a bad omen, which rose that year, and the facsimile shows the image of a flag (*Codex Aubin*, 1981, pp. 26, 126, 255). Chimalpahin (1963–5, vol. I, p. 137) also speaks of a *mixpanitl* which reached the center of the sky, but describes it as a black whirlwind. This is the only reference to the phenomenon as black, and I suspect that the author reached this conclusion when looking at an uncommented chronicle which shows it in a similar way as *Codices Vaticanus A* or *Telleriano Remensis* (Figure 22.7a, b).

For 5 Tochtil (1510) The *Historia de los Reynos* again refers to a *mixpamitl*, a 'cloud-flag', which arose in the east as a bad omen (Lehmann, 1938, p. 288). For this year Chimalpahin describes the *mixpanitl* as of great brightness and in its form similar to a rainbow (Chimalpahin, 1963–5, vol. I, p. 138), and the *Anales de Tula* refer to a *mispanitl* like fire (1979, p. 38). Ixtlilxochitl gives a more detailed description: according to him it was an '*esplandor*' which lasted many nights and arose in the east like a pyramid of flames (Ixtlilxochitl, 1965, vol. II, p. 313). The picture in *Codex en Cruz* (1981) could be an illustration of this phenomenon, because it shows a column which is more narrow towards the top; the latter, however, is turned a little to the left (Figure 22.8)[6]. The image of a pyra-

Figure 22.8 *Codex en Cruz* (1981, 5 Tochtli).

mid is also rendered by the illustrations in *Codex Mexicanus* (Figure 22.9) and the *Códice Florentino* of Sahagún (Figure 22.10), this time most convincingly. Sahagún has two descriptions in the *Códice*

Figure 22.9 *Codex Mexicanus* (1952, pl. 75).

Florentino (book VIII, fo. 11; book XII, fo. 1) and one in the manuscripts of *Madrid* (Sahagún, 1927, p. 110). He does not mention a specific year, but asserts that it happened 10 to 12 years before the arrival of the Spaniards.[7] He describes the phenomenon as a strong light which appeared in the east after midnight with a broad base and a narrow top like a pyramid, which disappeared at dawn; in this form it could be seen during the whole year.

Figure 22.10 Sahagún (1979, book 12 fo. 1r).

From the descriptions given we can easily conclude that it was neither a comet, nor a meteor. The description in pyramidal form and the appearance in the east clearly suggest an identification as zodiacal light, which has precisely this form in tropical latitudes (Herrmann, 1973, p. 138). The various references to an appearance in the east suggest that it was an observation before dawn around the autumn equinox. The information by Sahagún that it lasted a whole year should not be given too much weight, since he is the only one who gives this detail. When identifying the phenomenon as zodiacal light – which was already the conclusion of Lehmann (1938, p. 286) – there remains still one question: why is it reported specifically for the years 1508 to 1510, and especially for 1509 and 1510? Theoretically, in the tropics, it is a spectacular phenomenon every year around the equinoxes, in spring after dusk and in the autumn before dawn. March is not really a favourable month with regard to its visibility in Mexico because it is the dry season with a lot of dust, and furthermore it is the time when the burning of felled trees in the system of slash and burn agriculture begins, causing plenty of smoke. In September the main rainy season is over, but usually the sky is overcast on many days and there are also heavy rains as an effect of hurricanes in the Caribbean. Since the sources recorded point toward a phenomenon in the eastern sky, that is in September, the explanation may be that these were years without hurricanes around the autumn equinox.

In this context of other spectacular phenomena one more deserves mention. Again in codices *Vaticanus A* and *Telleriano Remensis*, for the year 7 Tecpatl (1512) there is the picture of the sky on top, the ideograph *tetl*, 'stone', at the bottom, and a line with branches on both sides connecting them (Figure 22.11a, b). According to the text of *Codex*

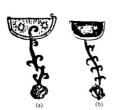

Figure 22.11 (a) *Codex Telleriano Remensis* (1899, fo. 42v). (b) *Codex Vaticanus A* (1979, fo. 85v).

Telleriano Remensis (1899, 42v), stones were smoking so much that year that it reached the heavens. This is somehow in accordance with the pictograph, but does not make much sense and would seem to be one of many hints of the fact that the Spanish text was not contained in the original, but was added independently at a later date by a more or less informed person. Aveni has suggested that this pictograph stands for a meteor (Aveni, 1980, p. 27). We may draw this more convincing argument further and suggest that it represents a meteorite, fallen from the sky. Should it be mere coincidence that the 'branches' of the line connecting heaven and earth point in the opposite direction to that of the *mixpanitl* (Figure 22.7a, b), which was described as rising from the earth to the sky? Another case of a meteorite could be depicted in *Codex Mexicanus* (1952, pl. 57) for the year 11 Acatl (1399). The entry shows two stars, one in the sky, the other touching the ground, and both conneced by a vertical line.

22.4 Meteors

The criterion by which meteors under the guise of the Spanish word *cometa* can be separated from real comets is that of a fast movement like running or falling. This is clearly a quality of meteors, not of comets. In Aztec there are two terms for meteors: *citlalin tlamina* and *xihuitl*. Both need a few words of explanation.

In the Spanish section of his dictionary, Molina uses the term *citlalin tlamina* as equivalent of *cometa que corre* (Molina, 1970, vol. 1, 27v), thus indicating that he refers to a meteor. Sahagún on the other had has caused some misunderstanding because of an unclear passage in his *Códice Florentino*. The literal translation of *citlalin tlamina* (Figure 22.12)

Figure 22.12 Sahagún (1950–69, book 7, fig. 21).

is 'star shoots' or more precisely 'star shoots arrow or spear'. He renders this translation, but adds the comment that the *inflamación de la cometa* was called by this term (Sahagún, 1979, book VII, fo. 8v). Anderson and Dibble as well as Garibay interpreted *inflamación* as the tail of a comet and added this word in parentheses to their respective translations (Sahagún, 1950–69, book VII, p. 64; 1969, vol. 2, p. 263). I think the idea of a comet's tail leads us on the wrong track. After all, the fast movement of shooting an arrow excludes the association with a comet.

Associating now *citlalin tlamina* with the phenomenon of meteors, the question remains whether this term refers to a particular star which shoots or to the projectiles. The image of 'star shoots' would rather point to the former. Already in 1910, however, Beyer rejected this idea, and he pointed to the parallel in English, where a 'shooting star' is also a meteor and not a distant star which shoots (Beyer, 1965, p. 284). This interpretation is supported by Muñoz Camargo, a chronicler of 16th century Tlaxcala: he also uses the word *cometa* when describing a meteor, and he explicitly says that these are the arrows of the stars (Muñoz Camargo, 1892, p. 13). Even more impressive are data recorded by Schultze Jena in 1929 in Zitlala, Guerrero, where the Nahuatl-speaking Indians have a concept of *citlaminalli*, 'star arrows', and these are nothing else than meteors (Sahagún, 1950, p. 64). After initial doubts we may safely conclude that *citlalin tlamina* refers to a meteor.[8] Interestingly enough, without commenting on the matter, Anderson and Dibble simply translate the term as 'shooting star' – with the meaning the word has in English – in another chapter of the *Florentine Codex* (Sahagún, 1950–69, book VII, p. 13).

The Aztec word *xihuitl* stands for a running *cometa* according to *Codex Telleriano Remensis* (1899, 39v) and Sahagún (1979, book VIII, fo. 12r), while Molina associates it with the idea of falling (Molina, 1970, II, 159v). These descriptions show that the phenomenon in question is not a comet but a kind of meteor.[9]

In Yucatecan dictionaries of the 16th century there are also a few entries under the headings of *cometa*, which rather seem to refer to meteors. The quality of fast movement is expressed by the Maya terms *halal ek*, 'running star', or *u halal dzutan*, 'running of witch', while *chamál dzutan*, 'cigarette witch', refers to a spark, like that of a falling cigarette butt (Álvarez, 1980, pp. 106f). The identification of this image with a falling star is supported by the Spanish gloss which describes the phenomenon as small *cometas*.

With regard to the effect of falling stars, only information from central Mexico could be detected. The Aztec considered them as dangerous, for, if that arrow hit an animal or man, an *ocuili*, a maggot or caterpillar, would be left in the wound.[10] Therefore, people protected themselves when walking at night and they refrained from eating animals wounded by a shooting star (Sahagún, 1979, book VII, fo. 8v). For Tlaxcala similar bad effects are reported by Muñoz Camargo (1892, p. 133): it was feared that game in the fields and woods was killed by these arrows.

There are two historical instances of meteors reported in the early sources from the center of Mexico. Since both refer to spectacular events, we may assume that these were not ordinary falling stars but strong showers of meteors. For the year 9 Calli (1489)[11] the *Codex Telleriano Remensis* (1899, p. 39v) reports: 'corrio una cometa muy grande que ellos llaman *xihuitli*', and it shows the picture of a snakeline being (Figure 22.13a), which also appears in *Codex Vaticanus A* (Figure 22.13b). If one observes the drawings meticulously, it

Figure 22.13 (a) *Codex Telleriano Remensis* (1899, fo. 39v). (b) *Codex Vaticanus A* (1979, fo. 92v).

becomes evident that there is no split tongue, which is the distinctive element for snakes in Mexican iconography. If not a snake, what kind of animal may there be depicted? On the grounds of its form it could well be a caterpillar, an *ocuili*, and, checking with *Códice Mendoza* (1979, pp. 10v, 34r), the animal in the hieroglyph for Ocuilan has indeed the same kind of head as the one under discussion. The spines along the body are common to several kinds of caterpillars in Mexico and may serve, therefore, as a further criterion to identify the animal as belonging to this category. Finally, the identification of the snakelike being as caterpillar is not really astonishing, since we know from Sahagún about the close relationship between such animals and meteors.

The second recorded appearance of an impressive meteor occurred some time between 1510 and 1519, and it is mentioned by several chroniclers as one of the bad omen which preceded the Spanish conquest and caused great fear. The most precise account is from Sahagún (1979, book VIII, fo. 12r), who even presents a picture of it (Figure 22.14).[12]

Figure 22.14 Sahagún (1979, book 8, fo. 12r).

He also uses the term *xiuitl* and describes it as a *cometa* which appeared before dusk and ran from west to east; it was split into three parts and gave the impression of showering sparks. The picture conveyed is clearly that of a shower of meteors. It would seem that the word *xiuitl* is used especially for such spectacular appearances of meteors, and indeed Molina's dictionary glosses it in another entrance as *cometa* in the form of a ball or great flame (Molina, 1970, I, 27v).

Let us now turn to the concepts of present-day Indians about meteors and the effects of their arrival on earth as meteorites. The data recorded show some instances of continuity of thought, but also offer further insight into indigenous ideas which are not contained in the early sources.

Of all the concepts of meteors encountered, their configuration as excrement of stars seems to be the most frequent. Up to now this idea has been reported for the Tlapanec in Guerrero (Schultze Jena, 1933–8, vol. III, p. 157), the Chontal in Oaxaca (Turner and Turner, 1971, p. 156), the Tzotzil (Gelwan, 1972, p. 9), Tzeltal (Slocum and Gerdel, 1971, p. 52; Gelwan, 1972, p. 9) and Tojolabal (Ruz, 1982, p. 52) in Chiapas, as well as the Quiché and Cakchiquel in Guatemala (Remington, 1977, p. 82). Apparently the respective words in these languages comprise both meteors and meteorites. Among the Tojolabal and the Tzotzil of Zinacantan the respective term for a shooting star stands as well for obsidian (Laughlin, 1975, p. 93; Ruz, 1982, p. 52) while other Tzotzil, as interviewed by Gelwan (1972, p. 9) and myself, are not that specific about the kind of stone which arrives on earth. Anyhow, also for them meteors and the objects which reach the earth are merely two aspects of one and the same thing.[13] Somehow akin to this school of thought are the ideas of certain highland Totonac who conceive meteors as the urine of the stars (Aschmann, 1962, pp. 2, 132).

For the Maya of Yucatan (Tozzer, 1907, p. 158)[14] and Chortí of Guatemala (Girard, 1966, p. 74) falling stars are the cigarette or cigar butts of the raingods. This idea has already been encountered in a Yucatecan dictionary of the 16th century, but while these leftovers were then attributed to an anonymous witch, the Indians – freed from early missionary endeavour – now seem to express freely what they really think.

By Yucatec Maya, Tozzer was also informed that meteorites are the arrowpoints of the raingods, and sometimes one finds them in the forest (Tozzer, 1907, p. 157). This concept is similar to that of meteorites in the form of obsidian among some Tzotzil and Tojolabal in Chiapas. The general idea is reminiscent of the Aztec visualization of meteors as arrows or spears, which is also reported for recent Nahua in Guerrero (Sahagún, 1950, p. 64) and the Totonac of Papantla, Veracruz (Aschmann, 1973, pp. 100, 120).

Another concept of shooting stars, up to now only recorded for the Huave in the Isthmus of Tehuantepec, is that of an iguana's tail (Lupo, 1981, p. 285; Stairs Kreger and Scharfe de Stairs, 1981, p. 113). A sole case is also the designation of *aerolitos* as *xopilli*, as I was informed by a Nahua from Ayutzinapa in Guerrero; according to Molina (1970, 2, p. 161), that word means 'toe'.

Among the Nahua of Los Reyes in the Sierra de Zongolica, state of Veracruz, I was confronted with ideas which bring together the Aztec notion of a close relationship between meteors and caterpillars with the widespread idea that meteors are the excrement of stars.[15] According to these Indians, where a *xihuitl*, the term applied both for meteor and meteorite, arrives on the earth, black caterpillars will appear, which are about 3 cm long and are piled together into a heap the size of a hand. These caterpillars are called *citlalocuile*, 'star-caterpillars', as well as *citlalcuitlatl*, 'starshit', and the heap looks indeed like excrement.

With regard to the ideas about the effects of falling stars, no uniform concept could be detected among present day Indians. Pessimistic, optimistic as well as ambivalent expectations have been encountered. An example for the first attitude was reported from southern Veracruz by Foster (1945, p. 187): 'The Popoluca knows better than to point at a falling star, for he knows that incurable eruptions will break out on his arm if this is done.' This is somehow reminiscent of the Aztec fear of being hit by a stellar arrow. Sickness, this time in the form of a swollen leg, is also reported for the Tzotzil of Zinacantan, but only if the shooting star is met before midnight (Laughlin, 1975, p. 284). Rather more pessimistic are the Jacalteca in the Highlands of Guatemala according to La Farge and Byers (1931, p. 129): 'If a falling star – *tahuwi'* – falls near a house, it is regarded as a sign of sickness. If it bursts over a house, someone will die.' Similar are the beliefs of the Huave who expect sickness and death in the surroundings of the place where a falling star has arrived on earth (Lupo, 1981, p. 285).

Positive expectations are scarce. The Tzotzil of Zinacantan believe that where a shooting star – seen after midnight – lands, treasure can be found (Laughlin, 1975, p. 232). This seems to be an idea imported from Europe. On the other hand, there is at least one example of an autochthonous belief in beneficial actions of shooting stars: that of the Totonac in the northern Sierra de Puebla. As reported by Ichon, certain stones, sent by stars, have therapeutic and other magic qualities; furthermore, there are four arrow-shooting stars in the corners of the sky which protect mankind (Ichon, 1969, pp. 98f).

The belief that meteors/meteorites may be

naguals, that is companion spirits of humans in the form of which the particular persons can act at their will, may be considered an ambivalent attitude towards these celestial phenomena, because the individuals in question may use such supernatural power to the benefit as well as to the detriment of other people or the community as a whole. Of healers with such a nagual one would rather expect beneficial actions, but not so of witches. This belief in meteors as naguals has been registered among the Tzotzil of Zinacantan (Laughlin, 1975, p. 284), and more recently among the Huave (Lupo, 1981, p. 296) and Mixe (Köhler, 1983, p. 30) in the Isthmus of Tehuantepec.

To this category of ambivalent expectations we may also count the ideas of Maya in Yucatan who informed Tozzer: 'Where a meteorite falls, there a lake will afterward be found, filled with alligators.' (Tozzer, 1907, pp. 157f). While we can only specu-

late about the feelings of these Indians towards the presence of alligators, we may assume that they would be rather happy to get such a source of water in their quite arid environment.

Tozzer just listed the general belief, but did not mention any concrete examples where that may have happened. Looking into much older sources, the 16th century *Relaciones de Yucatán*, there seems to be a reference, however, to a particular case. After mentioning the name Yocajeque [Yocah-Ek'], a place in the vicinity of Chuaca [Chauac-Ha], there is an explanation why the former is called that way: it received its name because there is a very deep lake, and the natives say that a star fell into it,[16] with massive rains (*Relaciones...*, 1983, vol. 2, p. 33).

Should this not be a case interesting enough to be seriously investigated by astronomers and geologists?

Notes

1 For this facsimile edition which has no throughgoing pagination, the form of quotation is altered. The information given here, about the respective book and folio, will be helpful, regardless of which edition of Sahagún's *Historia General* is at hand.

2 There is the picture of a smoking star, but above the glyph 1 Tecpatl (1532). For that year no other source from Mexico reports a comet, although two were seen in the Old World (Baldet, 1950, p. 46). The picture could refer to one of those, or – if placed a little too much to the right – to Halley's comet.

3 It is not necessary to give illustrations of all of them, because they follow the same pattern as Figure 22.4a, b and 22.5a, b.

4 This source is also known by the names of *Códice Chimalpopoca* and *Anales de Cuauhtitlan*.

5 The notion of 40 days is reminiscent of the duration of the Flood in the Old Testament, and seemingly this idea somehow found its way into the codex.

6 This form is given in the copies of León y Gama and Pichardo, while that of Dibble (*Codex en Cruz*, 1981) also shows a circle on top. Of the latter, however, nothing can be seen on the photograph of the original.

7 In two of the versions there is a brief notion that it began in 12 Calli. Interpreted as designation of a year this would be 1517. We know, however, from Sahagún himself as well as the many other sources that the phenomenon had already appeared toward

the end of the first decade. If not obsolete, the notion could refer to a particular day.

8 This interpretation does by no means deny the belief in stars which shoot. Such action was especially believed of Venus, as attested by illustrations in *Codex Dresdensis* (1975, pp. 46–50) or *Codex Cospi* (1968, pp. 9–11).

9 According to Molina, the word *xihuitl* has a variety of different meanings: 'año, cometa, turquesa e yerua' (Molina, 1970, II, 159v.), that is year, meteor, turquoise and more generally jewel, and herb. To a certain point this variety is due to the fact that the writers of the 16th century did not take into account the vowel length which is a prominent feature in Aztec. Modern authors present the difference between *xihuitl*, 'year', and *xi:huitl* as test-case to demonstrate the change of meaning, but they fell into the trap set by the early Spanish writers and wrongly translate the latter word as 'comet' instead of 'meteor' (Anderson, 1973, p. 9; Andrews, 1975, p. 4).

10 Anderson and Dibble translate *ocuillo*, in Spanish *gusano*, as 'worm' (Sahagún, 1950–69, book VII, p. 13), which does not seem to be correct. One may reach this conclusion when checking the meaning of *gusano* in modern Spanish dictionaries. On the other hand, Sahagún, in his 11th book which deals with animals, gives ample information on what kind of animals are ment by the terms *ocuili* (which is the same word as above) and *gusano*: it is a variety of caterpillars, maggots and larvae, and the only examples for worms mentioned are intestinal worms of

humans and dogs (Sahagún, 1979, book XI, fos. 103r–107r). Actually, even two kinds of *citlal ocujli*, 'star maggots or caterpillars', are mentioned. The first is a caterpillar which feeds on opuntia, and the other one is seemingly a kind of maggot or larva which breeds in the flesh of rabbits and other rodents. The second kind is also called *citlal mjtl*, 'star arrow or spear' (Sahagún, 1979, book XI, fo. 105v). This substantiates the belief that such animals in the flesh are the result of falling stars.

11 From the arrangement of the pictures the interpretation also seems possible that it occurred in 11 Tochtli (1490).

12 Although it looks almost like *metl*, 'agave', the plant at the right end could be a pictograph for *xihuitl*, 'herb', thus giving the name of its celestial homonym.

13 A more detailed account on ideas of the Tzotzil about falling stars can be found in a paper by Lamb (1984).

14 Tozzer uses the term 'comet', but from the context it is evident that the movement of meteors is described.

15 In Spanish the Indian informant also used the word *cometa*, but from his description it became entirely clear that he was not speaking about a comet.

16 *Yocah ek'* means 'star has pierced'.

References

Álvarez, C. (1980). *Diccionario Etnolingüístico del Idioma Maya Yucateco Colonial*, vol. 1: Mundo físico. México: UNAM.

Anales de Tecamachalco (1981). Crónica local y colonial en idioma Nahuatl, 1398 y 1590. Colección de Documentos para la Historia Mexicana, publicadas por Antonio Peñafiel. México: Editorial Innovación.

Anales de Tula (1979). Museo Nacional de Antropología (Cod. 35–9). Comentario de Rudolf A. M. van Zantwijk. Graz: Akadem. Druck- u. Verlagsanstalt.

Anderson, A. J. O. (1973). *Rules of the Aztec Language*. Salt Lake City: University of Utah Press.

Andrews, J. R. (1975). *Introduction to Classical Nahuatl*: Austin. University of Texas Press.

Aschmann, H. P. (1962). *Vocabulario Totonaco de la Sierra*. México: Instituto Lingüistico de Verano – SEP.

Aschmann, H. P. (1973). *Diccionario Totonaco de Papantla, Veracruz*. México: Instituto Lingüistico de Verano – SEP.

Aveni, A. F. (1980). *Skywatchers of Ancient Mexico*. Austin: University of Texas Press.

Báez-Jorge, F. (1983). La cosmovisión de los Zoques en Chiapas (Reflexiones sobre su pasado y presente). In *Antropología e Historia de los Mixe-zoques y Mayas. Homenaje a Frans Blom*, eds. L. Ochoa and T. A. Lee, jr., pp. 383–412. México: UNAM.

Baldet, F. (1950). Liste générale des comètes, de l'origine à 1948. *Annuaire du Bureau des Longitudes pour l'an 1950*, Paris.

Beyer, H. (1965). La astronomía de los Antiguos Mexicanos. *El México Antiguo* **X**, 266–84.

Brewer, F. and Jean, G. (1971). *Vocabulario Mexicano de Tetelcingo, Morelos*. México: Instituto Lingüistico de Verano – SEP.

Chimalpahin, D. de San Anton Muñón (1963–5). *Die Relationen Chimalpahin's zur Geschichte México's*, Parts I, II. Aztekischer Text, herausgegeben von Günter Zimmermann. Universität Hamburg: Abhandlungen aus dem Gebiet der Auslandskunde, vols. 68, 69. Reihe B. Hamburg: Cram, de Gruyter & Co.

Clark, D. and Stephenson, F. R. (1977). *The Historical Supernovae*. Oxford: Pergamon Press.

Codex Aubin (1981). Geschichte der Azteken. Der Codex Aubin und verwandte Dokumente. Aztekischer Text, übersetzt und erläutert von Walter Lehmann und Gerdt Kutscher, abgeschlossen und eingeleitet von Günter Vollmer. Quellenwerke zur Alten Geschichte Amerikas XIII. Berlin: Gebr. Mann Verlag.

Codex Cospi (1968). *Codices Selecti*, vol. 18. Graz: Akadem. Druck- und Verlagsanstalt.

Codex Dresdensis (1975). Sächsische Landesbibliothek Dresden (Mscr. Dresd. R 310). *Codices Selecti*, vol. 54. Graz: Akadem. Druck- und Verlagsanstalt.

Codex en Cruz (1981). Edited and comments by Charles Dibble, vols. I, II. Salt Lake City: Univesity of Utah Press.

Codex Mexicanus (1952). Bibliothèque Nationale de Paris, nos. 23–4. *Journal de la Societé des Americanistes*, NS **41**, pl. I–XLIV.

Codex Telleriano Remensis (1899). *Manuscrit mexicain du cabinet de Ch. M. Le Tellier, archevêque de Reims à la Bibliothèque Nationale* (Ms. mexicain no. 385). Transcription et commentaire de E. T. Hamy. Paris: Loubat.

Codex Vaticanus A (1979). *Codex Vaticanus 3738* (Cod. Vat. A, Cod. Ríos) der Biblioteca Apostolica Vaticana. *Codices Selecti*, vol. 65. Graz: Akadem. Druck- und Verlagsanstalt.

Códice Mendoza (1979). Colección de Mendoza o Códice Mendocino. Facsimile fototipico. México: Editorial Cosmos.

Durán, Fray D. (1967). *Historia de las Indias de Nueva España e islas de la tierra firme*, vols. I, II. ed. por Ángel Ma. Garibay K., Biblioteca Porrúa, vols. 36, 37. México.

Foster, G. M. (1945). Sierra Popoluca folklore and beliefs. *University of California Publications in American Archaeology and Ethnology* **42**, (2), 177–250.

Gelwan, E. M. (1972). *Some Considerations of Tzotzil–Tzeltal Ethnoastronomy (Summer Field Study Summary Report)*. Harvard Chiapas Project, Harvard University.

Girard, R. (1966). *Los Mayas. Su civilización, su historia, sus vinculaciones continentales*. México: Libro Mex Editores.

Hellman, C. D. (1944). *The Comet of 1577: Its Place in the History of Astronomy*. New York: Columbia University Press.

Herrmann, J. (1973). *dtv-Atlas zur Astronomie*. Tafeln und Texte. München: Deutscher Taschenbuch Verlag.

Hochleitner, F. J. (1974). Kometeninschriften in Maya-stelen. *Ethnologia Americana* **11**, (3), 565–72.

Hochleitner, F. J. (1979). Der Altar von Naranjo. Kometen als Warnsignale für Erdbeden? *Ethnologia Americana* **16**, (2), 918–21.

Hochleitner, F. J. (1985). Halleysche Komet in Mesoamerika. *Ethnologia Americana* **21**, (2), 1141–3.

Ichon, A. (1969). *La Religion des Totonaques de la Sierra*. Paris: CNRS.

Ixtlilxochitl, F. de Alva (1965). *Obras Históricas de Don Fernando de Alva Ixtlilxochitl*. Publicadas y anotadas por Alfredo Chavero, vols. I, II. México: Editora Nacional.

Köhler, U. (1983). Ethnographische Notizen zum Alter-Ego-Glauben und Nagualismus in Mexiko. *Mexicon* **V**, 30–2, Berlin.

La Farge II, O. and Byers, D. (1931). *The Year Bearer's People*. Middle American Research Series, publ. no. 3. New Orleans: Tulane University.

Lamb, W. W. (1984). *Tzotzil Starlore*. Paper read at the Conference on Ethnoastronomy, Washington, DC, September, 1983.

Laughlin, R. M. (1975). *The Great Tzotzil Dictionary of San Lorenzo Zinacantan*. Smithsonian Contributions to Anthropology, no. 19. Washington, DC.

Lehmann, W. (ed.) (1938). *Die Geschichte der Königreiche von Colhuacan und Mexico. Quellenwerke zur Alten Geschichte Amerikas I*. Stuttgart: W. Kohlhammer Verlag.

Lupo, A. (1981). Conoscenze astronomiche e concezioni cosmologiche dei Huave di San Mateo del Mar (Oaxaca, Messico). *L'Uomo* **5**, 267–314.

Molina, Fray A. de (1970). *Vocabulario en Lengua Castellana y Mexicana y Mexicana y Castellana*. Biblioteca Porrúa 44. México.

Muñoz Camargo, D. (1892). *Historia de Tlaxcala. Publicada y anotada por Alfredo Chavero*. México: Secretaría de Fomento.

Oppolzer, T. R. von (1962). *Canon of Eclipses [Canon der Finsternisse]*. Translated by Owen Gingerich. New York: Dover Publications.

Parsons, E. C. (1936). *Mitla, Town of the Souls, and other Zapoteco-Speaking Pueblos of Oaxaca, Mexico*. University of Chicago Press.

Redfield, R. and Villa Rojas, A. (1934). *Chan Kom. A Maya Village*. Carnegie Institution of Washington, publ. no. 448. Washington, DC.

Relaciones Histórico-geográficas de la Gobernación de Yucatán (1983). (Mérida, Valladolid y Tabasco), vols. I, II. *Fuentes para el Estudio de la Cultura Maya*, 1. México: UNAM.

Remington, J. A. (1977). Current astronomical practices among the Maya. In *Native American Astronomy*, ed. A. F. Aveni, pp. 77–88. Austin: University of Texas Press.

Ruz, M. H. (1982). *Los legítimos hombres. Aproximación antropológica al grupo tojolabal*, vol. 2. México: UNAM.

Sahagún, Fray B. de (1927). *Einige Kapitel aus dem Geschichtswerk des Fray Bernardino de Sahagún, aus dem Aztekischen übersetzt von Eduard Seler*. Stuttgart: Strecker u. Schröder Verlag.

Sahagún, Fray B. de (1950). *Wahrsagerei, Himmelskunde und Kalender der Alten Azteken. Aus dem aztekischen Urtext Bernardino de Sahagún's übersetzt und erläutert von Leonhard Schultze Jena*. Quellenwerke zur Alten Geschichte Amerikas IV. Stuttgart: W. Kohlammer Verlag.

Sahagún, Fray B. de (1950–69). *Florentine Codex. General History of the Things of New Spain*. Books I–XII. Translated and edited by Arthur J. O. Anderson and Charles E. Dibble. Santa Fé, New Mexico: University of Utah Press.

Sahagún, Fray B. de (1969). *Historia General de las Cosas de Nueva España*. Nueva edición con numeración, anotaciones y apéndices de Angel Maria Garibay K. vols. I–IV. Segunda edición, Biblioteca Porrúa 8–11. México.

Sahagún, Fray B. de (1979). *Códice Florentino*. Edición facsimilar, vols. I–III. México: Secretaría de Gobernación.

Schultze Jena, L. (1933–8). *Indiana* **I–III**. Jena.

Sedat S., G. (1955). *Nuevo diccionario de las lenguas K'ekchi' y Española*. Chamelco, Alta Verapaz: Instituto Lingüistico de Verano en Guatemala.

Slocum, M. C. and Gerdel, F. L. (1971). *Vocabulario Tzeltal de Bachajon*. México: Instituto Lingüistico de Verano – SEP.

Stairs Kreger, G. A. and Scharfe de Stairs, E. F. (1981). *Diccionario Huave de San Mateo del Mar*. México: Instituto Lingüistico de Verano.

Torquemada, Fray J. de (1969). *Monarquía Indiana*. vols. I–III. Biblioteca Porrúa 41–3. México.

Tozzer, A. M. (1907). *A Comparative Study of the Mayas and the Lacandones*. New York: Macmillan Company.

Turner, P. and Turner, S. (1971). *Chontal to Spanish–English Dictionary, Spanish to Chontal*. Tucson: University of Arizona Press.

23

Cosmic ideograms on petroglyphs of the Mesoamerican cultures of 'El Zape' region in Durango, Mexico

Alejandro Peschard Fernandez, Jaime Ganot Rodriguez and Jesus F. Lazalde Montoya   *Universidad Juarez de Durango*

The archaeological site of 'El Zape' was discovered very early by the friars during the conquest of Mexico. There are references to it in the 1604 and 1610 'Anuas', or yearly reports, that they were required to send to their superiors (Perez de Ribas, 1944).

Several scientific studies have been performed in the area; however, most of them have dealt with isolated aspects of its archaeology. If they are analyzed as a whole one may be able to establish affirmatively that a developed culture existed as El Zape with the following characteristics:

(1) The existence of large areas of stone constructions on the low, little hills near the Sextin River and on the high parts of the mountains (Tarayre, 1869).
(2) Corn, beans, and squash agriculture since early times, about 660 BC (Brooks, Kaplan and Cutler, 1962).
(3) Use of copper (Mason, 1971).
(4) Ceremonial polychrome pottery of the Mixteca–Puebla type (Ganot and Peschard, 1985).
(5) Intentional cranial deformation of the tabular oblique and tabular erect types (Brooks and Brooks, 1980).
(6) Burials in funeral urns (Larios, 1612).
(7) Spinning and weaving (Mason, 1966).
(8) Petroglyphs with designs similar to some glyphs of the Maya and Mixteca cultures (Mason, 1961; Peschard *et al.*, 1984, pp. 6–7).
(9) The use of a calendar to establish the seasons of the year (Peschard, Ganot and Lazalde, 1984, pp. 19–20).

The existence of these cultural elements in an area located geographically far away from the great ceremonial centers of Mexico, make El Zape a place of great importance in the archaeology of the north of Mexico, and it may be a fundamental link in interrelations between Mesoamerica and the cultures of the southwestern United States.

The existence in the area of a site with petroglyphs is of great importance. It was briefly described in 1936 by the American archaeologist J. Alden Mason (Mason, 1961), and in 1984 the authors of this paper reported that this place may have functioned as a solar observatory (Peschard *et al.*, 1984).

The 'observatory' was constructed through modifications made to a rhyolithic wall. This was accomplished by pulling out large flagstones, cutting down the walls to make them flat, widening the clefts into which wedges were put in order to keep them open (Figure 23.1), and by engraving figures in specific sites (Figure 23.2).

These modifications were probably made for determination of the pivots of the north and south maximum displacement of the sun on the horizon and to point to the place where the sun lies after half of its semiannual travel, by watching the sun during its decline through one of its clefts (Figure 23.3), or the projection of sunlight through another cleft and over the engravings (Figure 23.4). In this way they were able to determine the beginning and ending of the seasons of the year.

It is important to note that most of the engravings are geometrical, with different motifs in the interior, as if each of them was intended to express something

Figure 23.1 The observatory of El Zape.

Figure 23.2 The observatory of El Zape, showing petroglyphs.

Figure 23.3 Sun descending in cleft, El Zape.

different (Figure 23.5). Some are similar in design to Maya and Mixtec glyphs of cosmic significance (Figure 23.6). As a result of the advanced cultural development of El Zape, the Indians of nearby regions, perhaps less developed, may have been influenced in their customs. The petroglyphs are evidence in favor of this interpretation.

In 11 cases studied near the borders of the Sextin, Tepehuanes, and Santiago Papasquiaro rivers in the north of Durango and 120 km from El Zape (Figure 23.7), engravings like the ones described were found in the site of the solar observatory. Many others associated with figures of suns, moons, stars, plants, men, shamans, and hunting scenes also were noted (Figures 23.8–23.11).

A new type of petroglyph appears in which the designs of cosmic significance are incorporated into human and animal figures and, conversely, these figures form a part of the interior motifs of the geometrical engravings (Figures 23.12 and 23.13).

The association of figures such as the one of man and woman, sun and moon, and those with symmetrical designs, may signify the duality concept which is fundamental in Mesoamerican religion (Robelo, 1980; Fernandez, 1983) (Figures 23.11, 23.14 and 23.15).

Figure 23.4 Summer solstice, noontime, at El Zape.

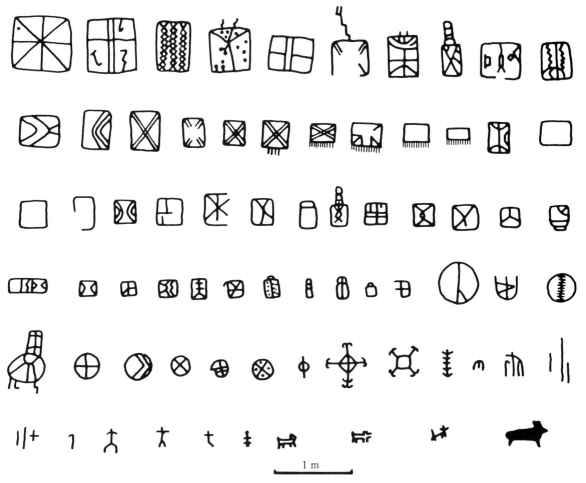

Figure 23.5 Forms of petroglyphs found at El Zape.

Figure 23.6 Petroglyphic motifs resembling Maya and Mixtec glyphs.

303

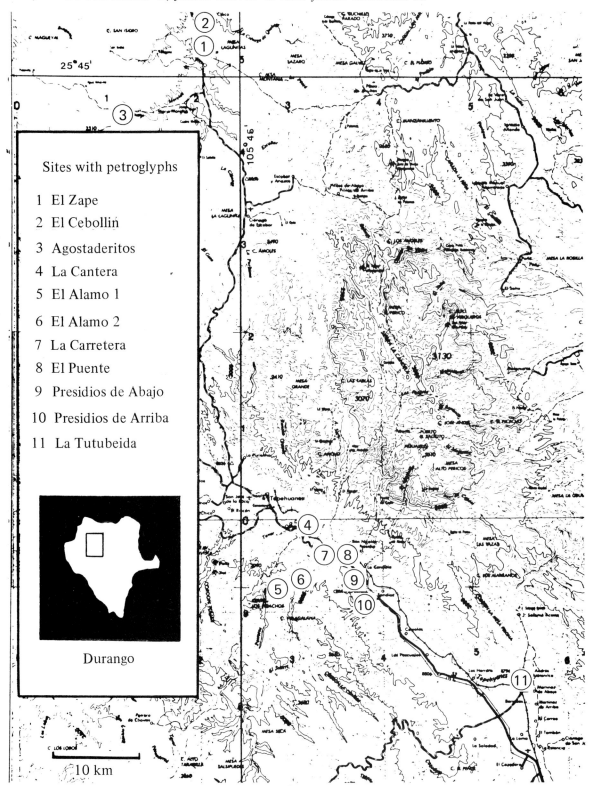

Sites with petroglyphs

1 El Zape
2 El Cebollin
3 Agostaderitos
4 La Cantera
5 El Alamo 1
6 El Alamo 2
7 La Carretera
8 El Puente
9 Presidios de Abajo
10 Presidios de Arriba
11 La Tutubeida

Durango

10 km

Figure 23.7 Sites with petroglyphs in the Mexican state of Durango.

Figure 23.8 Petroglyphs at El Alamo 1.

Figure 23.9 Petroglyphs at El Puente.

Figure 23.10 Petroglyphs at Agostaderitos.

Figure 23.11 Petroglyphs at El Alamo 2.

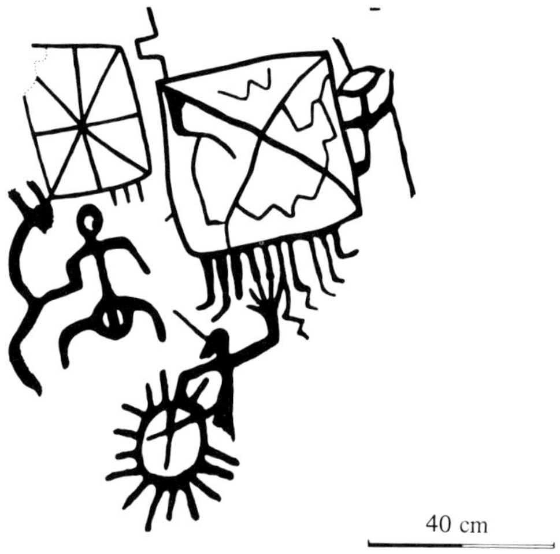

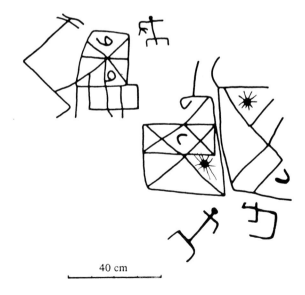

Figure 23.14 Human, animal and astral signs in the petroglyphs of La Tutubeida.

Figure 23.12 Human and animal motifs in petroglyphs at El Puente.

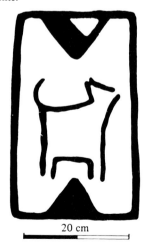

Figure 23.13 Human and animal motifs in petroglyphs at La Cantera.

Figure 23.15 Human, animal and astral signs in the petroglyphs of La Tutubeida.

The importance of observation of the celestial phenomena may have been associated with new religious concepts which played an important role in the interpretation of the cosmic events and therefore, in its representation in rock paintings.

Finally, other petroglyphs were found that were even more complex and for which we have no explanation at present (Figure 23.16).

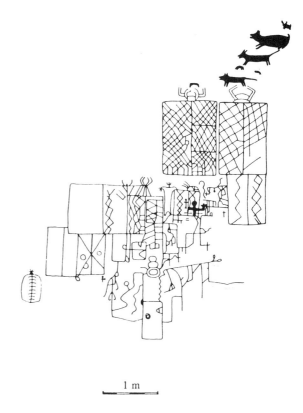

1 m

Figure 23.16 Petroglyphs at Presidios de Abajo.

References

Brooks, R. H., Kaplan, L. and Cutler, H. C. (1962). Plant material from a cave on the Rio Zape, Durango, Mexico. *American Antiquity* **27**, (3), 236.

Brooks, S. T. and Brooks, R. H. (1980). Cranial deformation: possible evidence of Pochteca trading movements. *Transactions of the Illinois State Academy of Science* **72**, (4), 8.

Fernandez, A. (1983). *Dioses Prehispanicos de Mexico*, pp. 9–11. Mexico: Panorama Editorial S.A.

Ganot, R. J. and Peschard, F. A. (1985). La Cultura Aztatlan, Frontera del Occidente y Norte de Mesoamerica en el Post-Clasico. *XIX Reunion de la Sociedad Mexicana de Antropología*, Queretaro, 10 pp.

Larios, D. (1612). *Anuas de 1612*. Archives of the Compañía de Jesús (Jesuit Company), Rome.

Mason, J. A. (1961). *Some Unusual Petroglyphs and Pictographs of Durango and Coahuila. Homenaje a Pablo Martinez Del Rio*, pp. 295–310. Mexico: Instituto Nacional de Antropologia e Historia.

Mason, J. A. (1966). *Cave Investigations in Durango and Coahuila.* Homenaje a Roberto Weitlaner, pp. 60–5. Mexico: Instituto Nacional de Antropologia e Historia.

Mason, J. A. (1971). Late archaeological sites in Durango, Mexico, from Chalchihuites to Zape, pp. 137–40. In *The North Mexican Frontier*, eds. B. C. Hedrick, J. C. Kelley and C. L. Riley, pp. 137–40. Carbondale: Southern Illinois University Press.

Perez de Ribas, A. (1944). *Historia de los Triunfos de Nuestra Santa Fe entre Gentes las mas Barbaras de Nuestro Orbe*, pp. 147–8. Mexico: Editorial LAYAC.

Peschard, F. A., Ganot, R. J. and Lazalde, F. J. (1984). Petroglifos de El Zape, Durango: un calendario solar en el norte de Mexico. Paper presented at the Simposio de Arqueoastronomia y Etnoastronomia. Ciudad Universitaria, Mexico, 1984.

Robelo, C. A. (1980). *Diccionario de Mitologia Nahuatl*, pp. 315–16. Mexico: Editorial Inovacion S.A.

Tarayre, G. E. (1869). *Archivies de la Commision Scientifique du Mexico*, pp. 183–6. Paris.

24

North American Indian calendar sticks: the evidence for a widely distributed tradition

Alexander Marshack   *Peabody Museum – Harvard University*

The analysis of an early 19th-century American Indian calendar stick (Marshack, 1985), made by Winnebago chief Tshi-zun-hau-kau around 1825 when his people inhabited the western shore of Lake Michigan, unexpectedly revealed the presence of a complex mode of lunar–solar observation and notation. The stick contained as its primary mode the notation of a lunar calendar year and a lunar month divided into three parts $(10 + 10 + 10)$, a division that was clearly not derived from late Mesoamerican calendars or from the Christian European calendar that had been brought into the region by traders and colonists. The analytic evidence for an *ad hoc* mode of lunar–solar intercalation suggested, likewise, that it was not derived either from Mesoamerican or European traditions.

Following analysis of the Winnebago calendar stick, a search was made for other possible calendar sticks in United States' museums and collections. Unknown and unpublished 19th-century examples from widely dispersed linguistic groups began to be found. Their presence and diversity suggested the dispersal of an already evolved tradition of North American origin in some period before European contact. At the same time ethnographic evidence began to accumulate reporting the use of calendar sticks among diverse linguistic groups that included both farmers and hunter-gatherers living east of the Rocky Mountains. A search for calendar record-keeping modes west of the Rocky Mountains has not yet been completed.

Widely dispersed and diverse traditions of North American Indian observational astronomy have in recent years been documented by workers in the still new discipline of archaeoastronomy. It has become increasingly evident that these traditions were part of and integrated with indigenous traditions of seasonal economic, ritual and ceremonial activity. While much of the recently published evidence relates to measurable solar and possible stellar alignments, it is becoming evident as well that these non-arithmetic observational 'calendars' often involved the corollary system of naming a sequence of lunar months after changing seasonal phenomena and human activities. There is a suggestion in the accumulating comparative evidence, ethnographic and analytic, that aspects of these observational 'astronomies' and 'calendar' traditions entered the New World with different peoples at various times as part of already developed systems of a northern latitude hunting–gathering adaptive lore.

Analysis of the first Winnebago calendar stick provided evidence for the presence of certain basic lunar–solar traditions of observation and for modes of notation and record-keeping which involved both individual idiosyncratic as well as traditional notational problem-solving strategies. Determination of the strategies and notational modes involved in the first stick made it possible to perform internal analyses of the other calendar sticks, each of which was found to be visually and structurally different, having been made by a local tribal 'sky-watcher' within a widely dispersed generalized tradition.

308

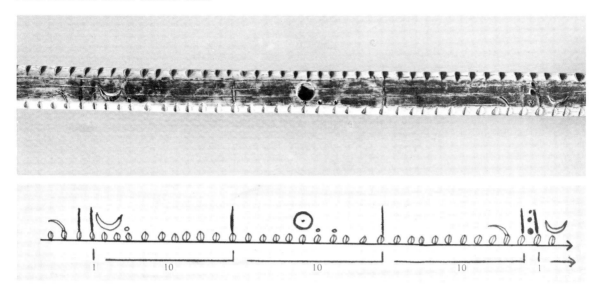

Figure 24.1 Close-up of the ninth month of the first year on the Winnebago calendar stick of Tshi-zun-hau-kau. The reading is from left to right and indicates the last crescent of the eighth month, the two days of invisibility that follow and the first crescent that comes after. Two days after the first crescent there is a dot over the day mark. The two days after the full moon sign in mid-month also have dots added above them. At right the ninth month ends with a sign for the last crescent. The two days of invisibility that follow have three dots that were added at different times since they are made by different points and pressures. The period is followed by the first crescent of the tenth month.

24.1 The first Winnebago calendar stick

The calendar of Tshi-zun-hau-kau[1] is a long (132 cm), thin, four-sided stick that contains six notated lunar months along each edge, recording two precisely notated lunar years. Each lunar month is *usually* divided into three parts of $10 + 10 + 10$, representing the periods of waxing, full and waning moon, but there is an occasionally variable last period of 7, 8, 9, 10 or 11, whose length was dependent on observation of the last crescent in the eastern morning sky. The months were separated by two long vertical strokes representing the two 'zero' days of invisibility. Eliminating these 'zero' days in the notated first year provided a sequence of 12 months whose lengths were: 26–28–29–26–29–26–28–28–29–27–28–27.

Each month ended with a symbol of the last crescent incised before the two days of invisibility, with a different symbol for the first crescent being incised after the days of invisibility:

The small scale of the marks and signs made it difficult to locate one's position within the year, so a subsidiary reading or cueing strategy was utilized, placing a large circle or 'full moon' sign at some point over the middle set of 10 in each month (Figure 24.1). Counting these circles placed one within a desired month. At particular points in the year dots were incised over certain days as though to indicate the position of special days of observation or of ritual or economic activity.

The first of the two years (Figure 24.2) proceeded in a serpentine or boustrophedon manner. It notated 360–361 days, but the first crescent of the first month was placed six days *after* the opening of that notated year. This suggested that the notation may have begun with a different type of astronomical observation, perhaps an equinoctial or solstitial sighting of the sun, the heliacal rising of a star, or the termination of a ritual period related to such observation. That placement of the first crescent six days after the notation began was not accidental is indicated by a count of the notated days from it to the last crescent of the 12th month. This count gives us 355 days, a near perfect observational lunar year. The second lunar year (Marshack, 1985) was notated right to left for each six-month period as though to differentiate this year from the preceding.

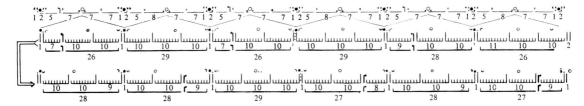

Figure 24.2 A schematic breakdown of the first lunar year notated on the calendar stick of Tshi-zun-hau-kau placed against an astronomically correct lunar model. Indicated are the serpentine mode of continuous reading, the decadal mode of dividing the lunar month into three periods of ten with occasional variation in the last third and the differences in the length of the months from crescent to crescent.

This year contained 356 days, but the last crescent of the 12th month fell three days after its proper astronomical position, suggesting that this terminating observation was delayed, perhaps by overcast skies. Except for these few anomalies, the two years provided a near perfect tally with an astronomically correct lunar year model, within the range of permissible variation that would occur in the use of an observational notation. The precision was obtained primarily by the strategy of using a variable third period at the end of the month.

Evidence that this two-year lunar calendar frame had been used for many years was found in the sets of marks that were added to the calendar later. At two places, in the middle of each six-month period and between the long vertical lines that mark the two 'zero' days of invisibility, there was an irregular accumulation of dots, each dot made by a different point and pressure, as though they had been added one at a time over a period (see Figure 24.1). In the second notated year (Marshack, 1985) dots were also accumulated within the 'zero' days of invisibility that fell at the beginning of each six-month period. These added dots, therefore, were placed before the first and seventh months and between the third and fourth, and the ninth and tenth months. Their placement suggested that they fell at significant calendric points that occurred at intervals of three months. Because of their placement, it seemed possible that they indicated an astronomical observation, perhaps equinoctial or solstitial, that fell in the lunar month *preceding* the marking, with the dots being placed in the following 'zero' days of invisibility. There is abundant ethnographic evidence among Indian groups that a named or ritual sequence of lunar months was often begun

with the first crescent to appear after a relevant solar observation.

An indication that corollary solar observation occurred during the lunar year and during use of the calendar is found at the bottom of the stick. Three sets or periods of notation, intentionally placed outside the precisely structured two year calendar frame, are engraved there. These added periods or 'months' contain 27, 28 and 23 days. That they represented subsidiary 'months' is indicated by the breakdown of the 28 into 10 + 10 + 8, in the mode of the main lunar notation (Figure 24.3).

Since the lunar year of 354 days is approximately 11 days short of the tropical solar year of $365\frac{1}{4}$ days, the only way that a continuing sequence of lunar years could operate in phase with the economically functional solar year was by the periodic addition of an extra 'month' every second or third year. How the notated two-year lunar frame would work with this mode of intercalation is indicated in the schematic model (Figure 24.4). In this model the two lunar years proceed in sequence with an intercalary inclusion each third year. The short intercalary 'month' of 23 days would, in this model, fit appropriately at the beginning of the long first month of the first year. The intercalary periods were apparently, then, not estimated by numerical count, but by observation within the lunar–solar system.

The Winnebago calendar stick therefore documents a number of different notational and astronomical problem-solving strategies that were used by Tshi-zun-hau-kau within what seems to have been an indigenous North American tradition. The tradition involved a counting by tens, rather than by 20s as in the vigesimal system of Mesoamerica;

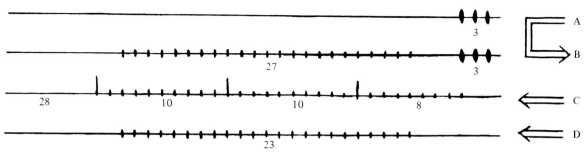

Figure 24.3 A schematic rendition of the added marks and sets at the bottom of the calendar stick of Tshi-zun-hau-kau. Rows A and B represent the direction of reading of the first calendar year, and C and D represent the direction of reading the second year. There are three intercalary periods incised along the edges and two sets of three marks which may be 'year' signs.

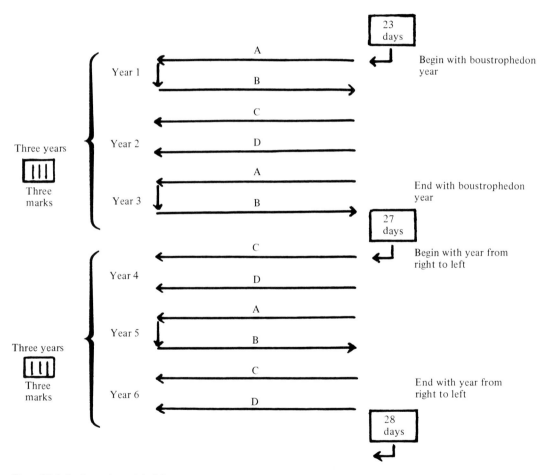

Figure 24.4 A schematic model of the manner in which the three engraved intercalary periods would work with a two year calendar frame by inserting one intercalary period at the end of each three years. (After Marshack, 1985.)

Figure 24.5 The decadal Zuñi calendar stick of Priest of the Bow, Nayuchi, as drawn by John G. Bourke in his diary (Bourke, 1884, 21 November).

it included a conceptual breakdown of the lunar month into the periods of waxing, full and waning; it included a non-arithmetical observational knowledge that the lunar and solar years were of different lengths; and it involved *ad hoc* observational and notational modes of intercalation. There is no evidence on this single stick of a standardized sum or count for the number of days in the lunar month or the lunar or solar year, though the potential for developing such counts, given the proper historical and contextual circumstances, was present in the notational and observational tradition and in a knowledge of the problems that were inherent in use of these variable periods.

24.2 The ethnographic evidence

The ethnographic evidence for a North American use of calendar sticks is as suggestive as the analysis of Tshi-zun-hau-kau's calendar. In the 19th-century Cushing (1941) lived among the agricultural Zuñi of New Mexico, was initiated into their priesthood, learned the language and described the activity of their calendar keepers or 'sky-watchers'. Cushing reported that the Sun Priest observed the rising sun and the alignment of the sun's shadow along a line extending from a mountain in the east to a solar monolith and additional pillars that were set up in Zuñi gardens. The Master Priest of the Bow, Nayuchi, 'responds as he cuts the last notch in his pinewood calendar and both hasten back to call from the housetops the glad tidings of the return of spring' (Cushing, 1941, pp. 116–17; Marshack, 1985).

Cushing gives no description of the calendar

stick. However, after I had published my analysis of the Winnebago calendar, M. Zeilik of the University of New Mexico called my attention to a 19th-century sketch of Nayuchi's Zuni calendar stick. In 1884, First Lieutenant John G. Bourke, a young officer and amateur ethnologist, was introduced to Nayuchi by Cushing. In his diary entry of November 21 Bourke describes the stick carried by Nayuchi, apparently as the symbol and sign of his status, much as chief Tshi-zun-hau-kau of the Winnebago had been portrayed early in the 19th century with his calendar stick in hand (Marshack, 1985). Bourke sketched one face of the calendar stick (Figure 24.5), documenting that it was lunar. He wrote that the Zuñi month was divided into three parts of ten days each, with 'dots' and 'signs' added to mark ritual and festival days. Bourke further stated that the Zuñi year began with a solar observation of the winter solstice and with the ritual Fire Festival that was held about ten days after that observation. According to Bourke one of the months on Nayuchi's calendar stick had the image of an ithyphallic male in ritual stance, apparently in the month or 'moon' of the summer solstice.

Analysis of the Winnebago calendar suggests that Bourke, who may have made his diary entry and comments after the meeting with Nayuchi, probably drew an idealized calendar based on what he saw and on what he had been told. The seven months drawn by Bourke depict a 10 + 10 + 10 breakdown with an elongated stroke between the months. Bourke indicates no variation in the counts. Such a sequence would put the months out of phase with the observed crescents and periods of invisibility after a number of months. It is possible, therefore, that there was some variation in the accumulation that was not noted by Bourke. Nayuchi, in fact, would most likely have given Bourke an explanation of the ideal model without mentioning the necessity of periodic adjustment, and Bourke saw and depicted the calendar in these terms.

Bourke also incorrectly states that the *notated* month begins with the full moon. While the ritual month may have begun at the full moon, a notational beginning at the full moon would have been impractical and awkward. Determination of the precise day of the full moon is difficult, and in a tripartite division of the lunar month a beginning with the full moon would place the more easily determined crescents and the days of invisibility in mid-month within a set of ten. It also contravenes the widespread North American concept of the lunar month as proceeding from the first crescent and period of waxing to full and then to waning.

The Zuñi of New Mexico were farmers who spoke a unique American language. They shared many cultural traits with the neighboring agricultural Hopi who spoke a Uto-Aztecan language, a language group that was widely distributed throughout the southwest. In 1977 McCluskey (1977) published a statistical analysis of the dates of the lunar–solar ceremonies and the practical agricultural cycle of the Hopi Indians of New Mexico. His analysis documented the presence of two synchronous and parallel cycles of observation and ritual and practical activity, one lunar and the other solar, with evidence of lunar–solar adjustment and intercalation every three years to bring the lunar and solar cycles into phase. The Winnebago Indians of the American heartland spoke a different language and inhabited a different woodland ecology. They spoke a Siouan language, and Siouan speakers inhabited a region encompassing some of the Mississippian drainage. Robert Hall, in commenting on the Winnebago calendar, noted that the 19th-century Winnebago may have derived aspects of their culture from the earlier Oneota culture, a climax farming culture in eastern Wisconsin near Lake Michigan. If so, the calendar tradition may have been related to calendar structures and frames in use by the earlier Mississippian agriculturalists. The evidence for astronomical calendric alignments in the structures of these early farming cultures is just beginning to be published and discussed.

After my presentation of the Winnebago calendar at an international meeting, Thor Conway, regional archaeologist for the province of Ontario, Canada, called my attention to the fact that one of his informants, an Algonquian-speaking Ojibwa named Fred Pine, in his 80s and one of the last surviving shamans of his people, recalled that his grandfather had been called 'He who walks with a calendar stick'. As a result of the Winnebago analysis, Conway inquired further and learned that Fred Pine's grandfather had stood at a particular spot on the lake shore to have a clear horizon view of the sun and moon. Fred Pine elaborated on the complex ritual involved in choosing a proper branch for the making of the shamanistically important calendar stick.

The Algonquians were primarily Eastern Woodland hunter–gatherers who had entered the fur trade after contact with European colonists. Where they came in contact with Indian farming groups, they occasionally acquired a partial farming mode, but they never gave up reliance on hunting–gathering. In Canada, in particular, hunting and gathering formed the economic base among the Algonquians. A fascinating 19th-century ethnographic account comes from the diary of an Algonquian-speaking Ojibwa (Chippewa) woman, Nodinens (Densmore, 1929):

> When I was young everything was very systematic. We worked day and night and made the best use of the material we had. My father kept count of the days on a stick. He had a stick long enough to last a year and he always began a new stick in the fall. He cut a big notch for the first day of a new moon and a small notch for each of the other days. I will begin my story at the time he began a new counting stick. After my mother had put away the wild rice, maple sugar, and other food we would need during the winter she made some new mats for the sides of the wigwam . . .

In the American heartland, therefore, among diverse, widely distributed linguistic groups, with differing ecologies and economies, the practice of observational astronomy, performed by an elite of specialized sky-watchers, was common. At least some of these sky-watchers kept notations of the year's progress, essentially a lunar sequence that

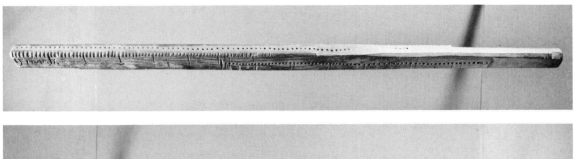

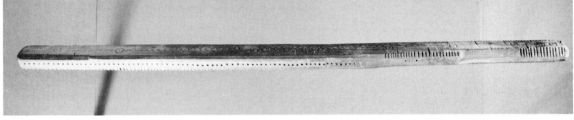

Figure 24.6 The two faces and edges of the second Winnebago calendar stick showing the sequence of dots on the faces, the bevelled edges and the sequence of strokes on these edges. The long strokes marking off sets of ten are indicated.

in some cases at least was periodically adjusted to the solar tropical year.

Within this tradition the decimal mode of counting by tens was common, as was the mode of making a long stroke to close out a sum of ten and the use of still longer strokes to mark out larger periods or sets of ten. The decadal mode was used by different North American Indian groups to count objects and persons. In his classic compilation of Indian graphic modes, Garrick Mallery (1898–9) illustrates the use of this decadal notational mode to count the number of persons killed in a tribal war and to count skins bartered in the fur trade. In his *Hopi Journal* A. M. Stephen (1936) illustrates the decadal notational mode in a tally of slain Apache and in a count of Apache women.

At many points, therefore, we have evidence that chief Tshi-zun-hau-kau had created his calendar stick within a widespread North American Indian 19th-century tradition which involved lunar and solar observation, calendric record-keeping, and the North American mode of counting and notating[2]. The simple notational strategy of completing a standardized numerical set of marks with a long stroke is documented as well in a calendar board kept by a contemporary shaman in Mesoamerica, but this time the accumulation was

in the Mayan 20 base vigesimal system (Marshack, 1974). It is this body of analytical and ethnographic data concerning record-keeping modes and strategies that made it possible to begin an internal, cognitive analysis of the other visually different North American calendar sticks and to place them within their historical and cultural context.

24.3 The second Winnebago calendar stick

By the mid-19th century the Winnebago had been battered by enemies and had been moved by the government in Washington westward to a reservation in Nebraska. Two Winnebago calendar sticks from the second half of the 19th century have been located, each visually different from the other and from the calendar stick of Tshi-zun-hau-kau. Each, nevertheless, derives from the same tradition. I present an analysis of the last Winnebago calendar stick, collected ethnographically in May of 1897 (Figure 24.6). The calendar is incised on a four-sided stick that has been marked with notation on the edges and the flat faces, primarily by accumulating sets of ten, the tenth stroke, as usual, being an intentionally longer line.

An initial study of the stick (Figure 24.7) presents us with what seems to be a random accumulation of idiosyncratic sets and sequences of different

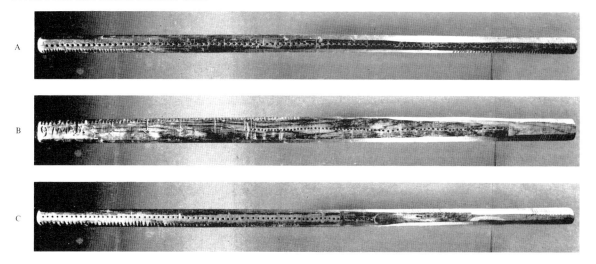

Figure 24.7 The three faces of the four-sided calendar stick of the Winnebago Indians which was collected in May, 1897. The faces contain sequences of dots, the edges are marked with sequences of notches or strokes. The dot sequence of Face C ends with an engraved lunar crescent. All the major sets are divided into decadal periods.

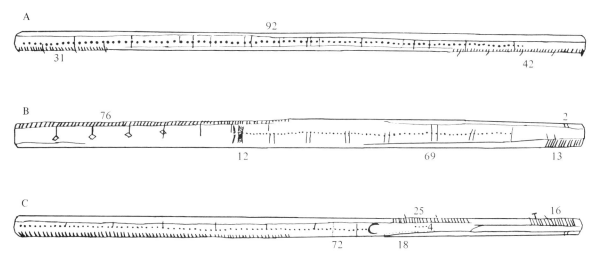

Figure 24.8 Schematic rendition of all the intentional sets and marks on the three notational faces of the main notation.

length, placed in an apparently haphazard manner. Some of the edges have been partially bevelled for a short distance at some time during the calendar's use, and these bevelled areas were then marked with additional sets. At first glance it seemed impossible to unravel this odd accumulation and organization of sets and groups of marks.

An insight into the strategies involved in the accumulation came by chance and through knowledge of the tradition that was acquired during

analysis of the first Winnebago calendar stick. On one of the faces (Figure 24.8, face C) there is a sequence of 72 dots ($7 \times 10 + 2$) which ends in mid-stick with an incised lunar crescent. Since calendar notations of this type almost always begin at one end of a stick and proceed horizontally for as long as deemed necessary, I assumed that the crescent terminated that line of notation and was either a first or last crescent.

It became apparent during the analysis that the

flat faces were usually marked with sets of dots accumulated in relatively long periods, whereas the edges were usually marked with strokes or notches in shorter periods. I assumed that because of this organization the long sequences on the faces and the shorter sequences on the edges represented different types of sets or sequences. On only one edge was there a relatively long sequence of notched strokes (Figure 24.8, face B), consisting of 76 marks ($7 \times 10 + 6$). The evidence for a bevelling of the edges, apparently after the faces were marked, and the subsequent marking of these 'new' surfaces with relatively short sets, suggested that these additions might have been similar to the intercalary sets added to the calendar frame by Tshi-zun-hau-kau.

Using these observations and assumptions a number of tests were conducted to determine the best lunar fit. I found that if I began with the lunar crescent as a terminal image and backtracked from it and placed the primary notation consisting of the longer sequences (each more than two months in length) against an astronomically correct lunar model, I achieved a near perfect lunar match. But this match for the backtracked lunar year occurred with some initial difficulty in positioning the crescent. I assumed, to begin with, that this crescent was either an observed last crescent or first crescent and that it ended the primary notation. In backtracking, however, I discovered that the notation formed a 'perfect' lunar match only by the insertion of one extra decadal group of ten marks into the notation. Insertion of this extra set into the initial test was called to my attention by the editor. This sent me to a microscopic re-examination of the engraved crescent. This revealed that, after it was made, three small marks or notches had been engraved upon and around the crescent and that a long thin line was then added, as though to close out the period (Figure 24.9). No such incised additions occurred at any other place on the calendar stick, though evidence of later forms of marking will be discussed below.

Placement of the crescent was therefore uncertain. However, because of the close tally of sets behind the crescent it seemed as though the crescent closing out the year had been marked on the stick within the last decade of the last month, before it was actually seen. Three days were then ticked off and a long thin terminating line was finally incised

at observation of the last crescent, closing out the year. Assuming such an ending with the observed last crescent, it was possible to backtrack the full year.

The reading was accomplished by turning the stick persistently away from me, or upward by one quarter turn, to present the next long sequence. Since I was backtracking, the completed notation (Figure 24.8) actually began where the backtracking ended, with face A.

The backtracking reconstruction proceeded in this manner. By turning the stick one quarter, from face C to face B, one is presented with a face that has *two* long sequences. Beginning at the left and incised as a set of linear strokes along the edge is a set of 76 marks, broken down in the decadal system ($7 \times 10 + 6$). The long linear strokes that close each set of ten descend onto the face so that no notation could thereafter be placed in this flat area of face B. The sequence of 76 terminates in mid-stick with an anomalous set of six. This sequence is connected in mid-stick to another sequence of marks engraved on the flat face as a set of dots that proceed to the right, ending before the sequence reaches the right end of the stick. The connecting element for these two long notational sequences is a 'ladder' consisting of 12 strokes (Figure 24.10). The dot sequence consists of 69 marks ($6 \times 10 + 9$). It is not the terminating set of dots, however, but the next-to-last seventh set that contains only nine marks. Because the earlier set of 76 edge marks had used up the space required for a notation on this face, I assumed that the set of 76 had begun the notation for face B, that the notation had then descended via the 'ladder' of 12 strokes and had continued in a rightward direction with the sequence of 69. In the backtracking reconstruction I therefore appended the 69 to the 72 (face C), then went to the 12 and then to the 76.

The stick was then turned one quarter to face A. This presented the longest sequence. It consisted of 92 dots ($9 \times 10 + 2$) that apparently began again at the left end of the stick and proceeded towards the right. It ended before reaching the right end with the anomalous count of two.

Placing this reconstructed sequence against an astronomically correct lunar model (Figure 24.11) gave one an almost precise observational match for a calendar of 11 lunar months. Surprisingly, this

Figure 24.9 Extreme close-up of part of the last crescent of the lunar year with three incised marks in and around the crescent. The last dot of the set of 72 marks is at left, the long terminating line of the primary notation is at right.

tally worked only because of the presence of the non-decadal counts and periods of two, six, nine and 12, which had been added to or were included among the decadal sets. The 'ladder' of 12, for instance, was a period that closed a six-month period. It terminated at invisibility. The next long sequence of 76, therefore, began with a first crescent. The first four decades of this period of 76 have appended to their long decadal strokes a rhomboidal 'diamond' sign, helping to fill the space in that area of the face and suggesting that these were probably 40 days of special significance.

These many anomalies, variations and inclusions within a lunar calendar that was essentially maintained in the decadal mode could function only within a well-known tradition of lunar observation, a division of the lunar year into culturally relevant periods and sequences and a knowledge of the dif-

ferent kinds of adjustments required to make the decadal mode function over an extended period. Winnebago chief Tshi-zun-hau-kau had devised his own strategies for adjusting the decadal mode to an observational process. Though arithmetic was used in counting relatively short periods, longer periods and the lunar year itself did not seem to have been arithmetically determined or structured.

There were other interesting aspects revealed by the reconstruction. The sequence of 72 which terminates the lunar notation ends with the period of last crescent, which was not apparently seen until four days after the 'closing' crescent was incised. The full 11 months begin with an observed *last* crescent. Normally we would expect a beginning with the first crescent, but this type of beginning occurred even in chief Tshi-zun-hau-kau's carefully planned and structured calendar stick (Figure 24.2). These seeming anomalies indicate that we

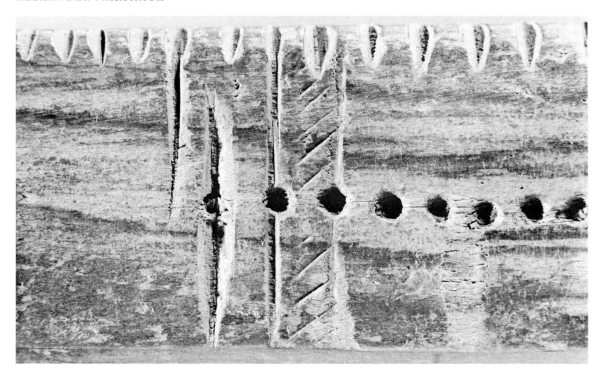

Figure 24.10 An extreme close-up of the 'ladder' figure with 12 incised strokes. The 'ladder' descends from the upper sequence of edge marks to the lower sequence of dots. A decadal tenth stroke descends from one edge mark and a terminal decadal tenth stroke crosses the dot to the left of the 'ladder'.

are dealing with observational modes of maintaining the notation. In such a system either the last or first crescent, or the days of invisibility, could be used as points for determining the beginning or end of the non-arithmetic, conceptual month. In a notation of this type it is possible that the calendar frame was intended to be used to determine relevant but variable periods that were as long as two and a half to three months or, as in the case of the sequence of 76, to mark a period that began with 40 special days and closed at the end of the next lunar month with a special period of 12 days. These longer periods were not used to structure a numerical year, but for prognostication of forthcoming ritual or economic periods and events.

As the complexities involved in the calendar stick tradition were unravelled, it became clear that the usual arithmetical modes of astronomical analysis

and measurement, used for determining astronomical periodicities and alignments, would not be able to resolve the problems inherent in a tradition which contained, besides astronomical observations, culturally relevant periods and variable, often *ad hoc*, notational problem-solving strategies. Cultural and cognitive as well as arithmetical and astronomical problems were involved and these required analysis at a number of levels.

Having reconstructed the main notational sequence of 11 lunar months, it became apparent that analysis of the calendar stick was not complete. Tshi-zun-hau-kau's calendar stick had documented the presence of intercalary periods that were apparently inserted to keep the lunar notation in phase with the functional solar year. Zeilik had pointed out that among the Zuñi the lunar year, which was notated by Nyachi in the decadal mode,

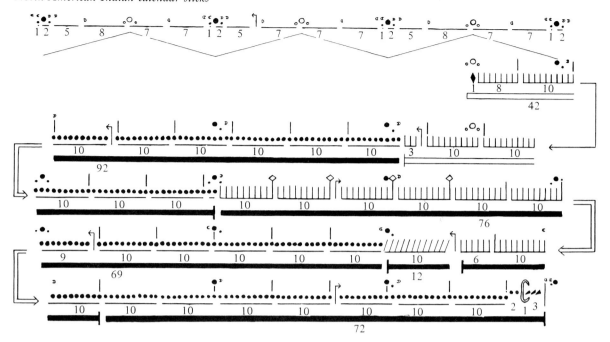

Figure 24.11 The 11 months of the main notation incised on the Winnebago calendar stick indicating a division of the lunar month into decadal periods, but containing anomalous additions and variations to make the calendar work as both an observational and a culturally relevant frame. The sequences of strokes are incised on two of the edges, the dot sequences on three faces. A one-and-a-half month period of 42 marks apparently proceeded the main lunar notation. A period of four or five days of waiting for observation of the last crescent closes out the lunar year.

began in the lunar month *after* the winter solstice and that there was, in addition, a ritual recognition of the full moon in the month or 'moon' of the winter solstice. Both concepts were found to be helpful as the analysis proceeded.

Among the shorter 'intercalary' periods of less than two months in length on the present stick was one incised along an edge of face A, the face that contained the opening sequence of 92 (Figure 24.8). The sequence was engraved at the bottom, far right, in a decadal mode, but it was unique in that it began with a deeply incised notch that formed a 'diamond' image at the end of the stick (Figure 24.12). Beginning with this notch, the sequence proceeded in the decadal mode to provide 42 marks (9 + 10 + 10 + 10 + 3). Unlike the other sequences, this one began at the right, and it ended with an anomalous set of three. Appending this one-and-a-half month long sequence to the opening major sequence of 92 and backtracking for the 42 notated days brought the 'diamond' notch to the day of the full moon, a month and a half earlier (Figure 24.11). With this addition, the year of 11 months had its

12th opening month with an indication that the observation and notation had begun in the month *preceding* opening of the lunar year, in a manner suggestive of that noted by Zeilik for the Zuñi. The anomalous set of three seems to have been the short observational period required to await the last crescent, which was then taken as an indication that the decadal notation of the next lunar month could begin. In light of this reconstruction, it is interesting that the long period of 92 itself ended with an anomalous set of two which fell in a period of invisibility.

If the initial engraving of this calendar stick did begin with this set of 42 and with an observation of the full moon in the month of a solar observation, then the other 'intercalary' periods incised on the calendar stick would necessarily have been appended at the end of that lunar year and the years following, to bring the lunar years into phase with the observed solar year. These shorter periods, incised on edges and occasionally on the faces, were of 31, 25, 18, 16, 13 and 13. Some smaller sets, of two, four, and six, were also incised on the stick.

319

Figure 24.12 An extreme close-up of the beginning of the edge-marked set of 42 that is incised between Face A and Face D. The deep diamond-shaped notch that begins this sequence is clearly seen. The beginning set has a count of nine if one includes the notch, the next set has a normal count of ten. The set of 16 edge marks below, intruding on Face D, has a decadal stroke over the tenth day and a 'T' sign over the 13th day, suggesting that it may possibly have had special ritual significance.

It did not seem, therefore, that intercalary periods were added regularly at three-year intervals. Some were apparently added at the end of one or two years, within an observational rather than an arithmetical frame.

That this calendar stick was used for a number of years, not only by the engraving of intercalary sets but by additions and 'corrections' made within the original engraving of the primary lunar year, is indicated by the many incised and painted marks that were placed over the notation at certain points within the lunar year, as though to indicate that a relevant day now fell at this point in one year and at another point in a different year (Figure 24.13). The calendar stick of Tshi-zun-hau-kau (Marshack, 1985) had its own set of periodic marks added to the originally incised calendar frame. Many of these additions had been placed within the 'zero' days of invisibility, but some had been incised within the month itself. The variable mode of marking such days during use of the calendar stick was, therefore, not culturally prescribed but represented a form of problem-solving and record-keeping that was devised by each sky-watcher and calendar keeper. The accumulating evidence suggests that there was no standardized arithmetical structure for the year or the month and no standardized or uniform system for maintaining a notation.

Such standardization apparently becomes useful and necessary with the creation of a well-established central administrative apparatus.

All the calendar sticks so far studied were made within a widespread tradition with a common 'informational' base. This not only included the decadal mode of notation, but a use of variable sets for adjustment as required by observation; it included a use of intercalary periods of irregular length. Each calendar stick so far studied is a variant within this system. The first Winnebago calendar stick studied, that of Tshi-zun-hau-kau, is the best organized and structured and would seem to represent an individual solution of the inherent problems invented by Tshi-zun-hau-kau himself.

There is a fourth face, face D (Figure 24.14) on the Winnebago calendar stick here being analyzed, and it presents us with a different set of analytical 'calendrical' and cultural problems. Instead of a set of notches or dots arranged in a decadal mode with long strokes to sum each set of ten, there is an unusual sequence of circles connected by festooned arcs (Figure 24.15). This represents a totally different type of image or iconography from any we have studied so far. Since the stick is calendric and none of the calendar sticks so far studied contain any evidence of 'decoration', I assumed that this sequence of circles

Figure 24.13 The closing group of marks of the terminal sequence of 92 with the engraved closing first crescent. Three of the dots have been marked with black paint, perhaps indicating that a last or first crescent fell at these points in different years. The crescent itself has been marked with two dots. Such indications of later marking occur throughout the notation.

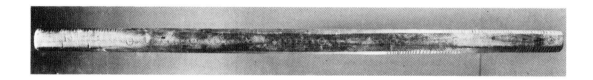

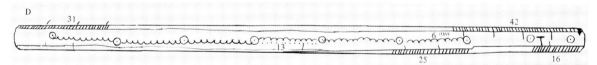

Figure 24.14 The fourth face, Face D, of the Winnebago calendar stick, marked with circles and festooned arcs. At far right are two circles without arcs, perhaps added last and without room for a set of arcs. The sequence seems to begin at the left. The first three sets are made with the arcs in one direction, the next three have the arcs in reverse, suggesting an intentional differentiation between each set of three 'years'. It is possible that this face contained the accumulation of months and years recording the period during which the calendar was used.

and arcs was related to the lunar year notation.

I assumed that the circles may have served, cognitively, as 'year' signs, much as the longer strokes of the decadal system had served as completion or summing signs for a set of ten. In that case, the sequence of arcs may have represented the months, kinesthetically representing the waxing and waning of each month. A search of Mallery's classic volume of Indian iconography and signs (Mallery, 1898–9) revealed that this abstract, schematic mode of iconographically visualizing the passage of time had in fact been used by Indian groups. The Sioux-speak-

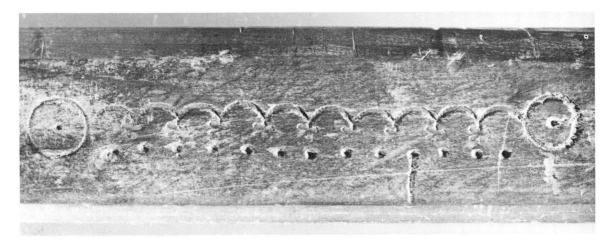

Figure 24.15 An extreme close-up of the fourth set of circles and arcs. The circles, the arcs and the star symbols at the ends of the arcs were made with metal stamps which were bought as trade items. A metal stamp also was used for the circles made on the first Winnebago calendar stick. Because of the standardiz- ation in the size of the arcs there was not room for a sequence of arcs among the last two circles stamped at far right. Below the arcs is a set of 13 dots broken down as 10 + 3, perhaps marking an intercalary period interposed at this point.

ing Dakota had imaged a sequence of days or years as a series of circles connected by lines:

O–O–O–O–O

The Athapascan Apache had used the same icono- graphic mode to image the passage of days. There was also ethnographic evidence that the circle could be used as a symbol both of the sun and the solar year.

Therefore, assuming that the circles and the sets of arcs may have indicated a higher order of struc- turing, the passage of months and years, and that the arcs were months while the circles indicated the completion or beginning of a new year, I attempted a count of the circles and arcs. This gave a sequence: $\odot$ + 11 arcs = 12; $\odot$ + 11 arcs = 12; $\odot$ + 10 arcs = 11; $\odot$ + 10 arcs = 11; $\odot$ + 12 arcs = 13; $\odot$ + 9 arcs = 10; and $\odot$. There were also two circles at far right that had not had arcs added to them, apparently because there was no longer room left for a full sequence of arcs. There was no certainty but, within the ethnographically known Indian tradition of naming and counting the number of 'moons' in a year, a sequence of months ranging from ten to 13 seemed possible. The suggestion would also have tallied with the *ad hoc*

mode of notating the lunar year. If the circles indi- cated 'years', the calendar stick carried the symbolic image of nine years. This would just about have included all the sequences of intercalation plus the notated primary year.

The extraordinarily complex accumulation of sets, periods, signs and symbols had, therefore, been 'read' as an accumulating sequence, a record which had been kept and had been continuously adjusted during a period covering approximately nine lunar years. Whether the analysis and interpre- tation were accurate in all respects could not at this stage be determined, but the fundamental cognitive mode and the nature of the problem-solving strate- gies utilized within this comparatively open tra- dition of calendrical notation had apparently been found.

Analysis of the two Winnebago calendar sticks has made possible the analysis of the third Winne- bago calendar stick, collected in 1893, and two lunar calendar sticks from the Sioux speaking Osage. Each is different from the two already published, but made within the same tradition. The analyses are being prepared for publication. If, as was sug- gested by R. Hall (Marshack, 1985), the Winnebago culture derives from a climax Mississippian farming culture, then current archaeological and archaeoastronomical investigation may be able to uncover aspects of an indigenous lunar–solar tradition among these early farming cultures that was not derived from Mesoamerican models. In addition, the suggestion that an early observational astronomy and hunter–gatherer's lunar calendar may have entered the Americas from Asia may provide the basis for an inquiry into the origins, development and dispersal of different North and South American Indian indigenous calendar traditions.

Notes

1 Tshi-zun-hau-kau engraved his calendar stick on the shore of Lake Michigan in the 1820s at the time that the Canadian blacksmith, Joseph Jourdain, was making and trading metal tools. Tshi-zun-hau-kau had used a metal punch, probably traded from Jourdain, to impress his lunar circles over the full moon periods. The two Winnebago calendar sticks from the end of the 19th century also evidence a use of metal trade tools. The circles on the calendar stick analyzed in this paper are made with a metal drill head and the arcs seem to have been made with a metal punch. Each month on face D is terminated by use of a tiny fleur-de-lys metal punch, perhaps originally made for decorating leather. It is probable that these metal trade items were curated by the Winnebago calendar keepers as specialized tools. The tradition of calendric notation, however, was indigenous and owed nothing to European influence. The presence of incised and painted signs throughout the accumulation indicates that the metal tools were merely adapted to the indigenous tradition.

2 Murray's paper (Chapter 25 of this volume) proposes a number of hypotheses based on the analysis of the first Winnebago calendar stick. He attempts to infer a 'bar-and-dot' system of notation derived from Mesoamerica. The use of sets of lines and dots in notational accumulation is so natural and widespread that no such inferences can be drawn from the presence of these modes. They appear in the Upper Paleolithic notations of Europe and in historic marking systems around the world. They appear on the wooden non-calendric tribal records of the Pima Indians of the historic period. Far more important as an indication of separate, indigenous origins is the widespread use of a decadal system on the calendar sticks and of a counting by fives and 20s in the Mayan bar-and-dot system. The North American notational tradition not only has a different trajectory of development but a different range of functional uses than the bar-and-dot counting found in Mayan calendrics.

Murray also hypothesizes eclipse predictions on the basis of the analysis of the first Winnebago calendar stick. That stick cannot be used as an example of a widespread tradition of eclipse prediction. For one thing, the stick is totally different in internal structure and breakdown from the other sticks so far studied, and on none of the other sticks is there an indication of the sums and counts that appear on the first stick. The sums and counts in the analysis were derived from the arithmetical mode of analysis. There is no evidence that they were of interest to the record-keeper. There is, in addition, no ethnographic evidence for eclipse prognostication or an arithmeticized calendar year in North America. As I have stated, it is theoretically possible that an arithmetical year and eclipse determination could have been derived from such a notational tradition given the proper cultural conditions. Unfortunately, the calendar sticks indicate individual problem-solving strategies within a widespread, but generalized tradition of lunar and solar observation. The formal regularities of a standardized, arithmetical calendar do not appear. In the Americas such standardized systems appear with the development of 'state' administrations. Despite the complexity of the Mississipian cultures and influences from Mesoamerica, there is no evidence for an administrative calendar or record-keeping of the Mesoamerican type. There is, also, no evidence for the artificial 260-day calendar cycle in the calendar stick tradition.

I suppose that it is natural to seek Mesoamerican influences in cultural developments that occurred outside of Central America. These influences, however, were felt and adopted by cultures and peoples with already evolved indigenous traditions. Traditions of wild plant harvesting and near domestication were already in process in parts of North America before the entry of maize. Observational calendar traditions and the decadal mode of counting were probably also present. The adoption of certain Mesoamerican cultural modes, therefore, was probably already culturally prepared for.

The complex problems in technological and symbolic dispersal and diffusion and in understanding how such elements are screened, used and changed by cultures at different levels of preparation cannot be explained by assumptions of wholesale export. The calendar stick traditions will have to be properly published before one can make final judgments concerning origins and influences.

References

Bourke, J. G. (1884). *The Diaries of J. G. Bourke.* The Special Collections Division, Library of the United States Military Academy, West Point, New York.

Cushing, F. H. (1941). *My Adventures In Zuni.* Santa Fe: Peripatetic Press.

Densmore, F. (1929). *Chippewa Customs.* Smithsonian Institution. U.S Bureau of American Ethnology, Bulletin 86, Washington, DC, pp. 119–23.

Mallery, Garrick. (1898–9). *Picture-Writing of the American Indians.* 10th Annual Report, Bureau of American Ethnology, Washington, DC.

McCluskey, S. C. (1977). The astronomy of the Hopi Indians. *Journal for the History of Astronomy* 8, 175–95.

Marshack, A. (1974). The Chamula calendar board: an internal and comparative analysis. In *Mesoamerican Archaeology: New Approaches*, ed. N. Hammond, pp. 253–70. Austin: University of Texas Press.

Marshack, A. (1985). A lunar–solar year calendar stick from North America. *American Antiquity* 50, (1), 27–51.

Stephen, A. M. (1936). *Hopi Journal.* New York: Columbia University Press.

25

A re-examination of the Winnebago calendar stick

William Breen Murray   *University of Monterrey*

In a recent article Marshack (1985) describes and analyzes a wooden calendar stick used into historic times by the Winnebago tribe of central Wisconsin. This and similar artifacts add a new dimension to our knowledge of Native American astro-calendrical counting, and should alert us to the growing complexity of the archaeological problem at hand, and to some tremendous new possibilities which may now be opening up. We begin to see the reflections of ancient man's mind in many different mirrors, and the task now is to integrate these reflections into a single, more coherent image. It is this question of integration which I would like to pursue, and I think the Winnebago calendar stick provides a very useful starting point.

The analysis of symbolism on ancient artifacts is always a risky business, even under the best conditions, but it is susceptible to certain kinds of hypothesis-testing, especially when we are dealing with astronomical phenomena whose fundamental nature is independent of cultural influences. These phenomena are real, and their variable rhythms are now known with great precision. This permits us to see more clearly in archaeological terms how culture has systematically transformed these phenomena in a given context into something which had meaning for man. The trick for anthropological archaeology is to find the cultural rules which govern this translation.

In the Mesoamerican high civilizations we have irrefutable evidence for the use of a unique bar-and-dot counting system. Its existence and association with calendrics can be documented at least to the late Preclassic period, but the origins of the

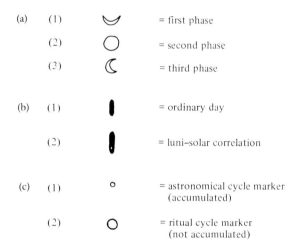

Figure 25.1 Symbolic vocabulary of the Winnebago calendar stick. (a) Lunar symbol system. (b) Solar symbol system. (c) Cycle marker system.

system are still largely a mystery. Nor can we be sure to what extent the system was shared by other Amerindian people. But both of these questions are open-ended and susceptible to new evidence, and I believe that the Winnebago calendar stick sheds new light on both, but particularly on the geographical diffusion of bar-and-dot counting.

The symbolic vocabulary of the Winnebago calendar stick is extraordinarily clear and consistent (Figure 25.1). It employs seven symbols, which can be grouped into three sub-systems, and deploys these symbols in a two-level parallel text fashion. Above, we have three symbols corresponding to the phases of the Moon. Along the edge of the stick,

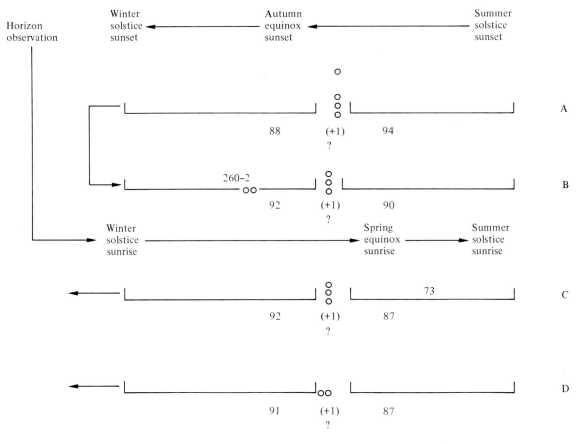

Figure 25.2 Astronumerical structure of the Winnebago calendar stick.

we find two kinds of slash marks which appear to be a solar day count: a short one, and a long one which extends into the level of the lunar symbols, and must represent some kind of correlation between the two. The third sub-system consists of two kinds of dot markings. The first kind are intercalated at the middle of each line, and on the CD side at the beginning as well.[1] Marshack's microscopic analysis (Marshack, 1985, p. 36) showed that these dots were made cumulatively, and this distinguishes them from another kind of dot marking which is carved directly above the solar day markings and is not accumulative. These occur at only two places: over the middle of the third month on line B (Marshack, 1985, Fig. 8), and over the full moon sign of the third month on line C (Marshack, 1985, Fig. 10). I think the dot symbol itself refers to cycle marking, and that the cumulative and non-cumulative dots record two different kinds

of cycles. Marshack suggests that the non-cumulative ones indicate ritual cycles, inferring that the cumulative dots were observational, and I think there is empirical support for this idea in the number pattern itself. This vocabulary accounts for all the symbols which are used on the stick, and one sees immediately that two-thirds of the symbols fall within the bar-and-dot tradition, even though they are not employed in accordance with Mesoamerican rules. Their numerical order shows a very precise correlation to luni–solar observational astronomy, however, which provides another way of getting at the underlying similarity.

If we examine the counting sequence (Figure 25.2) on the AB side of the stick, the serpentine reading order which Marshack suggests would divide the count into four elements, whose total number sequence would be: 94 + 88 + 92 + 90 = 364. This suggests immediately that the cumulative dots may represent the missing 365th day of the solar year, and that their accumulation might even permit some kind of leap year adjustment.

The numerical variability of the elements also raises the possibility that the variable length of the seasons was being intentionally recorded. Any attempt to compare the numbers to the modern season must take into account (1) the greater observational difficulty of pinpointing the solstices than the equinoxes to an exact day, and (2) the possibility that the stick's counting formula is old enough to be affected by long-term precessional changes. Nor can we say to which season the 365th day should be added. Nevertheless, if we compare the numbers to modern seasonal values, we can derive a best-fit sequence of Summer–Fall–Winter–Spring, with an error implied of three days in measuring the winter solstice, and one or two days in measuring the equinox. This is no mean achievement in horizon observation; it means that the users of the stick had empirically discovered the irregularity of the seasons, and were trying to reduce it to a codified number sequence. I doubt this could be done without many years of fairly systematic observation.

Moreover, the stick is so designed that, if both horizons are used, the counting sequence proceeds analogously to the sun's horizon movements. During discussions at the Oxford II conference, Ellegård pointed out that the stick could not have been used as a sighting device because all the notches are spaced approximately equidistant, whereas horizon observation would cluster the marks at the extremes, corresponding to the solstitial positions, and spread them out in the middle near the equinox position, when the sun's horizon movement is fastest. The counting by sections mirrors the perception of the horizon by quadrants, however, a cognitive linkage which allows us to confirm that the calendar stick was not simply an observational tool, but a true counting instrument.

If this numerical reconstruction is correct, then a much more important Mesoamerican indicator emerges along with it. The other dot cycle marker now indicates the 260th–262nd days after the beginning of the count. Marshack interprets this marking (correctly I think) as a reference to some kind of ritual event (Marshack, 1985, pp. 36–7), but does not establish its numerical parallelism to the Mesoamerican ritual year. Although bar-and-dot inscriptions have been identified on a figurine from

Hopewell, Ohio (R. L. Hall, 1984, unpublished manuscript: 'The cultural background of Mississippian symbolism'), to the best of my knowledge the Winnebago calendar stick constitutes the first solid evidence for the additional diffusion of the Mesoamerican ritual year into the Mississippian area. The calendar stick also tells us that among the Winnebago the 260-day cycle was definitely calendrical, and marked a period which might have run from the summer solstice to some time in early March just before the spring equinox. The ritual celebrated may have been anticipatory to the equinox in some way, as has been suggested by Zeilik and Elston (1983, p. 72) for the US southwest, but I am unable at this time to suggest any specific Winnebago ritual to which this could be tied.

Surprisingly enough, the presence of the 260-day ritual year among the Winnebago is quite consistent with what is known archaeologically, and inferred ethnologically, about the late prehistory of this region. R. L. Hall (1984, unpublished manuscript) has pointed out the considerable parallelism between Winnebago myths and Mesoamerican ones, and argues convincingly for various Mesoamerican influences in Mississippian culture. The site of Aztalan, near Lake Mills, Wisconsin, is generally considered to be a Mississippian intrusion, probably colonized directly from Cahokia (Illinois). Its material culture included Gulf Coast shell pendants, copper-covered ear spools, and long-nosed god masks, and maize–squash agriculture was practiced (Brose, 1978, p. 571). The 900 000 square foot stockaded palisade and plaza complex was constructed around AD 1100. It cannot be linked definitely to the historic Winnebago, but it is very close to their first known location. Moreover, Radin and others found evidence in the Winnebago kin/clan system that they had derived originally from a stratified society, and they are known to have depended more on horticulture than their neighbors (Lurie, 1978, pp. 690, 695), both of which would be consistent with an interpretation of Mesoamerican influences. Lurie also mentions that among the Winnebago 'great respect was accorded those individuals who were blessed by the Moon with the gift of prophesy and the requirement that they become berdaches' (Lurie, 1978, p. 695). The moon was a female deity, where as the sun was a primary male war deity, and may once have been the principal deity worshipped. It is not too hard to imagine the calendar stick in the hands of these prophets, and to see it even as the key to their prophetic prowess (cf. Callender and Kochems, 1983).

The prophetic aspect is better seen, however, on the CD side of the calendar stick, which works quite differently in numerical terms, and refers, I believe, to a different astronomical phenomenon. Marshack notes that the CD side is not read in serpentine form, but as parallel lines, and its dot indicators generate a completely different number pattern than on the AB side. Both lines include cumulative dots at the beginning, and both count to 87 in the first section. The second section varies by only one, giving a total of 179 on line C and 178 on line D. This variation stems from an additional long stroke added at the mid-point of line C, where the cumulative dots are also recorded differently on each line. The CD side seems to represent two closely related variants of the same count, and the numbers recorded strongly suggest that lunar eclipse intervals were being observed. I am not sure at this point what rules (observational or numerical) can account for the difference between the two versions. One obvious possibility is the dot cycle marker on line C, which falls on day 73, a reasonable approximation to two-and-a-half lunar months, or half of the five-month eclipse periodicity. Whatever the case, on both the C and D lines the five-month, five-and-a-half-month, and six-month eclipse intervals are all specially marked. In fact, the variant placement of the full moon symbol on each line, which Marshack noted (Marshack, 1985, p. 38), can be easily explained as a reference to the common day 162, the five-and-a-half-month eclipse interval. It seems fairly clear, then, that the counts could be connected with attempts at luni–solar eclipse prediction, a particularly significant exercise perhaps for berdaches who obtained their prophetic gifts from the moon.

Identification of lunar eclipse counting on the Winnebago calendar stick establishes other important links to Mesoamerican calendrics. The 'Lunar Eclipse' tables of the *Dresden Codex* and related counts in other codices are well-known expressions of the Mesoamerican interest in this periodicity, and a pecked cross symbol at Tepeapulco records it in petroglyphic form (Aveni, Hartung and Buckingham, 1978, table 1). Equally relevant are the petroglyphic representations of lunar eclipse counting in Nuevo León, which I reported at the first Oxford conference (Murray, 1982). Since then, additional counts have been discovered at Icamole, Nuevo León, one of which registers the missing five-and-a-half-month eclipse interval in vertical dot columns (Murray, 1985). The Nuevo León petroglyphs are especially important in that they, like the Winnebago calendar stick, show the bar-and-dot counting system separated into its two constituent components, and used under different rules from the Mesoamerican ones. The counts are also found at sites (Icamole and Boca de Potrerillos) where an equinox and solstitial horizon sighting system is strongly suspected, which shows that the problem solved on the AB side of the calendar stick might have been of interest to the petroglyph makers too. In fact, the Nuevo León sites are located just north of the Tropic at 26° N. Lat., precisely where zenith sighting would become impossible, and horizon observation becomes the only viable alternative for solar year record-keeping. Can it be that the ancient Mesoamerican tradition, like that of the ancient Mediterranean, travelled far enough to note these changes in the heavenly motions due to different latitudes? Aveni first raised this possibility at Chalchihuites (Aveni *et al.*, 1982); and I think the petroglyphs and the calendar stick taken together point strongly in that direction.

Another link in this chain are three portable bison scapulae found in a Late Prehistoric burial on the Texas Gulf Coast, near Corpus Christi bay (Hester, 1980, p. 80). The flat faces of these scapulae, which are unique in the Texas archaeological record, are engraved with three variant configurations of dots and slashes. According to one interpretation of their counting order (Murray, 1984), all three variants could relate to lunar eclipse counting. They show the transfer of the operation from a stationary rock wall to a new type of mobiliary artifact, more suited to a nomadic people like the Coahuiltecan buffalo hunters who are reported in south Texas at the time of contact. Might the wooden calendar sticks of the forest-dwelling Winnebago be an equivalent form within the same artifact family? Perhaps there are many other artifacts in between, whose existence has gone unnoticed for lack of an integrative model, just as had happened to the Winnebago stick before Marshack's report. Further investigations may tell us a great deal more about the evolution and permutations within the bar-and-dot counting tradition. Suffice to say that the problem of its origin and distribution has now become a North American one, rather than just a Mesoamerican one, as has been sustained traditionally.

None of these examples helps us much with the question of antiquity. The Winnebago calendar stick is an exceptionally clear ethnohistoric link to prehistory, but its time depth is purely speculative. While it is possible that some of the Nuevo León petroglyphs predate bar-and-dot counting in Mesoamerica, this cannot be proved, and the bison scapula counters are definitely Late Prehistoric. To date, all known examples of bar-and-dot counting outside Mesoamerica can be explained as radiations out from the high civilizations; none of them take us back necessarily to Archaic origins. Large numbers of petroglyphic dot configurations in the upper North American Great Basin (California, Nevada and Oregon) do hint that the trait could be more broadly Archaic, and predate Mesoamerica there by several thousand years, but again the petroglyphs cannot be dated absolutely, so the idea remains purely speculative at this point. The possibility that bar-and-dot counting can be linked to the Old World Paleolithic should not be ruled out, but it cannot be conclusively demonstrated at this time. Advances in petroglyphic dating techniques (Dorn and Whitley, 1983), could change that picture rapidly, however.

By the same token, I think it is premature to speculate on the meaning of subsidiary marks. Marshack has suggested that they are intercalary, and this is certainly a valid possibility. But until the ethno-archaeological context of the calendar sticks is better known, other possibilities cannot be ruled out by simple logic. They could be a parallel set of markings, or even a model set just as well. The growing sample of calendar sticks (see Chapter 24 by A. Marshack), certainly bodes well for an eventual solution to this question, however.

Finally, I think we must also assess what each piece of evidence teaches us about the broader features of the bar-and-dot counting tradition. Here the Winnebago calendar stick is especially important, in that it confirms several features previously noted on other artifactual evidence of counting. These features include:

(1) the counting of the lunar month in a variable set of unequal combinations, noted on the Lunar Count Stone at Presa de La Mula;

(2) the use of internal divisions of the lunar month which do not correspond to our Western system of quarters, also noted at Presa de La Mula;

(3) combination of dot-and-tally marking (binominal symbolism) within the same count, hypothesized for the Texas bison scapula counters; and

(4) the use of configurational spatial patterning to establish relevant divisions of the count, a feature noted on virtually all previous examples of New World counting.

This latter feature should warn us to take special caution. Each artifact identified so far is unique in place and time, and a wide variety of configurational patterns is represented. We must maintain open minds therefore in judging whether a given artifact represents intentional counting, because otherwise the relevant evidence will literally slip through our fingers. But we must also develop more stringent statements about the logical assumptions underlying any positive identification of numerical order. If not, the field will soon become cluttered with bracelets, bead necklaces, pottery decorations, and other assorted junk which probably have nothing to do with counting at all.

Notes

1 For an explanation of the faces A–D of the Winnebago calendar stick, see Chapter 24 by A. Marshack.

References

Aveni, A., Hartung, H. and Buckingham, B. (1978). The pecked cross symbol in Mesoamerica. *Science* **202**, 267–79.

Aveni, A., Hartung, H. and Kelley, J. C. (1982). Alta Vista (Chalchihuites), astronomical implications of a Mesoamerican ceremonial outpost at the Tropic of Cancer. *American Antiquity* **47**, (2), 316–35.

Brose, D. S. (1978). Late prehistory of the upper Great Lakes area. In *Handbook of North American Indians, 15*, pp. 569–82. Washington: Smithsonian Institution.

Callender, C. and Kochems, L. M. (1983). The North American berdache. *Current Anthropology* **24**, (2), 443–70.

Dorn, R. I. and Whitley, D. S. (1983). Cation ratio dating of petroglyphs from the western Great Basin, North

America. *Nature* **302**, 816–18.

Hester, T. R. (1980). *Digging Into South Texas Prehistory*. San Antonio: Corona Publications.

Lurie, N. O. (1978). The Winnebago. In *Handbook of North American Indians* **15**, 690–707. Washington: Smithsonian Institution.

Marshack, A. (1985). A lunar–solar year calendar stick from North America. *American Antiquity* **50**, (1), 27–51.

Murray, W. B. (1982). Calendrical petroglyphs of Northern México. In *Archaeoastronomy in the New World*, ed. A. Aveni, pp. 195–204. Cambridge University Press.

Murray, W. B. (1984). Numerical characteristics of three engraved bison scapulae from the Texas Gulf Coast. *Archaeoastronomy* **7**, 82–8.

Murray, W. B. (1985). Petroglyphic counts at Icamole, Nuevo León, México. *Current Anthropology* **26**, (2), 276–9.

Zeilik, M. and Elston, R. (1983) Wijiji at Chaco Canyon: a winter solstice sunrise and sunset station. *Archaeoastronomy* **6**, 66–73.

26

Navajo Indian star ceilings

Von Del Chamberlain *Hansen Planetarium, Salt Lake City*

26.1 Introduction

In another paper (Chamberlain, 1983), I have described Navajo constellations revealed in literature, on artifacts such as renderings of sandpaintings, masks, gourd rattles and on a rock art panel located in the Largo drainage in northern New Mexico. These indicate a consistent patterning which contrasts sharply in appearance with Navajo star ceilings, the subject of this chapter.

The star ceilings have been referred to as 'planetarium sites' (de Harport, 1951, 1953), a term still frequently used and one which is sometimes vaguely descriptively appropriate since some of the sites impart the feeling of being inside modern planetariums beneath simulated starfields. This is not the most appropriate term, however, since the root of the word 'planetarium' refers to planets, and also many of the pictographic panels are on small cliff overhangs rather than on large ceilings reminiscent of planetarium theatres. Furthermore, the word planetarium has the lineage of the history of classical science and the usage over time for reference to models and devices manufactured for the simulation of apparent celestial changes resulting from planetary motions. For these and other reasons I have mentioned previously (Chamberlain, 1974, p. 98, 1978, p. 82), the term seems inappropriate for the rock art panels decorating large and small ceilings along canyon walls in the American Southwest. At these sites, the pictographs occur on the undersurfaces of rock shelters and cliff overhangs, and the available information indicates that they do

represent stars. Therefore, the term 'star ceiling' is preferable and will be used in this paper.

Navajo star ceilings have been mentioned in literature, but they have not been studied in detail. Polly Schaafsma has referred to them in several publications (Schaafsma, 1963, 1966, 1972, 1980). Claude Britt (1975) has described examples in Canyon de Chelly, concluding that they are highly sacred places to the Navajo, that they date from about AD 1700 to 1864 and that at least some of them contain actual Navajo constellations. Campbell Grant (1978, pp. 219, 229–231) agreed with Britt that it is possible to distinguish star patterns on the ceilings. Finally, Stephen Jett (1984) has presented an analysis of how the ceilings might have been made.

My study (Chamberlain, 1983) of Navajo constellations convinces me that the ceilings do not reveal actual star patterns as Britt and Grant indicated. None of the ones I have studied clearly contain the patterns which are so consistently present in other Navajo art and which do depict known constellations. Indeed, some of them consist of rows of star pictographs. What are they? How and why were they made? The following is a descriptive summary together with interpretive comments, speculations and tentative conclusions.

26.2 Location and description of the star ceilings

I have examined 39 star ceilings and am aware of about a dozen additional ones. Two of these are now under water in the Navajo Reservoir. Thus there are at least four-dozen star ceilings which can presently be studied. Probably many others can

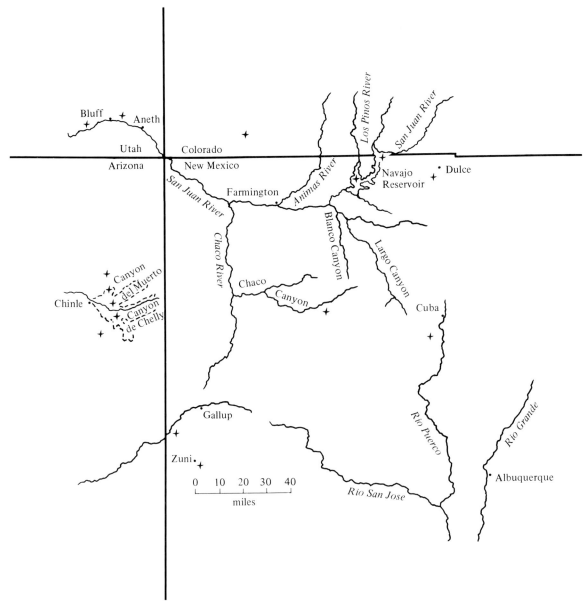

Figure 26.1 The southwest United States 'Four Corners' area showing approximate locations of Navajo Star Ceilings (+) which have been reported to or studied by the author at the time of this writing. Some of the sites contain multiple ceilings.

be found by a proper survey, which I hope to accomplish.

Figure 26.1 indicates the distribution of star ceilings which I have either studied or which have been clearly reported to me. Perhaps the oldest are the ones now located in the Navajo Reservoir on the San Juan River, within the area recognized as the

earliest homeland of the Navajo as a distinctive culture. Although Navajos lived there earlier, it has been suggested (Schaafsma, 1963, p. 5, 1966, pp. 1–11) that Navajo rock art, including the star ceilings, was made after 1698.

Schaafsma (1963, pp. 44, 49–51, 62) described two star ceilings, both located on rock overhangs which are now under the waters of the Navajo Reservoir (see Figure 26.1). She indicated that one of them was not completely recorded and that some star pictographs were scattered beyond the main groupings. Based upon other cases, I suspect there might have been additional star ceilings in the canyons of that region, but, alas, the opportunity to study any which were there is gone forever.

The pictographs on the Navajo Reservoir ceilings varied from simple, straight edged star crosses to petal-leaf forms. Schaafsma described them as mostly orange–red to red in color with a few which were dark gray, blue or black.

Not far to the east of the Navajo Reservoir, near the town of Dulce, New Mexico, there is a small star ceiling with red crosses, some of which are negative star figures (paint applied around a cross shape to leave a negative image).

About 60 miles south of Dulce, near the town of Cuba, New Mexico, is one of the larger star ceilings. Located in a good sized cave, some black crosses are found low where they easily could have been placed on the rock. Many of these resemble birds in flight (another example of a bird motif is found on a star ceiling in Canyon de Chelly), while others are the more common crosses. Among them are found tiny red figures of lance-carrying men on horseback. Higher up, some very high, are additional bird forms amid both black and red crosses and a few 'dragonfly' forms.

Brugge (1986, p. 111) recorded the existence of a small ceiling with several crude black crosses at a Navajo occupation site in Chaco Canyon. This is a borderline case, since the more traditional ones have not been found in this vicinity. I have examined it, and I think it belongs to the set of what I am referring to as star ceilings, but it is a very poor example. I have seen similar black sketched crosses immediately adjacent to unmistakable star ceilings.

Jane Young (private communication) has mentioned a star ceiling at Zuni which sounds very much like the others referred to in this chapter. Still another small one has been reported by Schaafsma (private communication) in Manuelito Canyon, southwest of Gallup, New Mexico.

Most known star ceilings are located in beautiful Canyon de Chelly, Canyon del Muerto and the smaller canyons nearby. Navajos have resided within and around these canyons for at least 300 years. Indeed, about 30 families still live in Canyon

de Chelly and Canyon del Muerto. Within these canyons star ceilings are quite abundant, varying from simple ones to a few which are spectacular. I have studied star ceilings in several of the canyons both within and outside of Canyon de Chelly National Monument and I continue to learn about new examples in the general vicinity.

A star ceiling overlooks a portion of the ruins in Step House at Mesa Verde in southern Colorado. The black star pictographs contrast with the other rock art at this ancient site. Polly Schaafsma has suggested (private communication) that the ceiling is of Navajo origin, noting that Navajos have been in the vicinity for a long time.

Two more star ceiling sites are indicated in Figure 26.1. Jett (1984, p. 38) indicated that Brugge observed a very simple one near Aneth, Utah.

Finally, I have studied a ceiling on the shore of the San Juan River southwest of Bluff, Utah. This one has three white star pictographs with pigment resembling that of the Anasazi rock art at the site. The lack of any apparent Navajo rock art here, combined with the similarity of pigment in the prevailing rock art, leads to the suspicion that this one might not be of Navajo origin. Pigment studies, requiring only minute samples, are among the significant things which could easily be done and which should help reveal the story of star ceilings.

When we are able to count up the recognizable star ceilings, it is likely we will find that those on small rock overhangs will prove to be most common (Figure 26.2). These vary from panels with a small number of star crosses located on little canopies to much more elaborate ones on larger overhangs. Some of these star panels are very high up, with no apparent way to get close to the surfaces upon which they are found.

Smaller in number, but considerably more impressive, are star panels found on large flat portions of deep rock shelters (Figure 26.3). Even here, we find considerable variation. Some such places have large numbers of stars, while others have only a few. One which I found has only a single star figure.

Some star ceilings are located in huge rock shelter alcoves, overlooking major Anasazi ruins. Within such places, the star panels may be concentrated on large flat surfaces, on narrow overhangs or both.

Of the 39 star ceilings I have studied to date,

(a)

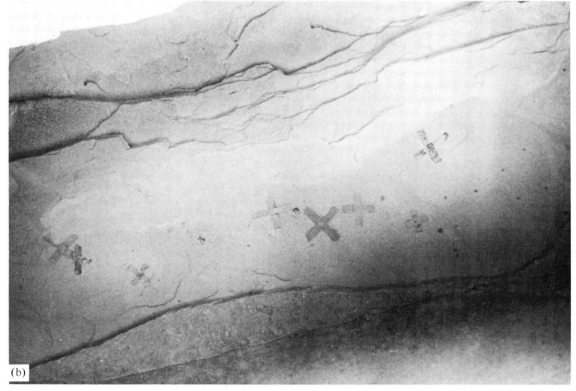

(b)

11 (28%) are found in deep 'planetarium-like' shelters, five (13%) are in large alcoves and 23 (59%) are on overhangs along canyon walls (four of the latter are quite large shelters, but not deep). Taken as a whole, the range is from very small overhangs, to shallow and deep rock shelters, to very large alcoves. Star panels are found low within easy reach to very high without apparent access.

The pictographs vary from thin to blocky crosses, a few dots, some negative crosses, to a small variety of unique figures. At least two ceilings contain crescents and an equal number have bird motifs. Many of the crosses have what appear to be tiny pits in the centers, and some have small, round, concentrated pigmented centers. One cross, for example, is composed of a white bar overlaying a black one, but having a black dot at the center.

The colors of the figures are most commonly black, frequently red or white, and occasionally gray. In a few cases we find individual star pictographs which are composed of more than one color. I have seen combinations of black and white and black and red.

26.3 Interpretation

Concerning how the star panels were made, my own observations, strengthened by statements from artist Harry Walters, Director of the Ned A. Hatathli Cultural Center at the Navajo Community College (private communication), lead me to agree with Campbell Grant (1978, pp. 218–19) and Stephen Jett (1984, p. 27) that most of the star pictographs were probably made by stamp techniques. It seems likely that the majority of them were made with tools consisting either of wood, leather or yucca leaves. These might have been dipped in paint and then applied to the ceilings. Walters suggested that some of the tools might have consisted of single blocks which were used to apply half of the cross, then turned 90 degrees to form the other component. Finally, some of the figures look as if they were painted by hand and others were almost certainly drawn using charcoal.

Figure 26.2 A star ceiling (a) on a small overhang located in the long horizontal shadow just above center of the photograph, (b). The crosses are black, gray and red and one of them consists of a black bar over a red one.

Jett (1984) convincingly showed that some of the pictographs located on high places were likely shot up on arrows, an idea supported by ethnographic evidence. The pits and concentrated centers of some star crosses confirm this hypothesis. However, some such pictographs are on low ceilings which could have been easily reached by other methods, and particular ceilings contain both images with and without these central features. This suggests the use of hand-held sticks, in addition to arrows. It also suggests that many of the ceilings were made over time, possibly by different people using somewhat different techniques. Additional experimental and ethnographic evidence, combined with detailed study of the pictographs, should lead to rather certain models of how the panels were made.

Most interesting is the question of *why* the star panels were made. Do the stars just overlook sacred or otherwise special places? Were they created as parts of ceremonies – perhaps ceremonies relating to the Navajo practice of 'star gazing' (Haile, 1947, pp. 38–40), or possibly to the Great Star Chant (Wheelwright, 1956)?

Star gazing is a method of divination employed by some Navajos for such things as finding lost or stolen objects, diagnosing illness, learning the source of misfortune or visualizing remote places. Crystals are used with the belief that they assist star gazers to receive communication from stars standing in the sky where they can see all things. It is conceivable that star ceilings might have some relationship to this practice.

The Great Star Chant involves a legendary trip into the sky where the hero received ritual knowledge from stars. In view of the bird motifs encountered on some star ceilings, it should be noted here that birds played a very important role in the Great Star story. Returned to earth, the ceremonies acquired in the star world were used to cure illnesses associated with evil (Wheelwright, 1956). Sometimes referred to as the Great Star Evil Chasing Ceremony, we find here the idea of protection from evil. Concerning this star ritual, we read in an important work on Navajo symbolism by Franc Johnson Newcomb, Stanley Fishler and Marry C. Wheelwright (1956, pp. 25–6):

> The Star Chant is one of the few Navajo ceremonies in which the sand paintings are made at night. This is necessary, for the stars made with colored sand must be lighted and

(a)

(b)

given spiritual power by the star shine of a particular star, which looks down through the opening in the roof of the medicine hogahn ... In the Star Ceremony there are not many sky maps such as we find in the Hail Chant and in the Shooting Chant, nor are there any real constellation groupings. Each medicine man chooses one or more that he believes he can influence, and then directs his ceremonial prayers and rites toward that particular one or group of stars.

In this same mode, the well-known scholar of Pueblo art, Barton Wright, suggested (private communication) that the stars on the ceilings might represent individual stars or groups of stars of special importance to those who made them. Each image might have symbolized some particular star which would be known only to the maker with no way to recover its identity. On the other hand, the pictographs might represent stars in general without any intent to symbolize particular ones. It seems possible, but remains unclear, that the star panels might relate to such stellar traditions.

Are the places watched over by painted stars shrines to be visited by singers (medicine men)? Were the stars placed there, like prayer-sticks, as prayer offerings to stellar deities? Britt (1975, pp. 96, 98) indicated that some Navajo prayer sticks have star crosses on them just like the crosses we find on star ceilings. Referring to prayer sticks as offerings to deities, Gladys A. Reichard (1974, p. 307) wrote: 'Offerings must be found by the deity to whom they are appropriate where he would look for them. Chanters give assistants explicit directions about the place and method of deposit.' It is hard to imagine more likely places to leave offerings to star deities than where we find the star panels in the Navajo canyon country.

Figure 26.4 illustrates the most likely concept known to the author of why the ceilings were made. The star panel is located on a small overhanging rock attached to a large crumbling rock standing over a pile of rocks which have fallen near a portion of a major ruin. The little group of star crosses is on a thin layer of rock which looks as though it might fall at any moment. With this example in mind, I will briefly relate three independent lines of evidence suggesting an interesting possible answer to the question of motivation for the star ceilings.

During the summer of 1985, a National Park Service official visited a star ceiling site with a Navajo medicine man who made the statement that the stars hold the sky together and the stars on rock ceilings hold the rocks together.

Then, toward the end of 1985, I had the pleasure of talking with a traditional Hopi practitioner, a medicine woman. Telling her something of my interests, including a description of star ceilings, I asked her if the Hopi had done anything like that and she answered, 'Not stars – they make prayer feathers and put them on ceilings so they won't fall on you.' Then she said that she thought the old people might have put the stars on ceilings to keep them from chipping and falling – 'to hold up the ceilings,' she said, 'like whoever holds up the universe.' We know that many Navajo traditions are remarkably like those of the Pueblos. Furthermore, we find Anasazi and Pueblo symbols on pottery and in kiva art which appear identical to the crosses on the star ceilings (Fewkes, 1973, pp. 155–6; Hibben, 1975, pp. 62–3).

Toward the end of Stephen Jett's article (Jett, 1984, p. 38), is information from National Park Service anthropologist David Brugge indicating that Navajo travelers camping in rock shelters had been known to draw charcoal crosses on ceilings to prevent their collapse. Communication with Brugge revealed that sometime during the first two decades of this century, probably in the teens, a Navajo informant's father had placed several charcoal star crosses in a cave located near Gallup, New Mexico. Brugge's information indicated that this was a small Blessingway prayer to keep the rock from falling.

These three sources, all seeming to be unrelated, suggest that the star panels might have been made with the belief that they would hold the rock together. This does not seem to be a sufficiently complete answer to explain all of the details of the painted star ceilings. There are so many appropriate places without star panels where rock threatens to fall, and some of the panels are on places which

Figure 26.3 A spectacular star ceiling located in Canyon del Muerto. A deep rock shelter (b) provides the theatre for the ceiling, only part of which is shown (a). White stars cover places blackened by smoke from fires which might have been lighted by Anasazi people; black stars are on lighter rock.

(a)

(b)

do not seem particularly endangered by crumbling rock. Yet, the rock fall idea does seem to me to be at least a partial answer of the correct kind. Perhaps it is the remnant of an older, more comprehensive concept.

According to one version of the Navajo emergence tradition, when the people came into the present glittering world, First Man raised the sky over the world like a giant hogan (Moon, 1970, pp. 167–83). Thus the sky is thought of as a shelter, protective in nature.

We have seen that most of the star pictographs on rock ceilings are black and red in color, with a fair number of white ones. These are the pigment colors most readily available for rock art, but the colors also may have significant symbolism. We have found color combinations of black and white, and black and red, forming individual star crosses. For the Navajo, black is a threatening color, or it can be used to protect from some threat (Reichard, 1974, p. 194). Red is the color representing danger, warning and threat, and it is also used for protection from danger (Reichard, 1974, pp. 197, 207). White represents perfect ceremonial control (Reichard, 1974, p. 206). The combination of red and black can indicate imminent danger (Reichard, 1974, p. 216).

All of this can be interpreted to suggest that the star panels might represent a protective concept. The ethnographic evidence cited above, together with the colors of the painted stars and the places where we find them, suggest symbolic protection from threat and danger. Danger or threat from what? From falling rock, of course, but from much more than this – from such things as the places themselves! Traditional Navajos stay away from the ruins unless they know the proper prayers which make them safe therein. The most extensive star panels overlook ruins. The rituals involved in putting the stars on ceilings might have been for the purpose of making people safe in these places.

Spending time in the deep, sheer rock-walled canyons makes one very much aware of the significance and danger of rock-fall. Every one of the star ceilings is, after all, located where rock *has* fallen and where additional rock-fall is not unlikely. The ceremonies, major or minor, which might have been associated with the ceilings could well have been directed to star deities in supplication to preserve and protect the canyons, its places of shelter and the people who dwell within them. If that is the answer, I not only applaud it, I add my fervent hope that the symbolic stars accomplish the end for which they were created.

Acknowledgments

Appreciation is given to Lloyd Jacklin who introduced the 'planetarium sites' to me in 1972. Thanks to Curtis and Polly Schaafsma for information, advice, directions to sites, suggestions and photographs. Mr and Mrs Harry Hadlock kindly escorted me to rock art sites in the Largo Canyon area of New Mexico and shared photographs of the Dulce star ceiling. Peggy Scott shared friendly dialogue, as did Barton Wright. Harry Walters accompanied me to a major star ceiling and shared ideas of importance. Joyce and Sidney Alpert treated me to a very special week in Canyon de Chelly. David Wilson, Chauncey Neboyia and others provided guide service. David Brugge willingly shared information and in other ways encouraged this work. Theadora Sockyma shared ideas through conversation. Jane Young, Roger Siglin, Russ Bodnar, Keith Franklin, Fred Blackburn and others informed me about star ceilings. At Hansen Planetarium, Nathan Gardner provided photographic services and Mark Brest van Kempen provided artwork for Figure 26.1. Finally, over the years, many National Park Service employees have made it possible for star ceilings to be studied in Canyon de Chelly. Without such assistance, the work reported here could not have been accomplished.

Figure 26.4 A small star ceiling located on crumbling rock in a large alcove sheltering a major Anasazi ruin. The star panel (a) is under the ledge at the center of the photograph, (b). In (a) note the faint star crosses on the large slab at the left as well as the concentrated group at the right.

References

Britt, C. (1975). Early Navaho astronomical pictographs in Canyon de Chelly, Northeastern Arizona, U.S.A. In *Archaeoastronomy in Pre-Columbian America*, ed. A. F. Aveni, pp. 89–107. Austin: University of Texas Press.

Brugge, D. M. (1986). *Tsegai: An Archaeological Ethnohistory of The Chaco Region*. Washington: US Department of the Interior National Park Service.

Chamberlain, V. D. (1974). American Indian interest in the sky as indicated in legend, rock art, ceremonial and modern art. *The Planetarian* 3, 89–106.

Chamberlain, V. D. (1978). Sky symbol rock art. In *American Indian Rock Art*. Papers Presented at the Fourth Annual American Rock Art Research Association Symposium, vol. 4, pp. 79–89.

Chamberlain, V. D. (1983). Navajo constellations in literature, art, artifact and a New Mexico rock art site. *Archaeoastronomy* 6 (1–4), 48–58.

Fewkes, J. W. (1973). *Designs on Prehistoric Hopi Pottery*. New York: Dover. Originally published in 1895 and 1898.

Grant, C. (1978). *Canyon de Chelly: Its People and Rock Art*. Tucson: The University of Arizona Press.

Haile, B. (1947). *Starlore Among the Navaho*. Santa Fe: Museum of Navajo Ceremonial Art.

de Harport, D. N. (1951). An archaeological survey of Canyon de Chelly: preliminary report of the field season of 1948, 1949, and 1950. *El Palacio* 58, (1), 35–48.

de Harport, D. N. (1953). An archaeological survey of Canyon de Chelly: preliminary report for the 1951 Season. *El Palacio* 60, (1), 20–5.

Hibben, F. C. (1975). *Kiva Art of the Anasazi at Pottery Mound*. Las Vegas, Nevada: KC Publications.

Jett, S. C. (1984). Making the 'stars' of Navajo 'planetaria'. *The Kiva* 50, 25–40.

Moon, S. (1970). *A Magic Dwells: A Poetic and Psychological Study of the Navaho Emergence Myth*. Middletown, Connecticut: Wesleyan University Press.

Newcomb, F. J., Fishler, S. and Wheelwright, M. C. (1956). *A Study of Navajo Symbolism*. Papers of the Peabody Museum of Archaeology and Ethnology, Harvard University, xxxii, no. 3. Reprinted 1974, Millwood, New York: Kraus Reprint Co.

Reichard, G. A. (1974). *Navajo Religion: A Study of Symbolism*. Princeton University Press. Reprinted 1983, Tucson: University of Arizona Press.

Schaafsma, P. (1963). *Rock Art in the Navajo Reservoir District*. Museum of New Mexico Papers in Anthropology, Number 7. Santa Fe: Museum of New Mexico Press.

Schaafsma, P. (1966). *Early Navaho Rock Paintings and Carvings*. Santa Fe: Museum of Navajo Ceremonial Art.

Schaafsma, P. (1972). *Rock Art in New Mexico*. Santa Fe: State Planning Office.

Schaafsma, P. (1980). *Indian Rock Art of the Southwest*. Santa Fe: School of American Research and University of New Mexico Press.

Wheelwright, M. C. (1956). *The Myth and Prayers of the Great Star Chant and The Myth of the Coyote Chant*, ed. D. P. McAllester. Navajo Religion Series, vol. IV. Santa Fe: Museum of Navajo Ceremonial Art.

27

A quipu calendar from Ica, Peru, with a comparison to the ceque calendar from Cuzco

R. Tom Zuidema *University of Illinois*

27.1 Introduction

Some years ago, Marcia Ascher informed me of the existence of a *quipu*, a Peruvian system of knotted cords used for counting, with a probable calendrical content (quipu AS 100, Ascher and Ascher, 1978, pp. 694–8). The quipu probably dates from Inca times (AD 1500 onwards) and was found in the coastal valley of Ica in southern Peru (see Figure 27.1).

According to the Spanish chroniclers who wrote about Andean culture in the 16th and 17th centuries, quipus played a central role in the administration of the empire that the Incas had conquered from their capital of Cuzco, also in southern Peru. Many quipus, including complete ones, have been preserved and published. This fact allowed R. and M. Ascher to analyze Inca methods of counting (Ascher and Ascher, 1981).[1] No quipu has been preserved together with the information that it stored. Nonetheless, in two particular cases we do possess an exact account of the information as it was read by an Inca specialist from a quipu and written down by a Spaniard. In the first case it was a tribute list of various quantities of goods given to the Spanish government (Murra, 1975, pp. 243–54). In the second case it concerned a calendar of interest to the political organization of the city of Cuzco and the valley in which it is located (Zuidema, 1977). The original quipu had recorded a list of 328 toponyms through which the Incas had mapped their interest in the territory of the valley. The toponyms corresponded to landscape features that, because of their role in the calendar, were worshipped as 'sacred', *huaca*. The 328 huacas were

Figure 27.1 Quipu from *Ica*. (Courtesy of Ascher and Ascher, 1981, p. 34.)

organized according to 42 directions, *ceque*, as viewed from the central temple of the Sun, *Coricancha*, in Cuzco towards the horizon. In one particular case, two directions were taken together as one ceque, so that, in fact, only 41 ceques were counted by the Incas. Some ceques as directions were used for making astronomical observations on the horizon of the sun and stars (Zuidema and Urton, 1976; Aveni, 1981; Zuidema, 1981, 1982a, 1982b). But independent of this function, each huaca and each ceque organizing a group of huacas were also given ritual attention in, respectively, their own day or period of days in the year. These days did not need to correspond to the time that a ceque was used for astronomical observation.

As I will make comparisons between the quipu calendar and the one of the ceque system, I want to elaborate somewhat on the latter. Two chroniclers mention the tradition that the population of the valley was divided into 12 localized groups, so that each group could take care of the rituals in one of the 12 months of the Inca calendar. With the help of this and other information we can reconstruct the calendrical organization integrating both the worship of the huacas and the ceques as well as of the monthly rituals (Zuidema, 1982a,b). Each of the 12 groups took care of a number of ceques, normally three, with their respective huacas. In the case of three groups, the number of associated ceques was higher so that all 41 ceques were accounted for. At the moment I am not concerned with the question of whether the calendrical order of the huacas on one ceque was read from the inside out or from the outside in. It is important to know the date when the calendrical role of one ceque was taken over by that of the next. The ceques were served in a clockwise order.

Figure 27.2 shows the ceque calendar with an explanation of the organization of the ceques, of the order in which they were mentioned in the Spanish chronicle (Cobo, 1956; Rowe, 1979) and of the dates related to each ceque as reconstructed by me. Especially the dates when each of the 12 groups entered service are of interest. Of these, ten were considered as being of Inca descent and were called *panaca*. They were divided into two groups of five and ranked politically in terms of their location and proximity to the king. The descending order is represented by the numbers 10–6 for the first group of five panacas and the number 5–1 for the second. The two remaining groups of the 12 were considered as being of pre-Inca origin in the valley and they were attached to the two groups of ceques where no panacas occurred. Thus the first pre-Inca group followed panacas 10–6 in the group of ceques III 3 a, b, c and the second one followed panacas 5–1 in the groups of ceques IVB 3 b, IVB 2 c, b, a.

For reasons of comparison with the quipu calendar, I want to elaborate upon the special role of the two pre-Inca groups in the calendar. The Incas managed, and were interested, to have in the case of these two groups the two roles of the ceques – one role of astronomical observation and the other of calendrical computation – combined into one. That is to say, the ceques in each group of ceques were used for an observation at the same time when the huacas on these ceques were also counted calendrically. Such a coincidence did not, and could not, occur in the case of the other ceques and groups of ceques. Thus the ceque system reflected in a particular way a type of organization that I could document also in other Andean examples (Zuidema, 1982c, 1986). It consisted of an organization of six ranked groups of which the last one played a particular role different from those of the other five groups.

Although we know that the ceque system was recorded from a quipu, we have not, as I said, preserved the latter. None of the chroniclers mention which particular panaca was related to which particular 'month'-like period in the ceque calendar. My reconstruction is based on evidence that needs to be reviewed critically against any new evidence that can be brought to bear upon the problem. Even though other quipus have been interpreted in the past as being of possible calendrical interest, in my opinion the one from Ica is the only one I know that includes a demonstrable calendrical organization and that can be confronted with our ethnohistorical information on Andean calendars. There are significant differences between the quipu as a calendar and the ceque system as such. But they both show important structural parallels that I will point out in this chapter. The quipu calendar allows us in particular to advance with the analysis of a problem that is crucial for an understanding of the ceque system as a calendar. Let me briefly describe the problem.

The ceque system, I argued, had 328 huacas and thus represented 328 days in the tropical year. In my reconstruction, these 328 days refer to an uninterrupted period in the year going from 9 June to 2 May inclusive. Thus, 37 days were not accounted for by the ceque calendar. This was a period when there were no ritual concerns with the

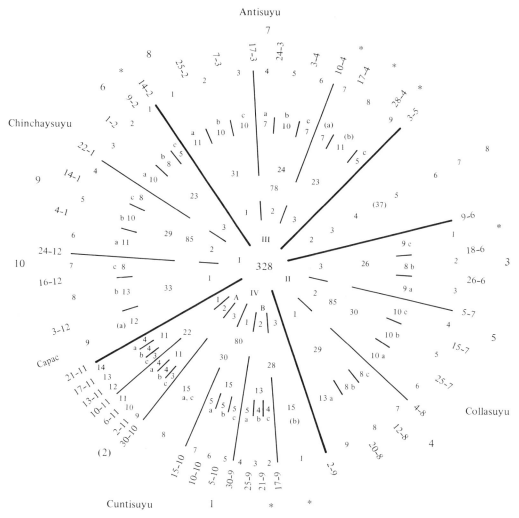

Figure 27.2 The ceque calendar. *Circles* (their order is enumerated in the 'empty' sector): (0) Total of *huacas* = 328. Each *huaca* represents one day and is given the space of one degree on the circle. The 37 days that are not accounted for by the *ceque* calendar are represented by the empty sector. (1) The four *suyus* ranked I, II, III, IV. (2) The groups of *ceques* in each *suyu* ranked 1, 2, 3. (3) The totals of *huacas* in each *suyu*. (4) The totals of *huacas* in each group of three *ceques*. (5) The *ceques* classified as *Collana* (a), *Payan* (b) or *Cayao* (c), and the numbers of *huacas* on each. The letters in brackets indicate that the *ceque* has a proper name replacing its generic name. Ceque I 1a is called *Capac* (royal). (6) The *ceques* in the sequence as given by Cobo. He enumerates the *suyus* in the sequence I, III, II, IV. (7) The calendar dates as reconstructed for the *ceque* calendar. (8) The numbers: the *panacas*, each in its association to one *ceque* and to one group of *ceques*; the location of (2) is hypothetical and not mentioned by Cobo. The asterisks: the autochthonous *ayllus* in the valley of Cuzco as named by Cobo; they occur only within the groups of *ceques* 3.

cultivated earth in terms of irrigation, plowing, planting or attending the plants in the field. We notice that the ceque calendar corresponded to a 'year'-like period – but 37 days shorter than the tropical year – and that the ten panacas and the two pre-Inca groups together were related to 12 'month'-like divisions of the year that all, or in part, also had to be shorter than the 12 solar months

of the tropical year.

In an article to be published (Zuidema, in press) I document the references in Inca culture that allude to another calendar that also existed in Inca culture of 12 solar months and I have argued for its origin in the pre-Inca cultures of *Huari* and *Tiahuanaco*. The organization of the ceque calendar seems to suggest that it was combined in Cuzco with the other calendar dividing the year into 12 solar months of 30 or 31 days each. I will argue that the quipu calendar from Ica represents another version of such a double calendar and that its analysis supports our understanding of the ceque calendar as part of a double calendar in Cuzco.

The ceque calendar is built up of 12 'month'-like periods, that I will refer to as 'months' and that vary considerably in length from each other, the shortest being of 22 days and the longest of 33 days. In order to compare the ceque calendar with the quipu calendar and to conclude how the ceque calendar might have been paired with a solar month calendar, I will suggest in the next section how the irregular ceque calendar might have been used by the Incas as a variation upon a more regular calendar, a variation probably used for describing exactly astronomically observed periods that are not dependent on, and cannot be described only in terms of, sequences of months, be these solar (30 or 31 days), synodic (29 or 30 days) or sidereal (27 or 28 days) months.

27.2 The ceque calendar: its possible background in a regular calendrical sequence

In this section I want to describe how the Incas lengthened and shortened certain 'months' in order to arrive at periods that were of astronomical interest to them. Perhaps the most important such period in Cuzco – and that I will compare with a similar period as reflected in the quipu from Ica – was the one of 107 days between the first (30 October) and second (13 February) passages of the sun through the zenith in the sky. The reconstruction of the regular calendar is guided by the observation that the ceque 'year' is divided, first, into two near-equal half-'years' of 165 and 163 days, respectively, (and not of 164 and 164 days, as we might expect) and, second, into four less equal 'seasons' of, respectively, 85, 80, 85, 78 days (and not of 82 days, as we also might expect).

As a first step towards the reconstruction, we can divide the series of 12 numbers of the 'months' into two groups: one of six larger numbers of days (30, 29, 30, 33, 29, 31), that have a total value of 182 days (equal to half a year) and an average value of $30\frac{1}{3}$ days, and one of six smaller numbers (26, 28, 22, 23, 24, 23), with a total value of 146 days and an average value of $24\frac{1}{3}$ days. The average value of all the 12 numbers is $27\frac{1}{3}$. This number, equal

to the length of a sidereal lunar month, suggests a sidereal lunar influence upon the ceque calendar. The six larger numbers alone, however, demonstrate that the Incas also were interested in periods derived from the solar month calendar.

We understand the meaning of the numbers 85, 80, 85 and 78 for the 'seasons' by regularizing the lengths of the 'months' to 30 days for the longer ones and to 25 days for the shorter ones, with the exception of the last short 'month', which remains one of 23 days. Ethnohistorical data suggest that the Incas were interested in representing their sequence of 12 months simultaneously as one of four seasons, each with three months, and as one of six double-months (Zuidema, 1966). By reversing in the ceque calendar in three instances the succession of a longer and a shorter 'month', we arrive at a regular alternation of long and short 'months' and to a series of six double-'months', of which five have 55 days each and the last 53 days (Figure 27.3, rows 1, 2, 4, 5). The double-'months' approximated each the length of a sidereal–lunar double-month of $2 \times 27\frac{1}{3} = 54\frac{2}{3}$ days. The excess of one-third of a day that occurred in the five 55-day periods was subtracted from the last ($54\frac{2}{3} - 5 \times \frac{1}{3} = 53$) in order to arrive at the exact length of a sidereal–lunar 'year' of 328 days.

The reason for the reversal in three instances of a longer and a shorter 'month' probably was not needed because of astronomical considerations, but of those of political hierarchy of the panacas and the two pre-Inca groups associated to the 'months'. Each reversal was done within the limits of a 'season' and thus it did not affect the length of the latter. In two 'seasons' two longer 'months' were combined with a short one, bringing the total of days to 85. In the case of the two other 'seasons', there were one longer 'month' and two shorter ones, leading to the totals of 80 and 78 days, respectively, (Figure 27.3, row 3).

Having arrived at this point of the reconstruction, I can present a preliminary hypothesis of how the ceque calendar was combined with a solar month calendar. Each double-'month' of 55 (or 53) days was paired with a solar double-month of 61 days and in one instance of 60 days ($5 \times 61 + 60 = 365$). The comparison with the quipu calendar will help us to argue how the pairing was done exactly (Figure 27.3, rows 5 and 6).

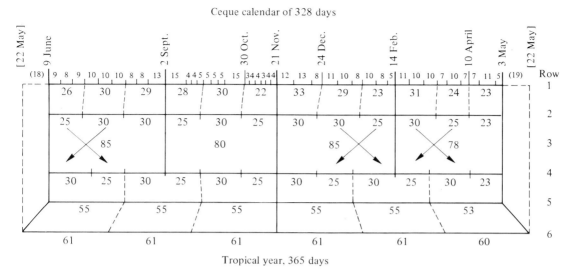

Ceque calendar of 328 days

Figure 27.3 Reconstruction of a regular sidereal lunar calendar from the ceque 'year' of 328 days. *Rows:* (1) ceque 'year' of 328 days with 12 'months'; (2) first step in reconstruction; (3) the four 'seasons' of the ceque 'year'; (4) second step in reconstruction; (5) the six sidereal, lunar double-'months'; (6) the six corresponding solar double-'months'.

27.3 A description of the quipu from Ica

The quipu, belonging to the Museum für Völkerkunde in Berlin and with a given provenience from Ica, is fully described by Ascher and Ascher (1978). From their description I will retain here only the information necessary for my analysis. To the main cord are attached 66 pendant cords. These are organized in groups, separated from each other by wider or smaller spaces on the main cord. Thus there are:

(1) one single pendant to which is attached another, subsidiary pendant;

(2) a group of six pendants, each also with its own subsidiary;

(3) six groups of pendants, each having, respectively, nine, ten, ten, nine, ten, eight pendant cords but without subsidiaries;

(4) a final group of three pendants. Unlike the other pendants and subsidiaries, these last three have no knots, and at the moment I do not need to take them into consideration.

The sum of the numbers represented by the knots on the six pendants of the first group is given also by the first single pendant and the sum of the six subsidiaries of the first group by the subsidiary of the single pendant. The totals of the next six groups of pendants are equal or very close to the numbers

of the six pendants in the first group. We can consider, therefore, the single pendant and the first group of six pendants as the first part of the quipu and the following six groups of pendants as its second part. I will analyze here only the totals, but I will give a list of the values on the individual pendants in the second part of the quipu and integrate them in Table 27.1. Figure 27.4 is a schematic representation of the quipu; the numbers in brackets above the main cord in the second part represent the numbers of pendants in each group, while the numbers below are the totals of each group.

The first part of the quipu

The calendrical character of the quipu is immediately given away by the number 365 on the subsidiary of the first, single pendant, this being the total of

Table 27.1.

	15	12	13	8	8	8	11	6	5	Total	86
0	8	4	5	4	3	2	4	3	6		39
5	4	2	2	2	1	4	3	2	6		31
	7	4	7	5	3	3	4	2	7		42
2	5	2	6	2	4	5	1	3	5		35
		3	9	1	3	5	8	5	2		36

Grand total 269

345

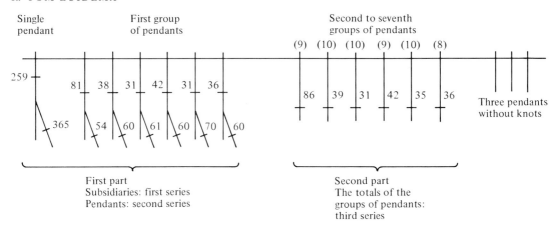

Figure 27.4 The quipu from *Ica*, with numbers represented by knots on pendant and subsidiary cords. The numbers in brackets above the main cord represent the numbers of pendant cords in each group of the second part.

the subsidiaries in the first group of pendants. The numbers as represented in the first part of the quipu are:

subsidiaries (first series): 54 60 61 60 70 60 (total: 365)
pendants (second series) 81 38 31 42 31 36 (total: 259).

While the numbers of the first series clearly seem to be of a calendrical nature, the calendrical context of the second series becomes acceptable only if we see both series as reflecting a type of social organization of 5 + 1 = 6 groups as mentioned before. In each of the two series, one number stands out from the rest as being most divergent; it might, therefore, refer to a divergent social role of that period in the calendar. By subtracting the divergent number from the total, we arrive in each series to a calendrically significant number; a number that becomes the basis for further analysis of the remaining five numbers. I should mention, moreover, that in early colonial times the town of Ica, located in the valley where the quipu had been found, was divided into six *ayllus* or local groups (Rossel Castro, 1964). The quipu may have been the calendrical reflection of a pre-Spanish organization of six ayllus, as the ceque system in Cuzco was one of 12 ayllus. The different local political and ecological conditions in the Ica valley could be the reason why people here also organized their calendar differently from the ceque calendar in Cuzco.

The first series of numbers The six numbers (54, 60, 61, 60, 70, 60) of the subsidiaries in the first part have two divergent numbers: 70, which is nine or ten more than four of the other five numbers, and 54, which is six or seven less. The operation of 365 − 54 = 311 does not lead, at this moment, to a number of calendrical interest. In the case of 365 − 70 = 295, however, this total is equal to that of ten synodic lunar months ($10 \times 29\frac{1}{2} = 295$). The quipu maker seems to have been interested in the observation of a period of ten synodic months and he measured its particular occurrence in a given year against the divisions of the solar year into double-months. The total number of days (241) by which the four double-months (60, 60, 61, 60 days) exceeded four synodic lunar double-months (236), was subtracted from another double-month, leading to the reduced size of 54 days (59 − 5 = 54) in the latter. The quipu maker could have chosen to represent all four solar double-months by periods of 61 days, instead of using only one such period (as 5 × 61 + 60 = 365). In that case, however, the period of 54 would have been reduced to 51 (244 − 236 = 8; 59 − 8 = 51), a number in which, apparently, he was not interested. The two other numerical series on the quipu explain the interest in the number 54. They explain also why the number 55 was not chosen (with a concomitant reduction of the number 61 to 60).

The second series of numbers The total (259) of the numbers on the six pendants (81, 38, 31, 42, 31, 36) also is given on the first single pendant. The number 81 stands out most clearly; subtracting it from 259 gives us 178 (259 − 81 = 178). The total of 178 is one unit more than that of six synodic lunar months ($6 \times 29\frac{1}{2} = 177$). Although the minor difference between 178 and 177 is not explained, the operation of subtracting the most divergent number from the total again leads to a promising calendrical result.

The numerical values of the three middle numbers (31, 42, 31), of the five that add up to 178, are all interesting calendrically. But, because the outer two of these three numbers are equal to each other, and because they are arranged symmetrically about the central one, we are led to consider further the total of the three middle numbers in the context of that of all five; thus:

38 + (31 + (42) + 31) + 36
total in parentheses: 104
38 + 104 + 36
total: 178.

The number 104 is of interest for comparison with a similar number in the ceque calendar. The most important period measured here is from first to second zenith passage of the sun (30 October–13 February), a period of 107 days. The length of the period is, of course, dependent on the latitude of the location where it is observed. The latitude of the Ica valley ($14\frac{1}{2}° S$) is about 1° further south than that of Cuzco ($13\frac{1}{2}° S$) and its period from first to second zenith passage is therefore 104 days. This number 104 may have served in Ica the same calendrical purpose as the 107 day period in Cuzco.

The third series of numbers, represented by the second part of the quipu

The second part of the quipu consists of six groups of pendants, ranging in number from eight to ten pendants for each group, with the numbers as in Table 27.1. The pendants have no subsidiaries. In this chapter I consider only the totals of the six groups.

The calendrical use of this third series of numbers is closely related to that of the second series, because only small additions are involved in three of the six numbers:

second series: 81 38 31 42 31 36
third series:

86(81 + 5) 39(38 + 1) 31 42 35(31 + 4) 36

If we begin the analysis by subtracting 86 from 269 (269 − 86 = 183), we arrive at the period of half a year (183 days), which is also equal to 3×61 days. The other five numbers, the three middle numbers and their totals are:

39 + (31 + (42) + 35) + 36
total in parentheses: 108
39 + 108 + 36
total: 183.

The period of 108 days approximates closely the 107-day interval in Cuzco from first to second zenith passage of the sun.

The numbers 54, 81 and 42 in the three series

In the three series of numbers of the quipu, the numbers 365 and 295 (365 − 70) of the first series, the number 178 (259 − 81) of the second series, and the number 183 (269 − 86) of the third series stand out most for their calendrical interest. If now we look at the individual numbers in the three series, we can consider those of 60, 61, 60, 60 in the first series, those of 31 and 31 in the second series, and that of 31 in the third series as being equal to either a solar double-month or a single month.

The numbers 38 and 36 in the second series and of 39, 35 and 36 in the third series do not seem to be of calendrical value by themselves. The numbers 81 and 42 in the second series and the number 42 in the third series might, however, be considered in a sidereal–lunar context. The number 81 is one unit less than 82 (three sidereal months), while 42 is one unit more than 41 (one-and-a-half sidereal months). We remember that in the ceque calendar the four 'seasons' were not represented by periods of 82 days, but by those of, respectively, 85, 80, 85, 78 days; this is because of the influence of the 55-day periods, each close to a sidereal–lunar double-month. We also remember that the Incas recognized 42 directions in the ceque system, consisting of 14 groups of three directions each, but that they reduced this total to that of 41 ceques in order to integrate it in the context of the 328 (= 8 × 41)-day calendar, a period equal to 12 sidereal–lunar months. We may suggest, therefore, that in the quipu calendar not only were the numbers 81 and 42 of sidereal–lunar interest, but so was the number 54 of the first series. One reason, suggested by the ceque calendar, why these numbers all deviate one unit from, respectively, the numbers 55, 41 and 82 may have been that they are divisible by three. Moreover, 81 (3×27) is paired with 54 (2×27) as a pendant and its subsidiary.[2]

The possible location of the three calendrical series in the year

The first part of the quipu calendar represented six pendant numbers (81, 38, 31, 42, 31, 36), the calendrical references of which were detected primarily because of the corresponding subsidiary numbers (54, 60, 61, 60, 70, 60) with calendrical significance. These last numbers (which belong to the first series) can only be considered in sequence, for together they refer to one full year of 365 days. One might think that such a requirement of a continuous sequence would not have been necessary for the numbers of the second and third series. However, because a calendrical meaning could be suggested for the numbers 178 and 183, the numbers 259 (178 + 81) and 269 (183 + 86) most likely also represented continuous periods in the year.

I suggested earlier that the quipu probably was the calendrical expression of a political organization of six groups. The combination of the three series within the year would help to codify the ritual obligations that all six groups had according to each of the three series. Moreover, the third series was a variation upon the second, and the first series was tacked onto the second. Therefore, the quipu makers seem to have considered the three series in relation to each other. Such a relationship was probably expressed by giving the second and third series overlapping locations in the year, represented by the first series.

One hypothetical argument for locating the three series in the year begins by taking the number 104 in the second series (31 + 42 + 31) as an expression of the period from first to second zenith passage of the sun at the latitude of Ica. With this assumption in mind, I have placed in circles 1, 2 and 3 of Figure 27.5 the first, the second and the third series of the quipu calendar in the tropical year. For comparative reasons I have placed the ceque calendar in circle 4.[3]

27.4 The quipu calendar and the ceque calendar compared

Having placed the two calendars in the tropical year, we become aware of certain structural similarities between them. One similarity is the way in which both calendars considered the interval of first to second zenith passage within the context of certain other calendrical periods; periods that, like the zenith passage interval, were distributed symmetrically around the date of the December solstice.

The quipu calendar measured the 104-day interval by employing a central 42-day period, equal to one-and-a-half sidereal–lunar months, in between two periods, each equal to a solar month. But then it bracketed the 104-day interval in between two periods, respectively, of 38 and 36 days, that themselves only served to arrive at the period of 178 days, equal to six synodic months. The third series considered this 178-day period in the context of a half-year (183 days) going from equinox to equinox.

The ceque calendar placed in the middle of its 107-day interval two 'months' that together were one day longer than a solar double-month (33 + 29 = 62). Here it was the almost equal shorter 'months' of 22 and 23 days, themselves of no calendrical interest, that bracketed the solar double-month.

The comparison of the quipu and the ceque calendars can lead us now to understand the significance of the 81-day period in the second series or the corresponding 86-day period in the third series. The argument departs from the observation that the quipu in its second part paid special attention to the 183-day period, breaking up its five constituent periods (39, 31, 42, 31, 35, 36) into many very short ones. The ceque calendar did something similar, although only during the periods from 17 September to 15 October and from 30 October to 21 November. Modern ethnographic parallels indicate a similar heightened ritual attention to this half-year (roughly going from September equinox to March equinox) of most intense agricultural activities. But the Incas introduced this period in Cuzco somewhat earlier, 2 September, and ended it somewhat later, 10 April. They marked these two dates – that are equidistant from the December solstice – by observing sunset in between two pillars set on the western horizon of Cuzco (Aveni, 1981; Zuidema 1981, 1982a).[4] The fact that the ritually important period in Cuzco (220 days) was 37 days longer than the comparable period (183 days) of the quipu calendar, and the additional fact that the ceque calendar had a stronger lunar character than the quipu calendar, may have been contributing fac-

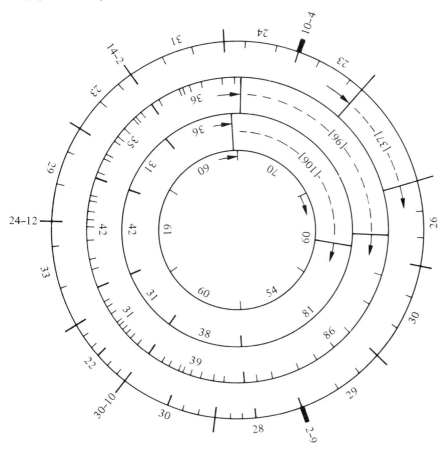

Figure 27.5 The quipu calendar and the ceque calendar compared. The circles, read from inside out, represent: (1) the first series of the quipu (subsidiary cords of first part); (2) the second series of the quipu (pendant cords of first part); (3) the third series of the quipu (groups of pendants of second part); (4) the calendar as represented by the ceque system. The calendars are located in the year recognizing in each the interval from first to second zenith passage of the sun. The dates for these passages in Cuzco are 30 October and 13 February. The rectangles on the outer circle represent the pillars on the Cuzco horizon where the sun set on 2 September and 10 April.

tors to the reason why the ceque calendar defined its 220-day sequence by four sidereal–lunar double-'months', and the quipu calendar its 183-day period of half a year by five periods ranging from 31 to 42 days. As a consequence of this argument we can suggest that the first 81-day period, in series 1, or the 86-day period, in series 2, of the quipu calendar had a function similar to that of the first 85-day 'season' (9 June–1 September) in the ceque calendar. In both calendars these periods played an introductory role. In Cuzco, this 'season' was dedicated primarily to preparing the ground by way of a first irrigation and a first plowing, activities with strong male connotations. After 2 September came planting, an activity with a strong female connotation.

Now, the different questions and problems discussed above may help us to propose more completely how the ceque calendar could have been related to a solar year calendar, just as the series 2 and 3 of the quipu calendar were related to series 1, also representing a solar year calendar. I will present my hypothesis on the basis of the regularized ceque calendar represented in Figure 27.3. My argument can be formulated in a simpler way and it remains valid for the ceque calendar itself.

We observed that the periods of 55 days were not broken down into two periods of 28 and 27 days each, as approximations to sidereal–lunar months, but into one period of 30 days (as an approximation, to a solar month), and a rest period of 25 days with no calendrical value of its own.

Although we can understand the periods of 328 days ($5 \times 55 + 53$) and of 220 days (4×55) as being of a sidereal–lunar interest, at no time in the year did the ceque calendar lose its connection with solar month periods. It is for this reason that I suggested in Figure 27.3, row 5, that the Incas calculated the sidereal–lunar periods (55 days) by subtracting six days from the corresponding solar month periods ($61 - 6 = 55$), or in the last case, by subtracting seven days from a 60-day period ($60 - 7 = 53$) day period. Our first chronicler, Polo de Ondegardo, who mentioned the solar month calendar, represented the seasons as consisting of one month before a solstice or equinox and of two months after (Zuidema, in press). He also mentioned, in a rather puzzling way, how the Incas came to 'months of 30 days or 31 days or 32 days according to the waning moon'. Thus, he may have been aware of an Inca method of correlating synodic months with solar months similar to that represented by series 1 in the quipu calendar. In Cuzco, the temple of the Sun, *Coricancha*, was aligned to a sunrise occurring one solar month before the June solstice (Zuidema, 1982a). Moreover, the two half-'years' of the ceque calendar were located in the year around the date of 21 November, one solar month before the December solstice date. Therefore, the solar month calendar in Cuzco probably began 18 days before the ceque calendar (22 May) and ended 19 days after.

The ceque calendar was used primarily for carrying out rituals from the time of planting to that of harvest. It was not used when agricultural interests were not dominant; and 37 days after it ended, it began again by establishing a new correlation between the solar, lunar and stellar movements. Methods similar to the one proposed here were known, for instance, to the Achenese of Sumatra (Snouck Hurgronje, 1893) and the Borana of Eastern Africa (Legesse, 1973). But, while in these last two cases a correlation was established between two movable lunar calendars, a synodic and a sidereal one, in the Inca case a solar year calendar was combined with another calendar, also with a fixed position in the year, but using periods of interest to a sidereal–lunar computation.

As a final remark, I want to observe that both the ceque and the quipu calendars show a remarkable freedom in combining periods of a solar and a sidereal–lunar character, placing these, by way of additional periods, in their desired locations in the year. Both calendars integrated certain important astronomical observations, other than those of the moon, into their respective systems of 'monthly' periods. It was for this reason that the ceque calendar allowed some of these periods (such as, for instance, the sidereal–lunar double-'months' of 55 days) to become a couple of days longer or shorter than their real value.

Notes

1 For the illustration of a quipu, with its main cord, pendant cords, subsidiary cords, color distinctions of cords and knots, as I will discuss these categories for the quipu of Ica, see Ascher and Ascher (1981, p. 34). (In the collection of the Museo Nacional de Anthropología y Arqueología, Lima, Peru.) See Figure 27.1.

2 It may be of interest to observe here that the total (108) of the three central numbers ($31 + 42 + 35$) in the third series is also equal to 2×54.

3 The place of the first series is perhaps less sure. My arguments in this respect are the following: (1) the period of 54 days should be placed within that of 81 days of the second series; (2) 61 is bracketed within the two equal numbers 60 just, as 42 is bracketed within the two equal numbers 31 in the second series; (3) the period of 181 ($60 + 61 + 60$) should coincide with those of 178 and 183 in the second and third series, respectively; (4) the period of 70 days should include the 11 extra days that are the difference between the solar and synodic lunar years ($365 - 354 = 11$) at the same place in the year where the ceque calendar places the 37 extra days.

4 Aveni and I argued originally that the two pillars indicated the antizenith sunsets on 25 April and 18 August, half a year apart from the zenith passages of the sun. Further research in the ethnohistorical sources force me to argue that the sunsets in between the pillars occurred at 10 April and at 2 September, both dates 15 days closer to the December solstice date.

References

Ascher, M. and Ascher, R. (1978). *Code of the Quipu Databook*. Ann Arbor: The University of Michigan Press.

Ascher, M. and Ascher, R. (1981). *Code of the Quipu. A Study in Media, Mathematics and Culture*. Ann Arbor: The University of Michigan Press.

Aveni, A. F. (1981) Horizon astronomy in Incaic Cuzco. In *Archaeoastronomy in the Americas*, ed. R. A. Williamson, pp. 305–18. Los Altos: A Ballena Press/Center for Archaeoastronomy Cooperative Publication.

Cobo, B. (1956). *Historia del Nuevo Mundo* (1653). Madrid: Biblioteca de Autores Españoles.

Legesse, A. (1973). *Three Approaches to the Study of African Society*. New York: The Free Press.

Murra, J. V. (1975). Las etno-categorías de un khipu estatal (1973). In *Formaciones Económicas y Políticas del Mundo Andino*. Lima: Instituto de Estudios Peruanos.

Rossel Castro, A. (1964). *Historia Regional de Ica*. Epoca colonial, vol. I. Lima: University of S. Marcos.

Rowe, J. H. (1979). An account of the shrines of ancient Cuzco. *Ñawpa Pacha* **17**, 1–80.

Snouck Hurgronje, C. (1893). *De Atjelers*, 2 vols. Batavia: Landsdrukkerij. Reprinted 1894 by E. J. Brill, Leiden. Translated as *The Achenese*, 2 vols, 1906. Leiden: E. J. Brill.

Zuidema, R. T. (1966). El calendario Inca. In *Actas del Congreso Internacional de Americanistas* vol. **II**, pp. 25–30. Seville.

Zuidema, R. T. (1977). The Inca calendar. In *Native American Astronomy*, ed. A. F. Aveni, pp. 219–59. Austin: University of Texas Press.

Zuidema, R. T. (1981). Inca observations of the solar and lunar passages through zenith and anti-zenith at Cuzco. In *Archaeoastronomy in the Americas*, ed. R. A. Williamson, pp. 319–42. Los Altos: A Ballena Press/Center for Archaeoastronomy Cooperative Publication.

Zuidema, R. T. (1982a). Catachillay: the role of the Pleiades and of the Southern Cross and Alpha and Beta Centauri in the calendar of the Incas. In *Ethnoastronomy and Archaeoastronomy in the American Tropics*, eds. A. F. Aveni and G. Urton, pp. 203–29. Annals of the New York Academy of Sciences 385.

Zuidema, R. T. (1982b). The sidereal lunar calendar of the Incas. In *Archaeoastronomy in the New World*, ed. A. F. Aveni, pp. 59–107. Cambridge University Press.

Zuidema, R. T. (1982c). Myth and history in Ancient Peru. In *The Logic of Culture: Advances in Structural Theory and Methods*, ed. I. Rossi, pp. 150–75. South Hadley, Ma: J. F. Bergin Publishers Inc.

Zuidema, R. T. (1986). *La Culture Inca au Cuzco*. Six lectures given at the Collège de France 1982–3. Paris: Collège de France/Presses Universitaires de France.

Zuidema, R. T. (in press). *Llama Sacrifices and Computation: Roots of the Inca Calendar in Huari–Tihuanaco Culture*. Acts of the First Congress of Ethnoastronomy, Washington DC, September, 1983.

Zuidema, R. T. and Urton, G. (1976). La Constelación de la Llama en los Andes Peruanos. *Allpanchis Phuturinqa* 9, 59–119.

III
Archaeoastronomy: an interdiscipline in practice

28

Lunar astronomies of the western Pueblos

Stephen C. McCluskey   *West Virginia University*

The Moon, who is our mother,
Yonder in the west waxed large;
And when standing fully grown against
the eastern sky,
She made her days. . . .
                    Sayataca's *Night Chant* (Bunzel, 1932a)

In the course of the Oxford II Conference the problem of intercalation and the lunar calendar appeared in various contexts. This paper approaches intercalation from a somewhat different perspective, by using historical data to determine the actual practice of intercalation among the Hopi and Zuni Indians of the Southwestern United States.

Stanisław Iwanisewski has suggested that archaeoastronomy is the study of astronomy in society (see Chapter 3). Traditionally historians of science have divided their work into internal history, which concerns the internal logic and development of a science, and external history, which studies a science within its broader intellectual, cultural and societal contexts.

The present analysis is concerned chiefly with internal questions of observational methods and observational precision, what we might call internal archaeoastronomy. Yet such internal concerns clearly impinge upon a wide range of external elements in Puebloan culture.

The motions of the Moon are complex; we therefore require a large amount of data if we are to understand lunar calendars quantitatively. The lack of adequate data for lunar ceremonials has heretofore frustrated efforts to understand the lunar

aspects of puebloan astronomies. In The summer of 1985 while in the Southwest I learned from a number of colleagues that a collection of dates for the Zuni ceremonial of Shalako had been assembled by Dr Luke Lyon, a retired chemist living in Los Alamos, New Mexico. Dr Lyon has graciously provided me with his data, which include an uninterrupted sequence from 1914 to the present and scattered dates extending back to 1879. Analyses of Lyon's data have led to a number of new insights into puebloan lunar astronomy, insights that would not have arisen without his data. I would like to give my heartfelt thanks to Dr Lyon both for his data and his suggestions.[1]

Lunar calendars are traditionally regulated by observations of the New Moon, and it is with the New Moon that I would begin my discussion. Among the Hopi, after the close of the Winter Solstice ceremonial of Soyal there are two lunar ceremonies, Powamu and Palulukonti, both of which are concerned with the growth of crops. Powamu begins at the Full Moon after the solstice with a ritual planting of beans in the kivas and culminates with a dance that occurs 18 days after a ritual observation of the New Moon (Voth, 1901; Stephen, 1936). Astronomically, Powamu is regulated by an observation of the Sun before the solstice, an intermediate period in which a fixed sequence of ceremonials is performed, and observations of the subsequent Full and New Moons; it thus provides an ideal example of a luni–solar festival.

Each year a given New Moon occurs 10.88 days earlier; after three years the Moon is running more than a month faster than the Sun, so an extra inter-

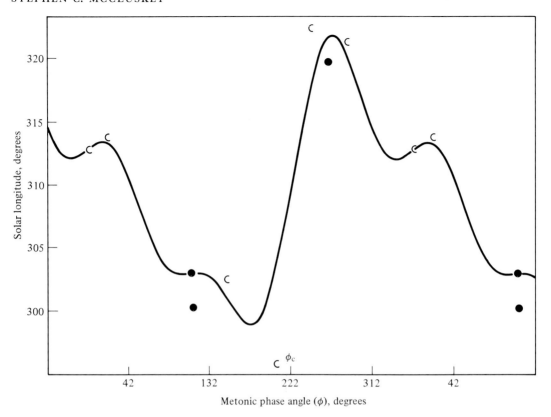

Figure 28.1 2.71 year intercalation periodicity; New Moon observation for Powamu Festival, Walpi Pueblo.

———— Fourier smoothed values;
○ Full Moon;
) Waning Crescent;
● New Moon;
(Waxing Crescent.

calary month must be inserted to restore the proper relation of the Moon and the Sun. To analyze luni–solar festivals I have developed a method based on the recurring 2.71 year periodicity of intercalations, which enables us to represent the recurring pattern of intercalations of a luni–solar festival. If we plot the date of the lunar observation preceding each year's festival (or more precisely, the longitude of the Sun on that date) against a phase angle φ, which represents the place of the year in the 2.71 year periodic cycle, we find those years in which any given lunar phase occurs on about the same date appearing side by side on the graph.[2]

Figure 28.1 reflects the pattern of New Moons by which the Powamu festival is determined. Notice the characteristic zigzag of the intercalation function. The critical phase angle (φ_c), which is indicated by the sharp break at the center of the graph, is a distinguishing characteristic to identify any

specific luni–solar calendar. A similar break will occur at the same critical phase angle for any festival that follows the same calendar.

The symbols indicate the phase of the Moon on the evening of the observation; note that for Powamu the Moon on the evening of observation was either a New Moon or an early crescent. It appears that in some cases the Powamu chief decided it was time for the festival before he actually saw the Moon. If we examine the data for consistency, we find that the phase of the Moon varies by only ±19°, corresponding to a change in the date of only ±1.6 days.

This indicates the precision of individual observations of the New Moon, but if we are to represent fully the pattern of intercalation, we need data from a full 19 years of the Metonic cycle. If we wish to find how consistently this pattern of intercalation was followed, we should have data extending over

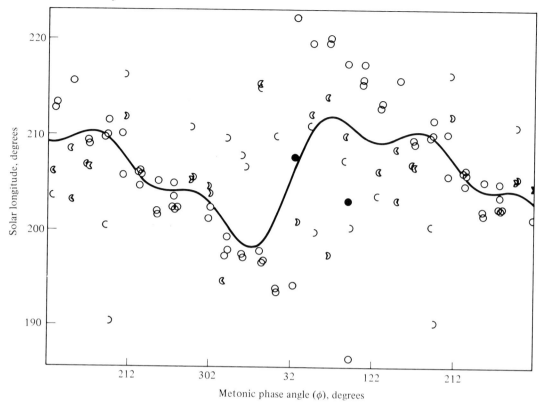

Figure 28.2 2.71 year intercalation periodicity; Tenth Full Moon prayer stick planting for Shalako, Zuni Pueblo. Symbols as in Figure 28.1.

several Metonic cycles. This is where the Shalako data come in.

At Hopi, the lunar ceremonials end in March with Palulukonti; at Zuni the Shalako Ceremony extends throughout the entire year. Shortly after the Winter Solstice Ceremony (*yatoika itiwonna kwin te'chikia*; the Sun's arrival at the Middle Place) the participants in the following year's Shalako are appointed. At the Full Moon following the Winter Solstice, and at every subsequent Full Moon, these men deposit prayer sticks at one of a series of shrines surrounding the village of Zuni.[3] If the Shalako impersonators are not appointed promptly, the first planting of prayer sticks may take place some days after the Full Moon. On the day after the prayer sticks for the tenth Full Moon are planted, the first of 49 knots is untied from two cords that had been prepared by the Pekwin to count the days of the Shalako ceremony. On the night of the 48th day,

when only one knot remains in the cords, the Shalako dance itself is performed; it is the record of these dances that provides the basis for analysis.[4]

This analysis (Figure 28.2) assumes that the observation takes place exactly 48 days before the festival. In fact, there are recorded instances when the dance was postponed for ten days. Lyon points out that the delay is announced shortly before the scheduled dance, specifically on the date that the 40th knot would have been untied (L. L. Lyon, personal communication, 19 October, 1985). The public reason given by the Zuni for postponing the ceremony was that the houses in which the dance was to take place were not ready, but there may be astronomical considerations as well.

Ten-day delays of this sort can be identified in about nine of the 76 years of data. It may be significant that many of these postponements occur when the tenth Full Moon, and consequently the Shalako

357

ceremony, would occur early in relation to the solar calendar. This suggests that in at least some cases the ten-day delay reflects an adjustment so that Shalako will mesh better with the subsequent solstice ceremony, particularly given the documented concern that the Pekwin schedule the subsequent Winter Solstice Ceremony at the Full Moon (Bunzel, 1932b).

How consistent is the Zuni pattern of intercalation? The scatter of data points around the critical phase angle indicates that, in the long run, intercalation was not as consistent as the idealized pattern suggests. Ten of the 77 dates show that Shalako occurred in the 'wrong' Moon; the band of phase angles encompassing half these deviations from the ideal is $\pm 36°$. That is, significant confusion about intercalation is likely in one year in five, while some confusion occurs over an even larger region.

In the third of the cycle where there is no confusion over intercalation ($195° \leqslant \varphi \leqslant 315°$) the inferred observations are almost all of Full Moons. We can take the scatter as representing the precision with which the Zuni could routinely observe the Full Moon, or more precisely the degree to which their determinations of the Full Moon agree with ours. For this data, the phase of the Moon varies by $\pm 40°$ (which corresponds to a variation of ± 3.3 days). If we consider that all the ten-day delays were intentional and omit them, the precision increases to $\pm 30°$, corresponding to ± 2.5 days, which we may take as an upper limit of precision for the Zuni determination of the day of Full Moon.

If we consider historical changes of precision, separate analyses of three groups of 25 festival dates (1879–1934, 1935–59, 1960–84) show that the more recent festivals follow the phases of the Moon more closely. Increasing precision represents something of a paradox, since the more recent festivals are all scheduled on weekends, which should tend to make them less precise astronomically. This suggests that as a compensating factor the Zuni came to rely upon the computations of the Full Moon that appear in calendars, newspapers and almanacs to aid their observations.[5]

The precision of Zuni observations of the Full Moon for Shalako is less than the corresponding Hopi observations of the New Moon for Powamu. In part this stems from the greater difficulty in observing the Full Moon; in part, however, it reflects a periodic change in Zuni observations of

the Full Moon that reflects the difference between Zuni observing techniques, with their implicit definition of the Full Moon, and our own.

28.1 The Full Moon

Using an analysis similar to that for the intercalary cycle, we can assign each year an arbitrary phase angle relative to the 18.61 year period of the lunar nodes to detect any periodic variation in the phase of the Moon (as measured by its elongation from the Sun) at the beginning of the Shalako festival. If we examine the Shalako Full Moon observations, we find significant indications of a periodic variation of the elongation of the Moon from the Sun.

For the data base as a whole (Figure 28.3) we find a distinct periodic signal for which there is a probability of only 0.037 that it occurred by chance. If we consider the earliest uninterrupted run of 56 years (Figure 28.4, corresponding to Hawkins' 56 year Stonehenge cycle), we find the same value of the critical phase angle, but the formal significance drops to the 0.127 level. Partitioning all the data into various subsets we find that only the most recent subsets exceed the 0.05 criterion for statistical significance, but that the critical phase angles for all subsets agree within 25° (Table 28.1).[6]

A cautionary note is in order here. Although the dates chosen for Shalako are modulated with an 18.61 year periodicity, there is no indication that the Zuni were aware of that periodicity. The Zuni say that the date on which the knots are tied in the string for Shalako is determined by the Full Moon; the 18.61 year period of the lunar nodes can appear in Zuni observations of the Full Moon without any awareness of this variation.

The appearance of this 18.61 year periodicity in the record raises questions about Zuni lunar obser-

Table 28.1 *Zuni Shalako Ceremony 18.61 year period analyses*

	φ_c (degrees)	p
Overall	78	0.037*
1879–1934	74	0.578
1935–59	72	0.141
1960–84	96	0.036*
1914–69	76	0.127
1914–41	80	0.283
1942–69	78	0.041*

* Statistical significance exceeds 0.05.

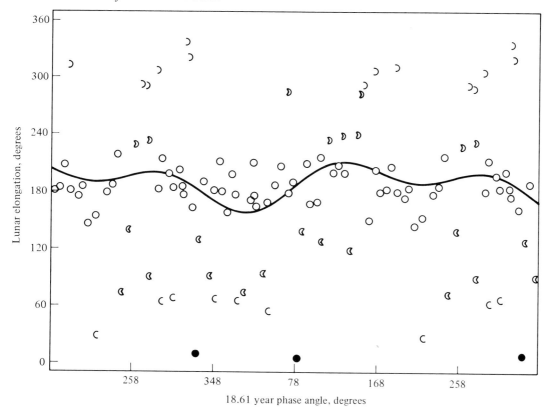

Figure 28.3 18.61 year periodicity; Tenth Full Moon prayer stick planting for Shalako, Zuni Pueblo (all available data, 1879–1984). Symbols as in Figure 28.1.

vations. Unlike observations of the Sun, there is no ethnographic description of observations for the Full Moon at Zuni. The only hint is in Sayataca's *Night Chant*, quoted at the beginning of this paper, which describes the Moon as 'standing fully grown against the eastern sky'. Since we know little about Zuni observational techniques, what can we infer about these techniques from such ethnographic hints and the quantitative data?

Since the declination, and consequently the azimuth, of a given Full Moon varies with this period, one might suspect that the Zuni were watching the azimuth of the Full Moon, or its shadow, against some fixed reference point. The parallels to the claims for lunar sites at Fajada Butte and in Great Britain seem obvious, and it is because of these claims that I routinely test festival dates for this period. In this case, however, the variation in dates of the Full Moon is so great that it rules out any

such observation in respect to a fixed azimuth.

Rather than looking simply at lunar azimuths, these results require that we take a careful look at how one observes the phases of the Moon. The first apperance of the lunar crescent is a distinct and readily observed phenomenon that is taken in most cultures as marking the beginning of the lunar month. The Full Moon is ambiguous and difficult to observe and does not play a major role in traditional calendars. If we intend to speak precisely, the very definition of the Full Moon is not at all obvious and requires careful examination. Just what do we – and other peoples – mean when we speak of a Full Moon?

Most simply we say that the Moon is full when it is opposite the Sun. Even within the well-defined tradition of European astronomy the exact time of opposition depends on whether we mean the time when the longitudes of the Sun and Moon differ

359

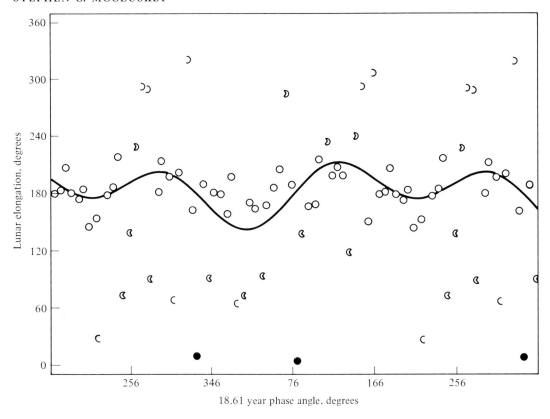

Figure 28.4 18.61 year periodicity; Tenth Full Moon prayer stick planting for Shalako, Zuni Pueblo (56 years, 1914–1969). Symbols as in Figure 28.1.

by 180°, when their right ascensions differ by 180°, or when they are at their maximum angular distance on the celestial sphere (which is generally somewhat less than 180°). Consideration of parallax, which can reach one degree in the case of the Moon, could be introduced to further complicate the situation; the meaning of the term 'Full Moon' is not as obvious as it seems.

When we move from an abstract definition of a Full Moon to the task of determining empirically when one has occurred, we find ourselves faced with greater difficulties. It is not easy to tell from its appearance whether the Moon is full or only nearly so; a gibbous Moon can fool even a trained observer. Only when we determine whether the Moon is opposite the Sun can we be certain that it is full. But how do we know when the Moon is opposite the Sun? The only time we can see a Full Moon and the Sun together is when one is

rising and the other is setting, which makes direct observation of opposition somewhat difficult. Modern astronomers would measure their position against some auxiliary reference points or in relation to some standard coordinate system.

The Shalako data indicate that the Zuni employ analogous astronomical techniques. Recall that puebloan astronomy is largely horizon based and that the system of four sacred directions, as determined by observations of the solstices, serves the puebloan peoples as a ritual and astronomical coordinate system. It is not surprising then, to see indications of the horizon coordinate system emerging in Zuni observations of the Full Moon.

We can define a number of different criteria for when a gibbous Moon on the horizon is actually a Full Moon. Consider the case where we watch the rising Full Moon in the evening. We can say the Moon is opposite the Sun either when it rises

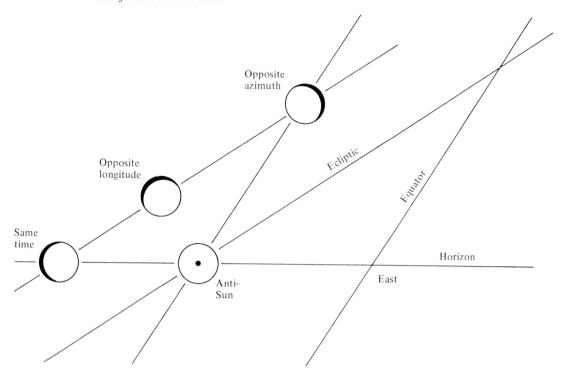

Figure 28.5 Alternative criteria for observation of the Tenth Full Moon.

at the same time that the Sun sets or when it rises at the point on the horizon directly opposite the point where the Sun sets. As Figure 28.5 shows, these criteria produce different results than the criterion that the Moon is opposite the Sun as measured in the geometrical abstraction of ecliptic coordinates. Two similar criteria could be defined by watching the setting Moon in the morning.

Since the time of moonrise or moonset changes on the average by almost an hour a day, this can fairly sharply determine the date of the Full Moon. This determination is less precise on autumn evenings in northern latitudes, when the ecliptic is roughly parallel to the horizon and the time of the Moon's rising is significantly influenced by the Moon's position north or south of the ecliptic. In autumn mornings, however, the ecliptic is almost perpendicular to the horizon so the time the Full Moon sets is scarcely affected by the Moon's distance from the ecliptic.

For the two criteria based on azimuth, where the Moon rises or sets at a point on the horizon directly

opposite the place of sunset or sunrise, the Sun and Moon must be at opposite declinations in order to be at opposite azimuths on the horizon. Thus, while the position of the Moon north or south of the ecliptic will influence its azimuth relative to the Sun, this effect does not change as we move from southern to northern latitudes or from moonrise on the eastern horizon to moonset on the western horizon; it does become more significant, however, near the time of the solstices.

If we consider the conditions for the tenth Full Moon at Zuni (35° N) when the date for Shalako was set, the different criteria will influence the dates differently (Figure 28.5). When the Full Moon is at its maximum northerly celestial latitude (5.15° north of the ecliptic) it will rise at the same time that the Sun sets 0.65 days after astronomical opposition; rise or set opposite the setting Sun 1.08 days before opposition; and set at the same time as sunrise 0.11 days before opposition. At the Full Moon's maximum southerly latitude the signs of these effects are reversed while their magnitude remains the same; at intermediate positions the effects vary linearly with lunar latitude.

The sinusoidal movement of the Moon north and south of the ecliptic reflects the rotation of the lunar nodes which also expresses itself in the major and minor lunar standstills, but since we are here concerned with maximum latitude of the Full Moon at a time other than the solstices, this oscillation will be out of phase with the cycle of the standstills. Nonetheless, this same 18.61 year periodicity, with northerly maxima at 1879, 1897, 1916, 1934, 1953 and 1972, should be reflected in a slight oscillation in the dates on which Shalako was performed.[7]

The analyses indicate that Shalako is observed slightly earlier in those years when the tenth Full Moon is near its northerly extreme than when it is near its southerly extreme (Table 28.1). This general pattern of early observations when the pre-Shalako Full Moon is north of the ecliptic is what we would expect if the Zuni defined the Full Moon in terms of the Moon rising or setting at an azimuth directly opposite that of the Sun.[8]

Considering this pattern more precisely, the value of the critical phase angle consistently obtained *for all subsets of the data* is most appropriate for a time slightly before the ninth Full Moon of the year. If this slight variation from the expected value is significant, it would indicate that the date of the planting of prayer sticks at the tenth Full Moon was determined, in part, by reckoning from observations of earlier Full Moons. Such a practice of using day counts to connect a sequence of lunar observations at Zuni would be fully consistent with the reckoning of lunar months revealed in Marshack's studies of calendar sticks, especially the calendar stick from Zuni reported by Bourke (see Marshack, 1985, and this volume, Chapter 24; Zeilik, 1986). We can see an analogous practice in the well-documented use of sequential solar observations at Hopi.

At first glance, observing the Full Moon in relation to the azimuth of the Sun seems to require that the Zuni used measuring instruments of some sort, which are not attested in the ethnographic literature. But given a clear and unobstructed horizon, instruments are not really necessary. Consider the dramatic shadows that mark the landscape at sunset. If our observer is standing, or is atop a building or a hill, or is even near a pillar (as Cushing describes the Pekwin observing the Sun), his shadow, and the shadows of objects near him, will point across the landscape directly opposite the setting or rising Sun.

If the Moon is already above the horizon, the observer can directly compare the Moon with the Sun's shadow. If the Moon has not yet risen he need only make a mental note of the place on the familiar horizon that was marked by the shadow of the setting Sun and wait to see if the Moon appears at that point. The horizon, the standard reference for solar observations, would thus serve as both measuring instrument and coordinate system for observations of the Full Moon. Since someone who regularly watches the Sun on the local horizon would know the point opposite which the Sun sets at a given time of the year, such resorting to shadows would not really be necessary.[9]

Observations of this sort would be fully consistent with the Puebloan peoples' widely documented use of the sacred system of four directions as the basic framework of their astronomies and extends their practice of observing the direction of the rising and setting Sun against the local horizon to include a concern with the direction of the rising Moon. In the case of the Moon, however, observations are not referred to a fixed landmark, the marker for the Moon is the continually shifting position of the Sun in his annual travels around the horizon.

28.2 Astronomy and environment

I would like to close by raising a few general considerations suggested by the differing roles of lunar observation at Hopi and Zuni. Studies of Zuni and Hopi culture have demonstrated the many elements of pueblo ritual and myth these close neighbors share. It is not surprising that we also find similarities between their astronomical practices. Both employ observations of the Sun along the horizon to fix their calendars; both observe the Moon after the solstices to relate the Sun and the Moon. Yet the emphasis upon these two aspects of astronomy is strikingly different.

The Hopi note that sometimes the Moon is ahead of the Sun, sometimes lags behind it, but for the Hopi the Moon remained a puzzle and there is no record of a serious effort to relate lunar and solar observations. At Zuni, however, there is the emphasis that the Pekwin's observations of the solstice should correspond to observations of the Full Moon.

Lunar observations and lunar rituals play a relatively minor part in Hopi astronomy and culture. Hopi agriculture, and most major ceremonials, are regulated by the observation of the Sun along the distant horizon. The Moon only regulates the ritual plantings of seeds in the kivas in ritual preparation for the planting of crops in the months following the Winter Solstice.

The prayer sticks planted for the Moon at Zuni after the Winter Solstice reflect a similar concern, but at Zuni these rituals continue throughout the entire year. It is clear that the Moon plays a greater role in Zuni ritual and calendrics than in that of the Hopi and that the Zuni are more concerned with the relationship of the Sun and the Moon than are their Hopi neighbors.

These differing emphases on shared elements of puebloan culture suggest a hypothesis on the relation of different environments upon astronomical practice. The villages of the Hopi are located atop mesas from which they can regulate their calendar precisely by direct observations of the rising and setting Sun against the distant horizon. Both the present village of Zuni and most of those abandoned in historic times were in the valleys of the Zuni River and its tributaries. While the landmarks against which the Hopi observe the solstice Sun are 45 to 127 km distant, Towa Yollanne behind which the Winter Solstice Sun rises at Zuni is less than 10 km from the village. Considered as an astronomical instrument, the horizon at Zuni is less suited for precise observations than is the horizon at Hopi.

If we take the slow annual movements of the Sun along the horizon as analogous to the hour hand on a watch, the monthly changes of the Moon's phases would provide something analogous to the minute hand. Observations of the Moon would help the Zuni overcome the limitations of solar observations against their nearby horizon. The concern that the Pekwin at Zuni relates the Sun and Moon could then be seen as reflecting his need to use the Moon to overcome the difficulties of keeping a purely solar horizon calendar from a location in a valley.

Since Puebloan astronomies are intimately linked to ritual, we should not be surprised by the greater influence at Zuni of those societies that embody lunar astronomy in their rituals. Further tests of this hypothesis by examination of lunar astronomy and ritual in other related cultures is needed.

Notes

1 For an early analysis of these data see Lyon (1985).

2 A more detailed discussion of this method is found in McCluskey (1977).

3 The Zuni practice of planting prayer sticks at the Full Moon after Winter Solstice is reminiscent of the Hopi planting beans in the kiva at the same time.

4 Parsons (1917) and Stevenson (1904) disagree on the date of knot tying, although the records documenting both prayer stick plantings and dances indicate that the interval from observation to dance is clearly 48 days. On this, see Lyon (1985).

5 Correlations of the festival dates with the 2.71 year intercalation cycle were statistically significant overall ($p < 0.001$), especially for the more recent data, i.e. 1935–59 ($p = 0.01$) and 1960–84 ($p = 0.002$); the correlation was not significant for data from 1879–1934 ($p = 0.1$). By inspection we can see that the recent improvement arose primarily through resolving those marginal cases where there was uncertainty whether a month should be intercalated.

These changes suggest that Zuni astronomical practices changed over the century to correspond more closely to what we would expect from observations of the Full Moon. Four hypotheses are suggested to account for these changes:

(a) Observations for more recent celebrations were influenced in one of several ways by the predictions of lunar phases available in newspapers, almanacs or calendars.

   (i) The reckoning of months could be regularized by their tabulation or by the model of the Gregorian calendar. This would decrease the ambiguity of intercalation

   (ii) Since it is hard to determine precisely when the Moon is full predictions could confirm, or even supplant, Zuni observations.

(b) Earlier celebrations were excessively confused as a result of events connected with the imposition of USA control on the reservation, including the move against pagan ceremonials of the 1920s.

(c) The observational criterion for Full Moon changed from some form of horizon observation to correspond more closely to the European concept of opposition.

(d) The Zuni have become more precise in observing the Moon for more recent ceremonies since they acquired a concern with atronomical precision from contact with Euro–Americans.

I prefer some form of the first hypothesis; the last two hypotheses are unlikely and are mentioned only for completeness.

6 Critical phase angles were computed for the subsets of data by iteratively applying a second order Fourier analysis after omitting outliers with a running median. The statistical significance was computed from a two-tailed t-test on the first Fourier coefficients computed from the original data including outliers.

The median value of the critical phase angle (78°) is what we would expect if the observations were made a bit more than a month before the tenth prayer stick planting for Shalako. This suggests that observations of this earlier Full Moon play a more important role than has heretofore been suspected and that the repeated monthly observations were connected by day counts of the intervening months.

Although not all the subsets are statistically significant, the general agreement on the value of the critical phase angle reflects the formal significance of the overall data. Since the partial data set for 1914–69 covers almost exactly three 18.61 year periods, it should be well-behaved statistically and give a reliable value for the phase angle despite its marginal statistical significance.

There are major gaps in the data prior to 1914. The sporadic record of Shalako dates due to bursts of intensive ethnographic field research makes results covering this period less trustworthy; further data are needed to fill these gaps.

7 The phase angle corresponding to maximum northerly declination of the Full Moon 48 days before Shalako can be obtained from a few simple considerations. The mean longitude of the Sun on the date of Full Moon before Shalako is 206.0°. When the Sun reached that longitude in the year 1900 (the base year for the phase angle) the longitude of the ascending lunar node was 243.8°. Thus we can compute the longitude of the ascending node (ω) at the time of pre-Shalako Full Moon as a function of the phase angle (φ):

$$\omega = 243.8 - \varphi. \qquad (28.1)$$

At its maximum northerly declination the Full Moon is opposite the Sun and 90° from the ascending node; for the pre-Shalako Full Moon the longitude is given by

$$L = 206.0 - 180, \qquad (28.2)$$

$$L = 90 + \omega. \qquad (28.3)$$

Combining (28.1), (28.2) and (28.3) we can readily see then that the maximum northerly declination of the pre-Shalako Full Moon occurs in years where φ is approximately 307.8°. Graphs of the dates should have a transition between late and early values (φ) of 18.61 year phase near 217.8° if northerly Full Moons are systematically late, or near 37.8° if northerly Full Moons are early.

8 It has been suggested that the Zuni may have set the date by observing the New Moon prior to the pre-Shalako Full Moon. Similar geometric analysis shows that north of the tropics New Moons tend to be late when the New Moon is south of the ecliptic (and the corresponding Full Moon is north of the ecliptic). Thus these results cannot be due to observations of the New Moon before the Shalako sequence begins.

9 This modified observing technique was suggested by Ray Williamson.

References

Bunzel, R. (1932a). Zuni ritual poetry. In *47th Annual Report of the Bureau of American Ethnology, 1929–1930*, p. 711. Washington.

Bunzel, R. (1932b). Introduction to Zuni Ceremonialism. In *47th Annual Report of the Bureau of American Ethnology, 1929–1930*, pp. 512–13, 534. Washington.

Lyon, L. (1985). Chronology of the Zuni Sha'lak'o Ceremony. In *Southwestern Culture History: Collected Papers in Honor of Albert B. Schroeder*, ed. C. H. Lange, Papers of the Archaeological Society of New Mexico 10, pp. 233–49. Sante Fe: Ancient City Press.

McCluskey S. (1977). The astronomy of the Hopi Indians. *Journal for the History of Astronomy* 9, 181–5.

Marshack, A. (1985). Lunar–solar year calendar stick from North America. *American Antiquity* 10, 27–51.

Parsons, E. C. (1917). Notes on Zuni. *Memoirs of the American Anthropological Association* 4, (3–4), 163–6, 176, 181.

Stephen, A. M. (1936). *Hopi Journal of Alexander M. Stephen*, ed. E. C. Parsons, vol. 1, pp. 159–60. New York: Columbia University Press.

Stevenson, M. C. (1904). The Zuni Indians: their mythology, esoteric societies and ceremonies. In *23rd Annual Report of the Bureau of American Ethnology*, 1901–1902, pp. 231–4. Washington.

Voth, H. R. (1901). *The Oraibi Powamu Ceremony*. Field Columbian Museum, Publication 61, Anthropological Series, vol. 3, no. 2, pp. 72–3. Chicago.

Zeilik, M. (1986). The ethnoastronomy of the historic Pueblos, part II. Moon watching. *Archaeoastronomy. (Supplement to the Journal for the History of Astronomy)* no. 10, suppl. to vol. **17**, pp. S1–S22.

29

The Great North Road: a cosmographic expression of the Chaco culture of New Mexico[*]

Anna Sofaer   *Solstice Project, Washington, DC*
Michael P. Marshall   *Cibola Research Consultants, Corrales, New Mexico*
Rolf M. Sinclair   *National Science Foundation, Washington, DC*

29.1 Introduction

The Great North Road is one of the most enigmatic constructs of the ancient Chaco culture of New Mexico. Efforts to establish strict utilitarian purposes for its construction do not explain certain unique features of it. We suggest it is a cosmographic expression of the Chaco culture.

The Chaco society, a prehistoric Pueblo culture, flourished between AD 950 and 1150 throughout the 80 000 km^2 of the San Juan Basin of northwestern New Mexico (Marshall *et al.* 1979; Powers, Gillespie and Lekson, 1983; Cordell, 1984; Marshall and Sofaer, 1988) (Figure 29.1). Chaco Canyon was the center of this culture. Here the Chaco people constructed multi-storied buildings containing 100 to 700 rooms (Lekson, 1984). These structures are noted for their planned, symmetric organization, massive core-veneer masonry construction, and numerous great kivas, the large ceremonial chambers of the prehistoric pueblo culture.

Descendants of the prehistoric Pueblo culture live today in the pueblos of New Mexico and Arizona. Ethnographic reports on the traditions of the historic Pueblo Indians suggest parallels between the historic and prehistoric and may provide insights into the general cosmological concepts of the prehistoric Chaco culture.

Astronomy played an important role in the Chaco culture. This is expressed in the cardinal alignments of the major axes of several large ceremonial structures at or near the center of the canyon (Williamson, Fisher and O'Flynn, 1977; Sofaer and Sinclair, 1986a), and in a complex set of solar and

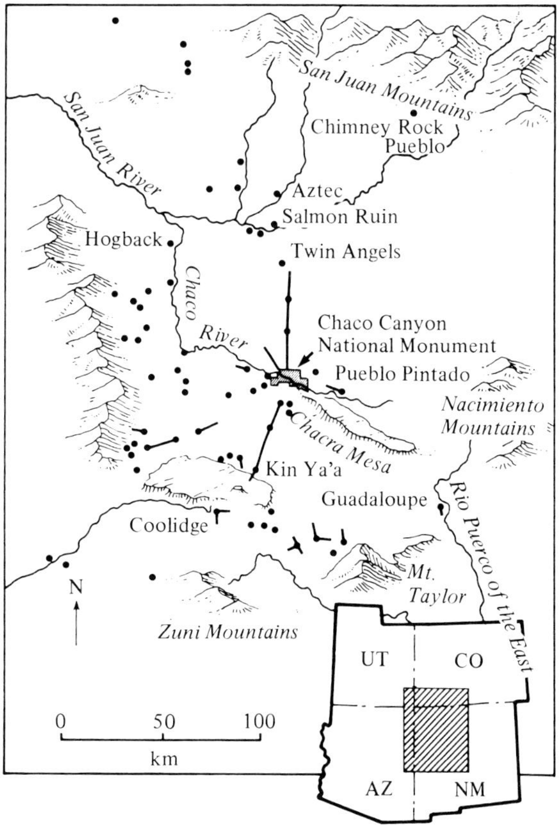

Figure 29.1 Map of the San Juan Basin, showing major Chacoan sites and roads. The inset shows this region within present State boundaries. —— prehistoric roads; ● outlying Chacoan communities. © Carol Cooperrider, 1986; additional data supplied by John Roney.

Figure 29.2 Portion of road on mesa above Pueblo Bonito. This unusually well-preserved section was formed by clearing to bedrock. © David L. Brill, 1985.

lunar markings on Fajada Butte, at the south entrance of the canyon (Sofaer, Zinser and Sinclair, 1979; Sofaer, Sinclair and Doggett, 1982; Sofaer and Sinclair, 1986a).

29.2 The roads

Roads also played an important role in the Chaco culture, judging from their extent and the effort required for their design and construction (Kincaid, 1983). In the late florescence of the Chaco civilization (*c.* AD 1050 to 1125) elaborate, formalized roads were constructed (Figure 29.2). No archetype for these roads appears to have existed in the region before their development by the Chacoans, and a recent inventory shows they were not used after the Chaco civilization's peak, about AD 1140. The Chaco roads have generally been interpreted as arteries connecting communities for trade, transportation of goods and materials, and movement of population. These explanations of the roads' functions have been premised on a model

of the Chaco culture that has envisioned Chaco Canyon as the political and economic center of a widespread trade and redistribution system extending throughout the San Juan Basin. (Vivian, 1983, provides a detailed summary of the various economic models which have been applied to the Chaco culture.)

The extensive religious architecture in Chaco Canyon suggests that the canyon may have served primarily as a ceremonial nexus for the outlying communities. Factors supporting this concept include: evidence from the middens of periodic intensive consumption of food at the large public structures (Judge, 1984); the dearth of burials and the presence of a few 'high status' burials (Akins and Schelberg, 1984); and possible large-scale ceremonial breakage of ceramic vessels (Toll, 1984). Consistent with this view of the religious function of Chaco Canyon, it appears that one of the Chaco roads – the Great North Road – and perhaps others, express religious considerations.

The Chaco roads have been noted for their great width and unusual linearity, and they have been described as 'extensively engineered' (Nials, 1983, p. 6.26). The roads were developed by excavation to a smooth, level surface, and some included masonry construction.

Approximately 300 km of roads, including the Great North Road, have been documented in the last 15 years by aerial photography and ground investigation in numerous intensive studies. A further archaeological investigation of the Great North Road was recently conducted by the authors. This involved the inspection of all structural sites and many kilometers of roadway. Numerous sites were mapped and sampled, and a technical report concerning this work is in preparation (Marshall and Sofaer, 1988).

29.3 The Great North Road – description

The Great North Road (Figure 29.3) has its origin in several routes which ascend by staircases carved into the cliff from Pueblo Bonito and Chetro Ketl in Chaco Canyon, which are the two largest structures of the Chaco region. These routes converge at Pueblo Alto, a large structure located close to the north rim of the canyon. From there the road runs 13° to the east of north for 3 km to Escavada Wash. It then heads within $\frac{1}{2}°$ of true north for 16 km, where it articulates with Pierre's Complex, an unusual cluster of small buildings on knobs and pinnacles. The road then heads close to 2° east of north for 31 km and ends at Kutz Canyon. It appears to terminate at three small, isolated sites, and a stairway recently located by the Solstice Project (Marshall and Sofaer, 1988) that descends from the Kutz Canyon escarpment to the canyon floor (see Figures 29.5 and 29.6).

From Pueblo Alto to Kutz Canyon, the road lies within one corridor, with no evidence of bifurcations. For much of its length, it exists as two, and occasionally four, closely spaced, parallel roads. The road's length and the complexity of its construction have led scholars to term it the 'Great North Road'.

The road traverses rolling, sagebrush country, where the only prominent natural features in view, and then only from rises, are the distant snow-capped mountains to the north. The only major topographic relief are the canyons at each end. With the possible exception of Pierre's Complex, there

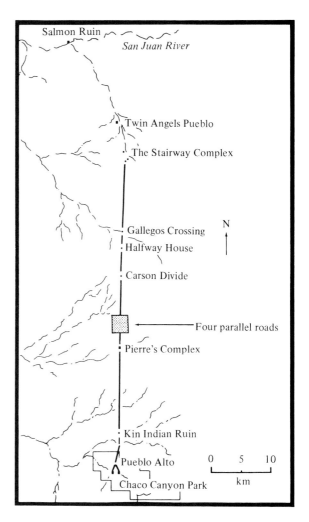

Figure 29.3 Map of the Great North Road. © The Solstice Project, 1986.

are no communities on the road's course from Pueblo Alto to Kutz Canyon. Two large complexes – Aztec and Salmon Ruin – lie to the northwest, 20 and 30 km beyond the road's terminus. Most of the outlying Chacoan communities are to the south, west and east of Chaco Canyon.

The road has been traced in numerous segments by aerial photography. On the ground, it has been intensively investigated from Chaco Canyon to Pierre's Complex and partially studied from there to Kutz Canyon. Its straight course and distinctive parallel segments have aided scholars in identifying and following it. Associated ceramic scatters and a number of unusual structural sites along its route have also aided its ground detection. Because earth and vegetation have refilled the road, only limited vestiges of it are visible on the ground today and sometimes only under particular lighting conditions.

Construction of the road involved primarily the removal of earth and vegetation. Extensive road cuts were made where the road crosses land elevations. Near the large community buildings of the canyon, several stairways were sculpted and large ramps were constructed. Several of the multiple roadways connecting these stairways and ramps with Pueblo Alto were curbed with masonry. Along one of these segments there is a curious linear groove cut into the bedrock. The stairway at Kutz Canyon, now largely collapsed, was built as a series of platforms which were supported by juniper posts and cross-beams and packed with earth (Marshall and Sofaer, 1988). The effort required for the road's construction testifies to the serious purpose that attended the decision to plan and execute it.

Considered from a utilitarian perspective, however, the road appears to be *overbuilt* and *underused*. Important features of the road – its extraordinary width and the redundancy of its routes – have no satisfactory functional explanation. The road averages 9 m in width – wider than a modern two-lane road and far wider than any of the other prehistoric roads or trails of the Southwest outside of the Chaco cultural region. The width is greater than required for draft animals or wheeled vehicles. Since this culture had neither, the width seems especially excessive in practical terms.

Redundancy occurs in the multiple stairways heading out of the canyon, the four routes that converge on Pueblo Alto, and most particularly where the Great North Road is expressed, for a good fraction of its length north of Pierre's Complex, as a set of two parallel roads. In addition, at one location, a set of four 'almost perfectly parallel' roads extending for 1.5 km is evident in aerial photography (Nials, 1983, p. 6.29) (Figure 29.4). Recent re-evaluation of the aerial imagery for the Solstice Project has revealed further portions of the road in previous gaps to the north of Pierre's Complex (G. Obenauf, 1986, unpublished report to the Solstice Project on re-evaluation of Bureau of Land Management aerial photography). Many of these segments consist of two parallel roads. (The new

portions lie on the straight line determined by the sections found earlier and thus further emphasize the overall linearity of the road.) There is no satisfactory functional explanation for these redundant features. Yet the effort devoted to achieving them indicates they are not casual expressions of the Chaco culture.

Viewed from a utilitarian perspective, we would expect the Great North Road to connect Chaco Canyon with other major population centers. An examination of the structures along the Great North Road and its destination, however, does not appear to support the earlier functional interpretation of its development and use. The road, after leaving the ceremonial complex of Chaco Canyon, traverses the least developed region of the Chaco cultural area. The structures along the road are small in comparison with other outlying Chaco structures, and minute in comparison with those in Chaco Canyon. All of the structures contain less than six rooms, and most of them contain less than three. Only Pierre's Complex suggests a possible community.

Earlier maps and reports of the Chaco cultural region have assumed that this road goes to Twin Angels Pueblo (Kincaid, 1983, Figure 4.1) and then extends at a NNW bearing, to one or both of the large San Juan River communities of Salmon Ruin (Powers *et al.*, 1983; Cordell, 1984) and Aztec (Morenon, 1977). There is, however, no ground inventory or aerial investigation that provides evidence that, in fact, the road goes to these pueblos. Moreover, efficiency for travel and transportation of goods to Salmon Ruin and Aztec would dictate a more direct and easier route from Chaco Canyon – one further to the west. Instead, the road goes north and descends a nearly impassably steep slope of Kutz Canyon.

Twin Angels Pueblo is located in the Kutz Canyon badlands, 6 km from the road's apparent terminus (Carlson, 1966). It is a relatively small pueblo of 17 rooms – less than one-tenth the size of Salmon Ruin or Aztec. (We note with interest that, although there is no evidence of the continuation of the road to or near Twin Angels Pueblo, that site lies only $\frac{1}{2}°$ east of north from the start of the road near Chaco Canyon. We cannot, at this point, rule out the possibility of a road relationship with this pueblo.)

A recent inventory of the Great North Road has produced no evidence that indicates extensive use for the transportation of economic goods (Stein,

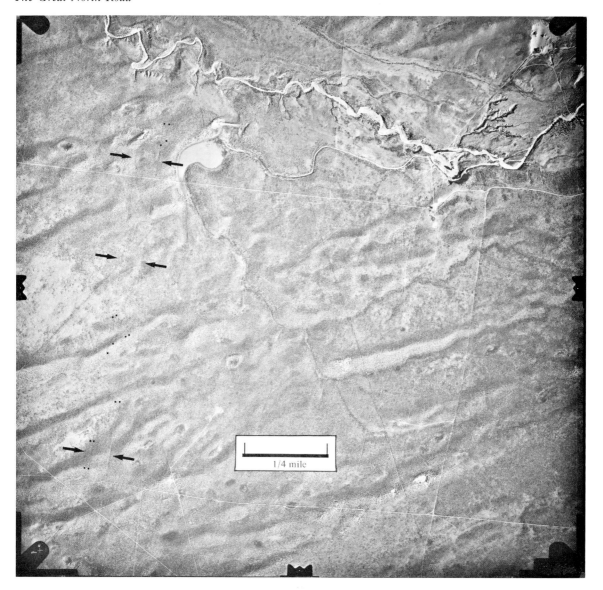

Figure 29.4 Aerial view of a 2 km section of the Great North Road north of Pierre's Complex. Arrows and dots indicate the road's four parallel segments. (Other linear features are modern roads.) Bureau of Land Management aerial imagery, 1981.

1983). It is estimated that only 10% of the ceramics found on the Great North Road are from the San Juan River communities (Stein and Levine, 1983), giving scant evidence of significant trade with them. The absence of hearths and ground or chipped stone in the road inventory suggests there was little encampment along the road.

To summarize, the road's great width and parallel routes, its ephemeral practical use, and apparent terminus at an isolated badlands canyon fail to justify, in functional terms, the effort entailed in its construction. The road apparently goes 'nowhere' and displays a level of effort far out of proportion to the meager tangible benefits that may have been

369

realized from it. In many important respects, the road appears to be its own reason for development – an end in itself.

29.4 The Great North Road – purpose

In the absence of a satisfactory functional explanation and practical destination for the Great North Road, and knowing what we do about Chaco and its interest in religious architecture, we posit that the primary purpose of the road may have been the expression of spiritual values. We will consider its direction to the north and its topographic direction with this in view. In addition, we will consider the sites along its course and their frequent location on distinctive land forms.

In the ceremonial architecture and astronomy of the Chaco culture the north–south axis is primary. Most of the great kivas have approximate north–south axes and the kivas generally have niches primarily located to the north (Reyman, 1976). The axes of two major ceremonial structures of Chaco Canyon, Pueblo Bonito and the great kiva, Casa Rinconada, are within $\frac{1}{4}°$ of north (Williamson *et al.*, 1977; A. Sofaer and R. M. Sinclair, 1984, unpublished survey). A bearing within $\frac{1}{2}°$ of north–south has been noted between two high ceremonial structures which are intervisible – Pueblo Alto and Tsin Kletzin (Fritz, 1978) – the former of which is itself aligned to the cardinals (Sofaer and Sinclair, 1986b). It is interesting to observe that Pueblo Alto and Pueblo Bonito are origin points in the canyon for the Great North Road. Just prior to the time of the road's construction, Alto was constructed; close to the time of the road's actual development, Bonito was greatly expanded and given its cardinal alignments (Lekson, 1984).

Seven noon–seasonal markings using shadow and light patterns on petroglyphs on Fajada Butte also involve the north–south axis. They occur within a few minutes of meridian passage of the sun, when the sun is due south, and thus involved a comparable interest in and knowledge of the north–south axis (Sofaer and Sinclair, 1986a).

The effort made by the Chacoans to construct the Great North Road with a bearing within $\frac{1}{2}°$ to 2° of true north is similar to the effort they made to involve the north–south axis in their large ceremonial constructions and the noon–seasonal markings on Fajada Butte. It is important to note that the road appears to deviate intentionally from astro-

nomic north after Pierre's Complex in order to arrive at the dramatic edge of Kutz Canyon. Clearly the people of Chaco had the capability of directing the road to within $\frac{1}{2}°$ of north, and as noted above they did so in a 16 km segment. For the next 31 km of the road they departed from this bearing and struck, with a rigorously straight course, their direction to a large mound on the edge of Kutz Canyon. The purpose of this deviation appears to have been a blending of astronomic north and symbolic use of topographic features in a cosmographic expression.

The road's 2° angle change directs it straight from the cone-shaped mound at Pierre's Complex, El Faro, to the large Upper Twin Angels mound. This mound (Figure 29.5) is located on the edge of the steepest slope of Kutz Canyon, where the stairway descends to the canyon floor. The mound stands out prominently above the deeply eroded slope of the canyon wall (Figure 29.6); from 10 km to the north, it is the only relief that extends above the southern horizon. These symmetrically shaped pinnacles, El Faro and Upper Twin Angels mound, while not very high, are the most distinctive prominences in the vicinity of the road corridor as it crosses the rolling northern terrain.

The straightness of the Great North Road has suggested that it was 'laid out as a single unit' (Morenon, 1977), and the 'chronological homogeneity' in the material culture associated with it has suggested 'that it can be viewed as a single construction event' (Kincaid, Stein and Levine, 1983, p. 9.76). The sites adjacent to the road were built at the time of its construction, apparently in association with its construction and use.

Five isolated structures along the road are small low-walled units located on distinctive land forms such as pinnacles or ridge crests (Kincaid, 1983; Marshall and Sofaer, 1988). They resemble shrines of the historic Peublo culture, which are similarly small, often in remote locations, and frequently on elevated land forms. Such a site was constructed on the top of the Upper Twin Angels mound.

At Pierre's Complex, almost all of the 27 structures are located on pinnacles, mesa tops, and steep ridge slopes (Powers *et al.*, 1983). While it is the largest development on the road and three of its structures are similar in scale to some small-to-medium outlying Chacoan pueblos, it is atypical. About a third of the structures are isolated rooms

Figure 29.5 Upper Twin Angels mound, looking north, seen from the edge of the canyon. A small shrine-like structure is located on top. The road terminus and stairway are just out of the picture to the left. Photograph by Anna Sofaer, © The Solstice Project, 1986.

or non-habitation sites. A recent report describes it as 'a constellation of special-function architecture', the location of which 'was probably predetermined during the engineering of the North Road' (Stein, 1983, p. 8.9). This report further states: 'indeed, arrangement of the major structures within the complex acted to preserve the bearing of the road and to "receive" it into the community.' These structures include a hearth construction on top of El Faro, from which there is extended visibility north and south.

Certain aspects of the ceramics associated with the Great North Road suggest the possibility of ceremonial activity on the road. There are several curious concentrations of shards along the road at locations isolated from structures and without evidence of nearby encampment. Unusually dense elongate ceramic scatters occur along the road

several kilometers south of Pierre's Complex (Kincaid *et al.*, 1983, p. 9.74). Along the isolated Kutz Canyon stairway, there is a concentration of ceramics (Marshall and Sofaer, 1988). The extensive quantity of broken ceramics at Pueblo Alto has suggested to some analysts the possibility of large ritual gatherings involving dispersal of food items and deliberate breakage of vessels (Judge, 1984; Toll, 1984). The ceramics along the Great North Road (Kincaid *et al.*, 1983) and at Pueblo Alto (Toll, 1984) have a significantly higher proportion of jars and non-utility ware than the ceramics at a typical Chacoan site. The road's enigmatic ceramic concentrations, the possibility of ceramic-related rituals at Pueblo Alto, and the character of the road ceramics suggest the possibility of ceremonial activity associated with ceramics on the road.

371

Figure 29.6 North slope of Kutz Canyon seen from the canyon floor. The arrow indicates the top of the stairway. Photograph by Anna Sofaer, © The Solstice Project, 1986.

29.5 Other Chaco roads

Many of the other Chaco roads also exhibit non-utilitarian features and suggest cosmological purposes. For instance, the other roads are as wide and straight, and show no evidence of frequent use. Long linear grooves were cut into the bedrock along certain roads. The principal road to the south, the South Road, has a segment of parallel roads. The ceramics on the other roads share the same non-utility ratio as the Great North Road. Small isolated structures resembling historic shrines have been found so frequently along the roads that they are now used as a means of predicting the presence of a road (Kincaid *et al.*, 1983, p. 9.16).

Recent road inventories have discounted several earlier postulated road connections between Chaco Canyon and major communities (Nials, Stein and Roney, 1988). Only a few such roads have been verified. Where certain roads do articulate with large communities, they appear to be only interconnecting avenues between nearby outlying communities or to link structures within the community, such as the large public building and the great kiva. Sometimes they appear to represent only a formalized entranceway to ceremonial locations in the community. Where the road articulates with public buildings, there is frequently evidence of large ceremonial earthen architecture: ramps, circular mounds, and platform mounds (J. R. Stein, 1983, private communication). The roads in this vicinity are usually wider and often curbed with masonry (see Figure 29.2). The pecked grooves and evidence of fire on ramps, burnt structures, elevated fire boxes, and fire pits warrant further investigation for possible ceremonial significance. At Pueblo Alto, very large firepits are located at the road's entrance points.

Some roads lead *only* to topographic features such as pinnacles, springs, or lakes. One major road, the Ashlislepah Road, which runs from Peñasco Blanco in Chaco Canyon 12 km to the northwest, connects with no other communities. It articulates, instead, with a group of cisterns, where there is a small, apparently non-utilitarian site, and then appears to terminate at now-dry Black Lake (Marshall and Sofaer, 1988). Another road, which runs from the community of Kin Ya'a and the southern terminus of the South Road 7 km to the south appears to terminate near the base of massive Hosta Butte, one of the highest and most prominent natural formations in the San Juan Basin (Nials *et al.*, 1988).

29.6 Ethnographic background

Historic Pueblo cosmology and ceremony may afford insights into the religious considerations underlying construction of the Great North Road and other Chaco roads. Here we find frequent symbolic use of straight roads, mythic and ceremonial journeys to and from the north and the 'middle place', and attention to prominent topographic features as elements of a spiritual landscape. There is even evidence of emblematic use of parallel roads and pecked grooves.

There are many symbolic uses of roads in Pueblo ritual and myth. 'Road' translates as 'channel for the life's breath' in Tewa, a Pueblo language (A. Ortiz, 1987, private communication). 'Life is a road; important spirits are ... keepers of the roads, the life roads. All spirits or sacrosanct persons have a road of cornmeal or pollen sprinkled for them where their presence is requested' (Parsons, 1939, pp. 17–18). These roads can represent the road travelled by the people to the middle place from the shipapu, the place where they emerged from the worlds below (Parsons, 1939, pp. 310, 363). Sometimes the road is for the spirits of the dead to return to the shipapu (White, 1942, p. 177).

For the Keresan Pueblos, north is where Iyatiku, the mother of all, resides at the shipapu. An account of Keresan cosmology describes the importance of north and the road to the north (White, 1960). When the people came out from the worlds below 'they stayed near the opening at Shipapu for a time, but it was too sacred a place for permanent residence, so Iyatiku told them they were to migrate to the south.' They moved south and stopped at a place where they lived for a long time.

When people died, their bodies were buried, but

their souls went back to Shipapu, the place of emergence and returned to their mother in the fourfold womb of the earth ... So every year, now, the souls of the dead come back to the pueblos of the living and visit their relatives and eat the food that has been placed for them on their graves and on the road to the north.

This 'road to the Shipapu' is described in another report as 'crowded with spirits returning to the lower world, and spirits of unborn infants coming from the lower world' (Stevenson, 1894, p. 67). This and other roads are frequently described as 'straight' (Stevenson, 1894, pp. 31, 41, 145).

When a person dies in the Keresan and Tanoan pueblos, the officiant takes offerings that represent that person's soul to the north and deposits them in a canyon or a mesa crevice (White, 1973, p. 137). Ceramic vessels are frequently broken in rituals related to the dead (Parsons, 1939, pp. 72, 77; Ortiz, 1969, p. 54). Sometimes a vessel containing food which is 'the last meal of the deceased' is put on the road to the north or sometime it is 'killed' (broken at the rim) and then thrown by the officiant 'out to the north, the direction in which the soul ... travels toward Shipapu' (White, 1942, p. 177).

Traditionally, the Pueblo people re-enact the creation and emergence events, especially at important solar times. As part of these ceremonies, they make ritual journeys to certain mountains, canyons, caves and lakes – places they regard as shipapu openings (Ellis and Hammack, 1968, pp. 31, 33; Ladd, 1983). These journeys may be as long as 500 km to and from the pueblo. Along the route, the ceremonialists leave offerings at shrines which are located on distinctive land forms, such as buttes, cone-shaped hills, ravines and springs. From a Keresan pueblo on the south edge of the prehistoric Chaco region, ceremonialists packed their burros with solar offerings and traveled north, stopping first at Chaco Canyon (Ellis and Hammack, 1968, p. 32). They made offerings at a shrine on the south side of the canyon and then travelled to a shrine at Jackson Butte and finally to the shipapu, a small lake or spring in the San Juan Mountains. In Pueblo culture, the mountains are where the cloud beings, the spirits of the dead, reside (Parsons, 1939, pp. 172, 173).

For Jemez, a Tanoan pueblo, north is 'spiritually indicative of the mythical and ancestral homeland' (Weslowski, 1981, p. 123), and the place of emergence is in the mountain range to the north of the pueblo. One of their most sacred shrines is located on its prominent peak. There the 'underworld chiefs make a pilgrimage... every June to begin the summer series of rain retreats and ceremonies' (Weslowski, 1981, p. 117). In the emergence and migration of this pueblo, the leader Fortease, upon emergence from the shipapu, chooses the direction 'towards the south' and then makes four roads for the people to travel on in search of their place of settlement (Parsons, 1925, pp. 137, 138). Fortease is reported to have made the roads by 'clearing away the brush'. Reference to two parallel roads occurs in Tanoan cosmology (Ortiz, 1969, p. 57):

> True to the underlying message of the origin myth... the Tewa do begin and end life as one people. The term they use for the life cycle is poeh, or 'path', after the two different migration paths the moieties followed after emergence. Thus, at the beginning of life there is a single path for all Tewa. ... it divides into two parallel paths and continues in that way until the end of life. At death the paths rejoin again and become one, just as the moieties rejoined in the myth of origin.

At the Zuni Pueblo, a pilgrimage is conducted every four years at summer solstice by 50 religious leaders to a lake, the Zuni 'village of the gods', the place where the spirits of all Zunis go after death (E. R. Hart, 1985, unpublished manuscript: 'The Zuni Indian tribe and title to *Kolhu/wala:wa* (Kachina Village)'). Fires are lit along the route by one of the participants, the Zuni Fire God. Another important pilgrimage, to the Zuni Salt Lake, is on roads that have been described as very straight and with shrine-like sites similar to those on the Chaco roads (Kelley, 1984). Although for the Zuni, these sacred lakes and the origin place are not located to the north, north is associated with the 'undermost' of the below worlds (Stevenson, 1904, p. 25) and has primacy in the ordering of ceremonial events and religious leadership (Cushing, 1979, pp. 188–90). In the prayers and chants telling of their emergence and migration to the middle (i.e. Zuni), reference is made to four parallel roads: 'Hither towards Itiwana (the middle) I saw four roads going side by side.' (Bunzel, 1932a, p. 717.) One Zuni ceremony includes breakage by the religious leaders of ceramic vessels throughout the pueblo (Cushing, 1979, p. 321).

Long linear grooves are cut into the mesas near two present-day pueblos. Contact with these grooves is reported to help diagnose sickness in curing ceremonies, to help people to regain their strength and to help persons 'to cease yearning for the dead or absent and to keep them from returning in their dreams' (Parsons, 1939, p. 449). There are many references to running in Pueblo ceremony and myth, sometimes on north–south roads (Parsons, 1925, p. 119) and sometimes symbolic of emergent events on north–south parallel courses (Dutton and Marmon, 1936, p. 12); and in certain instances on ritually swept east–west tracks to aid the journey of the sun (Parsons, 1939, p. 547).

In several reports, canyons or deep holes are seen in myths or in the actual topography as leading to the shipapu (E. R. Hart, unpublished manuscript). Symbolic ladders connect with the underworld (Bunzel, 1932b, pp. 589, 590; Cushing, 1979, p. 132). At one Keresan Pueblo, the shipapu is described as (Lange, 1959, p. 416) 'the Lagune... to the north, beyond the "Conejos"... very round and deep. Many streams flow into it, but it has no issue. Out of this Lagune came forth the Indians and in it dwells "Te-tsha-aa", our mother... The good ones return to her.'

The 'middle place', so important in Pueblo cosmology, is seen as the place of the convergence of the cardinal directions and the nadir and zenith directions (Ortiz, 1972, p. 142). Where these directions join is the sacred center for Pueblo people. This place is sometimes symbolically conveyed in the joining of ritual roads at the pueblo center (Goldfrank, 1962, p. 47). It is interesting to note that the cardinal directions in the Chaco architecture, and the north–south axis of the Great North Road, merge at a central ceremonial complex of the vast Chaco cultural province, Pueblo Bonito.

29.7 Conclusion

The Great North Road embodies many non-utilitarian aspects and has no clear practical destination. It displays a level of effort in its engineering and construction that is far out of proportion to any material benefits that could be realized from it. Its direction to the north and its linkage to the middle place of Chaco Canyon find echoes in much of the tradition of the historic Pueblos, where roads, and especially a road to the north, have intense symbolic value. We conclude that the Great North Road was conceived, in harmony with much of Chaco architecture, as a cosmographic expression uniting the Chaco world and its works with its spiritual landscape.

Acknowledgments

The Solstice Project investigations have relied heavily on the comprehensive Bureau of Land Management 'Chaco Roads Project Phase I' (Kincaid, 1983), which summarizes earlier pioneering studies of the Chaco roads, including the initial observations of the Great North Road by Pierre Morenon. We are grateful to Margaret Obenauf for her re-evaluation of aerial photography that led to further evidence for the Great North Road. We want to thank John Stein for sharing his early ideas regarding the possibility of cosmographic expression in the Chaco roads and his thorough knowledge of them. Conversations with Alfonso Ortiz, Fred Eggan, John Roney, Fred Nials and Chris Kincaid have been particularly helpful to this study. Our profound thanks go to a friend who edited this paper with remarkable insight and generosity, and who, most remarkably, insists on remaining anonymous.

References

Akins, N. J. and Schelberg, J. D. (1984). Evidence for organizational complexity as seen from the mortuary practices at Chaco Canyon. In *Recent Research on Chaco Prehistory*, eds. W. J. Judge and J. D. Schelberg, pp. 89–102. Albuquerque: US Department of the Interior, National Park Service.

Bunzel, R. L. (1932a). *Zuni Ritual Poetry*. 47th Annual Report, Bureau of American Ethnology, pp. 611–835. Washington: Smithsonian Institution.

Bunzel, R. L. (1932b). *Zuni Origin Myths*. 47th Annual Report, Bureau of American Ethnology, pp. 545–609. Washington: Smithsonian Institution.

Carlson, R. L. (1966). Twin Angels Pueblo. *American Antiquity* 31, (5), 676–82.

Cordell, L. S. (1984). *Prehistory of the Southwest*. Orlando: Academic Press.

Cushing, F. H. (1979). *Zuñi, Selected Writings of Frank Hamilton Cushing*, ed. J. Green. Lincoln: University of Nebraska Press.

Dutton, B. P. and Marmon, M. A. (1936). The Laguna calendar. *University of New Mexico Bulletin, Anthropological Series* 1, (2), 3–21.

Ellis, F. H. and Hammack, L. (1968). The Inner Sanctum of Feather Cave. *American Antiquity* 33, (1), 25–44.

Fritz, J. M. (1978). Paleopsychology today. In *Social Archaeology: Beyond Subsistence and Dating*. Orlando: Academic Press.

Goldfrank, E. C. (ed.) (1962). *Isleta Paintings*. Bureau of American Ethnology Bulletin 181. Washington: Smithsonian Institution.

Judge, W. J. (1984). New light on Chaco Canyon. In *New Light on Chaco Canyon*, ed. D. G. Noble, pp. 1–12. Santa Fe: School of American Research Press.

Kelley, K. B. (1984). *Historic Cultural Resources in the San Augustine Coal Area*. Draft on file at Bureau of Land Management, Socorro District, New Mexico, June 1984.

Kincaid, C. (ed.) (1983). *Chaco Roads Project Phase 1, A Reappraisal of Prehistoric Roads in the San Juan Basin*. Albuquerque: US Department of the Interior, Bureau of Land Management.

Kincaid, C., Stein, J. R. and Levine, D. F. (1983). Road verification summary. In *Chaco Roads Project Phase I, A Reappraisal of Prehistoric Roads in the San Juan Basin*, ed. C. Kincaid, pp. 9.1–9.78. Albuquerque: US Department of the Interior, Bureau of Land Management.

Ladd, E. (1983). Pueblo use of high-altitude areas: emphasis on the A: Shiwi. In *High Altitude Adaptations in the Southwest*, ed. J. C. Winters, pp. 168–76. Albuquerque: US Forest Service, Southwest Region Report no. 2.

Lange, C. H. (1959). *Cochiti: A New Mexico Pueblo, Past and Present*. Austin: University of Texas Press. (Reprinted in 1968 by Southern Illinois University Press, Carbondale.)

Lekson, S. H. (1984). *Great Pueblo Architecture of Chaco Canyon, New Mexico*. Albuquerque: US Department of the Interior, National Park Service.

Marshall, M. P. and Sofaer, A. (1988). *Report on the Solstice Project Archaeological Investigations in the Chacoan Province, New Mexico*. (To be published.)

Marshall, M. P., Stein, J. R., Loose, R. W. and Novotny, J. E. (1979). *Anasazi Communities of the San Juan Basin*. Albuquerque: Public Service Company of New Mexico.

Morenon, E. P. (1977). *A View of the Chacoan Phenomenon From the 'Backwoods': A Speculative Essay*. On file, Archaeological program of the Institute of Applied Sciences, North Texas State University, Denton.

Nials, F. L. (1983). Physical characteristics of Chacoan roads. In *Chaco Roads Project Phase I, A Reappraisal of*

Prehistoric Roads in the San Juan Basin, ed. C. Kincaid, pp. 6.1–6.51. Albuquerque: US Department of the Interior, Bureau of Land Management.

Nials, F. L., Stein, J. R. and Roney, J. R. (1988). *Chacoan Roads in the Southern Periphery: Results of Phase II of the Bureau of Land Management Chaco Roads Study.* Albuquerque: Department of the Interior, Bureau of Land Management. (In press.)

Ortiz, A. (1969). *The Tewa World: Space, Time, Being, and Becoming in a Pueblo Society.* University of Chicago Press.

Ortiz, A. (ed.) (1972). *New Perspectives on the Pueblos.* Albuquerque: University of New Mexico Press.

Parsons, E. C. (1925). *The Pueblo of Jemez.* Papers of the Phillips Academy Southwestern Expedition, 3. New Haven: Yale University Press.

Parsons, E. C. (1939). *Pueblo Indian Religion.* University of Chicago Press.

Powers, R. P., Gillespie, W. B. and Lekson, S. H. (1983). *The Outlier Survey, A Regional View of Settlement in the San Juan Basin.* Albuquerque: US Department of the Interior, National Park Service.

Reyman, J. E. (1976). The emics and etics of Kiva wall niche location. *Journal of the Steward Anthropological Society* 7, (1), 107–29.

Sofaer, A., Zinser, V. and Sinclair, R. M. (1979). A unique solar marking construct. *Science* 206, 283–91.

Sofaer, A., Sinclair, R. M. and Doggett, L. E. (1982). Lunar markings on Fajada Butte, Chaco Canyon, New Mexico. In *Archaeoastronomy in the New World*, ed. A. F. Aveni, pp. 169–86. Cambridge University Press.

Sofaer, A. and Sinclair, R. M. (1986a). Astronomical markings at three sites on Fajada Butte. In *Astronomy and Ceremony in the Prehistoric Southwest*, eds. J. Carlson and W. J. Judge. Albuquerque: Maxwell Museum Technical Series, University of New Mexico.

Sofaer, A. and Sinclair, R. M. (1986b). Astronomic and related patterning in the architecture of the prehistoric Chaco culture of New Mexico. *Bulletin of the American Astronomical Society* 18, (4), 1044–5.

Stein, J. R. (1983). Road corridor descriptions. In *Chaco Roads Project Phase I, A Reappraisal of Prehistoric Roads in the San Juan Basin*, ed. C. Kincaid, pp. 8.1–8.15.

Albuquerque: US Department of the Interior, Bureau of Land Management.

Stein, J. R. and Levine, D. F. (1983). Documentation of selected sites recorded during the Chaco Roads Project. In *Chaco Roads Project Phase I, A Reappraisal of Prehistoric Roads in the San Juan Basin*, ed. C. Kincaid, pp. C.1–C.64. Albuquerque: US Department of the Interior, Bureau of Land Management.

Stevenson, M. C. (1894). *The Sia.* 11th Annual Report of the Bureau of American Ethnology, pp. 3–157. Washington: Smithsonian Institution.

Stevenson, M. C. (1904). *The Zuni Indians.* 23rd Annual Report, Bureau of American Ethnology, pp. 3–634. Washington: Smithsonian Institution.

Toll, H. W. (1984). Trends in ceramic import and distribution in Chaco Canyon. In *Recent Research on Chaco Prehistory*, eds. W. J. Judge and J. D. Schelberg, pp. 115–36. Albuquerque: US Department of the Interior, National Park Service.

Vivian, R. G. (1983). Identifying and interpreting Chacoan roads: an historical perspective. In *Chaco Roads Project Phase I, A Reappraisal of Prehistoric Roads in the San Juan Basin*, ed. C. Kincaid, pp. 3.1–3.20. Albuquerque: US Department of the Interior, Bureau of Land Management.

Weslowski, L. V. (1981). Native American Land Use Along Redondo Creek. In *High Altitude Adaptation Along Redondo Creek*, eds. C. Baker and J. C. Winter. Albuquerque: University of New Mexico, Office of Contract Archaeology.

White, L. A. (1942). *The Pueblo of Santa Ana, New Mexico. Memoirs of the American Anthropological Association* 60, Menasha, Wisconsin.

White, L. A. (1960). The world of the Keresan Pueblo Indians. In *Culture in History: Essays in Honor of Paul Radin*, ed. S. Diamond, pp. 53–64. New York: Columbia University Press.

White, L. A. (1973). *The Acoma Indians.* Glorieta, New Mexico: Rio Grande Press.

Williamson, R. A., Fisher, H. J. and O'Flynn, D. (1977). Anasazi solar observatories. In *Native American Astronomy*, ed. A. F. Aveni, pp. 203–18. Austin: University of Texas Press.

The rise and fall of the Sun Temple of Konarak: the temple versus the solar orb

J. McKim Malville *University of Colorado*

30.1 Introduction

The small village of Konarak (19°53′N; 86°06′E) of eastern India contains the ruins of the largest and most impressive temple in India dedicated exclusively to the sun god, Surya. Although the tower of the main temple, once over 200 feet in height, is now in ruins, it is not a dead temple, as each day there is a small but constant flow of pilgrims amidst the bus loads of tourists who come to worship the three large statues of Surya (Figure 30.1) and perform *puja* to the nine planets of which the sun is one. The most impressive demonstration of the continuing grip which the sun has on worshippers is the annual festival held on the morning of the seven-day-old moon in the month of Magh (January–February) when approximately 50 000 pilgrims gather on the shore of the Bay of Bengal near the ruined sun temple to greet the sun at dawn and bathe in the ocean. The sunrise festival appears to pre-date the construction of the sun temple itself and has elements of non-iconic sun worship which is practiced today in some of the local Orissan villages.

In spite of the great prominence, both explicitly and symbolically, given to the sun in the Vedas and the subsequent sacred literature of India, no major temple dedicated to the sun has been constructed in India since the completion in approximately AD 1258 of the Temple of Konarak (Boner and Sharma, 1972). In no temple of the land, does Surya receive the attention by devotees comparable to that of the classic period of sun worship between AD 300 and 1200 (Pandey, 1971; Srivastava, 1972). Why

has temple-centered worship of the sun so faded over the past 800 years? Sun worship has been described as the real religion of India, yet where are the living temples now? Temples devoted to the sun as well as to other Hindu deities were destroyed in northern India by the Muslim invasion. Although temples to Shiva, Vishnu and Shakti rose again, those devoted to the sun never reappeared.

The Sun Temple of Konarak offers a number of clues for such decline of sun worship. There is a strange and ominous sense of unavoidable catastrophe associated with the temple. The ambitious edifice may have been built in part to celebrate the victory of the Orissan king's armies over the dark forces of the Muslim. Yet another battle appears to be enshrined in the fallen stones of the temple, fought not between human legions but between the sun and its iconic representations.

Expressions of devotion and gratitude to the natural sun have been, and continue to be, fundamental features of the ethos and cosmos of Hindu India for several millennia. For most of that time devotion to the sun has proceeded without the assistance of temples and icons. The *gayatri* mantra, repeated by every 'twice-born' man at his morning and evening devotions, calls upon the sun, *savitar*, as the supreme creative force in the cosmos. (*Om bhur bhva savat tat savitur varenyam bhargo devashya dimahi dhiyo yo nah pracodayat*: 'We meditate on that excellent light of the divine Sun; may he illuminate our minds'; Stutley and Stutley, 1977.) The *surya argha*, another important feature of morning devotions throughout India, expresses gratitude to the sun with an offering of water. The west bank

Figure 30.1 Mitra, a form of the Sun God, in a niche of the southern wall of the main temple. Mitra is friend and companion to man and 'he who awakens men at daybreak and stirs them to labour' (Rig Veda, III.59,1). The lotus is the first object to be touched by the sun on the morning of Magh Saptami.

of the Ganges at Varanasi serves the role of a vast open-air sun temple in which the rising sun at dawn is the major object of devotion for thousands who line the bathing ghats (Pandey, 1971; Srivastava, 1972; Eck, 1982; Malville, 1985).

The Sauras, exclusive devotees of the sun, were one of the six major philosophical schools of Hinduism (Saivism, Shiva; Vaisnavism, Vishnu; Shaktam, Shakti; Kaumaram, Skanda; Garnapatyam, Ganesh; and Saura, Sun) identified by Shankara, the great reformer of Hinduism in the eighth century AD (Bhandakar, 1983). Each of these schools

had deep historical precedence, but probably Saura worship was the most ancient. For the Sauras the sun was identified as *parabrahma*, the source and essence of the cosmos. In its original form the Saura system needed no temples; the natural sun was a sufficient symbolic and devotional object.

The alternate mode of devotion to the sun, that of employing icons, sun priests, and elaborate sun temples, was due in part to the foreign influence of the Magas, sun priests of ancient Persia, Sakadvipa, who accompanied the Scythian invaders, the Sakas, during the latter half of the first millennium

BC. As it spread from west to east, the new form of sun worship was absorbed into the Hindu culture and accepted by the mass of Indians as another form of devotion to the sun. In time the Magas were thoroughly Hinduized, becoming a separate caste, and continued to provide priests for the numerous sun temples which were constructed throughout India (von Stietencron, 1966; Bhandakar, 1983).

30.2 The troubled rise of the Sun Temple

Orissa state, in general, and the neighborhood of Konarak, in particular, had a tradition of sun worship predating the construction of the Sun Temple. According to local tradition, it was on the edge of the Chandrabharga River that Samba, the son of Krishna, was cured of leprosy after performing austerities for the sun for 12 years. It is to the Chandrabharga River that pilgrims today come to pray to the sun for the relief of illness. The original Chandrabharga River from which this story was borrowed was far to the west in the Punjab, now the Chenab. The temple of Multan, built around the fifth century AD and destroyed by the Muslims, may have been the location of the first example of a temple devoted to the sun in India (Mitra, 1976).

A local, Orissan story explaining the reason for the destruction of the great tower of the Sun Temple suggests an undercurrent of negative feelings toward the construction of a sun temple (Sriupendranayak, 1985, private communication):

> During the Golden Age, the Sattva Yuga, the gods were gathered at a meeting to which the god of love, Kama, and his consort came. In the middle of the gathering they commenced to make love. The Sun was outraged at such behavior in public and stormed out of the meeting. 'What airs the sun has acquired,' commented the other gods. The sun, in their estimation, had become too proud and too arrogant, thinking himself to be very powerful. 'Let us test the Sun and see who is more powerful, Surya or Kama,' said Brahma. All the gods created a daughter of the moon; named Chandravati, she was sealed in a box and set afloat on the ocean. The girl was found by a holy man living near the Chandrabharga river and was raised by him. One morning before sunrise, when she was sixteen, 'as lovely as a newly blooming flower in spring,' she was playing on the beach with a hoop while her step father was meditating. Just as the sun was rising, Kama shot an arrow into the sun and the reverse end of the arrow into the girl. Enthralled by her beauty, he went running toward the girl. But the frightened Chandravati ran from the sun, jumped into the ocean and drowned. Brought out of his meditations by the cries of his stepdaughter the rishi cursed the sun: 'You who have killed my daughter and have broken my meditation shall have any temple built for you in this location broken and destroyed.'

We have available the record of the details of construction of the Sun Temple due to the scholarship of Alice Boner and Sadisiva Rath Sharma. During eight years of search in villages near Konarak, Sharma discovered nearly 1000 palm leaf manuscripts which had been preserved in temples or as heirlooms in private houses. Two are of particular interest to us: the *Baya Cakada*, a detailed description of the construction of the Sun Temple and the *Padmakesara Deula Karmangi*, a manual of temple ritual. Both, written in Karani script and in an antiquated form of Oriya language, have been translated by Boner and Sharma (1972) and give us an almost unprecedented view of the process of temple design and construction in 13th-century India.

The original motivations for the construction of the Sun Temple in a remote area of Orissa at a time when sun worship was already declining in popularity were apparently manifold. The temple may have been constructed as a monument to celebrate the victories of the Orissan king, Narasimha Deva, a famous warrior who not only repulsed all attacks of the Muslims but also reconquered territory occupied by them in Bengal in battles of AD 1243 and 1245. As a result of his victories, Orissa state was able to preserve its traditions longer than the other parts of India, until its eventual occupation by the Muslims in AD 1568 (Boner and Sharma, 1972). The friezes of the temple display various war scenes glorifying the military victories of Narasimha Deva. The King himself is also found in various representations as both warrior and devotee to the sun god (Mitra, 1976). Narasimha Deva was known as a *langulia*, 'one having a tail'. It is possible that another of the reasons for construction of the great temple was an attempt to cure himself of a deformity of his spinal column.

The reign of King Narasimha lasted between

AD 1238 and 1264. Construction of the temple was started close to 1240, although plans apparently had been underway for six years prior to that time. According to the *Baya Cankada* (Boner and Sharma, 1972) the platform, or *pitha*, of the temple was constructed before the onset of the rainy season in 1242. Construction of walls and pillars began at the end of the rainy season in September, 1242. Building of the temple continued for approximately another 12 years, drawing heavily upon local resources and talents. Popular tradition relates that the revenue of the entire state was diverted into the construction of the temple. The total diversion of the revenue of the state into the temple is unlikely, but such belief is indicative of the popular conception of large resources which were expended on the massive temple and its extensive works of art.

Before the completion of the temple in AD 1258, there occurred the tragic death of the son of the chief stone mason (Mansimha, 1968):

> According to tradition some 1200 stone masons of Orissa state were ordered to work on the Konarak temple and not to leave until it was complete. But because of a defect in the design or construction, the finishing touch could not be accomplished; the heavy kalasa stone could not be put in place on the top of the temple. The son of the chief architect, who had been born after his father started work on the temple, came to see his father for the first time. The boy had constantly asked his mother about his father and although young had become something of a prodigy in architecture as he attempted to follow his father's profession. When he found that his father, after their initial joy of reunion, was dejected and sad, he learned about the inability of the builders to place the top stone of the temple. King Narasimha was becoming increasingly impatient with their inability and had ordered that if the temple was not completed by a fixed date all craftsmen would lose their heads. The talented son suggested a satisfactory solution to the problem and the top stone was put in place successfully. Although overjoyed at the completion of the tower, the masons were afraid that if the word reached the King that a mere boy had succeeded where 1200 masons had failed, the masons would lose their heads for incompetence. The son, overhearing these discussions, decided to save the lives of his father and his fellow masons by killing himself. He climbed to the top of the temple and jumped to his death.

The completed temple is a rich collection of astronomical symbols. The main temple, 220 feet from east to west, is built as the chariot of the sun carried on 24 large stone wheels drawn forward to the east by seven horses (Figure 30.2). Directly to the east, 30 feet in front of the stairway leading to the main temple, is the Natamandir, the Dancing Hall, consisting of nine chambers for each of the planets, the *navagrahas*. The roof of the Natamandir is missing but a stone lotus five feet in diameter has been found which may have formed the ceiling above the chamber dedicated to the sun. In the center of the lotus sits Surya holding a lotus in each hand pulled by seven horses. On the inner circle of eight petals and the outer circle of 16 petals are dancers and musicians, suggesting that the structure was at certain times associated with dancing and musical performances. The Natamandir is also known as the Bhoga Mandap, the Hall of Offerings. As we shall suggest, the rituals associated with the structure may offer a major clue for the nature of sun worship and its decline at Konarak.

Together with the lotus the wheel is supremely symbolic of the sun as the source of time, movement and change. Each of the 24 wheels, 9.75 feet in diameter, represented a particular fortnight of the year, either the waxing bright or waning dark portion of the moon. Each pair of wheels was dedicated to one of the 12 zodiacal constellations, starting in the east with Pisces, Mina, and ending in the west with Aries, Mesa. The wheels contain eight large and eight small spokes representing the eight divisions of night and day. (The period from 6.0 a.m. to 6.0 p.m. is divided into eight *praharas*; Kay, 1981.)

In Hindu mythology the sun is a powerful horse, white, golden, or red, Haridashva or Rohita, moving across the sky. The Vedic ritual of the horse sacrifice, the *asvamedha*, was directed toward the sun. The sun is a swift 'horse of light' adorned with seven colors and is visible in that form on the north side of the temple (Figure 30.3). The seven horses which pull the chariot across the sky are an amplification of the sun–horse metaphor. Furthermore, the number seven is a recurrent number in Hindu tradition and also refers to the holy meters of Sanskrit verse, the ruddy cows of dawn driven

Figure 30.2 South-east wheel representing Pisces.

before the sun, seven Rishis of Ursa Major, seven planets, seven Pleiades, seven rings of mountains around Mount Meru, seven days of the week, and seven colors of sunlight (O'Flaherty, 1975; Kramrisch, 1981; Mabbett, 1983).

The 'sun's birthday', the major sun festival of Orissa, occurs on Magh Saptami, the month of Magh when the moon is seven-days old. Magh Saptami follows the first new moon after Pongal. The festival of Pongal is celebrated when the sun enters the constellation of Capricorn, *makara*; the crossing from Sagittarius into Capricorn is known as *makara sankranti*, when according to tradition the sun is beginning on its northern course again, and thus appears to be the precessed winter solstice. Pongal, now on January 15, is considered to be the beginning of the year, and especially in south India is celebrated as a solar festival (Freed and Freed, 1964; Beck, 1976). It is a dramatic measure of the power of the number seven to have the sun's birth-day celebrated at Konarak, not when the year has just been born but at the time when the first moon of the year is seven days old.

The Sun Temple is a mixture of symbols of kingly self-aggrandizement, war, fertility attested by the abundant erotic sculptures, and sun worship. A miniature universe, the temple contains all: war, love, and transcendent spirit. Not just a symbol of the victory of Narashimha's forces of light over the forces of darkness on the field of battle, it is a vivid and explicit handbook for human sexuality, drawing upon human behavior for symbols of the fertilizing and creative powers of the sun. The erotic carvings affirm the Tantric motto: 'bhoga (delight) is yoga (religion)' (Zimmer, 1955), as well as demonstrating the 'descent' of the sky god to the role of fecundator (Eliade, 1967).

The Temple was a highly ambitious religious enterprise, involving extraordinarily elaborate rituals performed in and around it, as disclosed

Figure 30.3 Haridashva, the Sun God, riding a golden horse in a niche of the northern wall of the main temple.

by the *Padmaskeshara Deula Karmangi*. The presiding deity of the temple, Mahabhaskar, 'the creator of light', was hidden deep within the sacred center of the temple, its 'womb chamber' the *garbha griha* (Kramrisch, 1946). Although Mahabhaskar faced east, on only a very few days of the year could the light of the rising sun reach the sun god. From inside the *garbha griha* it is 160 feet to the eastern opening of the main temple and another 100 feet beyond to the eastern opening of the Natamandir, the Dancing Hall. There is a traditional belief that sunlight regularly struck the image of the sun god on the throne within the *garbha griha* of the main temple (Mishra, 1981). According to my measurements, however, the field of view of the eastern horizon from the *garbha griha* would have been only 2°15′, limited by a ten-foot wide eastward entrance to the Dancing Hall. Today the view of the eastern horizon is completely blocked by trees which stretch from the walls of the temple compound to the shore

of the Bay of Bengal some 3 km distant. Between the main temple and the Dancing Hall was a pillar bearing an image of Aruna, the charioteer of Surya. The pillar approximately two feet in diameter, which now stands in front of the main gate of the temple of Jagannath in Puri, would have further reduced the view of the eastern horizon by some 40′. According to the *Padmaskeshara Deula Karmangi* the entry of the light of the rising sun, even if it did occur, was not associated with any major temple festival. There were many festivals at Konarak, but none occurred at the vernal equinox. The major festival was then as it is now, that of Magh Saptami, the 'birthday of the sun'.

The temple records indicate that two small bronze statues of the sun god, *utsava murti* or *vijaya murti*, served as mediators between Mahabhaskar in the dark sanctuary and the sun in the sky. According to the detailed description of the rituals of the temple in the *Padmaskeshara Deula Karmangi* each morning before sunrise the movable image was placed in the center of the Natamandir where it received the light of the first rays of the sun.

My measurements of the Natamandir show that the movable image of the sun would not receive light from the rising sun on every day of the year if it were always in the center of the chamber of Surya. Only 21°50′ of the eastern horizon is visible from the center of the Natamandir. Measurements with a magnetic compass indicate that the east–west axis of the temple is rotated approximately 3° south of east, assuming a magnetic declination at Konarak of −1°. There are reports of a local magnetic anomaly near the temple, so these measurements must be treated with caution. Assuming for now that the axis of the temple is aligned to geographic east–west, only when the sun was within 10°10′ of the celestial equator would it have illuminated the bronze statue in the center of the Natamandir (Figure 30.4; point *a*). In order for the important first ritual of the day to work when the *utsava murti* had to be illuminated by the rising sun, the statue would have to have been displaced from the exact center of the Natamandir from the middle of April to late August and from late October to late February. A highly inauspicious situation must have arisen near solstices, as the movable image of the sun would have been placed in the next square to the east, that ruled by the demon Rahu, the seizer, the eternal enemy of the sun and moon. In the center of the square

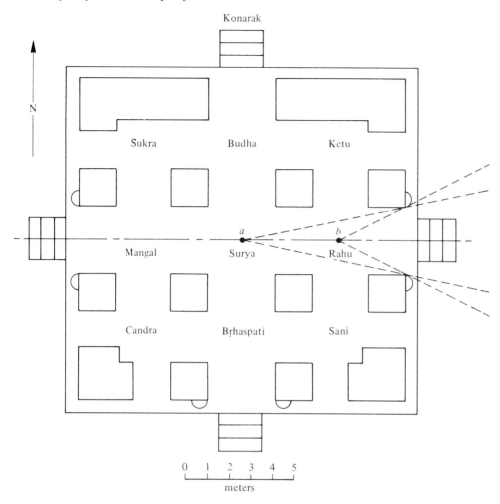

Konarak

N

Sukra   Budha   Ketu

a   *b*

Mangal   Surya   Rahu

Candra   Bṛhaspati   Sani

0   1   2   3   4   5
meters

Figure 30.4 Plan of Natamandir.

ruled by Rahu (Figure 30.4; point *b*), the maximum declination of an object visible on the horizon is approximately 22°50′ assuming an east–west alignment of the temple. Such a value is 42′ less than the obliquity of the ecliptic in the 13th century, 23°32′, and suggests that the chamber of Ketu may have been designed for the solstice sun. The bronze image may have been cycled throughout the year from the center of the Natamandir to the center of Rahu's chamber.

The arrangement giving Rahu such a prominent role in the temple dedicated to the sun seems odd and highly inappropriate. One reason for this structural design may well have been provided by the solar eclipse of AD 1242 which occurred just as the major construction of the temple was starting. The

center line of the total eclipse of September 26, AD 1242 was, according to Oppolzer (1962), some 100 km to the north of Konarak. An eclipse at Konarak, either total or partial, would have been viewed with great alarm by all associated with the construction of the temple as well as local villagers. In India an eclipse of the sun is a time when the sun is understood to be in pain, and it is filled with the potential for disaster to people, animals, and society (Lapoint, 1981). One can hardly imagine a less auspicious beginning for the temple.

Eclipses of sun and moon are associated with the shadow planets, Rahu and Ketu, who reside at the ascending and descending nodes of the moon's orbit. Rahu, the seizer, was the demon who appeared in the disguise of a god at the churning of the Ocean of Milk, and joined the gods in drinking the sacred *amrita*, the elixir of immortality. Surya, the sun, and Soma, the moon, informed Vishnu of the imposter's presence, who cut off Rahu's head with his golden discus, itself a sun symbol. Rahu had consumed enough *amrita* for at least his head to be immortal. The demon, a head without a stomach, remains in the sky, consumed by hunger which can never be satisfied and eternal hatred for the sun and moon. His counterpart, his lower half, Ketu, inhabits the descending node, and together at eclipse seasons they are responsible for seizing briefly the sun and moon. Vivid images of Rahu and Ketu together with those of the other planets were placed above the eastern doorway of the temple. The sun, interestingly, is also known at the 'great seizer', Mahagraha, as he is believed to seize the light of the moon during the day (Long, 1975; O'Flaherty, 1975; Stutley and Stutley, 1977; Dimmitt and van Buitenen, 1978).

During the recent total eclipse of 16 February 1980, shopkeepers of Delhi closed their stores and no Delphi Transport buses ran during the time of minimum light although only 58% of the disk was covered. People abstained from work to express sympathy for the sun in its plight and to avoid the pollution which was afflicting the sun. Many villagers retreated for safety inside their houses, and streets were largely deserted in many villages in both north and south India. Although a predictable and natural event, an eclipse is still considered to be highly dangerous (Gurumurthy, 1981; Lapointe, 1981).

Another explanation of the strange juxtaposition of the sun with Rahu may perhaps be found in the traditional use of the dancing hall in ancient India. Dancing halls of similar design, though normally constructed of wood, provided the space for dance performances in which the orchestra and performers occupied the three northern areas, those of Venus, Sukra, Mercury, Budha, and Ketu. The king and his retinue occupied the largest space in the center, that of the Sun God. The Brahmans occupied the space behind the Raja, that of Brihaspati, Jupiter, the Guru of the Gods. The Ksatriyas, the warrior caste, occupied the spaces of Mangal, Mars, and Rahu. The Vaishyas occupied the spaces of the Moon, Soma and Saturn, Sani (Boner and Sharma, 1972; Beck, 1976). Thus the Raja was seated under the sign of the sun, which rules and illuminates all. But, Narasimha Deva was also a warrior, and the design of the Dancing Hall may have been planned to include a special blessing for the Ksatriyas. The Sun Temple commemorated victories as well as honoring the sun, and it may have been viewed as entirely appropriate for the symbolic representation of the Sun God at the time of the consecration of the temple in January, AD 1256 to occupy the space of Rahu, granting blessings on the warrior King as well upon the Ksatriyas.

30.3 Local traditions of sun worship

Sun worship of an entirely different character occurs each year at Konarak on Magh Saptami, the seventh day of the bright moon of Magh. According to local newspaper estimates, some 35 000–50 000 pilgrims gather to bathe in the Chandrabharga River, watch the sun rise above the waters of the Bay of Bengal, circumambulate the Sun Temple, and perform *puja* to the Navagrahas. In the evening before Magh Saptami, pilgrims purchase black clay pots locally to cook their evening meal. After sunrise they offer rice on a palm leaf to the sun at the ocean's edge. According to my informants, the black pot is broken to signify that everything is being given to the sun and that nothing is held back. By noon, most of the pilgrims have departed for their villages, the food and supply stands have been torn down, and the ground where they camped and the beaches are littered with broken black pots.

Participants explain that viewing the sun rising above the ocean on that morning is like viewing the face of a god. Before the appearance of the sun the attention of the throng was focused on the faint glow on the east. Waves of high pitched cries mostly by women would sweep through the crowd, but otherwise people were silent and unmoving, watching intently for the first glimpse of the sun. When the crimson top of the sun broke above the sea, people leapt into the air with cries, threw rice and flowers toward the sun, ran into the ocean, and/or performed *namaskar* to the sun.

It is significant that the local tradition of sun worship in Orissa contains a form of non-personified, image-free sun worship which was practiced even by workers while engaged in the construction of the temple. In the inventory of the workmen's common room during construction of the temple there is an item identified as *dharma nirajana thakura*, which was a round white object resembling the shell of a tortoise, used for purposes of Dharma worship. Dharma worship is found in various parts of north India today and represents an austere and ancient form of sun worship (Dasgupta, 1962). An old cult, Dharma has been identified with *surya dharma* in the Vedas and signifies the supreme Truth and Law. In Buddhism, Dharma is equated with the Buddha himself, with the great void, *mahasunyata*, and with *nirajana*, the 'stainless' supreme reality. Dharmaraj, a related solar deity of potters of west India, and his horse are white (Bhattacharya, 1952).

In the Samba Ashram, adjacent to the Sun Temple, the void, *nirakar brahma*, is worshipped in the form of a stone containing an image of the sun and moon. According to an informant in every house of the local villages, Dharma is worshipped with no image. A corner is prepared with mud and cow dung, decorated with a *kolam* using rice flower, and a lamp is placed on the floor. No image is necessary. I was told: 'we already have the sun and moon. Dharma is the root of the gods.' In the local villages, the cobra is also a Dharma symbol; its spotted hood, swiftness, association with fertility, and dangerous aspect give it power as a sun symbol.

30.4 The decline of the Temple

The form of sun worship practiced at the Great Temple apparently did not have the effect of harmonizing and symbolically supporting the indigenous forms of sun worship already in place in Orissa, but appeared to have disruptive and psychologically disturbing features associated with it. The two folk stories of the deaths of the two youths, the beautiful daughter and the talented son, indicate some deep resentment toward the temple and a discontinuity between King Narashima Deva's form of official devotion to the sun and local tradition. I suggest a number of reasons for the failure of the Sun Temple at Konarak, each of which may have applicability to the wider puzzle of the general decline of temple-centered sun worship in India after the 13th-century.

Clifford Geertz (1973, pp. 142–68) suggests that systems of symbols may be models *of* reality as a map or they may be models *for* reality as in a set of instructions for human activity which serve to give meaning 'to social and psychological reality by shaping themselves to it and by shaping it to themselves'. The powerful symbols of our world possess the double aspect of being both symbols or models 'of' and 'for'. The traditional Hindu temple is a complex set of cosmological double symbols involving both models of and for reality. The temple is symbolic of the relationship between chaos and creation, darkness and light, unmanifest and manifest, and center and periphery. The center of the temple, the womb chamber, is symbolic of a dark cave in the sacred mountain as well as a fertile womb of the earth. As they enter the symbolic space of the temple, devotees experience it as a model for reality as they move into the dark and sacred center and then return to profane space having been touched by the power and purity of the center.

What is the symbolic meaning of the Sun Temple? It is a highly syncretic structure combining different symbolic traditions, perhaps not all well harmonized. Contained in the temple and its sculptures are symbols of war, eclipse, kingdom, sex, time, power, movement, primordial darkness, cosmic mountain, zodiac, and planets. A lack of coherence between the symbolic parts of the temple and between the temple and surrounding cultural traditions may have been seeds for its destruction.

Failure of the symbolic sun

At one level, the story of the death of the daughter of the rishi involves a failure on the part of the sun, as the god of fertility, to fulfil his symbolic role. The sun was punished because of his disdain for the public display of affection between Kama, the God of Love and his spouse, Rata. Such disdain is clearly inconsistent with the role of the sun as a god of fertility. His humiliation and the eventual destruction of his temple may be related to his rejection of his own nature. A prudish fecundator is absurd and illogical, but why should that issue be associated with the construction of the Great Temple? Perhaps in the eyes of the local inhabitants it was also absurd and illogical that the sun which rises above the horizon and which needs no other icon than itself should also require the elaborate ritual and extravagant architecture of the Temple. A god requiring such a temple may have been guilty of the sins of hubris. The sun had 'airs' as judged by Brahma, and he had to be punished for his self-conceit and priggishness. Even the erotic carvings on the sun's temple did not protect it from the consequence of the sun's rejection of Kama.

The suicide of the son of the master craftsman

of the Temple was another ominous portent of the destruction of the Great Temple. Perhaps this story is a similar admonition concerning the arrogance of both god and king to build such an extravagant temple, especially in the midst of a tradition which did not use or need such an approach.

The hiding of the light

If there was an incongruity between the natural sun and the temple sun, the discrepancy was even more intense as Mahabhaskar, the Great Creator of Light, was placed in the darkness of the *garbha griha*. The role of the Vedic sun god was to smite the darkness and destroy the enemies of light. Again and again in the Vedas light appears as the primary metaphor for order, knowledge, creation, and life itself. Light, both physical and spiritual, was the essence of the sun, and it seems highly inappropriate for the god of light to be locked in darkness.

The passive quality of the god of the temple who awaited the mediator image to return from the sunlight appears also inconsistent with the god who was *savitar*, the impeller, the initiator, and the energizer. Here perhaps is a case of a ritual which did not function properly. The traditional form of temple ritual in which the deity is maintained in the dark womb chamber was apparently copied without significant modification. While the darkness of the center of the temple is an evocative symbol of the chaos at the beginning of time, it is less appropriate as the house of the sun. The placing of Bhaskar in the *garbha griha* generated symbolic dissonance. The two symbols are mutually exclusive. The symbols and ritual, the meaning and the action were out of step.

Non-solar architecture

Not just the ritual of the temple, but its very architecture appears to be inconsistent with sun worship. The traditional location of the *garbha griha* in the great stone tower representing the cosmic mountain, generates a powerful set of mutually supporting symbols. It was out of the dark and fertile earth that creation emerged; the earth is connected to the cosmos by the axis of cosmic mountain. However, to attach 24 wheels to the cosmic mountain and the womb chamber and to pull it across the cosmos with seven horses is a mixture of incongruent symbols. The use of the mediator image

appears to be a contrived attempt at adaptation of sun worship to traditional temple design.

The design of the Natamandir and its use as a site for the moving image of the sun god to greet the sun at sunrise was strange, to say the least. The need to move the sun near solstices into the square ruled by Rahu may have been motivated by the memory of the terror of the eclipse which coincided with the start of the temple or may have been appropriate for the warrior king who wanted himself and his warriors blessed by the sun. But such specific design limited the symbolic meaning of the ritual and of the site and may have been disturbing and confusing to later generations.

Tradition of non-iconic sun worship

Set down in a culture with a tradition of non-iconic Dharma worship, the temple appears to have been an attempt to implant an alien religion or at least a new style of sun worship upon the local population. Such a symbolic discontinuity may have engendered the stories which anticipated the destruction of the temple and the eventual cessation of priestly ritual in the temple. The temple was built at a time when worship of Shakti, Shiva, and Vishnu were growing and worship of Surya had faded. Great temples to Shiva and Vishnu had been built within the two previous centuries in neighboring Bhubanesvar, the Lingaraja Temple, and in Puri, the Jagannatha Temple. The construction of the Great Temple at Konarak may have occurred because sun worship, although primarily of a different nature, was already a long standing local tradition.

It may have been part of the King's plan to establish his Sun Temple in a region in which the pre-existing devotion to Surya could be used to evoke and support the immense effort of human labor and creative talent necessary for the construction of the temple. Here is a striking example of an attempt to validate a new structure by an old sacred tradition. It appears to be a case of Roemer's Rule which may be stated: 'the initial effect of an evolutionary change is conservative in that it makes it possible for a previously existing way of life to persist in the face of changed conditions'. Rappaport's (1979) interpretation of Roemer's rule suggests that the 'threatening aspects of changed conditions can get somewhat neutralized by being incorporated into sacred tradition ... Sanctity allows the persistence of traditional forms in the face of "structural threats and environmental fluctuations"'. Within Mexico, and particularly in the Aztec and Toltec traditions, Carrasco (1982) has argued that Quetzalcoatl provided such sanctity for social systems and provided thereby social stability and system self-maintenance. In the case of the Sun Temple, the ancient traditions of sun worship in Orissa state may have provided the energy for the creative artistry and prolonged hard labor as well as the expenses demanded by such an ambitious temple at a time when temple-centered sun worship had passed its prime.

However, such an approach appears to have been ultimately counter-productive at Konarak, as it certainly was the case in the downfall of Tenochtitlan when Moctezuma believed Cortez to be Quetzalcoatl. The local folk stories of Konarak suggest the presence of an undercurrent of resentment toward the foreign and extravagant style of sun worship. The collapse of the Sun Temple was apparently neither violent nor sudden. The sources of funds needed to maintain the elaborate rituals, the large number of priests, and the structure itself may simply have dried up. Following, perhaps, a somewhat similar pattern, the collapse of the classic Maya civilization may have been initiated by the construction of elaborate temples, the consequent separation between commoners and elites, and the increased burden on the commoners (Culbert, 1973).

The priests of the Great Temple at Konarak apparently neglected to keep its roof in repair and a massive stone lion balanced on the eastern side of the tower may have become loose, bringing about the collapse of the structure some time in AD 1620 (Boner and Sharma, 1972). The center of the great stone temple did not hold and the golden orb of the sun at dawn returned to symbolic dominance.

Acknowledgment

This work was supported in part by a grant from the Council for International Exchange of Scholars, Indo-American Fellowship Program.

References

Beck, B. E. F. (1976). The symbolic merger of body, space and cosmos in Hindu Tamil Nadu. *Contributions to Indian Sociology* **10**, 213–43.

Bhandakar, R. G. (1983). *Vaisnavism, Saivism and Minor Religious Systems*. New Delhi: Asian Educational Services.

Bhattacharya, A. (1952). *The Dharma-cult*. Bulletin of the Department of Anthropology, 1. University of Calcutta.

Boner, A. and Sharma, S. R. (1972). *New Light on the Sun Temple of Konarak*. Varanasi: Chaukhambha Sanskrit Sansthan.

Carrasco, D. (1982). *Quetzalcoatl and the Irony of Empire*. University of Chicago Press.

Culbert, T. P. (ed.) (1973). *The Classic Mayan Collapse*. Albuquerque: University of New Mexico Press.

Dimmitt, C. and van Buitenen, J. A. B. (eds. and trans.) (1978). *Classical Hindu Mythology: A Reader in the Sanskrit Puranas*. Philadelphia: Temple University Press.

Dasgupta, S. (1962). *Obscure Religious Cults*. Calcutta: K. L. Mukhopadhyaya.

Eck, D. (1982). *Benares, City of Light*. London: Routledge and Kegan Paul.

Eliade, M. (1967). *Patterns in Comparative Religion*. New York: Meridian Books.

Freed, R. S. and Freed, S. A. (1964). Calendars, ceremonies, and festivals in a North Indian village: necessary calendric information for fieldwork. *Southwestern Journal of Anthropology* **20**, 67–90.

Geertz, C. (1973). *The Interpretation of Culture*. New York: Basic Books.

Gurumurthy, K. G. (1981). Eclipse ritual in South India. *Man in India* **61**, 346–55.

Kay, G. R. (1981). *Hindu Astronomy*. New Delhi: Cosmo Publications.

Kramrisch, S. (1946). *The Hindu Temple*. Calcutta University Press.

Kramrisch, S. (1981). *The Presence of Shiva*. Princeton University Press.

Lapoint, E. C. (1981). Solar contagion: eclipse ritual in a North Indian village. *Man In India* **61**, 327–45.

Long, J. B. (1975). Life out of death: a structural analysis of the myth of the churning of the Ocean of Milk. In *Hinduism: New Essays in the History of Religion*, ed. B. Smith. Leiden.

Mabbett, I. W. (1983). The Symbolism of Mt. Meru. *History of Religions* **23**, 64–83.

Malville, J. M. (1985). Sun worship in contemporary India. *Man in India* **65**, 207–33.

Mansimha, M. (1968). *Reflections on the Wonder and Enigma of Konarak in Konarak*. Bombay: Marg Publications.

Mishra, S. B. (1981). *Konarak*. Bhubaneswar: Biswajit Mishra.

Mitra, D. (1976). *Konarak.* New Delhi: Archaeological Survey of India.

O'Flaherty, W. D. (1975). *Hindu Myths.* New York: Penguin Books.

Oppolzer, H. von (1962). *Canon der Finsternisse.* New York: Dover.

Pandey, L. P. (1971). *Sun Worship in Ancient India.* Delhi: Motilal.

Rappaport, R. (1979). *Ecology, Meaning, Religion.* Richmond, CA: North American Books.

Srivastava, V. C. (1972). *Sun Worship in Ancient India.* Alahabad: Indological Publications.

Stutley, M. and Stutley, J. (1977). *Dictionary of Hinduism.* San Francisco: Harper and Row.

von Stietencron, H. (1966). *Indische Sonnenpriester (Samba and die Sakadvipiya Brahmanas).* Wiesbaden: Otto Harrasowitz.

Zimmer, H. (1955). The Art of Indian Asia, ed. J. Campbell. Princeton University Press.

31

Cognitive aspects of ancient Maya eclipse theory

Michael P. Closs   *University of Ottawa*

It remains questionable to what extent we can expect to derive significant information about complex non-Western systems of astronomy by focusing only upon the observational aspects of those systems. We must strive to comprehend the whole picture as seen through their eyes.

(Aveni, 1981, p. 13)

31.1 Introduction

This paper is concerned with determining the nature of eclipse phenomena as it was perceived by the ancient Maya. It approaches the problem by considering the linguistic information pertaining to eclipses and by exploring the traditional beliefs associated with the occurrence of eclipses among the postconquest Maya. These data yield a model of a native eclipse theory which is compatible with hieroglyphic and iconographic materials pertaining to the ancient Maya. Such a cognitive approach to Maya astronomy has been touched on in previous studies but only in a peripheral sense. Indeed, most studies of eclipse phenomena in the Maya area have concentrated on the interpretation of the eclipse table of the *Dresden Codex*, a Maya hieroglyphic book. These studies have been structural in nature and have drawn on data susceptible to arithmetic analysis. In a few instances attempts have been made to identify eclipses and eclipse glyphs elsewhere in the Maya inscriptions but little progress has been made in this direction. It seems likely that eclipses would have been recorded only if they happened to occur on a date which was being commemorated for some other event.

In considering sources for this work, I have attempted to be exhaustive in so far as material on the Yucatec, Lacandon, Mopan, Chol, and Chorti are concerned. Information from other groups is only dealt with selectively when it is felt that it sheds light on the matters being discussed. This approach is dictated by the conviction that the languages underlying Maya hieroglyphic writing are Yucatec and Chol. Terrence Kaufman (1976, pp. 108–12) asserts that linguistic geography and ethnohistory make it almost inescapable that the central Maya area was occupied by Cholan speakers. He estimates that the Cholan subgroup split into Chorti and Chol-Chontal about AD 600, at the end of the Early Classic, and that the date of separation of Chol and Chontal is somewhat later. In the northern Maya area the inhabitants spoke Yucatec. Lacandon and Mopan occupied an intermediate geographical position between Yucatec and Chol, although their linguistic affiliation is closer to Yucatec (Campbell, 1984, pp. 2–3). Together the northern and central Maya areas constitute the Maya lowlands and coincide with the region in which hieroglyphic writing was a principal feature of Maya civilization.

31.2 Linguistic data

In Yucatec, eclipses of the sun and the moon are described by a number of terms. The following list is selected from the *Diccionario Maya Cordemex* (Barrera Vásquez, 1980, pp. 33, 93, 824):

chi'bil k'in *eclipsarse el sol, porque en los eclipses la sombra o parte eclipsada es corva, como la señal que dejan los dientes en las mordidas o bocados quitados* [to be eclipsed the sun, because in eclipses the umbra or part eclipsed is arched, like the mark which is left by the teeth in things which are nibbled or bitten; the Maya term is literally 'to be bitten the sun']

chi'bil u *eclipsarse la luna* [to be eclipsed the moon; literally 'to be bitten the moon']

tupul u wich k'in *eclipsarse el sol* [to be eclipsed the sun; literally 'to be extinguished its face the sun']

tupul u wich u *eclipsarse la luna* [to be eclipsed the moon]

bala'an u wich k'in *eclipsado está el sol* [the sun is eclipsed; literally 'hidden its face the sun']

chi'ba k'in *eclipse de sol* [eclipse of the sun; literally 'bitten sun']

chi'ba u *eclipse de luna* [eclipse of the moon; literally 'bitten moon']

Cholan vocabulary is not as well documented as Yucatecan, and I have not been able to isolate eclipse terminology in the sources available to me. However, at the Tercera Mesa Redonda de Palenque held in Palenque, Chiapas in 1978, I had the opportunity to query a Chol informant about eclipses of the sun and moon. He responded that eclipses were referred to by the expression *woli k'uxbahlum* 'the jaguar is eating it'. The informant was unable to give a rationale for the origin of the expression. In this instance, the sun or moon is once again perceived as being eaten but, in addition, the jaguar is specified as the active agent.

31.3 Ethnohistorical data

There are many reports of folk beliefs concerning eclipses in the ethnohistoric literature. I begin the survey by considering an extract from an account of the expedition of Alonso Dávila to Chetumal. Francisco de Montejo had sent Dávila with 50 Spaniards and 16 horses to seek out gold mines in the vicinity of this Maya town. Dávila reached Chetumal in 1531 and there founded the town of Villa Real de Chetumal. The lives of Dávila and his men were filled with battles, skirmishes, and murders and they abandoned the town in 1532. A lunar eclipse occurred during Dávila's approach to Chetumal, and the reaction of the natives to it is described in an account published in Spanish by Manuel Rejón García. My translation of Rejón García's account, referred to as Text A, is given below. I subsequently present other reports from other sources without comment. For convenience,

these reports are labeled in alphabetic sequence as they are encountered.

Text A. Yucatec Maya of Chetumal, 1531 (Rejón García, 1905, pp. 10–11)

Dávila, meanwhile, was drawing near, becoming convinced of a rebellion of the Indians because he found the trails, which served as roads, closed. He arrived, without realizing it, to within a short distance of the fortified town.

As was said earlier, the Indian chief had ordered that they [i.e. his men] maintain absolute silence. This was observed until shortly before dawn on that beautiful night of full moon. In its last hours, a shadow began to be projected onto the Queen of the Night, a shadow which advanced slowly until it obscured the moon completely.

The order of the chief was definite and the discipline of his army strict. However, common tradition obliged them to bring themselves to the defense of Diana, who in their judgement called, soliciting the protection of mortals against the hard bite of the *xulab* which had hold of her [*contra la pica dura del xulab que la tenía*], unto the point of death, in the way it had superimposed itself. She was aided at once with a serenade able to deafen anyone who was less anxious than Dávila who made use of it to orient himself.

Those warriors set themselves to make all manner of noise in order to defend the moon. One was able to hear the raucus seashell, the monotonous cymbal (*tun kul*), the *lahte*, the *zacatán*, the *ddoroch ac*, the *chul*, a type of clay flute with a feather tip, all the instruments of their warlike music, and furthermore, shouts, laments, blows on the trees and mace blows on the stones. [It was] an infernal uproar, a noise so deafening that had it last longer than the immersion of the planet in the shadow, or not to know its cause, would have inspired fear in the spirit of the Spaniards.

The concert ended with the phenomenon, a little before dawn. The simple natives were left satisfied with having accomplished the moon's salvation, and the invaders much more, because knowing exactly the place to which they should direct themselves, they arranged the attack that took place on the morning of that day. They charged with bravery and ferociousness against the Indians. The latter followed the command to resist with tenacity although, as imprudently happened, they had not followed the command to be silent. They confronted the enemy until they became convinced that their families had evacuated the town. Then they withdrew in good order.

Text B. Yucatec Maya, 1639 (Sanchez de Aguilar, 1921, pp. 302–5)

During lunar eclipses they still believe in the tradition of their forefathers to make their dogs howl or cry by pinching them either in the body or ears, or else they will beat on boards, benches and doors. They say that the moon is dying, or that it is being bitten by a certain kind of ant which they call *xubab* [*sic*]. Once, while at the village of Yalcobá, I heard great noises during an eclipse of the moon which occurred that night, and in my sermon the next day I tried to make them understand the cause of the eclipse in their own language, according to the interpretation from the Philsopher: 'The lunar eclipse is the interposing of the earth between the sun and the moon with the sun on top and the moon in the shadow.' With an orange to represent the sphere of Sacrobosco, and two lit candles on either side, I explained to them plainly and at sight what an eclipse really was. They seemed astonished, and quite happy and smiling, cured of their ignorance and that of their forefathers. I gave orders to their chieftain (cacique) that he should punish in the future all those who made a noise on such occasions.

Text C. Yucatec Maya, 1861 (Mendez, 1921, pp. 186–7)

A pregnant Indian woman will not go outdoors during an eclipse, in order to avoid her child being born with spots or ugly birth marks on its body; nor do they visit women who have just given birth to a child, because it is their belief that the babies would become ill with pains in their bowels.

Text D. Yucatec Maya of Chan Kom, 1934 (Redfield and Villa Rojas, 1962, pp. 206–7)

Some of the people of the village begin to regard an eclipse as brought about by the motions of

the sun and moon. One suggests that the moon hits against the sun, knocking out its light; another says that every eighteen years the moon passes under the sun covering the earth with its shadow; a third proposes that there is probably a hole in the sun through which the moon has to pass at certain intervals. But these speculations have not seriously affected the general belief that periodically some evil animal seeks to devour the heavenly luminaries and that eclipses are occasions of great danger for mankind.

The name for eclipse is 'chibil kin' ('biting the sun'), or 'chibil *luna*'. Many cannot specify the animal which seeks to eat sun or moon, saying merely that it is 'a very bad thing' (kakazbaal); but several say that it is certain ants, either certain evil-smelling ants of a red color, or else the 'king of the leaf-cutting ants.' That is why one so often sees the ants carrying bits of leaf: as fodder for the ant king's horse. But one or two of the older men say that it is no ant, but an animal like a tiger that seeks to devour sun or moon. That is what the ancients taught by carving on stones at Chichen Itza a disk representing the sun and two tigers coming to eat it.

The moment of eclipse is one of threatened calamity, because, it is believed, should the sun or moon fail to reappear, then all the furniture and other objects with which people are surrounded would be changed into devils or beasts that would devour all living things. So the villagers strive to avert the cataclysm. They seek to frighten away the devouring animal by making as much noise as possible, beating on drums, cans and pails, and firing guns. If the danger of complete extinction appears imminent, they may also go to the *oratorio* and there begin a novena to San Diego.

Text E. Yucatec Maya of X-Cacal, 1932–6 (Villa Rojas, 1945, pp. 140B, 156A–7A)

A woman must be especially careful, after the first months of pregnancy, to avoid going out during an eclipse, for the influence of an eclipse is believed responsible for the fact that certain children are born with reddish or dark spots on the face or other parts of the body. According to some informants, however, the danger was not in exposing oneself to the eclipse, but rather in scratching the abdomen during it. Reddish spots are attributed to eclipse of the moon and dark ones to eclipse of the sun; the former are called *chibal-luna* (moon bite), and the latter *chibal-kin* (sun bite). Persons with such spots are rare. The only one I saw was a young woman who had on her right arm a black spot which she took great pains to conceal by trying to hide the arm in her clothing.

Eclipses are believed to occur when certain 'very bad animals' try to eat the sun or the moon. Thus, eclipses are called chibal-kin ('biting of the sun') or chibal-luna ('biting of the moon'). A few Indians express the idea that the 'bad animal' is the ant known as *xulab*. According to one version, the eclipse occurs because its surface is covered by an immense colony of these ants. According to a second version, the sun or moon is attacked by a monster similar to the queen of the xulab. The belief attributing eclipses to ants of this family seems to be of pre-Hispanic origin, for according to Sanchez de Aguilar, it was known among the Indians in the very earliest colonial times. Today, however, the idea is not well known in X-Cacal, for most of the Indians could tell me only that eclipses were due to an attack against the heavenly bodies by some 'very bad' animal or perhaps by the devil himself. [In a footnote to the text it is pointed out that the xulab is a large ant, black or reddish, which forms large colonies. It is feared because of its painful bite and expecially because of the destruction it works on trees and all sorts of plants. In the rainy season it causes great alarm when, in search of shelter, it invades the houses of natives. It is also said that there is among this family of ants one of greater size with a red head and much feared because its bite is severe.]

Eclipses are greatly feared because it is believed that a total eclipse of sun or moon would cause all the domestic instruments then to be transformed into living creatures and kill their masters, revenging themselves for the bad treatment they have suffered. Another belief is the one mentioned in preceding pages, namely, that Juan Tutul Xiu will destroy the earth on the day when the sun is covered with a black curtain. To drive away the enemy attacking the sun or moon, it is customary to fire off guns in the directions of the threatened luminary but no other noises are made, such as beating drums, as noted in other Maya regions. When an eclipse occurs, pregnant women are required to keep indoors so that their children may not be born with spots on their faces or bodies.

The fear inspired by eclipses is so great that any exceptional condition of sun or moon is noted with alarm. On two occasions when I was in Tusik the heavenly bodies were 'threatened' by eclipse. On the morning of April 3, 1936, just as the moon was about to set, the first symptoms were noted. These consisted of the progressive reddening of the moon and a sudden decline in its brilliance. The phenomena could have been easily explained by the fact that the burning of milpas, begun a few days before, had filled the air with smoke. But the natives immediately began invoking divine aid by Catholic prayers in the church, and for a long time the men were busy shooting off their guns toward the moon to save it from the threatening animal. The same activities took place in all the villages of the subtribe except the shrine village, where firearms could not be discharged because of the sanctity of the place. Some hours afterward the same symptoms appeared in the sun, but nothing was done to save it because 'what was done at dawn is enough'. That day the men went to their milpas as usual, but they had been considerably frightened and for several days the averted eclipse was the principal subject of conversation. It was generally thought that it presaged some calamity such as war or sickness.

Text F. Lacandon (Tozzer, 1907, pp. 96, 157)

The name of the other culture hero of the early natives of Yucatan, *Ququlcan* (written *Kukulcan*), is still retained among the Lacandones as the name of a mythical snake with many heads, living only in the vicinity of the home of *Nohotšakyum*. This snake is killed and eaten only at the time of great national peril, as during an eclipse of the moon and especially that of the sun.

The mythical serpent, *Ququlkan*, of the Lacandones is called *Ququikan* among the Mayas. It is described as a many-headed snake living in the sky. At intervals it comes to the earth to a place below the home of the red ants (*sai*).

Text G. Lacandon (Baer and Baer, 1952, p. 239)

The Lacandon fear the world will perish during an eclipse. At such times men chant to the god Ac Yanto, praying him to request Hachacyum to stop the calamity before the world comes to an end.

Text H. Lacandon (Tozzer, 1941, p. 138, n. 639)

The Chacs are found among the present Mayas and Lacandones. The latter have a god Noho Chac Yum, the Great Father Chac, who is at the head of the pantheon. He was supposed to be ill when there was a solar eclipse and his daughter sick in a lunar eclipse.

Text I. Lacandon (Bruce, 1979, p. 322)

ECLIPSE (*tak'äl*) OF SUN OR MOON is a fatal omen, foretelling the death of a man or a woman, respectively. The fatal significance of an eclipse, according to Maya cosmology, is so strong and clear that an eclipse may also be called by the alternate terms *xu'tan* or *xulik t'an/xulik tan*, which may be freely translated as 'the end of the world'. The recurrent destructions of the Eras of the world are said to begin with an eclipse. The end of the world is prophesied by vivid colors, such as a sunset, in a dream, thus tying the usually fatal prophecies of colors with the equally fatal omens of dreams of astronomical phenomena.

Text J. Lacandon of Lake Miramar, 1975
(Boremanse, 1981, p. 5)

A long time ago a moon eclipse occurred. They say that Our Ancestors were shooting arrows, waving macaw tails, praying to the gods, asking that Our Mother should be released. They looked and looked, and saw she was not there. She had been entirely swallowed up! Oh! Our Ancestors were really much worried. Thus they said: 'The jaguars will devour us! They will eat us!' The moon was not there ... Darkness. There were only stars in the sky. Not our Mother. She was not there. Don't you see, she had been swallowed up by an iguana. Indeed, an iguana had landed at the beginning of the arm [of Lake Miramar].

Long ago there were reeds there, and there went the iguana which had swallowed up Our Mother. Our Ancestors went to look for her. They were worried. 'Ah! It swallowed her up!' They went to shoot it, at the arm of Small Sea [Lake Miramar]. They said to their kinsmen: 'We have to look for Our Mother.' They went to look for her. They all took their arrows. Plenty of them went to shoot the iguana. But ... which iguana was it? ... That one? ... That one? ... The moon was not there ... Don't you see, the iguana which had swallowed up Our Mother would have its belly illuminated ... An iguana is really small, and that one should be big, enormous! ... 'That one did not swallow Our Mother' ... they said. Our Ancestors walked and looked for her ... Nothing ... 'Neither that one' ... Until they saw where she was, swallowed up. 'That one!' they said.

They saw a really big iguana. They saw one illuminated. They said: 'That one has swallowed up Our Mother!' They said to their kinsmen: 'What shall we do?' 'I don't know.' 'What do you think?' 'Maybe ... shoot it.' 'What if we shoot Our Mother?' 'You must shoot at its throat.' Our Ancestors hit the iguana exactly at its throat. They did not shoot at its body. Don't you see, there was Our Mother. They shot at its throat instead. The iguana would not die even though they shot arrows at it. They all shot at its throat, until their arrows cut it off. Then the iguana fell down. It was really heavy! It had swallowed Our Mother. It fell down. Our Ancestors cut it up right away. They cut it up, and out came Our Mother, She went up to the sky. So, there was Our Mother, as now, as always.

This is a very old tale, don't be afraid. If you see a moon eclipse, you will know it is because an iguana has swallowed her. Our Ancestors cut up the iguana, and up to the sky went Our Mother, and the sky was illuminated. If they had not shot the iguana ... Don't you see, these iguanas are celestial iguanas. They are not iguanas of the earth. They are iguanas from above, where is Our Mother.

[In a footnote, the author adds that the Lacandon believe that a total (solar or lunar) eclipse precedes the end of the world. When this occurs the heavenly jaguars will land on earth in darkness and will devour everybody.]

Text K. Mopan, 1935 (Thompson, 1972, p. 75)

On July 16, 1935, my compadre Jacinto Cunil wrote me in his somewhat shaky Spanish of events in the Maya village of Socotz, British Honduras: 'On the night of the 15th–16th of this month an eclipse of the moon was seen. All the people were frightened, saying the world would end that night. A lot of noise was made beating tin cans, boxes and bells. [There were] prayers and processions in the night at one o'clock. All were awake [believing] that that was the moment of the [last] judgment.'

Text L. Chol of Tumbala (Beekman, 1956, p. 262)

Many of them still adore the sun and the moon ... In case of an eclipse, they believe that the moon (which they call 'mother') is being attacked by a jaguar and that she needs help. Thus they shoot their guns, beat drums and other people sing, while still others just cry and shout in order to save her.

Text M. Chol (Whittaker and Warkentin, 1965, pp. 71–3)

When [the] moon is hidden [eclipse]. '[The] tiger is killing our holy mother. We will perish,' they say. 'Our holy mother has gone into [the] stomach of [the] tiger. He is coming to eat us now also,' they say. They shoot him with [the] gun. 'Bang,' it says when they shoot him. [The] tiger is afraid of [the] gun, it is said. They immediately bring corn shellers [pieces of iron used for shelling corn]. They gather them and old machetes and also machetes that are sharp. This is a figure of helping the moon, it is said, because they are stabbing the tiger, they say, because he is gathering the corn shellers, old machetes. They say [that the] tiger is eating the moon. [Ch'ay bajlum] is the name of the tiger. They wet them in water. They gather together there beside the old machetes. They pierce [the] gourds. They are very quiet. They are fearful. Some are crying because they say they are going to come to an end. They only do this at their own houses. They beat drums and tin cans so that our holy mother will be freed. They have a fiesta so that she will be freed. They drink much liquor. They dance there in [the] house.

On Sunday they go to put candles before [the] wooden god [a wooden cross]. 'We almost perished,' they say. 'Help us now,' they say, 'so that we still will not perish.'

Text N. Chorti (Wisdom, 1940, pp. 285, 399–400)

Pregnancy, however, is believed to be a ritually unclean condition, for which reason certain acts on the part of the pregnant woman are forbidden or frowned upon, as liable to cause harm to herself, her child, or another person or thing ... She is worried if an eclipse of the sun or moon occurs during her pregnancy, as this is said to cause great danger that her child will be born without one of its external body parts, such as a part of one of the extremities or a portion of the face or head. At such a time she is careful not to leave the houses.

Another deity, *ah kilis*, is associated vaguely with the sun and is said to eat the latter when angry, thus causing the sun's eclipse. [In a footnote, Wisdom adds that the derivation of *ah kilis* is unknown but that the deity is also called 'Sun-eater'. Several of his informants thought that this deity ate the moon also, thus causing its eclipse, but were not certain. John Fought (1972, p. 437) observes that *ah kilis* is an adaptation of Spanish *eclipse*.]

The moon deity is the patroness of childbirth and also has some connection with plant growth. She is said to lose partially her powers of fecundity during an eclipse, for which reason women fear an eclipse during pregnancy.

Text O. Chorti (Fought, 1972, pp. 428–9)

The Eclipse is feared by people when it lies in the sky. Because it is said that ... that eclipse, if given time by God, that eclipse would eat the sun or the moon. And so, they say that when the sun or the moon is grabbed by the eclipse, then the moon offers up mankind on the earth, so as not to be eaten by the eclipse. Because it is said that if one day passes, or two days, there will be no sun; it will become dark on the earth. All the animals there are will rise up in the peaks, in the seas, and all the spirits who are dead, just like those who are completely dead. And those spirits will come to life and eat the people who are alive on the earth.

Text P. Tzotzil of Zinacantán, 1970 (García de León, 1973: 307); [translation by author]

> People who don't know and who look at the moon during an eclipse think that the star *muk'ta k'anal* (Venus) is going to kill her because they are fighting.
>
> When they look they begin to shout, fire shotguns and bring out their flutes and drums in order to scare away the Devil, the Star; and the old people bring out sticks and strike themselves, and take *aguardiente* so as not to feel the moment of her death. They also think that with this they give help to the moon so that the Devil will not continue to strike her, and also (thereby) to frighten him away.

31.4 Synthesis of eclipse themes

The preceding linguistic and ethnohistorical data pertaining to eclipses repeatedly refers to the same ideas. It is clear that the material is a part of a single tradition which derives from ancient Maya roots. From the various sources, it is possible to extract the major themes which make up that tradition and thereby to reconstitute, at least approximately, the eclipse beliefs of the ancient Maya.

The dominant theme running through almost all accounts is that during eclipses of the sun and/or moon, the affected heavenly bodies are being eaten (Yucatec and Chol linguistic data; Texts A, B, D, E, J, L, M and O). In some of the references to lunar eclipses, it is said that the moon is being attacked (Texts E, L and P). In other references, the sun or moon is said to be sick or dying (Texts A, B and H). These statements form a common strand in which the sun or moon is attacked, bitten, being eaten, and in danger of dying.

In the accounts, eclipses are not natural occurrences but are caused by the intervention of an active agent. A variety of eclipse agents is specified in the various sources: the jaguar (Chol linguistic data; Texts D, L and M), the *xulab* (Text A), the ant called *xulab* (Texts B and E), certain red ants or the king of the leaf-cutting ants (Text D), a celestial iguana (Text J), an evil animal (Texts D and E), the deity *ah kilis* (= eclipse) (Texts N and O), the Devil (Texts E and P), Venus (Text P), and, by inference, the mythical snake Kukulcan (Text F). It is important to note that the eclipse agent involved in both solar and lunar eclipses was perceived to

be the same creature (Chol linguistic data; Texts D, E, N, O and, by inference, F). In addition, it may be noted that if these beliefs really stem from a common source, as is being advocated, then the eclipse agents which are enumerated here should be derivable from the same ancient deity. This problem is examined in detail in the next section.

The belief that eclipses were caused by a celestial creature devouring the sun or moon is much more than a metaphor. Indeed, this is amply demonstrated by the strong reaction which eclipses evoked among the common people and by the measures which they took to avert the impending death of the eclipsed body. The common tradition impelled the Maya to rush to the defense of the endangered sun or moon (Texts A, D, L, M and P). They attempted to frighten away the devouring creature by making all manner of noise (Texts A, B, D, K, L, M and P) and by discharging arrows (Text J) or firearms (Texts D, E, L, M and P). The people also sought divine intercession in their calamity through prayer (Texts D, E, G, J, K and M).

Eclipses aroused considerable fear among the common people (Texts E, K and O). Underlying this fear was the belief that eclipses were a great danger to mankind (Texts D, E and F) and portended the end of the world (Texts D, E, G, I, J, K, M and O). The sources specify three ways in which the destruction of man would be accomplished. One of these is that furniture and other inanimate objects would be changed into devils or beasts that would devour all living things (Texts D and E). The second is that celestial jaguars would descend to the earth and devour the living (Texts J and M). In the latter reference it is interesting to note that the celestial jaguar of the apocalypse is identified with the eclipse agent. The third appears to be a combination of the preceding two ideas: the moon offers up mankind to be eaten by the eclipse agent to secure her own safety and, at the same time, if the sun is eaten then the animals and spirits of the dead will arise to eat the living (Text O).

Finally, the sources attest that pregnant women were especially vulnerable during eclipses. It was believed that if a pregnant woman went outdoors during an eclipse then her child might be born with spots or birth marks (Texts C and E); reddish spots were attributed to lunar eclipses and dark ones to solar eclipses (Text E). It was also believed that such exposure might lead to the child being born with some external part of its body missing (Text N). In addition, an eclipse was a dangerous time to visit a woman who had just given birth, for the newly born child would then become ill with pains in the bowels (Text C).

31.5 The identity of the eclipse agent

In the previous section it was seen that the sources refer to a variety of eclipse agents, and it was suggested that if these derive from a common tradition then the various agents should coalesce into one. I believe that there is sufficient ethnohistorical data to achieve precisely this type of fusion. Text P (Tzotzil), the only non-lowland source among those which have been described, was included in the survey of ethnohistorical data because it specifically refers to Venus as a cosmological agent implicated in eclipse phenomena. It provides an explicit focal point for the contention that the eclipse agents are derived from a Venus deity.

Another explicit Venus reference occurs in Text F (Lacandon). It names Kukulcan as a mythical serpent which is killed and eaten during eclipses. Kukulcan is the feathered serpent, also known as Quetzalcoatl, and is a recognized manifestation of Venus as morning star. Text F makes sense and is an efficacious solution to the eclipse problem only if Kukulcan is regarded as the eclipse agent. That such a conclusion is warranted is confirmed by Text J (Lacandon), in which a celestial iguana is identified as the eclipse agent and the solution to the eclipse problem is obtained by killing and cutting up the iguana. Given that Text F is at least 68 years older than Text J, it is possible that the celestial iguana in the more recent text is a modern transformation of the mythical serpent appearing in the earlier text.

Text E (Yucatec) and Text P (Tzotzil) both identify the Devil as eclipse agent. The latter text makes it clear that the Devil is an alternative name for Venus. The identification of Venus with the Devil is an old theme in Mayan beliefs. It surfaces in the name *Hun Ahaw* (1 Ahau) which is the ritual day for Venus calculations in the Venus table of the *Dresden Codex*. Bishop Landa (Tozzer, 1941, p. 132) after giving a brief description of the Maya underworld, referred to as *Metnal*, writes: 'They maintained that there was in this place a devil, the prince of all the devils, whom all obeyed, and they

call him in their language Hunhau.' Barrera Vásquez (1980, p. 247) corrects the spelling to *Hun Ahaw*, and elsewhere (Barrera Vásquez, 1980, pp. 245, 351) notes *Hum Ahaw, Hum Haw, Kum Ahaw* and *Kum Haw* as other attested spellings for the name of Lucifer, the prince of devils.

The most detailed account describing *Hun Ahaw* as Lord of the Underworld and identifying him with Venus is found in the *Popul Vuh* of the Quiche Maya. Here it is related that *Hun Hunahpu* (1 Hunter), the Quiche calendrical equivalent of 1 Ahau, and his twin brother *Vuqub Hunahpu* (7 Hunter = 7 Ahau) were both sacrificed by the gods of death in the underworld. After his death the skull of 1 Ahau was placed in a tree which immediately bore numerous fruit. Eventually, the skull of 1 Ahau brought about the conception of the Hero Twins *Hunahpu* and *Xbalanque* in the womb of a young maiden. *Hunahpu* and *Xbalanque*, after reaching maturity, return to the underworld, defeat the gods of death, and reconstitute 1 Ahau and 7 Ahau. Edmonson (1971, pp. 143–4) translates the dénouement as follows:

And this then was the reassembling of their
    father by them,
    And then they reassembled 7 Hunter.
They went there to reassemble them
    in Dusty Court.
. . .
    And thus then they honored him
And left the heart of their father.
    'It will just be left at Dusty Court,
And here you will be called upon
    In the future,'
His sons then said to him.
    Then his heart was consoled.
'First will one come to you,
    And first also will you be worshipped
By the light born,
    The light engendered.
Your names will not be lost.
    So be it,'
They said to their father
    As they consoled his heart.

The Hero Twins then took leave of the underworld and rose to the sky as the Sun and Moon. In the portion of the *Popol Vuh* which has been cited, 1 Ahau, the father of the twins, has been reconstituted in Dusty Court, the ball-court of the underworld. His sons assert that he will remain

at Dusty Court and that there he will be called upon
in the future. It is further prophesied that 1 Ahau
will be the first to be worshipped in the new day
which is coming. This anticipates the apotheosis
of Hunahpu as Sun and of his father 1 Ahau as
Morning Star. The new reality can be seen in a
later passage of the *Popol Vuh* where Edmonson
(1971, p. 176) translates:

> They waited together
> For the coming up
> Of the Great Star
> Called the sun Passer.
> 'It will come up first,
> Before the sun,
> Then it will dawn,'
> They said.

When 1 Ahau was reconstituted and left behind
in Dusty Court he became effectively the Lord of
the Underworld. The previous gods of death had
earlier been destroyed by the Hero Twins and the
reconstituted 1 Ahau was left behind to rule in their
place. In addition to this, 1 Ahau was further
honored by his transformation into the Morning
Star. This was in fulfillment of the promise that
he would be the first to be worshipped in the new
dawn.

The eclipse agent specified in the Chorti sources
is personified as an eclipse beast called *ah kilis*, an
adaptation of Spanish *eclipse* (Texts N and O).
Thus, the Chorti name for the eclipse agent is of
modern origin and does not reflect on the ancient
identity of the creature. The generalized description of the eclipse agent in some of the texts as
an evil animal yields a similar inconclusive result
(Texts D and E).

However, Texts A, B, E and D (Yucatec) provide
more details on the name and identity of the eclipse
agent. Text A, the earliest of the sources, names
the eclipse agent as *xulab*; Text D identifies the
eclipse agent as an ant; and Texts B and E say
that the eclipse agent is the ant called *xulab*. These
references are very explicit and give a characterization of the eclipse agent which, once again, can
be referred to a Venus deity. Indeed, the term *xulab*
serves in a cosmological sense as a reference to
Venus as can be seen in the following lexical items:

Cholti: **xulab** *entrella* [sic; probably *estrella* 'star']
    (Moran, 1935, p. 260);
Lacandon: **xulab/xulaab** star, Venus (Bruce,
    1979, pp. 247, 285);

Mopan: **xülab** *estrella, planeta* [star, planet] (Ulrich
    and Ulrich, 1976, p. 248).

With respect to these items, it should be noted
that among the Maya the term 'star' could be used
as a proper appellation of the planet Venus. This
is indicated by the Lacandon example which has
been cited above. Elsewhere, I have presented other
linguistic and hieroglyphic evidence for this type
of usage (Closs, 1979, 147–8). Further information
on the same point is found among the folk beliefs
of the Kekchi, Kekchi-Chol and Mopan Maya of
Southern and Central Belize. Indeed, in this region,
Venus is called Lord Xulab. He is described as
the elder brother of the Sun and is specifically identified with the Morning Star (Thompson, 1930,
pp. 120, 125, 132).

The relationship between *xulab* 'ant' and *xulab*
'Venus' is not simply a linguistic one. In fact, there
is good ethnohistorical support indicating that the
ant itself is a manifestation of Venus. This appears
in the myth regarding the all important discovery
of maize. It has a widespread range in Mesoamerica
and must have ancient origins. A description of this
event, from the Mexican plateau, is given in the
Leyenda de los soles (Thompson, 1970, p. 348). The
source relates that Quetzalcoatl asked the ant, who
is credited with the initial discovery of maize, where
it obtained the maize which it was carrying. Upon
receiving the information, he transformed himself
into a black ant and accompanied his informant
to the deposit of maize secreted beneath a mountain.
He then took some grains and carried them to the
other gods in Tamoanchan. Eventually, the mountain was broken open by Nanahuatl and the grain
was secured.

Thompson (1930, p. 140) also gives another
description of this myth taken from the *Anales de
Cauhtitlan*. The only difference from the previous
one is that Xolotl, the god of Venus as Evening
Star, splits open the mountain rather than Nanahuatl.

The Mopan and Kekchi-Chol of Belize also
attribute the initial discovery of maize to the ants
(Thompson, 1930, pp. 132–4). The maize was hidden beneath a great rock and no one knew of it
except the leaf-cutting ants (*sai*). The ants supplied
the other animals with maize taken from beneath
a great rock but soon could not keep up with the
demand. The problem was overcome by Yaluk, the
greatest of the Mams (servants of Lord Xulab), who

hurled a thunderbolt at a spot on the rock indicated by a woodpecker and thereby shattered it.

The Bachajon-Tzeltal have a similar myth of the finding of the maize (Thompson, 1970, 351–2). It is a large black ant (*xolop*) who is first seen carrying maize. After some persuasion the ant reveals the source of the maize within a rock. Following some initial assistance from a woodpecker, the red Cha-huuc (one of the Tzotzil rain gods) split the rock open with lightning. In this instance, the large black ant implicated in the discovery of maize is called *xolop*, almost certainly a cognate of Yucatec *xulab* 'ant' and Cholti/Chol/Lacandon/Mopan *xulab* 'Venus'.

The Tzeltal of Tenejapa say that God took maize from the ants who had obtained it from the Anheles (Tzotzil rain gods) (Thompson, 1970, p. 352).

When these stories of the discovery of maize are considered as a group, it is clear that they are related in theme, of ancient origin, and attest to an identity or intimate relationship between Venus and the ant. This identity/relationship is echoed in the linguistic usage of the term *xulab* for both 'ant' and 'Venus' and in the ethnohistoric records which implicate both ants and Venus in eclipse phenomena. The identity/relationship is also productive in the sense that it suggests new insights into some of the material which has been discussed: cf. the role of the leaf-cutting ants called *sai* in the discovery of maize as related by the Mopan and Kekchi-Chol of Belize, the role of the red ants or the king of the leaf-cutting ants as an eclipse agent as mentioned in Text D (Yucatec), and the descent of the celestial Kukulcan to a place below the home of the red ants called *sai* referred to in Text F (Lacandon).

The Chol linguistic data, Text D (Yucatec), and Texts L and M (Chol) specify the jaguar as eclipse agent. The clearest reference to a jaguar aspect of Venus in the ethnohistorical sources is found in the augury for the day 8 Lamat in the books of Chilam Balam of Ixil, Tizimin, and Mani (Thompson, 1971, p. 300). Here one reads of 'jaguar-faced 1 Ahau with the protruding teeth'.

Thompson (1971, p. 218) has also argued that the augury for the day Lamat in the first list of the Book of Chilam Balam of Kaua refers to Venus. It reads: 'Drunkard, deformed dog is his prognostic. The head of a jaguar; the rear of a dog. A meddler, a prattler, dishonest is his speech, an experimenter in mutual hatred, a sower of discord.'

He notes that this description fits the Venus god Lahun Chan who is described in the book of Chilam Balam of Chumayel in the following terms: 'Mighty are his teeth ... Sin is in his face, in his speech, in his talk, in his understanding, and in his walk ... Forgotten is his father; forgotten is his mother ... He shall walk abroad giving the appearance of one drunk, without understanding ... There is no virtue in him, there is no goodness in his heart ...' (Roys, 1967, pp. 105–6). The initial emphasis on mighty teeth recalls a reference of Lopez de Cogolludo (cited in Roys, 1967, p. 101, n. 2) to *Lakunchan* as an idol with very ugly teeth and also recalls the description of 1 Ahau as having protruding teeth.

In summary, the ethnohistorical data on eclipses are consistent with the notion that Venus acted as a cosmological eclipse agent. In this role, the Venus deity was visualized as a jaguar and also in the form of the celestial serpent Quetzalcoatl. A number of sources also describe the agent as being in the form of an ant or ants. However, in these cases the accounts may have transferred the name of the Venus deity Xulab, remembered in some linguistic and ethnohistorical sources but not in others, to the name of a species of ant with which that deity was linguistically and mythologically identified.

31.6 The role of the other planets

The *Codex Pérez* observes that all the planets in the heavens cause earthquakes, thunder, and eclipses of the sun and moon (Craine and Reindorp, 1979, pp. 50–1). In the same section it discusses the significance of thunder on each of the seven days of the week and of eclipses on some of those days: Tuesday, Wednesday, Friday and Saturday. A similar association of thunder and eclipses to the days of the week appears in the Book of Chilam Balam of Chan Kan (Hires, 1981, pp. 248–51). In this case, eclipses are considered only on Wednesday and Friday. These comments occur within a Ptolemaic cosmology which assigns the seven 'planets' (Sun, Moon, Mars, Mercury, Jupiter, Venus, Saturn) to the seven days of the week (Sunday, Monday, Tuesday, Wednesday, Thursday, Friday, Saturday) as well as to seven corresponding angels. The entire scheme is of European origin as is made clear by the embedding of natural phenomena into the European week and its integration with a Ptolemaic cosmology. As a result, the idea that any planet could be an eclipse agent does not appear to be of ancient Maya origin.

31.7 The eclipse god Kolop U Wich K'in/Kolop U Wich Ak'ab

Additional information on cognitive aspects of ancient Mayan eclipse theory can be obtained through a consideration of the eclipse god Kolop U Wich K'in/Kolop U Wich Ak'ab (Snatcher of the Eye of Day/Snatcher of the Eye of Night). This deity has been poorly understood by previous investigators, and his position in the Maya universe has not been generally appreciated. The deity appears several times in the *Ritual of the Bacabs* and is referred to once in the colonial dictionaries. Barrera Vásquez (1980, p. 334) lists this dictionary entry as follows:

Kolop U Wich K'in *idolo mayor que tenian estos indios de esta tierra, del cual decian proceder todas las cosas y ser incorpóreo y por esto no le hacían imagen* [principal god which these Indians of this land had, from whom they said all things proceeded and who was incorporeal, and for that reason they made no image of him].

In his analysis of the *Ritual of the Bacabs*, Roys (1965, p. 145) considers that this important god is a solar eclipse god and that Kolop U Wich Ak'ab is a lunar eclipse god. However, there are several problems in the published work discussing these names. For example, Barrera Vásquez identifies both gods with Itzamna, an identification I find very dubious. Thompson (1970, p. 205) has expressed some doubt that Kolop U Wich K'in is an eclipse god and suggests that the deity, whose name he renders as 'Tears out Sun's Face', may be a creator god who has brought several creations of the world to an abrupt end. Roys (1965, p. 3) has noted that the name is coupled with K'in Chak Ahaw (Sun Great Lord) in a way which suggests that they are the same deity, but he believes that this is impossible. In addition, Roys has shown some ambivalence in translating the *k'in* component of the name, at times favoring 'sun' and at other times 'day'.

In an earlier work (Closs, 1981a), I have discussed the parallel couplet structure of the incantations in the *Ritual of the Bacabs* and the formulaic usage of contrastive word pairs. With the use of these concepts it is possible to resolve the above problems. For example, the deity pair Kolop U Wich K'in/Kolop U Wich Ak'ab employs the contrastive

word pair *k'in/ak'ah* (day/night). The same word pair appears in other couplets in the manuscript, and it is clear that the contrastive nature of the pair precludes *k'in* from being translated as 'sun'. (It may be noted that another common contrastive word pair in the manuscript is *k'in/u* (sun/moon); in this case, usage of the pair precludes *k'in* from being translated as 'day'.) Consequently, Thompson's translation of the name Kolop U Wich K'in is inappropriate and Roys' ambivalence in translating the name is not justified. In addition, I have argued (Closs, 1981a, pp. 11–13) that when deity names are written in couplet form with the members of a contrastive word pair distinguishing the members of the couplet, then the couplet does not represent two gods but only contrastive aspects of the same god. Thus, Kolop U Wich K'in and Kolop U Wich Ak'ab should not be viewed as distinct solar and lunar eclipse deities. Rather, they are contrastive aspects of one and the same eclipse deity; the solar eclipse agent is not to be distinguished from the lunar eclipse agent.

The couplet structure of the incantations sometimes permits one to derive supplementary data regarding the eclipse god. To exhibit this point, I consider below the beginning of Incantation I. The translation is faithful to that of Roys (1965, p. 3) but includes some minor changes in spelling and in the form of presentation.

1 Ahau, unique 4 Ahau
　4 Ahau would be the creation, 4 Ahau would be the darkness
　　when you were born.
Who was your creator? Who was your darkness?
　He created you, did K'in Chak Ahaw,
　　Kolop U Wich K'in, when you were born.

Roys (1965, p. 3, n. 3) considers that the '4 Ahau' in the fourth line is a scribal error for '1 Ahau'. However, I have left the language as in the original. The portion of the incantation presented above reveals a partial couplet structure as well as the use of some contrastive word pairs: creation/darkness (*ch'ah/ak'ab*), 1 Ahau/4 Ahau, K'in Chak Ahaw/Kolop U Wich K'in. The first of the pairs is among the most common in the *Ritual of the Bacabs*, the second is a contrast of the calendar names for Venus and the Sun, and the last is a contrast of the sun god and the solar eclipse god. Since the Sun is linked with the sun god, then the partial couplet structure suggests that Venus should be linked with the solar eclipse deity. Hence, once again, Venus is implicated as an eclipse agent.

This type of argument can be repeated in other incantations to yield additional data. For this purpose, I consider the following excerpt from Incantation XXII (Roys, 1965, p. 46) which has a strong couplet structure.

1 Ahau, 4 Ahau
 During his birth, during his creation
  Snake of creation, snake of darkness
 Who was his creator? Who was his darkness?
 He created him, did K'in Chak Ahaw
  Kolop U Wich K'in, Kolop U Wich Ak'ab
  In the heart of the sky, in the heart of
   Metnal
  During his birth, during his creation

In the above example, contrastive word pairs are 1 Ahau/4 Ahau, birth/creation (*sihil*/*ch'ab*), creation/darkness (*ch'ab*/*ak'ab*), sky/Metnal, and Kolop U Wich K'in/Kolop U Wich Ak'ab. The sequence of three deity names K'in Chak Ahaw, Kolop U Wich K'in, and Kolop U Wich Ak'ab is a triadic expression in which two of the members form a contrastive pair representing a single deity. This structure appears in a number of places in the *Ritual of the Bacabs*, and in such cases the contrastive pair can be reduced so that the triadic expression itself becomes a contrastive pair in which the deities are distinct (Closs, 1981a, pp. 12–13). In the present case, the triadic expression is equivalent to the contrastive pair sun god/eclipse god. With this understanding, one obtains a linkage of 4 Ahau (Sun), sun god, and sky and the contrasting linkage of 1 Ahau (Venus), eclipse god, and Metnal. Since Metnal is used here as the name of the Maya underworld, this text pertains to Hun Ahaw (1 Ahau/Venus) as both eclipse agent and Lord of the Underworld. The arguments are independent of those considered earlier, but lead to the same results.

There is an abstruse passage in Incantation X which Roys has suggested also refers to an eclipse god. With the exception of some minor changes (reading *wich* as 'face' rather than 'eye', or vice versa), Roys (1965, p. 31) translates the text as follows:

One day he is curled up at the door of his arbor, one day he turns about in the heart of the sky. There he spoiled the lintel of his father. One day, seated on the lintel of his father, he spoiled the face of the sun, . . . [also] the face of the moon. One day the eye of the sky was bitten by him. Then he was beaten on his tail, on his head-part.

If the passage is a reference to eclipses, which appears to be a good possibility, two deductions can be made. First, there is only one eclipse agent for both solar and lunar eclipses. Second, the eclipse agent is described as turning about in the heart of the sky and this suggests that the cosmological actor is a planet undergoing retrograde motion. In the latter case, there is a preference for identifying the planet as Venus since this planet and its motions are known to have been of far greater interest to the Maya than were the other planets and their motions. Thompson (1971, p. 220) mentions one ethnohistorical reference from outside the Maya area which also seems to refer to the retrograde motion of a planet. He refers to a remark of one commentator of the Telleriano-Remensis Codex that Itzlacoliuhqui is 'a star which they say goes backward'. He considers it 'virtually certain' that Itzlacoliuhqui is a god of Venus and identifies him with one of the Venus gods in the Venus table of the *Dresden Codex*. This datum, insofar as it appears to specifically attribute retrograde motion to Venus, supports the argument that the above excerpt from Incantation X is an oblique reference to Venus as eclipse agent.

31.8 Ancient Maya perspectives

I believe that the themes discussed in the earlier section synthesizing eclipse beliefs in the postconquest period are derived from an ancient tradition and can also be attributed to the Maya during the Classic period. To some extent, the cosmological identity of the eclipse agent has been obscured in the late survivals of the tradition but, nevertheless, enough information remains to implicate Venus as the eclipse deity. In addition, it has been seen that, in his role as eclipse agent, Venus is described variously as a celestial jaguar, a celestial serpent, or Lord of the Underworld. Finally, I recall the important theme that eclipses have an apocalyptic significance: an eclipse will portend the end of time and, subsequent to swallowing the sun or moon, the Venus jaguar will descend to the earth to devour mankind.

If the above model of ancient Mayan eclipse theory is accurate, it should be reflected in the hieroglyphic and iconographic sources of the ancient Maya. Without doubt, the most important environment for seeking evidence of such beliefs is the eclipse table of the *Dresden Codex*. This table is known to function as a prediction mechanism for finding dates on which solar or lunar eclipses may occur (Lounsbury, 1978, pp. 789–804; Bricker and Bricker, 1983). It is the product of Mayan astronomers who recognized the cyclical nature of eclipse phenomena and who sought to integrate that cycle with the Sacred Round, their ritual calendar of 260 days. The length of the eclipse table is $46 \times 260 = 11\,960$ days, the lowest multiple of the Sacred Round yielding a satisfactory eclipse period. The significance of the Sacred Round as a basic structural component of the eclipse table must not be underestimated.

The 260-day ritual calendar is the foundation on which the Maya system of divination rests. Every period of divination had to be a multiple of it. In particular, this is true of the astronomical tables appearing in the *Dresden Codex*. Indeed, solar and lunar eclipses, as well as heliacal risings of the planet Venus, were natural phenomena much feared by the Maya. The ability to predict the occurrence of such events and to take countervailing measures to avert potential disasters would have been highly valued activities in ancient Maya society. The divinatory nature of this type of activity can best be appreciated by a consideration of the Venus table of the *Dresden Codex*.

It has long been known that the canonical base of the Venus table is 1 Ahau 18 Kayab at 9.9.9.16.0. This is indicated by the Initial Series which records its Long Count position on *Dresden* 24, the introductory page to the Venus table, and by the associated ring number and companion number which tie it to a 1 Ahau 18 Kayab date at $-6.2.0$, preceding the zero point of Maya chronology. I have argued that this canonical base was selected for numerological reasons and that the true astronomical base of the table is a much later 1 Ahau 18 Uo at 10.15.4.2.0 (Closs, 1977, p. 97; Closs, 1979, pp. 157–8). Lounsbury (1983, pp. 12–16) has described an interesting heliacal rising of Venus with a simultaneous Venus–Mars conjunction on 1 Ahau 18 Kayab at 10.5.6.4.0 (assuming Thompson's 584285 correlation). He suggests that the Maya may have been so impressed by this unusual occurrence that they selected 1 Ahau 18 Kayab as a Calendar

Round base for the Venus table. He then proposes that the 1 Ahau 18 Kayab canonical base date is derived from the later date by numerological processes. I find the idea attractive since it serves to explain why this particular 1 Ahau date was chosen as a divinatory base rather than a 1 Ahau date which is closer to the zero point of Maya chronology. I note that it may be incorporated into my solution of the base dates of the Venus table since this is also compatible with Thompson's correlation (Closs, 1977, 1979, pp. 157–8; 1981b, table 2). I am uneasy about the solution of the base dates proposed by Lounsbury (1983) since it relies on a method of recycling which I find objectionable (Closs, 1977).

The 1 Ahau 18 Kayab date at $-6.2.0$ is the base date for the prefatory glyphic material on *Dresden* 24. This text, among other things, introduces a complex of glyphs including the direction east, the common verb appearing in the Venus table proper, and five deities marked with Venus titles. These glyphs are associated with heliacal rising of Venus as morning star in the Venus table (pages 46–50 of the *Dresden Codex*). Because the events in the preface take place in mythological time it is clear that 1 Ahau is a ritual day for heliacal risings of Venus as morning star. The divinatory nature of the date is further manifested by the fact that the actual dates in the table on which the phenomena are expected to occur vary widely. As noted earlier, Hun Ahaw (1 Ahau) is both a calendar name of Venus and the name of Venus as Morning Star and Lord of the Underworld. This underlies its divinatory importance in the Venus table and its role as an archetypal day for heliacal risings of Venus as morning star.

31.9 The day Lamat as a divinatory base of the eclipse table

The eclipse table in the *Dresden Codex* is introduced with prefatory material on *Dresden* 51a–2a. This includes a table of multiples of the eclipse cycle preceded by chronological data which anchors the table in the Long Count. The anchor is provided by an Initial Series of eight days, the smallest Initial Series on record, leading from a specified 4 Ahau 8 Cumku, the zero point of Maya chronology, to the date 12 Lamat (16 Cumku). To the latter date a distance number of 9.16.4.10.0, equal to 5434 Sacred Rounds, is then added, yielding the date

12 Lamat (1 Muan) at 9.16.4.10.8. The last date is a canonical base for the eclipse table in historical time. As was the case with the Venus table, I believe that this canonical base is selected for numerological reasons and not astronomical ones. In fact, the month positions of the 12 Lamat dates are not mentioned in this instance, a detail which serves to emphasize their divinatory nature. It may also be noted that a series of five Sacred Round dates (again without month positions) is recorded below each of the seven columns in the table of multiples. Each series of dates is the same, beginning with 12 Lamat and ending 60 days later with 7 Lamat. The dates in the series are separated by 15-day intervals, a potentially useful characteristic, since it is possible to have two solar eclipses a month apart with a lunar eclipse between them. The 60-day interval from 12 Lamat to 7 Lamat is the minimal interval with 15-day subdivisions which will accommodate this astronomical phenomenon and permit it to be bracketed with a Lamat date on either end. This feature suggests that it is the day Lamat and not specifically the date 12 Lamat which is of importance here. In fact, the date 12 Lamat may have been selected as the numerological base of the table simply because it is the closest day Lamat to the zero point of Maya chronology at 13.0.0.0.0.

I believe that the day Lamat was specifically selected as the numerological base of the eclipse table because of its underlying divinatory characteristics. These characteristics should pertain to eclipses and set a mood or tone for eclipse phenomena regardless of the actual dates on which they might occur. Lamat is one of the days for which the divinatory aspects are relatively clear. For the Classic period, this information can be recovered by a consideration of various glyphs for the day (Figure 31.1a–h). Thompson (1971, pp. 44, 77) considered the geometric form of the glyph for Lamat (Figure 31.1a–d) to be a variant of the full form Venus sign (Figure 31.1i). He reached this conclusion because of the underlying similarity in the forms of the two signs. Indeed, if the day sign cartouche is eliminated from certain Lamat glyphs (Figure 31.1c–d) one might well wonder whether the remainder is a Venus sign. Nevertheless, despite the similarity in form, the usual depictions of the Lamat glyph (Figure 31.1a,b) consistently differ from the Venus sign. However, it is apparent that there is some equivalence between Lamat and Venus signs. This is exemplified by variants of the Lamat glyph (Figure 31.1e–g) which employ the half form Venus sign (Figure 31.1j). (Once again, I remark that the glyphs which I refer to as Venus signs can be translated as 'star', but, in almost all determinable cases, the Maya employed these glyphs in specific reference to the planet Venus. Hence, if one wishes to reflect Maya usage, it is preferable to refer to them as Venus signs rather than star signs.) The head variant Lamat glyphs (Figure 31.1g,h) represent Venus in the form of a serpent with a Venus sign to the right of the head. The Venus variants of the Lamat glyph indicate that Lamat is under the influence of Venus. It can be inferred that in the Maya system of divination the day Lamat will be marked by Venus and its characteristics. Since the day Lamat is also the canonical base day of the eclipse table, it suggests that, in some sense, Venus will be implicated in eclipse phenomena.

More information on divinatory aspects of the day Lamat can be derived from almanacs in the books of Chilam Balam which give prognostications for the days of the Sacred Round. The following items are extracted from a summary of this data published by Thompson (1971, 300–1):

1 Lamat (Bad augury) Adhesion of leg of jaguar (?) [Tizimin, Perez 2, Perez 4]

8 Lamat (Bad augury) Jaguar-faced 1 Ahau with protruding teeth [Tizimin, Ixil, Perez 2–4]

12 Lamat (Bad augury) Encounter with Kisin [Kaua 2, Perez 2]

A number of prognostications for Lamat have been omitted in the above list. These are references to agricultural and environmental matters which are common in the almanacs and which are found with most days in the Sacred Round. By contrast, the items which are listed are atypical. In fact, the references to the jaguar are peculiar to the day Lamat. The reference to Kisin, the Maya god of death, is not unique to Lamat but is very limited, the only other appearances being in the prognostications for the days 1 Cimi and 8 Ahau.

The first augury, 'adhesion of leg of jaguar', is opaque. The second, 'jaguar-faced 1 Ahau with protruding teeth', has already been encountered. It refers to the Venus deity Hun Ahaw, Lord of the Underworld, who in his manifestation as a celestial jaguar devours the sun and moon during eclipses. The third augury, 'encounter with Kisin', is also appropriate for eclipses, since they portend not only the death of Sun or Moon but, potentially, the death of all mankind.

One may add an additional prognostication for Lamat which was not available to Thompson. This appears in the book of Chilam Balam of Chan Kan (Hires, 1981, p. 154) and reads:

5 Lamat its nature is dog, its tidings are jaguar.

This augury recalls another from the book of Chilam Balam of Kaua which was mentioned earlier: 'The head of a jaguar, the rear of a dog' (Thompson, 1971, p. 218). It was noted in the previous discussion that Thompson believed this augury to refer to the Venus deity Lahun Chan. That interpretation is consistent with the idea that prognostications for Lamat are dominated by the influence of Venus.

31.10 The picture and text of the 12 Lamat canonical base date

The eclipse table ends on a day 12 Lamat with a picture and an associated glyphic text (Figure 31.2). This is the tenth and final picture in the table and, unlike the other nine, is not preceded by a five-month period but comes in the middle of an eclipse season, after a six-month half-year. Because of this structural arrangement, the tenth picture is out of astronomical synchronization with the other nine insofar as eclipses are concerned. Yet, the tenth picture and its text prominently display paired eclipse symbols, solar and lunar, as do most of the others. As a result, Lounsbury (1978, p. 799) concludes that the tenth picture is an anachronism and is retained from an earlier version of the table. In his structural analysis, he further assumes that the day 12 Lamat is an implied node day; the structural constraints then entail that the other pictures are more or less maximally pre-nodal (Lounsbury, 1978, p. 796). Bricker and Bricker offer a different structural analysis of the eclipse table. In their interpretation, the assumption that 12 Lamat is at or near a node is unnecessary (Bricker and Bricker, 1983, p. 21). However, they do not discuss the problem raised by the 12 Lamat picture and its text. As noted above, these are similar to the other nine in terms of eclipse symbolism but dissimilar in terms of the astronomical alignment of 12 Lamat with respect to eclipse events.

In my interpretation of the eclipse table, the 12 Lamat base has a divinatory purpose. It was chosen because it is the closest day Lamat to the zero point of Maya chronology. The 12 Lamat base at 9.16.4.10.8 is a canonical base for the table in historical time but its essential purpose is still divinatory. One consequence of this notion is that the 12 Lamat picture and text (Figure 31.2), should be typical of all eclipses. Thus, the similarity of its eclipse symbolism with that in the other nine pictures and texts is to be expected, despite the fact that the astronomical alignment of the canonical base date with respect to eclipse events is unlike that of the dates associated with the other pictures. In this sense, the 12 Lamat picture and text serve as a preface, or better, a postscript, to the eclipse table as a whole. There is some internal evidence in support of this idea. Indeed, the 12 Lamat text is strongly differentiated from the other picture texts by the presence of chronological glyphs. The second glyph in the first line of the text records a period of 13 *tuns* (= 4680 days) and the first glyph in the fifth line of the text records a period of 1 *katun* (= 7200 days). Since each of these chronological counts exceeds the interval between any two picture dates, it is likely that the 12 Lamat text has a more global scope than do the other picture texts. The count of 13 *tuns* has both eclipse and Venus significance. Indeed, the length of the interval can be expressed as $4680 = 9 \times 520 = 27 \times 173 \cdot 3333$ days, that is 27 eclipse half-years (the average length of interval between lunar nodes is $173 \cdot 3096$ days, and also as $4680 = 8 \times 584 + 8$ days, that is eight average Venus cycles and eight days.

The picture accompanying the 12 Lamat canonical base depicts a creature plunging head first from solar and lunar eclipse symbols. The most striking feature of the diving figure is the Venus glyph replacing the head. This surely identifies him as a Venus deity. Immediately beneath the Venus-head is an apparent beard. Thompson (1930, p. 63) notes that the Kekchi-Chol of San Antonio in South-Central Belize consider the Venus god Xulab to be bearded and ugly. He adds that this accords with the depiction of his Mexican counterpart, Quetzalcoatl. The comparison is interesting since there is a sculpted diving figure with similar posture and beard from the Huastec area which has been identified as a Venus apparition, either Quetzalcoatl or Xolotl (Fuente and Gutiérrez Solana, 1980, pp. 368–71, pl. 361). A second unusual feature of the *Dresden Codex* figure is that he has an insect abdomen. This may reflect one of the names

(a) (b) (c)

(d) (e) (f)

(g) (h) (i)

(j) (k) (l)

of Venus in Yucatec, *xux ek'*, literally 'wasp star' (Barrera Vásquez, 1980, p. 957).

The imagery in the picture suggests that the Venus deity is plunging earthwards following a solar or lunar eclipse. Thompson (1971, p. 223) believed that this Venus deity could be explained by the Aztec belief in the Tzitzimime, monsters who dove earthwards from the sky during eclipses. Caso (1958, p. 37) writes that the Tzitzimime, or Tzontémoc, 'those who fell head first', are planets which will be transformed into jaguars on the terrible night at the end of the century and who will come down to earth to devour man. It has been seen that the Maya had similar beliefs associated with eclipses and the end of the world, although the only planet specifically implicated in these beliefs is Venus. In particular, Text M (Chol) observes that after the eclipse agent (described as a jaguar) had swallowed the moon, it was coming down to eat mankind. Bruce (1979, p. 236) refers to a probably related belief among the Lacandon who say that the *Ts'ulu'*, or *Nah Ts'ulu'*, mythical jaguars of the heavens and the underworld, come forth during the various destructions of the world to devour everyone who remains alive on the earth. There can be little doubt that the diving Venus deity in the picture is both the eclipse god and the agent of the apocalypse.

Finally, it may be recalled that the augury for 12 Lamat, 'encounter with death', is certainly pertinent to the apocalyptic vision represented by the 12 Lamat picture.

31.11 The Venus eclipse god as devil (Lord of the Underworld)

The first glyph in the 12 Lamat text (Figure 31.2) consists of a seated man in upside down position attached to a Venus glyph. The manikin, Glyph T227 in the Thompson (1962) catalog, has a peculiar darkened stump-like head without any facial characteristics. Because T227 is shown head down

Figure 31.2 The 12 Lamat picture and text from the eclipse table, *Dresden* 58b.

Figure 31.1 Glyphs for the day Lamat: (a) Naranjo Stela 24, A1; (b) Palenque Palace Tablet, M15b; (c) Piedras Negras Lintel 4, 01; (d) Uaxactun Fresco, Glyph 28; (e) Copan Stela J, Glyph 46; (f) Palenque Tablet of the 96 Glyphs, D4b; (g) Copan Hieroglyphic Stairway, Date 3; (h) Copan Hieroglyphic Stairway, Date 24. Glyphs for Venus: (i)–(j) *ek*, 'star'; (k)–(l) *chak ek*, 'great star'.

and is attached to the Venus sign, the glyph combination probably names or refers to the plunging Venus eclipse god illustrated in the corresponding picture. Such a notion is supported by the preceding picture and text in the eclipse table. In this instance,

(a)   (b)   (c)   (d)   (e)   (f)   (g)   (h)   (i)

Figure 31.3 Venus as Lord of the Underworld [T227 and T703]: (a) *Dresden* 57b; (b) *Dresden* 66b; (c) *Paris* 4b; (d) *Dresden* 24, C11; (e) *Dresden* 51b; (f) Palenque, Palace Tablet, F13; (g) Palenque, Tablet of the Foliated Cross, N12; (h) *Dresden* 35c; (i) *Dresden* 38c.

the text opens with the inverted T227 attached to a Venus glyph and the associated picture depicts a solar eclipse (Figure 31.3a). Further support is found in the *Paris Codex*, where on one occasion solar and lunar eclipse symbols are followed by an inverted T227 attached to a full form Venus glyph (Figure 31.3c).

On *Dresden* 24, the introductory page of the Venus table, the combination T172.227 (Figure 31.3d) appears as one member of a group of five sequential glyphs all having T172 prefixes. These five glyphs reappear in the five pages of the Venus table, in the same sequence, as successive victims of Venus at heliacal rising of morning star. Since these glyphs include names for God K, the maize god, the great jaguar, and another deity, it can be inferred that T227 is also a name. In all, T227

occurs five times in the Venus table, one of its most frequent environments. T227 also occurs in an almanac of the *Madrid Codex* having four sections, each beginning with a different directional glyph (*Madrid* 88c). In this almanac, T227 appears as one of a group of which the other members are a dog on a turtle, a monkey, and a bird on a Cimi (death) glyph. The entire almanac is presided over by a large figure of the Maya death god Kisin. Once again, the environment indicates that T227 is a name.

A different type of usage of T227 appears in an almanac of the *Dresden Codex* which gives various locations of the rain god Chak. In one of the sections (Figure 31.3b), the locative phrase is given by T552.227 immediately before the name of Chak. Thus, T227 should specify the location of the rain god who, in the associated picture, is shown seated on a bounded space marked with crossed bones. The location can only refer to the underworld. Thus, in addition to signifying a deity name, T227 should signify the underworld.

A widely travelled term for the underworld in the Maya area is *xibalba*. This is the name used in the *Popol Vuh* of the Quiche Maya and is also found in Moran's Cholti vocabulary (Moran, 1935, p. 38; Edmonson, 1971, p. 60). The same term is found in Yucatec, but there it means 'devil' (Barrera Vásquez, 1980, p. 930). In Chol, there is a cognate term *xib'a*, also signifying 'devil' (Aulie and Aulie, 1978, p. 136). In references to the world of the ancient Maya, a more appropriate rendition of the term would be 'Lord of the Underworld'.

In interpreting the *Dresden Codex* usage of T227, it is interesting to find that the Yucatec term *xibalba* fits the environmental requirements which have been noted. It can be interpreted as both a deity name (or title), 'devil', and as a place name, 'underworld'. In eclipse contexts, the inverted T227 and Venus combination may then refer to the Venus eclipse god as the '(plunging) devil star' or '(plunging) underworld star'.

Another context of T227 in the eclipse table is in the brief auguries or notes associated with the 69 eclipse stations. It occurs as a prefix of the sky glyph (Figure 31.3e) on four separate occasions. These cases can be interpreted as 'underworld sky', an appropriate augury for a potential eclipse date, given that the Venus eclipse god is also Lord of the Underworld. It may be noted that 'underworld sky' glyphs, like eclipse glyphs, also occur outside the eclipse table (Figure 31.3h). In one instance, a near synonymous usage 'underworld cloud' is also found (Figure 31.3i).

It has been noted earlier that the *Popol Vuh* describes how the hero twins set up their father 1 Ahau (*Hun Ahaw, Hun Hunahpu*) as lord of the underworld and also prophecies his apotheosis as the Morning Star. However, the *Popol Vuh* also speaks of difficulties which were encountered when the twins attempted to reconstitute the face of 1 Ahau. Edmonson (1971, pp. 143–4) translates this event in the following words:

And this then was the reassembling of their father by them,
    And then they reassembled 7 Hunter.
They went there to reassemble them

In Dusty Court.
But really his face wanted to exist,
    And was asked about the name of everything,
His mouth,
    His nose,
The socket
    Of his eye.
He first found its name,
    But very little more could he speak.
Only he couldn't name any longer
    The name of the lips of his mouth,
So he couldn't really speak.
    And thus then they honored him
And left the heart of their father.

The text gives a fairly lengthy description of problems encountered in forming the face of 1 Ahau, suggesting that it is of special significance. It appears that 1 Ahau was unable to describe the features of his face; he was able to name the features but was unable to say anything much about them. In the end, since he was unable to name the lips of his mouth he was also rendered speechless. I believe these events indicate that when 1 Ahau became lord of the underworld in Dusty Court he remained essentially faceless. I also think that Glyph T227 may represent 1 Ahau in Dusty Court as the hero twins left him and that this accounts for the stump-like head without facial characteristics. This identification would also give a rationale for the use of T227 as a reference to 1 Ahau, the Lord of the Underworld (devil, demon), and to the underworld itself.

John Justeson (1986, personal communication) has pointed out that there are two references in the inscriptions of Palenque which give support to this hypothesis. On two occasions, the name *Hun Ahaw* is spelled out (once in the form T1000a.168 and once in the linguistically equivalent form T1000a.747a) (Mathews and Justeson, 1984, pp. 208–9). In each case, the name of *Hun Ahaw* is immediately followed in the same glyph block by a crouching headless human figure (Figure 31.3f,g). The latter glyph, T703, is surely the inscriptional variant of the codical glyph T227 which has been discussed above. Consequently, its interpretation as a name or title of 1 Ahau as Lord of the Underworld is appropriate here. In fact, since the Palenque texts are very likely written in Chol, the glyphs in question may have been read as *Hun Ahaw xib'a*, '1 Ahau, Lord of the Underworld'.

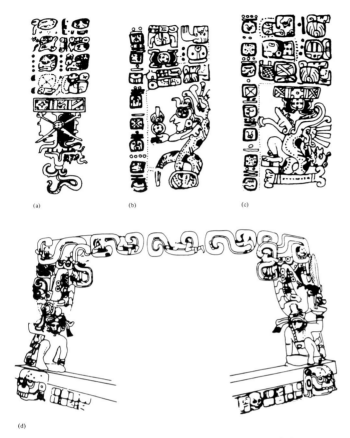

Figure 31.4 Venus as celestial serpent: (a) *Dresden* 56b; (b) *Madrid* 66b; (c) *Madrid* 67b; (d) the façade of the interior doorway of Temple 22 at Copan (courtesy of Linda Schele).

31.12 Venus as celestial serpent

The notion of Venus as a devouring celestial serpent in its role as an eclipse agent is depicted twice in the *Dresden* eclipse table (Figures 31.3a and 31.4a). In both instances, a serpent is shown devouring a solar eclipse symbol suspended from a celestial band. Since the pictures occur in the eclipse table itself, there is little question as to their significance and their interpretation.

On two other occasions, both in the *Madrid Codex*, one encounters similar iconography. In the first of these (Figure 31.4b), a serpent is shown devouring an anthropomorphic head, with crossed bands replacing the eye. The head is attached to a hand holding a **YAX-KIN**, 'new sun', glyph. In the legend above the picture, the third glyph is a solar eclipse glyph. In the second instance (Figure 31.4c), a serpent is devouring a solar eclipse symbol

attached to a celestial band. The serpent is also shown coiled around the death god atop a long bone. There is some degree of parallelism between the glyphic texts in these two almanacs. In both cases, the last glyph is identical, the fifth glyph begins with a locative prefix, the fourth glyph is a distance number, and the third glyph includes a sun sign. Since the third glyph in the first almanac is a solar eclipse glyph and since there are similarities in the texts and the iconography of the two pictures, it is possible that the third glyph in the second almanac is a head variant solar eclipse glyph.

Additional evidence attesting to Venus as a celestial serpent manifests itself in the sculptured two headed bicephalic dragons found in the Maya area. One of the finest examples is found on the facade of the interior doorway of Temple 22 at Copan (Figure 31.4d). The west head of this bicephalic dragon has an infixed Venus sign, while the east head has an infixed sun sign. The body of the monster is serpentine on either side of the doorway but over its central span consists of a series of S-shaped and reverse-S-shaped symbols in which small human figures are intertwined. Venus glyphs are attached to each side of the serpent body.

Another example of a bicephalic dragon functioning as a doorway and having one head marked with a Venus sign and the other with a sun sign is found on Palace House E at Palenque (Schele, 1976, p. 20). Examples of bicephalic dragons with skybands for bodies are found on Stelae 6, 11, 14, and 25 from Piedras Negras. These are also characterized by having the left head marked with a Venus sign and the right head marked with a sun sign (Kubler, 1969, p. 39).

It was noted earlier that in the *Ritual of the Bacabs* there is a repeated reference to the couplet 1 Ahau/4 Ahau which is a contrast of the calendar names for Venus and the Sun. This is precisely the type of dualism which is found in sculpted form in the bicephalic dragons discussed above. Thus, the ritual pairing of 1 Ahau and 4 Ahau in the *Ritual of the Bacabs* can be seen as the echo of an ancient practice still preserved in the stone monuments of the Classic Maya.

31.13 Venus as evening star and eclipse agent

The glyphic legend accompanying the 12 Lamat canonical base date of the eclipse table (Figure 30.2), features paired solar and lunar eclipse glyphs in its third row. These are followed in the fourth row by a second pair of glyphs T286.561.670 and T12.168.559. The second of these is a name glyph which may be read as *Ah Tsul Ahaw* (Closs, 1979, pp. 158–63). I do not wish to repeat my arguments from the earlier paper, but it may be noted that T559 has phonetic value **tsu** in the codices as can be seen from such phonetic collocations as **tsu-l(u)** for *tsul*, 'dog', and **ku-ts(u)** for *kuts*, 'turkey' (Kelley, 1976, p. 173). In addition, the sign is used iconographically as a rebus for *tsul*, 'dog', and also for the homonymous term *tsul*, 'spine' (Closs, 1979, fig. 8). I have argued that the glyph also has links to the deity *ah Zuli* appearing in the Chilam Balam of Kaua and that he may be identified with *Lahun Chan*, a god of Venus as the evening star who is depicted in the Dresden Venus table. The deity is the Mayan equivalent of the central Mexican god *Xolotl*, the god of Venus as evening star, who was

the twin brother of Quetzalcoatl, the god of Venus as morning star. It is also interesting to note that *Xolotl* was represented in both canine and skeletal forms.

The glyph which intervenes between the eclipse glyphs and the name glyph of *Ah Tsul Ahaw* is a sky-in-hand glyph which may function as a relationship glyph specifying that *Ah Tsul Ahaw* is the causative agent underlying eclipse events. Indeed, the glyph (with T125 replacing T286 as prefix) functions on at least one occasion as a relationship glyph linking a ruler to his mother (Kuna-Lacanha, Lintel 1, J1).

The four glyph sequence discussed here suggests that *Ah Tsul Ahaw*, a god of Venus as the evening star, is both eclipse agent and, given the apocalyptic content of the picture accompanying the glyphic text, agent of the apocalypse. This interpretation is supported by the repetition of the complete four glyph sequence in the text accompanying the first picture in the eclipse table (Figure 31.5a). In this instance, the picture shows the skeletal death god seated on a bone throne. The context once again relates eclipse phenomena, Venus (as evening star), and death.

The *Ah Tsul Ahaw* glyph is found elsewhere in the codices. Since it has been identified as a name glyph for Venus as evening star, it is not surprising that it appears in the introductory page to the Venus table and in the Venus table itself. In the first case, it appears paired with the sky-in-hand glyph which takes a numerical coefficient of 1 (Figure 31.5b). The two glyph sequence ends the opening sentence describing Venus events set in the mythological past. In the second case (with T172 replacing T12), it appears on the page of the Venus table depicting Lahun Chan (Figure 31.5c). The local context in which the glyph appears remains opaque. Other occurrences in the codices do not yield useful information for the present analysis. For example, the glyph appears twice in the *Madrid Codex* in the pages dedicated to the new year ceremonies. On *Madrid* 34d it follows the directional glyph for west and appears as the last glyph in the text. On *Madrid* 37d it follows the directional glyph for north and again is the last glyph in the text. The glyph also appears twice in the *Paris Codex*. On *Paris* 6b it appears in sequence with the sky-in-hand glyph; it then recurs on *Paris* 6d where it is followed by a death glyph. The complex of associations here

(a)

(b)

(c)

(d)

(e)

Figure 31.5 Venus as evening star (T12.168:559, T168:559, and T168:559.130): (a) *Dresden* 53a; (b) *Dresden* 24A; (c) *Dresden* 47f; (d) Palenque, Temple of the Inscriptions, East Panel, P10–O10; (e) Palenque, Temple of the Inscriptions, Middle Panel, H6–H9.

is not surprising but the content remains opaque.

The deity *Ah Tsul Ahaw* is also named three times in the inscriptions of Palenque, but in slightly simplified form as *Tsul Ahaw*. In these instances the contexts are very revealing. The name appears twice in a sequence of glyphs from the Middle Panel of the Temple of Inscriptions at Palenque (Figure 31.5e). I have previously shown that the sequence from G7 to G9 marked a greatest eastern elongation of Venus (Closs, 1979, pp. 163; 1981b). In this earlier work it was shown that the 'star over shell' glyph at G7 was associated with Venus events. Also, the directional glyph for east at H7, followed by *Tsul Ahaw* at G8, and the directional glyph for west at H8, followed by *Tsul Ahaw* at G9, gave a fairly precise description of the behavior of the evening star (*Tsul Ahaw*) at a greatest eastern elongation. At this moment, the planet would have been perceived as undergoing a change of direction; its motion changed from being eastwards and away from the sun to being westwards and towards the sun. Lounsbury (1982, p. 156, fig. 1d) later extended this sequence to include the skull glyph at H6 which he identified as a head variant glyph for Venus.

The other occurrence of the deity name *Tsul Ahaw* is found on the East Panel of the Temple of the Inscriptions at Palenque (Figure 31.5d). In this case, the glyph is preceded by the 'sky-in-hand' glyph with a coefficient of 1. As shown in my earlier work, this text also occurs at a greatest eastern elongation of Venus. It is 3 *katuns* [= 21 600 days], or five days less than 37 means Venus cycles of 583.92 days, prior to the greatest eastern elongation recorded on the Middle Panel of the Temple of the Inscriptions.

The sequence of glyphs in the Middle Panel of the Temple of the Inscriptions at Palenque which refers to a greatest eastern elongation of Venus can be extended to include the glyph at H9 (Figure 31.5e). Indeed, it has been suggested that the glyph consisting of a seated body with a stump-like head [T227 or T703] depicts the Venus god *Hun Ahaw* seated in Dusty Court as Lord of the Underworld. It was also noted that the Chol term *xib'a*, 'devil', probably corresponding to an old term signifying 'Lord of the Underworld', was a viable linguistic candidate for the reading of T227 or T703. With this in mind, the glyph at H9, T703.501.181, can be analyzed as **XIBA-(ba-ha)**, for *xib'ah*, 'Lord of

the Underworld', or possibly as **XIBA-ba-(ha)**, for *xibalba*, 'underworld'. In either case, the known phonetic reading of T501 as **ba** supports the tentative identification of T703 (or T227) as *xib'ah*. It is also interesting to note that the two glyphic expressions *Hun Ahaw xib'ah* (Figure 31.3f,g) and *Tsul Ahaw xib'ah* (Figure 31.5e) exhibit the same local syntactic and semantic structure.

It can be seen from the above that the codices implicate *Ah Tsul Ahaw* as an eclipse agent. The codices and the ethnohistorical sources also suggest that *Ah Tsul Ahaw* is a god of Venus as evening star identifiable with *Lahun Chan*. Finally, the inscriptions at Palenque provide independent confirmation that *Tsul Ahaw* represents Venus as evening star. The data of this section are consistent with the model of ancient Maya eclipse theory which has been outlined in the preceding pages.

31.14 Venus as 'jaguar-faced 1 Ahau'

In the discussion of the prognostications for the day Lamat, the divinatory base day for the *Dresden* eclipse table, it was mentioned that some of the books of Chilam Balam contain the augury 'Jaguar-faced 1 Ahau with protruding teeth' for the day 8 Lamat. It was also noted that the book of Chilam Balam of Kaua refers to the day lamat in the words: 'Drunkard, deformed dog is his prognostic. The head of a jaguar; the rear of a dog.' In addition, it was seen that the Chol linguistic data and some ethnohistoric sources in Yucatec and Chol (Texts D, L and M) specify the jaguar as eclipse agent. These jaguar attributes of 1 Ahau – god of Venus, eclipse agent, and Lord of the Underworld – are visible in some of the Classic depictions of God L. Perhaps, the clearest instance of this occurs on the Vase of the Seven Gods where God L is displayed in all his glory as Lord of the Underworld (Figure 31.6a). In this case, God L is seated on a very impressive jaguar throne and a 'star over earth' Venus bundle rests on the floor in front of him. God L is shown with a prominent jaguar ear and has his lower face covered with jaguar spots. It should also be emphasized that God L can be clearly identified as a Venus deity (Closs, 1979, pp. 149–52). The Vase of the Seven Gods specifically alludes to this in its association of the Venus bundle with God L. However, the identification is apprehended in a much more direct sense in the *Dresden* Venus table. Here, God L appears as the

(a)

(b)

first manifestation of Venus on *Dresden* 46 and *Lahun Chan* as the second manifestation on *Dresden* 47. More importantly, God L and *Lahun Chan* are singled out in the introductory page to the Venus table as the two principal manifestations of the planet (*Dresden* 24C, C1–C12). Hence, God L is jaguar-faced, is a principal Venus deity, and is Lord of the Underworld, just as is *Hun Ahaw* in the ethnohistoric sources. There can be little doubt that they are one and the same.

A consideration of the evidence which has been discussed also suggests that God L/*Hun Ahaw* is primarily a manifestation of Venus as morning star, while *Lahun Chan/Ah Tsul Ahaw* is primarily a manifestation of Venus as evening star. It is understandable that, at times, the distinction between these two pairs of related Venus deities may be somewhat vague since the morning star and evening star are the same planet.

Representations of God L as Lord of the Underworld are relatively common in the Classic period ceramics, although Venus symbolism is not usually overt. An interesting example of the head of God L/*Hun Ahaw* attached to a Venus sign is found on Stela 4 from Yaxchilan (Figure 31.6b). The head is characterized by a jaguar ear with additional jaguar spots below the ear ornament. In this case, and another like it on the same monument, the Venus

sign and attached head are suspended from a celestial band above the principal scene. The effect is reminiscent of the plunging Venus eclipse god in the 12 Lamat picture of the eclipse table (Figure 31.2). It has been proposed that Stela 4 from Yaxchilan has a dedicatory date at 9.17.5.0.0 (Riese, 1977, p. 8). This is five *tuns* after the *katun* ending at 9.17.0.0.0, when an annular eclipse of the sun was visible in the Yucatan according to the 584 285 correlation. It has been suggested that if any eclipse was recorded on the monuments, it ought to be this one since it occurred on a *katun* ending. However, in practice, monuments erected as part of a *katun* ending festival would necessarily have been carved before the dedicatory date, and it is unlikely to expect that what would have been only a potential eclipse date at the time of carving would be recorded in the glyphic commentary associated with the *katun* ending. On the other hand, if a visible eclipse did occur on a *katun* ending, the celestial event would have occurred on the seating day of the incoming *katun*. As a result, the event would likely be taken as an ominous portent of what the *katun* had in store for the Maya and so might well receive attention in a monument erected shortly thereafter. Therefore, I think it a reasonable possibility that the 'jaguar-faced 1 Ahau' attached to the Venus sign suspended from the celestial band on Stela 4 may reflect the fate of the current *katun* as determined by the eclipse of 9.17.0.0.0.

Figure 31.6 Venus as 'jaguar-faced 1 Ahau': (a) God L as Lord of the Underworld (from Coe, 1973, p. 109); (b) Venus sign with head of *Hun Ahaw* from Yaxchilan, Stela 4 (after Kelley, 1976, fig. 9).

Acknowledgment

This work has been supported by a research grant (410-84-0506) from the Social Sciences and Humanities Research Council of Canada.

References

Aulie, H. W. and Aulie, E. W. de (1978). *Diccionario Ch'ol – Español: Español – Ch'ol*. Serie de Vocabularios y Diccionarios Indigenas 'Mariano Silva y Acerves' 21. Mexico: Instituto Lingüístico de Verano.

Aveni, A. F. (1981). Archaeoastronomy in the Maya region: a review of the past decade. *Archaeoastronomy*. (*Supplement to the Journal of the History of Astronomy*) no. 3, suppl. to vol. **12**, pp. S1–S16.

Baer, P. and Baer, M. (1952). *Materials on Lacandon Culture of the Petha (Pelha) region*. Microfilm Collection, Middle American Cultural Anthropology, no. 34, University of Chicago.

Barrera Vásquez, A. (1980). *Diccionario Maya Cordemex, Maya–Español, Español–Maya*. Merida: Ediciones Cordemex.

Beekman, J. (1956). The effect of education in an Indian village. In *Estudios Antropológicas*, pp. 261–4.

Boremanse, D. (1981). A southern Lacandon Maya account of the moon eclipse. *Latin American Indian Literature* **5**, 1–6.

Bricker, H. M. and Bricker, V. R. (1983). Classic Maya prediction of solar eclipses. *Current Anthropology* **24**, 1–23.

Bruce, R. D. (1979). *Lacandon Dream Symbolism*. Mexico City: Ediciones Euroamericanas Klaus Thiele.

413

Campbell, L. (1984). The implications of Mayan historical linguistics for glyphic research. In *Phoneticism in Mayan Hieroglyphic Writing*, eds. J. S. Justeson and L. Campbell, pp. 1–16. Institute for Mesoamerican Studies, Publication no. 9. State University of New York at Albany.

Caso, A. (1958). *The Aztecs: People of the Sun.* Norman: University of Oklahoma Press.

Closs, M. P. (1977). The date-reaching mechanism in the Venus table of the Dresden Codex. In *Native American Astronomy*, ed. A. F. Aveni, pp. 89–99. Austin: University of Texas Press.

Closs, M. P. (1979). Venus in the Maya world: glyphs, gods and associated astronomical phenomena. In *Tercera Mesa Redonda de Palenque*, eds. M. Greene Robertson and D. C. Jeffers, vol 4, pp. 147–65. Monterey: Pre-Columbian Art Research, Herald Printers.

Closs, M. P. (1981a). Las palabras pareadas en el Ritual de los Bacab y las implicaciones para los estudios glificos. *Boletín de la Escuela de Ciencias Antropológicas de la Universidad de Yucatán* 8, (46–7), 2–27.

Closs, M. P. (1981b). Venus dates revisited. *Archaeoastronomy: The Bulletin of the Center for Archaeoastronomy* 4, (4), 38–41.

Coe, M. D. (1973). *The Maya Scribe and his World.* New York: The Grolier Club.

Craine, E. R. and Reindorp, R. C. (1979). *The Codex Pérez and the Book of Chilam Balam of Mani.* Norman: University of Oklahoma Press.

Edmonson, M. S. (1971). *The Book of Counsel: The Popol Vuh of the Quiche Maya of Guatemala.* Tulane University, Middle American Research Institute, publ. 35. New Orleans.

Fought, J. G. (1972). *Chorti (Mayan) Texts.* University of Pennsylvania Press.

Fuente, de la, B. and Gutiérrez Solana, N. (1980). *Escultura Huasteca en Piedra.* Universidad Nacional Autónoma de México.

García de León, A. (1973). Breves notas sobre la lengua Tzotzil: literatura oral y clasificadores numerales. *Estudios de Cultura Maya* 9, 303–12.

Hires, M. K. (1981). *The Chilam Balam of Chan Kan.* Transcription and annotated translation. Ann Arbor: University Microfilms International.

Kaufman, T. (1976). Archaeological and linguistic correlations in Mayaland and associated areas of MesoAmerica. *World Archaeology* 8, 101–18.

Kelley, D. H. (1976). *Deciphering the Maya Script.* Austin: University of Texas Press.

Kubler, G. (1969). Studies in Classic Maya iconography. *Memoirs of the Connecticut Academy of Arts and Sciences* 18.

Lounsbury, F. G. (1978). Maya numeration, computation, and calendrical astronomy. In *Dictionary of Scientific Biography*, vol. 15, supplement 1, pp. 759–818. New York: Charles Scribner's Sons.

Lounsbury, F. G. (1982). Astronomical knowledge and its uses at Bonampak, Mexico. In *Archaeoastronomy in the New World*, ed. A. F. Aveni, pp. 143–68. Cambridge University Press.

Lounsbury, F. G. (1983). The base of the Venus table of the Dresden Codex, and its significance for the calendar-correlation problem. In *Calendars in Mesoamerica and Peru: Native American Computations of Time*, eds. A. F. Aveni and G. Brotherston, pp. 1–26. BAR International Series 174.

Mathews, P. and Justeson, J. S. (1984). Patterns of sign substitution in Mayan hieroglyphic writing: the 'Affix Cluster'. In *Phoneticism in Mayan Hieroglyphic Writing*, eds. J. S. Justeson and L. Campbell, pp. 185–232. Institute for Mesoamerican Studies, publication no. 9. State University of New York at Albany.

Mendez, S. (1921). The Maya Indians of Yucatan in 1861. In *Reports on the Maya Indians of Yucatan*, ed. M. H. Saville, pp. 143–95. New York: Indian Notes and Monographs, vol. 9, no. 3. New York: Museum of the American Indian, Heye Foundation.

Moran, F. (1935). Vocabulario en lengua Cholti. Extract from 'Arte y diccionario en lengua cholti'. *Maya Society*, publ. 9. Baltimore.

Redfield, R. and Villa Rojas, A. (1934). *Chan Kom: A Maya Village.* With notes on Maya midwifery by Katheryn Mackay. Carnegie Institution of Washington. (Reprinted in 1962 by The University of Chicago Press.)

Rejón García, M. (1905). *Supersticiones y Leyendas Mayas.* La Revista de Mérida.

Riese, B. (1977). Yaxchilán (Menché Tinamit), Dokumentation der Inscriften. *Beiträge sur mittelamerikanischen Völkerkunde* **XIV**.

Roys, R. L. (1965). *Ritual of the Bacabs.* Norman: University of Oklahoma Press.

Roys, R. L. (1967). *The Book of Chilam Balam of Chumayel.* Norman: University of Oklahoma Press.

Sanchez de Aguilar, P. (1921). Notes on the superstitions of the Indians of Yucatan (1639). In *Reports on the Maya Indians of Yucatan*, ed. M. H. Saville, pp. 202–8. Indian Notes and Monographs, vol. 9, no. 3. New York: Museum of the American Indian, Heye Foundation.

Schele, L. (1976). Accession iconography of Chan-Bahlum in the Group of the Cross at Palenque. In *The Art, Iconography and Dynastic History of Palenque*, ed. M. Greene Robertson, pp. 9–34. Pebble Beach, CA: The Robert Louis Stevenson School.

Thompson, J. E. S. (1930). *Ethnology of the Mayas of Southern and Central British Honduras.* Field Museum of Natural History, publ. 274. Chicago.

Thompson (1962). *A Catalog of Maya Hieroglyphs.* Norman: University of Oklahoma Press.

Thompson, J. E. S. (1970). *Maya History and Religion.* Norman: University of Oklahoma Press.

Thompson, J. E. S. (1971). *Maya Hieroglyphic Writing: An Introduction.* Norman: University of Oklahoma Press.

Thompson, J. E. S. (1972). A Commentary on the Dresden Codex. *Memoirs of the American Philosophical Society* **93**.

Tozzer, A. M. (1907). A comparative study of the Mayas and Lacandones. *Arch. Inst. Am., Rep. Fellow Am. Arch. 1902–5.*

Tozzer, A. M. (1941). *Landa's Relacíon de las Cosas de Yucatan.* Translation, edited with notes. Papers of the Peabody Museum, vol. **18**. Harvard University, Cambridge, MA.

Ulrich, E. M. and Ulrich, R. D. de (1976). *Diccionario Bilingüe: Maya Mopan y Español y Maya Mopan.* Guatemala: Instituto Lingüistico de Verano, Apartado 74.

Villa Rojas, A. (1945). *The Maya of East Central Quintana Roo.* Translated from the Spanish by Burton Lifschultz, William McSurely, Isabel Sklow and Robert Redfield. Carnegie Institution of Washington.

Whittaker, A. and Warkentin, V. (1965). *Chol Texts on the Supernatural.* Summer Institute of Linguistics Publications in Linguistics and Related Fields, publ. 13. Norman University of Oklahoma Press.

Wisdom, C. (1940). *The Chorti Indians of Guatemala.* The University of Chicago Press.

32

The use of astronomy in political statements at Yaxchilan, Mexico

Carolyn Tate *University of Texas*

The inhabitants of all ancient Maya ceremonial centers shared cultural practices that made them uniquely Mayan: they spoke related languages and wrote them in hieroglyphs, they erected limestone stelae and corbel vaulted temples, their beliefs were symbolized by a system of pictorial images, and they observed and recognized the motion of the celestial bodies. Just as each ceremonial center had its own recognizable artistic style, so each manipulated astronomical knowledge in a manner reflective of the specific history and character of that site. Astronomical knowledge was incorporated into writing, art, architecture, and ritual in various ways at different centers. For example, a study by Floyd Lounsbury (1982) has shown the importance of Venus as Evening Star, in maximum eastern elongation, and on heliacal rising as Morning Star in the historical hieroglyphic records of several sites, including Bonampak, Dos Pilas, Copan, and Palenque.

The Mayas of Yaxchilan, on the other hand, although they must have been aware of the beliefs that prompted their neighbors to incite wars on these stations in the passage of Venus through the heavens, did not regularly capture victims on a Venus-related schedule. Of the over 17 captures recorded on Yaxchilan monuments, three fell upon the first appearance of Venus as Evening Star (recorded on stela 18 and the structure 20 steps). One of the Yaxchilan lintels uses the 'star-shell' verbal glyph that appears at other sites in situations of Venus warfare, but the capture commemorated did not occur on a first appearance of Venus. Instead, the Yaxchilan rulers incorporated

observations of the sun and of stationary conjunctions of Jupiter and Saturn into their monuments.

Integrated into the limestone structures at Yaxchilan were texts and images on stone lintels and stelae. The texts were commissioned by semi-divine, hereditary rulers and were the records of their ceremonial practices. The events I will discuss in this paper were recorded on monuments during the reigns of two kings of Yaxchilan, Shield Jaguar the Great, who reigned from 9.12.9.8.1 (October 19, AD 681) to 9.15.9.17.16 (June 22, AD 741), and his son and heir, Bird Jaguar IV,[1] who reigned from 9.16.1.0.0 (April 29, AD 752) to after 9.17.0.0.0 (January 20, AD 771).[2] I will also mention several women and male allies who were recorded on monuments by these kings. Shield Jaguar recorded the astronomical events of a woman who is now called Xoc, the Maya word for shark. Bird Jaguar documented the activities of his mother, Lady Wind Skull, the mother of his son, Lady Great Skull, and two ritual assistants, Lady Ah Pop Ik and Lady Ix.[3]

Yaxchilan is situated on the southwestern bank of the Usumacinta River, at 16° 57′ N latitude and 90° 58′ W longitude.[4] The ceremonial center consists of 88 mapped structures and over 125 carved stelae, lintels, and thrones. Over 30 structures line the edges of a 300 m long, low-lying plaza which runs approximately parallel to the river. Other structures were built on three acropoli, each on natural hills 35 to 110 m higher than the Plaza. These structures are all oriented to overlook the Plaza, and in Classic times, when the trees were cut, would have had clear views of the hilly eastern

416

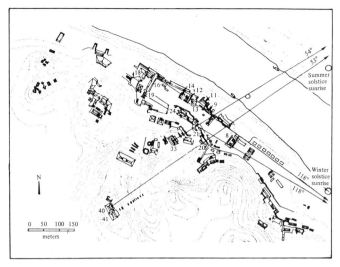

Figure 32.1 Map of Yaxchilan, after Heyden and Gendrop (1976), with directional lines and building numbers indicated by this author.

horizon across the river. The orientations of these hilltop structures form axes roughly perpendicular to that of the Plaza (Figure 32.1). To the west, behind the hilltop structures, a vista of the horizon is blocked by the nearby steep hills.

Orientations of the structures at Yaxchilan fall loosely into five groups:[5] 51–4°, 6–14°, 21–30°, 108–24°, with several structures facing exactly 118°, and a westerly facing group (see Table 32.1). I suspect that at least two of these groups of orientations were selected for astronomical reasons.

At Yaxchilan, on summer solstice, the sun rises in a pronounced notch on the eastern horizon at 63° east of north when viewed from Strs 33, 40, or 41.[6] The rising sun illuminates the interior of the group of buildings with 51–4° orientation, casting a patch of light in the form of the doorway on the floor of the chamber or onto the rear wall for a few minutes. In two of these buildings, I observed interesting interactions of the summer solstice sunrise light with the architecture.

Structure 41 is the earlier of these two solar–architectural hierophanies. Its entrances have step shaped tops (Figure 32.2). On summer solstice, the sun rises at 63° at 5.43 a.m. Faint light shines in the central door, down a long narrow entranceway. At 6.00 a.m. the sun appears next to the projecting flat stone that forms the stepped element in the right side of the doorway as one looks out. The sun illuminates a small portion of the rear wall, with a semi-

quatrefoil shaped patch of light created by the stepped-top doorway. This building was probably used as a solstice observatory by Shield Jaguar, and perhaps by rulers before him.

Structure 33 is the scene of a more striking solar–architectural hierophany. On summer solstice, as the sun rises (at 5.42 a.m. at 63° east of north in a notch next to the highest hill on the horizon approximately 1200 m distant from the temple), light passes through the doorways of Str 33. Inside the temple, in a niche created by transverse buttresses in the rear wall of the temple, is a crosslegged, seated statue of Bird Jaguar IV. The sun illuminates the statue of Bird Jaguar for approximately seven minutes. This hierophany occurs at least two days before and two days after the actual solstice.

Another group of alignments includes eight buildings oriented precisely to 118°. In Figure 32.1 it is clear that this is the orientation of the Main Plaza. These buildings were built over many generations. Structure 12 probably dates to about 9.5.0.0.0. Structures 16 and 24 were definitely built under the auspices of Bird Jaguar IV. Structure 24 is the obituary monument for his grandmother, father, mother, and aunt. The ancestral commemoration theme of Str 16 lintels will be discussed below. Structure 19 was probably built after 9.17.0.0.0, by Bird Jaguar IV's son. Every southeasterly hallway and the four front doorways of Str 19 are exactly 118° in orientation. The large pyra-

Table 32.1. *Chart of astronomical orientations of some major buildings at Yaxchilan.*

Measurements by this author, using a compass.

Structure	Orientation (degrees east of magnetic north)	Ruler
25	9	?
26	9	?
30	14	?
55	6	SJ2
39	20–1	BJ4
6	27	KEJ2?
7	30	?
23	29	SJ
36	26	?
51	30	?
20	53	SJ2
21	54	BJ4
33	54	BJ4
40	53	BJ4?
41	51	BJ3?
12	118	R10
13	118	BJ4?
14	118	?
16	118	BJ4
18	118–20	SJ?
19	118	BJ3?
24	118	BJ4
42	108–10	BJ4
67	124	?
11 + 74	220	?
71	318	?
1	345	BJ4

Key to rulers: KEJ2 = Knot Eye Jaguar II; BJ3 = Bird Jaguar III; BJ4 = Bird Jaguar IV. R10 = Ruler 10; SJ = Shield Jaguar the Great; SJ2 = Shield Jaguar II.

mid at the northwest end of the Plaza, Str 18, faces exactly 118° also. The ballcourt (Str 14) is situated so that the walls that form the narrow edges of the architectural bodies face 118° (although the walls sag now). Why do all these building face 118°?

On winter solstice the sun rises at 115–16° at Yaxchilan's latitude. Like the summer solstice orientation group, (Tate, 1986), the winter solstice buildings are set slightly outside of the path of the sun. Again, this allows for a few minutes of illumination of the interior of the buildings, providing the trees at the southeast end of the Plaza were kept cut. Note that all the winter solstice alignments are found on the Great Plaza. No tall buildings block the view of the sun along the Plaza on winter solstice.

The summer–winter solstice axes at the site must have been present from the Early Classic. Structure 41 dates to perhaps 9.10.0.0.0 or earlier, and Str 12 to 9.5.0.0.0. The view of summer solstice sunrise from Str 41 in the notch on the eastern horizon was an important facet in the selection of the location for monumental architecture at Yaxchilan. This notch or cleft in the eastern horizon is probably the source for Yaxchilan's Emblem Glyph, which is a sky glyph whose upper edge is cleft or split. If this view of summer solstice sunrise were not important, the site could have been located where the airstrip is today, and it would have been a simpler task to construct buildings at lower altitudes.

Other sites used architectural–solar alignments to measure the passage of summer and winter solstice, but not in the same manner in which Yaxchilan did. The solstitial sunrise observatory at Uaxactun is well known (Blom, 1924) and 12 additional examples of a similar structure exist at other sites, although they are only formal copies; they are not functional (Ruppert, 1940). Many other types of observatories are discussed by Aveni (1980, pp. 249–86). What is obvious is that the Mayas were aware of the annual cycle of the sun and used it ritually, even though there was not a method for recording it in the complex Maya calendar. Modern Maya groups are still acutely aware of the solar cycle as defined by the passage of the sun from north to south on the horizon (Gossen, 1974, p. 27). Gossen states, 'The whole cosmological system is bounded and held together by the paths of the Sun and Moon, who are the principal deities in the Chamula pantheon,' (Gossen, 1974, p. 22). In ancient times, like today, the solar year was probably observed on the horizon or with architectural assemblages designed to interact with the horizon.

The buildings at the ceremonial center of Yaxchilan were oriented to frame the annual path of the sun. The location of Strs 41, 40, 33, and perhaps some buildings on the Main Plaza, allowed the observation of the sun on the horizon all year. Evidence within the hieroglyphic inscriptions shows that the Yaxchilan astronomers and calendar priests related the horizon position of the sunrise to important events in the life of the city. If what was remembered about the nature of the day upon which an event occurred was the position of the sun relative

(a)

(b)

Figure 32.2 Diagrams of summer solstice phenomena. (a) Light passing into Structure 41, from the interior looking at the sunrise. Note the position of the sun relative to the notch in the distant hills. (b) A cutaway view of Structure 33, showing the summer solstice sunrise light illuminating the statue of Bird Jaguar IV. (Identity of the statue based on an *Ah Kal Bac* glyph on the rear of the head. This is one of Bird Jaguar's standard titles.) Drawings by Daniel Powers.

to an object on the horizon, then some error was to be expected.

The inscriptions at Yaxchilan do not explicitly state that one event is the solar year anniversary of another. Among the many Maya calendrical cycles, there is no known hieroglyph for counting periods of a solar year, 365.2422 days. In the Long Count the third place from the right is units of

Table 32.2. *Solstice events recorded in the inscriptions of Yaxchilan.*

Monument	Julian (AD)	Days from solstice	Calendar Round
L 26	6-20-720	+1	11 Chicchan 13 Yaxkin
L 23	6-22-726	+3	6 Caban 15 Yaxkin
St 16	6-20-735[a]	+1	6 Cimi[a] 19 Yaxkin
St 11	6-22-741	+3	12 Cib 19 Yaxkin
L 33	6-21-747	+2	5 Cimi 19 Yaxkin
L 9	6-16-768	−2	1 Eb, end of Yaxkin
L 50	no date recorded		

[a] One of several possible reconstructions

360 days, the *tun*. In the Calendar Round, the *haab* is a count of 18 months of 20 days, plus five days to equal 365 days. Except for the fact that annual anniversaries often fall within the same month, there is no indication in the hieroglyphs that an anniversary event is recorded. The presence of the annual anniversary only becomes clear when the Maya date is converted to a Christian date. Because the annual anniversaries do not fall at exactly 365.2422 day intervals, but are off by one to eight days, the anniversaries do not appear to have been calculated mathematically, but were more likely to have been determined by observation of the position of the sunrise relative to features on the horizon. In each instance, iconography or architectural alignment supports the argument that these solar year intervals were not mere coincidence.

A series of seven events recorded in stone at Yaxchilan occurred on summer solstice (see Table 32.2).

The earliest two events in the summer solstice series were performed by Lady Xoc. They are recorded on the lintels of Str 23, in unillustrated texts. The remaining summer solstice events are accompanied by illustrations of the king holding a staff with cutout quatrefoil shaped flaps (the flapstaff). The king always wears the same costume, the same headdress, the same jewelry on the summer solstice occasion (Tate, 1986). The most unusual items of the costume are a full length backrack and a chest ornament composed of a series of shrunken heads, sections of spondylus shells, and a shrunken torso (Figure 32.3). The flapstaff itself is an unusual feature of the monument. It was probably contrived with a frame of wood, to which was attached cloth. Quatrefoil shaped holes

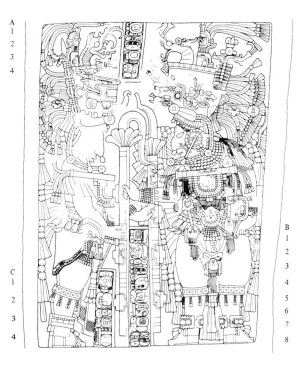

Figure 32.3 Lintel 9 of Structure 2. By Graham and von Euw (1977). © 1977 by the President and Fellows of Harvard College. Used by permission.

were cut on three sides from the cloth, and the cutout area flapped on the fourth side of the quatrefoil like a hinge, hence the term 'flapstaff'. Two of the monuments record the last or the next to last events in a king's reign. Those monuments show the king passing the staff to another individual.

Ethnographic data shows that for modern Chorti, the solstices demarcated a period of overlapping ritual responsibilities (Girard, 1966, pp. 121–6). As Girard describes them, the rituals concerning the seasons of the year function on several levels of the society: celestial, political, and agricultural. The rituals of summer solstice involve the positioning of a statue in one of two temples. The Chortis know that summer solstice is the day that the sun reaches its northernmost declination in the sky, and that on winter solstice the sun is furthest south. According to Girard, they duplicate this celestial division of the year through a ceremony in which an idol (obviously signifying the sun) is carried from a southerly temple to a northerly one on winter solstice and back again on summer solstice.

In the Chorti sphere of political/religious leadership, the solstices are the entry and exit points of service for the sacerdotes, each of whom serves an 18-month term. Entering service on winter solstice, the first year of his service, the sacerdote honors the chair of the idol which signifies the sun. When winter solstice returns, he begins a period of relative inactivity, during which he devotes himself not to interceding to the deity, but only to honoring the vacant seat of the idol which is ensconsed in the other temple. Summer solstice marks the day that he terminates his duties, having passed responsibility on to the new sacerdote who has had the benefit of a 'backup' sacerdote for the past six months. The ceremonies of the solstices begin the 20th or 21st of June and continue for two or three days when the chair and its statue arrive in the second temple.

The modern Maya concepts of overlapping periods of ritual responsibility are illustrated on the Classic period monuments at Yaxchilan. On St 11 and L 9, two individuals are shown exchanging the staff. On St 11, Shield Jaguar performs the last commemorated act of his life at approximately age 95 (Figure 32.4). His last recorded ceremony is passing the flapstaff to Bird Jaguar IV on summer solstice. Five days later, Bird Jaguar's mother performed a bloodletting and was called with the title 'Mah K'ina', (*k'in* is the word for sun). The ritual responsibility at Yaxchilan apparently overlapped between Shield Jaguar, Lady Ik Skull, and Bird Jaguar for ten years until the woman died and the son acceded. On L 9, Bird Jaguar IV is recorded performing his second to last official act – dressed as GI, he hands the staff to Lord Great Skull, who wears a solar headdress, on summer solstice. Lord Great Skull never acceded to the throne, but he remained important on the monuments of Bird Jaguar's son, Shield Jaguar II.

In addition to ethnographic evidence, tentative linguistic reconstruction of the verbal phrases provides a clue to the meaning of summer solstice rituals. The verbs on the summer solstice monuments are practically identical. Each verbal phrase consists of a T515b:103:683 auxiliary verb, and the *ti* locative plus verbal noun construction, T59:563:130:561. This might be pronounced 'mul-–taj–aj t(i)–aj+ ka wa' chan', written thus:[7]

In Ch'ol

| aj | kaw | wa' | chan |

means

| he | open | standing | sky |
| | | stopped | |

This phrase might refer to the swing of the sun, which lingers at its northernmost declination on the horizon for several days at summer solstice. For a group of people performing horizon observations of the sun, this is one of the most obvious stations of the year, when the movement of the sun slows to a standstill.

Yaxchilan St 11 and L 9 show the passage of ritual responsibility from an outgoing ruler to an incoming ruler on summer solstice. On L 33 (see Graham, 1979), Bird Jaguar wears a coiled serpent headdress which only appears on anniversary occasions (C. Tate, 1986, chap. 2 of doctoral dissertation presented at the University of Texas at Austin). That date is the six year anniversary of the summer solstice event by Bird Jaguar that was illustrated on St 11.

Several other groups of events at Yaxchilan were commemorated on solar year intervals. The following textual headings are designed to coordinate this commentary on iconography with the information in Table 32.3.

32.1 The Jupiter–Saturn conjunction

One of the ceremonies at Yaxchilan seems to have been modelled by the king after a grand ritual series performed by Chan Bahlum of Palenque to honor his deceased royal father, Pacal. At Palenque, Chan Bahlum organized a four-day ceremony to commemorate the 75th solar year anniversary of the accession of Pacal, and the eighth solar year anniversary of his own accession. This occasion fell during a six-week long stationary conjunction of Jupiter and Saturn, which were only 4° apart in the sky in July of AD 690. Shortly thereafter was the period ending 9.13.0.0.0, the dedication date

Figure 32.4 Stela 11, Structure 40. The flapstaff side faced the summer solstice sunrise and the river. Photo by Teobert Maler (1901–2, pl. LXXIV). © 1901–2 by the Peabody Mueseum of Archaeology and Ethnology, Harvard College. Used by permission.

of the Group of the Cross at Palenque.[8] Nineteen and a half years later, in October of AD 709, was the very next time Jupiter and Saturn were aligned and stationary. I suspect that Shield Jaguar, who was known to have been at or connected with Palenque in his youth, modelled the event celebrating the 80th year anniversary of his father and his 28th solar year anniversary during the Jupiter–Saturn conjunction after the event at Palenque. In addition to there being a parallel in the types of events that were commemorated, the verbs used for the ceremonies – bloodletting verbs and a rare God

N verb – were used both at Palenque and Yaxchilan. This ceremony appeared on Yaxchilan L 24 of Str 23 (Figure 32.5). The 5 Eb 15 Mac event (L 24; the Jupiter–Saturn conjunction) seems to later have been commemorated by various other types of rituals at Yaxchilan.

32.2 Bird Jaguar IV's supernatural ballgame event

On Step VII of the Hieroglyphic Stairway of Str 33, Bird Jaguar IV portrayed himself engaged in playing ball accompanied by two dwarfs (Figure 32.5). Step VII is the central and largest of the 13 carved step risers situated immediately below the entrances to Str 33. It textually links a ballgame played by Bird Jaguar with several mythological ballgames played by supernaturals in the distant past. Bird Jaguar locked the date, 3 Muluc 17 Mac 9.15.13.6.9, into eight larger cycles than the baktun, giving the date cosmic significance. The date was seven days short of 35 solar years since the 5 Eb 15 Mac event performed by his father, Lady Xoc, and purportedly by his mother, Lady Wind Skull, as well (L 32). The ballgame event occurred over seven years before Bird Jaguar's accession: in fact, it is the first event in which he is monumentally commemorated at Yaxchilan. Another event is related to this series. It occurs on three lintels, L 6 and L 7 of Str 1, and L 43 of Str 42. One hundred and sixty-eight days after his accession, Bird Jaguar dressed in underworld type garb[9] and performed a bloodletting ritual on the seven year anniversary of his ballgame event. The date 10 Lamat 16 Mac was the completion of 8×365.25 days – three days since the ballgame event. The ballgame event and the jaguar paw events bracket Bird Jaguar's accession, both temporally and spatially. The two jaguar paw lintels are in buildings that define Bird Jaguar's ritual arena at Yaxchilan, and an imaginary line drawn between them passes directly over Step VII.

32.3 Stela 11-solar year anniversary of a posthumous event

Stela 11 is the monument created by Bird Jaguar for his accession day that documents the transition of power from Shield Jaguar to himself during the ten year interregnum (Figure 32.4). One of the ceremonies recorded on that stela is a Hotun Ending on 9.15.15.0.0. The odd thing about this Hotun Ending is that Shield Jaguar, then deceased, is named as the protagonist, under the auspices of (*u cab*) Bird Jaguar.[10] On the same monument, Bird Jaguar impersonated Chac Xib Chac, a GI – Hero Twin avatar, on the fourth solar year anniversary of the posthumous event by his father. The Chac Xib Chac impersonation is illustrated on the temple side of the stela, while the passing of the flapstaff on summer solstice is illustrated on the side facing the sunrise.

32.4 Lady Xoc's posthumous event, an anniversary of Bird Jaguar IV's birthday

Lady Xoc is clearly said to have performed a 'fire' event about six years after her death. The event is recorded at the end of a continuous obituary inscription carved during the reign of Bird Jaguar (Str 24, L 28) (see Graham, 1979). The date 6 Caban 10 Yax was exactly 46 years after Bird Jaguar's birthday. It also seems to have been selected because it was even numbers of tzolkins distant from other events that seem to link Bird Jaguar and Lady Xoc. The posthumous event was 41×260 days since a summer solstice 'fire' event performed by Lady Xoc (L 23), and 5×260 days since Bird Jaguar captured Q, the prisoner whose capture was one of the important, and perhaps necessary, events prior to Bird Jaguar's accession.

32.5 Anniversaries of important ancestral women

The lintels of Str 16 each show a royal personage from the court of Bird Jaguar IV sitting crosslegged and holding a double headed serpent (Figure 32.6). Two women, Ladies Ix and Ah Pop Ik, are portrayed on the two outer lintels, L 38 and L 40, with fleshed serpents, while Bird Jaguar IV is shown on the central lintel holding a skeletal serpent. The date of L 38 is 3 Ix 7 Mol (9.16.12.5.14), which is 22 years minus three days since the day Lady Wind Skull let blood, and, I believe, assumed the royal power at Yaxchilan[11] on 4 Imix 4 Mol. The central lintel, L 39, is dated 4 Imix 4 Mol, and is an assertion made by Bird Jaguar 23 years after the actual date that he, too, participated in a bloodletting ritual on that day. L 40 shows Lady Ix Balam on a date I have reconstructed as 13 Ahau 18 Zip, 9.16.7.0.0 (C. Tate, 1986, appendix 1 of doctoral dissertation presented at the University of Texas at Austin). That date is nine solar years minus one day after

Table 32.3. *Chart of astronomically related periodic commemorations in the dates of Yaxchilan.*

Jupiter–Saturn conjunction and related accession anniversaries

Long Count	Calendar Round	Event	Actor	Str	Monu	Julian
(9.9.16.10.13)	12 Akbal 6 Yax implied	A	BJ3	20	St 6	9-5-629
9.12.9.8.1	5 Imix 4 Mac	A	SJI	23	L 25	10-19-681
9.13.17.15.12	5 Eb 15 Mac	Bl, N	SJI, X	23	L 24	10-23-709

$28 \times 365.25 + 4$ since accession SJ $80 \times 365.25 + 39$ since accession BJ3
Jupiter and Saturn aligned, stationary

Bird Jaguar IV's supernatural ballgame event

Long Count	Calendar Round	Event	Actor	Str	Monu	Julian
9.13.17.15.12	5 Eb 15 Mac	Bl, N	SJI, X	23	L 24	10-23-709

$28 \times 365.25 + 4$ since accession SJ $80 \times 365.25 + 39$ since accession BJ3
Jupiter and Saturn aligned, stationary

Long Count	Calendar Round	Event	Actor	Str	Monu	Julian
9.13.17.15.13	6 Ben 16 Mac	Bl	SJI, LIS	55	L 53	10-24-709
				13	L 32	
9.15.13.6.9	3 Muluc 17 Mac	Bl	BJ4	33	Sp VII	10-17-744
9.16.1.8.8	10 Lamat 16 Mac	God K	BJ4	1	L 7	10-14-752

7×365 after 3 Muluc 17 Mac ballgame

Stela 11 solar year anniversary of a posthumous event

Long Count	Calendar Round	Event	Actor	Str	Monu	Julian
9.15.15.0.0	9 Ahau 18 Xul	BL	SJI	40,	St 11	5-31-746
			?	19	A 1	
9.15.19.1.1	1 Imix 19 Xul	?	BJ4	40	St 11	5-31-750

4×365.25 since 9.15.15.0.0. St 11

Lady Xoc's posthumous event and the 6 Caban dates

Long Count	Calendar Round	Event	Actor	Str	Monu	Julian
9.13.17.12.10	8 Oc 13 Yax	B	BJ4	10	L 29–30	8-23-709
9.14.14.13.17	6 Caban 15 Yaxkin	Fire	Xoc	23	L 23u	6-22-726

summer solstice

Long Count	Calendar Round	Event	Actor	Str	Monu	Julian
9.16.0.13.17	6 Caban 5 Pop	C Q	BJ4	21	L 16	2-5-752

36×260 since L 23 Fire event 26×360 since L 23 Fire event

Long Count	Calendar Round	Event	Actor	Str	Monu	Julian
9.16.4.6.17	6 Caban 10 Yax (Zac)	Fire	L Xoc	24	L 28	8-23-755

41×260 since L 23 fire event 5×260 since capture of Q
46×365.25 since BJ4 birthday

Structure 16: anniversaries of women of the previous generation

Long Count	Calendar Round	Event	Actor	Str	Monu	Julian
9.15.17.15.14	3 Ix 17 Zip	D	L Xoc	24	L 59–28	3-30-749
9.16.7.0.0	13 Ahau 18 Zip	Bl	L Ix	16	L 40	3-29-758

$9 \times 365.25 - 1$ since Xoc's death

Long Count	Calendar Round	Event	Actor	Str	Monu	Julian
9.15.10.0.1	4 Imix 4 Mol	Bl	BJ4	16	L 39	6-27-741

summer solstice. 1 st app V as MS

Long Count	Calendar Round	Event	Actor	Str	Monu	Julian
			LIS	21	St. 32	
			LGS, BIL	20	L 14	
9.16.12.5.14	3 Ix 7 Mol	Bl	L Ah Ik	16	L 38	6-25-723

$22 \times 365.25 - 2$ since 4 Imix 4 Mol

Zenith passages

Long Count	Calendar Round	Event	Actor	Str	Monu	Julian
9.11.18.15.1	7 Imix 14 Zotz	4, Fire	SJI	44	SP IV	5-6-671

first zenith passage

Long Count	Calendar Round	Event	Actor	Str	Monu	Julian
9.16.0.0.0	2 Ahau 13 Zec	PE	BJ4	33	A 9	5-5-751

first zenith passage

Long Count	Calendar Round	Event	Actor	Str	Monu	Julian
9.16.1.0.9	7 Muluc 17 Zec	4	BJ4	22	L 21	5-8-752

first zenith passage 81×365.25 since Sp IV

Long Count	Calendar Round	Event	Actor	Str	Monu	Julian
9.16.4.1.1	7 Imix 14 Zec	C JS	BJ4 CoM	1	L 8	5-5-755

84×365.25 since Sp IV first zenith passage

Long Count	Calendar Round	Event	Actor	Str	Monu	Julian
		star	BJ4, L Ik	42	L 41	

Tzolkin and Tun commemorations of Str 44, step IV event

9.11.18.15.1	7 Imix 14 Zotz	4, Fire	SJI	44	Sp IV	5-6-671	
				first zenith passage			
9.12.9.8.1	5 Imix 4 Mac	A	SJI	23	L 25	10-19-681	
				accession			
9.14.11.15.1	3 Imix 14 Chen	N	Xoc	23	L 25e	8-1-723	
			compl 52 × 360 since Sp IV				
9.14.14.8.1	7 Imix 19 Pop	Anniv	SJI, X	23	L 23u	2-26-726	
		compl 45 × 360 reign 77 × 260 since Sp IV					
9.15.6.13.1	7 Imix 19 Zip	Fire	SJI, X	11	L 56	4-4-738	
		94 × 260 since Sp IV. 1st app V as MS					

Key to events: A = accession to rule; Anniv = stated anniversary; B = birth; Bl = bloodletting; C = capture; D = death; Fire = a variant of T 1035; God K = T 533.670; N = God N, T 1014a; 4 = 4 *te zotz*, T 87.756.
Key to actors: BIL = Lord Great Skull, BJIV's possible brother-in-law; BJ3 = Bird Jaguar III; BJ4 = Bird Jaguar IV; LIS = Lady Ik (Wind) Skull; LGS = Lady Great Skull; L Ix = Lady Ix; L Ah Ik = Lady Ahpo Ik; SJ = Shield Jaguar the Great; X or Xoc = Lady Xoc.

the death of Lady Xoc. So here, Bird Jaguar and his female ritual assistants perform ceremonies that commemorate important events in the lives of the two royal females of the previous generation, Lady Xoc and Lady Wind Skull. The iconography of the lintels supports the fact that these are ancestral celebrations, for the double headed serpents' mouths contain busts of God K, the god of royal lineage, and the serpent image is associated with ancestor portraits on the stelae of Yaxchilan and elsewhere (C. Tate, 1986, chap. 2 of doctoral dissertation presented at the University of Texas at Austin).

The presence of implied solar year anniversaries and summer solstice observations in the dates is supported by iconography and by city planning. Since the monuments referring to zenith passage are purely glyphic, I will leave the calendric data to stand on its own in Figure 32.3.

Discrete sets of political rituals have been shown to be linked by implied solar year and occasionally by tzolkin periodic commemorations at Yaxchilan. The links are between supernatural or astronomical type events in the past, and the documentation of Bird Jaguar IV's legitimacy during his reign. There must have been some ritual–political power in being able to coordinate the politics of the present with the celestial hierophanies observed by the ancestral royalty.

The kings of Yaxchilan created analogies between the monumental commemorations of their ritual actions and the motions of the sun. The site as it exists today was modified to suit the purposes of Bird Jaguar IV. Every temple he built was situated to clarify the summer and winter solstice sunrise axes. His buildings that are directed toward winter solstice sunrise contain information about his deceased ancestors: Str 24, the obituaries and Lady Xoc's posthumous event, Str 16, the commemoration of Lady Xoc's death and Lady Wind Skull's autosacrifice, and the Altar 1 of Str 19, which has the death date of Shield Jaguar. He oriented his accession commemoration structure, Str 33, so that the rays of the summer solstice sunrise struck his three dimensional portrait, and the image of his ballgame with the Lords of the Underworld. This ballgame event was temporally framed by solar anniversaries. It was linked to a prior event in the reign of his father (the Jupiter–Saturn conjunction) and to an event after his own accession (Ls 5, 6 and 43). Structure 40 also displayed a record of Bird Jaguar's accession on St 11, beneath the image of his father passing him the flapstaff of summer solstice. That stela faces summer solstice sunrise. The entire program of Bird Jaguar's legitimacy was documented by both summer solstice events and solar year anniversaries.

The growing body of data on the uses of astronomy at Classic Period Maya sites strongly suggests that Maya rulers selected specific astronomical phenomena to incorporate into the ritual–historical life of their ceremonial centers. The emphasis at Yaxchilan on solar rituals and the dual orientation of the site's axes toward the solstice sunrise points set Yaxchilan apart from other Maya ceremonial

(a)

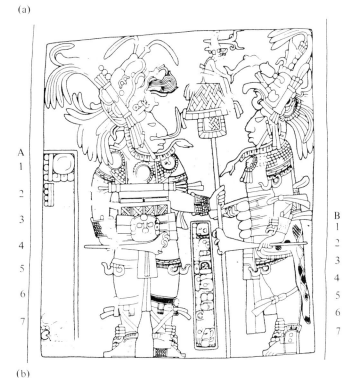

(b)

426

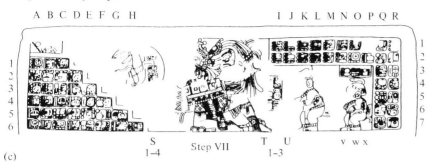

(c)

Figure 32.5 Events related to Bird Jaguar IV's supernatural ball-game. (a) Lintel 24 of Structure 23. (b) Lintel 6 of Structure 1. (c) Step VII of Structure 33. Lintels 24 and 6 from Graham and von Euw (1977). © 1977 by the President and Fellows of Harvard College. Step VII from Graham (1982). © 1982 by the President and Fellows of Harvard College. Used by permission.

Figure 32.6 Lintels 38, 39 and 40 of Structure 16: commemorations of ancestral females. From Graham (1979). © 1979 by the President and Fellows of Harvard College. Used by permission.

centers, and created a sense of local identity for the people of the place of the split sky.

Notes

1 For correlations between the Christian and Maya calendars, I use the 584285 correlation first proposed by J. Eric S. Thompson (1935) and re-evaluated by Lounsbury (1982). Christian dates in the Maya era are given in the Julian calendar.

2 The historical rulers Shield Jaguar and Bird Jaguar were first identified by Tatiana Proskouriakoff (1963, 1964).

3 For a discussion of the role of royal women at Yaxchilan, see Tate (1987).

4 These coordinates were taken from Carta Topográfica, Programación y Presupuesto de Coordinación General de los Servicios Nacionales de Estadistica Geografica Informacioncia, México.

5 I measured orientations with a compass, sighting along the interior walls of the three doorways and the exterior walls of the structures. This means that the number given corresponds to the direction the temple itself faces, or the direction one would face who had his back against the rear wall of the building and looked straight out. The orientations are expressed in degrees east of magnetic north.

6 Abbreviations used herein are: Str for structure, St for stela, and L for lintel.

7 Linguistic assistance has been rendered by Barbara MacLeod (1984, personal communication) and Nicholas Hopkins and Kathryn Josserand (1985, personal communication).

8 Linda Schele discusses the *u cab* glyph in Schele (1982, p. 73).

9 On these lintels, Bird Jaguar wears a jaguar skin, holds jaguar paws, and wears a water-lily jaguar headdress, associated with GIII of the Palenque Triad. His companion wore the headdress seen on the ballplayer steps of Str 33. See C. Tate (1986, chaps 2 and 3 of doctoral dissertation presented at the University of Texas at Austin) for more discussion of the implications of costume on monuments at Yaxchilan.

10 The Jupiter data at Palenque was discovered by Floyd Lounsbury, cited in Schele (1984) and was presented by Lounsbury at the Oxford II Conference (see Chapter 19).

11 The 4 Imix 4 Mol event is discussed in C. Tate (1987; and 1986, doctoral dissertation presented at the University of Texas at Austin). Lady Wind Skull, Lady Great Skull, Lord Great Skull, and Bird Jaguar IV are all recorded as having let blood on this day. The latter three individuals were recorded performing the bloodletting many years after the event occurred, and their motives for saying, 'Twenty or more years ago I participated in this event,' are unclear.

References

Aveni, A. F. (1980). *Skywatchers of Ancient Mexico.* Austin: University of Texas Press.

Blom, F. (1924). Archaeology: report of Mr Frans Blom on the preliminary work at Uaxactun. *Carnegie Institute of Washington Yearbook* 23, 217–19.

Girard, R. (1966). *Los Mayas.* Mexico City: LibroMex.

Gossen, G. H. (1974). *Chamulas in the World of the Sun: Time and Space in a Maya Oral Tradition.* Cambridge, MA: Harvard University Press.

Graham, I. (1979). *Corpus of Maya Hieroglyphic Inscriptions, vol. 3, no. 2. Yaxchilan.* Cambridge, MA: Peabody Museum of Archaeology and Ethnology, Harvard University.

Graham, I. (1982). *Corpus of Maya Hieroglyphic Inscriptions, vol. 3, no. 3. Yaxchilan.* Cambridge, MA: Peabody Museum for Archaeology and Ethnology, Harvard University.

Graham I. and Von Euw, E. (1977). *Corpus of Maya Hieroglyphic Inscriptions, vol. 3, no. 1. Yaxchilan.* Cambridge, MA: Peabody Museum for Archaeology and Ethnology, Harvard University.

Heyden, D. and Gendrop, P. (1976). *Pre-Columbian Architecture of Mesoamerica.* New York: Harry N. Abrams.

Lounsbury, F. G. (1982). Astronomical knowledge and its uses at Bonampak, Mexico. In *Archaeoastronomy in the New World*, ed. A. F. Aveni, pp. 143–68. Cambridge University Press.

Maler, T. (1901–2). Researches in the Central Portion of the Usumacintla Valley. Peabody Museum Memoirs, vol. 2, no. 2.

Proskouriakoff, T. (1963). Historical data in the inscriptions of Yaxchilan, Part I. *Estudios de Cultura Maya* 3, 149–61.

Proskouriakoff, T. (1964). Historical data in the inscriptions of Yaxchilan, Part II. *Estudios de Cultura Maya* 4, 177–201.

Ruppert, K. (1940). A special assemblage of Maya structures. In *The Maya and Their Neighbors*, eds. C. Hay, R. Linton, S. Lothrop, H. Shapiro and G. Vaillant, pp. 222–31. New York: D. Appleton-Century Co.

Schele, L. (1982). *Maya Glyphs: The Verbs.* Austin: University of Texas Press.

Schele, L. (1984). *Notebook for the Maya Hieroglyphic Workshop at Texas.* Austin: Institute of Latin American Studies.

Tate, C. (1986). Summer solstice events performed by Bird Jaguar IV of Yaxchilan. Paper presented at *Simposio de Arqueoastronomia e Etnoastronomia en Mesoamerica,*

Mexico, D. F., September, 1984.

Tate, C. (1987). The royal women of Yaxchilan. Paper presented at the *Primer Simposio Internacional de Mayistas*, Mexico, D. F., 1985.

Thompson, J. E. S. (1935). *Maya Chronology: The Correlation Question.* Carnegie Institute of Washington publ. no. 456, contrib. 14, Washington, DC.

Further studies on the astronomical orientation of Medieval churches in Austria

Maria G. Firneis and Christian Köberl   *University of Vienna*

33.1 Introduction

This chapter is a summary of various research activities predominantly published in German. The fact that a pronounced sun orientation might be imminent in the axis-orientation of Medieval churches in Central Europe was investigated. The idea had, of course, been formulated earlier. The very first hints of it were put forward in a treatise by Thijm (1858) and independently by Charlier (1902), but due to some erroneous results the concept later had come under severe attack.

Early Christian communities followed the antique tradition considering the orientation towards the east as sacred, as is known from the patristic literature of Origines, Clemence of Alexandria and Tertullianus, who state that prayers usually were said turning in an eastward direction. The fact that in the first half of the fifth-century pope Leo the Great (AD 440–61) issued a decree against the veneration of the rising sun shows that the sun only had to be taken as a symbol of the Christ entering his church and not as a deity itself (Firneis and Ladenbauer, 1978b). While Pope Vigilius (AD 537–55) confirmed the east orientation of churches put forward by Bishop Athanasius during the Council of Nicea (AD 325) it was ordered that a priest specifically had to turn eastward during prayer in church (as specified in the Rationale Divinorum Officiorum V.2.57) (Guzsik, 1978), this direction still was considered sacred by tradition during the age of the Renaissance as can be seen from the figure (on p. F III v.) given in Apianus' *Astronomicum Caesareum* printed in 1540 (Figure 33.1).

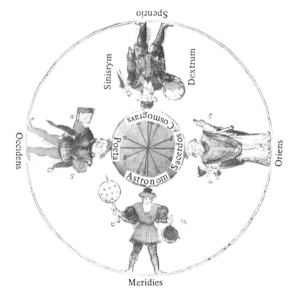

Figure 33.1 Four professions related to the cardinal directions are depicted: the priest to the east, the cosmographer to the north, the poet to the west and the astronomer to the south. Taken from Apianus (1540, p. F III v).

Though early churches show this east orientation in two ways (either the altar is positioned in the eastern main apse while the entrance door points to the west, or the door might be in the east, so that the beams of the rising sun could illuminate the altar then positioned in the west) only very few churches actually point to the cardinal directions. Deviations may be attributed to various sources (e.g. as given in Guzsik, 1978). From Firneis, Köberl and Göbel (1981b) and Köberl (1984), the fact can be deduced that, as far as Central Europe is concerned, the solar rising azimuth on the patron saint's day or another important date related to the foundation history of the specific church played an eminent role for the setting out of its axis. Orientation studies of this kind always have to be related to the Julian calendar, in use up to 1582. The orientation habit only can be found for Romanesque churches (Firneis and Ladenbauer, 1978a), or in a few cases for the early Gothic style after which epoch it obviously ceased. In order to investigate the fact that an axis-orientation actually is related to the building concept, thorough investigations on the history of each specific building as well as its internal decoration have to be carried out. It is reasonable to collaborate at this stage with archaeologists, historians and art historians.

33.2 Methodological suggestions

As magnetic orientation measurements may be affected too much by external influences, usually a survey with a well shaded theodolite relating the measurements to the solar limb to obtain the azimuth with sufficient accuracy is considered appropriate. Usually at least 20 measurement points along the older parts of the main nave walls or the main apse are considered adequate to yield the azimuth directions.

The fit of the direction of the church axis to the rising or setting sun is computationally an inverse interpolation. A straightforward procedure is given here.

A trivial reasoning in spherical trigonometry relates the solar rising aximuth A to the solar declination d and the latitude φ of the observing site

$$\cos A = -\frac{\sin d}{\cos \varphi},$$

where A is measured from north to east. The natural horizon is usually elevated by the angle h_0 above the mathematical horizon (omitting the influence of the atmosphere). The azimuth A thus changes by the value ΔA according to

$$\Delta A = \arcsin\left(\frac{\tan h_0}{[(\cos d/\sin \varphi)^2 - 1]^{\frac{1}{3}}}\right).$$

If the effect of refraction near the horizon is included, a simple interpolation formula is no longer adequate. Now the atmospheric pressure and the temperature at sunrise have to be taken into account and two second-order polynomials can be set up in the following form:

$$a = p(0.1594 + 0.0196 h_0 + 2 \times 10^{-5} h_0^2),$$
$$b = (273 + \delta_C)(1 + 0.5050 h_0 + 0.0845 h_0^2),$$

with p being the atmospheric pressure and δ_C the temperature in Celsius (Duffet-Smith, 1985).

For a rough approximation plausible mean values for p and δ_C have to be inserted. When the season has been specified, the procedure can be repeated with the appropriate seasonal values. With these polynomials the apparent height h_{app} of a celestial body – including the sun – can be formulated:

$$h_{app} = h_0 + \frac{a}{b}.$$

In practice, another source for frequent reiteration of the procedure has to be considered. The term 'rising sun' may indicate the sun's upper or lower limb as well as the bisected solar disk. Yet from previous investigations it turned out that by 'rising' (and 'setting') really the first (last) beams of the solar limb have to be considered – which is in accordance with the practised religious concept.

The formulas for a and b, however, can only be applied if h_0 for a celestial body is known. While the inverse procedure from h_{app} to h_0 seems only to be accessible via inverse interpolation, the substitution by a direct one can be considered favourable. Solving the refraction correction term for h_0 when h_{app} is known results in the cubic equation

$$h_0^3 + a_1 h_0^2 + a_2 h_0 + a_3 = 0$$

with the coefficients

$$a_1 = 5.9763 - h_{app} + 2.3669 \times 10^{-4} \frac{p}{\delta_K},$$

$$a_2 = 11.8343 - 5.9763 h_{app} + 0.2320 \frac{p}{\delta_K},$$

$$a_3 = -11.8343 h_{app} + 1.8864 \frac{p}{\delta_K},$$

where $\delta_K = \delta_C + 273$. Transforming this cubic equation into its reduced form by the linear transformation

$$h_{app} + \frac{a_1}{3} = y$$

431

results in

$$y^3 + py + q = 0$$

with

$$p = a_2 + \frac{a_1^2}{3}$$

$$q = \frac{a_1}{3}\left(\frac{2a_1^2}{9} - a_2\right) + a_3,$$

where the discriminant

$$D = \left(\frac{q}{2}\right)^2 + \left(\frac{p}{3}\right)^3.$$

The elevation angle of the natural horizon then can be calculated:

$$h_0 = \left(-\frac{q}{2} + D^{\frac{1}{2}}\right)^{\frac{1}{3}} + \left(\frac{q}{2} - D^{\frac{1}{2}}\right)^{\frac{1}{3}} - \frac{a_1}{3}.$$

Thus, from h_{app} the true h_0 can be recalculated. If we no longer consider a rectangular spherical triangle but a general triangle, the following formula, which relates the measured azimuth A_{meas} of the church axis to the true angle h_0 and implicitly to the solar declination d and the latitude φ, is obtained:

$$180° - A_{meas} = \arcsin\left(\frac{\tan h_0}{[(\cos d/\sin \varphi)^2 - 1]^{\frac{1}{2}}}\right)$$

$$- \arcsin\left(\frac{\sin d}{\cos \varphi}\right).$$

Though there exists an explicit formula for the difference of two arcsines, it is computationally advantageous to solve this equation for d by an iteration procedure.

Since the same solar declination value is observed twice a year, there still exists an ambiguous solution for the dates corresponding to a specific azimuth. From the historical background related to the church under consideration one possibility has to be excluded for a reasonable explanation. It is recommended to the astronomer to search the historical literature for some hidden hints and countercheck with a historian in order not to overlook facts of importance.

If the building accuracy is low Tuckerman's tables (Tuckerman, 1964) or the less well known (and less accurate) tables of Neugebauer (1922, 1929) help in handling the orientation problem quite fast, especially if an accuracy of the result within $\frac{1}{3}°$ is considered sufficient as related to the limited alignment precision of the walls of old buildings.

33.3 Samples in certain areas

As a first object of discussion St. Mary's church in Unterfrauenhaid in the Austrian province of the Burgenland, which was set up in AD 1222, was investigated. Details can be found in the literature (Firneis and Köberl, 1982). The fact that during excavations in 1982 a clearly marked alignment dislocation by 1°8 along the Gothic apse between the foundations and the uprising walls could be found is a good indication that possibly a slight error of the apse-orientation was corrected at the last minute. While the Romanesque nave walls can be attributed to the first building epoch, the Gothic apse dating back to AD 1450 (which shows an axial line differing by 5°2 from that of the existing nave) would yield a corresponding azimuth for the rising sun on March 25th and August 31st with an accuracy of 0°25. In this special case the true horizon could be substituted by the mathematical one as it was completely flat in the eastern direction.

March 25th – Annunciation Day – is a very important festival, as during the Middle Ages it marked the beginning of a new year. No equivalently important festival is known for August 31st. From the written records it can be deduced that the church was dedicated to St. Mary from the beginning (with no change in dedication). It also possessed on one of the side altars a painting showing the angel's annunciation. This is another example that apse-orientation and building concept were closely linked. For comparison, another example is given in Firneis and Ladenbauer (1978b). No reasonable explanation for the nave-orientation leading to a solar rising azimuth on March 17th and September 11th about the year AD 1200 (equinox falling on March 13th and September 15th) can be given (Figure 33.2).

Situated two valleys further south of Unterfrauenhaid the remains of another Romanesque church in Pilgersdorf – also built around AD 1200 – were discovered during earthwork for an inundation construction. In 1975 these remains were

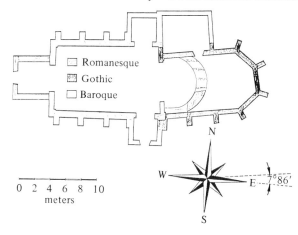

Figure 33.2 Plan of the various style elements of the church foundations at Unterfrauenhais/Burgenland.

destroyed to a greater part by road building machines, but photographs permitted a theoretical reconstruction. As the preserved wall remains are excessively thick (1.8 and 2.5 m, respectively), this church clearly was used as a shelter during wartime. However, the building was burnt down in the second half of the 13th century (as deduced from coin findings). The event can be attributed to an uprise of some Hungarian magnates against the Habsburgs in AD 1289/90, and has also led to the devastation of Unterfrauenhaid. When in the beginning of the 14th century a new parish church was set up 300 m to the south-east of the old location, material was taken from the old building, which by AD 1560 clearly was forgotten when a cemetery was set up on top of it as well as later a secondary school (Figure 33.3).

Because the eastern horizon is obscured by a small mountain ridge right behind the investigated location and the valley opens to the west, an orientation calculation was carried out also for a west-orientation (taking into consideration the natural horizon and the fact that a correction of 1°6 between the foundations and the uprising walls had actually been carried out). An azimuth corresponding to March 25th and September 4th (AD 1200) resulted. Here again the importance of the beginning of the Medieval year intentionally was put before the eyes of the congregation much in the same way as in Unterfrauenhaid.

From an unpublished result (M. Firneis, unpublished manuscript: 'Orientierung der Pfalzkirche von Tilleda am Kyffhäuser', 1981) it is known that

a church within the imperial fortress of Tilleda on top of the Kyffhäuser mountain (GDR) built about AD 1000 also points in the direction of the solar rising azimuth on March 25th.

Another interesting example for this orientation theory is Vienna's main church: St. Stephen's Cathedral (Firneis, 1984). The year of the ceremony of the laying of the foundation stone is known to be AD 1137.

In 1945 the church had suffered severe damage from the Second World War. When the reconstruction started, excavations were carried out to a greater extent. Then it turned out that this Romanesque church already had very large dimensions. Three parallel naves (83 × 26 m) were conceived for a congregation of several thousand people, which was far more than could be anticipated to come to services. The idea underlying this impressive concept must have been to set up a cathedral rivalling the ones of Salzburg, Regensburg and Augsburg which were the seats of bishops (under imperial influence) – something which the dukedom of Austria did not possess at this time.

The axis-orientation turned out to be A_{Steph} = 125°03'18" (measured from north). With a solar rising azimuth on St. Stephen's Day (December 26th in the Julian calendar system) of $A_{\odot}$ = 124°41'35", this is considered as a remarkable coincidence underlining the fact that the patron saint's day was considered to be of importance already in the building concept. To investigate the sensitivity of such an orientation from an intended orientation

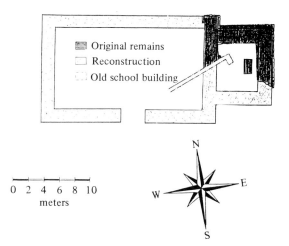

Figure 33.3 Plan of the partially reconstructed church foundations at Pilgersdorf/Burgenland.

towards the winter-solstice point a confidence interval for the slope of the linear regression with a covering probably $\gamma = 70\%$ (corresponding to the 1σ value of a Gaussian distribution) was set up. This confidence interval can be determined from the computed slope k, the deviation s_k and the fractile of a t-distribution by

$$[k - t_{n-2;\,\gamma/2} \cdot s_k, \quad k + t_{n-2;\,\gamma/2} \cdot s_k].$$

This slope allows a confidence interval for the azimuth angle to be expressed as

$$[124°38'05'', \quad 125°24'39''].$$

It can be shown that for the winter solstice which occurred on December 14th, AD 1137, the solar rising azimuth was $A_\odot = 125°34'56''$, a value already outside the confidence interval. This is another hint that St. Stephen as patron saint was already incorporated into the foundation azimuth (Figure 33.4).

Celtic tribal god who – as deduced from Roman writings – might have corresponded to some 'Celtic Mars'.

When using measurement points along the outer wall of the nave falling in the same axis direction as the apse (strictly parallel to the old temple walls) the azimuth was found to be $A_{\text{Law}} = 60°12'18'' \pm 38'25''$. The large uncertainty corresponds to the actual variations of these points along the wall. For the construction time under consideration this azimuth corresponds to the rise of the sun's upper limb (as in all previous cases) on May 14th and July 31st, AD 200. Now indeed the great festival of Lug (known from the Calendar of Coligny) was celebrated on August 1st in Celtic regions (Norton-Taylor, 1974) and would have given a reasonable explanation for the axis-orientation of the temple (Figure 33.5).

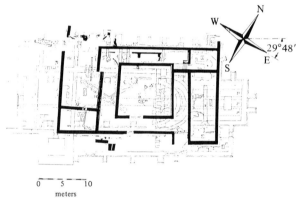

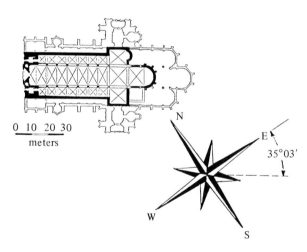

Figure 33.4 Dark outline shows Romanesque parts of St. Stephen's Cathedral in Vienna.

Figure 33.5 Dark outlines represent the primary structure superimposed on the plan of the church at Enns-Lorch.

As a final example, an investigation of the impressive parish church of St. Lawrence at Enns-Lorch in Upper Austria was carried out. Here the pitfalls of the astronomical method became evident.

Excavations at the site starting in 1960 revealed the striking concept of an old bishopric church with its main apse enlarged and burnt down three times. In the literature (Eckhardt, 1980) a hint was given that this church had been built on top of the remains of a Gallo–Roman temple dedicated to the major

Different results of the same excavation campaign (Harl, 1985) revealed that the primary building probably never had been a temple, but a Roman 'villa' – which had no pronounced orientation whatsoever, but was set up parallel to some existing road.

So for the astronomer the problem of overinterpretation exists when, as in the present case, archaeological concepts are not fully developed or rival theories are put forward.

33.4 Results

A comparison of the results on the 19 objects investigated by the present authors in the Central European region with the respectable collection of Guzsik (1978) shows that for the majority an astronomical explanation for their orientation can be given. As far as true equinoctial orientation is concerned (in the cardinal east–west direction) we could not find one in Austria. However, we are able to explain (M. Firneis and C. Köberl, manuscript in preparation) one orientation according to the Easter festival in the foundation year of the church of Königsbrunn/Burgenland (AD 1075), which is in accordance with some of the results available in Hungary. The new fact that the present authors want to stress is that the beginning of the Medieval year (March 25th) quite obviously was specially marked out too.

A generalization of the results presented here, however, has to be treated with care. In an area so closely linked together by cultural bonds as Austria, Hungary and Czechoslovakia (Köberl, 1984), architectural habits can be regarded as similar. In Poland, J. Dobrzycki (1984, private communica-

tion), who surveyed churches consecrated to St. Mary, could not attribute specific feasts to the proper church history, although the azimuthal swing of the orientation axes coincides with the extremes of the solar declination. As far as Medieval English churches are concerned, there is an absolute lack of coincidence with specific feasts.

However, the work on church axis orientation can be considered as a real link between history of astronomy and archaeoastronomy. When starting out with the measurements we found astronomical alignments and then started looking for some written records *ordering* axis orientation. Actually they can be traced in the old writings. For instance, some monastic orders like the Franciscans and the Cistercians had definitive rules according to which churches had to be oriented toward solar positions on data related to the life of their monastic founders.

References

Apianus, P. (1540). *Astronomicum Caesareum*. Ingolstadt: Iodocus Nass.

Charlier, C. (1902). On the Orientation of Early Christian Churches. *Vierteljahresschrift d. A.G.* **37**, 229–31.

Duffet-Smith, P. (1985). Astronomy with Your Personal Computer, pp. 87, 88. Cambridge University Press.

Eckhardt, L. (1980). Die Kontinuität in den Lorcher Kirchenbauten. *Denkschriften d. Österr. Akademie d. Wissenschaften, phil.-hist Kl.* **145**, 23ff.

Firneis, M. (1984). Untersuchungen zur astronomischen Orientierung des Domes von St. Stephan (Wien). *Sitzungsberichte d. Österr. Akademia d. Wissenschaften, math.-nat. Kl* **193**, 549–56.

Firneis, M. and Köberl, C. (1982). Untersuchungen zur astronomischen Orientierung der Kirchen von Unterfrauenhaid und Pilgersdorf (Burgenland). *Sitzungsberichte d. Österr. Akademie d. Wissenschaften, math.-nat. Kl.* **191**, 479–94.

Firneis, M., Köberl, C. and Göbel, E. (1981). Zur astronomischen Orientierung der 'Virgil'-Kapelle. *Anzeiger d. Österr. Akademie d. Wissenschaften, phil.-hist. Kl.* **118**, 240–53.

Firneis, M. and Ladenbauer, H. (1978a). Studien zur Orientierung der Kirchen von St. Ulrich in Wiesel-

burg und St. Ruprecht in Wien. *Forschungsberichte zur Ur- und Frühgeschichte* **10**, 124–6.

Firneis, M. and Ladenbauer, H. (1978b). Studien zur Orientierung mittelalterlicher Kirchen. *Mitteilungen d. Österr. Arbeitsgemeinschaft für Ur- und Frühgeschichte* **23**, 1–14.

Guzsik, T. (1978). Sol Aequinoctialis: Zur Frage der äquinoktialen Ostung im Mittelalter. *Periodica Polytechnica* **22**, 191–213.

Harl, O. (1985). Zum gallo-römischen Umgangstempel in Österreich. *Archäologisches Korrespondenzblatt* **15**, 217–28.

Köberl, C. (1984). On the astronomical orientation of St. Vitus' Cathedral and St. George's Church in the Castle of Prague. *Bulletin of the Astronomical Institutes of Czechoslovakia* **35**, 216–20.

Neugebauer, P. V. (1922). *Tafeln zur Astronomischen Chronologie*, vol. III. Leipzig: Hinrichs'sche Buchhandlung.

Neugebauer, P. V. (1929). *Astronomische Chronologie*. Leipzig: Walter de Gruyter & Co.

Norton-Taylor, D. (1974). *The Celts*. Amsterdam: Time-Life.

Thijm, A. (1858). *The Holy Line*. Amsterdam.

Tuckerman, B. (1964). *Planetary, Lunar and Solar Positions AD2–AD 1649*. Philadelphia: American Philosophical Society.

34

Some archaeoastronomical problems in East–Central Europe

K. Barlai *Konkoly Observatory, Budapest*

The tropical solar year plays a prominent role in the archaeoastronomy of East–Central Europe. The phenomenon of the heliacal rising of the stars is not as conspicuous there as it is in the tropical regions. Zenith passage of the sun does not take place either. Thus, equinoxes, solstices, sunrise and sunset are the most spectacular events.

Maps of numerous prehistoric cemeteries excavated so far in this region show clearly at first glance an orientation of graves in an E–W direction, suggesting that the burial cult might be connected in some way to the path of the sun (Bognár-Kutzián, 1972). The problem has been dealt with mathematically. It can be determined, even in the case of cemeteries having data too poor to be evaluated statistically, whether one single preferred direction (solstice, equinox) prevailed in the orientation of the graves or whether the actual sunset or sunrise on the day of the burial influenced the orientation of the grave pits (Barlai, 1980).

In Figure 34.1, the half solar angle span ($\alpha/2$), the deviation between the directions at horizon of the sunrises at summer solstice and at the equinoxes, is plotted versus φ, geographic latitude (continuous line). After a flat start, the slope of the curve becomes steeper. While in the tropical region this angle is small and its increase is negligible, the steepness grows considerably, approaching the latitude of the Carpathian Basin, and shows a rapid increase towards the Polar Circle.

Examining the formula

$$\sin \frac{\alpha}{2} = \frac{\sin \delta_\odot}{\cos \varphi},$$

that describes the dependence of the solar angle on the geographic latitude, one can see that this behavior seems quite natural. The declination ($\delta_\odot$) of the sun at summer solstice, a constant value, is being divided by continuously diminishing values of the cosine function proceeding toward 90°. So the formula shows immediately why solstices are wide-ranging phenomena at higher latitudes.

Neolithic and copper age cemeteries of different geographic latitude are represented by crosses in Figure 34.1. In these cemeteries, the E–W oriented graves fan out in the same angle which is between the directions of solstitial sunrises at the given latitude.

The width of the latitude belt studied so far is about 20°. It is conspicuous, however, that this distance is very unevenly filled with cemeteries. The major part of the archaeological material being at my disposal comes from the Carpathian Basin. It would be desirable to cover this region more evenly with international cooperation.

A strong relation to the sun has been manifested in the orientation of early and Medieval Christian churches. Saint Athanasius, bishop of Alexandria, writes about orientation in the fourth century: '*ecclesiarum situs plerumque talis erat. ut fideles facie altare versa orantes orientem solem, symbolum Christi, qui est sol iustitiae et lux mundi intuerentur.*' ('. . . the location of the churches was generally such that the faithful, in praying, with their faces turned toward the altar, would be looking toward the rising sun, the symbol of Christ, who is the sun of justice and the light of the universe.') This rule of orientation had been accepted by the first synod at Nicea (AD 325).

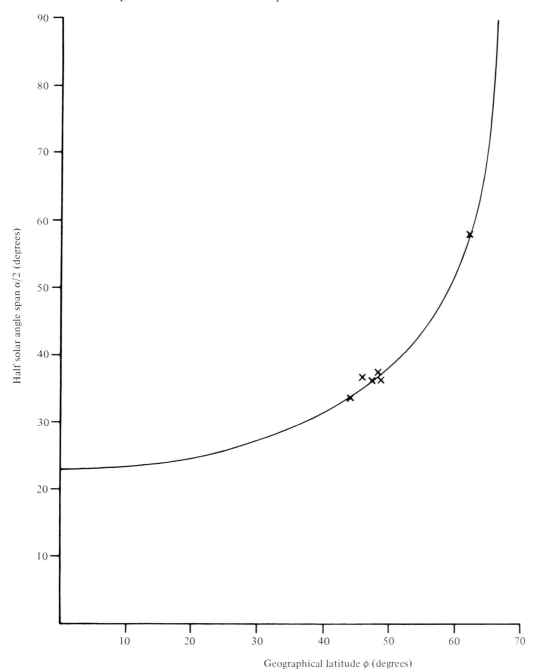

Figure 34.1 From south to north – towards increasing geographic latitudes – the crosses represent cemeteries as follows: Cernica (Cantacuzino and Morintz, 1963), Romania; Jászladány (Patay, 1944–5), Bodrogkeresztur (Bella, 1923), Tiszapolgár (Bognár-Kutzián, 1963), Hungary; Vel'ké Raskovce (Vizdal, 1977), Slovakia; Olenij ostrov (Gurina, 1956), Soviet Union.

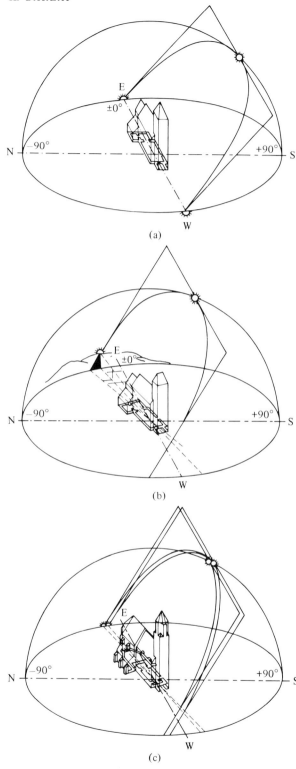

Figure 34.2 (a) Equinoctial orientation. (b) Orientation with a hill behind the church. Courtesy of T. Guzsik. (c) Reorientation of a church.

But what does *'versa solem orientem'* ('towards the rising sun') mean? The orientation of about 800 churches has been investigated so far in East Central Europe, from Czechoslovakia and Poland down to Bulgaria (Erdei and Kovacs, 1964; Guzsik, 1978; Sasselov, 1980; D. Sasselov, private communication, 1985). In Figure 34.2, some usual ways of orientation are presented. Figure 34.2a shows equinoctial orientation carried out in a flat region. The rising sun shines through the longitudinal axis of the church on days of equinoxes. The sun's path has been chosen in the figure at 47° N geographic latitude, e.g. as in Hungary. Solstitial orientation has also been accepted, although it was not carried out as often as the other versions.

There are numerous examples of orientation which are neither equinoctial nor solstitial but which were carried out on the feast day of the patron saint, on the day of the laying of the foundation of the church, or on the day the building was actually begun. Figure 34.2b shows an orientation with a hill behind the church. The sunrise actually observed takes place later than the true one because of the elevated horizon in the background. Thus, churches devoted to the same patron saint have deviations in their orientation depending on whether they are situated in a plain region or surrounded by mountains or hills.

Churches have been demolished in many cases by fire or wars – not a rare phenomenon in European history. When it came to rebuilding, the axis of the reconstructed church often deviated from the axis of the former one due to the inconsistency between the old Julian Calendar and the tropical solar year. In Figure 34.2c the reorientation of a church built originally in the 12th century AD and rebuilt in the 15th century has been depicted very schematically. In the course of about three centuries the name day of the patron saint shifted back in time by about three days. This resulted in a slight but measurable difference of angle between the former axis and the new axis of the church.

Let us consider real examples. There is a small chapel devoted to Saint George in the castle district of the Hungarian town Veszprém, a Medieval royal residence. This interesting and important remnant of Hungarian Romanesque architecture shows a good example of drastic reorientation (see Figure 34.3).

Relatively small churches (several meters in dia-

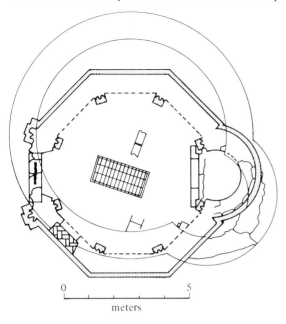

Figure 34.3 The reoriented St. George chapel in Veszprém, Hungary. Courtesy of K. H. Gyürky.

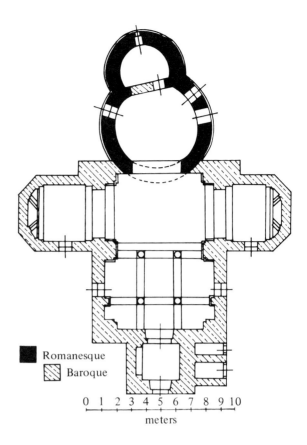

Romanesque
Baroque

Figure 34.4 Baroque church with rotunda at Keresztúr. Courtesy of K. H. Gyürky.

meter) having rounded nave and semicircular apse (the so-called rotundas) had appeared from the ninth century in East–Central Europe and they had been built until as late as the 13th century. They spread over the territory from present-day Yugoslavia up to Poland and Czechoslovakia (Gervers-Molnár, 1972).

The lower chapel with circular walls represents a common type of Central-European rotunda. Archaeological and historical evidence suggests that the rotunda was built not later than the end of the tenth century. It served as a royal chapel attached to a royal palace, the remnants of which are being excavated today. Its orientation is roughly equinoctial. Prince Emericus, the son of the first Hungarian king, Saint Stephen, had prayed in that chapel many times during the night until the morning came, when, as the 'Legenda Sancti Sephani regis maior' claims, '... *subito lumen cum ingenti claritate totum ecclesiam circumfulsit edificium...*' (Gyürky, 1963). If so, the orientation of the rotunda left nothing to be desired concerning the illumination of the building.

Still, in the middle of the 13th century, a new octagonal chapel was built upon its demolished walls, directed north-west with almost solstitial orientation, Saint George having been retained as

patron continuously. Thus, the reorientation must not have taken place for astronomical or iconographic reasons. There are a great variety of individual stories connected to orientation, all of which are worthy of study.

Another rotunda of Medieval Hungary in the village Keresztúr (now Keresztúr nad Váhom, Slovakia), built in the 13th century to the patron Holy Cross, was attached later to a bigger baroque church. The new church was erected in the late 18th century and has kept the medieval rotunda as its organic part locating the choir within it (see Figure 34.4). However, a slight reorientation was needed due to the local conditions.

The rule of orientation was valid throughout the medieval ages, both in eastern and western churches, and has never been abolished. In later centuries, however, the architectural structure of the crowded settlements, villages, and capitals rendered it impossible to keep this rule with regard to newly built churches.

It is of the utmost interest to know whether the Christian churches built from the 16th century in the New World continued to keep this rule of the orientation or ignored it, or, as a consequence of the clash between two cultures, some heathen sighting practice influenced the actual orientation of those churches.

References

Barlai, K. (1980). On orientation of graves in prehistoric cemeteries. *Archaeoastronomy* **3**, 29.

Bella, L. (1923). A bodrogkeresztúri aeneolith-kori temetö (The aneolithic burial site at Bodrogkeresztúr). *Régészeti és történeti értesítö (Archaeological and Historical Bulletin)* **1**, 7. (In Hungarian.)

Bognár-Kutzián, I. (1963). *The Copper Age Cemetery of Tiszapolgár-Basatanya*. Budapest: Publishing House of the Hungarian Academy of Sciences.

Bognár-Kutzián, I. (1972). *The Early Copper Age Tiszapolgár Culture in Carpathian Basin*. Budapest: Publishing House of the Hungarian Academy of Sciences.

Cantacuzino, G. H. and Morintz, S. (1963). Die jungsteinzeitlichen Funde in Cernica (Bucharest). *Dacia* **7**, 27.

Erdei, F. and Kovács, B. (1964). A váraszói templom (The Váraszó church). *Az egri muzeum (Yearbook of the Museum in Eger)*, vol 2, p. 1. Eger, Dobó István Múzeum. (In Hungarian with summary in German.)

Gervers-Molnár, V. (1972). *A középkoi Magyarország rotundái (Romanesque Round Churches of Medieval Hungary)*. Cahiers d'histoire de l'art, no. 4. Budapest: Publishing House of the Hungarian Academy of Sciences. (In Hungarian with summary in English.)

Gurina, N. N. (1956). Neoliticheskiy mogil'nik na Yuzhnom Olen'em ostrove (Neolithic burial site on the South Deer island). *Materialy i Issledovaniya po Archeologiy* **47**, 1. (In Russian.)

Guzsik, T. (1978). Sol aequinoctialis – Zur Frage der aquinoktialen Ostung in Mittelalter. *Periodica Politechnica – Architecture* **22**, 191.

Gyürky, K. H. (1963). Die St. Georg-Kapelle in der Burg von Veszprém. *Acta Archaeologica Acad. Sci. Hungariae* **15**, 431.

Patay, J. (1944–5). Les trouvailles archéologiques de cimetière de l'âge du cuivre à Jászladány. *Archeologiai Ertesíto*, series III, vol. **V–VI**, p. 11.

Sasselov, D. (1980). La coupole de l'église 'St Jean Baptiste' de Nessebre. *Nessebre* **2**, 191.

Vizdal, J. (1977). *Tiszapolgarske pohrebisko vo Vel'kych Raskovciach (Tiszapolgár Style Burial Site in Vel'ké Raskovce)*. Kosice: Vychodoslovenské Vydavatelstvo. (In Slovak.)

Uaxactun, Guatemala, Group E and similar assemblages: an archaeoastronomical reconsideration

A. F. Aveni   *Colgate University*

H. Hartung   *University of Guadalajara*

35.1 Introduction: historical background on Group E, Uaxactun

The site of Uaxactun was first reported in the literature by Sylvanus Morley (1916) who named it after the combination of two Maya words, 'uaxac', meaning eight, and 'tun', meaning stone. The name was suggested by the Baktun 8 date found on Stela 9, then the stela with the earliest recorded date: 8.14.10.13.15 (=AD 327).

Ricketson (1928, p. 222) tells us that, 'In 1924 ... Frans Blom made an accurate plane-table survey of the city. Although no excavation was possible at that time, Mr Blom drew attention to the fact that the arrangement of buildings, mounds, and stelae in the easternmost plaza, called Group E on his map, suggested its probable function as a solar observatory.' Excavation was begun in 1926 under the direction of Ricketson, who selected Group E for investigation. After five years the center of activity shifted to Group A (excavated between 1931 and 1937), but also included surrounding smaller groups.

The Group E complex is the easternmost group of buildings on the Uaxactun map (Smith, 1950, fig. 143), a section of which is reproduced in Figure 35.1. It rests on an elevated 200×100 m terrace aligned north–south that is 'artificial throughout, including the fill beneath the plaza' (Ricketson and Ricketson, 1937, p. 44); however, there is a limestone bedrock layer beneath E-VII. On the east side is an elongated terraced mound atop which lie three west-facing structures (E-I, E-II, E-III; see Figure

35.1 and the enlargement in Figure 35.2). Overlooking these buildings across the open plaza about 50 m to the west is pyramid E-VII sub, the earliest structure that has been covered by at least one other pyramid, E-VII. According to the excavators, this pyramid had 'never supported any building, either wooden or stone' (Ricketson and Ricketson, 1937, p. 43). Structures E-VIII-IX-X on the north and E-IV-V-VI on the south complete the enclosure of the small plaza. Four stelae, numbered 18, 19, 20 and E-I also were found: no. 20 at the base of the stairway of E-VII, no. 19 at the base of the stairway of the platform of EI-III, and nos. 18 and E-I flanking no. 19 (consult Figure 35.2 for their exact positions). The whole complex, with its clear access to the eastern horizon, is aligned almost precisely true north, while, by contrast, the main groups A and B are skewed a few degrees east of north. Groups E, D and F, not shown in Figure 35.1, lie even farther off the cardinal points.[1]

Evidently Blom's attention to astronomical motives was attracted by both the N–S axiality of the grouping and the possibility that the three eastern mounds could have marked the extremes and midpoint of the rising sun's course along the horizon throughout the year as viewed from somewhere on E-VII. Using a Brunton compass, he made several trial measurements from various points on the latter building (Ricketson and Ricketson, 1937, p. 42). After the excavations were complete, measurements were made by his successors with a theodolite. In both cases, the magnetic readings were corrected to true north with the use of tables supplied by the US Government Department of Terrestrial

Magnetism. In no case was an astronomical bearing ever taken directly in the field,[2] a factor which has been demonstrated to be necessary for astronomical orientation studies (Aveni, 1975, p. 164).

In his pre-excavation plan, Blom proposed that the sun had been viewed rising over the eastern mounds from Stela 20, which is not a very convincing location as we see it now. Ricketson, who excavated Group E in 1926–31, rectified this scheme and presented another plan (Ricketson, 1928, p. 223). He proposed three possible points of observation: in front of Stela 20 (as had Blom); in the middle of the stairway ('fifteen feet above the plaza level'); and at the end of the stairway. These are labelled A, B and C, respectively, in Figure 35.2 and in the plan and section of E-VII shown in Figure 35.3. Note that there is only one stairway drawn in the plan and it goes straight up to the top (cf. Ricketson, 1928, p. 218, fig. 3).[3]

Some of Ricketson's selected reference points on Temples I and III and on E-VII sub are not convincing on 'architectural' grounds. Why, for example, would one observe from the middle of a stairway to the extreme edges of the northern and southern buildings? Evidently, Ricketson was well aware of the stages of construction of Group E; for example,

he tells us that Floor III 'is the period of primary construction' (while E-VII sub stands on Floor II) with the East Platform 'surely surmounted by primary Temple E-II and possibly by Temples E-I and E-III, though the evidence is negative for these latter two' (Ricketson and Ricketson, 1937, pp. 134, 135). Curiously, both Ricketson and Blom seem to have become pre-occupied with making the best fit of solar orientations to the alignments regardless of whether the alignments made much sense architecturally.

Although in 1938 an airfield was established in Uaxactun for the exportation of chicle, a development which facilitated access to these remote ruins, no investigations known to the authors have been carried out since, except for some reconstruction work on Structure E-VII sub by several groups of volunteer amateur archaeologists under the direction of Edwin Shook in the summer of 1974.

While Shook's reconstruction was well done, he failed to duplicate the original finish of brilliant white mortar that once covered the whole structure and also the plaza floor in front (cf. Figure 35.4a with b and c). A. V. Kidder, who visited the excavation in 1928, had remarked: '... we were to keep our eyes on the ground until we were told to look

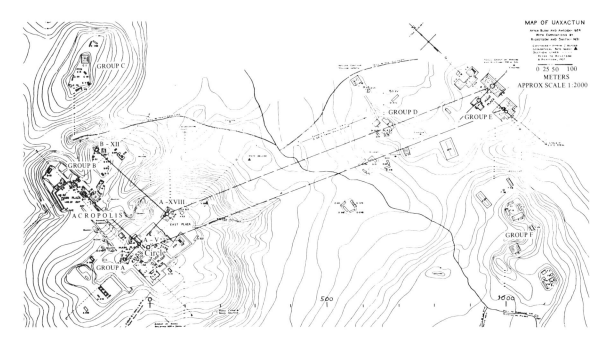

Figure 35.1 Map of Uaxactun (after Smith, 1950, Figure 143), with authors' suggested orientations.

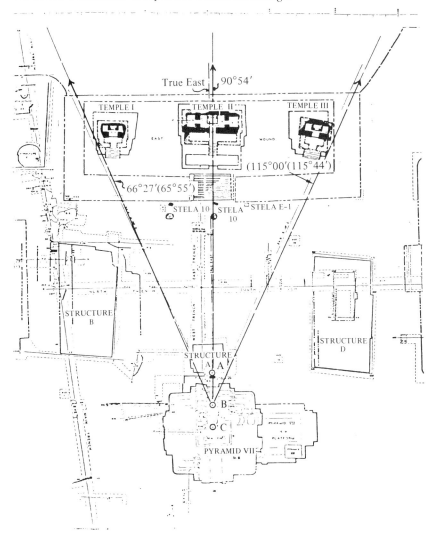

Figure 35.2 Plan of Group E (after Ricketson and Ricketson, 1937, Figure 197) with authors' measurements added.

up.' Thus viewing the facade of E-VII sub, he stated: 'There, snowy white against the deep green wall of the jungle was a terraced pyramid ... It was one of those moments one doesn't forget.' (Kidder, quoted in Smith, 1950, p. 3). A model in the Peabody Museum (Figure 35.5) suggests what the building might have looked like originally. Although E-VII sub was restored to its original condition, it appeared as a rather drab construction when we saw it a few years after Shook's Earthwatch project.[4]

Though Group E has been discussed in other contexts (cf., e.g., Rathje, 1973; Hammond, 1974;

Rathje, Gregory and Wiseman, 1978), our interest in this particular complex was motivated by its frequent appearance in the archaeoastronomical literature as the archetypal Maya solar observatory and the general recognition in the literature that the Group E form was widespread throughout the Petén. But we decided to visit and survey these structures when we realized no one had made accurate measurements of the alignments. Finally, we were interested in other features of Uaxactun architecture that might have been employed to relate astronomy and calendar, among them the three Teotihuacan-type pecked crosses which had been

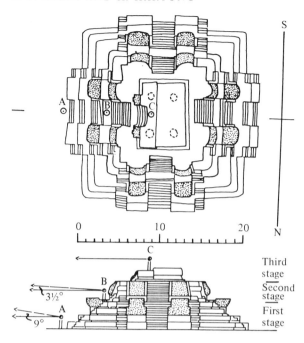

Figure 35.3 Plan and North façade of E-VII sub (after Ricketson and Ricketson, 1937, Figures 33 and 34).

reported at the site (Smith, 1950, pp. 21–2, figs. 15a, 60, 61, 62). Consequently, in January 1978, the authors traveled from Tikal to Uaxactun, a distance by rough track of approximately 30 km, with a group of students from Colgate University. Our goal was to make measurements of alignments with a theodolite on the existing architecture, and particularly on Group E. In the next section, we report on the analysis of the data we collected.

35.2 Group E, Uaxactun as an astronomical 'observatory'

The problem we proposed to solve can be stated quite simply: Does the sun as viewed from various selected positions on E-VII sub rise over the three buildings to the east (EI–III) at the June solstice, the equinoxes, and the December solstice, respectively?

We present in Table 35.1 alignments measured during January, 1978, from the aforementioned hypothetical points of observation along the center-line (east stairway) of E-VII sub (again, see Figures 35.1 and 35.2): (A) at the base;[5] (B) at the end of the stairway of the first stage of construction (four small platforms, height about 3.30–3.50 m); and (C)

at the top of E-VII sub (the third stage of construction, height about 7.50–8.00 m) and terminating either over (i) the centers of the doorjambs, or (ii) the extreme outer edges of the eastern temples. In each case, the measurement to the central temple (Temple II) passes over the center of the doorjamb. Table 35.1 also provides columns for computed ('true') solar orientations based upon the observed altitudes of the visible horizon and (also tabulated) differentials in azimuth between the measured alignments and the true solar orientations.[6]

Judging from the size of the tabulated differentials in the last column of Table 35.1, the best fit to a functioning solar observatory[7] corresponds to case IIB. For this case, the observer would have been situated about 5 m above ground level, provided we remember to add about 1.5 m for the height of the observer to the 3.3–3.5 m height of the platform from which he made his sightings.[8] In Figure 35.6 we depict this view. Looking east across the open plaza, the observer's eyes would have been about level with the top of the platform of the East Mound ($h = 4.6$ m); the tops of the individual terraces would have lain about $3\frac{1}{2}°$ above both the top of the East Mound and the natural horizon. Owing to dense overgrowth, only the steps of Temple II are visible in our photograph. Figure 35.7 depicts the present view looking the other way, while in Figure 35.8a,b we offer closeup looks at the present condition of the stairway and the doorway, respectively, of Temple II. In Figure 35.9 we employ our measurements to depict what the view from point B would have disclosed with the eastern platform and its contents restored, the vegetation trimmed away, and the sun permitted to move along the horizon on its annual course.

The fit between the astronomical and architectural data for case II appears to be quite satisfactory. The June and December solstice sunrises each deviate in azimuth by about one full sun disk from the edges of the two temples, but they do so in the sense that each would have been visible, the June solstice sun appearing to the left of the northern edge of E-I, the December solstice sun fully visible to the right of the southern edge of E-III. These deviations correspond to a distance along the terraces of these buildings of only 0.5 m, but because the sun advances very slowly along the horizon around the time of the solstice,[9] we must caution against concluding that, though the

Table 35.1. *Uaxactun Group E-VII. Building alignments for several trial solar observatories obtained from measurements taken along the center-line of E-VII sub*

Measurement made from	Measurement made to	Altitude of horizon	Measured azimuth of alignment	True azimuth of center of sun at June solstice, equinoxes, December solstice	Azimuth difference (measured minus true)
Case 1A. Base of E-VII sub	Dwy E-I	9°	68°34'	67°44'	0°50'
(from Stela 20)	Dwy E-II	9°	90°54'	90°54'	0°00'
(point A in Figures 35.2 and 35.3)	Dwy E-III	9°	112°53'	117°30'	−4°37'
Case IB. Middle of first platform	Dwy E-I	$3\frac{1}{2}$°	70°31'	65°55'	4°36'
(point B in Figures 35.2 and 35.3)	Dwy E-II	$3\frac{1}{2}$°	90°54'	90°54'	0°00'
	Dwy E-III	$3\frac{1}{2}$°	110°56'	115°44'	−4°48'
Case IC. Top	Dwy E-I	<0°	71°30'	64°50'	6°40'
(point C on Figures 35.2 and 35.3)	Dwy E-II	<0°	90°54'	89°50'	1°04'
	Dwy E-III	<0°	109°55'	114°38'	−4°43'
Case IIA. Base of E-VII sub	N Edge E-I	9°	64°30'	67°44'	−3°14'
(from Stela 20)	Dwy E-II	9°	90°54'	92°48'	−1°54'
(point A in Figures 35.2 and 35.3)	S Edge E-III	9°	116°57'	117°30'	−0°33'
Case IIB. Middle of first platform	N Edge E-I	$3\frac{1}{2}$°	66°27'	65°55'	0°32'
(point B in Figures 35.2 and 35.3)	Dwy E-II	$3\frac{1}{2}$°	90°54'	90°54'	0°00'
(The 'best' observatory)	S Edge E-III	$3\frac{1}{2}$°	115°00'	115°44'	−0°44'
Case IIC. Top	N Edge E-I	<0°	67°26'	64°50'	2°36'
(point C in Figures 35.2 and 35.3)	Dwy E-II	<0°	90°54'	89°50'	1°04'
	S Edge E-III	<0°	114°01'	114°38'	−0°37'

Dwy = doorway.

architecture appears to enframe these key sun positions rather neatly, the Group E complex in any sense offered a precise means for determining the solstitial dates.

What sense, architecturally, can be made of these alignments? In fact, the location of the viewer for the 'best observatory' (case IIB) seems the most logical starting position, because, as we already stated, it lies on the same level as the surface upon which EI–III were constructed. From this vantage point, then the top of E-VII sub, construction of the East Mound, as well as of EI–III upon it, could have been closely supervised, both the height and the width of the platform being judged by the elevation of the horizon beyond it and the extremes along that horizon attained by the sun. Thus deriving the alignments of the sun at the two solstices and its midpoint (true east) from the landscape, the builders could have transferred them to ceremonial–architectural space. Then the buildings E-I and E-III could have been added in such a way that the solstice sunrises would impinge upon

their outer edges. In this scheme, the equinox sunrise strictly could not have been viewed in the completed architecture, for building E-II would then have stood in the way. Still, this orientation, though no longer constituting a visual line, would have been incorporated within the design rather accurately (note that the value of the corresponding azimuth difference in Table 35.1 is 0°). It may be significant that while the equinox sun, though blocked by E-II when it first appeared over the natural horizon, nevertheless would have been seen minutes later perched atop the building just to the right of center. Unfortunately, because of the ruined state of these buildings, we cannot make estimates of their original heights. Consequently, Group E should be regarded as a functioning (though not precise) *solstice* observatory only and not as an equinoctial one.

In Figure 35.10 we sketch out our proposed building sequence for the complex, which we believe is consistent with both the archaeological and the astronomical data. In the first phase of construction (Figure 35.10a), let us suppose that E-VII

(a)

(b)

Figure 35.4 (a) E-VII sub when excavated (from Benson, 1967, p. 43). Photo permission: Peabody Museum, Harvard University. (b) E-VII sub in 1978, North façade (photograph by H. Hartung). (c) E-VII sub in 1978, East façade (photograph by H. Hartung).

was built at least to the level of point B, from where one views the sun rising over the natural horizon (alt. $= 0°$). The East Mound is built to precisely the proper height to accommodate the natural sunrise and is of the proper width to encapsulate the annual solar course along the horizon. Thus commences the transferral of the alignment from the natural environment to the architecture. Next (Figure 35.10b) the individual platforms and then the buildings E-I, II, and III are added, the equinoctial alignment becoming non-functional once E-II is placed upon its platform.

When the later E-VII sub construction phases proceeded (Figure 35.10c), the height of the pyramid was extended from 3.30–3.50 m up to about 8 m (for the top of the third phase).[10] The top of the secondary pyramid (E-VII) lay 13–15 m above the plaza floor. Thus, over time the astronomical perspective would have changed considerably. By this later stage of development, all of the sunrise

events would have taken place along a natural horizon that lay well above the level of the platform and its three buildings. It is likely that by this time the complex could not have functioned as a solar observatory in any sense.

Thus, these architectural transformations suggest that a dimunition of astronomical order and symmetry in the architecture likely accompanied an abandonment of its original, intended function. This astronomical decontextualization that took place will be discussed in a later section in the light of historical evidence relating to the demise of Group E-type structures in the Petén in general.

35.3 Group E, Uaxactun: inter-site and infra-site alignments

We have argued that Group E, among the earliest complexes at Uaxactun, was most closely aligned to the cardinal directions, and that operationally it would have functioned quite well as a solar

447

Figure 35.5 E-VII sub model (Peabody Museum, Harvard University).

observatory. (Alignments of some of the other buildings at Uaxactun are listed in Table 35.2.) We also noted the existence of the pecked cross motif carved in the floor of one of the buildings of neighboring Complex A. All of these facts raise the question whether Group E might have played a role in the determination of the placement and orientation of the other buildings, specifically those in Complex A, where we find the very artifact with respect to which a case for astronomical orientation already has been made at other Mesoamerican sites (cf. Aveni, Hartung and Buckingham, 1978; Aveni, Hartung and Kelley, 1982).

Table 35.2. *Alignments of selected buildings at Uaxactun based on measurements made on Blom/Ricketson map*[a]

1	N–S bisector of EVII–EII line in Group E (measured with the transit)	00°54′
2	D-II (west side)	1°
3	A-XII (west side)	5½°
4	Group A (west side of south plaza)	5½°
5	A-V (east side)	6½°
6	B-VIII (east side)	7°
7	A-XVIII (west side)	349½°
8	F-VI (east side)	21°
9	C-I (south side)	46°

[a] Corrected to transit measures made by Aveni and Hartung at A-V and E-VII.

Figure 35.6 E-VII sub view to Temple II, from point B (photograph by H. Hartung).

Figure 35.7 E-VII sub view from East, Temple II to E-VII sub (photograph by H. Hartung).

Accordingly, we turn next to an analysis of the possible relationships between Group E and other buildings at Uaxactun as well as at neighboring Tikal, which lies about 19 km almost due south.

These relationships take the form of certain lines connecting points of significance in the architecture (Hartung, 1969, 1971, 1975), which suggest that primary alignments in the Group E buildings may have been transferred to the main acropolis (Groups A and B).

In Figure 35.1 we note the following *parallel lines* connecting Group A to Group E:

(1) A line from the central doorway of E-II to the central doorway of A-XVIII (at the east end of that group), which coincides with the DSSR–JSSS[11] line. With the vegetation cleared, these structures, about 1000 m apart, should be intervisible; Group E is on the 16 m contour and A-XVIII on the 38 m contour of the map, but the latter is situated on a high platform and probably contained the proposed reference point, an upper-level central doorway 15 m above local ground level. Also note that there

is an inner stairway leading from the ground floor to the upper story (Smith, 1950, figs. 83–8).[12]

On archaeological grounds, Smith (1950, pp. 67–8) has argued that the decline of the importance of Group E was accompanied by the expansion of Group A so that by the end of the Classic Period, when E was a minor center at Uaxactun, A had developed into the 'largest and most important precinct in the entire city'. We suggest that the builders of the latter complex took at least some prior knowledge concerning calendar and orientation from their predecessors.

(2) The line from the Altar E-I in front of E-V (the main structure fronting the Group E plaza on its south side) to the pecked circle (so labelled in Figure 35.1) in the floor of A-V also coincides with the DSSR–JSSS direction and is exactly parallel to line (1) and as close to parallel as one can read from the revised map with the DSSR alignment that we measured in Group E. In the present case, we offer the

suggestion that the pecked circle might have been employed by the architects of A-V in order to establish a solstitial reference point. An identical usage has been proposed at Alta Vista (Chalchihuites) by Aveni *et al.* (1982).

That later builders wished to transfer the N–S axis of Group E to both Groups A and B is supported by the fact that the line connecting the doorways of A-XVIII and B-XII runs south–north. The buildings (A-V, A-XVIII and B-XII) which figure in these N–S and solstice alignments previously found in Group E are among the major ones in their respective groups, as one may discern by consulting the map. But there are still other coincidences of alignments between buildings that preserve the cardinal directions in these otherwise deviated buildings. As one can see in Figure 35.11, an enlargement of the appropriate section of Figure 35.1, another south–north line connecting the A and B groups passes from the central doorway of the west building of the A-V complex to the center

Figure 35.8 (a) Stairway to Temple II. (b) Temple II, doorway. (Photographs by H. Hartung.)

of the Ballcourt (B-V). A perpendicular to the latter line, taken from its point of origin exactly westward, passes to the doorway of A-II. An extension of the same line to the east strikes the upper central doorway of A-XVIII.

In Figure 35.12, another enlargement of a section

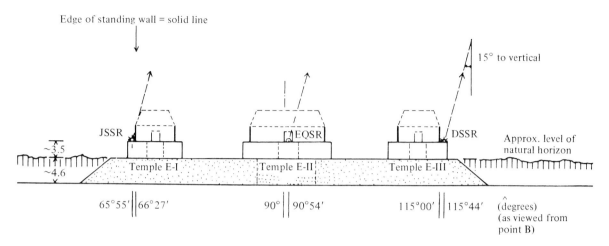

Figure 35.9 View from point B on E-VII sub to Temples E-I, E-II, E-III showing sun positions.

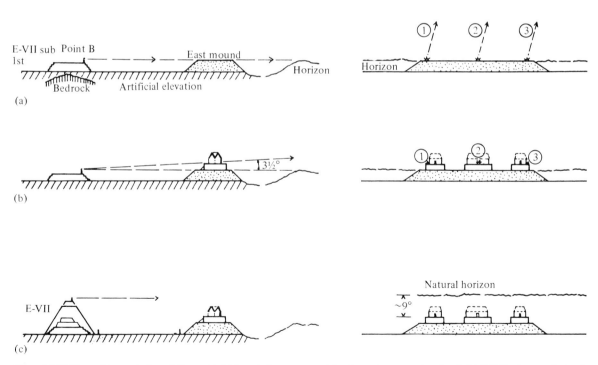

Figure 35.10 Plan of the evolution of Group E. (a) Level alignment in landscape transferred to artificial platform. Observer at point B on E-VII sub. (b) Platforms and buildings added so that desired positions of sun (1, 2, 3) fit the architecture; position 2 now becomes functional. (c) E-VII sub extended and E-VII added later; perspective changes. Right side shows view from point B on E-VII in each case.

451

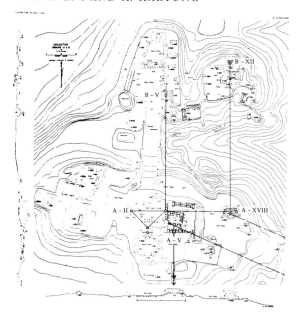

Figure 35.11 Map of Groups A and B (after Smith, 1950, Figure 143) with authors' measurements added.

of Figure 35.1, extending one of these building alignments a bit farther in distance, we wish to raise the possibility that, as in the case of Uxmal-Nohpat (Aveni, 1975, 1980; Hartung, 1975; Hartung and Aveni, 1982), wherein two ceremonial centers at kilometric distances may have been connected by an astronomical sight line, a deliberate alignment also may have connected Uaxactun with Tikal. In this case the line, a north–south one, which seems to be so preponderant in the architecture of Uaxactun, passes almost due south into the center of Tikal (see Figure 35.13). According to A. L. Smith (1950, p. 13): 'During the Early Classic Period all important buildings in Group A faced south, in the direction of Tikal ... only 19 km distant; with the jungle cut down, it would have been visible from Group A.' We note that the last addition to A-V aligns most closely with E–W, perhaps intended to confront Tikal. Puleston (1983, fig. 20) also alludes to a Uaxactun–Tikal N–S visual line on a map of the environment, but because this map may not be entirely faithful to the topography (see Puleston, 1983, pp. 7–8) no precise possible reference lines between Uaxactun and Tikal can be given. Yet another reference to visual sightings between Uaxactun and Tikal appears in Puleston (1983, p. 21), which states that: 'From the crest, S of Str. EP2-

151, one can clearly see the temples of Tikal without climbing a tree.'

35.4 Group E as a special assemblage in Maya architecture in general

The familiar form of Group E can be recognized on site maps of other Maya ruins, but did the architecture function in the same way as that at Uaxactun? If not, what were the differences? Is it possible to determine where, when, and under what conditions these types of structures occur? And, does their situation bear any relation to other structures at the sites where they occur? Because the relevant data in most cases are rather sparse and uncertain, we entertain these engaging questions only very briefly in this section.

The widespread occurrence throughout the Petén of complexes resembling Group E was first reported by Ruppert (1934). In a 1940 review, he listed 13 sites where the Group E form is unmistakable and six additional sites where the resemblance is less clear. In the latter instances, either the three eastern buildings do not rest on a platform or they do not lie in a straight line. In a number of instances, a given complex closely resembles the archetype at Uaxactun, except that it is misaligned from the cardinal directions. However, most of the alignments taken at sites other than Uaxactun are based on magnetic compass measurements or are derived from highly conventionalized maps. Table 35.3 is a collection of all the alignment data we were able to gather on Group E structures; it incorporates all of the data from Ruppert (1940, table II), as well as our own transit and map measurements. Possible Group E-type structures not noted in Ruppert's Table 2 appear in italics.

We follow Ruppert's conventions of listing the deviation of the central east mound from true east as viewed from the top center of the west mound (column a), and the axis of the east mound from true north (column b). In addition, we tabulate (in column c) the deviation of the northernmost and southernmost corners of the outer eastern buildings from the true east as well as from the center mound, the latter data appearing in parentheses.

Consulting these data, we draw the following general, tentative conclusions:

(1) Uaxactun, the first site at which the observationally functional Group E assemblage was pointed out, also seems to be the only one

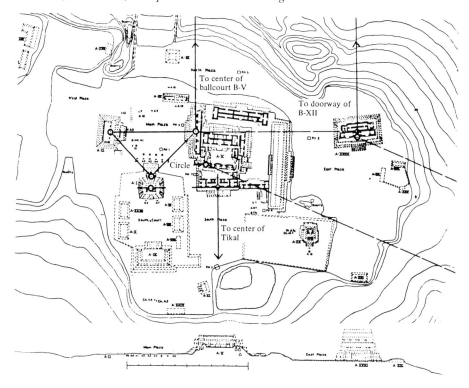

Figure 35.12 Map of Group A (after Smith, 1950, Figure 143) with authors' measurements added.

to possess an accurate observatory. However, recently we have begun a series of measurements at Dzibilchaltun, which lies well outside of the Petén area, and where we find a 3°50′ deviation from cardinal in the E–W axis and a reasonably good fit to solar observations when one takes account of elevation differences. Also, we are currently investigating similar possibilities at Kabah and El Mirador, which we include in Table 35.3 provisionally and for the sake of completeness.

(2) The directions to the extreme corners of the eastern buildings reckoned from true east at nearly all the sites are too far off the mark to register the solstice even approximately. Possible exceptions with respect to one of the two corners may be noted in the table. Yet, in many cases, the angles measured from the central east building to the outer edges of the companion buildings are of the correct magnitude. This correct relative proportion but incorrect absolute orientation leads one to wonder whether most E-complexes might have

been non-functioning copies of the astronomically operational archetype at Uaxactun.

(3) The deviation of the axis of the east-supporting platform from true north is practically always east of north, in the same sense as most Maya buildings (cf. Aveni and Hartung, 1986). Rarely does one of the Group E-type complexes deviate W of N or in a different sense from the other buildings at a given site.[14] Uaxactun provides the exception, as Table 35.2 illustrates.

(4) Practically all of the sites exhibiting Group E assemblages lie in the northern Petén and seem to focus on Uaxactun. We illustrate this principle in Figure 35.14, which updates Ruppert's (1940) fig. 15.[15]

35.5 Group E structures in historical context

It is simple enough to suggest from the study of alignments that a building complex was deliberately oriented astronomically. However, there are questions of far greater significance triggered by such

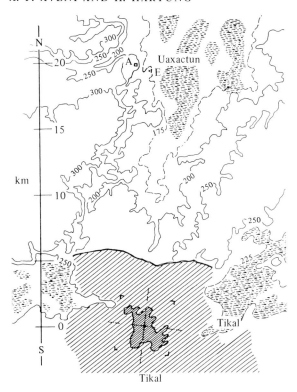

Figure 35.13 Map of Uaxactun – Tikal zone. (Redrawing of part of Figure 20 of Puleston, 1983.)

an investigation: Is the observational astronomy in Group E structures consistent with our knowledge of Maya history? For what purpose was the architecture actually used by the people who built it? What role did it play in the complex and changing Maya social structure that existed at the time of the Early Classic Period? In this section, we deal with these more anthropologically oriented inquiries.

Utilizing iconographic and inscriptional data from the northern Petén, Coggins (1983, pp. 38–41) has argued that certain ideological changes caused by the Mexican invasion of the Maya region about the fourth century AD resulted in a different way of perceiving and utilizing the calendar. Prior to the invasion, the Maya appear to have isolated their esoteric elite way of marking time by the Long Count from the empirical, less abstract, way of reckoning it by marking the position of the sun. Thus, up until about AD 400, Maya astronomers were constructing specialized observatories like the Group E structures specifically for marking the positions of the sun. With the arrival of the foreigners, things changed: for example, carved stelae before the end

of the fourth century celebrate dynastic events, while inscriptions emphasize the completion of pure time cycles. The establishment of katun-completion ceremonies in this area coincides with the demise of the practice of erecting the Group E observatories. Especially at Tikal, there is good evidence that twin-pyramid groups replaced Group E-type structures, at least as far as ritual function was concerned. Coggins believes the motive for this calendar reform was conditioned by the Mexicans' appreciation of a similarity between the structural principles of their 260-day calendar and the Maya katun round of 13 20-tun periods, a similarity that would have made it easier to propagate their own concepts of calendric ritual among their new subjects.

The Uaxactun Group E structures had been the only ones of this type to have been significantly excavated until very recently, when A. Chase (1983, unpublished dissertation, University of Pennsylvania, and 1985) reported on archaeological work undertaken at two (possibly three) hitherto unrecognized Group E-type buildings (Cenote, Paxcaman, and, later, Tayasal) by the University of Pennsylvania group. He views these specialized architectural groups, with their accompanying stone monuments and particular ceramics, as signs of the formation of a new Maya society with the beginning of the Early Classic Period.

In addition to offering a wealth of archaeological data on these sites, Chase attempts to assess the importance of the Group E assemblages in the development of Maya society. He defines two Group E 'styles': Cenote and Uaxactun, and he offers a variant of the Cenote category. The most visible difference is the much longer (and smaller) eastern mound in the Cenote type; also the central temple with its pyramidal platforms is larger and higher than the lateral ones, which are on attached 'wings' and not on the same long platform (cf. the isometric drawing of Cenote in A. Chase (1983, unpublished dissertation, p. 1302, and 1985, p. 39). On the basis of burials with tetrapod vessels found in caches within it, Chase dates the complex to the Pre-Classic/Classic transition (*c.* AD 300), or earlier than Uaxactun Group E, although the formulation of the different platforms generally indicates a later construction than the simple form one finds at Uaxactun (which probably was a bit more complicated than the drawings show). Also, there was an older structure below the long east mound at Uaxactun

(A. Chase, 1983, unpublished dissertation, p. 191) that may be older than Cenote. We do note that the north and south temples C-108 and 109 at Cenote have their stairways on the east, like the corresponding buildings at Uaxactun, but on the other hand the southern temple is 9 m more to the south, so that the observation from the pyramid to the west surely cannot mark the same distances or arcs to the north and south of equinox as was the case at Uaxactun. The west pyramid in its two stages has stairways to the *west* side; but this does not rule out observation points on the eastern side. Chase is careful not to attribute specific astronomical functions to the Cenote example, but then he has not conducted a precise study of the alignments. While he 'does not rule out a potential astronomical significance for the Cenote group ...', his discussion does not focus upon the astronomical problem to the extent that he can really hazard a definitive opinion on the issue (A. Chase, unpublished dissertation, 1983, pp. 188, 189).

At the later period sites such as Tayasal, Chase suggests that the Group E form might have given way to a so-called Plaza Plan 2 arrangement, which consisted of three buildings employed around a plaza, the middle structure on the east being constructed over a burial. A very different class of artifacts has been unearthed from this type of complex. Chase (1985, p. 38) speculates that the ritual activity associated with each of these types of buildings might have undergone considerable change, evolving from 'group-oriented' to family or ancestor-oriented rituals. He views these changes having taken place from within the Lowland Maya culture. It will be very important to obtain accurate alignments on both the Group E and Plaza Plan 2-type structures at these sites in order to see whether and to what extent astronomical orientations might have been carried forward with the alteration of the activity that took place there.

Finally, in an interesting unpublished work, A. Schlak has drawn attention to human remains by examining Chase's (unpublished dissertation, 1983) and Ricketson's earlier data on burials associated with Group E structures. Citing several examples of caches in Uaxactun Group E that can be interpreted to consist of decapitated remains, he also employs inscriptional material to argue the hypothesis that these solar observatories were employed specifically for decapitation rites. Caches

with similar contents and arrangement at Tikal, accompanied by inscriptions with calendrical dates relating to equinoxes and solstices, help bolster his claim.

35.6 Summary

We have attempted to re-evaluate the arguments of Ricketson and Ruppert that Group E, Uaxactun, served as a device to mark the positions of the sun at the equinoxes and solstices. Utilizing our own transit measurements, the first taken at the site based upon an astronomical fix, we found that the most accurate fit between the alignments and solar orientations corresponds to an observation point located at the top of the first structure of E-VII sub, on a level with the top of the platform supporting E-I, II, III. We developed an evolutionary construction scheme for the Group E complex that included a stage during which the completed group actually served, by design, to register the sun at the solstices, but not at the equinoxes. Also, we attempted to relate the positions and alignments of certain other later buildings at the site to Group E, which emerges as the only truly cardinally aligned complex at Uaxactun. Among these later buildings is Str. A-V, which contains a Teotihuacan-type pecked cross, here intended perhaps to transfer an astronomical (solstitial) line from one part of the site to another.

Regarding other sites exhibiting the Group E form, we have extended Ruppert's listing to 27 possible members, mostly Petén sites, at least 18 of which unmistakably resemble the archetype at Uaxactun. Based on measurements taken on extant site maps, many of which, unfortunately, may be highly conventionalized, we find that none of these Group E-type complexes appear to have functioned either as precise or approximate observatories that registered both solstice extremes and/or the equinoxes; however, in some instances, other solar positions could have been indicated. Many of the complexes appear highly distorted. Having examined nearly all the plans, we support Ruppert's hypothesis that at these highly distorted complexes more attention had been given to ritual and ceremony and less care had been paid to the rigid and rigorous sort of calendric concern derived from the orientation scheme we find in Uaxactun.

Whether the buildings were astronomically oriented or not, the widespread occurrence of the Group E style in the Petén may be taken as a general

Table 35.3. *Alignments of possible group E-type structures*

	(a) Variation from E–W	(b) Bearing of center E. platform	(c) Bearing to corners of N–S mounds of E. platform	Structures forming group E	Structures forming group W	Map, ref.	Notes
Balakbal	10°15'S	9°03'E	17°, 37° (27°), (27°)	VIIIa, b, c	VI	Morley (1937–8, pl. 218)	After Ruppert (1940)
Benque Viejo (Xunantunich)	5°03'N	2°23'W	26°, 11½° (21°), (16½°)	II, III, IV	VIII	Morley (1937–8, pl. 191a) (Chase,[a] p. 1301)	Published scale incorrect; after Ruppert (1940)
Cahal Pichik	15°54'S	14°30'E		D, E, F	13	Thompson (1931, Fig. 4)	After Ruppert (1940)
Calakmul	9°00'S	12°25'E		IVa, b, c	VI	Marquina (1951, p. 572. Lam. 176) (after Ruppert, 1940)	Reliable map not available
Caracol							
Cenote	4½°S	11½°E	27½°, 45½° (32°), (41°)	Court A1 C3	C1	(Chase, p. 1301) (Chase), Chase (1985)	Suggested by Chase Suggested by Chase
Comalcalco	5°S	5°E	3½°, 13° (8½°), 7½°	1	8, 9, 17	Andrews (1975)	Not measured
Dzibilchaltun (east Group E looking W)	3°50'N	3°20'E	33°, 37° (40°), (33°)	Str 1-sub	Str 4–5, 6–7, 8–9	Andrews V, (1974); Stuart et al. (1979) Transit measurements by authors	Study in progress
Dzibilchaltun (east Group E looking E)	3°50'S	3°20'E	19°, 20° (15°), (24°)	Stela 3	Str 4–5, 6–7, 8–9	Kurjack (1979)	
El Mirador				3D3-1, 2, 3	4D3-3	Matheny (1980); Dahlin (1984)	Study in progress
Hatzcab Ceel	9°05'S	5°18'E	24½°, 41½° (33½°), (32½°)	J, F, E	A	Thompson (1931, Fig. 7)	After Ruppert
Ikkun	6°13'S	7°08'E	18°, 29° (24°), (23°)	X	A	Maudslay (1889–1902, vol. 2, pl. 67)	After Ruppert (1940)
Ixtutz	3°30'S	6°30'E	19° 28½° (22½°), (25°)	A9, 10, 11	A2	Greene Robertson (1972)	Also suggested by Chase
Kabah							
La Muñeca	6°15'N	–	27½°, 11½° (21°), (17½°)	1B4-a, b, c VI, VII, VIII	1B3 IX	Pollock (1980) CIW unpublished	Study in progress Conventionalized map. E buildings not on platform; W building not a pyramid
Naachtun	2°21'S	5°24'E	11½°, 15½° (14°), (13°)	XXIIIa, b, c	XX	Morley (1937–8, pl. 206)	After Ruppert (1940). At N end of Group A complex
Nakum	4°40'S	N	14°, 23° (19°), (18°)	I, A, 2	C	Tozzer (1913, pl. 32) Hellmuth (1976a)	After Ruppert (1940). On N side of main acropolis B19 on north.
Naranjo	2°30'S	2°W	35½°, 37° (38°), (34½°)	B20	B18	Graham (1975, vol. 2(1), pp. 6–7)	Stairway of B18 faces west.

Site	Latitude	Longitude	Azimuth	Structure	No.	Reference	Notes
Oxpemul	8°37'S	11°21'E	12½°, 29½° (21°), (21°)	Va, b, c	II	CIW unpublished	Group appears aligned more closely to true N than other buildings
Paxcaman	8°S	6½°E	Mounds not present 10	10	11	(Chase), Chase (1985)	After Ruppert (1940). In the middle of the principal grouping
Rio Bec II	7°40'S	7°36'E		I	III	CIW unpublished	Map not available
San Jose	0°20'S	?		A2, 3, 4	A-7	Thompson (1939, Fig. 1) Marquina (1951, p. 567)	Conventionalized map. Map not available
Tayasal				T93–96	T99?	(Chase), Chase (1985)	Group E: str. not obvious on plan. Suggested by Chase (n.d. 1301). See note 13.
Tikal				5D-84, 86, 87	5C-54	Tozzer (1911, pl. 29). Carr and Hazard (1961, pp. 11, 19). Marquina (1951, p. 536)	Also suggested by Chase (n.d. 1301).
Uaxac Canal	13°44'N	10°00'W		Temple Grp B		Seler (1901)	Conventionalized map. Map not available
Uaxactun	0°54'S	0°00'	23°33', 25°00' (24°27'), (24°06')	E1-II-III	EVIIsub	Transit measurements by the authors Aveni (1980, pp. 277–81). Marquina (1951, pp. 518–35)	See Table 35.1 for details
Ucanal	E?	N?	26°, 26°	Grp B		Merwin (unpublished)	After Ruppert (1940) Conventionalized map
Uxul	3°25'S	0°10'E	9°, 22½° (12½°), (19°) (Other possibilities exist; see map)	XIIIa, b, c	XI	CIW unpublished	After Ruppert (1940). On easternmost eminence of site (like Uaxactun). Horizon falls off sharply to E
Xultun	13°34'N	N?		VIII	XI	Morley (1937–8, pl. 100)	Scale incorrect Conventionalized map. Map not available
Yaxha	(a) 11°S	11°E	14°, 36° (25°), (25°)	150, 152, 154	157	Hellmuth (1972, p. 148 1976b)	
	(b) 7°30'S	11°E	9°, 32½° (20°), (21½°)	119, 120, 123	116, 117	Hellmuth (1972, p. 148, 1976b)	

[a] A. Chase (1983), unpublished dissertation, University of Pennsylvania.
CIW = Carnegie Institute of Washington.

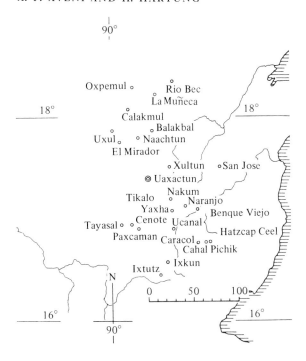

Figure 35.14 Map of Petén Region with Group E sites.

indication of social cohesion in that area and the (probably) earlier Cenote Group E variant may signal the onset of that reunification process, as A. Chase (1983, unpublished dissertation, pp. 1250–1) has suggested.

In Uaxactun Group E we find the epitome of the inclusion of precise time–space measurements in sacred architecture that we might anticipate in an advanced theocracy. If we could establish a more detailed chronology of Group E-type complexes,

we might be able to trace the evolution of the role of timekeeping in Maya society further. For example, we might hypothesize that the loss of order in the other Group E complexes, was related to the removal of the attending ritual from its astronomical context. Such a process might have resulted in the greater degree of randomness apparent in the alignments of the other E-type structures. The transformation of the astronomically operational Group E, Uaxactun, to other purely symbolic such complexes possessing different power and meaning may have been occasioned by cultural contact, but conflict among various rulerships or other reasons cannot be ruled out.

Given these considerations, we propose that transit measurements of the type we made at Uaxactun also should be carried out at these other sites (many of which unfortunately still lie totally in ruin) before we can conclude that all of them served no astronomical function. Furthermore, we urge that efforts be made to chronologically document these sites. Finally, the search for other Group E complexes should continue. Particular attention should be paid to whether all sites that contain the Group E-type also possess other buildings that are aligned close to cardinal or slightly E of N. In this regard, it may be significant that no such arrangement is found either at Piedras Negras or Yaxchilan where the skew from cardinality is rather large.

Acknowledgment

The authors thank the OSCO Fund for its enduring support of the Colgate University archaeoastronomy field research program.

Notes

1 Concerning structural detail, some oddities about the trio of small temples and about the E-complex in general may be relevant in the context of the forthcoming discussion of the possible astronomical use of the buildings. We note that Temple III, while situated on an east–west aligned platform, is skewed out of line with the base of the platform. According to Ricketson and Ricketson (1937, p. 57), it faces 7°42′N of W, while its companion buildings face exactly west. This skew, which is very noticeable in Figure 35.2, appears to direct the facade of E-III toward point

B (to be discussed later) on the eastern side of E-VII sub and could have been intended to focus attention upon a viewer who stood on the latter temple. We know of no other reason for this skew, which seems far too large to be attributed to sloppy construction work. Incidentally, the base of E-VII sub also is slightly cardinally out of line. It faces 87°43′ or about 2°N of E.

2 Ricketson and Ricketson (1937, p. 106) specifically state that, though the latitude of the site was determined astronomically, the alignments of the buildings were fixed with the magnetic compass, then corrected using the Department of Terrestrial Magnetism tables.

3 A similar plan can be found in Ricketson and Ricketson (1937, p. 107, figs. 68 and 197) that shows the observation point on step 2 of the third of five flights, a place that might have been chosen to fix sunrise alignments on the solstice days.

4 In part, the tarnished appearance may have resulted from the use of cement in the mortar. Very recently it was disclosed that another structure may lie buried within E-VII sub (Valdéz, 1986).

5 This choice of a backsight adjacent to Stela 20 by the early investigators makes little sense chronologically. It should be noted (see plan of Ricketson and Ricketson, 1937, p. 135) that there actually was no stela present in the Floor V stage when the Group E complex was fully developed; the stelae appeared only in the Floor VI stage, by which time E-VII sub already had been covered over by E-VII.

6 Our measurements of the relative positions of the temples are in agreement with the published maps, except we conclude that Temple III should be shifted about 1 m to the north.

7 We use the word 'observatory' to suggest that the event of the sun passing over a particular place in the architectural scheme was actually observed; thus, regardless of the end served by it, that is whether it was used as a time marker in a calendar count or as the harbinger of a ritual enactment, or both at the same time, the proposed orientation would have consisted of a *visual line* incorporating an astronomical event. Later, we shall have occasion to refer to a *relation line*, by which we shall imply an alignment intended to associate two presumably significant points in architectural space which, though it does not in its final state imply an act of direct observation, nevertheless was intended to preserve a direction in space held to be significant and originally derived from a visual line. Though it might seem rational to us to associate precisely oriented alignments with calendrical procedures and imprecisely oriented ones with ritual behavior, we believe that any attempt to connect the matter of precision with the issue of intention or purpose on the part of the builders can be misleading.

8 The supposed observation point B on E-VII sub also is supported by the stylized serpent heads on both sides of the end of the stairway (to the first platform), which also are slightly higher than the level of the first platform. This reminds one of the serpents that define the space in front of Temple 22 at Copan, noted as a particular reference point (Hartung, 1971; also visible in Closs, Aveni and Crowley, 1984, fig. 4).

9 At these times of the year the sun would cover a distance equal to its own diameter in about ten days.

10 That the final eastern platform was built before the final western pyramid also is supported by the study of human remains at Uaxactun. Dedicatory burial caches in EI–III are classified as earlier than those found connected with the later constructional stages of E-VII.

11 We adopt the following abbreviations: JSSR = June solstice sunrise; JSSS = June solstice sunset; DSSR = December solstice sunrise; DSSS = December solstice sunset; ZP_1 = first solar zenith passage (sun moving north); ZP_2 = second solar zenith passage (sun moving south).

12 The difference between the alignments we measured on our corrected version of the Blom–Ricketson map and the relevant orientation of the sun at horizon that we actually computed is about 2° (measured azimuth = $296\frac{1}{2}°$ v. computed azimuth = 294°24′ over a horizon elevated by $2\frac{1}{2}°$ for JSSS; measured azimuth = $116\frac{1}{2}°$ v. computed azimuth = 114°39′, horizon = 0° for DSSR.

13 The disposition of the structures on Tozzer's (1911) map looks not unlike a Group E structure, particularly because 5C–54 (= Tozzer's no. 66) was alleged to have four stairways and no temple on top. Coe (1967) wrote: 'Standing close to 100 feet high, it is square in plan with a stairway on each side. Each stair is flanked by terraces and gigantic masks employing masonry blocks over six feet long. This pyramid is Late Pre-Classic in date and was one of the greatest structures of its time in all of Mesoamerica.' We draw our measurements from the Carr and Hazard (1961) 1:2000 map.

14 This result may possess some relevance in the context of practical attempts to orient buildings astronomically. Suppose one wished to incorporate the equinox sunrise into the design of a building by requiring that the sun be viewed rising over the top of that building as seen from another structure. One would need to take the third dimension into account, especially if the rising position of the sun is elevated well above eye level at any stage of the construction process, as we have demonstrated to be the case in the Templo Mayor of Tenochtitlan (see Aveni and Gibbs, 1976). For a northern hemisphere observer, as the sun rises it moves not only upward but to the right (south). Thus, unless the builders took special care when transferring an equinox sunrise observation from a relatively flat landscape to the vertical architectural environment of the ceremonial center, they might encounter considerable difficulty in their attempts to fit the sun's annual course into their architecture. Evidently the builders of Uaxactun knew enough and cared enough to make the relationship work. But do the generally southerly deviations we

find at the other sites suggest to us that, in other cases, the builders made only half-hearted attempts?

15 Some inconsistencies and overlaps: Ruppert includes Las Palmas and El Paraiso in his map of sites exhibiting the Group E-type assemblage but he does not list them in his table. In the absence of concrete information, we have decided to exclude them. Uaxac Canal, which is included in our table, is too far to the west to be included in our Figure 35.14, and Kabah and Dzibilchaltun, of course, lie well to the north of the Petén. A. Chase (1983, unpublished dissertation, table 44) excludes Uaxac Canal as well as La Muñeca, San Jose and Xultun, but he adds Nakbe, Cerro Ortiz, Chachaclun and Caracol. Also, he recognizes three structures at Yaxha rather than the two listed in our table.

16 However, A. Chase (1983, unpublished dissertation, pp. 1241–4) points out that Thompson's (1931) excavations at Cahal Pichik and Hatzcab Ceel were relatively thorough and often have been neglected.

References

Andrews, G. (1975). *Maya Cities: Placemaking and Urbanization.* Norman: University of Oklahoma Press.

Andrews, V. E. W. (1974). Some architectural similarities between Dzibilchaltun and Palenque. In *Primera Mesa Redonda de Palenque, Part I*, ed. M. Greene Robertson, pp. 137–47. Pebble Beach: R. L. Stevenson School.

Aveni, A. F. (1980). *Skywatchers of Ancient Mexico.* Austin: University of Texas Press.

Aveni, A. F. (1975). (ed.) Possible astronomical orientations in Ancient Mesoamerica. In *Archaeoastronomy in pre-Columbian America*, pp. 163–90. Austin: University of Texas Press.

Aveni, A. F. and Hartung, H. (1986). Maya city planning and the calendar. *Transactions of the American Philosophical Society* **76**, (1), 1–87.

Aveni, A. F., Hartung, H. and Kelley, J. C. (1982). Alta Vista (Chalchihuites), astronomical implications of a Mesoamerican ceremonial outpost at the Tropic of Cancer. *American Antiquity* **47**, 316–35.

Aveni, A. F., Hartung, H. and Buckingham, B. (1978). The pecked cross symbol in ancient Mesoamerica. *Science* **202**, 267–79.

Aveni, A. F. and Gibbs, S. (1976). On the orientation of Pre-Columbian buildings in Central Mexico. *American Antiquity.* **41**, (4), 510–17.

Benson, E. (1967). *The Maya World.* New York: Crowell.

Carr, R. and Hazard, J. (1961). *Tikal Reports.* Philadelphia: University of Pennsylvania Museum.

Chase, A. (1985). Archaeology in the Maya heartland. *Archaeology* **38**, (1), 32–9.

Closs, M., Aveni, A. and Crowley, B. (1984). The planet Venus and Temple 22 at Copan. *Indiana* **9**, 221–47.

Coe, W. (1967). *Tikal: A Handbook of the Ancient Maya Ruins.* Philadelphia: Museum of the University of Pennsylvania.

Coggins, C. (1983). The stucco decoration and architectural assemblage of the Str. 1-sub, Dzibilchaltun, Yucatan, Mexico. *MARI*, publ. 49.

Dahlin, B. H. (1984). A colossus in Guatemala: the Preclassic Maya city of El Mirador. *Archaeology* **37**, (5), 18–25.

Graham, I. (1975). *Corpus of Maya Hieroglyphic Inscriptions*, vol. 2, parts 1 and 2. Cambridge: Peabody Museum.

Greene Robertson, M. (1972). Notes on the Ruins of Ixtutz, Southeastern Peten. *Contributions of the University of California, Archaeological Research Facility (Berkeley)*, no. 16, 89–104.

Hammond, N. (ed.) (1974). The distribution of Late Classic Maya major ceremonial centres in the central area. In *Mesoamerican Archaeology: New Approaches*, pp. 313–34. Austin: University of Texas Press.

Hartung, H. (1969). Consideraciónes sobre los trazos de centros ceremoniales mayas. *Centro de Investigaciones Historicas y Esteticas, Universidad Central: Boletin 11*, pp. 127–37.

Hartung, H. (1971). *Die Zeremonialzentren der Maya. Ein Beitrag zur Untersuchung der Planungsprinziplien.* Graz: Akademische Druck- und Verlagsanstalt.

Hartung, H. (1975). A scheme of probable astronomical projections in Meso-american architecture. In *Archaeoastronomy in Pre-Columbian America*, ed. A. F. Aveni, pp. 191–204. Austin: University of Texas Press.

Hartung, H. and Aveni, A. F. (1982). El Palacio del Gobernador en Uxmal. Su trazo, orientación y referencia astronómica. *Bol. Esc. Ciencias Antrop. U. de Yucatan*, no. 52, pp. 3–11.

Hellmuth, N. (1972). Excavations begin at Maya site in Guatemala. *Archaeology* **25**, (2), 148–9.

Hellmuth, N. (1976a). Maya architecture of Nakum, El Peten. Guatemala *FLAAR Progress Reports* **2**, (1).

Hellmuth, N. (1976b). Maya archaeology: travel guide. Guatemala *FLAAR*, p. 51.

Kurjack, E. (1979). Map of the ruins of Dzibilchaltun, Yucatan, Mexico. *MARI*, publ. 47.

Maudslay, A. (1889–1902). *Biologia Centrali-Americana: Archaeology*, 5 vols. London: R. H. Porter & Dulau.

Marquina, I. (1951). *Arquitectura Prehispanica.* Mexico: INAH.

Matheny, R. T. (1980). *El Mirador, Peten, Guatemala. An Interim Report.* Provo, Utah: New World Archaeological Foundation.

Morley, S. (1916). Report. *Carnegie Institute of Washington Year Book*, no. 15, 337–41.

Morley, S. (1937–8). *The Inscriptions of Peten* CIW publ. 427 (5 vols). Washington.

Pollock, H. (1980). *The Puuc.* Memoirs of the Peabody Museum, vol. 19. Cambridge, MA.

Puleston, D. (1983). *The Settlement Survey of Tikal.* Tikal Report no. 13. The University Museum, Philadelphia.

Rathje, W. (1973). Trade models and archaeological problems: the Classic Maya and their E Group Complex. *XL Cong. Int. Americanists*, vol. 4.

Rathje, W., Gregory, D. and Wiseman, F. (1978). Trade models and archaeological problems: Classic Maya examples. *Papers of the New World Archaeological Foundation*, no. 40, pp. 147–75.

Ricketson, O. G. (1928). Astronomical observatories in the Maya area. *Geographical Review* 18, 215–25.

Ricketson, O. and Ricketson, E. (1937). 'Uaxactun, Guatemala, Group E, 1926–31.' *Carnegie Institute of Washington*, publ. 477.

Ruppert, K. (1934). Explorations in Campeche. *Carnegie Institute of Washington Year Book.* Annual Report 1933–34.

Ruppert, K. (1940). A special assemblage of Maya structures. In *The Maya and their Neighbors*, eds. C. Hay, R. Linton, S. Lothrop and H. Shapiro, pp. 222–31.

New York: Appleton.

Seler, E. (1901). *Die alten Ansiedlungen von Chaculá im Distrikte Nenton des Departements Huehuetenango der Republik Guatemala.* Berlin: Dietrich Reimer.

Smith, A. (1950). Uaxactun, Guatemala: excavations of 1931–1937. *Carnegie Institute of Washington*, publ. 588.

Stuart, G., Scheffler, J., Kurjack, E. and Cottier, J. (1979). Map of the ruins of Dzibilchaltun, Yucatan, Mexico. *MARI*, publ. 47.

Thompson, J. E. S. (1931). Archaeological investigations in the Southern Cayo District, British Honduras. *Field Mus. Nat. Hist., Anthr. Ser.* 17, (3).

Thompson, J. E. S. (1939). Excavations at San Jose, British Honduras. *Carnegie Institute of Washington*, publ. 506.

Tozzer, A. (1911). A preliminary study of the ruins of Tikal, Guatemala. *Mem. Peabody Museum* 5, (2).

Tozzer, A. (1913). A preliminary study of the ruins of Nakum, Guatemala. *Mem. Peabody Museum* 5, 137–201.

Valdéz, J. (1986). Uaxactun: recientes investigaciones. *Mexikon* 8, (6), 125–8.

461

36

Inca observatories: their relation to the calendar and ritual

D. S. P. Dearborn   *Lawrence Livermore National Laboratory*
R. E. White   *Steward Observatory, University of Arizona*

36.1 Introduction

The Inca creation myth involved descent from the sun god, and the superiority implied from such a heritage was used as a method to control subject populations in the empire. For this reason, Pachacutic ordered temples of the sun to be built throughout the empire (Cobo, 1983). Also, the main religious structure of the empire, the Coricancha in Cuzco, was reported to contain temples to the sun, moon, stars, lightning, and rainbows (Garcilaso de la Vega, 1960). Clearly astronomical objects held a fundamental place in Inca cosmology. The skies were watched not only for basic calendric information, but for portents (Cieza de Leon, 1551). The work of Urton (1981) showing such behavior among modern Quechua-speaking peoples suggests that at least some of this interest in astronomy is not just Inca, but Andean.

Our work in the Cuzco area has involved searches for the physical remains of structures used for astronomical observations. We have found structures capable of accurately determining the dates of the solstices and the zenith passages (Dearborn and White, 1983). These structures demonstrate the interest of the Inca in making precise astronomical observations, and their desire for accurate observations in places other than Cuzco. Since we have published detailed descriptions of these structures elsewhere, we will give only a summary account of them here.

The identification of the 'solstice observatories' that we describe in the first part of this chapter is firmly supported by statistics and ethnohistoric

accounts. We have not, however, found any structure for observing the equinoxes. The equinoxes were important in the Inca year as discussed by Zuidema (1979). In the remainder of the chapter, we will discuss an alternate hypothesis of how the year was quartered. Modern astronomy divides the year into four seasons based on a geometrical definition of the sun's position along the path that it appears to traverse annually through the sky. While the solstices are easily determined from horizon observations of sunrise and sunset, alignments to equinoctial rise and set points cannot be so precisely defined. If instead the year were quartered with an enumerative (day) counting system, the year could be accurately and consistently quartered without constructing equinoctial alignments. Moreover, if such alignments were constructed they would not point precisely to the celestial equator. While the physical data that we have on the equinoxes is by no means compelling, it serves as a basis for a working hypothesis which subsequent field work may confirm or deny.

36.2 Inti Raymi – the June solstice

Because of its relatively pristine condition, we began our studies at Machu Picchu. The large proportion of fine Inca stone work suggests it had a special function either as an administrative center, or perhaps as an estate of one of the Incas. Among the finest structures in the site is the one called the 'Torreon' (Figure 36.1). Bingham (1948) had noted similarities between this building and the Coricancha, and suggested it might be a temple dedicated to the sun.

462

Figure 36.1 The dominant feature of the interior of the Torreon is the stone platform which contains a single sharp edge. This edge is precisely aligned to the June solstice.

We studied this structure in 1980, and found that it functioned as a precise astronomical observatory for determining the June solstice and the zenith passage dates. Subsequent work in 1982 in the 'Intihuatana Barrio' at Pisac revealed a structure with similar capabilities. The Pisac structure was in much poorer condition, but, like the Torreon, was founded on a rock prominence which had a clear view to the east. Additionally, a solstitial alignment was carved into the foundation rock of both structures.

While the Torreon was capable of being used to determine the solstice without anything appended to the window (the solstitial alignment in the central stone was accurate to approximately 2 arc-minutes, a precision comparable to the best naked eye astronomical measurements), simply sighting along

the edge would be awkward at best. The edge is situated in a manner that one cannot sight the horizon directly along it. It is necessary to raise your eye to the horizon after sighting along the edge. It would be much easier and more precise to use a feature that is unique in the Torreon among all of the buildings at Machu Picchu. The two windows in the curved wall had pegs projecting from the exterior corners. As one can see in Figures 36.2a–c, the pegs are quite distinct from the projections that occur on many Inca structures. Furthermore, they are located specifically at the corners of the windows instead of randomly along the wall. The south window has an additional set of pegs at the same level as the lower window pegs.

Unlike the more common projections which

463

Figure 36.2 (a) The pegs at the corners of the windows are longer than the projections commonly found at the lower edges of large stones. Also they are flattened on top and are located specifically at the window corners. (b) The southeasterly window has pegs at the corners, and an additional pair of pegs level with the bottom of the window. (c) A more common projection located at the base of a large stone in the Principal Temple at Machu Picchu.

people often speculate were used for levering the stone into place (such projections are flat on the bottom), these pegs are flat on the top, and appear to be placed there to support something. The shadow cast by an object supported on these pegs (Figure 36.3) would not only clearly identify the solstice (when the shadow is parallel to the edge), but could be used to provide dates for a range of time around Inti Raymi, and the zenith passage dates. While the suggestion that these pegs formed part of a shadow casting mechanism to measure the angle at which sunlight enters the window is speculative, it is consistent with Poma de Ayala's (1583–1615) description of the type of observations done by the Inca.

> *lo ciguin en el sembrar la comida en que mes y q.*
> *dia y en que hora y en que punto por donde anda*
> *el sol lo miran los altos serros y por la manana de*
> *la claridad y rrayo q. apunta el sol a la ventana por*
> *este rreloxo cienbra y coxe la comeda del ano en este*
> *rreyno.*

In the sowing of the crops, they follow the month, the day, the hour, and the point where the sun moves; they watch the high hills in the morning, the brightness and the rays that the sun aims at the window; by this clock they sow and harvest the crops each year in this domain.

The shadow casting experiments that we conducted demonstrate that individual days could be distinguished from early May to early August. Zuidema (1982) has argued that the ceque system in Cuzco was used as a calendar beginning with

465

Figure 36.3 The shadow cast by a plumb-bob supported by the upper window pegs becomes parallel to the stone edge on the June solstice.

the first morning rise of the Pleiades on 9 June. His belief in this start date is based in part on the statement by Molina (1573) that 'they started to count the year in the middle of May'. The system counted through 328 days and was followed by a 37-day intercalation period in which observations were made to restart the calendar. Because of the difficulty in making a consistent first morning rise observation, Zuidema suggests that solar observations were actually used to initiate the new count. This adjustment period falls in the time when the Torreon window is suitable for determining the date. This provides a link between the capabilities designed into the Torreon, and recorded actions in Cuzco.

Additionally, there is an association between the Torreon and Coricancha. The Coricancha is aligned to the rise point of the Pleiades. We found that the window through which the solstice is observed at Machu Picchu is not aligned on the solstice, but on the 15th-century rise point of the Pleiades.

Finally, the Torreon does not stand alone as a structure capable of observing the solstice. The temple of the sun in the Intihuatana Barrio at Pisac is a structure of similar construction and orientation. It is also located on a large rock prominence, with an edge carved into the rock aligned to the June solstice. The precision of the alignments in these two structures, when coupled with the known interest of the Inca in the sun, make the likelihood that these alignments become very small. The zenith passage date detection depends entirely on the position of the windows in the Torreon.

The walls of the Pisac structure have been destroyed to a point where we cannot determine if such windows existed in this structure (beyond stating that at least one window would have been necessary for using the solstitial feature if the original wall height was the same as the present day entrance).

36.3 Capac Raymi – the December solstice

Whereas Inti Raymi was a major festival celebrated by everyone, Capac Raymi (associated with the December solstice) was a festival for the Incas themselves. This was the festival at which the young nobles were initiated into manhood, and the ritual culminated on the solstice by watching the sunrise and receiving their ear plugs. It is therefore not surprising that the structure for observing the December solstice should be separate from that for observing the June solstice.

The structure, 'Intimachay' (Dearborn, Schreiber and White, 1988), that we believe was associated with Capac Raymi is located on an upper terrace, just below the southeastern portion of Machu Picchu. It consisted of a cave whose entry was formed from fine masonry, and whose interior had a wall of coursed masonry with two niches. The most significant feature of this structure was a window which was carved through 2.2 m of solid rock. The full view of the window contained only 2° of horizon. The view from the interior of the cave was further limited by a stone which baffled the window from inside of the cave, resulting in a view of approximately 10 arc-minutes. This window was ineffectual for the normal function of a window (illumination), but was precisely aligned with the rising of the sun on the date of the December solstice. Because of the finite diameter of the sun, and the view of the window, light from the rising sun was capable of entering the cave for about ten days before and after the solstice. Again, shadow casting would be able to determine the precise date, and, given that the December solstice occurs during the rainy season, the ability to function over a range of dates is a benefit. Even neglecting the possibility of shadow casting, the view of the window was remarkably well aligned and collimated, and was constructed with some considerable effort.

As with the Torreon, the alignment found in Intimachay is known to correspond to a culturally

important event, and is of a precision that makes a coincidental alignment unlikely. The ability to construct alignments to such precision is not at all mysterious, and is in fact relatively easy when observing solstices.

36.4 Quartering the year in Tawantinsuyu

Before discussing the requirements that must be placed on an astronomical alignment for observing equinoxes, we must consider the different ways in which the year may be quartered. In modern astronomy, we divide the year into four seasons based on the position of the sun along its annual path. The path of the sun when viewed on the sky is spatially divided into four equal quarters, one for each season. Determining the precise path of the sun with naked eye is quite difficult, and involves constructing detailed maps of the sky. If the observer has a level horizon, this geometrical quartering can be accomplished without a map by observing the positions of the solstices, and bisecting the angle between them. However, Andean horizons are seldom level. Consider, for example, a flat horizon with constant altitude of 5° above level. Bisecting the angle between the solstices on this horizon leads to a placement error in the alignment of 6 arc-minutes north of the true equinox position. A horizon which varies in altitude will have some error dependent on the shape of the local horizon. The equinoctial date determined by this method could then vary depending on the local horizon.

As stated above, the modern convention for dividing the year does so by evenly quartering the path through which the sun appears to move. Because the earth's orbit is elliptical, the sun's apparent motion along this path is uneven, and the length of each season varies. In AD 1450, the season lengths were: autumn = 93.15 days, winter = 93.40 days, spring = 89.46 days, summer = 89.23 days (southern hemisphere seasons). The longitude of perihelion passage changes by 1.7° per century causing the current values to be different by about half a day.

An alternative way to quarter the year is to divide it evenly in time. The time between solstices can be easily and accurately determined from observations, and the year may be accurately quartered by dividing this intra-solstice interval in half. Such a counting or enumerative scheme would obtain the same date for the 'equinoxes' regardless of the local horizon, and is consistent with the counting scheme proposed by Zuidema (1982) for the ceque system of Cuzco.

If an enumerative scheme were used to quarter the year, no equinoctial observatory would be necessary. Nonetheless, an alignment might be constructed for ceremonial purposes. In the 15th century, there were 182.38 days between the December and June solstices, and 182.86 days from the June solstice back to the December solstice (the times between solstices were more even in the 15th century than they are today: 181.77 and 183.47 days).

The mid-point in time between the solstices was offset from the time of the true equinox by approximately 2.1 days. The declination of the sun at this time is approximately 48 arc-minutes north of the celestial equator. This is the position that the sun occupies two days after the March equinox, and two days before the September equinox. Finding a set of alignments to this celestial position would indicate the existence of an enumerative system but such alignments could not have the precision that we found in the solstice observatories.

The reason that solstitial sunrise and sunset positions can be precisely defined is that the daily declination change of the sun (and so the change of its position on the horizon) becomes very small near the times of the solstices. The solar motion at the time of an equinox includes a declination change of nearly 24 arc-minutes per day. At the latitude of Cuzco this leads to a typical daily change in the azimuth of 25 arc-minutes. Because the length of the year is not an integral number of days, sunrise (or sunset) on the day of the equinox will vary from the actual moment of the equinox by a substantial fraction of a day. The horizontal position of the rising (or setting) sun will then vary through a range of approximately 25 arc-minutes from year to year. An equinoctial alignment (or one offset from the equinox by a couple of days) would therefore either have to be approximate, or would have to define a range of rising (or setting) positions which could only be defined by a series of observations over a period of several years.

As we stated in the introduction, we have found no alignments which point with precision to an equinoctial rising or setting position, but as explained in the preceding paragraph, a horizon alignment to the rise or set point of the sun on a date near the equinox is not possible with the precision found in the solstitial alignments. We have, however, examined some westward alignments.

While working at Ollantaytambo, R. Randall pointed out a set of three irrigation canals which appeared to radiate from a point within the 'Fortress'. The central canal was a few degrees north of true west, and marked a point on the horizon where a small platform was located just below a ceremonial structure. This structure contained a double niche door, and large (human size) niches with holes penetrating the stones on the sides of the niches (similar to the 'Stocks' at Machu Picchu). Tom Zuidema identified this structure for us as Cachicalla.

Such an alignment was at least superficially reminiscent of the ceque system in Cuzco, which Zuidema (1982) has suggested was used in quartering the year. When accounting for the altitude of the horizon, and refraction, the declination of objects setting at this position was +20 arc-minutes (north), offset towards the position which the sun would have if the year were quartered in time (and 30 arc-minutes north of the point that would have been chosen by bisecting the positions of the solstices). This alignment by itself is not precise enough to argue that it was intended to be equinoctial. Its direction may well have been chosen for convenience, or intended as a generally western alignment. Furthermore, beyond crudely defining the extreme points on the horizon at which the moon sets, the other two canals are even less well associated with any significant setting positions (aligning to declinations −30.0 and +29.13). However, if close examination of other 'generally western' alignments shows a systematic bias to the north, it could argue that the Inca practiced an enumerative quartering of the year instead of a geometric one.

36.5 Summary

We have found structures at Machu Picchu and Pisac which function as precise astronomical observatories. This demonstrated that the Inca interest in the sun went beyond ritual observation to true astronomy. Finally, we have suggested that the lack of equinoctial observatories could be associated with the difficulty in constructing a precise structure, and that an enumerative procedure might have been used to quarter the year.

Acknowledgments

We gratefully acknowledge the sponsorship of 'Earthwatch', and the many people whose help made this project possible. Furthermore, we wish to thank the Peruvian National Institute of Culture for the opportunity to study the sites mentioned in this chapter.

References

Bingham, H. (1948). *Lost City of the Inca*, p. 148. New York: Duell, Sloan & Pearce.

Cieza de Leon, P. (1551). *La Cronica del Peru*, chap. 65. Edicion de la Revista Ximenez de Quesada, 1971.

Cobo, B. (1983). *History of the Inca Empire*, transl. R. Hamilton, p. 134. Austin: University of Texas Press.

Dearborn, D. S. P. and White, R. (1983). The Torreon at Machu Picchu as a solar observatory *Archaeoastronomy. Supplement to the Journal for the History of Astronomy* **14**, Suppl. no. 5, S37–S49.

Dearborn, D. S. P., Schreiber, K. J. and White, R. E. (1988). Intimachay: a December solstice observatory at Machu Picchu. *American Antiquity* (in press).

Garcilaso de la Vega (1960). *Inca, Obras Completas del Inca Garcilaso de la Vega II*, p. 133. Madrid: Biblioteca de Autores Espanoles.

Molina, C. de (1573). *Fábulas y Ritos de los Incas.* Reprinted in 1943 by D. Miranda, Lima.

Poma de Ayala, F. Guaman (1583–1615). *El Primér Coronica y Buen Gobierno*. Reprinted in 1980, eds. J. V. Murra and R. Adorno. Mexico: Siglo XXI.

Urton, G. (1981). *At The Crossroads of the Earth and the Sky*. Austin: University of Texas Press.

Zuidema, R. T. (1977). The Inca calendar. In *Native American Astronomy*, ed. A. F. Aveni, pp. 219–59. Austin: University of Texas Press.

Zuidema, R. T. (1982). Catachillay: the role of the Pleiades and of the Southern Cross and Alpha and Beta Centaun in the calendar of the Incas. In *Ethnoastronomy and Archaeoastronomy in the American Tropics*, eds. A. F. Aveni and G. Urton, pp. 203–29. New York Academy of Sciences vol. 385.

37

The year at Drombeg

Ronald Hicks   *Ball State University*

37.1 The nature of the problem

Having never been convinced that the design of
the stone circles of Ireland and Great Britain incorporated the sort of astronomical precision claimed
by Alexander Thom, I have not been surprised by
the rising tide of skepticism about all astronomical
claims concerning the circles except the simplest.
However, I continue to believe that these sites were
meant to mark astronomically significant directions,
although only in an imprecise way important to the
builders merely for its symbolism. Stonehenge is,
after all, oriented toward midsummer sunrise, or
midwinter full moon rise, depending on your interpretation. Examples of simple agricultural societies
paying close attention to the heavens are abundant.
Further, astronomical symbolism has always struck
me as the likeliest explanation for the circular shape
these monuments commonly take (with either surveying error or a tendency toward multiple layers
of symbolism like that evident in early Celtic literature, probably accounting for the rest, it being, after
all, a simple step from a circle to the shape of a
moon that is not quite full or that of an egg). And
even though it may date from a period too late to
apply to the stone circles, we do have Caesar's comment that the druids were as inclined to discuss
astronomy as theology. (Whether the druids did or
did not have anything to do with the stone circles
is an aspect of the more general question of the
degree of cultural continuity from the Neolithic
through the Bronze Age and into the late Iron Age
in the British Isles and is thus a question for another
time, though my belief is that the continuity is very
considerable indeed, at least in ritual matters.)

What seems to me to have been slighted in all
this discussion is the question of exactly *how* the
circles were designed to reflect astronomy. At
Stonehenge, the most obvious astronomical marker
is the line from the center through the entrance.
But it seems clear that other alignments were
marked there in other ways. When Gerald Hawkins
was carrying out his calculations for *Stonehenge
Decoded*, he chose to consider the positions of
'stones, stone holes, other holes, mounds' as well
as each 'singular geometric point, like the center
and the archway midpoints' (Hawkins, 1965,
p. 104). Of the resulting two-point alignments,
those that were astronomically significant were
limited primarily to the station stone alignments (i.e.
with the short sides of this rectangle pointing toward
the midsummer sunrise/midwinter sunset while the
long sides indicate the major lunar standstill positions) plus alignments from the central *area* (not
just the exact center) of the monument through the
gaps presented by each of the trilithons and nearby
pairs of stones in the surrounding circle (Hawkins,
1965, pp. 108–10). Alexander Thom (1967, p. 94)
considered the alignment of a slab (i.e. its long axis),
two stones separated by some distance, two circles,
a circle and a stone outlier, or a natural (horizon)
foresight identified by some simple indicator (and
viewed from the center of the ring) to be possible
identifiers of the astronomically important direction. The emphasis in his work, however, tended
to be on the last of these possibilities, the presence
of a horizon foresight. In my initial report on the
orientation of Irish henge earthworks (Hicks, 1981,
p. 349), I assumed the relevant sightlines were from
the center of the monument along lines bisecting

the entrances, but suggested that the builders might have had other possibilities in mind. With slight variations the works of others have made assumptions similar to these examples.

It seems very likely that not only these possibilities for marking alignments but also others were known and used by the builders of the stone circles. We have no reason to suppose that only one technique was used. Indeed, if one looks at the monuments in the field, one is immediately struck by the fact that there is more to their designs than just the shape of the ring and the presence or absence of outliers. The size, shape, and placement of the stones themselves all seem factors worthy of consideration. In the literature, however, our attention has been drawn to this primarily by Aubrey Burl in his work on the recumbent stone circles of Scotland (e.g., Ruggles and Burl 1985, p. S28). Here, one is presented with the likelihood that design attributes related directly to characteristics of individual stones may be of primary significance in determining how the circles were used to mark astronomical events. This emphasizes a fact that needs frequent repetition: the statistical method used alone, without reference to other types of data, can lead us astray, or blind us to actual astronomical elements.

If, in fact, size, shape, and placement of the stones in the rings are important clues to interpretation of the astronomical use of the sites, then the commonly seen plan views produced by surveys, however carefully done, must omit vital data. Certainly such surveys are the most efficient way to study a site's design, since one can do them at any convenient time without worrying about what is actually going on in the sky. I would question, however, whether any survey, even if profile drawings of the site elements are also produced, can provide all the information one needs. Owing to greatly reduced scale and possible errors in determination of true north or other measurements, as well as the simple impossibility of viewing the site as a whole from different perspectives, the results of such surveys are in my opinion only mediocre substitutes for actual observations in the field at times suspected of being significant, i.e. solstices, equinoxes, cross-quarter days (the days midway between the solstices and equinoxes known to have been important in the ancient Celtic calendar), and times of the major and minor lunar standstills. There may, in other words, be clearly marked astronomical sightlines that are not at all obvious from survey data.

Unfortunately, field observations are not easy to carry out. They are time consuming and subject to disruption by uncooperative weather, and they must be carried out at unreasonable times of the day. In the case of the moon, there is the further complication of only being able to make major standstill observations at 18.6-year intervals. Nonetheless, during a sabbatical in 1983–4, I took advantage of the opportunity to carry out field observations of sunrise and sunset on appropriate dates over several months at a recumbent stone circle in Drombeg Townland in County Cork, in the southwest of Ireland. The results confirmed my suspicions that the best way to determine how a stone circle may have been used for ancient astronomy is to try to use it. I offer the description that follows as an example in support of the points I have made in the preceding paragraphs.

37.2 Drombeg

The Drombeg stone circle (Figure 37.1), locally known as 'The Ring' or 'Druid's Altar', was first brought to the attention of archaeologists in 1903 in a brief note by Denham Franklin in the *Journal of the Cork Historical and Archaeological Society* (Franklin, 1903). Very shortly thereafter, in 1909, Captain (later Vice Admiral) Boyle T. Somerville published the first archaeoastronomical study of the site, concluding on the basis of his survey that the monument incorporated sightlines for not only winter solstice sunset and summer solstice sunrise and sunset but also three separate alignments for sunrise at Beltaine, the cross-quarter day at the beginning of May, a date also prominent in Geoffrey of Monmouth's account of the building of Stonehenge (Somerville, 1909). Of these, the most convincing was the alignment for winter solstice sunset, for which one stood in the entranceway to the circle and looked across the recumbent on the opposite side toward a notch in the hill that rises just to the west.

Somerville reaffirmed his conclusions about the winter solstice alignment in a later paper that also considered other stone circles in the vicinity (Somerville, 1930, p. 74), and it is this alignment that has claimed the attention of later researchers (Fahy, 1959; Barber, 1973; Patrick and Freeman, 1983). When E. M. Fahy excavated the site in the

Figure 37.1 Drombeg stone circle, viewed through the entrance.

1950s, he did in fact take the trouble to double-check Somerville by observing and photographing the winter solstice sunset. John Barber and Jon Patrick both considered only the axis of the monument from the center of the entrance to the center of the recumbent – the solstice line – as part of surveys of the Cork/Kerry stone circles as a group.

The stones

While he does not relate them to astronomy, Fahy did pay close attention to design elements in the circle, and concluded on the basis of an array of evidence that the shapes and positioning of at least some of the stones – eight of the 17 in the circles are discussed – was deliberate. The results of Fahy's excavation reassure us that the stones as they appear today are positioned essentially as they were when the circle was built. I believe that in several cases the design features of the circle noted by Fahy, plus others, do relate to astronomy. Let us look at the circle stone by stone.

In contrast to the Aberdeenshire recumbent stone circles, in those of County Cork (Figure 37.1) the tallest stones are not normally those flanking the recumbent but rather those flanking the entrance. Figure 37.2 shows the stones as viewed from inside the circle. Using Fahy's numbers, the two portal stones are 17 and 1. Note that Stone 1 – the one to the right, or south, of the entrance – is not only the tallest in the circle but also has its height emphasized by coming to a point. Note also that that point is not over the center of the stone's base, the location that would be considered important in an analysis of survey data.

Stones 2 and 3 do not have any characteristics that at first glance set them apart, although Stone 2 does have a top considerably lower on one side than the other, and Somerville claimed that a line from the center of the recumbent to Stone 2 marked the line of Beltaine/May Eve sunrise. That line does *not* cross the center of the circle. Stone 3 had fallen and was re-erected in its original socket by Fahy.

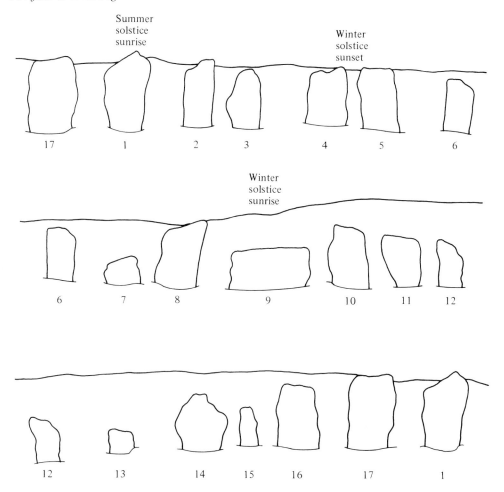

Figure 37.2 The stones of Drombeg, and the surrounding horizon, as viewed from the center of the circle. To make clear the relationships between the stones, the three sections of the illustration have been drawn to overlap. At this greatly reduced scale, the hills and notches along the horizon are not as apparent as they are in the field.

Stones 4 and 5, on the other hand, are distinguished by being spaced noticeably closer together than any others in the circle, by being slightly taller than their neighbors which are set noticeably apart, and by having the sea for a horizon behind them.

Stone 6 has nothing to distinguish it except being somewhat narrower, or more peglike, than most, and by being closest to the south point of the circle.

Stone 7 is missing. The stump shown in the drawing was set in place by Fahy to mark its position. From Fahy's plan of the socket (Figure 37.3), the original stone must have had a very narrow base, comparable to those of Stones 12 and 15.

Stone 8 is the left flanker of the recumbent. Like Stone 1, it has a slanting top, the slant in this case rising toward the recumbent.

Stone 9 is the recumbent. Apart from being on its side with a very flat top and being astride the axis of the circle directly opposite the entrance, it is distinguished by having on its top a petroglyph (Figure 37.3) – a cup mark with a pear-shaped ring enclosing it and with another cup mark just to the west.

Stone 10, the right flanking stone, like its mate across the recumbent, has a top that slants up toward the recumbent side. This stone stands out because of the prominent quartz vein marking it (Figure 37.4). The presence of one stone that is markedly

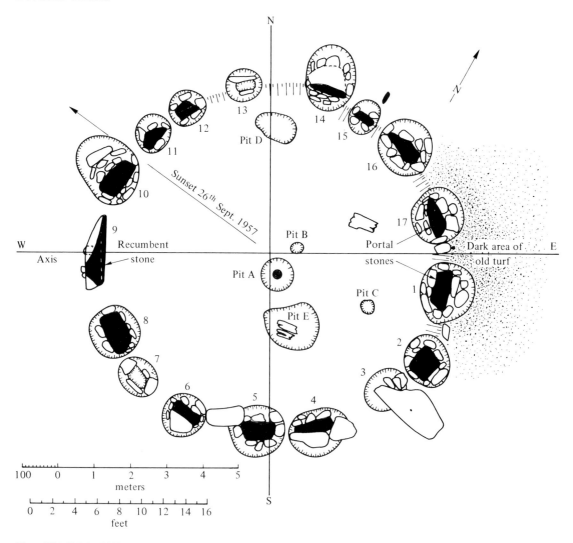

Figure 37.3 Fahy's (1959) excavation plan of Drombeg, showing the stone sockets. (Courtesy of the Cork Historical and Archaeological Society.)

whiter than the others, usually because of the presence of quartz, is a widespread characteristic of Irish stone circles. However, I have so far been able to discover no consistency to the placement of the quartz stones so it is possible that they are meant to represent some deity rather than to indicate an alignment. That does bring up the interesting point that in various Medieval accounts (Stokes, 1895, pp. 35–6; Gwynn, 1924, pp. 18–23; Duignan, 1940) the stones of one presumed circle (no longer extant), at Magh Sleacht in Co. Cavan, are described as having been sheathed in copper and gold images of the gods in Patrick's time. Stone 10 is also note-

474

worthy because of one other sign of care in its positioning. The socket had evidently been dug too deep, so it alone of the stones in the circle has a pedestal under its base to raise it to the correct height.

In Figure 37.2 you can see that Stones 11 and 12 continue the slope downward to the right started by the top of Stone 10. In both cases the sockets had been dug just deep enough to produce this effect and it is clearly deliberate, as though intended to direct the eye to ... something.

Just what that something was we may never know, at least not entirely, because Stone 13 was also

Figure 37.4 The petroglyph on the top of the recumbent. As viewed from the center of the circle, it lies toward the right end of the stone.

among the missing ones. Once again, the stump you see was set in place by Fahy to mark the location of the socket of the missing stone.

The next stone, 14, is as distinctive as the recumbent. It has a lozenge or diamond shape also found on petroglyphs in the Irish passage graves and at various monuments in Great Britain. The very peg-like character of its neighbor, Stone 15, is reminiscent of the pairing of lozenge and peg found in the avenue at Avebury, the combination having been taken to be female and male symbols, respectively. While I feel the lozenge is once again symbolic of a deity, I do not find Stone 15 to be particularly phallic and suspect it may instead merely be intended to mark the north end of the circle. That the lozenge shape of Stone 14 was intended is supported by the fact that the stone was inserted in its socket with the heavy end up. In other words, it is quite pointy. If it were reversed, you would see a triangle, not a lozenge. This arrangement took some considerable planning and effort by the builders. (As an aside, the presence of stones with a distinct zigzag or dogleg along one edge at the neighboring stone circles of Brogar and Stenness in the Orkneys is another case where I believe a stone has been chosen to represent a particular deity, but I have noticed no examples of that type in Ireland.)

The lozenge lies too near the north point of the circle to figure in any solar alignment involving a circle diameter. However, according to Somerville, the long axis of the lozenge itself provides a second May sunrise alignment. May cross-quarter day alignments in connection with stone circles were also referred to by Sir Norman Lockyer (1909). Aside from the comment by Geoffrey of Monmouth regarding Stonehenge, it is not clear why Somerville and Lockyer should have chosen to emphasize Beltaine, the May cross-quarter day, over Lughnasa, the festival at the August cross-quarter day, for which the same alignment works. Lughnasa practices have been much more thoroughly documented in the literature, thanks to a magnificent, and monumental, study by Maire MacNeill (1962). Lughnasa refers to so-called 'funeral games' supposedly established by the god Lugh at various sites around Ireland to honor dead goddesses, who presumably should be equated with the grain or the earth's fertility, which has just 'died' for the year, to be resurrected in the spring, quite in parallel with the classical legend of Ceres and Persephone. The principal characters involved in the legends associated with this festival are Lugh (literally 'light' or 'brightness'), the Celtic god of all skills, to whom the leprechauns and fairy shoemakers seem to have been answerable, and Crom Dubh, a bull god credited in his human form with having brought grain to Ireland (MacNeill, 1962). Crom Dubh's name literally means 'the black stooped [or bent] one'. There is some reason to suspect that Crom Dubh may be the summer aspect of the sun god, though why he should then be called 'black' is a very large question.

Lugh (in later versions St. Patrick) wrests the harvest, or a bull, from the humanized 'king' Crom Dubh for the people. Crom Dubh, or at least Crom Cruaich who is probably the same deity, is mentioned as the principal god associated with the Magh Sleacht stone circle mentioned earlier, where he alone is said to have been clothed in gold. This is a considerably simplified explanation, but it should be enough to indicate that the conflict, as the names of the gods indicate, is on one level between light and dark. All this information may or may not be relevant to the interpretation of Drombeg. Whether it is relevant once again depends on the degree of continuity of belief from the Bronze Age into Iron Age and later times. Available evidence suggests there was considerable continuity, but one must always bear in mind the danger of extrapolating folkloristic or ethnohistoric evidence this far back in time.

Finally, we have Stones 16 and 17, which have no particular distinguishing characteristics, except that the latter flanks the entrance and is one of the circle's tallest stones.

Field observations

Now let us turn to the results of my field observations. To make the observations, I stood a few feet outside the circle. While the exact distance outside is relatively unimportant, variations will alter slightly the time at which sunrise or sunset will be seen. The height of the observer is another factor affecting timing somewhat. Since my contention is that the builders were not concerned with high precision but rather with symbolic alignments, these variations have little practical effect. A more important factor is that if one stands either too near or too far from the stones, it is difficult to see the relevant alignments.

I will start with the winter solstice, since this is the date of the one generally recognized alignment at the site. However, rather than the sunset, I will start with the sunrise, which no one has referred to previously, because this provided my first surprise. While viewing conditions were not perfect on the days I was able to visit the site, they were good enough to determine that the brightest spot at sunrise was between the closely spaced Stones 4 and 5 as viewed across the center of the circle. Diametrically across the circle from the paired stones is the site of the missing Stone 13. It would be of considerable interest to know what the original stone in that socket looked like. Fahy's excavation showed the socket to be nearly the same size as that for the peglike Stone 15 and a bit shallower, so it cannot have been large (Figure 37.5).

The weather was even less cooperative at winter solstice sunset than at sunrise. Three tries produced no satisfactory photographs. Fortunately Fahy

Figure 37.5 Stone 10, the right flanker of the recumbent, showing the quartz vein.

(1959) published his photograph, which shows the sun setting just to the south of the deepest part of the notch in the hill crest behind the recumbent.

The next date likely to be of consequence is the cross-quarter day at the beginning of February, celebrated as Imbolc or Oimelg in late pre-Christian times, the sightlines for which would also work for the Samhain cross-quarter day at the beginning of November, the old Celtic year end celebration. This time the clouds were more cooperative, but there seemed to be nothing to mark the alignment, which as viewed from between Stones 12 and 13 across the center of the circle had the sun rising slightly to the north of Stone 4.

Sunset that day once again saw less cooperative weather. There was only a sufficient glow in the sky for me to have some idea where the sun was setting, which as viewed from the gap between Stones 3 and 4 across the center of the circle appeared to be near Stone 11, in the middle of the sloping group of three to the north of the recumbent. This did not strike me as being sufficiently clearly marked to be of any significance.

The weather at the vernal equinox, the third of the solar dates being considered, was equally

Figure 37.6 Beltaine sunrise. As viewed across the center of the circle, sunrise appears to occur just to the right of the telephone pole on the horizon, as evidenced by the bright glow in the break in the clouds just above that point. Note the prominence of the sloping tops to the stones as viewed from this angle.

uncooperative. For sunrise, my best estimate for the sightline is from somewhere in the vicinity of the second of the sloping stones, Stone 11, to the gap between Stones 2 and 3. With slightly more confidence – again based on noting where the glow of the sun through the clouds was centered – I would place equinox sunset on a line from the center of the Stone 3/Stone 4 gap to the north edge of Stone 12. Neither of these equinox lines strikes me as being sufficiently well marked to have been deliberate, though both are possible. It is also just possible that the line for sunrise should run from the point of Stone 11, or just to its south, to the south edge of Stone 2, which would be somewhat more believable.

Next came the observations of Beltaine, which would also have served for Lughnasa. This proved to be more interesting because for sunrise one looks across the quartz-marked Stone 10 (Figure 37.6). One in fact looks across the point of Stone 10 toward the spot on the horizon indicated by a continuation of the slope of Stone 1. The slopes of the two stones match, a point I had not noticed until reviewing my slides for this paper, and it is possible that one is intended to align them exactly. Since a narrow

band of clouds along the horizon made determination of the precise point of sunrise impossible, this is not clear.

Beltaine sunset requires one once again to look through the rather wide Stone 3/Stone 4 gap, this time toward the gap between Stones 12 and 13. This again is an alignment that seems to lack any clearly defined markers. However, discovery of the matching slopes of Stones 1 and 10 has led me to look at photos taken from other angles, and it is clear that the slopes of Stones 11 and 12 can both be aligned with those of 2 and 4 *if* one disregards the center of the circle. This question of the sloping tops of the stones needs to be examined in more depth – which simply means that more field observations need to be made.

The cross-quarter and equinox alignments all involve the three stones with the carefully planned slope to their tops – 10, 11 and 12. This is such a distinctive element in the design of the circle that it must have been important, and most probably in an astronomical context. Just how remains to be determined, but the use of three stones may itself be significant given the prominence of that number in later Irish mythology, when deities often occur

Figure 37.7 Summer solstice sunrise.

in triplets, with for example, each of three sisters linked to a different season (MacNeill, 1962, p. 273).

The final solar event to be considered is summer solstice. Somerville claimed that a line from the center of the recumbent across Stone 1 marks this sunrise, which turns out to be very nearly right. The actual alignment, however, is a bit more interesting, and more convincing. Remember the petroglyph? It lies near the north end of the recumbent. If one looks across the petroglyph toward the point of Stone 1, the south flanker to the entrance (Figure 37.7), one sees a summer solstice sunrise whose place on the horizon has clearly been marked with considerable care. We cannot be sure that the petroglyph is contemporary with the construction of the monument, but there seems to be no reason for suspecting that it was added at any very much later time.

Solstice sunset (Figure 37.8), alas, produces nothing nearly so interesting. The sun sets near a line along which observation would cause Stone 4 and the missing Stone 13 to appear to abut. Once again the form of the missing stone is a bit of important but lost data. Somerville's map shows solstice sunset as occurring slightly further to the north along a line from the center of Stone 4 to the gap between Stones 13 and 14. A large bush on the

close-by horizon adds to the complications in analyzing this alignment.

Inspired by the discovery that at Drombeg the summer sun rose over the point of the tallest stone as viewed across the circle, during the next few days I visited two other stone circles, both in the Wicklow Mountains south of Dublin. At Boleycarrigeen (Figure 37.9) the solstice sunrise is quite neatly framed by the two shortest stones in the circle, looking across the center to the peak of the tallest. At Castleruddery (Figure 37.10) a bush interfered with the view and the high grass of the site made determination of the center difficult, but the same phenomenon appears to be present. Castleruddery is another circle where I think further investigation might prove very interesting since its entrance – around the arc to the south of the photo – is flanked by two massive quartz boulders. (See Leask, 1945, for a description of the site.)

Another point of interest in these three circles is the fact that in all three cases a hill provides a raised horizon on the summer solstice sunrise sightline. This is reminiscent of Ruggles and Burl's (1985) findings in Aberdeenshire. If nothing else, it means one would not have to get up quite so early to watch the sunrise.

It is worth noting here that in one case – the

Figure 37.8 Summer solstice sunset.

Figure 37.9 Summer solstice sunrise at Boleycarrigeen stone circle, Co. Wicklow, a sight that will soon be obscured by the growing trees.

Figure 37.10. Summer solstice at Castleruddery stone circle, Co. Wicklow, not long after sunrise. The crest of the hill on the horizon is obscured by the trees.

Grange Stone Circle near Limerick (Figure 37.11), which I did not have a chance to check for a solstice alignment – the tallest stone, which *is* to the north-east, is called Rannach Chrium Dhiubh (which has been translated as Crom Dubh's Peak or Promontory) (Ó Ríordáin, 1951, p. 42). Crom Dubh is often associated with hilltops, and an alternative name for the deity at Magh Sleacht seems to mean 'head of the hill' (MacNeill, 1962, pp. 28, 346).

While I feel uncomfortable about attempting to interpret prehistoric symbolism, it seems quite reasonable to postulate that the sun was considered male (as seems evident from all the recorded European mythology). This being the case, perhaps the tall stone marking the summer solstice in these circles was interpreted by the builders as being phallic (allowing another translation of Rannach Chrium Dhiubh). Similarly, the narrow slit through which the midwinter sun penetrates the circle may have been seen as female, providing midwinter symbolism quite in parallel with that evident at Newgrange, where the midwinter sunlight penetrating the central chamber seems clearly meant to represent the solar deity impregnating the earth to give rise a few weeks later to the rebirth of spring.

37.3 Summary and conclusions

The two primary points I have tried to make in this chapter are that (1) in at least some cases, the shapes and relative placements of the stones in a ring – data not considered in most analyses – are important design elements relevant to archaeoastronomy, and (2) one element in the design at Drombeg, the use of the tallest stone in the circle as a pointer for midsummer sunrise, is present in at least two other Irish stone circles.

While my initial assumption proved correct – that there is a level of design sophistication present in some stone circles that will not be apparent from a plan of the monument – at present the only alignments at Drombeg that I am completely convinced were intended by the builders are those for midwinter sunrise and sunset and midsummer sunrise. These are all clearly marked in the design of the circle – by the axis from the entrance to the recumbent and the notch in the hill, by a line across the center and between the paired stones in the southeast quadrant of the circle, and by a line from the petroglyph to the point of Stone 1, tallest point in the circle, respectively. While alignments to the

Figure 37.11 The interior of the henge-like Grange Stone Circle in Co. Limerick, showing the tallest stone. Another stone circle can just be made out in the field beyond.

other likely sunrises and sunsets are not completely ruled out, proof of their existence awaits a more detailed field study of the role of the sloping tops of some of the stones.

The presence of alignments other than those for the solstices, and perhaps the May/August cross-quarter days, seems to me unlikely given what my study revealed about viewing conditions. Weather conditions today are not thought to differ significantly from those existing at the time of the monument's construction. Nor was the weather during the period over which I carried out my observations atypical. Yet cloudiness made the necessary observations very difficult except for the summer alignments – Beltaine/Lughnasa and the summer solstice. The length of the winter solstice period makes it likely that adequate observations could be made then as well. But one's chances are likely to be very low at the equinox and only slightly better at Imbolc or Samhain, the February and August cross-quarter days.

Finally, at least one additional set of observations needs to be made at Drombeg, those for the major and minor lunar standstills. And the hypothesis that a tall pointer stone marks the direction of mid-summer sunrise should be tested at other Irish stone circles as well.

Acknowledgments

The work on which this paper is based was made possible through an American Council of Learned Societies fellowship, faculty research grants from Ball State University, and a faculty travel grant from the Ball State Alumni Foundation.

References

Barber, J. (1973). The orientation of the recumbent stone circles of the southwest of Ireland. *Journal of the Kerry Historical and Archaeological Society* **6**, 26–39.

Duignan, M. (1940). On the medieval sources for the legend of Cenn (Crom) Croich of Mag Slecht. In *Feil-Sgribhinn Eoin Mhic Neill*, ed. Rev. J. Ryan, pp. 296–306. Dublin: Three Candles.

Fahy, E. M. (1959). A recumbent-stone circle at Drombeg, Co. Cork. *Journal of the Cork Historical and Archaeological Society* **64**, 1–27.

Franklin, D. (1903). Stone circle near Glandore, Co. Cork. *Journal of the Cork Historical and Archaeological Society* **9**, 23–4.

Gwynn, E. (1924). *The Metrical Dindshenchas.* Royal Irish Academy Todd Lecture Series 11, Part IV. Dublin: Hodges, Figgis.

Hawkins, G. (1965). *Stonehenge Decoded.* Garden City: Doubleday.

Hicks, R. (1981). Irish henge orientation. In *Archaeoastronomy in the Americas*, ed. R. A. Williamson, pp. 343–50. Los Altos: Ballena Press.

Hicks, R. (1984). Stones and henges. In *Archaeoastronomy and the Roots of Science*, ed. E. C. Krupp. Boulder: Westview Press.

Leask, H. G. (1945). Stone circle, Castleruddery, Co. Wicklow. *Journal of the Royal Society of Antiquaries of Ireland* **75**, 266–7.

Lockyer, Sir N. (1909). *Stonehenge and Other British Stone Monuments Astronomically Considered.* London: Macmillan.

MacNeill, M. (1962). *The Festival of Lughnasa.* Oxford University Press.

Ó Ríordáin, S. P. (1951). Lough Gur excavations: the Great Stone Circle (B) in Grange Townland. *Proceedings of the Royal Irish Academy* **54(C)**, 37–74.

Patrick, J. and Freeman, P. R. (1983). Revised surveys of Cork–Kerry stone circles. *Archaeoastronomy (UK)* **5**, S50–6.

Ruggles, C. L. N. and Burl, H. A. W. (1985). A new study of the Aberdeenshire recumbent stone circles, 2: interpretation. *Archaeoastronomy (UK)* **8**, S25–60.

Somerville, B. T. (1909). Notes on a stone circle in County Cork. *Journal of the Cork Historical and Archaeological Society* **15**, 105–8.

Somerville, B. T. (1930). Five stone circles of West Cork. *Journal of the Cork Historical and Archaeological Society* **35**, 70–85.

Stokes, W. (1894). The prose tales in the Rennes Dindsenchas. *Revue Celtique* **15**, 272–336, 418–84.

Stokes, W. (1895). The prose tales in the Rennes Dindsenchas. *Revue Celtique* **16**, 31–83, 135–67, 269–312.

Thom, A. (1967). *Megalithic Sites in Britain.* Oxford University Press.

38

Star walking – the preliminary report*

Claire R. Farrer *California State University at Chico*

And then He,[1] the Power, . . . lined up all His creation. And then He told the Sun, He said, 'You I have created so that you will be My representative. You will be that which Man sees.' And then the Moon, He told the Moon, 'You will be their eyesight at night.' And the Stars, He said, 'When they travel, by you they will know the directions to guide them.'
From the Mescalero Apache creation story as told by Bernard Second to the author.

In this, the preliminary report of continuing work concerning timing and location through celestial phenomena among the Mescalero Apache, I raise questions of how terrestial navigation is, and was, accomplished without contemporary instruments. Ethnohistory, history, myth, and a long term familiarization with the culture suggest directions to follow to answer the questions. While complete answers cannot yet be given, there are interesting points to consider at this, my present, level of understanding.

We should bear in mind that the system of knowledge I am attempting to describe is one that is taught through years of nightly observation. The training begins in childhood and continues for as long as is necessary for the individual to become competent in the system. Thus, an individual possessing the knowledge has quite literally made the information part of his persona.

It is not like Western knowledge systems where we have a fixed, codified corpus to be learned in sequence. Rather, an individual learns when the sky is right, when the social conditions are met,

when there is sufficient time (increasingly difficult to manage in these days of acculturation), when the proper ritual observances have been observed. The information is learned slowly and becomes so ingrained as to be a natural part of the individual. It is akin to what Hall (1983, pp. 211–12) terms 'primary level culture': once assimilated, it is basic to understanding and behavioral operations, but is difficult, if not impossible, for people to articulate.

My learning the system, then, is compounded by not only being an adult trying to learn what is properly taught to children but also by being an outsider, an Anglo, and a woman. This knowledge is men's knowledge and is usually kept within the ritual realm. While women are not prohibited from gaining the knowledge, they will not be directly taught as a man would teach another man or would teach a male child. However, once a woman, or a girl, gains some of the knowledge, men will discuss it with her and correct her where necessary. Women are free to observe the night sky, should they so choose. They are free to navigate by the stars. But Apache women don't do these things. Once I learn an aspect of the system, I may question; but I may not question to learn. An Anglo woman who seeks this male-realm knowledge must exercise patience if she is to learn how Apaches practice star walking.

How did our own remote ancestors find their way through the world? Did we always blaze scout-like trails to lead us back to our camps? We can surmise that getting *to* a place was not particularly difficult – simply follow one's nose. The problem

483

is in returning back to one's fellows, going yet again, and repeating the process many times.[2]

Biblical accounts of star following are familiar to most of us. And Gladwin (1970) detailed how South Pacific Islanders use stars as a navigational technique that they combine with astute observations of waves and winds. Of course, our own Western European derived backgrounds are rich in technological discoveries, using the celestial sphere, that improved naval navigation. Is it possible hunters and gatherers also used sophisticated techniques and, if so, how did they work? It is this puzzle that currently perplexes me. Hunting/gathering populations, it is generally accepted, followed animals; but this does not account for large scale migration over hundreds or thousands of miles. The Athabaskan speakers, to which the Mescalero Apache belong, migrated over 3500 miles from far northwestern Canada to their present location in south-central New Mexico. They were within 50 miles of their present-day site when they were first contacted by Europeans in the 16th century.

Imagine what it must have been like for the early Apaches on the vast plains of the Llano Estacado, the staked plains of what is now West Texas, Eastern New Mexico, and the adjacent portions of Kansas and Oklahoma. The Apache forebears were in this area when first encountered by Spainards in the 16th century (Winship, 1896). There are no natural landmarks on the Llano Estacado other than deep chasms that suddenly fall away from the level surface of the plains. The grasses, until quite recently, grew to the level of a man's eye or even over his head. How did people find their way through this grass desert and back to their home camp? And how did they find their home when the camp was quite likely to move to exploit other vegetal resources or to follow the bison herds? We know from the reports of the Spaniards, as well as later chroniclers, that the men from the plains hunters and gatherers groups were often gone for months at a time hunting bison or other game. The place where they left their families might well be abandoned by the time their hunt was finished. How did people find each other? For that matter, how did they find the game? How did they find other tribes when they went on trading expeditions? How did they get back home? Or, when not following game, how did they get simply from place A to place B, whatever the motivation for beginning the journey?

We have no time machine that allows us to travel back to ask such questions of those no longer with us; therefore, we must rely upon our own constructions. Or, it we are fortunate, as I am, we work with people who have not forgotten the old traditions even while they are busily assimilating the new traditions of the Anglo-Europeans. While it may be difficult to articulate the sytem, it is, nonetheless, intact.

Today, among the Mescalero, there are professional rememberers, those who carry the cultural knowledge in their heads and who share it at proper times and through proper requests. Perhaps their knowledge was once common; perhaps at one time everyone knew the things now held to be almost sacred. But today the knowledge is kept alive by only a select few, usually men who are Singers of Ceremonies.

One such man is my primary consultant at Mescalero, Bernard Second; we have worked together for over ten years now. The system he uses is frustratingly incomplete from my perspective, but it is an active system, portions of which I have described elsewhere (Farrer and Second, 1981; Farrer, 1988). I am still trying to learn the system and he is unable to articulate it in ways that make sense to me. His statements of 'I *know* where I am by looking at the stars', or 'You just feel where you are', are not considered sufficient. But they are true. The Mescalero say, 'It's a funny thing; Apaches don't get lost.' And that, too, is true.

Consider the two stories below as indications of travel and location. The first, The Prophecy, is paraphrased and edited from several of Second's tellings over the years while the second is my report of an event from 1982.

The Prophecy: He was already an old man, they say, when he came in from hunting and lay down on his bed opposite the tipi entrance. All he said to his wife was that he was tired. But when next she looked at him, he had the look of death on him. She began the ritual wailing. And, since he was an important and well respected man in the tribe, the medicine man came immediately when the death cry was heard. But it was too late; even his powerful medicine could not help now.

He lay there for four days and nights, as was the custom in those days and is still the custom today. She cut her long hair, for her husband was dead. The people came to comfort her, to cook for her, to help her prepare him for his journey; they sharpened their knives in preparation for cutting through the west side of the tipi to release his spirit. The Singer, who would sing him onto the first part of his journey to The Land of Ever Summer, began to shake the deer hoof rattles as he sang of this shadow life and of the real life the man was about to enter.

Then, suddenly, there was the quiet of death itself as the man raised his head and asked for water. His voice was weak, but it was his own voice – not the voice of a spirit but the voice of a man. He told of being dead for four days and four nights, of what he had seen, of what was to come, of what would happen to his people. He told of things of joy and he told of things of sorrow. He told of being reunited with the animal they called nełį́į́yé (horse)[3] whom they had lost in their previous travels, of going to a new place that would be theirs, of knowing it by particular drawings on rocks in the area, of finding much food. And he told of the four winds that would eventually destroy their new home, of the strange pale people they would see who would conquer them, of the times of no food when the children would sicken and die. Truly he brought messages from the Real World, the world of the Creator, of spirits, of the beginning time linked to our time.

But that was a very long time ago, they say. And when he did finally die, instead of a raspy death rattle, pollen came from his mouth, they say. They say these things because they are true.

All this happened, the grandparents say, at the Shell River where the Flint Rock Dig is, where the Mud That Boils is, Where a Mountain Ridge Began. Today, the Mescalero Apache people say that it happened in the Yellowstone country. They can tell you how to get there, although they have never been there themselves, nor were their grandparents or the grandparents of their grandparents. But they know the way, the landmarks. They know where the man who breathed his last breath in pollen gave this prophecy.

Contemporary Mescalero Apaches have place names for many places they have never been; they also have instructions for how to reach those places. These have been handed down in oral tradition for many, many generations. When a contemporary Apache travels to one of the sites mentioned in the oral literature, he or she invariably describes a sense of joy and déjá vú, knowing full well it is a new place but yet knowing what to expect and where to find significant landmarks. They know how long it should take to get there by foot or by pickup truck, the kind of weather to expect in the place, what is edible there, and they know what the sky will look like. Alfonso Ortiz chronicles similar events for his Tewa relatives (Ortiz, 1969, pp. 148–9).

But there is a different level of knowledge in simultaneous operation as well. This level involves actual location by celestial observation. While it is referred to as knowledge from stars, stars should here be understood in a generic sense. Surely stars are of primary importance but also important in location are other celestial phenomena and especially the moon.

In the Fall of 1982 Bernard Second was a visiting scholar for a three week period at the University of Illinois in Urbana-Champaign where I was teaching. He left south-central New Mexico and the Mescalero Apache Reservation in late afternoon for a bus ride to Albuquerque, NM, a four hour trip. After dark he caught an airplane in Albuquerque and landed in central Illinois just before midnight. We talked briefly before retiring.

It is Bernard's custom to pray at sunrise each day, but upon arising in Illinois he found he was confused. As he described it, with chagrin, 'The sun came up in the wrong place!' He attributed his confusion to travelling inside the airplane at night and to landing in a place with overcast skies, so that he was unable to ascertain immediately his bearings. Nonetheless he was acutely embarrassed because, 'Apaches don't get lost.' And, most particularly, Singers of Ceremonies who have an intimate knowledge of the skies do not look in the wrong direction for sunrise.

He spent a goodly part of the next few evenings walking around the flat Illinois prairie and commenting upon how high the corn grew. On the fourth day of his stay, he told me, 'Sis, I guess I'm about one-thousand miles from home.' I was astounded: the shortest distance by car is 1168 miles. When I asked him how he determined the distance, he stated that he just walked around looking at the stars change throughout the night. He also maintained that he knew how far he walked at home, and how long it took him, before the stars took on a somewhat different appearance. Thus, he avered, all he did was walk and look and figure how long it would have taken him to have walked far enough for the stars to be as they were in Illinois compared to what he would see at home in Mescalero.

Bernard is drawing upon a knowledge base that he cannot adequately describe to me and that I do not understand well enough myself to be able to eludicate in its finer points. I cannot account for the longitudinal problem other than to note that through telephone conversations and questioning me he knew before he arrived in Illinois that my local time was an hour in advance of his local time in Mescalero. Perhaps he merely did simple calculations of time on the airplane that, coupled with his knowledge of a two time zone difference, allowed him to extrapolate to distance. But this begs the question of how. He does walk great distances, on the order of hundreds of miles, on some of his forays. This practice gives him a practical knowledge of walking and noting celestial phenomena. He maintains that despite what scholars may think, his people took their bearings at night and walked, as much as was possible, at night, making better time, of course, when the moon was brightest. He still walks at night, whenever possible, when he is on serious and long trips, maintaining that not only is it cooler but also it is easier. In his words,

I'm always outside at night and I'm always looking up so I know the changes with the seasons. Even in the wintertime I get up and walk around outside at night and watch them so I'm familiar with it, 'cause of the seasonal changes, ... to use the sky as a guide ... It's still the same *except* for the seasonal change; you have to remember that it's changed ... We *always* look at the Dipper no matter what season. I don't know how to explain it. ... See, these ones [stars/planets he focuses on other than the Dipper] will change but the Dipper doesn't change: it's still there ... I just know it ... whether its in December or July, I can gauge whereas things change. Star-That-Does-Not-Move and the Morning Star are always constant, dependable, we always see them. As long as we see them, it doesn't matter what angle they're

at. When we see them, we guide ourselves by them.[4]

While Bernard is one of only a handful of people who now know how to use the stars to determine position and direction, I cannot help but suspect that at one time this was a more generalized knowledge – especially since archaeological and linguistic evidence, as well as the Apachean verbal traditions, maintain that they travelled from the northwestern parts of Canada into their present homelands.

We, as followers of a Western European derived life style, find it difficult to believe that people can navigate without sophisticated instruments, without ways of shooting the sun or measuring celestial angles. We do not attend to our own biological clocks, preferring to rely upon alarms and appointment books to get where we should be. Our notions of time and science have served us well. But they are not the only possible ways to live and live well in our universe.

Consider that just a short time ago our lives were not clock-bound but rather responded to Matins (about midnight), Lauds (sunrise), Prime (around 6 a.m.), Terce (about 9 a.m.), Sext (noon), None (approximately 3 p.m.), Vespers (sunset), and Compline (about 9 p.m.) Mescalero indigenous time is similarly arranged, with fixed times dependent upon celestial events and relative times dependent upon divisions between the fixed celestial events. Those celestial events, in combination with a sophisticated biological time using the body as referent, allow one to have a measure of time that is certainly sufficiently accurate to allow travel.

Further, many Mescalero people still walk great distances and are fully aware of how long it takes to walk over various kinds of terrain in particular kinds of weather. This walking is during the day as well as at night, when people walk by the stars.

During the summer of 1985 a teenager attached to the household in which I was staying walked home from a ceremonial; this is not an unsual event until one considers that she had to walk over two mountain ridges and through three canyons, a distance of approximately 13 miles, at night. I last saw her at the ceremonial grounds at about 10:30 p.m. when she declined to accompany me home, perferring to stay to dance; upon arising the next morning at 6:20 a.m. I found that she was at the door – she said she'd gotten a ride from 'the road', an intersection about two-and-a-half miles from the house where we were staying. Her feat was not unusual at Mescalero, despite the fact that she also was pregnant: if one does not have a car, one walks, unless a ride is offered. And when walking, usually unencumbered by a watch, a device that most Mescalero eschew, one learns the ways of the body in conjunction with the ways of the universe. Further, children are specifically taught to pay attention to their bodies and to their surroundings. Simply because we do not know where we are without scientific instruments does not mean that others require the same devices to place themselves in space and time.

When first encountered on the Llano Estacado by Coronado's 1540 expedition, the Apaches were described as a people with a simple tool kit, tipis, dogs and travois to carry bundles, vast numbers of bison skins, and products manufactured from bison parts. They were described by Casteñada, a chronicler of the expedition, as tall, intelligent, clean, and curious: traits that can be used today as well. The Llano Estacado was a formidable barrier to that expedition; Coronado lost soldiers and horses to its vastness – those that went out never to return again through the sea of grass. The Spaniards were properly impressed by people who managed not only to live on the plains but also to move freely within their boundaries and beyond them, as well as guiding others through them (Winship, 1896).

Perhaps we need to re-examine some of our own presuppositions and realize that a simple tool kit and a simple way of life, devoid of much technology, do not necessarily mean a simple mind or simple thought processes and simple calculating ability. If we can accept that premise, then there is a logical extension of it that bears examining.

Let us assume that hunters and gatherers in general had sophisticated knowledge of celestial phenomena and terrestial navigation using the stars as maps. Most of those who provide us with reconstructions of past life times and ways believe that hunting and gathering preceded an agricultural way of life. If we accept this, then it is logical to assume that when people became agricultural, and sedentary, they *already* had a complete system of knowledge concerning stars and other celestial phenomena. Having such a system would mean that, when planting time was approaching, people already had the calendar in mind. The Apache native calen-

dar, as an example, begins with the summer solstice (*sha sizi*, sun standing still [with connotation of straight up and down]); the entire calendar round is termed *dateʔe naańtʔa*, one growing season (of wild plants). Hunters and gatherers, who depend upon wild plant foods for all their nutritional needs, other than meat protein, must have an elaborate knowledge of what can be expected to ripen when as well as how to process the foodstuffs available in their environment. They must know how to 'read' the seasons and how to predict their round. It is not poetry that caused the Mescalero to name one of the winter months 'famine time'.

Embedded within the system of a seasonal calendrical round is a knowledge of star movement. This suggests, then, that when people in prehistory began their monumental architecture, the buildings were *confirmatory* constructions having orientations to particular astral or calendrical phenomena that were well known. That they may also have been used as sight lines would only re-confirm what was already known.

I am suggesting that the elegant pre-Columbian architecture of the Americas reflected just this kind of system and that the buildings were constructed to take advantage of particular events rather than to observe them. Events keyed to the sky, coming when priests knew they would, surely would have been ever so much more spectacular to an uneducated lay people when, for example, the sun lit an inner chamber on the solstice or a star set over a particular constructed peak, confirming a priest's prediction. I suspect when the first stone was laid in the first pre-Columbian ceremonial center, it was laid under the direction of a man similar to Bernard Second who knew precisely where it should be placed to take advantage of natural phenomena visible from that site, the natural phenomena that the buildings on the site had been placed to confirm. Knowing where and how to place structures depends upon a thorough understanding of the sky; we accord this knowledge postdictively to the pre-Columbian city builders. I wonder why we are chary of according the knowledge as well to hunters and gatherers?

Astronomical knowledge was and is significant for agricultural peoples. But it is no less significant for hunting and gathering people who not only needed to know where they were, how to go to another specific place (as, for instance, a salt source), and how to return home but also who needed to know when to expect certain foods to be available, when animals would be in rut and their meat taste badly, when it would be turning cold, when it would be the beginning of a new year with the summer solstice, or when the winter solstice would occur so judgments could be made for food-stuffs to last through the famine time ahead.

I find it interesting that the star names and planet names for the Mescalero are based upon a summer sky;[5] and we must keep in mind that until a bare 100 years ago, these people were still pursuing a hunting and gathering way of life. Many, on a reservation with too few jobs for the population, find that hunting and gathering is an important supplement in today's world as well. And many walk vast distances that we with our technology would not consider without mechanical transport of some kind.

The knowledge of the sky among contemporary Mescalero Apaches is based upon long term obser-vations and oral history that is often encoded in myth. The Twin War Gods figure prominently in these stories; the Twins are usually represented iconographically by triangles. It is the Twins who are credited by the Apache with showing the people how to move through space without getting lost. I suspect that as this research continues, I will have to look again, with newly informed eyes, to the Twins and their iconography.

We no longer have the opportunity to talk with many who follow a traditional life style, unencum-bered by the vestiges of modernity. But those few who still live such a life or who are only a generation or two removed from it can tell us a very great deal about perceptions and knowledge utilized in the seemingly simple life. What may seem to have come out of nowhere, such as the sophisticated monu-mental architecture of pre-Columbian America, I believe was rooted in countless generations of pre-agricultural people, people who were not very dif-ferent from the Mescalero Apache. It behooves us to remind ourselves that elegant technical and scientific instruments are a very recent addition to the tool kit of humankind; we need to remember to lay aside our ethnocentric biases that the sky can only be made to reveal its secrets through technolo-gical manipulation. It remains my task to continue the investigation of precisely how the Mescalero became, and how they maintained their mastery of star walking.

Acknowledgments

I am indebted to Ray Williamson and Gene Am-marell for their willingness to assist me with the astronomical points in this paper. Earlier versions were read by them as well as by Trudy Griffin-Pierce. Their comments, along with those made during the discussion section of the Mérida meet-ings, have been incorporated here. I, of course, remain responsible for errors of omission or com-mission; it cannot be assumed this version accords with arguments presented by the readers or the dis-cussants.

Notes

1 'He' is necessary and conventional in English. How-ever, Mascalero Apache does not indicate gender in nouns or pronouns. There is no attribution of maleness to the Creator in the Mescalero Apache languge; rather the Eternal Power, Creator, is beyond the human terms and notions of gender and sexuality. Bernard Second is here following an English language conven-tion of referring to the deity in English with a male pronoun.

2 Bernard Second, in reading a draft of this paper, dis-putes this statement. He maintains I have it precisely opposite: it is difficult to find what one seeks but very easy indeed to return home, since one knows where one began and can follow one's own trail back to the starting point. It is, he avers, in the getting to where one wishes to be with an economy of time and effort that requires precise sky watching and star walking.

3 I find this an intriguing idea. If we suppose that the Athabaskan speakers were at one time contiguous to the Tibeto-Burman speakers, as most linguists concede by their classification of the Athabaskan lan-guages within the Tibeto-Burman stock, then might we also grant that they may, indeed, have had the horse prior to its re-introduction on the North American con-tinent by the Spaniards. The term, netįįyé, refers to one which carries burdens; today it is considered the ancient term for 'horse', and the now proper word for 'dog', an idea consistent with the dog-powered travois

Coronado's expedition described. The contemporary-term for 'horse', *łįį*, is found in the root of the word. The sequence, then, if this story and related comments by Singers are to be believed, is the term originally referring to horse, then being transferred to dog when horses were no longer a part of life for whatever reason(s), and now returning to the root for the contemporary term for horse. Today most people will insist that the word for dog is 'chúúne', a term that can also mean 'friend'. It is only Singers and the very old who recall '*nełįįyé*' in both its horse and dog connotations.

4 This quotation is from 1984 when Gene Ammarell worked with me at Mescalero. We are grateful for support from the American Council of Learned Societies.

5 We must keep in mind here that, since I have been interested in pursuing ethnoastronomy seriously, I have been unable to spend a full year at Mescalero; thus, there can well be other stars and constellations consonant with a winter sky that I am unaware of because I am not there when they are important and thus when they would be discussed. Nonetheless, the year is said to begin at the summer solstice, giving *de facto* primacy to the summer sky.

References

Farrer, C. R. (1988). Mescalero Apache ceremonial timing. In *Ethnoastronomy: Indigenous Astronomical and Cosmological Traditions of the World*, eds. J. B. Carlson and V. D. Chamberlain (in press).

Farrer, C. R. and Second, B. (1981). Living the sky: aspects of Mescalero Apache ethnoastronomy. In *Archaeoastronomy in the Americas*, ed. R. A. Williamson, pp. 137–50. Los Altos: Ballena Press.

Gladwin, T. (1970). *East Is A Big Bird*. Cambridge, MA: Harvard University Press.

Hall, E. T. (1983). *The Dance of Life: The Other Dimension of Time*. New York: Doubleday and Company.

Ortiz, A. (1969). *The Tewa World*. University of Chicago Press.

Winship, G. P. (1896). *The Coronado Expedition 1540–1542*. Annual Report of Bureau of American Ethnology 1892–1893. Reissued in 1964 by Rio Grande Press, Chicago.

IV
Additional abstracts submitted

Myth, Environment, and the Orientation of the Templo Mayor of Tenochtitlan

A. F. Aveni, E. E. Calnek, and H. Hartung

In the light of the recent excavations of the Templo Mayor in downtown Mexico City, we explore the problem of the role of astronomy and calendar in the design and orientation of the building and of the city in general. We employ the ethnohistoric data relating to the foundation myth of Tenochtitlan as a means of generating hypotheses concerning astronomical orientation that can be tested by reference to the archaeological record.

We find that eastward-looking observations (implied in dismantling and reconstructing the myth) that took place around the time of the equinox may have been related to an attempt to transform a true east orientation from the natural environment into the architecture via a line that passed through the center of the temple of Huitzilopochtli (the more southerly temple of the pair constituting the top of the Templo Mayor). It is also possible that the deliberately skewed notch between the twin temples served a calendrical/orientational function.

Evidence is presented to support the view that the mountain cult of Tlaloc, represented in the environment on the periphery of the Valley of Mexico by Monte Tlaloc, also may have directly influenced the orientation of the building. In this connection we discuss the association between the Templo Mayor and an enclosure containing offertory chambers atop Mt. Tlaloc, which is located on a line extended to the visible horizon 44 km east of the ceremonial center. The ethnohistoric record implies that this place had been used for sacrifices to the rain god after whom the other of the twin temples of the Templo Mayor was named. A horizon calendar consisting of the marking of solar divisions at 20-day intervals is postulated. Full text published in *American Antiquity* **53**: 287–309 (1988).

La Gran Tabla de Venus en el *Codice de Dresden*

W. Brito Sansores

En la tabla de Venus, como en el contenido total de los treṣ Códices Mayas, pueden distinguirse principalmente 3 clases de textos: el calendárico, el pictórico y el glífico o explicativo.

Desde luego, esos tres aspectos, consideramos que están relacionados entre sí; por lo que el presente trabajo tiene como mira principal lo siguiente:

(A) Destacar algunas partes del texto calendárico que a nuestro juicio, desde luego no somos expertos en astronomía, tienen bas tante interés para el estudio de la cultura Maya.

(B) Buscar la relación que pueda haber entre el texto calendárico ya anotado y los dos aspectos pictórico y explicativo.

(C) Las páginas de la gran tabla de Venus, a las que nos estamos refiriendo, podemos dividirla, siguiendo la manera tradicional, en tres partes: A, B, C contando de arriba para abajo.

En el trabajo que presentaremos en el IV Simposio de Arqueoastronomía hacemos un estudio de las relaciones que hay entre el texto pictórico y glífico correspondiente al contenido de las páginas tales como:

(I) Descripción de todos y cada uno de los personajes e in tentos de identificación de ellos, etc.

(II) La representación gráfica de Venus con el signo Chac-Venus entero o medio. Posición del afijo Chac (rojo) a la izquierda o arriba del signo del planeta.

(III) Identificación de los puntos cardinales o lados del mundo.

(IV) Identificación o relación de los jeroglíficos o deidades o personajes.

(V) Interpretación de algunos glífos explicativos y las relaciones que pueden tener.

Significant dates of the Mesoamerican agricultural calendar and archaeoastronomy

Johanna Broda

The 'Fiesta de la Santa Cruz' (3 May) is celebrated today in all important Indian peasant communities of Mexico and Guatemala. Particularly interesting ceremonies full of Prehispanic reminiscences take place in the Nahua area of northeastern Guerrero and were observed by the author. Modern ethnographic data are compared with ethnohistorical information on the Aztecs as well as archaeological evidence from Cerro Tlaloc situated on the southeastern limits of the Valley of Mexico. It is shown that there existed a close relation between astronomical observation (of the zenith passage of the sun, the annual disappearance of the Pleiades), the beginning of the rainy season, sowing time in the dry-land cycle of maize, and the agricultural rites taking place at the Feast of the Holy Cross. Modern ethnographic reminiscences as well as 16th-century historical data point to the observation of the annual cycle of the Pleiades in Prehispanic Mesoamerica which was closely connected to the solar year, to climatological phenomena and the agricultural cycle. In this paper it is argued that the significance of astronomical observation and knowledge for Prehispanic society should be analyzed in this rather complex cultural and social context, and that the concept of ancient astronomy should be broadened to include the observation of the natural environment in more general terms.

Cataclysmic coincidence: the archaeoastronomic context of Cortes' Entrada and the conquest of Mexico–Tenochtitlan

R. David Drucker

Major Mesoamerican events, dated by native participants, may be examined for astronomic significance in a rigorously systematic fashion perhaps familar to the Mesoamericans themselves.

Central to the enterprise is the 'scroll' – a table of all the days of a single solar year generated by pairing days of equal solar declination in the region of the year when the sun's geographic positions are equal to the latitudes of Mesoamerica and by filling in the rest of the year completely and systematically.

Next, the Mexica calendar is regressed using the modified Thompson (584,283) correlation. European dates are kept in the Gregorian rather than Julian calendar.

With the aid of the scroll it is possible to examine specific dates for solar significance (zenith passage, solstices, equinoxes) by inspection. Obliquity changes for changes in epoch may be noted readily.

A check of the method is provided by projecting the Venus tables of the *Dresden Codex* forward from base date H in the modified Thompson correlation staying in the Gregorian calendar.

Using the method it is possible to see that key Tenochcan events concerning the founding and development of Mexico–Tenochtitlan have calendric and solar parallels to fourth century AD Teotihuacan and that these may have been interpreted differently by Moteuczoma Xocoyotzin and the regular priesthood of Tenochtitlan.

El Bisiesto Mesoamericano: the Middle American leap year

M. S. Edmonson

This paper will summarize the results of a comparative study of Middle American calendrics leading to conclusions about the accuracy of the native computation of the solar (tropical) year and the manner and probable date of the attainment of this computation. The data offered should clarify the reasons for the lack of intercalary leap year days in the native calendars, and will answer some questions and generate others about the dates at which particular calendars may have been instituted.

Tres Zapotes Stela C: a Mesoamerican eclipse record

J. A. Fox

Tres Zapotes Stela C and its complement, the Covarrubias Stela, constitute the second oldest dated inscription in the New World. The original calculation of the date was in error by several months. The correct date fell on or at most six days after the annular eclipse of the sun on August 31, 32 BC, the path of annularity of which passed almost directly over the site. The iconography of the monument prefigures what later became the Maya eclipse glyphs. The striking coincidence of the eclipse with the recorded date and iconography suggests that the monument commemorated the eclipse. This eclipse record has important implications for our understanding of calendrical correlation, the earliest writing in Mesoamerica, and Protoclassic intellectual activity.

Astronomy and the patterns of five geometric earthworks in Ross County, Ohio

N'omi Greber

The present project is part of a long term study which focuses on five major Hopewell sites, located near present-day Chillicothe, Ohio. This set of sites is defined by a number of distinctive cultural features, one of which is the design of the earthen embankments. Part of this design includes a square shaped element, which is uniquely defined, among Ohio sites, by the number and positioning of the line segments and mounds forming the square and by its size. Originally the walls were near human height and extended over 330 m on a side. The majority were easily traced and surveyed in the 1890s. By ground survey today, at best, some can be seen as surface stains or as very low ridges in cultivated fields. Data from 1847 surveys indicates that the squares, as a set, appear to note the complete annual solar cycle. This was accomplished by a consistently defined siting line which marked a different part of the cycle (*c.* AD 250) in each square. Sighting lines towards significant horizon positions of the moon and Venus were apparently incorporated into the pattern differently for each square. Archival aerial photographs and records, modern aerial photographs, traditional land survey, and geophysical remote sensing instruments are being used to obtain additional data needed to check and to refine the model.

Archaeoastronomical investigations of prehistoric stone pilings in Oregon

Roberta L. Hall and Phillip C. Green

Aboriginal cultures of the northwestern part of the United States provide a great challenge to archaeoastronomy. Occupied by a diverse set of peoples, the area was thrown into turmoil in the mid-19th century by disease and relocation. As a result, people left the areas where their natural history – both terrestrial and heavenly – was based, and lost their native languages. In 1980 Hall collected and tested a story about celebration of winter solstice by natives on the mid-coast, utilizing a natural marker. This was followed by a systematic investigation in 1984–5 of Oregon sites with stone constructions. The most promising site included a double line of stacked rock piles, averaging 15 inches high and three paces apart, which extend one-and-a-quarter miles across a high plateau in south central Oregon. Investigations in the summer of 1986 indicated the rock row was built in the 19th century as a base for a crude juniper fence, now decayed, to control sheep on the open range. To date no constructions indicating unequivocal astronomical alignments have been found in Oregon, though the oral tradition indicates knowledge of astronomical features and observance of winter solstice by means of natural markers.

The talayotic culture of Menorca

M. Hoskin

The small Mediterranean island of Menorca contains some 1600 archaeological sites, which have been little studied. Most of the remains are structures of stone, including the prominent towers or *talayots* and the remarkable *taulas*, which consist of a huge vertical slab on which is balanced an equally huge horizontal slab to form a letter T. The purpose of the taulas, which are located in horse-shoe shaped precincts, is much disputed, but three-quarters of them face between SSE and SSW, and none faces the northern half of the horizon. The paper discusses the taulas, and also the isolated menhirs of rectangular cross-section that are found some distance from taula/talayot sites.

Centipedes, rainbows, and power mountains: the ethnoastronomical investigation of a Gila River site in SW Arizona

T. Hoskinson

AZ:Y:3:1 is a prehistoric/historic archaeological site which occupies two small volcanic mesas and their associated drainages on the south bank of the Gila River in southwestern Arizona. The site contains a large quantity of prehistoric rock art, which includes petroglyphs, rock alignments and geoglyphs. The site is an integral part of the approximately 1747 acre Sears Point Archaeological District (SPAD). A variety of cultures, including the Desert Archaic, Patayan, and Hohokam cultures, dating between 10 000 BC to approximately AD 1450, are believed to have utilized the site area. A large historic/prehistoric village site is located on the north bank of the Gila River immediately across from AZ:Y:3:1. This village was occupied by the Hokan speaking Kaveltcadom for some time previous to the first recorded Hispanic contact on the Lower Colorado River in the 16th century, through the early part of the 19th century.

At AZ:Y:3:1 there is a 23 m linear ground figure that forms a solstice line which appears to connect two far horizon mountains. Study of the local and distant horizons led to the discovery of petroglyphs, cairns and rock alignments associated with the solstitial and equinoctial horizon points. Specific motifs were identified which have unique light and shadow interactions with the sun at the solstices.

This paper describes the site and some of its rock art. Ethnographic information for the area is presented which deals with shamanic transformations involving power mountains, rainbows, centipedes, rock art, calendrics, mythic Coyote, and mythic Coyote as calendar keeper. Hypothetical intepretations are offered for some of the rock art which attempt to synthesize images, astronomy and myth.

The golden mean in Mesoamerica and its astronomical significance

A. Ponce de Leon

The apparent motions of the sun and Venus as seen in the sky, especially at the horizon, reflect a kind of 'golden mean' which is expressed in spatial and temporal concept by the number of days and the angular values of the different signified positions. The spatial–temporal registration of these celestial bodies employs the geometry of the buildings and urban–geographical spatial planning in prehistoric Mexico, and I believe that some manifestations of the 'Golden mean' are inherent in them.

This establishes the hypothesis that the origin of the 'golden mean' in Mesoamerica lies in observations and spatial expressions of the apparent positions of at least those two celestial bodies, and that it led to the more abstract mathematics, of the middle rate and of the extreme concept.

In this work, the usual descriptive–geometrical analysis of charting the sun's movement with three planes of the structures is employed utilizing in addition, the images of apparent motions of the planets and the solar course.

Ambiguity, change and what astronomical symbols *do*: an example from the Tabwa of Zaire (Central Africa)

Allen F. Roberts

This presentation was adapted during the conference to react to points raised by colleagues. Leon Portilla finds that the celestial order recognized by Mesoamerican peoples confirms social order; their 'conquest of time', in Munro Edmonson's words, legitimizes political *status quo*. Conference speakers neglect the paradoxes, ambiguities, perplexities and indeterminacies of the lives of those studied. The multireferential symbols people use as vehicles to comprehend and cope with the irregularities of life must accommodate dilemma and change, as well as expectation. A discussion of the use of astronomical concepts in the preparation of magic by the Tabwa of Zaire allows consideration of how symbols provoke action and transformation in moments of misfortune, when the usual order of things is threatened. For Tabwa, the Milky Way is of a named conceptual set including lines and principles of demarcation and definition. Through a logical progression of metaphors, a substance is obtained for use in active magic that represents the definition but also the inherent change of the Milky Way's swing. The replacement for the Milky Way is used as an activating agent for a powerful medicine bundle used to protect from and apprehend evil sorcerers. This bundle is packed into a horn and set to swing on a pivot (in a way similar to the Galaxy's turning about the South Celestial Pole); it will 'point to' sorcerers. Action can be initiated thereafter to resolve conflict and restore order, albeit perhaps a *new* order.

Medicine wheels: testing the astronomical theory

J. H. Robinson

Attempting to remedy certain imperfections in J. Eddy's astronomical treatment of medicine wheels, I assumed that the heliacally rising stars were observed about 2° higher than their extinction angles. The revised procedure led to the following discoveries: (1) At both the Bighorn and Moose Mountain Medicine Wheels, the alignment Eddy had associated with the star Rigel was more accurate for Spica. (2) If we substitute Spica for Rigel, then all of the star-alignments at the Moose Mountain Medicine Wheel were accurate around AD 800. (3) At the Bighorn Medicine Wheel, if we associate the rising of Aldebaran with the E–B alignment rather than the F–A alignment – the latter indicating instead the rising of the Pleiades, then all of the star-alignments there were accurate in the era AD 1000–1100. (4) In spite of substituting some different star-targets, at both of these medicine wheels the heliacal risings of the stars remained limited to the warmer part of the year – a period of four to five months, from May or June through mid-October (including summer solstice).

The new approach was tried on the medicine wheel near Fort Smith, Montana. The results were: (1) If we assume that the observer stood in the little circle of stones at the center and looked out in the direction indicated by each spoke (which is my hypothesis), then five of the spokes were accurately aligned toward places where bright stars rose or set between AD 1800 and 2000 (and the only other spoke indicated the direction of winter-solstice sunset). (2) This range of dates is consistent with Crow Indian legend about the origin of the Fort Smith Medicine Wheel. (3) At no other time between 1000 BC AND AD 2000 were all five star-spokes accurate for bright stars. (4) If one assumes that the observer stood at the ends of the spokes and used the center of the wheel as a foresight (as others have hypothesized), then there was no time between 1000 BC and AD 2000 when all spoke-alignments were simultaneously accurate for bright stars (or solstice sun).

Thus the revised astronomical theory of medicine wheels successfully explains the one near Fort

Smith as well as the Bighorn and Moose Mountain Medicine Wheels; in the case of the Fort Smith Medicine Wheel, it discriminates between rival hypotheses about where the observer stood; and for each of these three medicine wheels it yields a reasonable date for astronomical usage. Furthermore, because of the stricter criterion that all alignments must function accurately at the same time, the probability of alignments pointing to bright stars just by chance is greatly reduced.

Sun tables of Star-Oddi in the Icelandic sagas

Curt Roslund

The old Icelandic sagas are a great source of information on the cultural achievements of the Norsemen. An interesting compilation of astronomical and calendrical knowledge is found in what is now known as the Rím-book from the middle of the 12th century. It contains the story of a learned man called Stiornu-Oddi, or Star-Oddi, who made extensive observations of the sun.

His table of the sun's declination is of particular interest. It gives in a novel and original way the sun's noon altitude at weekly intervals, using a simple arithmetic progression. During the first week after the winter solstice the sun's altitude is said to increase by its own semi-diameter, by two semi-diameters in the second week and so on by another semi-diameter for every new week until the spring equinox. This simple rule seems to have no earlier precedent in classical or Medieval literature, and despite its intriguing simplicity it gives a fairly accurate description of the variation of the sun's declination.

There has been considerable speculation about the origin of Oddi's declination table. It is here suggested that Oddi was familiar with an approximate method of dividing the semicircle, later used by Italian Renaissance shipwrights and by them called the mezzaluna-theorem: if a semicircle is divided into $2n$ equal parts, then the segments parallel to the diameter and defined by the divisions on the circumference will increase in width by ratios close to $1:2:3: \ldots :n$ from the top to the base of the semicircle. This theorem can be extended without great loss of precision to the lune formed on a sphere by two great circles of moderate inclination. If one of the great circles represents the celestial equator and the other the ecliptic, then the sun's declination can be expected to increase by the ratios close to $1:2:3: \ldots :n$ for equal time divisions between solstices and equinoxes as a natural consequence of the sun's almost uniform motion along the ecliptic. By pure coincidence, the unit of weekly increase nearly equals the sun's semi-diameter which gives added usefulness to Oddi's method.

The mezzaluna-theorem may have had a wide application. Computer analysis of a table of directions on the horizon for the first glimpse of dawn and the last shimmer of dusk makes it plausible that Oddi also used this theorem in his other calculations for the sun. In another example in the Rím-book, this time not attributed to Oddi, it is stated that the sun's declination changes by 4° in the first month after the winter solstice, by 8° in the second month and by 12° in the third, that is by ratios $1:2:3$ as in the mezzaluna-theorem.

Star-Oddi's sun tables show the extent of the dispersal of astronomical knowledge of Medieval times and its assimilation into the culture of that far outpost of the world.

Heliacal rise phenomena

B. Schaefer

The heliacal rising of a star or planet is when the object first becomes visible after the yearly period of invisibility caused by the glare of the moving sun. Ethnoastronomy offers many examples where heliacal phenomena were featured prominently in the calendrics of ancient civilizations (e.g. the Egyptians, Mayans, and Moslems). Many other examples are known or claimed from ethnoastronomy or archaeoastronomy. As such, it is important to have an understanding of heliacal phenomena. Unfortunately, as an *astronomical* problem, *no* observations appear in the literature, and the last theoretical work was due to Ptolemy. This paper presents, for the first time, a large data base of observations and a complete model. Primary attention is paid to the application of the model to many archaeo-astronomical and ethnoastronomical claims.

An interpretation of a unique petroglyph in Chaco Canyon, New Mexico

Anna Sofaer and Rolf M. Sinclair

The design of a petroglyph on Fajada Butte in Chaco Canyon is a half circle surmounted by a spiral. This pattern corresponds closely to the ground plan of Pueblo Bonito, a massive semicircular structure that is the largest construction in the prehistoric US southwest, and to this building's relationship to the solar cycles. The recumbent 'D' shape of the petroglyph, unusual in Chacoan rock art, is the same shape as the outline of this pueblo. Two lines of the petroglyph (the diameter and the perpendicular radius) appear to represent two major walls of Pueblo Bonito that are accurately aligned north–south and east–west. In addition, the location of a drilled hole in the petroglyph corresponds to the location of the primary kiva of the pueblo. The spiral design above the 'D' appears to represent the sun's daily and seasonal passage over the pueblo and to refer to the pueblo's cardinal alignments and solar orientation. The carving also shows the pueblo design as bow-and-arrow, a symbol associated with the sun in the mythology and ritual of the successor historic pueblos. The use of this petroglyph to express solar phenomena is consistent with the use of other petroglyphs near the top of the same butte, where sunlight forms a number of noon–seasonal markings at the solstices and equinoxes on spirals and other shapes. This recognition of the petroglyph as a solar symbol of Pueblo Bonito is the first reading of a petroglyph as an explicit statement of cosmology and architecture.

Hawk migration, meteorology, and astronomy in Quiché-Maya agriculture

B. Tedlock

Farmers everywhere are dependent on large-scale seasonal climatological patterns and local ephemeral meteorological phenomena; consequently, attempts are made to predict both of these variables. Mayan farmers in highland Guatemala compute seasonal climatological patterns by careful astronomical observation of the movement of the sun, individual stars, constellations, and asterisms, including Regulus, Castor and Pollux, Orion, the Pleiades and Hyades, the Southern Cross and a seven-star asterism in Sagittarius (Sigma, Phi, Delta, Gamma, Lambda, and Epsilon). The phenomena observed include heliacal risings and settings, transits, oppositions, and zenith passages. Observations of the biannual migratory passages of Swainson's Hawks, which are keenly attuned to both climatological and meteorological patterns, are combined with these astronomical observations and recorded in 260-day almanacs.

The astronomical dimension of Quiché Maya mythology

D. Tedlock

In the commentaries to my recently published translation of the *Popul Vuh*, I have argued that the mythological portions of this ancient book contain allusions and direct references not only to the sun and moon, but to all five visible planets, various constellations, and the Milky Way as well. In the present paper I will pursue these arguments further, attempting to draw out the general principles of Mayan astronomical thought as expressed in the *Popól Vuh*. Special attention will be paid to the discrepancies between the Mayan Venus calendar and the observable heliacal risings and settings of

Venus. In attempting to solve this problem, I will consider such phenomena as human pregnancy, agricultural cycles, the synodic and sidereal perio- dicities of the moon and Venus, and the Mesoamerican 260-day divinatory calendar.

Archaeoastronomy and archaeological preservation

R. A. Williamson

Archaeoastronomical research has shown that in planning sites and building structures, traditional historic and prehistoric cultures around the world have incorporated astronomical alignments of calendrical or sacred ritual importance. Alignments to the solstices, the equinoxes, and to the major standstills of the moon and bright planets have been demonstrated or suggested. Stellar alignments have also been shown to have been important for some cultural groups.

The confidence archaeoastronomers place in the accuracy of structural alignments they measure depends in many cases on the efforts of archaeologists and others directly engaged in the preservation, stabilization, and reconstruction process to reflect accurately the orginal state of the structure. Archaeoastronomers therefore have a major stake in understanding the methods by which archaeologists gather and analyze their data. They also have an interest in conveying to the archaeological community the importance of maintaining structural alignments when sites are stablized or reconstructed.

At the request of the Committee on the Interior of the US House of Representatives, the Office of Technology Assessment (OTA) is now engaged in a study of *Technologies for Prehistoric and Historic Preservation*, and is seeking input from a wide variety of individuals interested in the preservation and interpretation of cultural resources. This paper briefly describes the OTA study, and summarizes the various technologies upon which the archaeoastronomer depends for accurate analysis and interpretation. it also analyzes preservation issues important to the archaeoastronomer.

Recent archaeoastronomical research in Davis Canyon, UT

R. A. Williamson

Davis Canyon, on the east side of Canyonlands National Park, Utah, lies along the northernmost boundary of the Anasazi, the prehistoric Pueblo peoples of the Southwest. It is the site of several unusual sandstone structures that were apparently built and utilized during the latter part of the 14th century. At least one of these structures, the so-called South Ruin, located on a promontory east of the South Six Shooter peak, may have been intentionally aligned to the winter solstice.

This paper presents results of research on these structures carried out during 1984 and 1985. It describes the structures and examines their possible function in the context of the local ecology. Because these structures were built comparatively late in the Anasazi occupation, when they had begun to move from their traditional areas as a result of prolonged drought and other factors, and because they occur at the boundary between the Anasazi and the Fremont peoples to the north, they are of particular interest in understanding the relationship of these prehistoric peoples to the land and to the sky.

Index

501